Structure of Turbulence and Drag Reduction

International Union of Theoretical and Applied Mechanics

A. Gyr (Editor)

Structure of Turbulence and Drag Reduction

IUTAM Symposium Zurich, Switzerland
July 25 – 28, 1989

Springer-Verlag
Berlin Heidelberg New York
London Paris Tokyo Hong Kong

Dr. Albert Gyr
Institut für Hydromechanik
und Wasserwirtschaft
ETH Zürich
8093 Zürich – Hönggerberg
Schweiz

ISBN 978-3-642-50973-5 ISBN 978-3-642-50971-1 (eBook)
DOI 10.1007/978-3-642-50971-1

Library of Congress Cataloging-in-Publication Data
IUTAM Symposium (1989 : Zurich, Switzerland)
Structure of turbulence and drag reduction : IUTAM Symposium,
Zurich, 25.-28.7.1989 / A. Gyr, editor.
(IUTAM symposia)
At head of title: International Union of Theoretical and Applied Mechanics.
ISBN 978-3-642-50973-5 (U.S. : alk. paper)
1. Turbulence--Congresses.
2. Frictional resistance (Hydrodynamics)--Congresses. I. Gyr, Albert.
II. International Union of Theoretical and Applied Mechanics.
III. Title.
IV. Series: IUTAM-Symposien.
QA913.I97 1990
532'.0527--dc20 90-9675

Softcover reprint of the hardcover 1st edition 1990

2161/3020 5 4 3 2 1 0 – Printed on acid-free paper

Scientific Committee

A. Gyr, Switzerland (Chairman)
G.I. Barenblatt, USSR
B. Gampert, FRG
J. Hinch, United Kingdom
G. Leal, USA
A.E. Perry, Australia
J.-M. Piau, France
H. Sato, Japan
L. van Wijngaarden, Netherlands

Symposium Sponsors

International Union of Theoretical and Applied Mechanics (IUTAM)

Federal Institute of Technology Zürich, Switzerland.
Institute of Hydrodynamics and Water Resources Management.

Preface

In 1976 a similar titled IUTAM Symposium (Structure of Turbulence and Drag Reduction) was held in Washington . However, the progress made during the last thirteen years as well as the much promising current research desired a second one this year.

In Washington drag reduction by additives and by direct manipulation of the walls (compliant walls and heated surfaces) were discussed. In the meantime it became evident that drag reduction also occurs when turbulence is influenced by geometrical means, e.g. by influencing the pressure distribution by the shape of the body (airfoils) or by the introduction of streamwise perturbances on a body (riblets).

In the recent years turbulence research has seen increasing attention being focused on the investigation of coherent structures, mainly in Newtonian fluids. We all know that these structures are a significant feature of turbulent flows, playing an important role in the energy balance in such flows. However their place in turbulence theories as well as the factors influencing their development are still poorly understood. Consequently, the investigation of phenomena in which the properties of coherent structures are altered provides a promising means of improving our understanding of turbulent flows in general.

The mechanisms responsible for the phenomenon of turbulent drag reduction produced by the addition of polymers, surfactans, fibres, flakes, needles, algae, gas bubbles etc. deserve particular attention. A more detailed knowledge of coherent structures would also enable a better understanding of the influence of modified wall characteristics, in particuar the drag reducing effect produced by special wall roughness (riblets) and by flexible or heated walls.

The intention of the Symposium was to review recent research results from studies of turbulence structures in Newtonian as well as in non-Newtonian fluids and in particular from drag reduction studies which include details of the structure of the flow. Emphasis has been placed on the available knowledge about coherent structures to improve the understanding of drag reduction.

This review process provided at the same time an opportunity for analyzing the quality and completeness of available results on coherent structures. Such a critical analysis helped to define the fundamental questions more precisely, thus stimulating further progress in the field of coherent structures and drag reduction.

At a IUTAM Symposium on drag reduction held at the University of Essen in 1984 attention was restricted to polymer additives and mainly to the influence of their physico-chemical parameters (molecular structure, molecular weight, molecular weight distribution) as well as rheological aspects, such as the elongational viscosity. This emphasis reflected the current research interests at this time. However, even at the Symposium in Essen it was suggested that aspects excluded due to the limited available time, such as the influence of turbulence structures (in particular coherent structures) and the multiplicity of other additives, should be the subject of a meeting in the future .

A thematical structure was provided to the large topic of the Symposium by nine invited lectures which reviewed the various aspects of coherent structures, rheological properties of the additives, the different drag reducing effects and aspects of the stability of a turbulent flow. These state of the art reviews were supplemented by 23 orally presented contributions and 20 papers presented as posters. Fortunately practically all contributions could be accepted for publication in the proceedings. This is mainly the merit of the members of the scientific committee and some external reviewers who preselected the contributions from about a hundred based on the submitted abstracts.

The achieved scientific progress can be summarized as follows:
- Mainly it became evident that drag reducing effects are caused more by specific interactions of all size of structures rather than by an interaction of only selective parts of the turbulence spectrum.
- The models explaining drag reduction are still controversal. The main contribution of the Symposium therefore seems to be the idea that drag reduction should not be discussed only in terms of supressing mechanisms but also in terms of topological explanations based on specific anisotropies of the turbulent flow field, e.g. high helicity.
- Drag reduction by riblets have a lot in common with drag reduction by additives and should be describable in similar terms.
- The audience was very sceptic whether drag reduction by LEBUs exists or not.

Last but not least I would like to thank the members of the Scientific Committee who at all stages during the preparation of the conference, maintained a very close rapport. In addition to the members of the Scientific Committee we would like to thank especially Prof. T.Dracos and Drs. J. Bühler, A. Müller and X. Studerus (ETHZ), Dr H.-W. Bewersdorff (Univ. of Dortmund) and Prof. A. Tsinober (Tel Aviv Univ.). Thanks are also given to D.Fruman and B. Gampert and his wife for their contributions which were not part of the official program. Thanks go also to my wife Edith who organized the ladies program and to my two daughters Nadia and Regina for their help in the organisation work.

The ready and efficient cooperation of Herr von Hagen and the editorial staff of Springer Verlag during the preparation for the printing of the Proceedings is specially acknowledged.

It is our hope that this book will stimulate further research efforts in this exciting field of fluid mechanics.

Zürich, November 1989 Albert Gyr

List of Participants

John B. Anders
NASA Langley Res. Center
Mail Stop 163
Hampton, Virginia 23665-5225
USA

Theodor R. Anderson
Dept. of the Navy
New London Lab. Bld. 80
New London, CT 06320
USA

Fabien Anselmet
Universite d'Aix-Marseille I M S T
12 Avenue General Leclerc
F - 13003 Marseille
France

Ram S. Azad
University of Manitoba
Dept. Mech. Eng./Ind. Eng. Progr.
Winnipeg, Manitoba R3T 2N2
Canada

S. Banerjee
Dept. Chem. & Nucl. Eng.
University of California
Santa Barbara,CA 93106
USA

S.P. Bardakhanov
Inst. Pure & Appl. Mechanics
USSR Academy of Science
Novosibirsk, 630090
UDSSR

Dietrich W. Bechert
DLR, Abt. Turbulenzforschung
Müller-Breslau-Str. 8
D- 1000 Berlin 12
Germany

Neil S. Berman
Dept. Chem, Bio & Mater. Eng.
Arizona State University
Tempe, Arizona 85287-6006
USA

Arild Bertelrud
NASA Langley Res. Cent.
M/S 163
Hampton, Va. 23665
USA

Horst Bestek
Univ. Stuttgart/Inst. f. Mechanik
Pfaffenwaldring 9
D- 7000 Stuttgart 80
Germany

Marco Bettelini
Inst. f. Fluiddynamik
ETHZ
CH-8092 Zuerich
Switzerland

Hans-Werner Bewersdorff
Univ. Dortmund, Chemietechnik
Postfach 500500
D-4600 Dortmund 60
Germany

Ron F. Blackwelder
University of Southern California
Dept.Aerosp.Eng./University Park
Los Angeles, CA 90089-1191
USA

J.P. Bonnet
C.E.A.T.
43, rue de l'areodrome
F- 86036 Poitiers Cedex
France

An. A. Borisov
Inst Thermophysics
USSR Acad. Sc.
Novosibirsk
UDSSR

Kwing-So Choi
British Maritime Technology
1 Waldegrave Road, Orlando House
Teddington, Middlesex TW11 8LZ
England

David G. Clark
Queen Mary College/Univ. London
Mile End Road
London, E1 4NS
England

Eric Coustols
Dep. d'Etudes en Aerothermodyn.
2 Avenue Edouard Belin
F - 31055 Toulouse
France

Nicholas Daish
Topexpress Limited
Poseidon House/Castle Park
Cambridge CB3 0RD
England

Antonio Delgado Rodriguez
ZARN Univ. Bremen, FB4
Postfach 330440
D-2800 Bremen 32
Germany

J. Deville
C.E.A.T.
43, rue de l'Aerodrome
F- 86036 Poitiers Cedex
France

J. Dickinson
Dep. of Mechanical Eng.
Univ. Laval, Cite Universitaire
Quebec, G1K 7P4
Canada

Lyazid Djenidi
Unite Mixte Univ. I M S T
12 Av. du Gen. Leclerc
F- 13003 Marseille
France

Themistocles Dracos
IHW
ETHZ-Hoenggerberg
CH-8093 Zuerich
Switzerland

Regis Dumas
Univ. d'Aix-Marseille II/IMST
12 Av. du Gen. Leclerc
F- 13003 Marseille
France

Arthur Dyment
ONERA- IMFL
5 bd Paul Painleve
F- 59000 Lille
France

Chaim Elata
Ben Gurion University of the Negev
P.O.Box 653
Beer-Sheva 84105
Israel

Marcel Escudier
University of Liverpool/ Mech. Eng. Dep.
P.O. Box 147
Liverpool L69 3BX
England

Bob E. Falco
Dept. of Mechanical Engineering
Michigan State University
East Lansing, MI 48824
USA

Pandelis Fanourakis
Univ. of Cambridge/Dept. of Eng.
Trumpington Street
Cambridge CB2 1PZ
England

B.E. Forestier
Inst de Mechan. des Fluids
1, Rue Honnorat
F- 13003 Marseille
France

Rainer Friedrich
TU Muenchen, Lehrst.f.Strm.
D-8000 Muenchen 2
Postfach 20 24 20
Germany

Daniel H. Fruman
ENSTA Centre de l'Yvette
Chemin de la Huniere
F- 91120 Palaiseau
France

Mohamed Gad-el-Hak
Dept. of Aerospace & Mechanical Eng.
University of Notre Dame
Notre Dame, IN 46556
USA

Bernhard Gampert
Universitaet Essen GH
FB 12 Stroemungsmechanik
D - 4300 Essen 1
Germany

Spyros Gavrilakis
Inst. M.H & Mec. des Fluides
EPFL, ME-Ecublens
CH-1015 Lausanne
Switzerland

Hakan Gustavsson
Lulea University of Technology
Div. Fluid Mechanics
S-95187 Lulea
Schweden

Albert P. Gyr
Inst. of Hydromechanics
ETHZ
CH-8093 Zürich-Hoenggerberg
Switzerland

Thomas J. Hanratty
Dept.Chem.Eng. Univ.of Illinois
1209 W.California,Box C-3,113 RAL
Urbana, Illinois 61801
USA

Tomiichi Hasegawa
Fac. Eng./ Niigata University
Ikarashi-2
950-21 Niigata City
Japan

Reinhard Hess
Inst.f.Fluiddynamics
ETHZ
CH-8092 Zürich
Switzerland

Sebastian Hirschberg
Inst. f. Fluiddynamik
ETHZ
CH-8092 Zürich
Switzerland

Chih-Ming Ho
University of South. California
Dept. Aerospace Engineering
Los Angeles, CA 90089-0192
USA

Hui-Chang Hou
Guangdong Res.Inst.of Hydraulic Eng.
Sougoulin, Saho
Guangzhou 510610
CHINA

Jack W. Hoyt
Dept of Mechanical Engineering
San Diego State University
San Diego, Ca. 92182-0191
USA

Osamu Inoue
Tohoku Univ.7/Inst. of Fluid Sc.
2-1-1 Katahira
Aoba-ku, Sendai 980
Japan

David F. James
Dept.Mech.Eng. Univ.of Toronto
5 King's College Road
Toronto, Ontario M5S 1A4
Canada

John Kim
Mail Stop 202A-1
NASA-Ames Research Center
Moffett Field, CA 94035
USA

Harald Klein
TU Muenchen/Lehrst.f.Stroemm.
Arcisstrasse 21
D- 8000 München 2
Germany

Stephen Kline
Mechanical Engineering
Stanford University
Stanford, CA 94305
USA

Köster
Univ. Dortmund/ Chemietechnik
Postfach500500
D-4600 Dortmund 60
Germany

Michat Krol
Universitaet Kaiserslautern
Postfach 3049/Fb Masch.w.
D- 6750 Kaiserslautern
Germany

Marten Landahl
Dept. of Aeronautics & Astronautics
M.I.T. Room 37-451
Cambridge, Mass. 02139
USA

Shaoping Li
UMIST
PO Box 88
Manchester,M60 1QD
England

Michael List
Universitaet Essen
Postfach 103 764
D- 4300 Essen 1
Germany

Harrison T. Loeser
Consultant
1 Connshire Drive
Waterford, Connect. 06385
USA

Earl Logan
Arizona State University
Dept. Mech. & Aerosp. Eng. 102718
Tempe, Arizona 85287
USA

T.B. Lynn
Hermann-Foettinger Inst.
TU Berlin
D-1000 Berlin 12
Germany

Mehrdad Mansour
Univ.College London/Civ.Eng.
37a Penwith Rd.
EArlsfield London SW18-4PU
England

T. Maxworthy
University of Southern California
Dept. Mech. & Aerosp. Eng.
Los Angeles, CA 90089-1453
USA

Jose Meyer
Inst.of Aerospace Engineering
TECHNION
32000 Haifa
Israel

Richard Meyer
Haldenstrasse 11
8967 Widen
Switzerland

B.P. Mironov
Inst. of Thermophysics
Prospekt Akademika, Lavrentzeva 1
630090 Novosibirsk
UDSSR

Hiroshi Mizunuma
Dept.Mech. Eng./Fac. Techn.
Tokyo Metropolitan University
2-1-1,Fukazawa,Setagaya-ku,Tokyo 158
Japan

Tatsuo Motohashi
Dep.Aerosp.Eng./Nihon Univ.
7-24-1 Narashinodai
Funabashi, Chiba
Japan 274

Andreas Mueller
Inst. for Hydromechanics
ETHZ-Hoenggerberg
CH-8093 Zuerich
Switzerland

Hirosuke Munakata
1-408 Marukodori,Nakahara-Ku
Kawasaki-Shi
Kanagawa, 211
Japan

Richard H. Nadolink
Dep. of the Navy/Naval Underwater Sy.Cen
Newport Lab/Code 821
Newport, RI 02841-5047
USA

Hassan M. Nagib
Illinois Inst. of Technology
IIT Center/ Armour Coll. of Engin.
Chicago, IL 60616
USA

Shin-Ichi Nakao
M.I.T.I.
1-1-4,Umezono
Tukuba-City,305
Japan

Dieter Neumann
Max-Planck Inst. f. Stroem.forsch.
Bunsenstrasse 10
D - 3400 Goettingen
Germany

Vinh Duy Nguyen
Dep. de Genie Mecanique
Universite Laval/Cite universitaire
Quebec G1K 7P4
Canada

D.E. Nikitopoulos
Louisiana State University

Mech. Eng. Dept.
Baton Rouge, LA 70803-6413
USA

Stavros Nychas
Univ. of Thessaloniki
Univ. Box 453
GR- 54006 Thessaloniki
Greece

Volker Oles
Universitaet Dortmund FachB CT/EPT
Postfach 500500
D-4600 Dortmund
Germany

Dimitris A. Papantoniou
Inst. for Hydromechanics
ETHZ
CH-8093 Zuerich-Hoenggerberg
Switzerland

Andrew Pollard
Queen's University
Dept. Mech. Engineering
Kingston, Ontario K7L 3N6
Canada

Lucio Pompeo
Inst. f. Fluiddynamik
ETHZ
CH-8092 Zuerich
Switzerland

Krishna K. Prasad
Eindhoven Tech. Univ. /Phys Dep.
P.O. Box 513
5600 MB Eindhoven
The Netherlands

Philippe Pulvin
Inst. de Ma.Hydr. et Mec.des Fluides
EPFL
CH-1015 ME-Ecublens/Lausanne
Switzerland

Laurel W. Reidy
1153 Balour Dr.
Encinitas, CA 92024
USA

Michael M. Reischmann
U.S.Office of Naval Research
800 N. Quincy Street (Code 1132)
Arlington, VA 22217-5000
USA

Oliver Riccius
ABB- Forschungszentrum
CRB.P2
CH- 5405 Baden- Daettwil
Switzerland

Stephen K. Robinson
NASA Ames Res. Cent.
Moffet Field, CA 94035
USA

Jim Rohr
Naval Ocean System Center
San Diego, CAL 92152-500
USA

Jason B. Roon
Univ. Southern California
Dept. Aerospace Engineering
Los Angeles, CA 90089-0192
USA

Inge L. Ryhming
Inst. Mach. Hydr./Mec. des Fluides
EPFL
CH- 1015 ME-Ecublens Lausanne
Switzerland

Mark A. Savill
University of Cambridge/Dept. of Eng.
Trumpington Street
Cambridge CB2 1PZ
England

Adriana D. Schwarz-van Manen
Eindhoven Tech. Univ./Phys. Dept
P.O. Box 513
NL- 5600 MB Eindhoven
The Netherlands

Charles R. Smith
Dept. of Mechanical Engineering
Lehigh Univ./Building #19
Bethlehem, PA 18015
USA

Karl Strauss
Universitaet Dortmund/FB Chem.Tech
Postfach 500500
D-4600 Dortmund
Germany

Itiro Tani
National Aerospace Laboratory
1880 Jindaij,Chofu
Tokyo
Japan 182

Hans Thomann
Inst. f. Fluiddynamik
ETHZ-Zentrum
CH-8092 Zuerich
Switzerland

William G. Tiederman
School of Mechanical Eng.
Purdue University
West Lafayette, IN 47907
USA

Trong Vien Truong
Swiss Federal Inst.of Tech. EPFL
Inst Mach. Hydr. & Mech. d. Fluides
CH- 1015 Lausanne- ME-Ecublens
Switzerland

Arkady Tsinober
Tel-Aviv University
Dept. of Fluid Mech. & Heat Tran.,Eng.
Tel- Aviv 69978
Israel

Hiromoto Usui
Yamaguchi University/Dept. Chem. Eng.
2557 Tokiwadai
Ube, Yamaguchi-ken 755
Japan

Venkatesa I. Vasanta Ram
Ruhr-Univ. Bochum (ITh&M)
Postfach 102148
D- 4630 Bochum 1
Germany

P.S. Virk
Mass. Inst. of Technology
Dep. of Chemical Engineering
Cambridge, Mass. 02139
USA

Claudia Vossmerbaeumer
Univ. Duisburg/FB Maschinenbau
Lotharstrasse 1
D- 4100 Duisburg 1
Germany

D.L. Wagger
MIT/ Dep Chem. Eng.
BLDG 66
Cambridge, MA 02139
USA

J.D.A. Walker
Dept. of Mech.Eng.& Mech.
354 Packard Lab#19,Lehigh Univ.
Bethlehem, PA 18015
USA

Cadance Wark
Illinois Inst. of Technology
IIT Center/Fluid Dy. Res. Cent.
Chicago, Ill. 60616
USA

Marvin Weiss
Nova Husky Res. Corp.
2928 16th Street N.E.
Calgary, ALB. T2E 7K7
CANADA

Kenneth C. Wilson
Dept. Civil Engineering
Queen's University
Kingston,Ont.,K7L 3N6
Canada

Thomas Winkler
Universitaet Essen
Postfach 103 764
D- 4300 Essen 1
Germany

I.J. Wygnanski
Dept. Aerospace & Mech. Eng.
University of Arizona
Tucson, AZ 85721
USA

I.J. Wygnanski
Tel-Aviv University
School of Engineering
Ramat- Aviv, Tel- Aviv 69978
Israel

Jacques L. Zakin
Ohio State Univ., Dept. Chemical Eng.
121 Koffolt Lab., 140 West 19th Ave,
Columbus, Ohio 43210-1180
USA

Final Program

SECOND IUTAM SYMPOSIUM ON STRUCTURE OF TURBULENCE AND DRAG REDUCTION

FEDERAL INSTITUTE OF TECHNOLOGY, (ETHZ) ZUERICH SWITZERLAND 25th to 28th JULY 1989

Tuesday July 25, 1989

Opening Address

STRUCTURE OF TURBULENCE

Session 1 (Chairman: R.F. Blackwelder)

S.J. Kline & S.K. Robinson
(Stanford University, NASA-Ames Res. Cent., USA)
Review: given in two parts.

S.J. Kline: Turbulent boundary layer structures: Progress, status, and challenges.

S.K. Robinson: A review of vortex structures and associated coherent motions in turbulent boundary layers.

C.R. Smith, J.D.A. Walker, A.H. Haidari & B.K. Taylor (Lehigh Univ., USA). Hairpin vortices in turbulent boundary layers: The implications for reducing surface drag.

M. Gad-el-Hak (Univ. Notre Dame, USA) Flow control by suction.

D.E. Nikitopoulos (LSU Baton Rouge,USA) & J.T.C. Liu (Brown University, USA) Non-linear coherent mode interactions and the control of shear layers.

Session 2 (Chairman: T. Maxworthy)

J.D.A. Walker (Lehigh Univ., USA) Modes of wall-layer eruptions in turbulent flows.

R.E. Falco, J.C. Klewicki & K. Pan (MSU East Lansing, USA)
Production of turbulence in boundary layers and potentials for modification of the near wall region.

R.A. Antonia (Univ. Newcastle, Australia), R. Dumas & L. Fulachier (Univ. d'Aix-Marseille, France) Visualisation of the organized motion in a turbulent boundary layer.

T.R. Anderson (Dept Navy NUWSC, USA) Wavenumber-frequency spectral densities of turbulent wall pressure and wall shear fluctuations.

Session 3 (Chairmans: I.J. Wygnanski & T. Dracos)

Poster A discussion

(1) R.S. Azad & B. Doell (Univ. of Manitoba, Canada) Behaviour of separation bubble with different roughnes elements at the leading edge of a flat plate.

(2) J. Meyer, A. Sevrain, H.C. Boisson & H. Ha. Minh (IMFT Toulouse, France) Organized structures and transition in the near field of a plane jet.

(3) O. Inoue (Tohoku Univ., Japan) Artificial control of turbulent mixing layers.

(4) A. Müller (ETHZ, Switzerland) Quadrant analysis and instantaneous momentum transport - a critical review.

(5) H.T. Loesser (USA) Analysis of experimental data on boundary layer pressure fluctuations in turbulent pipe flow.

1. Panel Discussion "Coherent structures in turbulence"
Conductors: C.M. Ho, R.E. Falco & C.R. Smith

Wednesday July 26, 1989

DRAG REDUCTION BY DILUTE POLYMER SOLUTIONS & INHOMOGENEOUS INJECTION OF CONCENTRATED SOLUTIONS.

Session 4 (Chairman: M.M. Reischman)

L.G. Leal (CALTEC, USA) The review was given by S. Banerjee UCSB, USA)
Review: The rheological behavior of dilute polymer solutions

W.G. Tiederman (Purdue Univ., USA)
Review: The effect of dilute polymer solutions on the viscous drag and turbulence structure.

P.S. Virk (MIT, USA) Aspects of mechanisms in type B drag reduction.

A. Gyr (ETHZ, Switzerland) & H.-W. Bewersdorff (Univ. Dortmund,F.R.G.) Changes of structures close to the wall of a turbulent flow in drag reducing fluids.

B. Gampert & C.K.Yong (Univ. of Essen, FRG) Influence of polymer additives on the coherent structures of turbulent channel flow.

R.H. Nandolink (NUWSC, Newport,USA) Interaction of molecules and turbulent flow in dilute polystyrene solutions.

Session 5 (Chairman: P.S. Virk)

H. Usui (Yamaguchi Univ., Japan)
Review: Drag reduction caused by the injection of a polymer solution into a pipe flow.

N.S. Berman (ASU Tempe, USA) Large eddies and polymer strings.

V.N. Mamonov, B.P. Mironov & S.V. Panov (USSR Acad. Sc., Novosibirsk,USSR)
Drag reduction at injection of polyethyleneoxide solution into turbulent boundary layer through perforated section or a slot.

Session 6 (Chairmans: C. Elata & R.H. Nadolink)

Poster B discussion

(1) H. Mizunuma, H. Kato & T. Kurita (Tokyo Metrop. University, Japan)
Λ - shaped vortices in dilute polymer solutions.

(2) AN.A. Borisov, B.P. Mironov, B.G. Novokov & V.D. Fedosenko
(ITP USSR Akad. Sc., Novosibirsk, USSR) Wake flows in dilute polymer solutions.

(3) H.C. Hou (Guangdong Res. Inst., Guangzhou, China)
Hydrodynamic stability of laminar sublayer and drag reduction.

2. Panel Discussion "Drag reduction by dilute and heterogeneous polymer solutions"
Conductors: B. Gampert, W.G. Tiederman & M.M. Reischman

Thursday July 27, 1989

DRAG REDUCTION BY OTHER MEANS

Session 7 (Chairman: D.H. Fruman)

H.-W. Bewersdorff (University of Dortmund, FRG)
Review: Drag reduction in surfactant solutions.

K. Abe, A. Matsumoto, H. Munakata (Nihon Univ.) & I. Tani (NAL, Japan)
Drag reduction by sand grain roughness.

M. Krol (Univ. Kaiserslautern, FRG) Inertial interaction of spherical and fibre like solid particles with turbulent flowing liquids.

H. Gustavsson (Lulea Univ. of Techn., Sweden) The effect of three-dimensional surface elements on boundary layer flow.

Session 8 (Chairman: S.K. Robinson)

A. Tsinober (Tel-Aviv University, Israel)
Review: Turbulent drag reduction versus structure of turbulence.

COMPUTATION AND STABILITY

M. Landahl (MIT, Cambridge Mass., USA)
Review: Hydrodynamic instability and coherent structure in turbulence.

S.L. Lyons, T.J. Hanratty (Uni. ILL Urbana) & J.B. McLaughlin (Clarkson Uni. NY.) Relation of turbulence production to eddy structure in wall turbulence

J. Kim & P. Moin (NASA Ames Res.Cent., Moffet Field California,USA) Active turbulence control in a wall-bounded flow using direct numerical simulations.

Session 10 (Chairmans: N.S. Berman & R.H. Nandolink)

Poster C discussion

(1) C. Vossmerb%umer & G. Schweiger (Univ. Duisburg, FRG)
Determination of regular structures in turbulent mixing processes.

(2) F. Anselmet*, R.A. Antonia (Univ. Newcastle, NSW, Australia), T. Benabid* & L. Fulachier* (*Univ. Aix-Marseille, France)
Effect of wall suction on the transport of a scalar by coherent structures in a turbulent boundary layer.

(3) J. Deville, J.P. Bonnet (CEAT/CNRS, Poitiers, France) & J. Lemay (Univ. Laval, Quebec, Canada)
Analysis of the wake of an outer layer manipulator.

3. Panel Discussion "Drag reduction by surfactants implication of stability, computation of turbulence making use of the concept of coherent structures"
Conductors: N.S. Berman, J. Kim & J.L. Zakin

Friday July 28, 1989

DRAG REDUCTION BY PASSIVE MEANS

Session 11 (Chairman: I.L. Ryhming)

A.M. Savill (University of Cambridge, England)
Review: Drag reduction by passive devices - a review of some recent developments.

C.E. Wark & H.M. Nagib (IIT, Chicago, USA) Relation between outer structures and wall-layer events in boundary layers with and without manipulation.

J.B. Anders (NASA Langley Res. Cent., Hampton, VA, USA) Boundary layer manipulators at high Reynolds numbers.

Session 12 (Chairman: M. Gad-el-Hak)

R. Friedrich & H. Klein (TUM, M¸nchen, FRG) Large scale turbulence structures in a manipulated channel flow.

R.F. Blackwelder (USC, Los Angeles, USA) The effects of longitudinal roughness elements and local suction upon the turbulent boundary layer.

D. Bechert (DFVLR, Berlin, FRG) Turbulent drag reduction by nonplanar surfaces. A survey on the research at TU/DLR Berlin.

Session 13 (Chairmans: R. Friedrich & R. Dumas)

Poster D discussion

LEBU's

(1) S.P. Bardakhanov, V.V. Kozlov & V.V. Larichkin (USSR Acad. Sc.,Novosibirsk, USSR)
Influence of an accoustic field on the flow structure behind a LEBU in a turbulent boundary layer.

(2) M. Stanislas & M.C. Hoyez (IMFL-ONERA, Lille, France)
Analysis of the structure of a turbulent boundary layer, with and without a "LEBU", using light sheet smoke visualizations and hot wire measurements.

RIBLETS.

(3) K.N. Liu, C. Christodoulou, O. Riccius & D.D. Joseph (UM, Minneapolis MN, USA)
Drag reduction in pipes lined with riblets.

(4) E. Coustols & J. Cousteix (ONERA/CERT, Toulouse, France)
Experimental investigation of turbulent boundary layers manipulated with internal devices: "riblets".

(5) L. Djenidi, J. Liandrat, F. Anselmet & L. Fulachier (UM Univ., Marseille, France)
A possible explanation for drag reduction over riblets in turbulent boundary layer.

(6) K.-S. Choi (BMT, Teddington, England)
Drag reduction mechanisms and near-wall turbulence structure with riblets.

(7) A.D. Schwarz- van Manen, J.H.H. Thijssen, C. Nnieuwvelt & K.K. Prasad (Eidhoven Univ. of Techn., Netherland)
The bursting process over drag reducing grooved surfaces.

(8) Ph. Pulvin & T.V. Truong (EPFL, Lausanne, Switzerland)
Riblets in internal flows with adverse pressure gradients.

4. Panel Discussion "Drag reduction by passive means"
Conductors: D. Bechert, H.M. Nagib & A.M. Savill

Summary (Chaired by the Scientific Committee) & Closing session.

Contents

1 Structure of Turbulence

Reviews

3 Drag Reduction by other means

Reviews

Contributions to drag reduction by other means

4 Stability and Computations

Review

Riblets

Part 1
Structure of Turbulence

Turbulent Boundary Layer Structure: Progress, Status, and Challenges

S. J. Kline
Department of Mechanical Engineering
Stanford University, CA 94305-3030, USA

S. K. Robinson
NASA-Ames Research Center
Moffett Field, CA USA 94035

ABSTRACT

The current community-wide cooperative study reviewing structure in the turbulent boundary layer has allowed construction of a list of the various forms of structure observed in laboratory experiments and provided a number of detailed features of each structure. The study also has revealed a lack of consensus on three matters: spatial relationships among the forms of structure; temporal relations in creation, evolution and decay of structures; a complete model of the important structure(s).

An ordering of the various known structures in terms of relative importance to the creation of turbulent Reynolds stresses and in the production and dissipation of turbulent kinetic energy is tentatively suggested. This ordering suggests a central role of vortices in boundary layer turbulence structure. Recapitulation of some essential concepts needed to understand the relationships of vortices to other structures is therefore given. The validity of a central role of vortices is then tested by examining two questions: (i) Are the vortices proximate in space to other known structures and features of importance in the flow? (ii) Are the vortices of sufficient strength so that they could induce other observed features and do they have proper orientation and direction of rotation to do so?

Since direct Navier-Stokes simulations allow inspection of a sample of all vortices throughout a volume of a simulated turbulent layer, it becomes possible to examine individual realizations of vortices as well as various types of statistics concerning the forms of vortices observed. The sample of individual realizations reveals a wide variety of vortex element orientation and also of shapes of more complex vortical structures.

Some features for which we still need further information are noted, and steps which may aid in moving toward increased agreement on the overall picture of structure and its use in forming useful predictive models in wall-bounded turbulent shear flows are suggested.

SPECIAL NOMENCLATURE AND DEFINITIONS

DNS = Direct Navier Stokes: Used in connection with numerical simulations; assumed to have sufficient numerical accuracy, but may be open to questions about sufficiency of grid resolution.

Ejection = Velocity perturbation with positive v' and negative u'.

Foot = The part of a leg in a vortical structure farthest upstream and nearest the wall.

Head = A transverse vortex element either alone or as part of a vortical structure

A. Gyr (Editor)
Structure of Turbulence and Drag Reduction
IUTAM Symposium Zurich/Switzerland 1989

Inrush = A special class of sweep with high negative v′, and which emanates from the outer portion of the layer and moves to, or very near, the wall.
Leg = A quasi-streamwise vortex either alone or as part of a vortical structure.
Neck = Portion of a vortical structure turning from transverse to quasi-streamwise.
Sweep = A motion with positive u′ and negative v′.
TKE = Turbulent Kinetic Energy
TRS = Turbulent Reynolds Stress, $\rho\, \overline{u'v'}$
Vortex, vortices, vorticity lines: see section titled, "Vorticity Lines, Vortices and Their Observation".
Vortex element, vortex structure, vortical structure: See section titled "Vortex Elements and Vortical Structures."

INTRODUCTION

If we take the term "Structure" to denote "those motions (or events, or eddies) which convert mean motions into Turbulent Reynolds Stresses and/or produce and dissipate TKE, then the structure is, by definition, the center of the physics of turbulence. If we accept this idea, we do not need further justification for the study of structure provided only we are willing to follow the guide of the history of science. That guide, if it tells us anything with high assurance, tells us that understanding the physics cannot hurt, and, in fact, nearly always aids us in the long run in predicting and controlling physical processes.

Statistical representations fill the important needs of providing *quantitative* information about turbulent flows and providing numbers usable in Reynolds-averaged equations. However, there has been considerable difficulty in relating the quantitative statistics to the underlying structures as observed in visual studies. Part of the reason for these difficulties is that statistics of the fluctuations may or may not provide information equivalent to the full details of the structure, depending on the "completeness" of the statistics given. Moreover, it is not a priori obvious what statistics are needed to be complete in this sense. Thus one of the things we can learn from study of structure is a better understanding of what "completeness" in this sense requires in the statistics which need to be measured in order to determine structure using methods like that of Aubrey et al (1988), for example.

Since a primary purpose of using statistical representations is to capture enough of the structure to represent the physics well and at the same time provide a description simple enough for use in practical predictive models, a second thing we can learn from an accurate picture of the structure is what conditions to put on the statistics so that we can use procedures such as that of Adrian (1988) to learn more about structure quantitatively in efficient ways.

Beyond these gains, one hopes that agreement within the research community of what constitutes the structure will prove of direct aid in both modeling and control of turbulent flows. This is not to suggest that using structure information to guide either modeling or control will be an easy task, for both are likely to be difficult. It is to suggest that lack of an agreed complete picture of the structure in the research community has almost certainly hampered use of information about structure in both model formation and control of turbulent flows.

THE COOPERATIVE, COMMUNITY-WIDE REVIEW: SOME GENERAL RESULTS

When the cooperative review project began in late 1986, there was little agreement on the appropriate model of the structure features in the boundary layer. There also appeared to be some lack of knowledge of the results of other laboratories within each particular research group. The central purpose of the cooperative study was therefore to create a data base of "truth assertions" which would do two things. First, distinguish results that were founded on reliable data on the one hand from results which involve speculative extensions which pass well beyond the data on the other. Second, from this data-base we had hoped to distill a complete, or nearly complete, model of structure on which at least most of the research community could agree. As part of this process, we also have carried out one or two day discussions with many of the leading researchers on structure in wall-bound shear layers.

As we had anticipated, some groups suggested that hairpin-like (or other forms) of vortical structures explain what occurs; other groups emphasized the effects of sharp shear layers or 'fronts'. Still other groups believed the structure near the wall was largely the result of the passage of outer large-scale-motions (also called 'bulges') and what we now call 'backs' which are narrow regions of high vorticity observed on the upstream face of the large-scale-motions*. Other groups had additional differing ideas concerning a complete model of structure. There is no intent in this paper to review all these model ideas. We are, however, collecting the model ideas for discussion in a later community-wide meeting. Nor is an exhaustive list of references on boundary layer structure given in this paper since that now runs to several hundred entries. The intent of this section of the paper is rather to make two points as clear as possible: (i) there is good agreement in the research community on the existence of a number of structure features in the boundary layer and many of the details about each feature; (ii) there is a distinct lack of consensus on three matters: (a) the spatial relations of the various structures; (b) the time evolution of various structure features; (c) a complete picture of the important structure(s). Both items (i) and (ii) need comment.

Based on the cooperative study, a list of eight of the major types of structure was given by Kline and Robinson (1988). As the 1988 paper shows, quite a lot is known about each of these structures; all of them are at least reasonably well documented by experiments. In a section below some improvements to list of the eight types of structure are given. An ordering of the eight types of structure with regard to which are probably more important in creation of TRS and creation and dissipation of TKE is also provided since a mere listing of all of these structures provides an undifferentiated eclecticism which does little to aid our understanding. This ordering is tentative and will hopefully be subjected to intensive discussion.

---

* The 'backs' have been called 'fronts' in much of the literature. However, the word 'front' had also been widely used by various groups to describe the far more numerous sharp regions of high vorticity observed below $y+ = 100$ or so, which we now call 'near-wall shear-layers'.

Despite a long history of excellent contributions by many members of the research community, the cooperative study showed a good many questions remained open concerning spatial relationships among the eight kinds of structures. The review also showed relatively little was agreed upon about the temporal order in which the individual structures appear, evolve, and disappear. If, as seems to be the case, a number of the structures act sequentially in the central dynamics of the production and dissipation of TRS and TKE, then information about them one-by-one, or even in pairs, lacking information on how they behave over time is not sufficient for formation of a complete model.

Given these residual uncertainties and lack of knowledge about the spatio-temporal relationships, it is not surprising that various researchers and research groups filled in details of a complete model of structure in different ways. Since a complete model is the central goal of structure research, and is needed even to guide strategy on further structure research, it seemed appropriate to seek the missing spatial and temporal relationships via study of the DNS generated data bases available at NASA-Ames. These simulated turbulent flows are the work of several people, notably Parviz Moin, John Kim, Philippe Spalart, Bob Moser and co-workers. A great deal has already been contributed to the goal of understanding structure via study of the simulations by a number of workers. Here only some results which are particularly important for the current paper are summarized; a more complete review of these results appears in Robinson (1989b).

Particularly notable is the pioneering work of Moin and Kim using both LES and DNS simulations primarily of channel flows (Kim and Moin 1979; Moin and Kim 1982; Kim 1983; Moin 1984; Moin and Kim 1985; Kim and Moin 1986; Moin 1987; Kim, Moin and Moser 1987). Some particularly important results of these papers include not only confirmation of many structure features which had been measured in the laboratory but also improved understanding of the orientation of vortices. Agglomerated vorticity lines were often seen with tilted or "quasi-streamwise" orientation; the tilt angle gradually increased along the vortex element with increasing $y+$ up to 45°; however truly streamwise agglomerations of vorticity lines were very rarely observed even near the wall. Another particularly important result is given by Moin (1987); Moin reported regions of high $u'v'$ were associated with every observed quasi-streamwise vortex in the wall region, and concluded that relatively short, single vortices are fundamental structures associated with regions of high turbulence production. The studies also confirmed that the wall-proximate streaky structure had the well-known mean-transverse-scale of 100 wall units; however, the low-speed streaks were observed to be far longer than the quasi-streamwise vortices. Moin (1984) also confirmed that a large fraction of TKE was captured in a first eigenfunction similar to that found by Bakewell and Lumley (1967). Kim and Moin (1985, 1986) also showed that agglomerations of vorticity lines often appeared as "hairpin-like" shapes. These vortex-line bundles were interpreted as true vortices; and the shapes observed included both downstream-leaning upright horseshoe like shapes associated with high $u'v'2$ and upstream-leaning inverted horseshoe-like shapes associated with high $u'v'4$. Moin and Kim also showed that one-sided shapes with a head and one leg were more common in the agglomerated regions of compact vortex-lines than symmetrical horseshoe shapes in which a head had two attached legs.

Spalart (1988) observed delta-scale bulges in the outer part of the boundary layer with sharp turbulent/non-turbulent interfaces on the contorted outer face of the bulges. Spalart also reported interspaced deep valleys of irrotational flow between bulges. These features had been reported by many others in earlier laboratory measurements, notably Kovasznay et al (1970).

In 1987 the present authors began to investigate some of the remaining open questions focused by the cooperative study. The DNS simulation of the flat plate at Re_θ = 670 by Philippe Spalart (1988) was chosen as a first flow for investigation. Emphasis was placed on study of the spatial and temporal relationships among the various known forms of structure. In this work, vortices were made visible by inspecting iso-contours of low static pressure in the flow field utilizing the idea that the Euler n-equation specifies the existence of low pressure cores in true vortices, and thus distinguishes true vortices from regions of distributed vorticity which do not contain vortices. This has led to a significant amount of information about the spatial relationships among the structures reported in Robinson et al 1988a, 1988b. It has also led to increased knowledge about the shape of vortex elements and vortical structures and about their orientations in the flow. Work over the past year has revealed relationships between the vortices and dissipation of TKE which is reported in Robinson (1989b). These results taken together with the vast amount of earlier results suggest a central role of true vortices in boundary layer structure. This suggestion is investigated more fully below.

SOME ESSENTIAL CONCEPTS

As part of the work of the cooperative study, the project coordinators constructed a nomenclature list and circulated it for comment to interested members of the research community. After two rounds of comments and revisions, this list is reasonably complete, and seems agreed upon by at least most members of the research community. Construction of this list seemed necessary since various terms had been used in conflicting and overlapping ways in the literature. As a result there was considerable misunderstanding in the community from miscommunications. One hopes these communication difficulties are reduced for the structure problem by the nomenclature list just mentioned. This list will be supplied by the authors on request to any interested worker.

However, there is a second level of the use of words, namely as concepts, that also needs clarification. In particular, with regard to the problem of boundary layer structure, observations in both the laboratory and in the DNS simulated flows have employed a variety of concepts. The relationships among these concepts is readily misinterpreted in some cases. In order to think clearly about the structure results, we therefore need to be as clear as we can about the concepts involved. Hence a discussion of three concepts is given which addresses this task, in part. Patience on the part of the reader already entirely familiar with this conceptual territory is solicited.

The three sets of concepts are:

- Vorticity lines/Vortices
- Vortex elements/Vortical structures
- Average model/Realization

VORTICITY LINES, VORTICES, AND THEIR OBSERVATION

The phrase "vorticity line" is used to denote what most textbooks call a "vortex line", that is, *a line everywhere parallel to the direction of the vorticity vector*. We use the phrase "vorticity line" (or "line of vorticity") rather than the more conventional phrase "vortex line" in order to emphasize the distinction between vorticity lines and what we will call a vortex or vortices.

Vorticity lines move at particle speed in the absence of viscosity, as we know from Helmholtz' theorem. Even in viscous flows, we can assume with good accuracy that *vorticity lines move at local particle speed* for short distances in the flow.

There seems to be no agreed definition of what constitutes a vortex. In this paper, the word "vortex" will denote a structure for which an *observer moving with the structure sees* a circular (or near circular) pattern of the direction of the velocity vectors (or instantaneous streamlines) in the plane normal to the center (core) of the structure. Such a pattern of vectors represents either closed (near) circular streamlines or a spiralling motion.

Several points about vortices need noting. First, all vortices contain lines of vorticity, but not all bundles of vorticity lines constitute vortices. There are two cases which need to be considered.

The first and simpler case is vortices normal to the direction of flow, in either the transverse or wall-normal direction. Consider for example a planar, laminar Couette flow with x the flow direction, y the wall-normal, and z transverse. In this flow there are z-oriented (transverse) vorticity lines everywhere. However, until instabilities arise, there are no transverse vortices whatsoever since dv/dx nowhere has a value the same order of magnitude as du/dy, and hence the necessary circular pattern of vector directions cannot be present in an xy-plane regardless of the frame of reference. Similar remarks apply to wall-normal vorticity lines and vortices. So long as the vorticity lines or the vortex is normal to the flow direction there is not much likelihood of confusion between vortices and lines of vorticity.

The second case concerns streamwise vorticity and vortices. In this case, there is considerably more likelihood of confusing lines of lines of vorticity with true vortices because a strong streamwise vortex will induce motions which create loops (or hairpin-like shapes) in the vorticity lines on each side of the vortex, but outside the region of circular motions. If this occurs in a flow with a mean strain the loops of vorticity lines will become tilted as time proceeds. The kinematics are illustrated in Robinson (1989a). Moreover, the marker particles will make visible true vortices only if the markers are introduced at points lying inside the closed streamlines of the vortex unless very rapid vicious diffusion occurs, and we normally try to arrange markers so that viscous diffusion is not rapid. These remarks also apply to any vortex with a significant component of streamwise orientation of the vortex axis.

These same ideas apply to the speed of vortices. *Vortices can move at any speed in the direction of the vorticity lines.* Since the making (and unmaking) of a vortex comes from crowding together (or spreading) of the vorticity lines *normal* to the direction of these same vorticity lines, the process can go faster or slower than local particle speed along the direction of the vorticity lines. However, the speed of motion of the vorticity lines in the direction normal to the vorticity lines must be at approximately particle speed locally by virtue of Helmholtz' theorems. As a result of this difference, nearly streamwise vortices (legs) can behave quite differently from essentially transverse vortices (heads), and in fact appear to do so when observed in the DNS simulated flat plate flow.

Finally, vortices isolate regions of low pressure in their core from the surrounding flow; this is necessary by virtue of the Euler n-equation in inviscid, or nearly inviscid, motion. Lines of vorticity, which are not vortices, do not have this property. Observations of the flat plate flow via DNS show z-oriented vortices have a surprisingly long persistence, not infrequently thousands of wall units. Given these facts, and the preceding paragraph, it can be seen that it would be easy to interpret the passage of successive z-oriented vortices past a probe as a wave motion rather than a quasi-coherent structure moving in space but with considerable persistence over time. It is equally clear that output from a probe stationary in lab coordinates could suggest that the important structural events are more intermittent than is actually the case since the persistence observed is over time, and is therefore not observable except in an appropriate convected frame, a frame which is not known a priori.

VORTICAL ELEMENTS AND VORTICAL STRUCTURES

For reasons that will become more evident in the descriptions which follow, we make a distinction between vortex elements and vortical structures (Both are called vortices in the definitions and in the preceding sections).

The phrase "vortex element" will denote *a vortex with a single (or very nearly single) orientation in space.* Thus a streamwise vortex will be called a vortex element. So will a transverse vortex or a quasi-streamwise vortex which is tilted and has either a constant or slowly varying inclination to the wall.

The phrase "vortical structure" will denote *a linked set of vortex elements of several orientations possessing a single connected core threading through all the vortex elements of the particular structure.* For example, a head element connected to one or two necks, or a head connected to a neck and then to a leg will be called vortical structures.

This distinction is important not only because both solitary vortex elements and vortical structures are frequently observed in the boundary layer, but also because we have only a few studies that show complete vortical structures in the flow rather than vortex elements, notably those reported by Clark and Markland (1971), Head and Bandyopadhyay (1981); and C. R. Smith and his co-workers (e.g. Acarlar and Smith, 1987). Each of these experimental studies is limited in some ways which prevent obtaining a completely delineated picture of the spatio-temporal features of the various structures. Since the identification of a vortex requires measurement of two derivatives of velocity over space, it has been difficult to document even vortex elements, quantitatively. This is particularly true of spanwise vortex elements because the necessary frame of reference is not known a priori. Moreover, it has been extremely difficult, or impossible, to measure the instantaneous shape of a complete vortical structure with probes; this has restricted observations to visual studies.. These difficulties make it easy to understand why there has been a lack of consensus concerning a complete model of the vortical structures within the boundary layer.

AVERAGE MODELS/REALIZATIONS

The distinction between one realization of a flow, and an average over many realizations (ensemble average) is obvious to workers in the turbulence community. However, certain aspects of the distinction need to be re-emphasized in order to be fully clear about coherent structures in the boundary layer.

The literature abounds with statistics that are limited to planes or other two-dimensional representations of flow structures which are intrinsically four-dimensional. These statistical measurements are essential in providing quantitative knowledge about turbulent fluctuations. However, they need to be interpreted with considerable care. Specifically, the ensemble average representations, unless interpreted with great care, can give the impression that there exist in the flow stationary structures which move with the flow in a convective sense. As discussed below, many of the vortical elements and many sections of the vortical structures do not move with the local convective speed of the flow, and the shapes of the vortical structures change over time as elements are created, evolve and decay.

Another quite surprising result has arisen from the cooperative, community-wide study. Several important structures are created in more than one way, each of which appears to be statistically relevant to one or more processes that are important in the dynamics. (See also further comments in the section discussing evolution of structure.) These facts provide another reason why establishing the full model of the structure in the boundary layer from ensemble average measurements has been difficult.

GENERAL DESCRIPTION OF OBSERVED VORTEX ELEMENTS AND STRUCTURES

Before we discuss ordering of structures, it will be useful to note several features which are observed when one looks at a large sample of vortices in an instantaneous realization of a large volume of flow in Spalart's (1988) simulated, flat-plate boundary layer by examining a surface defined with a contour of negative p'. An example is shown in Figure 1. This sample covers a volume extending through the layer in y, 1200 wall units in z, and over 4000 wall units in x. One observes many isolated vortex elements with various orientations, and also many vortical structures of various shapes. There is no single shape which dominates the observed type of vortices by its frequency of appearance. Also, the shapes of vortical structures vary more continuously than discretely. The contour level used in Figure 1 was $-4.2\,\rho\, U_\tau^2$. However, variation of the contour level over a wide range (covering all the appropriate values) does not alter the qualitative nature of the descriptions given.

Because of the variety and continuous variation in shape observed in figure 1, it is hard to decide on a set of categories to describe the vortex elements and structures. However one shape does occur more commonly than most others, and can be used to delineate some distinctions and associations. This relatively common vortical structure is illustrated in Fig 2.

Fig. 2 shows a hook-like shape which includes three vortex elements designated: Head (spanwise element), leg (near-streamwise element) and neck (a curving structure which connects the head and leg). The wallward (upstream) end of the leg is called the foot. The inboard side toward the head and the outboard side are marked in Figure 2 for reference. One observes many more hook-like vortical structures than full hairpin-like structures with two legs and two necks.

One also observes a significant number of what appear to be piled up "complexes" of vortical elements that are tangled together in a variety of shapes. These "complexes" of vortices must be observed from more than one perspective since when observed in plan view they may appear to touch, but when viewed observed in side view, are sometimes seen to be separate vortical structures lying at different y+ locations. Not infrequently as many as four or five separate layers of vortices are seen in a side view. This description stands in considerable contrast to many attempts to describe a single characteristic vortex structure. It suggests that we may need to deal with a distribution of vortex elements rather than a single structure or even single form of structure; this point needs further study.

Despite these complications, some clear associations between various types of vortex elements and other structures appear to be characteristic, and can be delineated. These associations are discussed, in part, and specific references cited on specific points below.

A PRELIMINARY ORDERING OF THE IMPORTANCE OF STRUCTURE ELEMENTS WITH RESPECT TO CREATION OF TRS AND CREATION AND DISSIPATION OF TKE

The list which follows was obtained by creating a series of interconnectivity diagrams showing how many other features preceded (inputs) and followed (outputs) each of the eight known structures in time (Robinson, 1989b). Structures with more inputs and outputs have then been taken as probably more central to the processes. The results of this input/output study were checked by considering which structures appear to actively influence the flow, and are most closely associated in space with volumes of high Reynolds stresses and/or high dissipation of TKE. The list is presented below. It is highly preliminary; we expect and hope it will be subjected to discussion by many others.

A TENTATIVE PRIORITY OF STRUCTURES

MOST CONNECTED AND APPARENTLY MOST ACTIVE

- Vortices-- Vortex elements and vortical structures

PLAYS AN IMPORTANT ROLE

- Ejections (including lifted Low-speed streaks)
- Sweeps (including inrushes)
- Near wall shear-layers

PLAYS SOME ROLE

- Bulges
- Backs
- Pockets
- Wall-attached low speed streaks

These are the same eight elements listed by Kline and Robinson (1988), but, in addition, include two improvements and provides a tentative ordering of importance. The improvements are: (i) lifting of low-speed-wall-streaks is a included as a special case of ejections; (ii) Inrushes are included as a special case of sweeps, where 'inrush' is used in the sense of Grass (1971) and of Praturi and Brodkey (1978); that is, an inrush denotes: motions containing a large negative v' component and also emanating from the outer region of the layer and moving to, or very near to, the wall. Not all sweeps are inrushes since many sweeps move only short distances in the y direction and/or have a low angle of inclination to the wall, as Praturi and Brodkey (1978) explicitly noted. Sweeps which are not inrushes are known to play a significant role in the near wall region. We have delineated the concept of inrushes because the observations suggests inrushes play a significant role in the *inward* interaction between the outer and inner layers. However, it is noteworthy that the *outward* interaction between the inner and outer layers appears to have a different character. Details will be reported separately.

Of these elements of structure, the most important, as many others have suggested earlier, seem to be the vortices. However, most earlier studies have suggested one single form of vortical structure. Observations of the DNS data base for the canonical plate flow suggest a variety of vortex elements (each with varying orientations) and vortical structures of a variety of shapes all play a role.

If this suggestion is taken as a hypothesis to check, then we can examine it by asking two questions:

- Do the other seven forms of structure occur close enough to vortex elements or structures so that the vortices could play a central role in what occurs?

- Are the strengths of the vortices sufficient and the directions of rotation such that the observed motions could be induced by the vortices?

The answer to both these questions appears to be, "Yes!" Some specifics follow; more details are reported elsewhere as noted in context.

THE ASSOCIATION OF VORTICES WITH OTHER ELEMENTS OF STRUCTURE

The following forms of structure from the list above appear most commonly immediately adjacent to vortices (for pictures see Robinson et al 1988a):

- ejections
- strong sweeps at the wall
- inrushes
- many (but not all) near-wall shear layers
- many (but not all) bulges
- many (but not all) backs
- attached low-speed wall streaks. (These occur wherever a leg vortex element is observed near the wall; however the low-speed wall streaks persist longer than the leg vortex elements; the streaks therefore are also observed when there is no proximate leg vortex element.)
- pockets (observed beneath intense near-wall sweeps)

In addition to these associations, a number of other observations also point to vortex elements and structures as central to the structure of the boundary layer.

As noted above, Moin (1987) reported from study of a DNS channel flow that regions of high $u'v'$ occurred adjacent to every observed quasi-streamwise vortex element. Robinson et al (1988a) reported observations of high values of $u'v'2$ primarily in two places, along the inboard side of leg vortex elements where low-speed streak lifting was observed, and underneath and upstream of vortex head elements. These observations held regardless of whether vortex elements were observed alone or as part of vortical structures. They also reported that high values of $u'v'4$ were observed primarily in two places: outboard of necks and outboard of leg vortex elements. Moreover, the $u'v'2$ inboard of leg elements, and the $u'v'4$ outboard of legs occur very close to the wall, and thus include the region where production of TKE is known to be a maximum.

There is also an observed association between vortices and regions of high static pressure in the flow. Specifically regions of high pressure are observed wherever high speed fluid overtakes slower moving fluid downstream. This occurs primarily in two places: upstream of backs which lie just upstream of vortex head elements and downstream of inrushes near the wall, motions which are associated with neck vortex elements. These results are reported in Robinson et al (1988b). In the same paper, it is noted that large regions of high pressure are observed just upstream from the piled up vortex elements. These regions of high pressure are of considerable extent both normal to the wall and spanwise. Alfredsson et al (1988) report high pressure regions upstream of lifted-low-speed streaks near the wall. Here also one observes high speed fluid overtaking low-speed.

Finally, observations of the complete dissipation term for TKE in Spalart's (1988) DNS simulated boundary layer (Robinson, 1989b) show that high values of dissipation occur along the leg of vortices, particularly toward the foot end of the leg and thus in the very near wall region which is the region of highest dissipation as indicated by extrapolation from the measurable elements of dissipation for example by Klebanoff (1955) for a boundary layer and Laufer (1952) for a channel flow.

In summary, this section indicates there are close spatial relationships between all the known important structures and some form of vortex element.

THE STRENGTH OF VORTEX ELEMENTS

In a study of Spalart's DNS simulated boundary layer Robinson (1989a) reports the strengths and other detailed properties of vortex leg and vortex head elements for isolated vortex elements and vortical structures. The results show that the vortex elements, both legs and heads, are clearly strong enough so that one can think of them as playing a major role in the dynamics of the nearby flow field via induction.

In summary, the data show three things about vortices in the boundary layer:

- No one shape, size or orientation of vortices is sufficiently common to be entirely characteristic; rather a variety of shapes, sizes, and orientations are observed;

- The vortex elements, whether solitary or in vortical structures, are associated directly in space with the other elements of structure, and have both the strength and direction necessary to induce, or at least augment, the observed motions;

- Head and leg vortex elements play somewhat different roles in the physics; however, this matter needs further study.

A FEW REMARKS ON TIME EVOLUTION OF STRUCTURES

When one observes low-speed streak-lifting either along the inboard side of leg vortex elements or beneath and behind head vortex elements, whether alone or in vortical structures, one sees the formation of the near-wall-shear-layers following from the streak-lifting. No sweep or dynamic action is needed for this creation of a near-wall-shear-layer since the lifting of the low-speed-streak creates a volume of low speed fluid at a distance from the wall of a surfboard-like shape or of a triangular wedge-like shape. This volume of low-speed fluid then lies away from the wall and downstream from a region of significantly higher speed fluid which already existed at this distance from the wall, virtually by construction. The near-wall-shear-layer then exists by virtue of kinematics alone; it is the interface between the lifted-low speed volume of fluid and the following higher speed fluid. No strong negative v' motion is needed for these things to occur, nor is such a motion typically observed; this has been reported by Brodkey and his co-workers (Corino and Brodkey, 1969; Praturi and Brodkey 1978), and is confirmed in the study of DNS simulation of the plate flow.

When one follows this near-wall-shear-layer farther in time after its formation, the observations show a significant fraction of the lifted near-wall-shear-layers roll-up, and create new spanwise vortex elements (heads) usually with one or two necks attached; see Robinson et al 1988a, 1988b. This roll-up appears to begin with a perturbation of the near-wall-shear-layer which creates rapid further mutual induction of the vorticity lines, and thus appears to be a local instability of the near-wall-shear-layer. Such roll-ups occur from both lifted streaks which form alongside legs and those which form behind and beneath heads as reported by Acarlar and Smith (1987) for perturbed laminar layers and as observed in DNS simulations of the plate. The vortex head elements formed in this way tend to persist in the flow for long distances downstream, in many instances farther than can be tracked in the available data-bases. In some instances the vortex head elements have been observed migrating, relatively slowly, to higher values of y+. The near-wall-shear-layers which do not form heads, appear to break-up and disappear over significantly shorter times. The precise fraction of near-wall-shear-layers which roll-up into persistent vortex head elements as contrasted with those which break-up relatively rapidly has not yet been measured; however each fraction is significant.

When the vortex heads formed by roll-up of the near-wall-shear-layers are followed still further in time, one observes, not uncommonly, a rapid growth of a leg from the open end of a neck which moves inward and ends in the very near wall region. In some instances the observed growth of the vortex leg element moves directly to the wall, and in others it appears to run upstream. In either case it is elongated, often rapidly, over time since the neck region moves faster than the foot region of the leg.

In many cases ones sees two or three lifting sections of the low-speed-streak closely spaced in x lying inboard of a vortex leg element. Each lifting section typically moves into a volume of high u'v'2. We interpret this as representing the several ejections in a burst as documented by Bogard and Tiederman (1987). The observations then suggest that a "burst" in the sense used by Kim et al (1971) is associated with the passage of a vortex element; either a leg or a head.

Lifted streaks have spanwise dimensions of the order of approximately 20-80 wall units; sharp interfaces with high values of dU/dz exist along the xy faces on the sides of the lifted-low-speed-streaks. The formation of solitary leg vortex elements with varying orientations have been observed along the sharp interfaces on the xy faces of lifted low-speed streaks. High values of dU/dz in the near wall region have been documented by many observers, notably Blackwelder and Eckelmann (1979). Blackwelder also stressed the possibility and potential importance of instabilities occurring as a result of the high gradients of both dU/dy and dU/dz, of roughly equal magnitude, in a number of informal workshops and conversations over the years.

The preceding paragraphs describe a sequence of events consisting of: low-speed-streak-lifting; formation of near-wall-shear-layers; roll-up of some of the near-wall-shear-layers forming spanwise vortices (heads); and finally growth of a leg vortex element from the head with the leg often extending essentially to the wall. However, little was said about how lifting of low-speed-streaks is initiated. Some remarks on the initiation of low-speed-streak lifting are now added.

Data by R. E. Falco, communicated privately, show simultaneous plan and side views with laser sheet marking. Lifting of low-speed-streaks observed in these two views often occurs when a strong spanwise motion impinges on a low-speed streak lying adjacent to the wall. Thus one often observes a strong spanwise kinking of the low-speak-streak when observed in plan view. Such kinks of low-speed-streaks in plan view were commonly observed in the earliest pictures of streaks, Kline and Runstadler (1959), and have been confirmed in numerous later visual studies. However, the implications of this kinking of the low-speed streaks in the plan view had remained unknown up to the time Falco's two simultaneous views became available. In a quantitative study Alfredsson et al (1988), using VISA statistics in a DNS simulated wall flow, showed that the ensemble average lifted-low-speed-streak and the associated high values of $u'v'2$ lying downstream from the lifted streak, both have a strongly spanwise skewed orientation at $y+ = 15$. Alfredsson et al did not track the results to the wall; they suggested the triggering source may be the pressure perturbation which they also mapped. In this paper, and in Robinson et al (1989b) we have suggested a different source and different role for the pressure perturbations. Thus we need also to suggest a different source of the spanwise kinking of low-speed streaks commonly associated with streak-lifting. The descriptions in Robinson et al 1988a, 1988b suggest the following possibility. The spanwise kinking of low-speed-streaks associated with lifting may arise from motion induced by the foot of a quasi-streamwise leg vortex with its underside lying essentially at the wall. In visual study of the DNS simulations, one can observe the resulting spanwise motions as part of the lifting of low-speed-streaks on the inboard side of vortex legs. This includes motions generated by inrushes around a vortex neck as well as those arising from a vortex foot. This suggestion needs more detailed studies to determine statistical relevance.

WHERE DO WE GO -- RESEARCH CHALLENGES

Let us first recapitulate where we stand in summary. Two main facts stand out. First, the sum of three decades or so of work in the laboratory has provided us with documentation of eight major structure features one-by-one, and a number of possible models of ensemble averages of the structure. Recent study of DNS data bases made accessible by workers at NASA-Ames have begun not only to confirm many characteristics of the eight structure features measured by laboratory experiments but also to provide several kinds of added information concerning: regions of high and low pressure; the location of regions of high dissipation of TKE; details of a variety of shapes of vortex elements and vortical structures and their creation, evolution and decay over time; considerable information about the size, distribution in $y+$ space, circulation, and intensity of vortex elements in the flow. Perhaps most important, in terms of constructing a complete model of the structure, these studies are providing considerable information about the spatio-temporal relations among the eight kinds of structure and regions of high and low pressure. All these results, from both laboratory and DNS studies, taken together suggest that vortices, which move through the flow in various shapes and orientations, and often persist for long distances measured in wall units, are the central features of the structure.

However, the available DNS data bases cover only a few canonical flows, are limited to very low R_θ values, and have thus far only been studied for significant times by a relatively small number of workers. Can we take these results from DNS to be reliable pictures of the structure of near-wall turbulence? What questions remain open, and how can we best approach them?

Let us deal with the problem of restriction to low values of R_θ first since that is in some ways the easiest to deal with. The question of the effect of R_θ has been much discussed; it constitutes a real problem since we cannot expect DNS to manage much higher R_θ for at least a long time. Nevertheless, we believe it would be easy to overemphasize the importance of this question for boundary layers. Kline and Robinson (1988), summarize data from many sources which show that the large fraction (of the order of 80%) of the creation of TRS and production of TKE occur in the near wall region. In this region, effects are known to scale, or at least nearly so, on U_τ. U_τ in turn is known to scale as the minus one-tenth power of R_θ. Moreover, for the canonical plate flow we do not see changes in slope on the curves of either Ln(Cf) or Ln(St) vs Ln(Re). Experience, covering many cases in viscous flows, tells us when the slope remains constant in this type of non-dimensional correlations, we ought not expect *qualitative* changes in the flow structure (or in what we sometimes call flow regime). Thus in so far as the structure of the near-wall region is concerned, we do not expect to see much effect of Reynolds number. After all, 100 to the minus one-tenth power is 0.63, and R_θ of 67000 is large in terms of most applications. Thus we expect what is found at $R_\theta = 670$ for the near wall region to be a reasonable qualitative picture for flat plate layers in general. Moreover, we also know the law of the wall is surprisingly tenacious; it applies when properly used to rough wall flows, to pressure gradient cases; to curved wall flows, and other applications. It is possible to create flows which do not obey the law of the wall, but one has to work at it rather hard. This suggests that the qualitative features of structure observed in the inner layers of boundary layers for the canonical plate flow ought to provide guidance in a much wider class of flows albeit the idea needs firmer proof via laboratory and DNS investigations.

In so far as the inward and the outward interactions between the inner and outer layers are concerned, we must expect changes as R_θ increases. Changes in the shape of the space-time correlations as presented for example by Antonia et al (1988) verify changes in the outer layer. Moreover, the ratio of outer to inner scales increases along a flat plate as x to the 0.7 power. Hence, we can anticipate, other things being unchanged, that the direct connection between the outermost part of the layer and the innermost regions of the layer will weaken as R_θ increases. The saturation of profile at $R_\theta = 5000$ given by the Coles (1968) tends to support this idea. So do recent data by Anders (1989) investigating the effect of LEBUS as a function of Reynolds number. Thus some changes are likely to be seen in the interactions between the inner and outer layer at least up to $R_\theta = 5000$. However, these interactions account for the smaller portion of the creation of RNS of the order of 20% for canonical plate flow. The inward interactions do not dominate production in the canonical flat plate layer albeit they are not without effect right to the wall. Moreover, even in the outer region of the flow, if we find, for example, that the vortex head elements are central to the structure, we ought to assume, in the absence of contrary information, that this will also be *qualitatively* true at higher R_θ since this is far more likely than the contrary assumption. Also, the results of Head and Bandyopadhyah (1981) suggest a linkage of the heads to the wall layers through an outward interaction with an aspect ratio that increases as R_θ increases.

Thus there is a need for investigation of structure at higher Reynolds number, but the authors believe this is less important as a research priority than investigation of many other effects which we know occur when we depart from canonical flat plate conditions. These include roughness, free-stream fluctuations, pressure gradients, three-dimensional layers, wall curvature. compressible cases, and body-force effects. Many of these effects will not be available for study in DNS data bases for some time to come, if ever. Hence there remains much to do experimentally. And we need to keep in mind the connections between the statistical measurements and the structure picture suggested by the comments on concepts and the results indicated above; these problems of connections between the statistics and the structure and the associated questions are unlikely to change significantly from one flow to another.

Let us move on then to a discussion of what we can do in utilizing experiments and DNS to move toward a more complete, generally agreed picture of structure in the boundary layer.

Let us begin by enumerating some open questions that appear important.

The comments in the section "Association of Vortices With Other Elements of Structure" suggests a long list of structure features for which we need to gather improved samples in order to have reliable statistical means and distributions. The availability of DNS simulated flows now makes this feasible.

Some questions which probably cannot be studied by using the existing DNS data bases include: the details of the outward interaction between the inner and outer layers; the origin of the large vortex head elements frequently observed at the center of bulges and why these heads are much longer spanwise than heads near the wall. What are the origin and role of the large "piled-up complexes" of vortex elements observed in the layer and their relation to high pressure regions. It may also be difficult to determine the dynamics of how leg vortex elements grow from heads with available data bases and software.

There is also a need for extensions of the DNS data bases to other flows, to cover the cases mentioned above. A good many workers are proceeding with such studies, and we have little doubt that the results will prove important in advancing our understanding of structure for a larger class of wall-bound shear flows over the next decade or so. These studies need to be linked to, augmented, and confirmed or disconfirmed by parallel experimental work.

If vortices are the central element of structure then several questions assume importance. How can we provide improved methods for observing vortices (in contrast to visually observed streaklines and VITA detected shear-layers) in the laboratory? What is the nature and the details of the formation of vortices? Is it as suggested above an unstable roll-up of near-wall-shear-layers with dU/dy and/or dU/dz orientation? Is this also true for the vortices under bulges far from the wall?

The central questions which remain seem to be, "How do we combine the results from study of DNS data bases with the experimental results to approach consensus on a complete model of structure?" and, "How do we capture the essence of such a model in a simple enough way so that it becomes useful in creating predictive models and controlling turbulence?"

This question leads naturally to a further question. If we assume that we know the distribution of location, sizes and strengths of head and leg vortex elements not only for the plate but also for other cases such as flows with pressure gradients, can we use that information to form a "Statistical-structural model" which has more physics built in than available Reynolds Stress Averaged models, and still is simple enough to provide practical engineering predictions? At the present state of knowledge this question seems worthy of investigation.

Finally, it seems very desirable for a wide based group of researchers to discuss the existing experimental results together with the emerging results from study of the DNS data bases in order to see to what degree consensus can be obtained on a complete model picture of boundary layer structure. Since this discussion requires in-depth examination, and that in turn requires a significant amount of time, a workshop type meeting open to all members of the research community seems indicated.

REFERENCES

Acarlar, M.S. and Smith, C.R., "A Study of Hairpin Vortices in a Laminar Boundary Layer. Part I: Hairpin Vortices Generated by a Hemisphere Protuberance," *J. Fluid Mech.*, vol. 175, p.1, 1987.

Acarlar, M.S. and Smith, C.R., "A Study of Hairpin Vortices in a Laminar Boundary Layer. Part II: Hairpin Vortices Generated by Fluid Injection," *J. Fluid Mech.*, vol. 175, p. 43-1987.

Adrian, R.J., "Linking Correlations and Structure: Stochastic Estimation and conditional Averaging," *Near Wall Turbulence: 1988 Zaric Memorial Conference*, Hemisphere, 1989.

Alfredsson, P.H., Johansson, A.V. and Kim, J., "Turbulence Production Near Walls: The Role of Flow Structures with Spanwise Assymmetry," *NASA Report CTR-588, "Studying Turbulence Using Numerical Simulation Data Bases - II"*, 1988.

Anders, J.B., "LEBU Drag Reduction in High Reynolds Number Boundary Layers," *AIAA-89-1011*, 1989.

Antonia, R.A., Bisset, D.K. and Browne, L.W.B., "Effect of Reynolds Number on Space-Time Correlations and Topology of Large Scale Motions in a Turbulent Boundary Layer," *Near Wall Turbulence: 1988 Zaric Memorial Conference*, Hemisphere, 1989.

Aubrey, N., Holmes, P., Lumley, J.L., and Stone, E., "The Dynamics of Coherent Structures in the Wall Region of a Turbulent Boundary Layer," *J. Fluid Mech.*, vol. 192, p. 115, 1988.

Bakewell, H.P. and Lumley, J.L., "Viscous Sublayer and Adjacent Wall Region in Turbulent Pipe Flow," *Phys. Fluids*, vol. 10, no. 9, p. 1880, 1967.

Blackwelder, R.F. and Eckelmann, H., "Streamwise Vortices Associated with the Bursting Phenomenon," *J. Fluid Mech.*, vol. 94, part 3, p. 577, 1979

Blackwelder, R.F. and Swearingen, J.D., "The Role of Inflectional Velocity Profiles in Wall-Bounded Flows," *Near Wall Turbulence: 1988 Zaric Memorial Conference*, Hemisphere, 1988.

Bogard, D.G. and Tiederman, W.G., "Characteristics of Ejections in Turbulent Channel Flow," *J. Fluid Mech.*, vol. 179, p.1, 1987.

Clark, J.A. and Markland, E., "Flow Visualization in Turbulent Boundary Layers," J. Hydr. Div. ASCE, vol. 97, p. 1653, 1971.

Coles, D.E., "The Young Person's Guide to the Date," Computation of Turbulent Boundary Layers--1968 AFOSR-IFP-Stanford Conference, eds: D.E. Coles & E.A. Hirst.

Corino, E.R. and Brodkey, R.S., "A Visual Investigation of the Wall Region in Turbulent Flow," *J. Fluid Mech.*, vol. 37, part 1, p. 1, 1969.

Grass, A.J., "Structural Features of Turbulent Flow over Smooth and Rough Boundaries," *J. Fluid Mech.*, vol. 50, p. 233, 1971.

Head, M.R. and Bandyopadhyay, P., "New Aspects of Turbulent Boundary Layer Structure," *J. Fluid Mech.*, vol. 107, p. 297, 1981.

Kim, H.T., Kline, S.J. and Reynolds, W.C., "The Production of Turbulence Near a Smooth Wall in a Turbulent Boundary Layer," *J. Fluid Mech.*, vol. 50, part 1, p. 133, 1971.

Kim, J., "On the Structure of Wall-Bounded Turbulent Flows," *Phys. Fluids*, vol. 26, p. 2088, 1983.

Kim, J. and Moin, P., "Large Eddy Simulation of Turbulent Channel Flow--ILLIAC IV Calculation," *Turbulent Boundary Layers--Experiments, Theory, and Modelling*, AGARD Conf. Proc. no. 271, 1979.

Kim, J. and Moin, P., "The Structure of the Vorticity Field in Turbulent Channel Flow. Part2: Study of Ensemble-Averaged Fields," *J. Fluid Mech.*, vol. 162, p. 339, 1986.

Kim, J., Moin, P., and Moser, R.D., "Turbulence Statistics in Fully-Developed Channel Flow at Low Reynolds Number," *J. Fluid Mech.*, vol. 177, p. 133, 1987.

Klebanoff, P.S., "Characteristics of Turbulence in a Boundary Layer with Zero Pressure Gradient," *NACA TN 3178, or Report 1247*, 1954.

Kline, S.J. and Robinson, S.K., "Quasi-Coherent Structures in the Turbulent Boundary Layer: Part I. Status Report on Community-wide Summary of Data," *Near Wall Turbulence: 1988 Zaric Memorial Conference*, Hemisphere, 1989.

Kline, S.J. and Runstadler, P.W., "Some Preliminary Results of Visual Studies on the Flow Model of the Wall Layers of the Turbulent Boundary Layer," *J. Appl. Mech., Trans. ASME*, Ser. D, vol. 26, no. 2, 1959.

Kovasznay, L.S.G., Kibens, V. and Blackwelder, R.F., "Large-Scale Motion in the Intermittent Region of a Turbulent Boundary Layer," *J. Fluid Mech.*, vol. 41, part 2, p. 283, 1970.

Laufer, J., "The Structure of Fully Developed Pipe Flow," *NACA TN 2954, or Report 1174*, 1953.

Lian, Q-X., "A Visual Study on the Coherent Structure of the Turbulent Boundary Layer in Flow with Adverse Pressure Gradient," to be published in *J. Fluid Mech.*, 1989.

Moin, P., "Probing Turbulence via Large Eddy Simulation," *AIAA-84-0174*, 1984.

Moin, P., "Analysis of Turbulence Data Generated by Numerical Simulations," *AIAA-87-0194*, 1987.

Moin, P. and Kim, J., "Numerical Investigation of Turbulent Channel Flow," *J. Fluid Mech.*, vol. 118, p. 341, 1982.

Moin, P. and Kim, J., "The Structure of the Vorticity Field in Turbulent Channel Flow. Part 1: Analysis of Instantaneous Fields and Statistical Correlations," *J. Fluid Mech.*, vol. 155, p. 441, 1985.

Pearson, C.F. and Abernathy, F.H., "Evolution of the Flow Field Associated with a Streamwise Diffusing Vortex," *J. Fluid Mech.*, vol. 146, p. 271, 1984.

Praturi, A.K. and Brodkey, R.S., "A Stereoscopic Visual Study of Coherent Structures in Turbulent Shear Flow, *J. Fluid Mech.*, vol. 89, part 2, p. 251, 1978.

Robinson, S.K., Kline, S.J., and Spalart, P.R., "A Review of Quasi-Coherent Structures in a Numerically Simulated Turbulent Boundary Layer," *NASA TM-102191*, 1989.

Robinson, S.K., Kline, S.J. and Spalart, P.R., "Quasi-Coherent Structures in the Turbulent Boundary Layer: Part II. Verification and New Information from a Numerically Simulated Flat-Plate Layer," *Near Wall Turbulence: 1988 Zaric Memorial Conference*, Hemisphere, 1988a.

Robinson, S.K., Kline, S.J. and Spalart, P.R., "Spatial Character and Time Evolution of Coherent Structures in a Numerically Simulated Boundary Layer," *AIAA 88-3577*, 1988b.

Robinson. S. K., "A Review of Vortex Structures and Associated Coherent Motions in Turbulent Boundary Layers," *Second IUTAM Meeting on Structure of Turbulence and Drag Reduction*, Zurich, Switzerland, July 25-28, 1989a.

Robinson. S. K. (1989b), "Kinematics and Dynamics of Coherent Motions in Turbulent Boundary Layer," Ph.D. Dissertation, Stanford University, 1989.

Smith, C.R. and Schwartz, S.P., "Observation of Streamwise Rotation in the Near-Wall Region of a Turbulent Boundary Layer," *Phys. Fluids*, vol. 26, p. 641, 1983.

Smith, C.R., "A Synthesized Model of the Near-Wall Behavior in Turbulent Boundary Layers," *Proc. of 8th Symp. on Turbulence*, University of Missouri, Rolla, 1984.

Smith, C.R., "Visualization of Turbulent Boundary Layer Structure Using a Moving Hydrogen Bubble Wire Probe," *Coherent Structure of Turbulent Boundary Layers*, AFOSR/Lehigh University Workshop, p. 48, 1978.

Spalart, P.R., "Direct Simulation of a Turbulent Boundary Layer up to $Re_\theta = 1400$," *J. Fluid Mech.*, vol. 187, p. 61, 1988.

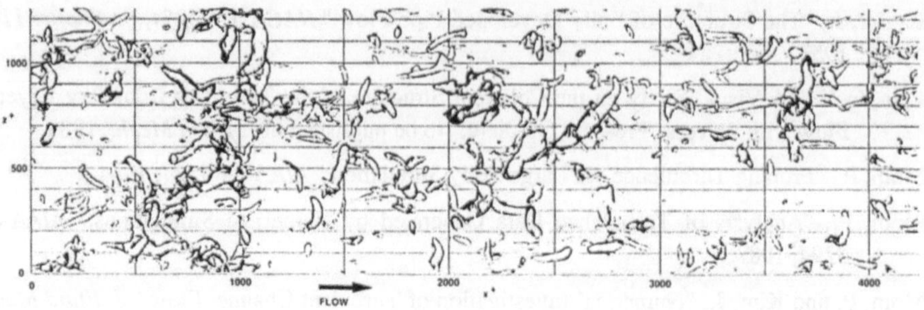

Fig. 1. Top-view of instantaneous three dimensional low-pressure structures in numerically-simulated turbulent boundary layer. Isobaric surfaces computed for $p' = -4.2\rho u_\tau^2$.

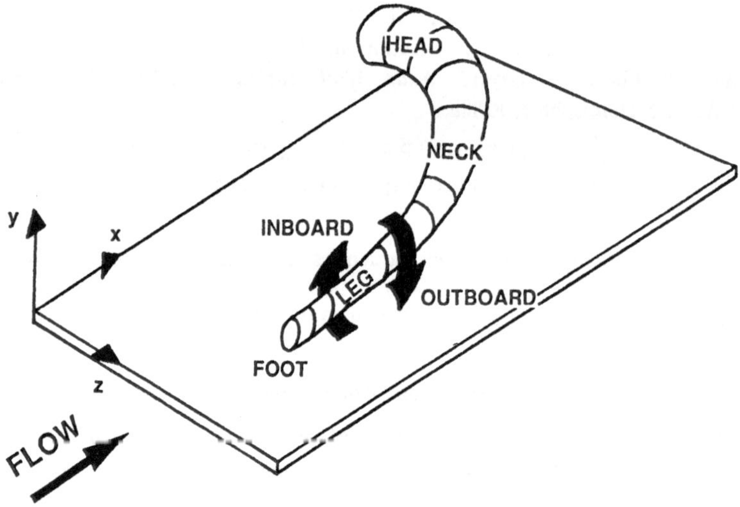

Fig. 2. Schematic diagram of one-sided, hook-shaped vortical structure.

A Review of Vortex Structures and Associated Coherent Motions in Turbulent Boundary Layers

Stephen K. Robinson - NASA Ames Research Center, Moffett Field, CA, USA, 94035

ABSTRACT

The experimental and computational evidence for the existence and role of vortices in turbulent boundary layers is briefly reviewed. Quasi-streamwise and transverse vortices are considered, and various published conceptual models for horseshoe-like vortical structures are compared. The causes for upright and inverted horseshoe-shaped vorticity lines are discussed, and the distinction between vorticity lines and vortices is demonstrated. Finally, results from a numerically-simulated turbulent boundary layer are used to compute distributions of diameter, height, and strength for quasi-streamwise and spanwise vortices. These results confirm that quasi-streamwise vortices are clustered near the wall, while spanwise vortices are distributed throughout the layer. The variation of spanwise vortex core diameter with distance from the wall is found to be consistent with the mixing-length distribution for a boundary layer.

INTRODUCTION

Objectives

The concept of vortical motion is intrinsic in the study of coherent structures in turbulent flows. Even in the term "eddy," we find an implied vortical motion. A review of 40 years of turbulence structure literature uncovers a mass of information, much of which has vortices and their effects as a common denominator. The motivation for investigating vortices is their potential to function as "pumps" which transfer mass, momentum, and heat while extracting energy from the mean flow. It is the intent of the current effort to briefly review this important concept of imbedded vortical structures as it pertains to turbulence physics in boundary layers.

The oldest and most common idea for coherent vortical structures in the turbulent boundary layer is that of the horseshoe or hairpin vortex. Therefore, attention will be focussed upon conceptual models that rely upon such vortical structures. Kinematical and dynamical aspects of the various published models are compared and experimental and computational evidence for their existence and importance is examined.

Motivations

Turbulence structure research does not generally aim to replace traditional, Reynolds-averaged modeling concepts. Instead, the motivation is to open a window upon the physics responsible for the statistics we are trying to model, in the hope that a deeper physical insight will provide guidance for improvements in Reynolds-averaged modeling and in turbulence control. The eventual objective is a class of models in which structural elements with modeled characteristics will provide the statistics of engineering interest. A leading candidate for the core element of such a modeling concept is a family of vortex structures, hence the topic of the current paper.

DEFINITIONS AND DISTINCTIONS

One of the hindrances to the study of vortex structures in turbulent boundary layers has been the lack of a rigorous, widely-accepted definition of a vortex for unsteady, viscous flows. For the present effort, the following working definition is employed: a vortex exists when instantaneous streamlines mapped onto a plane normal to the vortex core exhibit a roughly circular or spiral pattern, when viewed from a reference frame moving with the center of the vortex core. This definition requires an a-priori method for identifying vortex cores, and the process of choosing a reference-frame velocity may be iterative.

It is useful to distinguish between vortical elements and vortical structures (Kline and Robinson, 1989). In the present paper, a vortical element is defined as a vortex or vortex segment with a single dominant orientation. Examples are "leg," "neck," and "head" vortex elements of which a hairpin

A. Gyr (Editor)
Structure of Turbulence and Drag Reduction
IUTAM Symposium Zurich/Switzerland 1989

vortex is composed (Figure 1). A vortex structure is defined as any combination of elements, generally forming a complex three-dimensional shape. Vortex structures considered within this paper are hairpins (with extended trailing legs) and horseshoes (without well-defined legs). Horseshoe vortices possess a width/length ratio of approximately unity (as in Theodorsen's model, Fig. 2) and hairpins are longer in the streamwise direction than they are wide (as in Head and Bandyopadhyay's high-Reynolds number vortical structures, Fig.8c). This distinction is often unclear or unnecessary, however, and in those cases "hairpin/horseshoe" will be used. Horseshoe-shaped vortical structures will also be referred to as "arches".

The term "quasi-streamwise vortex" will be applied to any vortical element with a predominantly streamwise (x) orientation, although it may be tilted at a significant angle to the x-axis. Brodkey's (1987) term for streamwise vortices with an upward tilt is $\omega_{x/y}$-vortices, which is also a useful nomenclature.

A distinction must be made between vortices and vorticity. In the turbulent boundary layer, the association between regions of strong vorticity and actual vortices seems to be rather weak (Robinson et al, 1989). Thus additional methods are necessary for vortex identification. Visual techniques have been moderately successful in the laboratory (eg. Smith and Lu, 1988), and vorticity lines (Moin and Kim, 1985) and the pressure field (Robinson et al, 1988) have been useful for detecting vortices in numerical simulations.

Sweeps and ejections are defined here as $(u'v')_4$ (or Q4) and $(u'v')_2$ (or Q2) motions, respectively, in accordance with Wallace et al's (1972) $u'v'$ quadrant-splitting scheme. There are other interpretations of the terms, but the present usage is the most common, and has been chosen for its strong association with the Reynolds shear stress.

REVIEW OF EVIDENCE AND MODELS FOR VORTICAL STRUCTURES

The presence of Reynolds shear stress ($-\rho\overline{u'v'}$) in a boundary layer implies the existence of cross-gradient mixing; that is, transport of relatively low-momentum fluid outward into higher-speed regions and of high-momentum fluid wallward into lower speed regions. Since any vortex with an orientation other than exactly wall-normal will induce such transport, vortices are natural candidates for major (and perhaps dominant) producers of Reynolds shear stress in the boundary layer. A significant portion of the boundary layer structure literature is devoted to investigating the role and quantitative character of vortical elements and structures. Although the comments presented here are necessarily brief, an extensive discussion of boundary-layer vortex statistics and dynamics can be found in Robinson (1989).

Experimental and Computational Evidence for Quasi-Streamwise Vortices

The ubiquitous presence of an elongated low/high-speed streak pattern in the near-wall region of turbulent boundary layers has prompted suggestions that the streaks are fluid accumulated between counter-rotating pairs of near-wall streamwise vortices. These theories, along with the observed appearance of streamwise vortices during violent turbulence production events has continued to motivate new research into near-wall quasi-streamwise vortices.

Vortices with a major streamwise (x) component are generally identified in experimental and computational results with single cross-stream (y-z) planes of marked fluid or computed velocity vectors. This method may not allow for differentiation between purely streamwise vortices and those tilted with respect to the wall. Thus, much of the data reviewed in this section could apply to both leg and neck vortices (in the nomenclature of Fig. 1), so the vortex elements are referred to here as "quasi-streamwise." Both purely streamwise and tilted quasi-streamwise vortices are capable of momentum transport across the velocity gradient, but the vortex dynamics are obviously affected by the orientation, since the wall-normal velocity gradient provides rapid stretching of the tilted vortices only.

Table 1 lists some of the many articles that make reference to quasi-streamwise vortices in turbulent boundary layers, either by flow-visualization in the $y-z$ plane, by probe measurement, or by numerical simulation. The A-F categorization shows that about half of the results are from visual studies, and half include quantitative data. Some studies provide both by using particle or bubble displacement techniques. Nearly a third of the references include some kind of simplified predictive model for near-wall streamwise vortices in the turbulent boundary layer.

Quasi-streamwise vortices were observed early in the history of turbulent structure research. Kim et al (1971) noted the common appearance of quasi-streamwise vortices in conjunction with the oscillation phase of the turbulence-generating bursting process. Grass (1971) also observed quasi-streamwise vortices, generally during near-wall "inrush" and "ejection" events. Clark and Markland (1971) made careful observations of relatively long quasi-streamwise vortices (often counter-rotating pairs) with a 3 to 7 degree upward tilt in the wall region of a turbulent water channel.

Perhaps the most extensive direct information concerning quasi-streamwise vortices has come from the end-view hydrogen-bubble visualization studies of Smith and Schwartz (1983) and Kasagi, Hirata, and Nishino (1986). These studies confirmed the common occurrence of quasi-streamwise vortices in the near-wall region, including frequent observation of counter-rotating pairs. In the simultaneous top and end views by Smith and Schwartz, counter-rotating vortex pairs in the near-wall region were always associated in space and time with low-speed streak formation. Further recent results by Kasagi (1988) suggest that solitary quasi-streamwise vortices are more common than vortex pairs in the near-wall region, and that the vortical structures are not as long as the near-wall low-speed streaks.

Additional visual evidence of the existence and character of quasi-streamwise vortices can be found in Kastrinakis et al (1978) (who re-analyzed the film of Corino and Brodkey, 1969), Praturi and Brodkey (1978), and Lian (1987). Probe-based (or quantitative flow-visualization) results from which vortex behavior is inferred are presented in Willmarth and Lu (1972), Blackwelder and Eckelmann (1979), , Utami and Ueno (1987), Kreplin and Eckelmann (1979), and Nakagawa and Nezu (1981).

The apparent association between near-wall quasi-streamwise vortices and both the generation of low-speed streaks and turbulence production has motivated a number of streamwise vortex models with at least some predictive abilities. These include Bakewell and Lumley's (1967) proper orthogonal decomposition of experimental data, which showed that most of the near-wall Reynolds shear stress and turbulence kinetic energy could be represented by a dominant eddy structure which consisted of a streamwise vortex pair. These results were confirmed and extended by Herzog (1986), and more recently by Aubrey et al (1988) with much more detailed data. Additional streamwise vortex models of have been proposed and developed over the years by a number of groups. Representative references of recent results are Hanratty (1988), Ersoy and Walker (1986), Pearson and Abernathy (1984), and Jang et al (1986).

Recent numerical simulations of turbulence have provided important new tools for investigating the structure of low Reynolds number wall-bounded flows, and have been used extensively to study the nature of embedded vortices in simulated channel flows and boundary layers. Although most of the attention has been given to three-dimensional vortex structures, quasi-streamwise vortices have been found in both instantaneous realizations and in conditionally-averaged results from the simulations.

Kim (1983) used VITA-type conditional averages to educe tilted streamwise vortex pairs in the near-wall region of a Large-Eddy Simulation (LES) of a fully developed turbulent channel flow. Kim's average near-wall vortices were considerably shorter than the low-speed streaks in the simulation. In an extensive study of the vorticity field of an LES channel flow, Kim and Moin (1985, 1986) only rarely observed quasi-streamwise vortices (bundles of largely streamwise vortex lines). The vortices that were detected were of generally limited longitudinal extent ($\Delta x^+ < 100$), in agreement with Kim's (1983) earlier conclusions from the same data.

In the results of a direct simulation of a channel flow, Moin (1987) observed that, unlike the u' field near the wall, the v' and w' contours do not show significant streamwise elongation and that the regions of large v' tend to occur in side-by-side inward/outward pairs, suggesting quasi-streamwise vortices. Moin found these vortices occurred most commonly singly, rather than in counter-rotating pairs of equal strength, in contrast to conditionally averaged results (e.g. Blackwelder and Eckelmann, 1979; Herzog, 1986; Kim, 1983; Moin, 1984). Moser and Moin (1984), Robinson et al (1989), and Guezennec et al (1989) have also concluded that single quasi-streamwise vortices are statistically more common than equal-strength pairs in direct numerical simulations of turbulence.

Moin's (1987) quasi-streamwise vortices were only 100 to 200 viscous units long, but retained their coherence while travelling several channel half-widths downstream. Moin reported regions of high $u'v'$ adjacent to every observed quasi-streamwise vortex in the wall region, and concluded that relatively short, single vortices are the fundamental structures associated with regions of high turbulence production.

Statistical decompositions of the simulated turbulence velocity fields have provided additional insight into the nature and statistical significance of quasi-streamwise vortices (Moin, 1984; Moin, Adrian, and Kim, 1987).

Given the available evidence, there is no doubt that quasi-streamwise vortices exist in numbers sufficient to play an active role in near-wall turbulence dynamics. Some association between quasi-streamwise vortices and both ejections and low-speed streaks seems certain, although the details have not been clear. The obvious potential for outward pumping of low-speed fluid by single and paired quasi-streamwise vortices has been confirmed, but the apparent violence of the ejection phase of bursting may signify a more complex and transient vortex behavior. In addition, the formation of near-wall shear layers are apparently associated with the upward-rotating sides of quasi-streamwise vortices (Stuart, 1965; Robinson et al, 1988). There is reasonable agreement in the literature on the size and location of quasi-streamwise vortices: diameters from $15\nu/u_\tau$ to $50\nu/u_\tau$, with centers occurring predominantly between $y^+ = 20$ and $y^+ = 70$. Although early papers proposed quasi-streamwise vortex extents on the order of the streak lengths ($\approx 1000\Delta x^+$), it now appears to be a consensus that quasi-streamwise vortices are about an order of magnitude shorter than the longest sublayer streaks. Quasi-streamwise vortices are generally considered to occur both singly and in counter-rotating pairs, and, in the near-wall region, to tilt upwards from the wall at a shallow (3 to 7 degree) angle. In the outer region, vortices more commonly make angles of approximately 45 degrees with the wall, which corresponds to the direction of maximum vorticity production due to stretching by the mean gradient. There is little information available on the strength (circulation) of quasi-streamwise vortices in turbulent boundary layers.

The formation mechanisms and evolution of the quasi-streamwise vortices are poorly understood. The most popular theory is that they are the trailing, stretching legs of hairpin vortices (see later discussion). However, Falco (1982, 1983) suggests that quasi-streamwise vortices are created from the passage of a ring vortex eddy over the sublayer. It has also been proposed that quasi-streamwise vortices arise due to local flow curvature and an accompanying Görtler instability (Brown and Thomas, 1977), and Acarlar and Smith (1987) propose yet another means by which quasi-streamwise vortices may be generated (see Fig. 12). None of these theories are fundamentally contradictory, and in fact multiple formation mechanisms of quasi-streamwise vortices seems probable.

Experimental and Computational Evidence for Spanwise Vortices

In the spanwise (x-y side-view plane) of a turbulent channel or boundary layer, the detection of a vortex in the velocity field is obviously dependent upon the motion of the observer's reference frame. The mean velocity gradient combined with the turbulent unsteadiness make experimental detection of transverse vortices in the boundary layer difficult and often ambiguous. Cautions against potentially deceptive illusions created by reference frame choice and streaklines in unsteady flows have been published by Hama (1962) and again recently by Kurosaka and Sundaram (1986).

As a result of these experimental difficulties, relatively few references discuss transverse vortices specifically, although they are usually an element of three-dimensional vortex structures. Some of the exceptions are listed in Table 2. For the present purposes, the table does not include the many papers on the large, relatively weakly rotational δ-scale motions, such as those described in, for example, Kovasnay et al (1970), Brown and Thomas (1977), and Antonia et al (1988).

Most of the evidence for strong local transverse vortical motion is from side-view flow-visualization studies such as Kim et al (1971), Clark and Markland (1971), Nychas et al (1973), Praturi and Brodkey (1978), Lian (1986), and Smith and Lu (1988). Probe-based data which imply transverse vortices are included in Nakagawa and Nezu (1981).

The most common reference to transverse vortices is in regard to near-wall hydrodynamic instability (e.g. Einstein and Li, 1956; Kline et al, 1967; Black, 1968; Smith, 1984), in which a local instability occurs at the shear-layer interface of high and low-speed fluid, resulting in a transverse vortex rollup.

Perhaps the earliest extensive description of transverse vortices is included in Clark and Markland (1971), who found them to be the predominant vortex element in the $y^+ > 70$ region. Clark and Markland observed that the streamwise lifetime of transverse vortices increases as distance from the wall increases, and that the diameter grows but the spin-rate slows during the vortex lifetime.

Kim et al (1971) found that transverse vortical motion occasionally appeared during the oscillatory phases of what they defined as the near-wall bursting process, but not as commonly as streamwise vortical motion or "wavy" motion.

Nychas et al (1973) described large-scale transverse vortices (of both rotational signs) that appeared to roll up at the shear-layer interface of high and low-speed fluid for $y^+ > 70$. These vortices were suggested to be the cause of the outer interface bulges described by Blackwelder and Kovasznay (1972), and others. In addition, a close spatial association was observed between the passage of transverse vortices and the occurrence of near-wall ejections of low-speed fluid. Outer-region transverse vortices were therefore suggested by Nychas et al to be the key structural element which connects the near-wall activity with the outer-flow large eddies.

Nychas et al's results were basically confirmed and greatly extended in stereo visualizations of the three-dimensional boundary layer structure by Praturi and Brodkey (1978). These results showed inflows of free-stream potential fluid (entrainment) in the vicinity of the large outer-flow transverse vortices. Contrary to the earlier speculations of several groups, it was stated by Praturi and Brodkey that large-scale bulges in the outer turbulent/non-turbulent interface are not caused by unusually high-momentum ejections of fluid from the near-wall region, but instead by the outer-region transverse vortical motions.

The presence of transverse vortices at the interface of high- and low-speed fluid in the outer-layer was also emphasized by Nakagawa and Nezu (1981). Formation of transverse vortices for $y^+ < 100$ was described by Robinson et al (1988, 1989) as a rollup of near-wall shear layers consisting of locally concentrated spanwise vorticity.

It is fairly well-established that transverse vortices are common in the outer region of turbulent boundary layers, and less so in the near-wall region. It remains unclear whether the common mode is for transverse vortices to form locally in the outer region (Nychas et al, 1973) or in the buffer region and then migrate outward (Smith, 1984).

Hairpin/Horseshoe Vortex Conceptual Models

Horseshoe or hairpin-shaped vortical structures dominate the proposed conceptual models for boundary layer turbulence. Notable alternative (but not necessarily conflicting) concepts are the inclined roller-eddy model of Townsend (1976) (see also Guezennec, 1986), and the vortex ring models of Falco (1983, 1988) and of Kobashi and Ichijo (1986). (See Table 3). In the interest of brevity, we will review only the horseshoe/hairpin models here. Excellent reviews of such models have already been published by Wallace (1982, 1985), so the current section may be considered an update of Wallace's work.

In a landmark paper, Theodorsen (1952) proposed the first vortex model for turbulence production throughout the boundary layer. Theodorsen's model may be considered the ancestor of most vortex structure models proposed in the last 35 years. This model was developed to satisfy the vorticity transport form of the Navier-Stokes equations and consisted of vortices bent into an inclined horseshoe shape (Fig. 2). The vortex structures were described as "tornadoes" which were inclined head-downstream at 45 degrees, with spanwise dimensions proportional to the distance from the wall. This vortical model was proposed as the fundamental structure of transitional and fully turbulent boundary layers, being responsible for both the production and dissipation of turbulence energy. It is notable that extended quasi-streamwise vortex elements (legs) did not play a major role in Theodorsen's model.

Theodorsen also described the outward and downstream growth of a horseshoe vortex after its birth "embracing a region of low velocity medium adjacent to the boundary." The streamwise distance travelled by a mature vortex structure was postulated as approximately equal to the distance from its head to the wall. A useful synopsis of Theodorsen's paper is given in Head and Bandyopadhyay (1981).

Willmarth and Tu (1967) used their space-time correlations between the wall pressure and all three velocity components near the wall to devise a model for the average eddy structure of the near-wall region (Fig. 3). The model describes the deformation of initially two-dimensional transverse vorticity lines into three-dimensional hairpin shapes sloped downstream at about 10 degrees from the wall, with the dominant element being vorticity lines with a streamwise component. Although Willmarth and Tu (1967) proposed their hairpin vortex-line model for the near-wall region only, Willmarth and Lu (1972) suggested that near-wall hairpin vortices may evolve to a larger scale, producing the intermittent bulges in the outer edge of the boundary layer and providing an outward interaction mechanism between the inner and outer regions.

A concept by Lighthill (1963) was invoked by Kline et al (1967) to explain the formation of the sublayer streaky structure. In this idea, any fluctuating velocity normal to the wall stretches (for wallward movement) or compresses (for outward movement) the near-wall spanwise vorticity lines. Since the spanwise vorticity is due mainly to $\partial u/\partial y$, this stretching and compressing would lead to spanwise variation in the near-wall value of u. Kline et al also drew on Stuart's (1965) vortex-stretching concepts from transition research to explain the formation of intense local shear layers above lifted low-speed elements, which are a precursor to the oscillation and breakdown phases of bursting. Kline et al's paper concludes with a diagram of an initially spanwise vorticity line being lifted and stretched into a loop (Fig. 4).

Black (1968) used a simple instability argument to propose a flow model based upon horseshoe vortices which are "shed" from a near-wall instability. In their early formation stages, the vortex structures were described as closed loops, or rings (Fig. 5a). As the outer portions of the vortex evolve outward and downstream, the wallward transverse element of the original ring is left behind to decay in the viscous sublayer, leaving a horseshoe-shaped vortex. The heads of these discrete vortices move outward from the wall, thereby stretching the trailing legs and inclining the horseshoe vortex. The vortex structures induce an inviscid outflow of low-speed fluid from within the vortex loop, creating motions which are seen as sharp, intermittent spikes of Reynolds shear stress by a stationary probe. Instead of individual horseshoe vortices, Black proposed a structure comprised of several horseshoe elements in various stages of growth, which share a common front-like trajectory in space. The vortex structure is maintained for much longer periods than the lifetime of the component vortex elements by the continuous creation of new elements which replace the older members (Fig. 5b).

In a review of boundary layer structure concepts, Hinze (1975) attempted to relate the known coherent elements of near-wall turbulence production to the dynamics of horseshoe-shaped vortices (Fig. 6). In his scenario, Hinze suggests that fluid lifted between the legs of the vortex loop gives rise to a locally unstable shear layer, which then violently breaks down (bursts) into a "blob of fluid of high turbulence intensity," apparently destroying the parent vortex structure in the process. Wallward inrush motions were suggested to be initiated by the tip of the vortex loop on its downstream side, and later aided by pressure waves created during the sudden vortex/shear-layer breakdown.

Offen and Kline (1975) also attempted to synthesize most of the known visual features of near-wall boundary layer structure with a lifted and stretched vortex loop which was essentially the same as that of Kline et al (1967). Offen and Kline describe how the three kinds of oscillatory motion observed during the near-wall bursting process by Kim et al (1971) may be related to the passage of a horseshoe vortex (Fig. 7). Pairing of aligned vortex structures and violent interaction of non-aligned vortices are also postulated.

The most extensive and influential experimental evidence for the existence of loop-shaped vortical structures in turbulent boundary layers was that of Head and Bandyopadhyay (1981). These authors' flow-visualizations of boundary layers over a broad Reynolds number range ($500 < Re_\theta < 17500$) provided images of hairpin-shaped structures virtually dominating the boundary layer. These structures were interpreted as vortices by the authors (Fig. 8). At high Reynolds numbers, the vortices were elongated and hairpin-shaped, forming a characteristic angle of 45° with the wall. Large-scale structures were observed to consist of agglomerations of hairpin-shaped vortices. At low Re_θ, the vortices were less elongated, and more horseshoe-shaped, and the large-scale features were composed of just one or two vortices. Although Head and Bandyopadhyay's visual evidence for the existence of hairpin/horseshoe vortices is compelling, the dynamics that underlie their evolution as well as their contribution to turbulence production and dissipation remain unclear.

Head and Bandyopadhyay's work helped to inspire Perry and Chong's (1982) analysis of a model for the mechanism of wall-bounded turbulence (Fig. 9a). In this model, the boundary layer is represented by a forest of potential-flow Λ-shaped vortices, which were introduced as a candidate form for Townsend's (1976) "attached-eddy" hypothesis (Fig. 9b). Biot-Savart calculations of a geometrical hierarchy of such vortices gave promising reproductions of the mean profile, Reynolds shear-stress, turbulence intensities, and spectra for a turbulent boundary layer, lending further credibility to the idea of vortical loops as the dynamically dominant boundary layer structure.

Perry, Henbest, and Chong (1986) extended both the attached-eddy hypothesis of Townsend (1976) and the Λ-vortex model of Perry and Chong (1982) to include the entire turbulent boundary layer rather than just the wall (log) region. The updated Perry et al model for wall turbulence is based upon the existence of hierarchies of attached coherent eddies. The first hierarchy of attached eddies forms at the outer edge of the sublayer, then stretch and grow with a fixed orientation to the wall (e.g. 45° for hairpin vortices). Eddies that do not die through viscous diffusion or vorticity cancellation merge to form eddies of a larger length scale, which comprise the second hierarchy. This continual process creates a "hierarchy of geometrically similar hierarchies" of attached eddies, which are responsible for the mean vorticity, Reynolds shear stress, and most of the energy-containing motions.

Perry et al propose that the attached eddies are immersed in a soup of detached isotropic small-scale motions which are responsible for the Kolmogoroff spectral region and most of the turbulent energy dissipation. Thus, in the model, energy is extracted from the mean flow by coherent motions and dissipated into heat by incoherent, small-scale motions. The model involves energy flow to low wave numbers through eddy-merging, and energy flow to high wave numbers through the unattached, dissipative motions. The assumptions involved in Perry et al's model lead to a logarithmic law of the wall, a constant Reynolds shear stress region, and an inverse power law u' spectrum near the wall for a variety of attached eddy shapes and distributions. This indifference of the statistics to the geometry of the eddy structure tends to de-focus the attention being paid to the exact form of coherent eddy structure dominant in the boundary layer.

Falco (1982) observed the formation of short-lived, near-wall hairpin vortices on the downstream edge of sublayer "pocket" modules, which are created from the impact of relatively high-speed fluid upon the sublayer. These hairpin vortex structures were found to be associated with outward ejections of low-speed fluid, in agreement with the hypotheses of Kline et al (1967), and others.

Wallace (1982; updated in 1985) convincingly reviewed the quantitative evidence for hairpin/horseshoe vortices in boundary layers (Fig. 10). To explain the birth of the horseshoe vortices, Wallace invokes the Navier-Stokes equations at the wall, which show that local wall-pressure gradients are equivalent to an outward diffusion of vorticity from the wall. Although the equations predict the generation of strongly kinked vorticity lines near the wall, the concept is not necessarily applicable to the the formation of true vortices, which can be quite distinct from vorticity lines. This issue will be revisited in the following section.

Working from the many vortex models in the literature as well as from his own extensive visualization studies, Smith (1984) described the most complete conceptual model yet proposed for hairpin-shaped vortices in the wall region ($y^+ < 100$) (Fig. 11). The model describes both the kinematics and dynamics of hairpin vortices and their relations to low-speed streaks, the bursting process, near-wall shear layers, ejections, and sweeps. Smith proposes that the "bursting" of a low-speed streak is the visual and probe signature of vortex roll-up (one or a packet) in the unstable shear layer formed on the top and sides of the streak. Once formed, a vortex loop moves outward by self-induction and downstream due to the streamwise velocity gradient. The trailing legs of the loop remain in the near-wall region but are stretched, forming counter-rotating quasi-streamwise vortices which serve to pump fluid away from the wall (ejection) and to accumulate low-speed fluid between the legs. Coalescence of the stretched legs of multiple "nested" hairpins is postulated as a mechanism by which low-speed streaks are preserved or redeveloped during the bursting process, leading to observed streak lengths considerably greater than the streamwise extent of any particular hairpin vortex. The streamwise array of vortices which comprises a burst grows outward and may agglomerate into large-scale rotational outer-region bulges.

Although elements of Smith's scenario may be traced to the literature cited above, his model is the most complete in its schematic of vortex structure evolution and its description of the relationship of hairpin vortices to the stages of the bursting process.

Acarlar and Smith (1987a,b) extended the investigation of hairpin vortex dynamics with a pair of papers in which vortices were generated in a laminar boundary layer by shedding from a wall-mounted hemisphere or by rollup on an artificial low-speed streak at the wall (Fig. 12). The results support the concept of three-dimensional vortex formation through the rollup of an unstable shear layer wrapped over the top and/or sides of a low-speed streak. Low-speed regions in the laminar layer were observed to lift, oscillate, and break into small-scale motions during the passage of a hairpin vortex.

Smith and Lu (1988) have applied digital image processing techniques to detect hydrogen-bubble patterns in side-views of a turbulent boundary layer, using pattern-recognition templates obtained from bubble patterns surrounding artificially-generated hairpin vortices in a laminar boundary layer. The resulting distribution of detected hairpin patterns is skewed toward the wall, with a peak in the distribution centered at $y^+ \approx 40$.

Since direct experimental evidence for the existence of horseshoe/hairpin vortices in turbulence has been so limited, Moin and Kim have analyzed their Large-Eddy and direct numerical simulations of turbulent channel flow to locate, identify, and characterize hairpin vortices in the simulated turbulent channel flow.

In the first of two papers, Moin and Kim (1985) analyzed the vorticity field, employing vorticity vector angle histograms, two-point vorticity correlations, and vorticity line tracing in individual flowfields. A hairpin vortex was defined as "an agglomeration of vortex lines in a compact region (with higher vorticity than the neighboring points) that has a hairpin or horseshoe shape." (A definition of this form can be utilized only by numerical simulation researchers, who have complete access to the three-dimensional vorticity vector field.)

In their results, Moin and Kim showed that for the outer region of the flow, the vorticity vectors tend to be inclined at about 45 degrees to the wall. Two-point velocity and vorticity correlations in the 45 and 135 degree planes provided additional evidence for 45 degree vortical structures. These data described only single, inclined vortical structures, with unknown connection to hairpins themselves. To investigate the three-dimensional nature of the vortical structures, instantaneous vorticity lines (everywhere parallel to the vorticity vector) were traced through the flow, and were commonly found in horseshoe shapes, though usually asymmetric (Fig. 13). The horseshoe-shaped vorticity lines appeared to coalesce from deformed vortex sheets, and generally did not exhibit elongated streamwise legs. From these results, Moin and Kim concluded that 45 degree, hairpin-shaped vortices are statistically relevant features of turbulent channel flow structure.

In the follow-on paper, Kim and Moin (1986) applied variants of the VITA and $u'v'$ quadrant-splitting technique to detect turbulence-producing events in the LES channel flow. Although the objective was to isolate the structures associated with the near-wall bursting process, the detection points were placed in the outer regions of the flow ($y^+ = 100, 200, 300$) to avoid triggering by near-wall sweep motions. The resulting conditionally-averaged fields were then visualized by vorticity line tracing. The results exhibited horseshoe-shaped vorticity lines in both the conditionally averaged and selected instantaneous fields. The bunched instantaneous vorticity line structures were interpreted as true vortices (with pronounced circular motion about the axis). Upright horseshoe-shaped vorticity lines were found to be associated with ejection ($(u'v')_2$) motions, while inverted horseshoes were found in conjunction with sweep ($(u'v')_4$) motions. Since the horseshoe vortex structures were detected well away from the wall, it was argued that horseshoe vortices are a result of only vortex stretching, and are thus characteristic of all turbulent shear flows, whether or not there is a wall. (The argument that vortex stretching is a sufficient condition for the generation of horseshoe vortices was also the fundamental premise of Theodorsen's paper). Quasi-streamwise vortices were rarely found in the simulation results, and were generally of limited ($\Delta x^+ < 100$) streamwise extent. The major results of this paper were also confirmed in a new direct simulation without a subgrid model.

In a general review of coherent structures in turbulent flows, Fiedler (1986) included a sketch of hairpin vortices which suggests a phase relationship between the vortex structures and the large-scale motions associated with bulges in the outer interface of the boundary layer (Fig. 14). The figure depicts hairpins with 45 degree heads and nearly horizontal near-wall legs, and shows the hairpins residing on the backs of the outer-flow large-scale motion.

A dramatic example of the advantages of modern digital image-processing techniques as applied to coherent structure research is the work of Utami and Ueno (1987). These authors used successive pictures of particles in $x - z$ cross-sections at several y-values to obtain instantaneous distributions in the $x - z$ plane of the three components of velocity and vorticity, and of various associated spatial statistics. (This work has probably come the closest to producing experimental data approaching the detail provided by direct numerical simulations.) The results were interpreted by the authors with vortical structures in mind, and a horseshoe vortex model with coalesced legs (reminiscent of Smith, 1984) is proposed which exhibits causal relationships to low- and high-speed streaks, ejections and sweeps (quadrant 2 and 4 $u'v'$ motions, respectively), longitudinal vortices, and internal shear layers (Fig. 15).

Adrian (1988) has suggested that the topological distinction between near-wall hairpin vortices and a ring vortex plus two trailing quasi-streamwise vortices may not be necessary (Fig. 16). This proposal (as well as Black's, 1968) provides a conceptual link between hairpin vortices and the vortex ring structures documented by Falco (1983, 1988).

Kasagi (1988) presented a model of the near-wall region which featured alternating streamwise vortices with accumulated regions of low-speed fluid. Included in the paper was a figure illustrating the connections between streamwise vortices in hairpin shapes (Fig. 17).

Robinson et al (1988, 1989) analyzed Spalart's (1988) numerically-simulated boundary layer to educe the character and evolution of three-dimensional vortical structures, as well as their relationships to other coherent motions. Vortices were identified in the simulation by their elongated low-pressure cores. The results showed a strong spatial correspondence between vortical structures and both $(u'v')_2$ and $(u'v')_4$ motions (Fig. 18). Although hairpin-shaped vortices were occasionally sighted, long trailing legs were observed to be rare and fleeting, compared to the relatively common occurrance of long-lived vortical arches without extended legs.

Literature Summary

The existence of vortical elements in turbulent boundary layers is well established. However, experimental and even computational detections of horseshoe/hairpin vortical structures are outnumbered by the theoretical arguments for their existence. Most critically, the role and statistical relevance of vortical structures as a dynamical element of boundary-layer turbulence is mostly hypothesis. Quasi-streamwise vortices are the most thoroughly documented of the vortex elements, but distributions of the diameters, distance from the wall, strength, and population of all types of boundary-layer vortices are scarce.

In some ways, this lack of information may not be surprising, since we are in need of both a general definition and a set of useful detection techniques for vortices. Nevertheless, there is almost universal agreement that some form of vortical structures play a key role in the creation and maintenance of turbulence in boundary layers.

THE USE OF VORTICITY LINES

The use of vorticity lines (often called vortex lines) in numerically-simulated turbulence has proven useful for detecting vortical structures in the simulation databases (e.g. Moin and Kim, 1985; Kim and Moin, 1986). However, vorticity lines can be misleading unless the distinction between vortex and vorticity is maintained. In the literature of coherent vortical motions, this distinction has not always been made clear. Therefore, as a reminder, this section briefly outlines the causes and implications of horseshoe-shaped vorticity lines in turbulent shear flows. A more thorough discussion of the use of vorticity lines is included in Moin and Kim (1985).

A vorticity line is defined as a line everywhere parallel to the instantaneous vorticity vector, and its location in space is defined by

$$\frac{d\vec{x}}{ds} = \frac{\vec{\omega}}{|\vec{\omega}|} \tag{1}$$

where $\vec{x}$ is the position vector of the vorticity line, s is the distance measured along the vorticity line, and $\vec{\omega}(\vec{x})$ is the vorticity field. In the absence of viscous diffusion, vorticity lines cannot end within the flow, and must travel with the fluid particles.

In a shear flow, it is a kinematical necessity that upright, downstream-leaning, loop-shaped vorticity lines surround a region of fluid which is lifted upwards from the lower speed region into the higher speed flow. If we consider outward movement of a low-speed fluid parcel of limited spanwise extent (like a $(u'v')_2$ ejection), and ignore viscous diffusion effects, initially spanwise vorticity lines will be lifted outward in the region of the upward-moving fluid, and carried downstream by the higher speed flow above. The result is a sloping upright loop in the vorticity lines, without the existence of a vortex. This is illustrated in Figure 19, in which a vorticity line has been integrated through a localized region of lifting, low-speed $(+v', -u')$ fluid within an otherwise two-dimensional, linear shear flow. Similar arguments describe the formation of inverted, upstream-leaning vorticity-line loops for parcels of high-speed, wallward-moving fluid (sweeps).

Vortices consist of "bundles" of vorticity lines, with rotational motion about the bundle axis, so single vorticity lines cannot be construed as vortices. Thus, arrays of "horseshoe"-shaped vorticity lines must always accompany ejections and sweeps, but the association of vortices with these motions is not necessary from a kinematic point of view. To carry the point a bit further, any shear flow with a mean Reynolds shear stress (and hence $(u'v')_2$ and $(u'v')_4$ motions) must possess horseshoe-shaped vorticity lines, whether or not vortices are present.

When vorticity lines are traced in the vicinity of true vortices, the results can be surprisingly misleading. Consider, for example, a streamwise vortex in a shear flow (a common occurrence in the near-wall region of turbulent boundary layers). Unless the starting point for the integration of equation (1) is chosen almost precisely within the vortex core, the vorticity lines will trace out well-defined upright hairpins on the outward-rotating side of the streamwise vortex, and inverted hairpins on the wallward-rotating side. This is demonstrated in Figure 20, which is schematic drawing of vorticity lines computed by the author on both sides of a near-wall quasi-streamwise vortex found in Spalart's (1988) numerically-simulated boundary layer.

These points are not made to denigrate the method of vorticity line tracing, but to raise a warning that horseshoe or hairpin-shaped vorticity lines are vastly more common than similarly shaped vortices. This is a crucial dynamic issue, since isolated vorticity lines play a different role than true vortices with regard to induced motions and propagation of pressure disturbances.

Other than direct observation of rotational motions, no vortex detection technique has yet been found that is free of ambiguities of interpretation. For instance, the static pressure field available in the numerical simulations has been found to be useful for identifying the low-pressure cores of vortices (Robinson et al, 1989). Most low-pressure regions in the simulated boundary layer are observed to be elongated, and every elongated low-pressure region checked so far has corresponded to a vortex, under the definition set forth earlier in this paper. However, non-elongated low-pressure regions in the simulation apparently are not vortices, and are created through other types of local motions. Unambiguous vortex detection awaits improved experimental and computational detection methods as well as a more widely accepted definition of a vortex.

VORTICES IN THE NUMERICALLY-SIMULATED BOUNDARY LAYER

As part of a larger-scale research project, Spalart's (1988) direct Navier-Stokes simulation of a flat-plate, zero pressure gradient boundary layer has been analyzed extensively to clarify the kinematics of all detectable forms of coherent motions (Robinson et al, 1988, 1989; Robinson, 1989). In this section, new statistical results concerning the size, locations, and strength of vortices in the numerical boundary layer are presented.

The results presented here are for a momentum thickness Reynolds number of approximately 670. This is very low, so the results are directly indicative of the character of a very low Reynolds number boundary layer only, with unknown extensibility to higher Reynolds numbers.

Spalart's spectral code has been used to compute and save 104 time-steps ($\Delta t^+ = 3$) beyond the initial 1200 timesteps required to achieve statistical equilibrium. These 104 time-steps represent nearly 54 gigabytes of turbulence information, and comprise the database utilized for all of the analyses in the project. Each time-step is computed on a 384 x 288 x 85 grid, comprising 9.4 million nodes. At each node, for each time-step, pressure and all three components of velocity (and thus vorticity) are available. Grid resolution is is 12.8 viscous lengths in the streamwise (x) direction, and 4.3 viscous lengths in the spanwise (z) direction. Resolution in the wall-normal (y) direction varies from 0.03 to 16.0 viscous lengths, with 14 grid points between the wall and $y^+ = 10$. The grid spacing for the $Re_\theta = 670$ case gives a computational domain with streamwise, spanwise, and wall-normal dimensions of 4900, 2500, and 1100 viscous units, respectively. The boundary layer is approximately 300 viscous units thick at this Reynolds number. Further details concerning the computational method and the results are included in Spalart (1988). (The simulation used for the present study has been computed with a finer grid than the results reported in Spalart, 1988.)

In accordance with the working definition of a vortex stated above, vortices in the end-view ($y - z$) and side-view ($x - y$) planes have been identified by plots of instantaneous streamlines or vectors. For quasi-streamwise vortices, streamlines were plotted in 25 $y - z$ planes with dimensions $\Delta z^+ = 418$ by $\Delta y^+ = 250$. For spanwise vortices instantaneous vectors constructed with the fluctuating velocity (u', v') components were found to display all vortices identifiable from any moving reference frame. (This suggests that transverse vortices travel downstream at velocities not too different from the local mean velocity.) Dimensions of the 15 $x - y$-plane views used for the transverse vortex counting were $\Delta x^+ = 845$ by $\Delta y^+ = 250$. All data-planes were chosen at widely different and unrelated spatial and/or temporal locations in the simulation database. Examples of the $y - z$ and $x - y$ planes are given in Figure 21, in which vortices are clearly visible in the streamlines, but less obvious in the vectors.

For each vortex visually identified in the plots, the diameter, distance from the wall, core area, and approximate circulation was recorded. For the streamlines in the end-view planes, the diameter was estimated by measuring the extent of the bounding streamline of a group of closed concentric streamlines, and the area was estimated as the area of a circle with that diameter. Transverse vortices were more difficult to identify in the side-view vector plots, although strong (high circulation) vortices with significant perturbation velocities tended to be fairly obvious. For this reason, there may be a bias towards stronger transverse vortices in the data presented below, and this effect is accentuated near the wall. Though tedious, hand-counting of the "visual vortices" was considered preferable to more sophisticated, but less reliable vortex identification methods, at least for a first set of control data.

Quasi-Streamwise Vortex Results

Statistics were computed for a sample of 229 quasi-streamwise vortices identified visually in the simulated boundary layer. Recall that vortices visible in end-plane ($y - z$) views need not be (and usually are not) strictly streamwise but may possess significant spanwise and/or wall-normal components, and are thus referred to as "quasi-streamwise."

The distribution of distances of the vortex centers to the wall is shown in Figure 22. In agreement with experimental results, quasi-streamwise vortices are more commonly found in the wall region, with 72% of the total occurring for $y^+ < 100$. The diameter distribution in Figure 23 shows that 73% of the visual quasi-streamwise vortices have diameters between 10 and 40 viscous lengths, and 90% between 10 and 60 viscous lengths. The average diameter of the quasi-streamwise vortices in the sample is 34 viscous lengths.

The variation of quasi-streamwise vortex diameters with distance from the wall is plotted in Figure 24. Larger vortices generally reside further from the wall, but the data is broadly scattered between the $d^+ = 2y^+$ line and the lower bound of $d^+ \approx 10$.

Since visual identification of vortices cannot easily determine their potential as significant momentum transport "pumps," the circulation of each visual vortex was also computed. Following Pearson and Abernathy (1984), the circulation is described in terms of a vortex Reynolds number defined as

$$R_V \equiv \Gamma/2\pi\nu, \tag{2}$$

where Γ is the circulation computed by integrating the streamwise vorticity ω_x over the vortex area. As a reference value for evaluating the strengths of vortices, the vortex Reynolds number of a δ by δ box in the $x-y$ plane may be computed. For the numerically simulated boundary layer, this reference circulation is

$$R_{V_\delta} = \frac{U_e \delta}{2\pi\nu} \approx 955. \tag{3}$$

Since Γ is computed as an area integral, very large vortices are likely to contain high values of circulation. To estimate the intensity of the vortices, the vortex Reynolds number R_V may be divided by the area of the vortex, non-dimensionalized by viscous units. Using the δ by δ box, a reference intensity for the simulated boundary layer is ≈ 0.011, which may be taken as a mean circulation intensity for the boundary layer.

The vortex Reynolds number and intensity distributions for visually-identified quasi-streamwise vortices are shown in Figures 25 and 26, respectively. The average value of R_V is 22, but the distribution has a long tail toward much higher values. However, the intensity distribution (Figure 26) shows that about 90% of the quasi-streamwise vortices have a higher intensity than the boundary layer reference value of 0.011. The most intense quasi-streamwise vortices have intensity values up to nearly 10 times the reference value, and nearly all of the very intense vortices occur for $y^+ < 75$. Vortices with higher intensities than the average circulation intensity in the boundary layer may be expected to have sufficient strength to induce significant momentum transport through induction.

Spanwise Vortex Results

The sample used for transverse vortex statistics was 85 "visual vortices." From the wall-distance distribution in Figure 27, it is seen that transverse vortices tend to be located outside the near-wall region, with over 80% of the population occurring for $80 < y^+ < 180$, in a boundary layer with $\delta^+ \approx 300$. The broad, outer-region distribution of transverse vortices differs significantly from the near-wall preference of quasi-streamwise vortices (Figure 22).

Transverse vortices also tend to exhibit larger diameters than quasi-streamwise vortices. The distribution in Figure 28 shows 74% of transverse vortex diameters measuring between 30 and 70 viscous lengths, with an average value of 51. The variation of transverse vortex diameter with distance from the wall (Figure 29) exhibits wide scatter, but the data is fit reasonably well by a $d^+ = \kappa y^+$ line, where $\kappa = 0.41$ is the Karman constant. This apparent agreement with the boundary-layer mixing-length distribution is not surprising, and suggests that transverse vortices do indeed play a statistically significant role in determining the average statistics within a turbulent boundary layer.

Since transverse vortices are embedded in a mean velocity gradient, their circulation was computed with the fluctuating component $(\omega_z'(x,y) = \omega(x,y) - \overline{\omega}(y))$ of spanwise vorticity. This approach may be criticized on the grounds that the transverse vortices may themselves be the mean vorticity, but it was desired to compare the circulation intensities between quasi-streamwise and transverse vortices.

The distribution of vortex Reynolds number (Figure 30) shows that transverse vortices tend to have higher circulation values than quasi-streamwise vortices (comparing again to the boundary-layer reference value of 955), which is to be expected given their generally larger diameter. The R_V distribution peaks at about 30 for transverse vortices, and at about 12 for streamwise. About 17% of the transverse vortices were found to have positive values of circulation, i.e. containing lower total circulation than the local mean, which is negative.

The circulation intensity (R_V/A^+) distribution for transverse vortices is plotted in Figure 31. Transverse vortices are significantly less intense than quasi-streamwise vortices, with the peak in the distribution appearing at just above the reference intensity value of 0.011. However, occasional intense transverse vortices do occur and are generally found below $y^+ = 75$, as in the case of quasi-streamwise vortices.

If circulation intensity is assumed to determine the potential of a vortex for producing contributions to $-\overline{u'v'}$, the clustering of high-intensity quasi-streamwise vortices near the wall, and the distribution of low-intensity transverse vortices in the outer flow are consistent with the shape of the turbulence production ($-\overline{u'v'}\frac{\partial U}{\partial y}$) profile in a boundary layer. Further statistics of vortices in Spalart's simulated boundary layer may be found in Robinson (1989).

Vortical Structure Results

To identify three-dimensional vortical structures in the numerically-simulated boundary layer, advantage is taken of the fact that the pressure is low within the cores of vortices with significant circulation. When isobaric surfaces of constant low pressure are plotted, elongated tubular structures emerge which exhibit circular streamlines (from a reference frame moving with the low-pressure region) in any cross-section. As mentioned above, not all low-pressure regions are elongated, and those which are not apparently do not correspond to vortices. An example of a subvolume from the simulation database is shown in Figure 18, in which several vortical structures are clearly evident. Figure 18 contains a large, assymetric loop-shaped structure with a single long, trailing leg, as well as a smaller horseshoe-shaped structure with a secondary transverse vortex visible just upstream. Regions of significant $(u'v')_2$ and $(u'v')_4$ are seen to have a close spatial association with the vortical structures in Figure 18 (Robinson et al, 1989).

The low-pressure regions in the simulated boundary layer exhibit a wide variety of shapes and sizes, as seen in the top-view of the computational domain in Figure 32. In this picture, the low-pressure isobaric surfaces corresponding to $p'/\rho u_\tau{}^2 = -4.2$ are plotted in three dimensions in a volume measuring 4900 by 1225 viscous lengths in the streamwise and spanwise directions, respectively. (The subvolume of Figure 18 is in the upper, far-right portion of Figure 32.)

The most striking impression gained from Fig. 32 is the wide variety of shapes of the elongated low-pressure regions. A number of arch-like structures are visible, with spanwise dimensions similar to the height of their heads from the wall. (This was determined through stereo top-views combined with selected side-view slices.) Hairpins with two elongated legs are rare, as are quasi-streamwise vortices longer than 200 viscous lengths; this observation is not dependent upon the pressure value chosen for the contour surfaces. The stereo versions of Figure 32 also confirm that the most common near-wall vortical structure has a mostly streamwise orientation, while most outer-region structures are arches with well-defined transverse elements. Although many arches can be found without trailing legs, it also true that many quasi-streamwise vortices exist without a clear connection to arches. This suggests that although a symmetric hairpin may be a reasonable idealized conceptual model for the vortical structures, predictive modeling might be successfully approached by treating the transverse and streamwise vortex elements separately. This scheme should be sufficient for a kinematic/statistical picture of the flow, but would not be appropriate for a dynamically accurate model.

For vortices with significant circulation, the radial pressure gradient at the edge of the core is relatively strong. As a result, small changes in the pressure contour level used to compute the isobaric surfaces in Fig. 32 do not significantly affect the overall picture of the vortical structures. The most noticeable effect of varying the contour level is to alter the topological connectivity between adjacent low-pressure regions.

CONCLUSIONS

From the brief literature review presented, the following general conclusions may be drawn regarding vortices as coherent motions in turbulent boundary layers:

The major vortex elements (quasi-streamwise "legs," transverse "heads," and "necks") all exist within the boundary layer. However, vortex structures of a wide variety of shapes also apparently exist, and the evidence for the prevalence of particular forms (horseshoes, hairpins, rings, etc.) is not yet conclusive.

The near-wall region ($y^+ < 100$) is dominated by quasi-streamwise vortices with an outward tilt. These vortices appear to be kinematically associated with $(u'v')_2$ and $(u'v')_4$ motions, and are probably related to the near-wall streaky structure, although the specific cause-and-effect relationships for the latter remain unclear. Spanwise and 45-degree vortices are more commonly found in the outer region ($y^+ > 100$). These vortices have a consistent kinematical relation to a large fraction of the Reynolds stress production outside the buffer zone.

Horseshoe/hairpin vortex models and simulations are consistent with many forms of experimental data, but actual evidence of such vortical structures existing in turbulent boundary layers is limited. As a result, much of the kinematic and dynamic behavior of horseshoe/hairpin vortices is postulated or inferred from measurable statistics. For example, the following issues (a partial list) concerning boundary-layer vortical structures may all be considered unresolved from the standpoint of a community consensus:

- Formation: shear-layer instability or vortex stretching alone?
- Growth: self-induction, circulation lift, or wall-normal pressure-gradient?
- Destruction: short or long life? dissipation or destructive instability?
- Regeneration: pairing? secondary rollup? bifurcations?
- Contribution to the gross statistics: rare, significant, or dominant?

The arguments and evidence for the existence of horseshoe/hairpin vortices, at least in the near-wall region, are convincing (e.g. Wallace, 1985). The relationship of such vortices to the "bursting" process is quite unclear, however. Among the vortex-structure interpretations of bursting are:

1) Rollup of a locally unstable shear layer atop a low-speed streak into a vortex (e.g. Kim et al, 1971; Smith, 1984; Blackwelder and Swearingen, 1989)
2) Instability breakdown of lifted fluid between the legs of a hairpin vortex (e.g. Hinze, 1975).
3) Direct induction during passage of a relatively long-lived vortex structure past a measuring or observing station (e.g. Black, 1968; Acarlar and Smith, 1986; Robinson et al, 1989).
4) Indirect influence from the passage of an outer-region transverse vortex (e.g. Nychas et al, 1977; Praturi and Brodkey, 1979).
5) Violent eruption due to local pressure fields generated by near-wall vortical motion (e.g. Walker and Herzog, 1987).

Clearly, a considerable body of work needs to be done before the kinematics, dynamics, statistical contribution, and Reynolds number dependence of vortical structures in turbulent boundary layers is agreed upon. The recent development of direct numerical turbulence simulations should help accelerate the learning pace by suggesting new experimental techniques as well as by providing direct information on low Reynolds-number vortical structures. Cautions are in order for the simulated turbulence too, however, since their spatial resolution is finite and thus may be insufficient to capture all viscous interactions between vortices. For example, the "pinch-off" of a hairpin vortex loop into a ring-shaped vortex is dynamically possible (Falco, 1983; Moin, Leonard, and Kim, 1986), but has not yet been observed in channel-flow (Kim, Moin, Moser, 1987) or boundary layer (Spalart, 1988) simulations. In addition, vorticity-line tracing as well as low-pressure marking in the simulations must be interpreted with care, as discussed in the present paper.

The obvious need is for the development of more robust experimental methods for vortex identification in turbulent flows. For this reason, recent advances in multi-sensor hot-wire anemometry (Antonia et al, 1988), scanning two-component laser-doppler anemometry (Williams and Economou, 1987), particle-displacement imagery (Landreth et al, 1988), and digital image pattern recognition (Corke, 1984; Smith and Lu, 1988) may hold the promise of the most productive era yet in turbulence structure research.

ACKNOWLEDGEMENTS

The work reported herein has been jointly funded by NASA Ames Research Center, The Air Force Office of Scientific Research, and the Office of Naval Research. This project is under the supervision of Prof. S.J. Kline, to whom the author is grateful for invaluable insight and advice. The author also thanks Dr. P. Spalart of Ames for the use of his numerical simulation code and data, and Dr. D. Stretch of the NASA/Stanford Center for Turbulence Research for helpful review comments. Data analysis assistance was provided by L. Portela, O. Manickham, and M. Bauer.

REFERENCES

Acarlar, M.S. and Smith, C.R., A Study of Hairpin Vortices in a Laminar Boundary Layer. Part I: Hairpin Vortices Generated by a Hemisphere Protuberance, *J. Fluid Mech.*, vol. 175, p. 1, 1987.

Acarlar, M.S. and Smith, C.R., A Study of Hairpin Vortices in a Laminar Boundary Layer. Part II: Hairpin Vortices Generated by Fluid Injection, *J. Fluid Mech.*, vol. 175, p. 43, 1987.

Adrian, R.J., Linking Correlations and Structure: Stochastic Estimation and Conditional Averaging, *Near Wall Turbulence: 1988 Zaric Memorial Conference*, Hemisphere, 1989.

Antonia, R.A., Browne, L.W.B., and Bisset, D.K., Effect of Reynolds Number on the Organised Motion in a Turbulent Boundary Layer, *Near Wall Turbulence: 1988 Zaric Memorial Conference*, Hemisphere, 1989.

Aubrey, N., Holmes, P., Lumley, J.L., and Stone, E., The Dynamics of Coherent Structures in the Wall Region of a Turbulent Boundary Layer, *J. Fluid Mech.*, vol. 192, p. 115, 1988.

Bakewell, H.P. and Lumley, J.L., Viscous Sublayer and Adjacent Wall Region in Turbulent Pipe Flow, *Phys. Fluids*, vol. 10, no. 9, p. 1880, 1967.

Black, T.J., An Analytical Study of the Measured Wall Pressure Field Under Supersonic Turbulent Boundary Layers, *NASA CR-888*, 1968.

Blackwelder, R.F. and Eckelmann, H., Streamwise Vortices Associated with the Bursting Phenomenon, *J. Fluid Mech.*, vol. 94, p. 577, 1979.

Blackwelder, R.F. and Kovasznay, L.S.G., Time Scales and Correlations in a Turbulent Boundary Layer, *Phys. Fluids*, vol. 15, no. 9, p. 1545, 1972.

Blackwelder, R.F. and Swearingen, J.D., The Role of Inflectional Velocity Profiles in Wall-Bounded Flows, *Near Wall Turbulence: 1988 Zaric Memorial Conference*, Hemisphere, 1989.

Brodkey, R.S., personal communication, 1987.

Brown, G.L. and Thomas, A.S.W., Large Structure in a Turbulent Boundary Layer, *Phys. Fluids*, vol. 20, no. 10, part 2, p. 5243, 1977.

Chu, C.C. and Falco, R.E., Vortex Ring/Viscous Wall Layer Interaction Model of the Turbulence Production Process Near Walls, *Exp. Fluids*, vol. 6, p. 305, 1988.

Clark, J.A. and Markland, E., Flow Visualization in Turbulent Boundary Layers, J. Hydr. Div. ASCE, vol 97, p. 1653, 1971.

Corino, E.R. and Brodkey, R.S., A Visual Investigation of the Wall Region in Turbulent Flow, *J. Fluid Mech.*, vol. 37, part 1, p. 1, 1969.

Corke, T.C., Digital Image Filtering in Visualized Boundary Layers, *AIAA J.*, vol. 22, no. 8, p. 1124, 1984.

Einstein H.A. and Li, H., The Viscous Sublayer Along a Smooth Boundary, *J. Eng. Mech.*, A.S.C.E., vol. 82, no. EM 2, 1956.

Ersoy, S. and Walker, J.D.A., The Boundary Layer due to a Three-Dimensional Vortex Loop, *AIAA J.*, vol. 24, p. 1597, 1986.

Falco, R.E., A Synthesis and Model of Wall Region Turbulence Structure, *Structure of Turbulence, Heat, and Mass Transfer*, p. 124, Hemisphere, 1982.

Falco, R.E., New Results, a Review and Synthesis of the Mechanism of Turbulence Production in Boundary Layers and its Modification, *AIAA 83-0377*, 1983.

Fiedler, H.E., Coherent Structures, *Advances in Turbulence*, Springer-Verlag, p. 320, 1986.

Grass, A.J., Structural Features of Turbulent Flow over Smooth and Rough Boundaries, *J. Fluid Mech.*, vol. 50, p. 233, 1971.

Guezennec, Y., Documentation of Large Coherent Structures Associated with Wall Events, Ph.D. Dissertation, Illinois Institute of Technology, 1985.

Guezennec, Y., Piomelli, U., and Kim, J., On the Shape and Dynamics of Wall Structures in Turbulent Channel Flow, *Phys. Fluids*, vol. A1, 1989.

Hama, F.R., Progressive Deformation of a Curved Vortex Filament by its own Induction, *Phys. Fluids*, vol. 5, p. 644, 1962.

Hanratty, T.J., A Conceptual Model of the Viscous Wall Region, *Near Wall Turbulence: 1988 Zaric Memorial Conference*, Hemisphere, 1989.

Head, M.R. and Bandyopadhyay, P., New Aspects of Turbulent Boundary Layer Structure, *J. Fluid Mech.*, vol. 107, p. 297, 1981.

Herzog, S., The Large-Scale Structure in the Near-Wall Region of Turbulent Pipe Flow, Thesis, Cornell University, 1986.

Hinze, J.O., *Turbulence*, McGraw-Hill, 1975.

Jang, P.S., Benney, D.J., Gran, R.L., On the Origin of Streamwise Vortices in a Turbulent Boundary Layer, *J. Fluid Mech.*, vol. 169, p. 109, 1986.

Kasagi, N., Hirata, M., and Nishino, K., Streamwise Pseudo-Vortical Structures and Associated Vorticity in the Near-Wall Region of a Wall-Bounded Turbulent Shear Flow, *Exp. Fluids*, vol. 4, p. 309, 1986.

Kasagi, N., Structural Study of Near-Wall Turbulence and its Heat Transfer Mechanism, *Near Wall Turbulence: 1988 Zaric Memorial Conference*, Hemisphere, 1989.

Kastrinakis, E.G., Wallace, J.M., Willmarth, W.W., Ghorashi, B., and Brodkey, R.S., On the Mechanism of Bounded Turbulent Shear Flows, *Lecture Notes in Physics* 75, Springer-Verlag, 1978.

Kim, J. and Moin, P., The Structure of the Vorticity Field in Turbulent Channel Flow. Part 2: Study of Ensemble-Averaged Fields, *J. Fluid Mech.*, vol. 162, p. 339, 1986.

Kim, H.T., Kline, S.J., and Reynolds, W.C., The Production of Turbulence Near a Smooth Wall in a Turbulent Boundary Layer, *J. Fluid Mech.*, vol. 50, part 1, p. 133, 1971.

Kim, J., Moin, P., and Moser, R.D., Turbulence Statistics in Fully-Developed Channel Flow at Low Reynolds Number, *J. Fluid Mech.*, vol. 177, p.133, 1987.

Kim, J., On the Structure of Wall-Bounded Turbulent Flows, *Phys. Fluids*, vol. 26, p. 2088, 1983.

Kline, S.J. and Robinson, S.K., Turbulent Boundary Layer Structure: Progress, Status, and Challenges, *2nd IUTAM Symposium on Structure of Turbulence and Drag Reduction*, Zurich, Switzerland, 1989.

Kline, S.J., Reynolds, W.C., Schraub, F.A., Runstadler, P.W., The Structure of Turbulent Boundary Layers, *J. Fluid Mech.*, vol. 30, p. 741, 1967.

Kobashi, Y. and Ichijo, M., Wall Pressure and its Relation to Turbulent Structure of a Boundary Layer, *Exp. Fluids*, vol. 4, p. 49, 1986.

Kovasznay, L.S.G., Kibens, V., and Blackwelder, R.F., Large-Scale Motion in the Intermittent Region of a Turbulent Boundary Layer, *J. Fluid Mech.*, vol. 41, part 2, p. 283, 1970.

Kreplin, H.-P. and Eckelmann, H., Propagation of Perturbations in the Viscous Sublayer and Adjacent Wall Region, *J. Fluid Mech.*, vol. 95, part 2, p. 305, 1979.

Kurosaka, M. and Sundaram, P., Illustrative Examples of Streaklines in Unsteady Vortices: Interpretational Difficulties Revisited, *Phys. Fluids*, vol. 29(10), 1986.

Landreth, C.C., Adrian, R.J., and Yao, C.S., Double Pulsed Particle Image Velocimeter with Directional Resolution for Complex Flows, *Exp. Fluids*, vol. 6, no. 2, p. 119, 1988.

Lian, Q.X., Coherent Structures of Turbulent Boundary Layer in Flows with Adverse Pressure Gradient, in *Sino-U.S. Joint Fundamental Experimental Aerodynamics Symposium*, Hampton, VA, 1987.

Lighthill, M.J., in *Laminar Boundary Layers*, Clarendon Press, Oxford, p. 99, 1963.

Moin, P. Probing Turbulence via Large Eddy Simulation, *AIAA-84-0174*, 1984.

Moin, P. Analysis of Turbulence Data Generated by Numerical Simulations, *AIAA-87-0194*, 1987.

Moin, P., Adrian, R.J., and Kim, J., Stochastic Estimation of Conditional Eddies in Turbulent Channel Flow, *6th Symposium on Turbulent Shear Flows*, Toulouse, France, 1987.

Moin, P. and Kim, J., The Structure of the Vorticity Field in Turbulent Channel Flow. Part 1: Analysis of Instantaneous Fields and Statistical Correlations, *J. Fluid Mech.*, vol. 155, p. 441, 1985.

Moin, P., Leonard, A., and Kim, J., Evolution of a Curved Vortex Filament into a Vortex Ring, *Phys. Fluids*, vol. 29, no. 4, p. 955, 1986.

Moser, R.D. and Moin, P., The Effects of Curvature in Wall-Bounded Turbulent Flows, *J. Fluid Mech.*, vol. 175, p. 479, 1987.

Nakagawa, H. and Nezu, I., Structure of Space-Time Correlations of Bursting Phenomena in an Open-Channel Flow, *J. Fluid Mech.*, vol. 104, p. 1, 1981.

Nychas, S.G., Hershey, H.C., and Brodkey, R.S., A Visual Study of Turbulent Shear Flow, *J. Fluid Mech.*, vol. 61, p. 513, 1973.

Offen, G.R. and Kline, S.J., A Proposed Model of the Bursting Process in Turbulent Boundary Layers, *J. Fluid Mech.*, vol. 70, part 2, p. 209, 1975.

Pearson, C.F. and Abernathy, F.H., Evolution of the Flow Field Associated with a Streamwise Diffusing Vortex, *J. Fluid Mech.*, vol. 146, p. 271, 1984.

Perry, A.E. and Chong, M.S., On the Mechanism of Wall Turbulence, *J. Fluid Mech.*, vol. 119, p. 173, 1982.

Perry, A.E., Henbest, S., and Chong, M.S., A Theoretical and Experimental Study of Wall Turbulence, *J. Fluid Mech.*, vol. 165, p. 163, 1986.

Praturi, A.K. and Brodkey, R.S., A Stereoscopic Visual Study of Coherent Structures in Turbulent Shear Flow, *J. Fluid Mech.*, vol. 89, part 2, p. 251, 1978.

Robinson, S.K., Kline, S.J., and Spalart, P.R., A Review of Quasi-Coherent Structures in a Numerically Simulated Turbulent Boundary Layer, *NASA TM-102191*, 1989.

Robinson, S.K., Kline, S.J., and Spalart, P.R., Spatial Character and Time Evolution of Coherent Structures in a Numerically Simulated Boundary Layer, *AIAA 88-3577*, 1988.

Robinson, S.K., Kinematics and Dynamics of Coherent Motions in Turbulent Boundary Layers, Ph.D. Dissertation, Stanford University, 1989.

Smith, C.R. and Metzler, S.P., The Characteristics of Low-Speed Streaks in the Near-Wall Region of a Turbulent Boundary Layer, *J. Fluid Mech.*, vol. 129, p. 27, 1983.

Smith, C.R. and Schwartz, S.P., Observation of Streamwise Rotation in the Near-Wall Region of a Turbulent Boundary Layer, *Phys. Fluids*, vol. 26, p. 641, 1983.

Smith, C.R., Visualization of Turbulent Boundary Layer Structure Using a Moving Hydrogen Bubble Wire Probe, in *Coherent Structure of Turbulent Boundary Layers*, AFOSR/Lehigh University Workshop, p. 48, 1978.

Smith, C.R., A Synthesized Model of the Near-Wall Behavior in Turbulent Boundary Layers, in *Proc. of 8th Symp. on Turbulence*, University of Missouri-Rolla, 1984.

Smith, C.R. and Lu, L.J., The Use of a Template-Matching Technique to Identify Hairpin Vortex Flow Structures in Turbulent Boundary Layers, *Near Wall Turbulence: 1988 Zaric Memorial Conference*, Hemisphere, 1989.

Spalart, P.R., Direct Simulation of a Turbulent Boundary Layer up to $Re_\theta = 1410$, *J. Fluid Mech.*, vol. 187, p. 61, 1988.

Stuart, J.T., The Production of Intense Shear Layers by Vortex Stretching and Convection, *AGARD Rept.* 514, 1965.

Swearingen, J.D. and Blackwelder, R.F., The Growth and Breakdown of Streamwise Vortices in the Presence of a Wall, *J. Fluid Mech.*, vol. 182, p. 255, 1987.

Theodorsen, T., Mechanism of Turbulence, in *Proc. 2nd Midwestern Conf. on Fluid Mech.*, Ohio State University, Columbus, Ohio, 1952.

Townsend, A.A., *The Structure of Turbulent Shear Flow*, Cambridge University Press, 1976.

Utami, T. and Ueno, T., Experimental Study on the Coherent Structure of Turbulent Open-Channel Flow Using Visualization and Picture Processing, *J. Fluid Mech.*, vol. 174, p. 399, 1987.

Wallace, J.M., Eckelmann, H., and Brodkey, R.S., The Wall Region in Turbulent Shear Flow, *J. Fluid Mech.*, vol. 54, part 1, p. 39, 1972.

Wallace, J.M., On the Structure of Bounded Turbulent Shear Flow: A Personal View, in *Developments in Theoretical and Applied Mechanics*, XI, University of Alabama, Huntsville, p. 509, 1982.

Wallace, J.M., The Vortical Structure of Bounded Turbulent Shear Flow, in *Lecture Notes in Physics* 235, Spinger-Verlag, p. 253, 1985.

Walker, J.D.A. and Herzog, S., Eruption Mechanisms for Turbulent Flows Near Walls, *Proc. 2nd Int'l Symp. on Transport Phenomena in Turbulent Flows*, Tokyo, Japan, 1987.

Williams, D.R. and Economou, M., Scanning Laser Anemometer Measurements of a Forced Cylinder Wake, *Phys. Fluids*, vol. 30, no. 7, p. 2283, 1987.

Willmarth, W.W. and Lu, S.S., Structure of the Reynolds Stress Near the Wall, *J. Fluid Mech.*, vol. 55, p. 65, 1972.

Willmarth, W.W. and Tu, B.J., Structure of Turbulence in the Boundary Layer Near the Wall, *Phys. Fluids*, vol. 10, p. S134, 1967.

Categories:
A: Conceptual model (description of physics only)
B: Analytical model (predictive in some sense)
C: Probe data evidence (quantitative)
D: Visual evidence (qualitative)
E: Laminar boundary layer simulation
F: Numerical simulation (LES or DNS)

Table 1: Quasi-Streamwise Vortex References

Authors	Year	Category
Bakewell and Lumley	1967	B,C
Kline et al	1967	D
Kim et al	1971	C,D
Grass	1971	D
Clark and Markland	1971	D
Willmarth and Lu	1972	C
Brown and Thomas	1977	A,C
Kastrinakis et al	1978	C,D
Praturi and Brodkey	1978	D
Kreplin and Eckelmann	1979	C
Blackwelder & Eckelmann	1979	A,C
Nakagawa and Nezu	1981	A,C
Falco	1982	A,D
Falco	1983	A,D
Smith and Schwartz	1983	D
Kim	1983	C,F
Pearson and Abernathy	1984	B
Moin	1984	B,C,F
Moin and Kim	1985	C,F
Herzog	1986	B,C
Jang et al	1986	B
Kasagi et al	1986	B,C,D
Kim and Moin	1986	C,F
Ersoy and Walker	1986	B
Swearingen & Blackwelder	1987	C,D,E
Acarlar and Smith	1987	A,D
Lian	1987	D
Utami and Ueno	1987	C
Moin	1987	C,F
Moin, Adrian, Kim	1987	C,F
Aubrey et al	1988	B,C
Hanratty	1988	B
Kasagi	1988	B,C,D
Guezennec et al	1989	C,F
Robinson et al	1989	D,F

Table 2: Spanwise Vortex References

Authors	Year	Category
Clark and Markland	1971	D
Kim et al	1971	C,D
Nychas et al	1973	D
Praturi and Brodkey	1978	D
Nakagawa and Nezu	1981	A,C
Lian	1987	D
Smith and Lu	1988	C,D
Robinson et al	1988	D,F
Robinson et al	1989	D,F

Table 3: Vortex Structure References

Authors	Year	Category
Theodorsen	1952	A
Willmarth and Tu	1967	A,C
Kline et al	1967	A,C,D
Black	1968	A,B
Hinze	1975	A
Offen and Kline	1975	A
Townsend	1976	A,B
Smith	1978	A,D
Head & Bandyopadhyay	1981	A,D
Wallace	1982	A
Perry and Chong	1982	B
Falco	1982	A,D
Falco	1983	A,C,D
Smith and Metzler	1983	A,C
Smith	1984	A
Wallace	1985	A
Moin and Kim	1985	C,F
Kim and Moin	1986	C,F
Guezennec	1986	A,C
Kobashi and Ichijo	1986	A,C
Fiedler	1986	A
Perry et al	1986	B
Acarlar and Smith	1987	A,C,D,E
Utami and Ueno	1987	A,C
Chu and Falco	1988	A,D,E
Smith and Lu	1988	C,D
Kasagi	1988	A,B,C
Robinson et al	1988	D,F
Robinson et al	1989	D,F

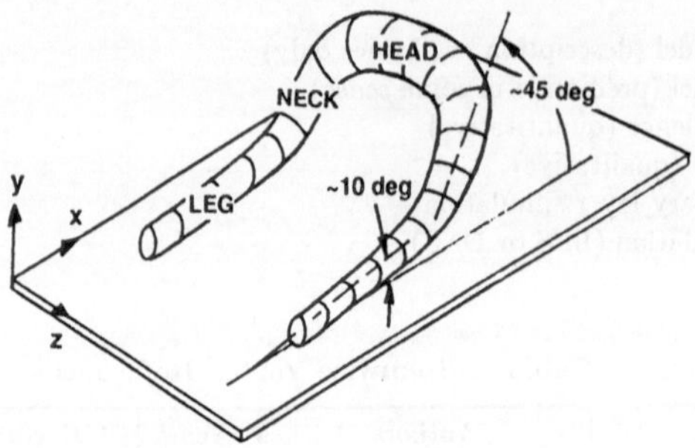

Fig.1. Nomenclature for schematic hairpin vortex.

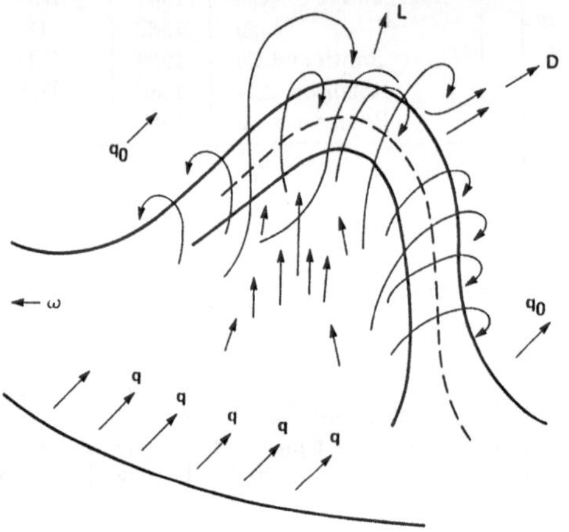

Fig. 2. Theodorsen (1952). "Primary structure of wallbound turbulence."

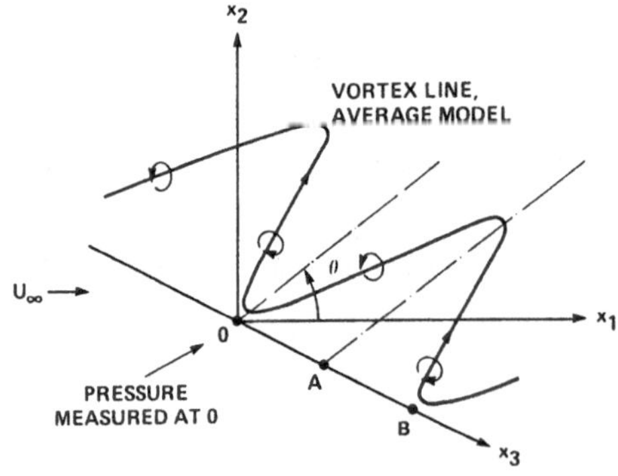

Fig. 3. Willmarth and Tu (1967). "Structure of an average model of vortex line near the wall..."

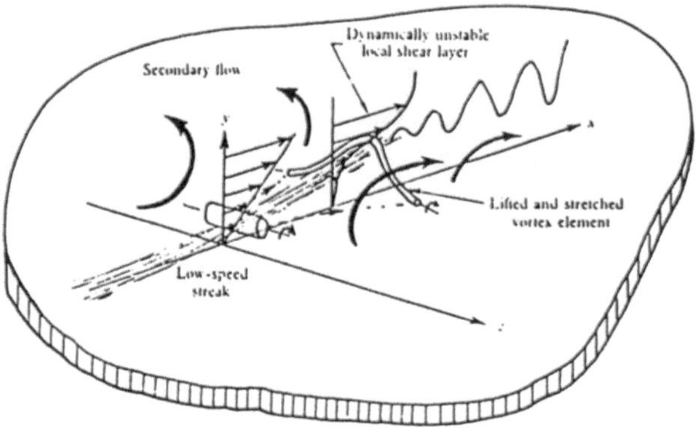

Fig. 4. Kline et al (1967). "The mechanics of streak breakup."

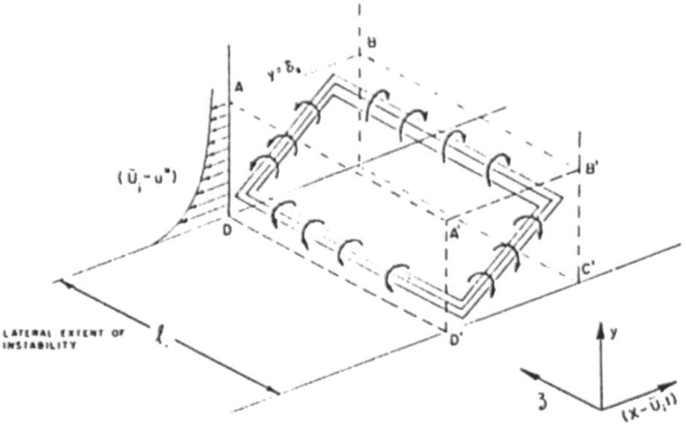

Fig. 5a. Black (1968). "Generation of ring-vortices by instability in actual shear layer."

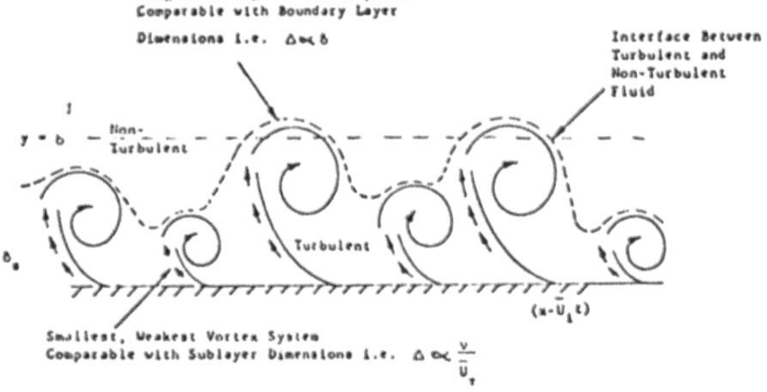

Fig. 5b. Black (1968). "Intermittency explained by random variation in strength of consecutive vortex systems."

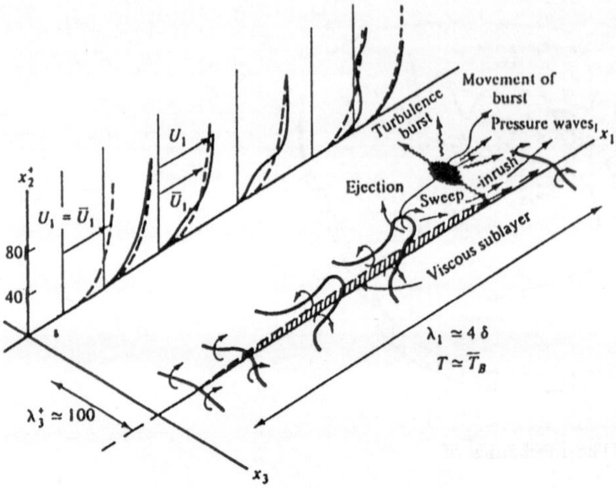

Fig. 6. Hinze (1975). "Conceptual model of the turbulence near the wall during a cyclic process, with average spacings λ_1 and λ_2.

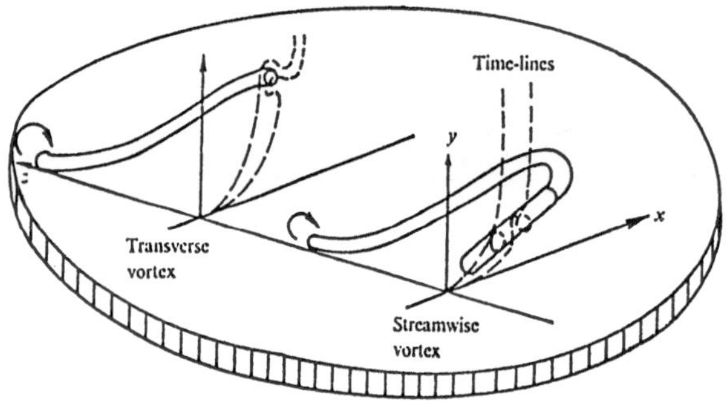

Fig. 7. Offen and Kline (1975). "Time-line patterns at different locations of a lifted and stretched vortex element."

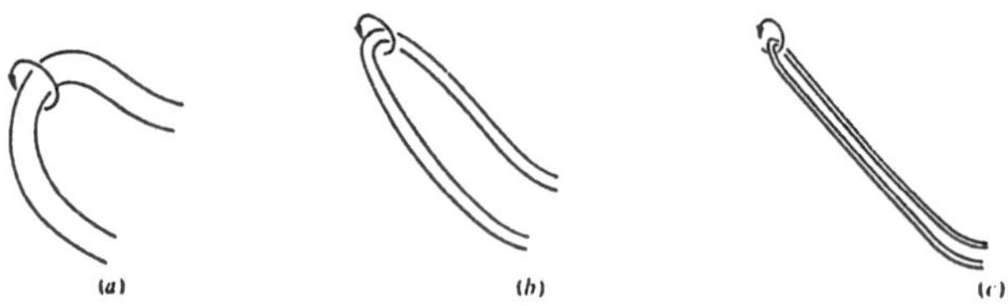

Fig. 8. Head and Bandyopadhyay (1981). "Effect of Reynolds number on features composing an outer region of turbulent boundary layer. (a) *Very* low *Re* (loops); (b) low-moderate *Re* (elongated loops or horseshoes); (c) moderate-high *Re* (elongated hairpins or vortex pairs).

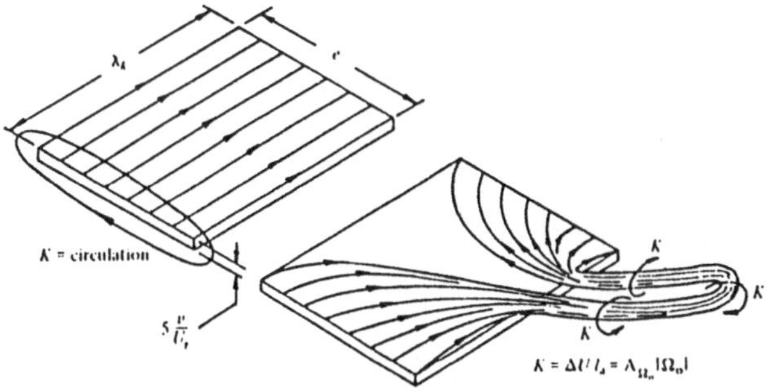

Fig. 9a. Perry and Chong (1982). "Carpet of vorticity being wrapped into vortex (schematic)."

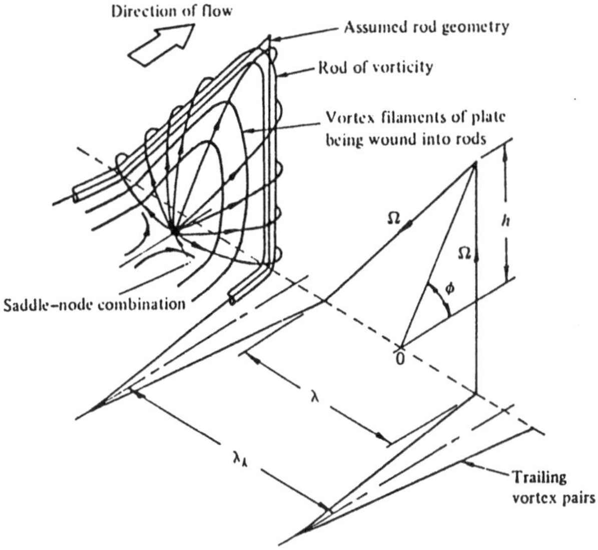

Fig. 9b. Perry and Chong (1982). "Λ-vortex configuration".

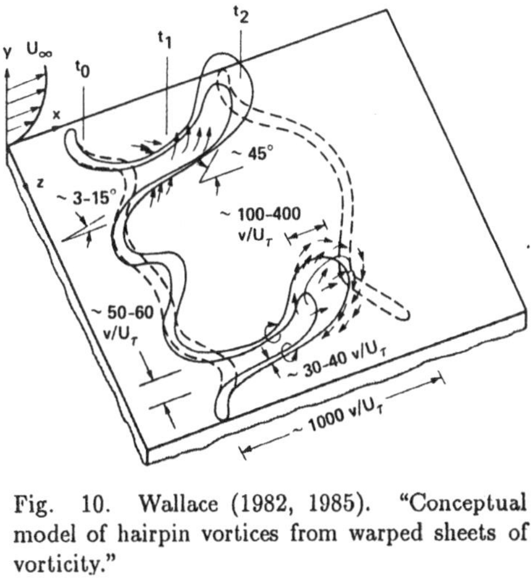

Fig. 10. Wallace (1982, 1985). "Conceptual model of hairpin vortices from warped sheets of vorticity."

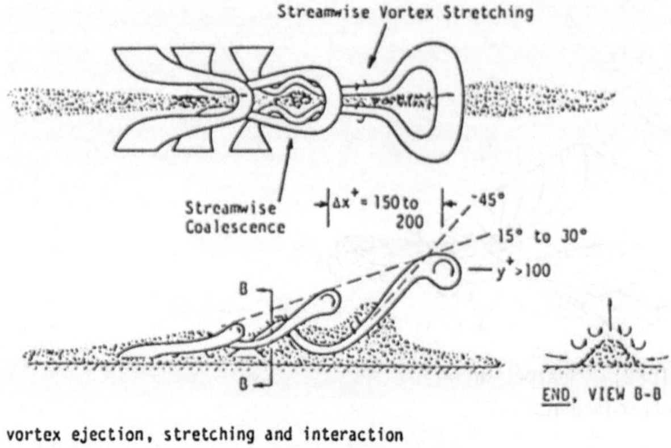

d) vortex ejection, stretching and interaction

Fig. 11. Smith (1984). "Illustration of the breakdown and formation of hairpin vortices during a streak bursting process. Low-speed streak regions indicated by shading."

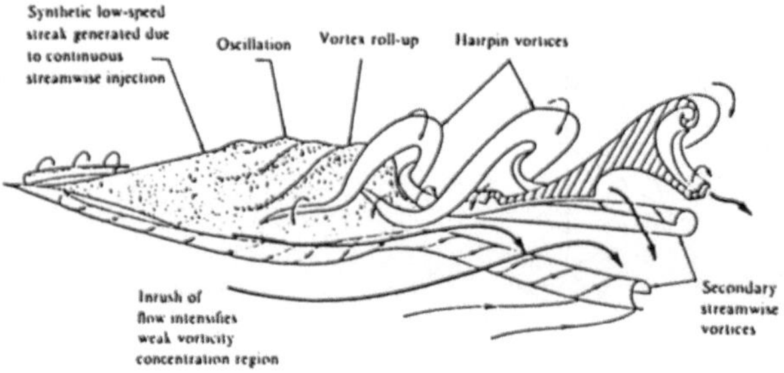

Fig. 12. Acarlar and Smith (1987b). "Schematic of breakup of a synthetic low-speed streak generating hairpin vortices. Secondary streamwise vortical structures are generated owing to inrush of fluid."

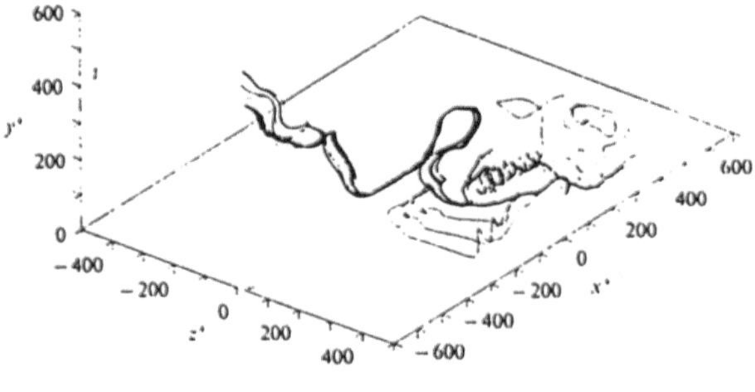

Fig. 13. Kim and Moin (1986). "Vortex lines showing an instantaneous structure detected by QD-2."

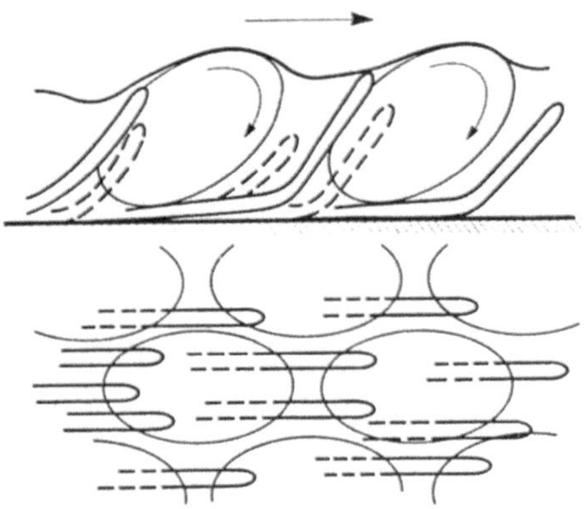

Fig. 14. Fiedler (1986). "Boundary layer structure."

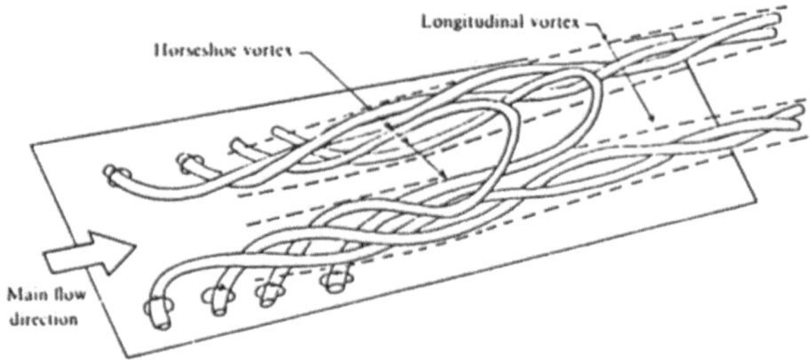

Fig. 15. Utami and Ueno (1987). "Conceptual model representing the overall structure of turbulence in the wall region in the fully developed stage. Solid lines denote vortex tubes."

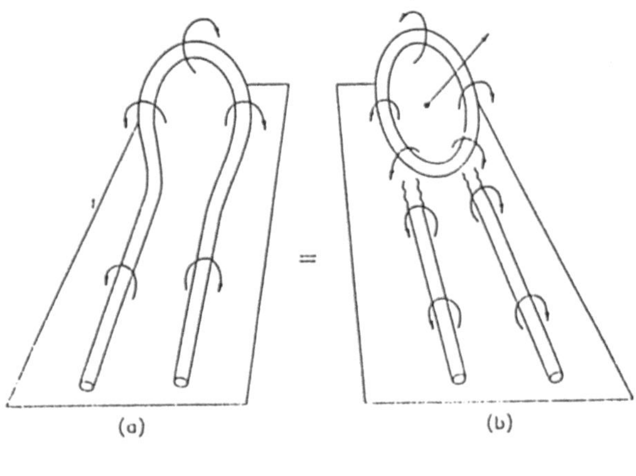

Fig. 16. Adrian (1988). "(a) Hairpin vortex close to the wall, (b) decomposition of a wall hairpin into a vortex ring plus mean shear plus two streamwise vortices."

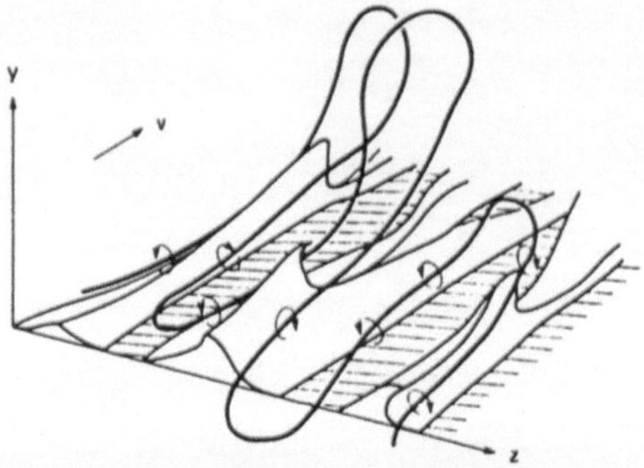

Fig. 17. Kasagi (1988). "Conceptual model of the wall-layer structure."

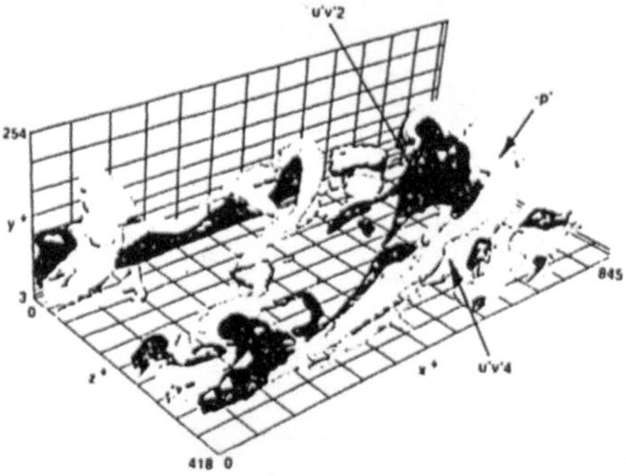

Fig. 18. Robinson et al (1989). "Spatial relationship between elongated low-pressure regions, strong ejections ($u'v'_2$), and strong sweeps ($u'v'_4$)."

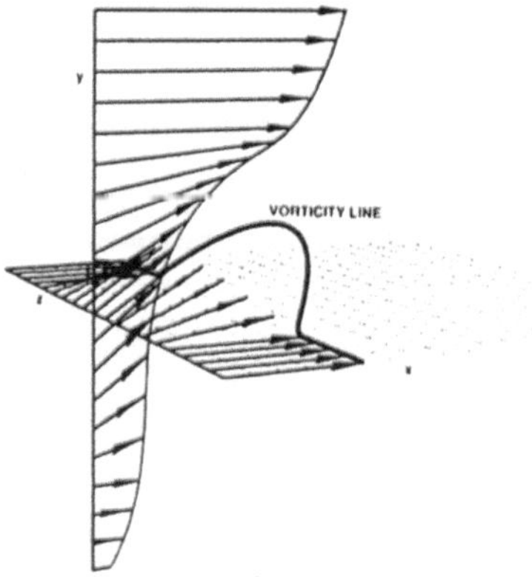

Fig. 19. Vorticity line computed for a localized region of lifting, low-speed ($+v'$, $-u'$) fluid within an otherwise two-dimensional, linear shear flow.

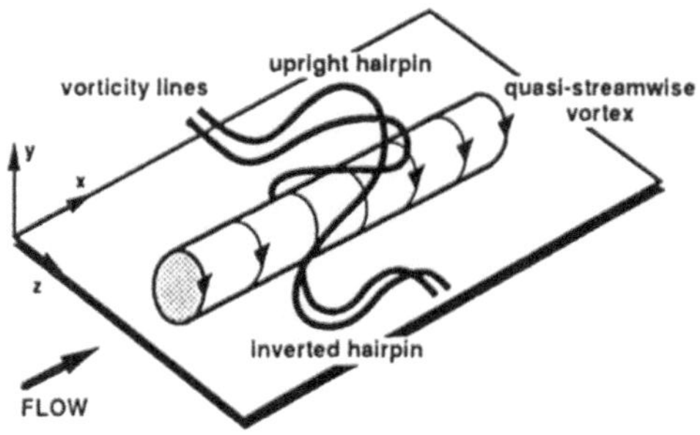

Fig. 20. Vorticity lines traced from either side of a quasi-streamwise vortex in a boundary layer, showing upright and inverted hairpin shapes.

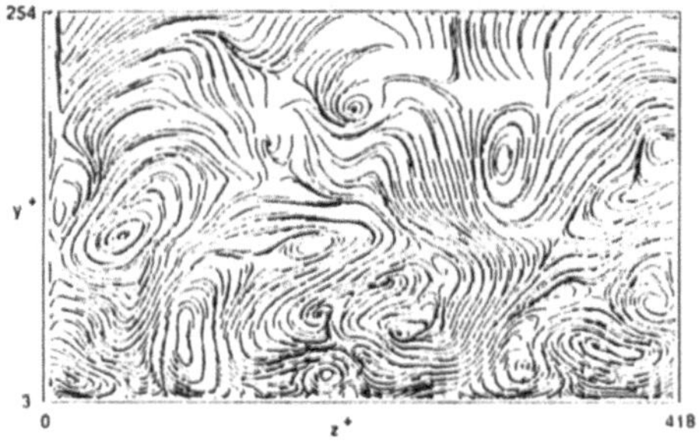

Fig. 21a. Example from Spalart's (1988) numerically simulated boundary layer: instantaneous streamlines in an end-view $(y - z)$ plane.

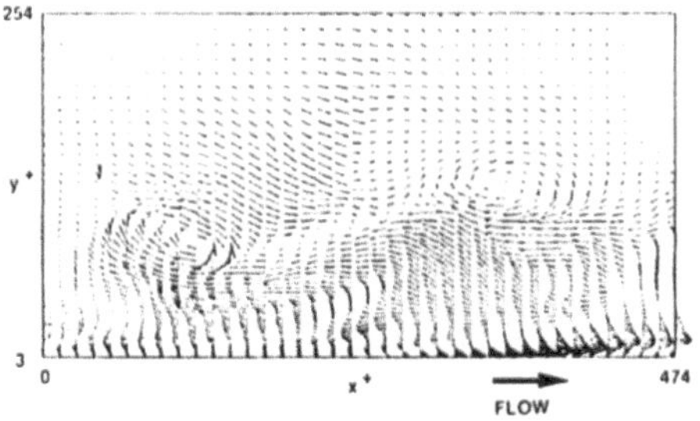

Fig. 21b. Example from Spalart's (1988) numerically simulated boundary layer: instantaneous $u'v'$ perturbation velocity vectors in a side-view $(x - y)$ plane.

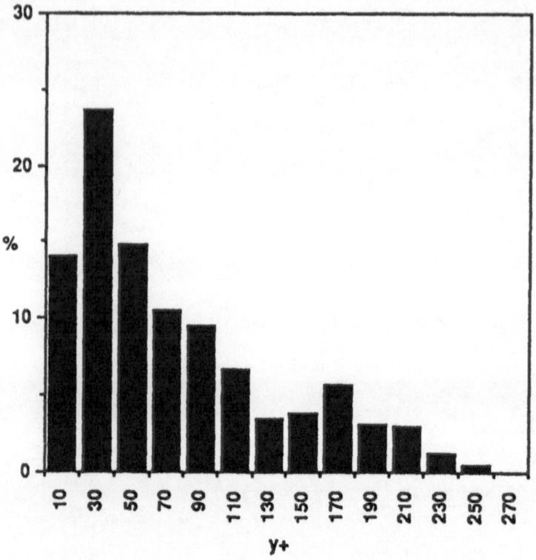

Fig. 22. Distribution of distances from the wall for visually-identified quasi-streamwise vortices.

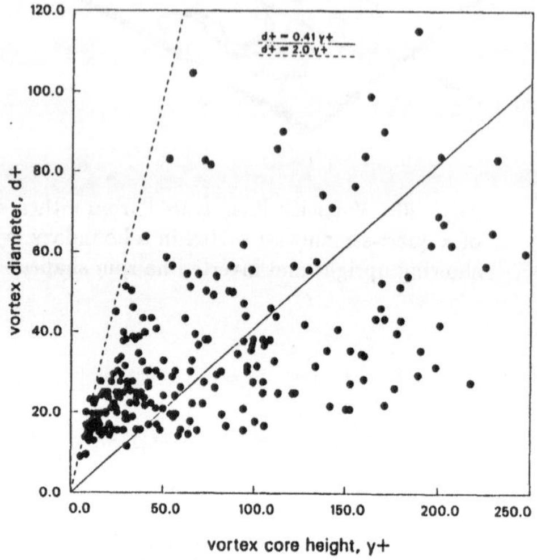

Fig. 24. Variation of vortex diameter with distance from the wall for visually-identified quasi-streamwise vortices.

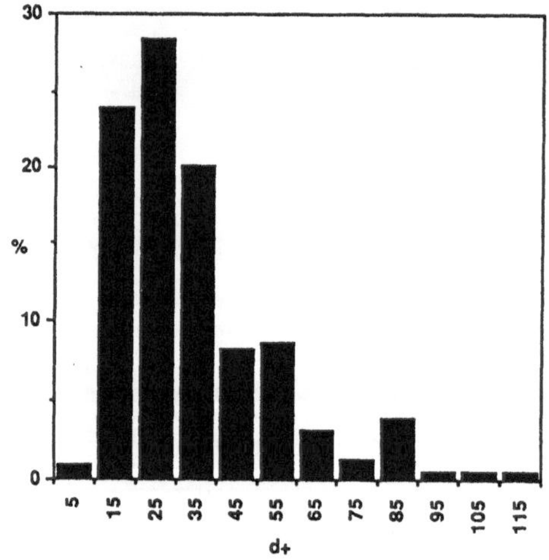

Fig. 23. Distribution of vortex diameters for visually-identified quasi-streamwise vortices.

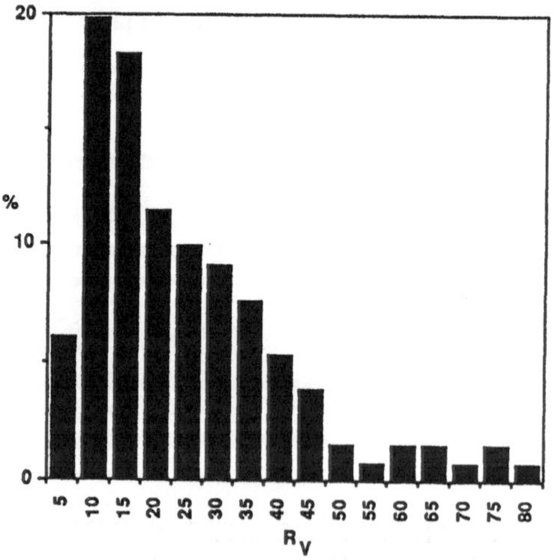

Fig. 25. Distribution of vortex Reynolds number R_V (circulation) for visually-identified quasi-streamwise vortices.

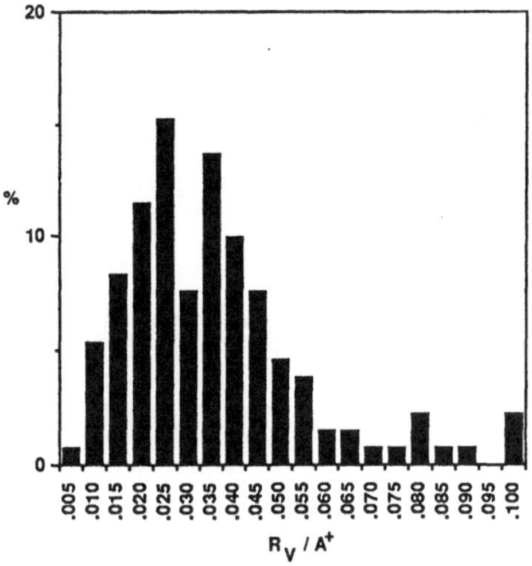

Fig. 26. Distribution of vortex intensity (R_V/A^+) for visually-identified **quasi-streamwise vortices.**

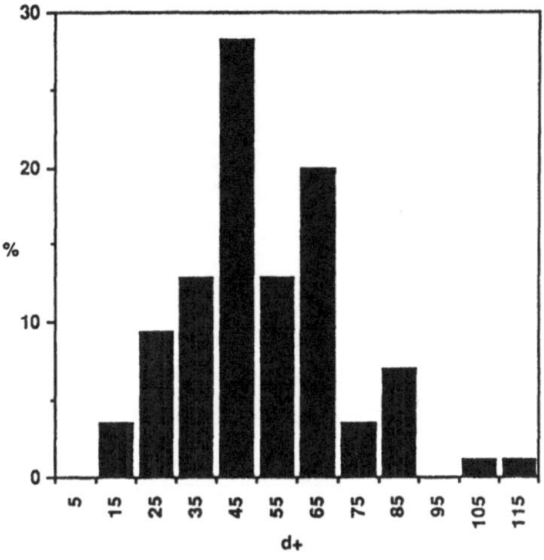

Fig. 28. Distribution of vortex diameters for visually-identified **transverse vortices.**

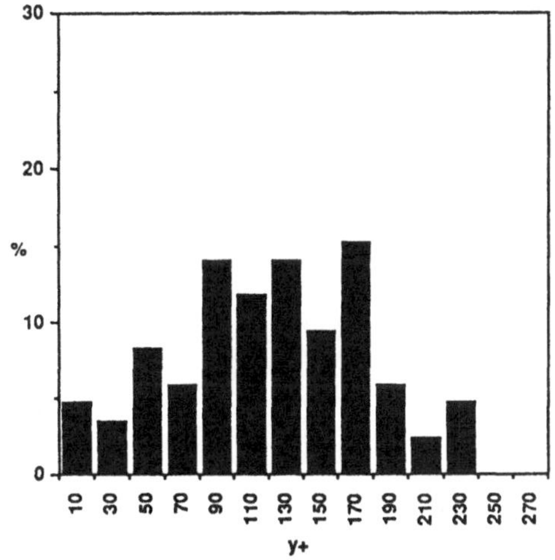

Fig. 27. Distribution of distances from the wall for visually-identified **transverse vortices.**

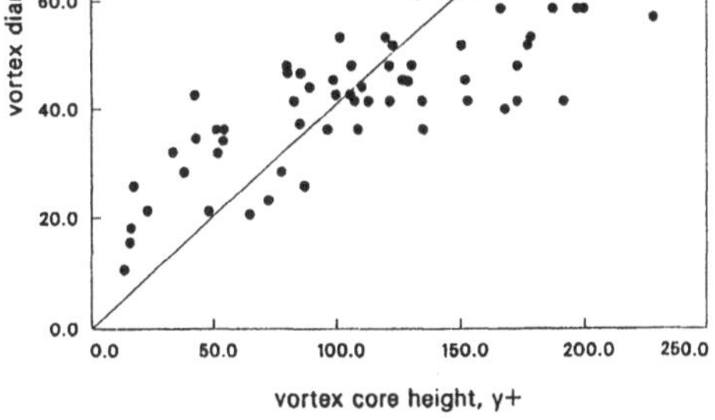

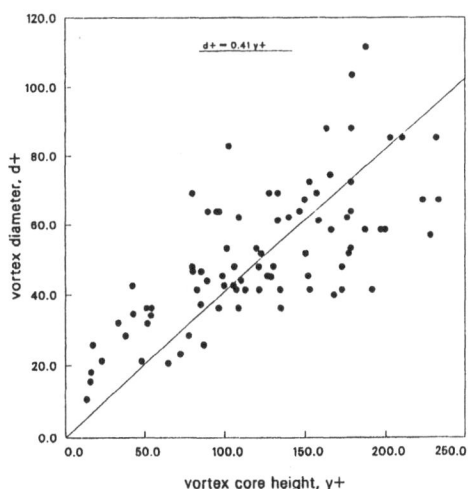

Fig. 29. Variation of vortex diameter with distance from the wall for visually-identified **transverse vortices.**

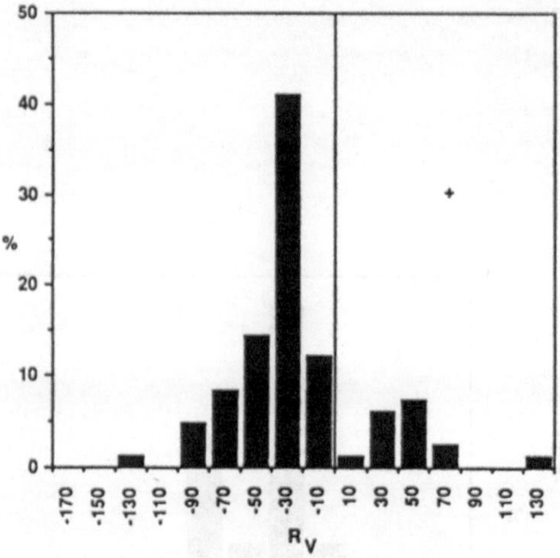

Fig. 30. Distribution of vortex Reynolds number R_V (circulation) for visually-identified **transverse vortices.**

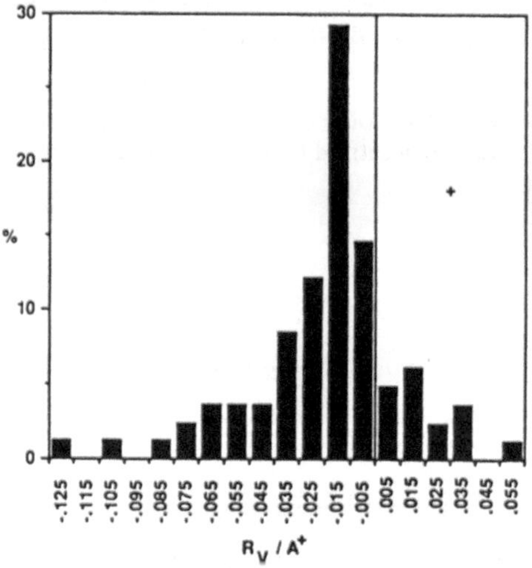

Fig. 31. Distribution of vortex intensity (R_V/A^+) for visually-identified **transverse vortices.**

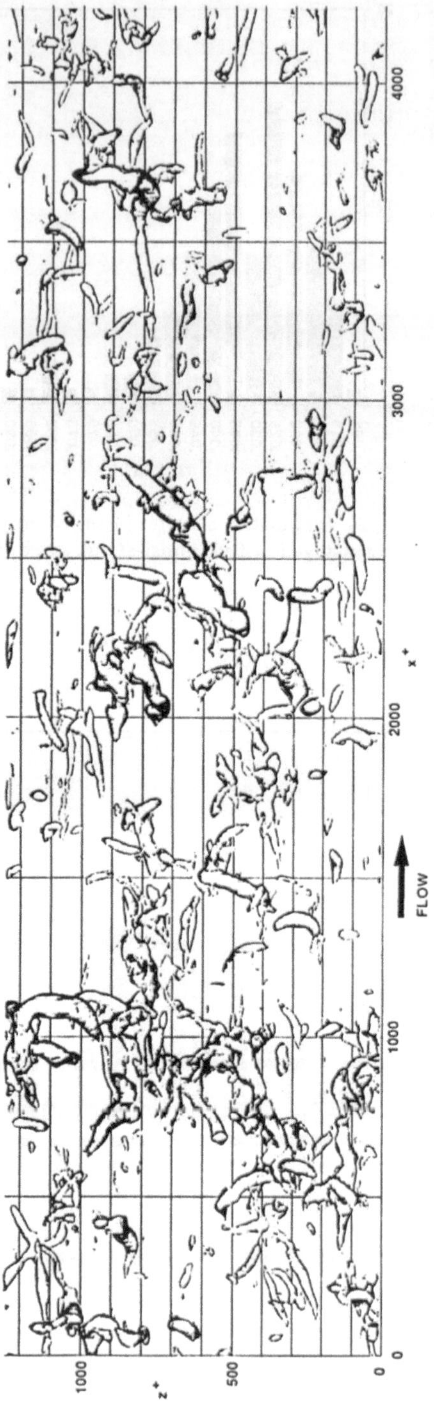

Fig. 32. Top-view of instantaneous three dimensional low-pressure structures in numerically-simulated turbulent boundary layer. Isobaric surfaces computed for $p' = -4.2\rho u_\tau^2$.

Hairpin Vortices in Turbulent Boundary Layers: The Implications for Reducing Surface Drag

C.R. Smith, J.D.A. Walker, A.H. Haidari, B.K. Taylor
Department of Mechanical Engineering & Mechanics
Lehigh University
Bethlehem, PA 18015

ABSTRACT

Ultimately, control of turbulent shear flows must center around the selected modification of the flow structure of the turbulence, or the conditions which gives rise to the flow structure. Consequently, in order to develop rational methods of control, it is necessary to understand the basic dynamical elements of the turbulent boundary layer. In this paper, a model of the turbulent boundary layer ,based upon a series of complimentary computational and experimental studies, is described in which the hairpin vortex is suggested to be a basic flow structure of boundary-layer turbulence. The paper discusses the development of single hairpin vortices, how such vortices interact with a shear flow and with one another, and how they interact with the flow near a wall to produce new vortices through a viscous interaction. It is demonstrated how key features of turbulent flow structure (e.g. low-speed streaks, bursting, pockets, etc.) can be explained in terms of the motion and effects of convecting hairpin vortices. The implications of this model in reducing drag are also discussed.

BACKGROUND

It is commonly believed that vortex motions play an important role in the dynamics of boundary-layer turbulence. Several different vortex motions have been proposed as basic models; however, the most commonly hypothesized vortex model is the hairpin vortex, which has been suggested as a basic flow structure of boundary-layer turbulence by Theodorsen [1], Willmarth and Tu [2], Offen and Kline [3], Head and Bandyophadyay [4], Perry and Chong [5] and Smith [6], to name a few.

Theodorsen[1], using analytical arguments, first suggested that hairpin flow structures are the most efficient structure for transfer of energy within a shear gradient, and thus an important feature in turbulent boundary layers. Based on a series of visual studies, Head and Bandyophadyay [4] suggest that turbulent boundary layers consist of "forests" of hairpin vortices. Perry and Chong [5] conjecture that a turbulent boundary layer is dominated by a systematic hierarchy of hairpin vortices, which grow and die due to vortex reinforcement and cancellation effects. Recently, Smith [6] has suggested that the turbulent boundary layer is controlled by a process of both hairpin evolution and nesting of multiple hairpins, with the hairpins giving rise to the low-speed streak structures by viscous interaction, and the streaks breaking down to form the hairpins. More recently, Acarlar and Smith [7] examined the behavior of streets of hairpin vortices in a laminar boundary layer and established that these flow structures give rise to both visualization patterns and velocity statistics which closely mimic the characteristics of the wall-region of turbulent boundary layers.

Despite the attractiveness of hairpin vortices as key structures of turbulent boundary layers, most of the hairpin vortex hypotheses are speculative. Until recently, little examination and direct comparison of this potential flow structure with the flow structure and behavior of turbulent boundary layers had been done. The present paper recounts the key findings of a series of complementary computational and experimental studies which have

A. Gyr (Editor)
Structure of Turbulence and Drag Reduction
IUTAM Symposium Zurich/Switzerland 1989

examined the detailed behavior of hairpin vortex flow structures. Based on the results of these studies, it is illustrated how the behavior and growth of hairpin vortices appropriately model both the kinematics and dynamics of the near-wall region in a turbulent boundary layer. Using this model, the implications in the reduction of surface drag are discussed.

OVERVIEW

In order to observe basic flow behavior, single hairpin vortices were generated in a laminar boundary layer in a low-speed water channel using computer-controlled fluid injection through a streamwise-oriented slot (similar to previous continuous injection studies[7]). Figures 1 shows a typical dye visualization of a single hairpin [8]. Detailed observations and quantitative measurements of the behavior and effects of the hairpin vortex were carried out using a combination of flow visualization, image processing, and phase-averaged probe measurements (Haidari [8], Taylor [9],Smith and Lu [10]). Computational studies of the motion and effects of similar hairpin vortex structures have also been done by Hon and Walker [11], who examine both the kinematic deformation of a single vortex in a shear flow into a hairpin vortex, and the dynamic viscous-inviscid interaction induced near a boundary by the hairpin vortex. Viscous-inviscid interactions induced by counter-rotating vortices have also been examined computationally in the studies of Ersoy and Walker [12,13].

The experimental and computational results indicate that a hairpin vortex in a shear layer grows by generation of similar hairpin structures. An interaction with the background shear flow results in the evolution of subsidiary vortices to the sides of the original vortex, which leads to a lateral spreading of the initial hairpin vortex. A second process of growth and regeneration occurs through a viscous-inviscid interaction with the flow near the surface in which an eruption of the surface fluid culminates in the production of secondary vortices [11]. The experimental studies clearly show that this growth process proceeds in a rapid, but very organized manner, evolving into a turbulent spot-like disturbance [8,9].

HAIRPIN GROWTH PROCESSES

It is impossible to summarize the details of the entire hairpin evolution and regeneration process in the present paper. Here, a selected review of hairpin vortex kinematics and dynamics will be done. For a more detailed review, the reader is referred to references cited in the overview section.

In three-dimensional flow situations, vorticity evolves in ways which are not easily understood and which cannot readily be adduced through intuitive arguments. To gain an appreciation of the development of hairpin vortices in a shear flow near a wall, Hon and Walker [11] carried out numerical simulations of the evolution of a hairpin vortex, including deformation in a background shear through Biot-Savart induction and the interaction that develops with the viscous flow near a surface. Their kinematic simulations show that a hairpin vortex developing in a strong shear layer deforms in the streamwise direction, with the vortex head lifting away from the wall, bending backward, and rising through the shear layer. An example of a typical computation is shown in figure 2. This same general type of behavior has been observed experimentally [7], most recently for single hairpin generation [8]. In addition, the line vortex which evolves into a hairpin vortex (figure 3) also develops symmetric, lateral deformations referred to as subsidiary hairpin vortices which are located outboard of the original hairpin.

The spreading of the hairpin vortex in the spanwise direction and the evolution of subsidiary hairpin vortices may be regarded as regenerative, since the interaction of the hairpin vortex with the shear stimulates the development of lateral hairpins. This process has been observed both computationally [11] and experimentally [7,8,9]. Note that the symmetric hairpin is regarded here as the simplest conceptual model which captures the main features of turbulent dynamics near the wall. In a turbulent boundary layer, there are considerable three-dimensional background disturbances, and it is evident that most hairpin vortices in this environment will not be symmetric. However, the general characteristics of symmetric hairpins are expected to apply since the local, self-induced velocities on each part of a three-dimensional vortex dominate its subsequent motion.

Dynamically, a hairpin moving over a wall induces an unsteady, three-dimensional pressure distribution on the flow near the wall. As the hairpin translates with the local flow, a region of deceleration develops upstream of the vortex; this is immediately followed by a region of acceleration and adverse pressure gradient which develops behind the vortex head. For large vortex Reynolds numbers, the adverse pressure gradient induced by the moving vortex has a strong effect on the viscous flow near the wall, eventually causing a rapid viscous-layer growth in the region behind the head and near the trailing portion of the hairpin vortex legs. This growth promotes a local, rapid thickening of the near-wall flow and development of a local inflection in the velocity profile, which precipitates a local breakdown. Observations suggest that an inflectional profile is the precursor of a strongly interactive event and the roll-over into a vortex structure. Figure 4 is a scene from a side-view hydrogen bubble visualization illustrating a single, primary hairpin with the initial vestiges of a secondary hairpin forming in its wake. This overall development of a secondary vortex is very "eruptive" in nature, and manifests the characteristics of a turbulent "burst" [10].

Several points bear upon the viscous-inviscid interaction of hairpin vortices with near-wall fluid. First, a moving hairpin vortex will induce an inflow toward the symmetry plane which strengthens as the hairpin legs move toward the surface. Secondly, the streamwise extent over which significant inflow toward the symmetry plane occurs will increase with both time and vortex stretching. The flow in the cross-flow plane is similar to that induced by a pair of counter-rotating vortices [12,13]; however, the strongest part of the interaction may not occur on the symmetry plane. The strongest response is expected off the symmetry plane where the hairpin vortex legs approach the surface [13]. This zone is indicated schematically in figure 3; the appearance of these dual growth regions in a cross-section plane of the hairpin legs is show in figure 5a. These regions of viscous growth are what are perceived as the low-speed streaks in a turbulent boundary layer, which are in turn the regions which break down to form other secondary hairpins. Depending on the proximity of the hairpin legs, these streak-like regions can form either centrally between the legs, or adjacent to *each* leg, a fact that has been recently demonstrated [8] [see figure 6a,b].

When the orientation of the vortex pair is not symmetric with respect to the wall as shown in figure 5b, similar rapid viscous growth effects occur with the exceptions that (1) the induced separations in the boundary layer are more complex than in the symmetric case [12] and (2) the vortex closest to the wall always has the dominant effect. Thus, the observed wall-layer streaks can result from either a pair of counter-rotating legs, or from one dominant leg of an asymmetric hairpin.

SIMILARITIES WITH TURBULENT BOUNDARY LAYERS

The visual similarities of hairpin vortices with typical visual structures observed in turbulent boundary layers are illustrated in figure 6, which shows the downstream growth of a single hairpin structure and its effect on a horizontal hydrogen bubble-sheet [8]. Figures 6-a through 6-d are plan-view pictures obtained with a hydrogen bubble-wire located parallel to the boundary at $Y/\delta=0.4$, and at streamwise distances of $\Delta X/\delta=1$, 5, 10, and 30 from the point of hairpin initiation. Figures 6-a' through 6-d' are obtained at the same corresponding $\Delta X/\delta$ locations with the wire located near the surface at $Y/\delta=0.1$. Note that in order to capture the continually growing patterns, the fields-of-view of the photographs in figure 6 were increased with streamwise distance (i.e. the field-of-view in figure 6-a,a' is 4.3cm x 2.9cm; the respective linear scale in the subsequent photographs is 1.2, 1.37, and 1.58 times that of figure 6-a,a').

The pictures shown are a taken from a typical sequence, and clearly illustrate both the significant hairpin growth/interaction characteristics discussed above, as well as demonstrating strong similarities to characteristic patterns commonly observed in turbulent boundary layers.

Figure 6-a illustrates a pattern characteristic of the passage of a single HPV (hairpin vortex). The counter-rotating legs of the primary vortex are marked by faint tubes of of bubbles which trail from right to left below the bubble-sheet, extending down toward the plate. The clear region in the picture has a pocket-like (a common feature of turbulent boundary layers [14]) appearance; this pattern actually represents a saddle point in the flow caused by the action of the counter-rotating legs of the HPV. Closer to the wall, at the same streamwise location, a narrow bubble concentration appears, caused by the viscous-inviscid interaction of the legs of the vortex (note the similarity to a turbulent boundary-layer streak).

The initiation of a secondary HPV (labeled as S) is apparent in figure 6-b. The structure is just beginning to rise through the bubble sheet and thus is not completely visualized. Note the continued development of a streak-like pattern near the surface in figure 6-b', with a streak associated with each hairpin vortex leg. As the secondary vortex develops and moves away from the wall, the head of the vortex appears as a well-defined ridge in figure 6-c. The secondary vortex follows directly behind the head of the primary vortex, which is out of the picture to the right in figure 6-c .

Figure 6-c also illustrates the development of the subsidiary vortices (labeled SU), which appear laterally removed from the centerline of the primary vortex. The subsidiary vortices are symmetrically located about the center-line, with the same initial shape and sense of rotation as the the primary vortex. With the appearance of the subsidiary vortices, the pattern near the surface in figure 6-c' begins to reflect the initiation of an additional pair of low-speed streaks.

As the disturbance proceeds downstream, the continuous interaction between newly formed and existing vortices, and the evolution of new vortex structures results in the systematic development of a turbulent spot-like flow structure which is remarkably organized and symmetric about the center-line. Figure 6-d, and 6-d' illustrate this turbulent-like appearance at $X_s/\delta=30$. In particular, note in figure 6-d' that the elongated streamwise patterns have developed into essentially four very definitive turbulent boundary layer-type streaks (indicated by the arrows). Beyond this streamwise location the spot-like structure continues to spread both streamwise and laterally. The further development of this organized turbulent spot is discussed in detail elsewhere [8,9].

FURTHER RATIONALE FOR HAIRPIN VORTICES IN TURBULENCE

It is clear that once hairpin vortices are present in a shear flow near a wall, they are able to kinematically multiply themselves in the spanwise direction and dynamically create new vortices through an interaction with the viscous flow near the wall. Thus, once the process starts, it feeds on itself and becomes regenerative. The process of growth to a fully turbulent boundary layer can be explained by the proximity of multiple hairpins in different phases of development, which creates a condition conducive to coalescence, or three-dimensional vortex amalgamation. This process is demonstrated in simulations [11] that show that two vortices generated in proximity to each other will intertwine and interact to yield essentially one stronger hairpin of a somewhat larger scale. Recent studies [8,9], have experimentally observed the development of just such an amalgamation process, carefully tracking the controlled evolution of a single hairpin vortex into a multi-hairpin, turbulent spot-like structure. This clearly suggests how the outer region of a turbulent boundary layer evolves from hairpin structures and that the large rollers observed in the outer region of turbulent boundary layers are essentially large amalgamations of initially smaller hairpin vortices. The outer part of the boundary layer may thus be regarded as a "graveyard" for vorticity, where the remnants of the hairpins pass through a complicated process of dissipation, diffusion and mutual cancellation.

Clearly, the hairpin vortex growth process described above is consistent with many of the kinematic and dynamic processes associated with turbulent boundary layers. However, there are several general points which solidify the argument for such structures being key elements of turbulent shear flows.

First, turbulent flow requires an increased exchange of momentum with the wall in order to sustain the elevated shear stress levels in excess of a laminar, diffusion dominated flow. This requires an active mechanism within the boundary layer which will augment the removal of low-momentum fluid from the wall. Three-dimensional vortex concentrations, such as the hairpin vortex, provide such a mechanism, acting as both a pressure gradient pump and a transport vehicle to remove low-momentum fluid from the wall. The suggested creation of hairpin vortices during a bursting event clearly provides the necessary mechanism for transfer of low-momentum fluid away from the wall and into the outer region.

Second, three-dimensional vortex stretching by the mean velocity gradient is necessary to sustain the energy transfer from the mean motion to the turbulent fluctuations. As suggested by Theodorsen[1], the formation of hairpin vortices provides the logical mechanism to accomplish this energy transfer process in the near-wall region, with the mean flow stretching process providing the energy input to the hairpins.

Third, the stretching of the legs of the hairpin also provides a clear mechanism for elevated energy dissipation in the near-wall. During rapid stretching by the mean velocity gradient, the angular momentum in a leg of a hairpin (proportional to ωr^2) will be essentially conserved, while the energy (proportional to $\omega^2 r^2$) will grow significantly as the vortex tube narrows. The velocity gradients within a leg will also increase dramatically, which strongly increases viscous dissipation (proportional to the square of the velocity gradient in the legs) in the vicinity of the wall.

SURFACE DRAG CONSIDERATIONS

The reduction of turbulent surface drag requires a reduction in the level of momentum exchange at the wall, which in turn requires a reduction of the bursting activity near the wall. In principle this may be accomplished either by reducing the number of low-speed streaks (i.e. "burst" sites) adjacent to the wall or by increasing the cycle time for the momentum exchange bursting process. Considering the hairpin dynamics described above, it is clear that to accomplish this, one must generally inhibit the viscous-inviscid interaction induced in the near-wall flow by the convected hairpins. Hairpin vortices produce the streaks and eventually provoke an eruption to produce new hairpin vortices; this process defines a cyclical momentum exchange process with the surface flow which can be interrupted in at least two ways. The first is to provide mechanisms which inhibit the viscous-inviscid interaction by interfering with the lateral movement of low-speed fluid at the wall. The second is to maintain the streaks in a stable state for a longer period. The first of these approaches is clearly the mechanism implemented by riblets and fences, which have been shown to be effective in reducing surface drag up to 10%. The riblets inhibit lateral flow near the surface and thereby act to make the adverse pressure gradients induced by the convecting hairpins less effective at provoking an eruptive response from the near-wall flow; the development of low-speed streaks is significantly affected, as evidenced by the increase in streak spacing [15]. Modification of surface topography to interfere with the viscous-inviscid interaction is therefore a viable approach for reducing surface drag. The recent success with streamwise fences and shark scale-like surface modifications are examples of this technique.

The maintenance of streak stability is a more tenuous approach since this entails a delicate balance between the inherent stability of the low-speed streak and the amplitude of the destabilizing pressure perturbations in the outer flow. The addition of polymer to the boundary layer is an example where streak stability is increased by addition of an external additive, as evidenced by wider streak spacing and reduced bursting activity. It is speculated that polymer addition may either affect the streak stability directly, by inhibiting lateral concentration of fluid by the hairpin vortices, or indirectly, by providing a region which is locally more viscous, which (1) more effectively damps external perturbations, thus preventing streak breakdown, and (2) dissipates the energy in the hairpin vortices generated by breakdowns, thus weakening the vortex strength of the hairpin vortices and inhibiting their effectiveness in the viscous-inviscid regeneration process.

ACKNOWLEDGEMENTS

We gratefully acknowledge the continuing support of the U. S. Air Force Office of Scientific Research under contract number AFOSR-89-0065.

REFERENCES

1. Theodorsen, T., "Mechanism of Turbulence," Proc. Second Midwestern Conference of Fluid Mechanics, Ohio State University, Columbus, Ohio, 1952.
2. Willmarth, W.W. and Tu, B.J., "Structure of Turbulence in Boundary Layers," Phys. of Fluids, Suppl. 10, p.3134, 1967
3. Offen, G.R. and Kline, S.J., " A Proposed Model of the Bursting Process in Turbulent Boundary Layers" J. Fluid Mech. , Vol. 70, p. 209, 1975
4. Head, M. R., Bandyopadhyay, P., "New Aspects of Turbulent Boundary Layer Structure," J. Fluid Mech., Vol. 107, p. 297, 1981.
5. Perry, A. E. and Chong, M.S., "On the Mechanism of Wall Turbulence," J. Fluid Mech., Vol.119, p.173, 1982..
6. Smith, C. R., "A Synthesized Model of the Near Wall Behavior in Turbulent Boundary Layers, "Proceedings of 8th Symp. on Turbulence, G. K. Patterson, J. K. Zakin, ed., Dept., of Chem. Engrg., University of Missouri-Rolla, Rolla, Missouri, 1984.
7. Acarlar, M. S., Smith, C. R., "A Study of Hairpin Vortices in a Laminar Boundary Layer. Part 1. Hairpin Vortices Generated by Hemisphere Protuberances," and ". Part 2. Hairpin Vortices Generated by Fluid Injection,"J. Fluid Mech., Vol.175, p.1 and p.43, 1987.
8. Haidari, A.H., "The Generation of Single Hairpin Vortices," PhD Dissertation, Lehigh University, Bethlehem, PA., 1989.
9. Taylor, B. K., "The Effects of Pressure Gradient on the Generation, Development, and Break-up of Synthetically Generated Hairpin Vortices," PhD Dissertation, Lehigh University, Bethlehem, PA., 1989.
10. Smith, C.R. and Lu, L.J., "The Use of a Template Matching Technique to Identify Hairpin Vortex Flow Structures in Turbulent Boundary Layers," Proc. of Zaric Int'l Seminar on Wall Turbulence, S.J. Kline, ed, Hemisphere, i989.
11. Hon, T. L., Walker, J. D. A., "Evolution of Hairpin Vortices in a Shear Flow," NASA Tech. Memo.100858, ICOMP-88-9, 1988. Also to appear in Computers in Fluids.
12. Ersoy, S. and Walker, J.D.A., "Flow Induced at a Wall by a Vortex Pair," AIAA Journal, Vol.24, No. 10, p. 1597, 1986.
13. Ersoy, S. and Walker, J.D.A., "Viscous Flow Induced by Counter-Rotating Vortices, " Physics of Fluids, Vol.28, No.9, p.2687, 1985.
14. Falco, R. E., "Structural Aspects of Turbulence in Boundary Layer Flows, "Proceedings of 6th Symp. on Turbulence, G. K. Patterson, J. K. Zakin, ed., Dept., of Chem. Engrg., University of Missouri-Rolla, Rolla, Missouri, 1981.
15. Bacher, E.V. and Smith, C.R., " A Combined Visualization-Anemometry Study of the Turbulent Drag Reducing Mechanisms of Triangular Micro-groove Surface Modifications," AIAA paper 85-0548, 1985.

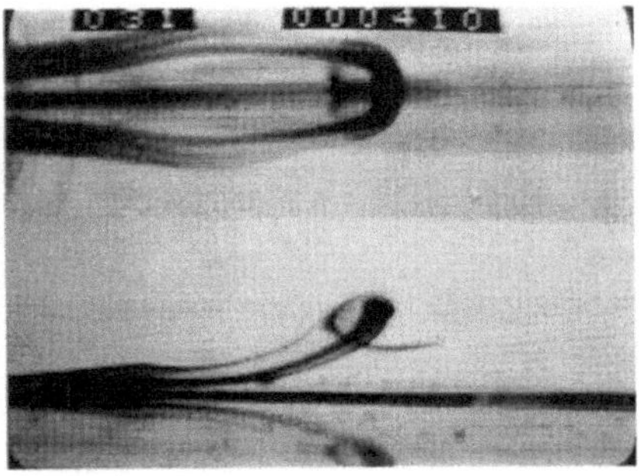

Figure 1. Dual view of dye marked single hairpin vortex.

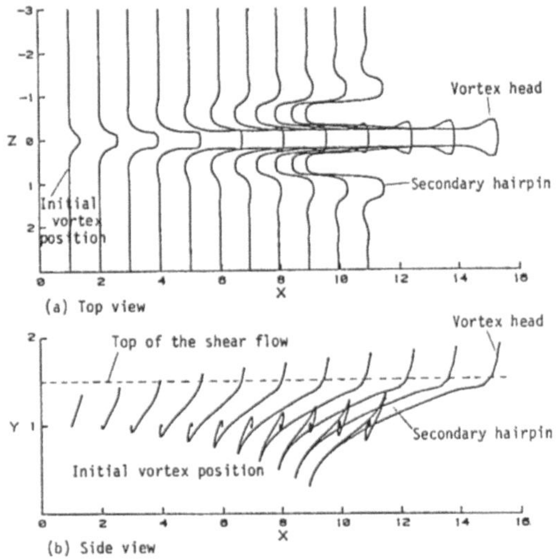

Figure 2. Temporal development of a hairpin vortex from a small distortion in a two-dimensional vortex convected in a shear flow.

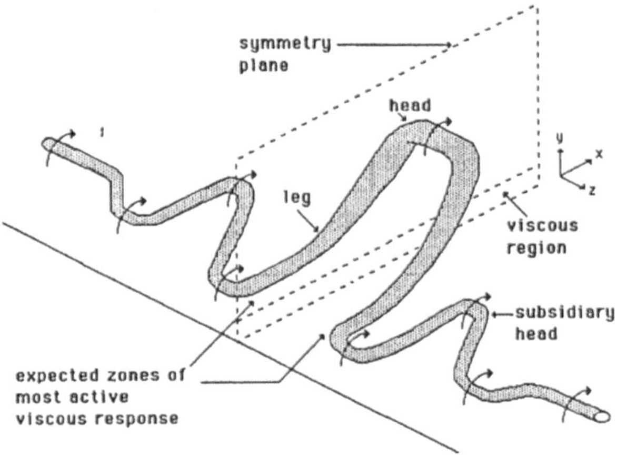

Figure 3. Schematic diagram of a hairpin vortex convected above a wall.

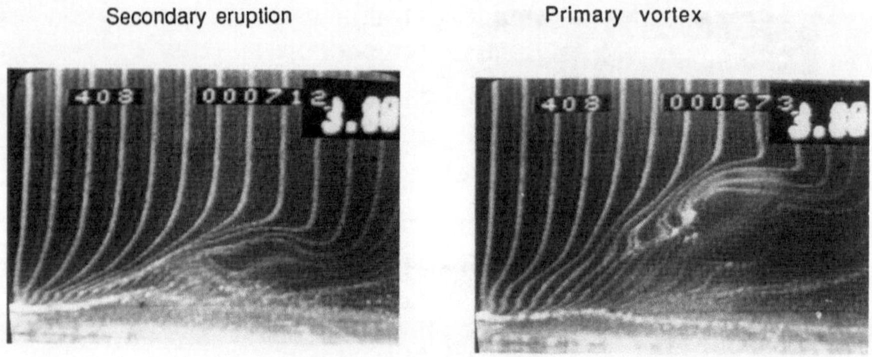

Figure 4. A sequential side-view hydrogen bubble pictures illustrating the passage of a primary hairpin, followed by a secondary eruption.

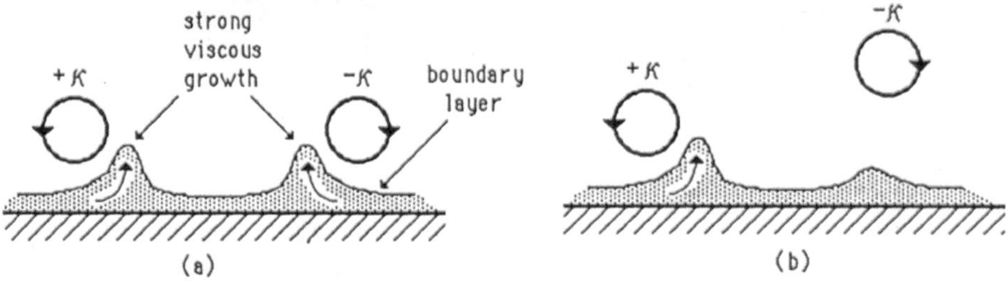

Figure 5. Schematic showing the general location of vortex-induced viscous growth for counter-rotating pairs. Cases: (a) Symmetric pair; (b) Asymmetric pair.

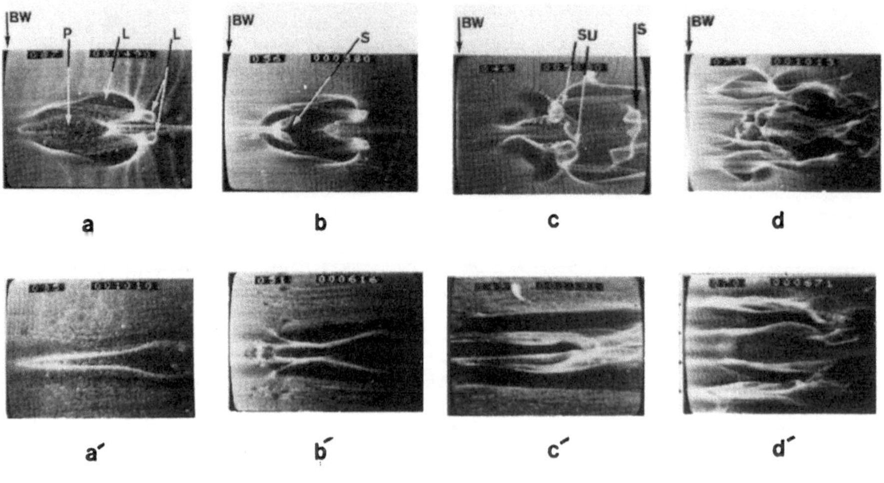

Figure 6. Dual-level, plan-view hydrogen bubble patterns generated by passage of a developing hairpin vortex structure.

Production of Turbulence in Boundary Layers and Potential for Modification of the near Wall Region

R. E. FALCO, J. C. KLEWICKI and K. PAN

Turbulence Structure Laboratory
Department of Mechanical Engineering
Michigan State University
East Lansing, MI 48824

Important relationships between known wall region coherent motions have been found, and an integrated overall model is put forth. Visual information has been used to qualitatively clarify the complete spatial/temporal response of the near wall region to perturbations originating in the outer region. Furthermore, two-point spanwise vorticity correlations have been used to characterize the outer region events initiating the near wall response. These findings lead to the result that a substantial integration of detail in the wall region is essential to uncover the underlying relationship of the organized features that are responsible for the production of turbulence near a wall. An important aspect of the integrated picture is the discovery that strong ejections of fluid occur as a result of specific spatial-temporal relationships. The correlation experiments statistically identify the presence of strong vortical motions associated with the proposed Overall Production Module (OPM). It was found that the vortical motions away from the wall have an average scale of 25 wall units, and rotate in the opposite sense of the mean vorticity. The present results suggest means by which drag can be reduced by modifying the wall so as to reduce the amplification of strong vortical motions created near the wall.

1. INTRODUCTION

Flow visualization studies of the wall region are complicated by the "history" of the flow, and thus are not reliable in determining the temporal evolution of the flow, or in quantifying the instantaneous spatial correlations and/or spatial-temporal dynamics for events that have length scales longer than the order of a hundred wall units.

The present set of experiments extend both visualization experiments and hot-wire techniques to obtain instantaneous and statistical spatial-temporal information. Visual information is extended by using a single "refresh" of wall visual information provided by a second slit. In addition, long time averaged two-point spanwise vorticity correlation data, with a fixed wire at the edge of the sublayer, are presented. These two sets of data provide complimentary information that together with outer region visual data and previous experimental simulations strongly support the cause and effect relationships proposed to exist between the outer flow and the organized motions in the wall region, and between the various

A. Gyr (Editor)
Structure of Turbulence and Drag Reduction
IUTAM Symposium Zurich/Switzerland 1989

wall region motions. Understanding the sequence of events that comprise the OPM is seen as the best means of discovering ways to interfere with the process.

2. EXPERIMENTS

Experiments were performed in a water channel and a boundary layer wind tunnel. The water tunnel was used to obtain two (and at times three) orthogonal views of two or three differently colored dyes being emitted from wall slits. In these experiments a special parallel wall slit was constructed that enabled the essentially tangential introduction of two colors of dye through slits separated by about 270 wall units in the streamwise direction. This double-dye-slit enabled a "refresh" of the information, thus minimizing the "history" effect as well as enabling some sequencing of visualized wall layer information. The spanwise vorticity measurements were made in the 17.1 meter low speed continuous flow visualization boundary layer tunnel in the Turbulence Structure Laboratory at Michigan State University. These measurements were made using four-wire hot-wire probes. The facility, the probe design and its calibration, and details pertaining to the accuracy of the measurements are described in Klewicki [1].

3. RESULTS

Extensive laser sheet visualization results from smoke marked air boundary layers and dye marked water boundary layers have been integrated to form what is referred to here as the wall region subset of the OPM. This subset of the OPM integrates long streaks of passive contaminant [2,3], pockets [4,5,6,7], streamwise vortices [8,9], hairpin vortices of different origin [10,11,12], and the pocket vortex [4,5,10]. Furthermore, this subset of the OPM incorporates information pertinent to the cause and effect relationships between the above coherent motions [13,14,15]. The physical simulations of Chu and Falco [16] and Chu [17] provide additional details pertinent to the OPM.

3.1. THE WALL REGION SUBSET OF THE OVERALL PRODUCTION MODULE (OPM)

In experiments designed to examine only the wall region, observations indicate that the production sequence starts with the formation of a pair of sublayer streaks. Using dye from the two slits (separated by $x^+ = 270$) illuminated by both streamwise and cross-stream laser sheets and plan view flood lighting, observations indicate that after the streak pair forms (both later in time and at the downstream end of the streak pair) a pocket forms between the streak pair. The pocket often appears after the pair of streaks has been forming for several hundred wall units in the streamwise direction. Figure 1 shows an example of observations in the plan view. The red dye emanating from the upstream slit has been rearranged into a pair of streaks (arrows). Of course there is additional dye rearrangement due to either parts of similar events or interactions with the current event. The pair in Fig. 1 has an average spacing along its length of $z^+ \cong 60$. The average streak spacing [2,3] is $z^+ = 100$. Without spatial information pertaining to the evolution combined with a temporal resolution the order of a wall time scale, it is difficult to ascertain that the pair formed in a highly correlated way. In the first few frames of Fig. 1 the pair of streaks has already developed over the distance between the two slits ($x^+ = 270$). By the fourth frame it is seen that the streak pair is now marked by the dye coming out of the second slit. Thus, the mechanism responsible for the streak pair formation is still active. By the fifth frame a pocket can be seen opening up, and by the sixth frame, i.e. a little more than $\Delta t^+ = 15$, a fully formed pocket is evident.

Figure 2 shows another sequence in which we focus on the flow development in the neighborhood of the pocket. This sequence begins with the formation of a pair of streaks just downstream of the upstream slit (not in this field of view of the print). The first frame in the sequence starts just after the pocket has formed. In subsequent frames secondary hairpins associated with the streak pair develop. By the second frame the dye indicates that these hairpins have moved over the edges of the pocket. In the next two frames these hairpins are being rotated towards the center of the pocket. In the last two frames the twisted hairpins are induced back towards the wall into the center of the pocket. The pocket vortex cannot, of course, be seen using dye. However the very rapid evolution (the entire sequence took only $t^+ = 5$) suggests that a strong vortex is present within the boundary of the pocket Falco [4]. The observation of lift-up and lateral movement towards the pocket centerline and then movement back down towards the wall have been confirmed by simultaneous end view experiments.

Repeated observations of the above kind have resulted in the conceptual model shown in plan view Fig. 3. It represents the present picture of the complete near wall response to a single outer region perturbation. This figure, however, does not include the motion that initiates the near wall response. The sequential evolution is as follows. First, a *pair of streaks* is formed. Then, a *pocket* forms at the downstream end, positioned between the streak pair. The formation of the pocket is often accompanied by a hairpin forming at its downstream end. Next in the evolution the streak pair undergoes an instability that results in hairpins forming over them [18]. The hairpins which form first are over the oldest portion of the streaks, and thus are furthest downstream and closest to the pocket. These hairpins are observed to lift up and be rotated by the pocket vortex into the center of the pocket, and are subsequently convected towards the wall. As the hairpins rise up they move a little to the inside or outside of each streak, and are induced towards the center of the streak pair by the pocket vortex. This process gives an impression of streak waviness in the plan view [8].

The hairpins which form over the pockets are called *secondary hairpin vortices*, and the one forming downstream of the pocket is called the *primary hairpin* vortex. The vortex which exists within the pocket is called the *pocket vortex*. (See Fig. 3).

Consideration of the different viscous and inviscid mechanisms acting on the pocket vortex, versus any of the hairpin vortices, reinforces the results of the experimental simulation measurements [16]. Specifically, the pocket vortex results from a local deceleration of the existing convected spanwise vorticity in the near wall region. The initially small upstream kink is amplified through self induction resulting in further bending of the vorticity filaments and moving them closer together. These bent vorticity filaments move upstream relative to the convecting disturbance, and induce themselves closer to the wall [19]. Thus, these motions tend to maintain their position in the region of maximum shear. The sharp upstream bend, or tip, of what is now identified as the pocket vortex is in the sublayer whereas the legs are further out. This causes intense vortex stretching (the viscous mechanism). Furthermore, as the tip of the pocket vortex induces itself towards the wall the inviscid impermeability constraint (mirror image mechanism) is amplified, since it is inversely proportional to the distance from the wall. The inviscid condition acts to further decelerate the pocket vortex, primarily in the neighborhood of the tip, and thus adds to the stretching of the pocket vortex.

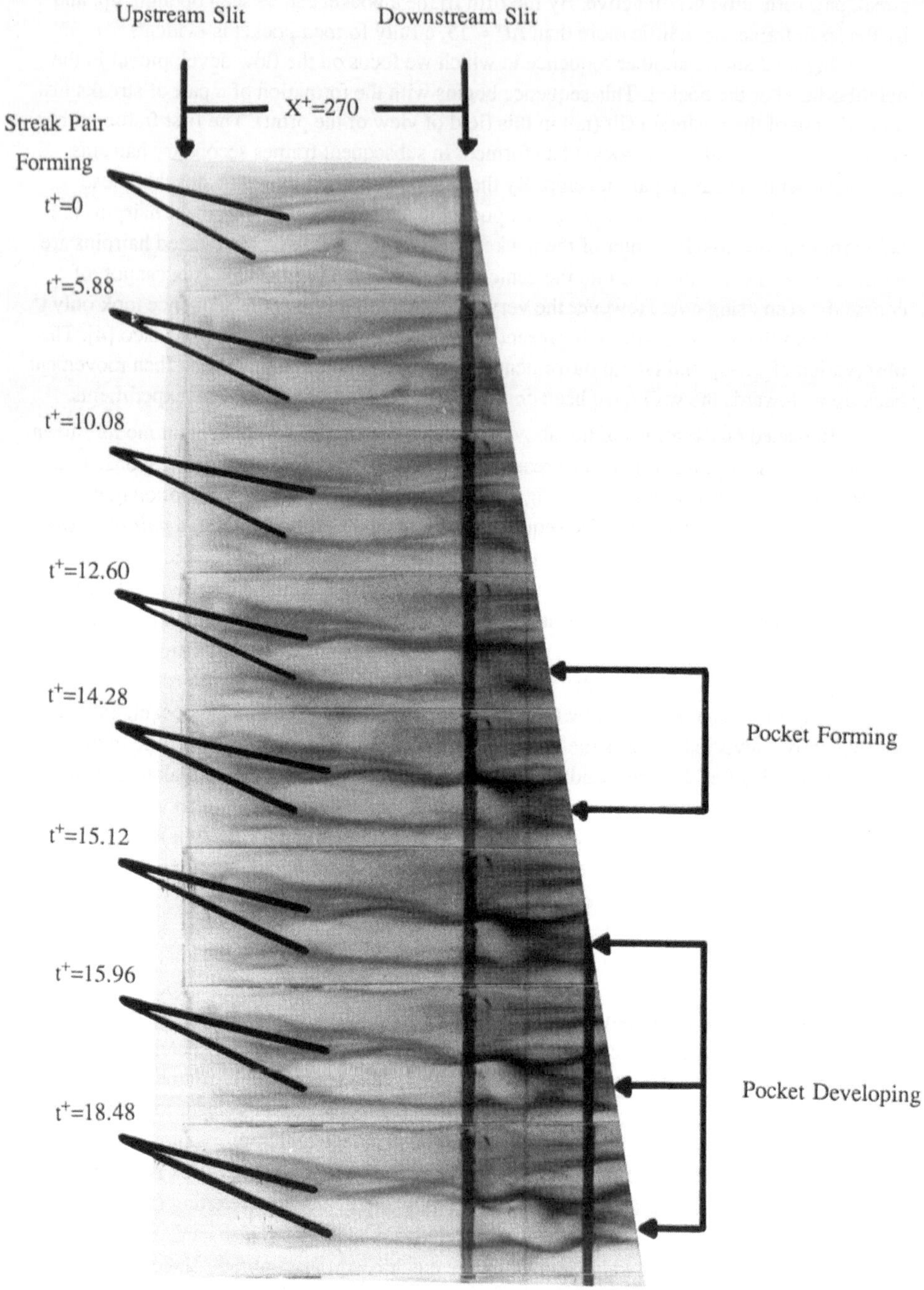

Figure 1. Formation of a streak pair and the downstream development of a pocket. Flow is from left to right.

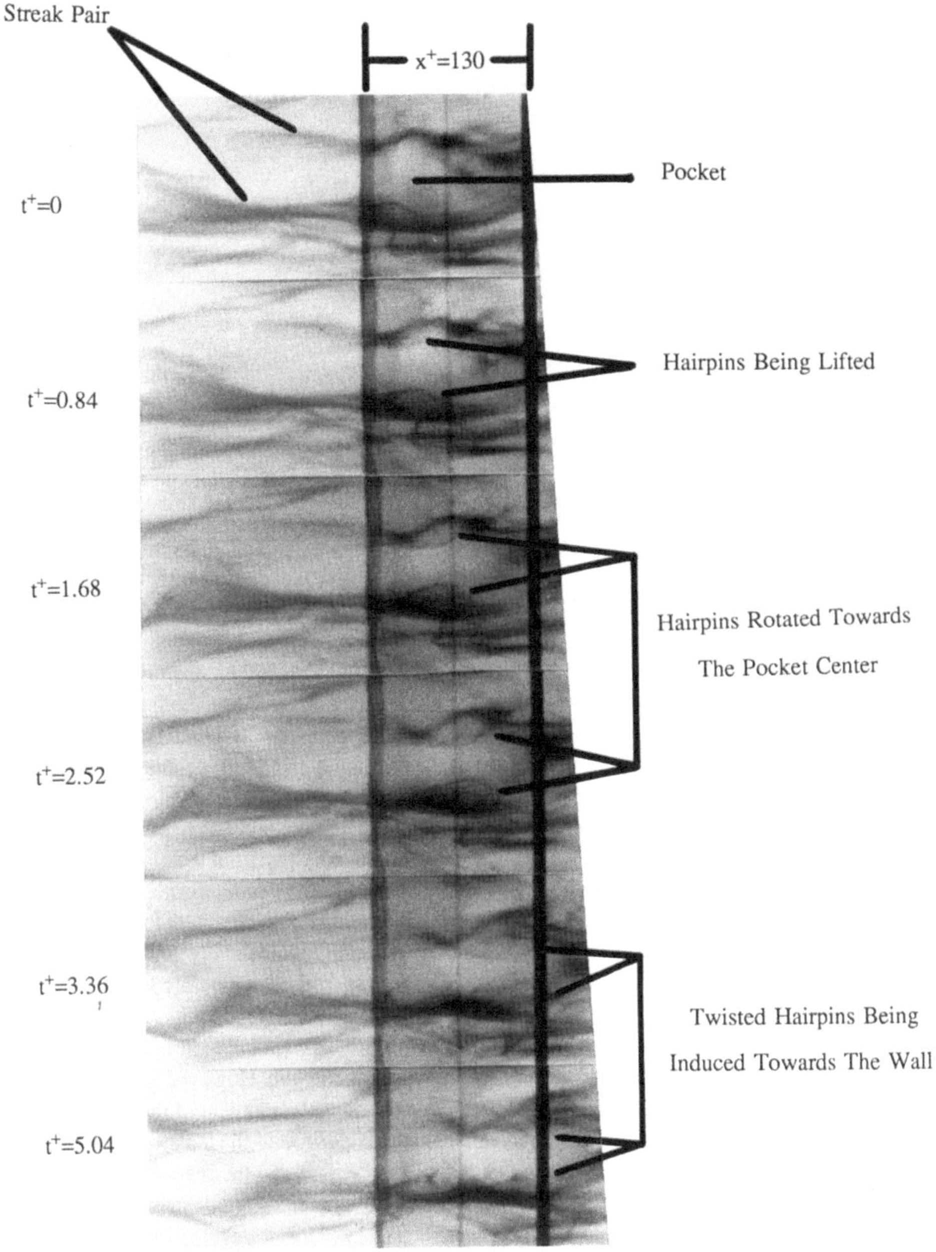

Figure 2. Pocket evolution resulting in the engulfment of secondary hairpins that have developed on the streaks.

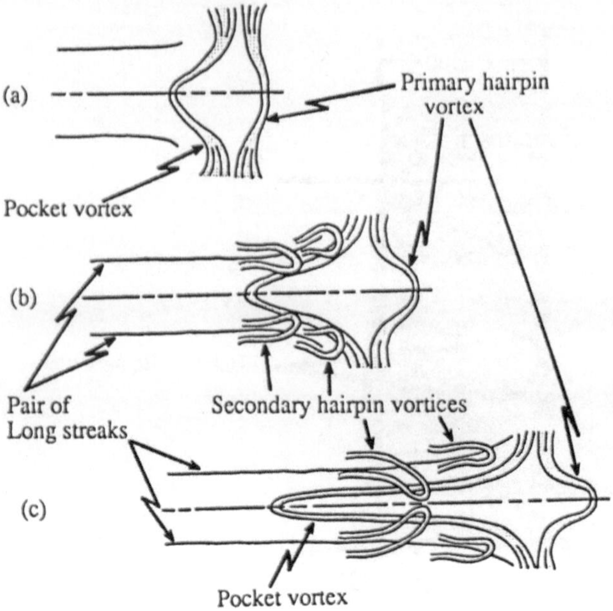

Fig. 3. A plan view showing three sequences in the evolution of the wall region subset of the OPM. It starts with the formation of a pair of streaks (a), followed by the formation of vortices in the pocket and over the streaks (b), followed by the interaction of the pocket vortex and the hairpin vortices (c).

The hairpins are stretched by the viscous shear. The magnitude of the shear across the secondary hairpins (that form further out over the long streaks), is on average less than the shear across the pocket vortex. The shear across the primary hairpin [20] generally tends to be intermediate to that experienced by the pocket vortex and the secondary hairpins. For all the hairpins, the inviscid condition generally acts to reduce stretching, and hence relatively weaken them. However, because the hairpins are self induced away from the wall the effects of the inviscid mechanism are less important than they are for the pocket vortex. The net result is that the pocket vortex is the strongest vortical motion associated with the wall portion of the OPM. As a result of vortex stretching, its scale becomes very small, becoming less than the local Kolmogoroff scale (approximately 1.5 viscous units) before loosing its identity through viscous dissipation.

During the evolution of the various features just described, each feature contributes a portion to the total turbulence produced by the overall flow module. However, important spatial/temporal relationships associated with the formation and alignment of the features results in significantly more transport than would occur as a result of the isolated evolution of each of the parts. The additional production results from the fact that the secondary hairpins which form over the long streaks first occur at the downstream end of each streak in the pair -- exactly where the pocket forms. Thus, as these secondary hairpins develop they lift themselves above the streaks to a position very close to the pocket vortex. The pocket vortex, as a result of reorientation and stretching, has a strong streamwise component. Consequently it induces the secondary hairpin vortices to move around it, resulting in lateral motion and twisting of these vortices, which subsequently brings them back down to the wall (see Fig. 3b,c). When modeled in experimental simulations [16], an important fraction of the measured Reynolds stress was attributable to this twisted hairpin/pocket vortex interaction. The lifted hairpins (both primary and secondary) constitute ejections of low speed fluid, as does the lifted fluid from the hairpin/pocket vortex interaction. The multiple ejections observed in [13,21,22] associated with a single burst can be seen to be a natural consequence of the wall region subset of the OPM.

3.2. CORRELATION OF SPANWISE VORTICITY ACROSS THE LAYER

As an alternative to constructing statistically significant samples from the required three dimensional visual data bases, two point spatial correlations of spanwise vorticity have been used (see Klewicki [1]). Correlation coefficients for a fixed probe at $y^+ = 7.5$ and positions further out in the layer are shown in Fig. 4. This experiment provides information relevant to the statistical importance of the relationship between the Typical eddy and the pocket vortex, and the Typical eddy and the primary hairpin vortex. Chu and Falco [16] have indicated that a wide range of possibilities exist in the instantaneous details of the evolution. However, the occurrence of a negative ω_z peak at the lower probe and a positive ω_z peak at the upper probe is predicted from their visualization studies. Thus, one would expect that if these events constituted a significant fraction of the events important to the OPM, then the correlation should reflect this by being negative.

The correlation coefficient of Fig. 4 is strongly negative for the closest measured separation of $\Delta y^+ = 14.4$ and remains negative for separations at least as large as $y^+ = 120$. By definition this correlation coefficient must go to 1.0 for zero probe separation. The strong negative correlation indicates that on average there exists fluctuating vorticity of the opposite sign at the two probes. The increasingly negative correlation at the closest separations indicates that the interaction between the opposing sign vortical fluctuations becomes most intense as the outer vortical motions get closer to the edge of the sublayer.

It is of interest to know whether the lower probe measures positive *or* negative vorticity fluctuations when the upper probe measures the corresponding opposite sign fluctuation. Furthermore, the determination of the absolute sense of rotation, or total vorticity (i.e., $\varpi_z = \Omega_z + \omega_z$), at each of the probes was desired. Finally, it is of interest to determine the length scale of the initiating motions. These measures allow correspondences to be made between vorticity probe data and flow visualization information.

Examination of a two-dimensional probability distribution function (pdf) from the instantaneous two-point vorticity correlations allows the instantaneous sense of rotation to be determined. An example of these pdf's is shown in Fig. 5 for the case where the separation is $y^+ = 14.4$ (the fixed probe is at $y^+ = 7.5$). Figure 5 includes both the pdf of the fluctuating vorticity (about axes labeled $\omega_z|_{lower}$, $\omega_z|_{upper}$), and of the total vorticity (about the axis displaced by the magnitude of the local mean shear $\Omega_z|_{lower}$ and $\Omega_z|_{upper}$).

From the point of view of the fluctuating vorticity, it is clear that the peak in the pdf lies in the fourth quadrant (i.e. upper probe sees positive fluctuations while the lower probe sees negative fluctuations; Ω_z is negative). Furthermore, the motions represented by all four quadrants exist.

From a total vorticity point of view virtually the entire pdf lies within the third and fourth quadrants. Thus, in an absolute sense the probability of observing motions associated with quadrants one and two is very small. A distinct feature of this pdf is that about one quarter of the area lies in the fourth quadrant, indicating that the upper probe is measuring ϖ_z with a sense of rotation opposite to that of the Ω_z. Counter-rotation in an absolute sense only occurs with positive ϖ_z at the upper probe.

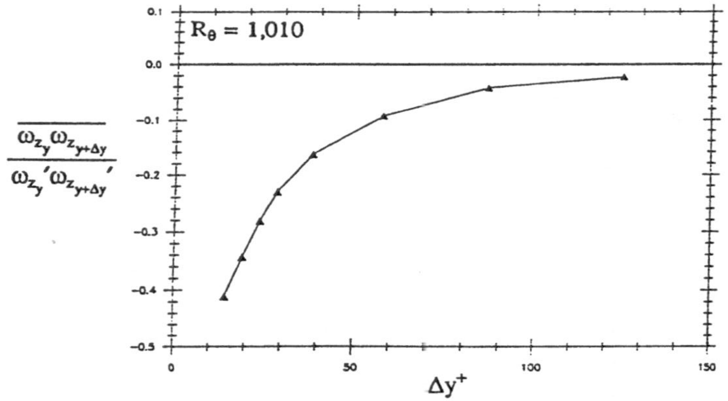

Fig. 4. Results of the two-point fluctuating ω_z correlations for positive Δy separations and the stationary probe at $y^+ = 7.5$.

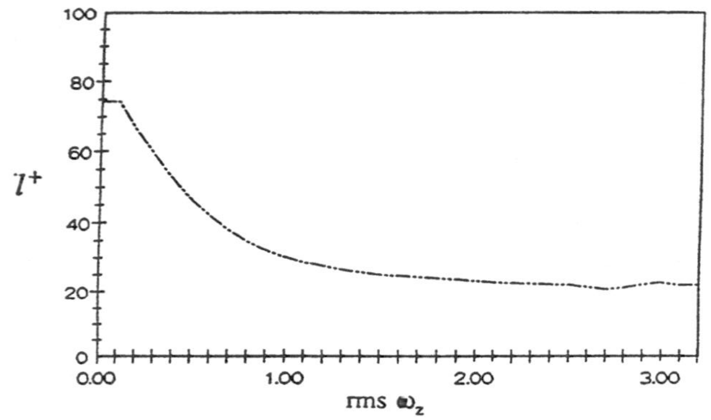

Fig. 6. Length scales characteristic of positive ϖ_z for the upper probe positioned at $y^+ = 22$.

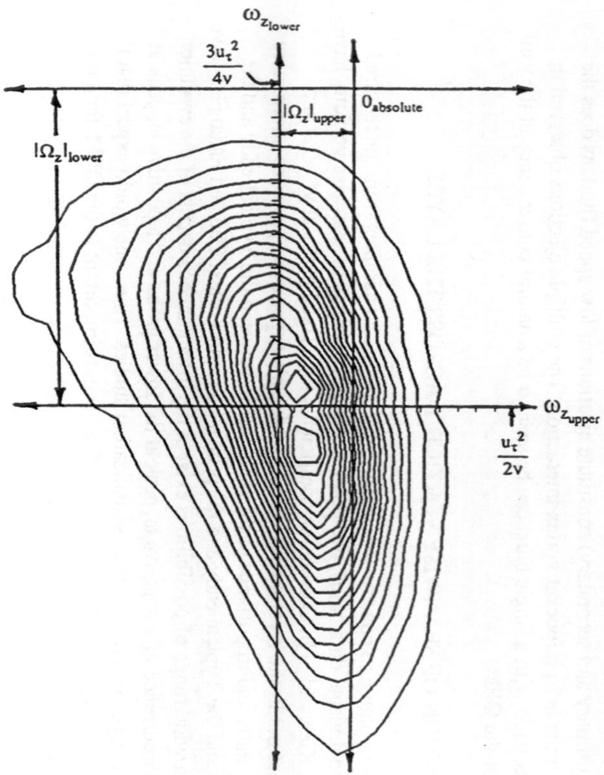

Fig. 5. Two dimensional probability contour map as derived from the ω_Z signals contributing to the data point of closest separation in Fig. 4.

In order to gain a further understanding of the motions represented in quadrant four, the durations characteristic of positive vorticity fluctuations at the upper probe were determined, and converted to length scales using $0.6U_\infty$ as a convection velocity. Figure 6 shows the average lengths of positive vorticity fluctuation events as a function of thresholds (based upon multiples of the rms ω_z at the upper probe). For thresholds above ≈ 1rms, the length scale takes on a characteristic value of about 25 wall layer units. Examination of the ratio of ω_z' to $|\Omega_z|$ shows that for $y^+ > 22$, $\omega_z' > |\Omega_z|$ [1]. This suggests that vortical motions which have rotation of sign opposite that of the Ω_z have a preferred scale.

The variability of the length scales of Fig. 6 was investigated by calculating individual probability distributions of the length scales for each value of the threshold. For low thresholds there is a wide spread indicating no preferred scale. By 1rms the mode of the distribution reaches an essentially constant value of approximately 20 wall units, and the width of the peaks are also constant for all higher thresholds. This reinforces the result that the positive ϖ_z motions have a well defined characteristic length scale.

4. DISCUSSION

The above statistical results characterize features essential to the motions initiating the wall region subset of the OPM. These results indicate that the motion is characterized by: a) strong ϖ_z, b) rotation of sign opposite that of Ω_z, and c) a length scale of about $25\nu/u_\tau$. The negative ω_z correlation is consistent with the presence of a Typical eddy at the upper probe (which would result in positive ϖ_z at the upper probe during its passage), and a strong negative ω_z at the lower probe (which could either be the pocket vortex or the primary hairpin). Additionally, the scale of the core of the Typical eddy is the order of $100\nu/u_\tau$ [14,23]. Thus, the extent of the negative correlation (≈ 120 viscous units) is also consistent with Typical eddy interactions. The joint pdf shows counter-rotating contributions to the correlation occur only when positive ϖ_z is measured at the upper probe. This length scale of ≈ 20 viscous units embraces interactions involving either the lower lobe of the Typical eddy and/or twisted hairpins that have been inverted.

The importance of the presence of an ordered sequence of coherent events and their interactions is that it offers a number of places to impart a control mechanism to break the chain and interrupt the production process. With respect to drag reduction the pocket vortex appears to be the most important link in the production chain that can be influenced by wall region modifications. This is because its amplification occurs within the sublayer and buffer layer by both viscous and inviscid mechanisms. Experiments have been suggested Falco [24] that will directly interfere with its formation and amplification.

5. CONCLUSIONS

Based upon qualitative flow visualization using a marker "refresh" scheme, it was shown that a sequential evolution exists connecting essentially all of the important events of the wall region which together comprise the wall region subset of the Overall Production Module (OPM). An important aspect of the integrated picture is the discovery that strong ejections of fluid occur as a result of specific spatial-temporal relationships between distinct vortical motions. The OPM includes vortical motions found in the outer region that interact with the near wall region and initiate these events. Two-point ω_z correlations were used to statistically characterize the spatial interactions between the inner and outer region motions. Results show the existence of a significant occurrence of vortical motions with sign opposite that of the mean outside of the sublayer. Furthermore, a strong correlation exists between these motions and intense vortical motions in the sublayer with ϖ_z of the same sense as Ω_z. A preferred length scale of about $25\nu/u_\tau$ was found for the positive rotating outer motions. These visual and statistical results provide additional support for, and more detail about the model developed in references [10, 14, 15, 16 and 22].

We want to thank Chuck Gendrich for writing the programs for calculation of the histograms and length scales. This work was supported by AFOSR. Dr. Jim McMichael was the contract monitor.

REFERENCES

1. J. C. Klewicki: Ph D Thesis, Dept. of Mech. Engr. Michigan State University, 1989.
2. S. J. Kline, W. C. Reynolds, F. A. Schraub and P. W. Runstadler, J. Fluid Mech. *30*, 741 1967.
3. C. R. Smith and S. P. Metzler, J. Fluid Mech. *129*, 27, 1983.
4. R. E. Falco, in *Turbulence in Liquids*, ed. Patterson and Zakin, Univ. Missouri-Rolla, 1, 1980.
5. R. E. Falco, AIAA Paper 80-1356, 1980.
6. M. Gad-el-Hak, in *Proceedings of the Third International Symposium on Flow Visualization*, Ann Arbor, 1983.
7. J. Kim, P. Moin and R. D. Moser J. Fluid Mech *177*, 133, 1987.
8. H. T. Kim, S. J. Kline and W. C. Reynolds, J. Fluid Mech. *50*, 133 1971.
9. R. F. Blackwelder and H. Eckelmann, J. Fluid Mech. *94*, p577, 1979.
10. R. E. Falco, in *The Structure of Turbulence, Heat and Mass Transfer*, ed. Z. Zaric', 124, Hemisphere, 1982
11. M. R. Head and P. Bandyopadhyay, J. Fluid Mech. *107*, p297, 1981.
12. C. Smith, in *Proc. 8th Symp on Turb*, Univ of Missouri-Rolla, 1984.
13. G. R. Offen and S. J Kline, J. Fluid Mech. *70*, p209, 1975.
14. R. E. Falco, Phy Fluids *20*, S127, 1977.
15. R. E. Falco, AIAA Paper 83-0377, 1983.
16. C. C. Chu and R. E. Falco: "Vortex ring/viscous wall layer interaction model of the turbulence production process near walls", Exp. in Fluids 6, 305 (1988).
17. C. C. Chu, Ph.D. Thesis, Department of Mech. Engineering, Michigan State Univ., 1988.
18. M. S. Acarlar and C. R. Smith, J. Fluid Mech. *175*, 1, 1987
19. R. E. Falco, in *Coherent Structure of Turbulent Boundary Layers*, ed. C. Smith and D. Abbott, 448, Lehigh Univ., 1978.
20. R. E. Falco, in *Flow Visualization IV*, ed. Verlin, Paris, 1986
21. E. R. Corino and R. S. Brodkey, J. Fluid Mech. *37*, 1, 1969.
22. D. G. Bogard and W. G. Tiederman J. Fluid Mech. *162*, 389, 1986.
23. R. E. Falco, AIAA Paper 74-99, 1974.
24. R. E. Falco, AIAA Paper 89-1026, 1989.

Visualisation of the Organized Motion in a Turbulent Boundary Layer

R. A Antonia[*], R. Dumas[+] and L. Fulachier[+]

* Department of Mechanical Engineering, University of Newcastle, N.S.W., 2308, Australia

+ Institut de Mécanique Statistique de la Turbulence, Université d'Aix-Marseille, UM-CNRS 38, 13003 Marseille, France

ABSTRACT

Visualisations of dye streaklines in a turbulent boundary layer show features that are consistent with the existence of a number of aspects of the organised motion and in particular hairpin-like vortices that originate close to the wall.

1. INTRODUCTION

Much of the research on wall-bounded turbulent flows in the last thirty years or so has concentrated on the sequence of events, sometimes known as the bursting phenomenon, which occur near the wall. In a recent review of some of the knowledge concerning coherent structures in the canonical turbulent boundary layer, Kline [1] proposed a taxonomy which included eight categories of quasi-coherent structures. The aim of the present paper, modest in comparison with Kline's taxonomy or Robinson et al's [2] classification based on numerical simulations of a turbulent boundary layer, is to discuss a series of photographs and film sequences, taken in the I.M.S.T. water tunnel, in the context of the above categories. As a result, we believe that further physical insight into the formation of structures which make up the organised motion has been obtained.

2. EXPERIMENTAL ARRANGEMENT/CONDITIONS

The flow visualisation was carried out in the I.M.S.T. closed-circuit water tunnel, described in Dumas et al [3]. The boundary layer develops on one of the sidewalls of the 20 cm x 20 cm vertical working section of about 1.2 m height. The visualisations were made for two different types of trips (fig. 1).

A. Gyr (Editor)
Structure of Turbulence and Drag Reduction
IUTAM Symposium Zurich/Switzerland 1989

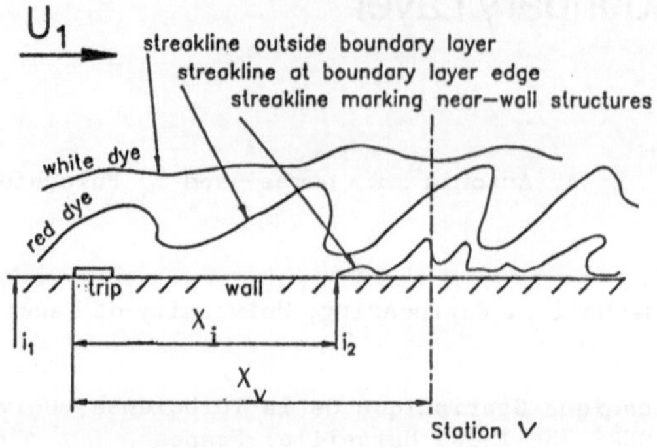

Fig. 1 - Schematic arrangement

For case A [4] the boundary layer is tripped with two pebble roughness strips (average height 5 and 3 mm) each 28 mm wide and separated by 29 mm in the flow direction. Rhodorsil (a white silicone oil emulsion) was injected through a narrow slit i_2 (streamwise length = 0.2 mm, spanwise length = 18 cm) flush with the wall.

In case B, the boundary layer is tripped with a flat strip (thickness ≈ 4 mm) which spans the full width of the working section. Dye was injected through 1 mm holes at locations i_1 and i_2. Green (water-diluted fluorescein) dye was injected at i_1, which corresponds approximately with the beginning of the working section, to mark the complete boundary layer transition. Rhodorsil was injected at i_2. Red dye (water-diluted rhodamine) was also injected, through hypodermic needles, in the contraction upstream of the working section, to mark the edge of the boundary layer inside the potential flow field. In the same manner a second streakline (white rhodorsil) was used at a larger distance from the wall. The dye flow rate was quite small, however even moderate injection rates can lead to strong instabilities in the dye immediately downstream of the dye exit. Although great care was exercised when machining the dye slot, a few imperfections remained, acting as nucleation sites for streaks.

The following table shows the main characteristics of the experiments.

	X_i (mm)	X_v (mm)	U_1 (cms^{-1})	U_τ (cms^{-1})	δ (mm)	R_θ
Case A	610	740	16.0	0.78	50	620
Case B	705	899	17.6	0.83	28	≃450

Despite the low Reynolds numbers, we verified that the flows have the characteristics of classical boundary layers in the inner part.

Photographs of the dye views in the (x,y) and (x,z) planes were taken using a flash synchronised with illumination from a narrow sheet of light, at station V. This station was a short distance downstream of i_2 to diminish the possible integration or memory effect when visualising flow structures.

3. RESULTS

3.1. Plan View of Dye Injection through a Wall Slot (Case A)

Low-speed streaks are highlighted in Figure 2. The average spanwise separation between streaks was found to be about 100 wall units, at a location of about 780 walls units downstream of the slit. Extremely narrow streaks can also be seen, often in the neighbourhood of the concentrated streaks; it is thought that the narrow streaks arise from the clinging property of rhodorsil [4].

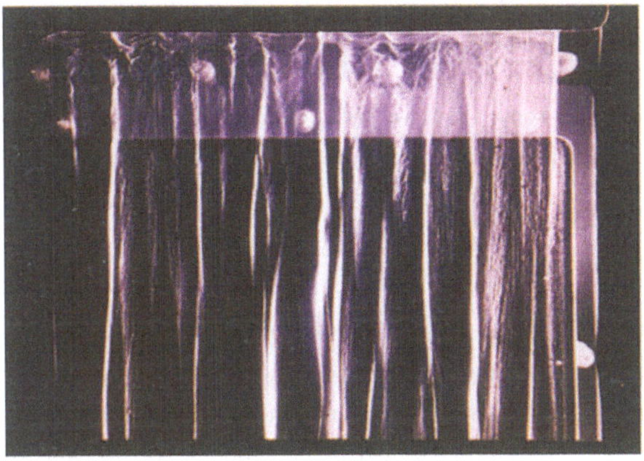

Fig. 2 - Low-speed streaks for case A. Flow direction is top to bottom.

Continuous film sequences (not shown here) of plan views of slot-injected dye have indicated that low-speed streaks are often drawn together with occasional coalescence, as previously reported by Nakagawa and Nezu [5] and Smith and Metzler [6]. Sometimes, adjacent streaks diverge rather suddenly, possibly due to the impingement of relatively high velocity fluid. This phenomenon is arguably related to the near-wall pocket [1].

3.2. Elevation Views of Dye Injected through a Hole in the Wall

A selection of photographs of views in the x-y plane of dye injected through a hole in the wall is shown in Figure 3, for case B. Most of the views in Figure 3 show that the dye tends to be lifted away from the wall, initially at a relatively small angle, typically less than 10°, but as the distance from the wall increases, the inclination can sometimes reach values in excess of 45°. The lifted dye is often in the form of a loop; one example can be seen in Figure 3.b, where the extremity of the loop is bent over and the inclination of the outer part of the loop is of order 60°. Quite often, the loop develops a kink as it rises from the wall. The part of the loop downstream of the link is often pushed back towards the wall (e.g. Fig. 3.c, 3.a) so that the kink assumes a cusp-like character. In Figure 3.d, the section of the loop downstream of the link is nearly parallel to the wall. If one assumes that the dye marks a vortex, for example a hairpin-like vortex, the explanation given by Acarlar and Smith [7] seems plausible: the legs of the vortex are drawn together by mutual induction, hence causing vorticity cancellation.

The extremity of the loop is then deflected towards the wall by the action of the wall shear. Another possibility for the deflection may be the event known as the sweep [8]. Sometimes the extremity of the dye loop reaches a significant proportion of the boundary layer thickness in a relatively short time, however this occurence is not as frequent as the deflection of dye towards the wall. The lifted dye often undergoes rapid oscillations as indicated in Figure 3.e (reminiscent of the description by Kline et al. [9]). A rough estimate of the wavelength of these oscillations is about 150 wall units. Occasionally, there is a large concentration of dye which moves away from the wall en masse (Fig. 3.f).

It is difficult to establish a direct connection between the large scale bulges in the outer layer and the creation, close to the wall, of structures. This difficulty is more evident when viewing a film sequence instead of photographs. Another, possibly more important, difficulty pertains to the strong three-dimensionality of the flow [3].

The sequence in Figure 4, taken from an 8 mm movie, captures the time history of a single dye ejection for cases A (X_i = 577 mm, no injection through the wall slot). Prior to lift-up, there is a significant agglomeration of dye on

- a -

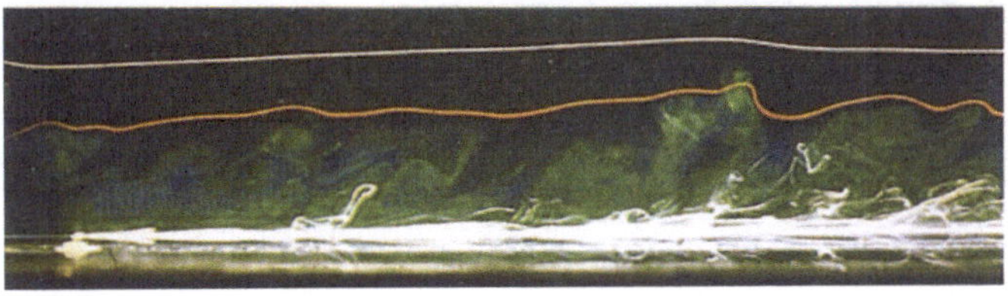

- b -

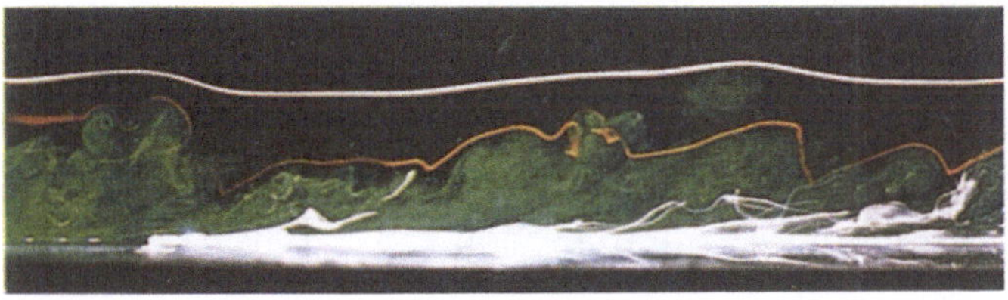

- c -

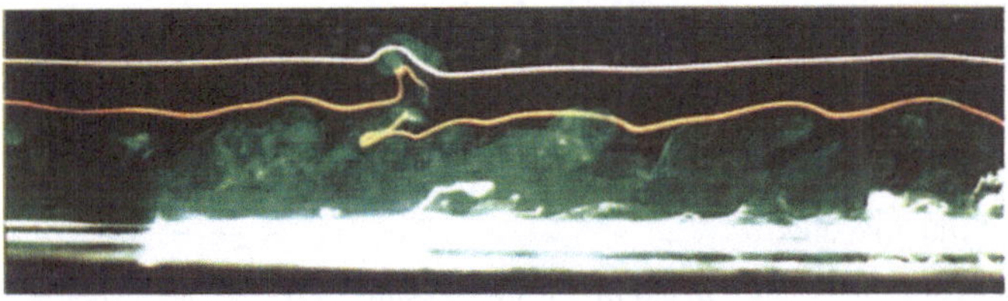

- d -

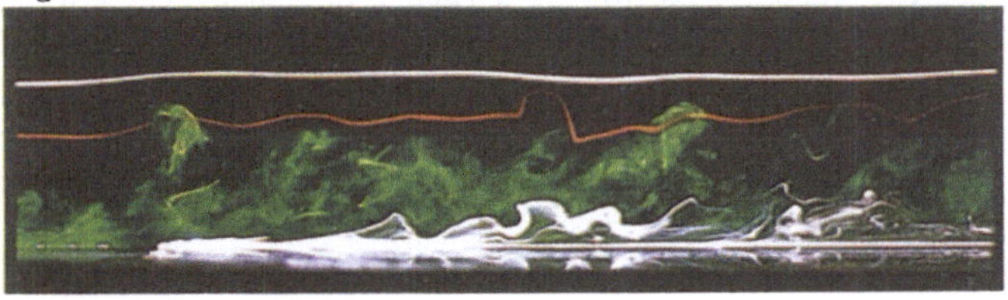

- e -

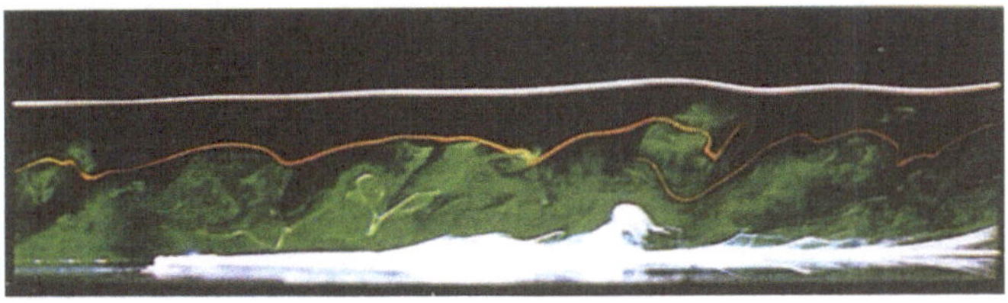

- f -

Fig. 3 - Elevation views for case B.

the wall streakline. Almost as soon as lift-up occurs, the downstream extremity of the dye develops a first kink, the tip eventually almost reaching the wall. In the following views the dye lifts up through the boundary layer. The discontinuous lines which appear can be explained by kinks in the planes normal to the view plan.

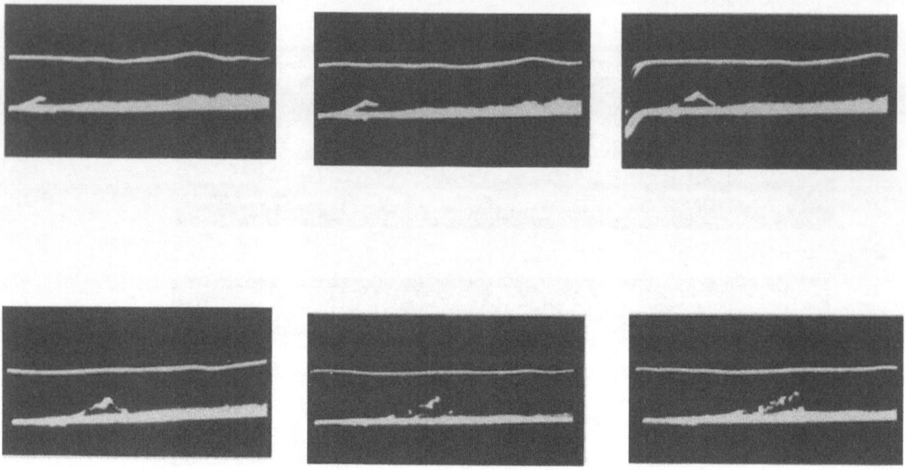

Fig. 4 - Sequence, taken from a film, of elevation views for case A. Flow direction is left to right.

4. A FEW CONCLUDING REMARKS

A recurrent feature in the present flow visualisation is the wallward deflection ejected dye loops, possibly as a result of sweep-type motions near the wall or large-scale motions in the outer layer. The loops which form at the wall can extend to a wide range of heights from the wall: this would be consistent with the similarity requirement that wall eddies scale with the distance y from the wall.

REFERENCES

1. KLINE, S. J. : 1988. Quasi-Coherent Structures in the Turbulent Boundary Layer : Part I. *Proc. Zaric Memorial International Seminar*, Amsterdam [to appear].

2. ROBINSON, S. K., KLINE, S. J. and SPALART, P. R. : 1988. Quasi-Coherent Structures in the Turbulent Boundary Layer : Part II. *Proc. Zaric Memorial International Seminar*, Amsterdam [to appear].

3. DUMAS, R., BONMARIN, P. and FULACHIER, L. : 1982. Visualisations of Turbulent Structures of Wakes and Boundary Layers, in Z. P. Zaric (ed.) *Structure of Turbulence in Heat and Mass Transfer*, New York, Hemisphere, 551-561.

4. ANTONIA, R. A., FULACHIER, L., KRISHNAMOORTHY, L. V., BENABID, T. and ANSELMET, F. : 1988. Influence of Wall Suction on the Organised Motion in a Turbulent Boundary Layer, *J. Fluid Mech.*, 190, 217-240.

5. NAKAGAWA, H. and NEZU, I. : 1981. Structure of Space-Time Correlations of Bursting Phenomena in an Open-Channel Flow, *J. Fluid Mech.*, 104, 1-43.

6. SMITH, C. R. and METZLER, S. P. : 1983. The Characteristics of Low-Speed Streaks in the Near-Wall Region of a Turbulent Boundary Layer, *J. Fluid Mech.*, 129, 27-54.

7. ACARLAR, M. S. and SMITH, C. R. : 1987. A Study of Hairpin Vortices in a Laminar Boundary Layer. Part 2. *J. Fluid Mech.*, 175, 43-83.

8. BRODKEY, R. S., OGUNDE, K. and CHANG, L-K : 1988. Visualization and Simulations of Fluid Interactions Between Large Scale Motions and the Streaky Structure in a Turbulent Boundary Layer Shear Flow, *Proc. Zaric Memorial International Seminar*, Amsterdam [to appear].

9. KLINE, S. J., REYNOLDS, W. C., SCHRAUB, F. A. and RUNSTADLER, P. W. : 1967. The Structure of Turbulent Boundary Layers, *J. Fluid Mech.*, 30, 741-773.

REFERENCES

1) KLINE, S. J., 1968. Quasi-Coherent Structures in the Turbulent Boundary Layer. Part I. Proc. [illegible] International Seminar, [illegible] (to appear).

2) ROBINSON, [illegible] and STRATFORD, [illegible] 1968. Quasi-Coherent Structures in the Turbulent Boundary Layer. Part II. Proc. [illegible] International Seminar, [illegible] (to appear).

3) COMAS, E., BORMANN, [illegible] and LEVACHER, [illegible] 1962. Visualisation of Turbulent Structure of Wakes and Boundary Layers. In [illegible] (ed.) Structure of Turbulence in Heat and Mass Transfer, New York: [illegible], 111-301.

4) [illegible]

5) [illegible]

Analysis of Experimental Data on Boundary Layer Pressure Fluctuations in Turbulent Pipe Flow

Harrison T. Loeser
Consultant

ABSTRACT

Experimental data on pressure fluctuations and convection velocity of turbulent air flow in a pipe are analyzed to measure the velocity, spacing, location and approximate size of the principle 'eddy' structures.

INTRODUCTION

Recent improvements in computing power have made it possible to compute accurate solutions to the Navier-Stokes equations for three dimensional unsteady flows. J. C. R. Hunt states in 1) that the results of these direct simulations are sufficiently accurate to have the status of experimental data. However, despite this capability, Hunt states that there are few signs of any convergence in turbulence research. In effect, individual researchers, attempting to understand the nature of this baffling phenomenon, continue to attack their separate problems using available data whether computed or experimental.

While researching the structure and dynamics of turbulent boundary layer flow ref 2) was reviewed. The experimental data on turbulent pipe flow in that report permits, a measurement of the main structures of the flow. The following discussion is based principally on this data.

THE EXPERIMENTS

The experiments used a 8.89 cm diameter pipe through which fully turbulent air flowed. The range of Reynolds numbers was from 100,000 to 300,000 based on pipe diameter.

A. Gyr (Editor)
Structure of Turbulence and Drag Reduction
IUTAM Symposium Zurich/Switzerland 1989

Fluctuating pressure measurements in the pipe wall were taken in the frequency range from 212 Hz to 20,000 Hz. Great care was taken to eliminate all vibration in the experimental equipment in that frequency range.

In addition to determining the velocity profile, and frequency spectrum, space-time correlation measurements of the pressure fluctuations at the wall were made from two similar transducers at several streamwise separations.

Cross-correlation maxima in the longitudinal direction were used to determine the rate at which the pressure fields for six octave bands with center frequencies from 300 to 9600 Hz were convected downstream.

MODEL OF TURBULENT PIPE FLOW USED IN ANALYSIS

Many available photographic depictions of turbulent boundary layer flow, see Schlichting for example, make evident that the principal features of the flow are eddy structures of varying size. This is supported by other experimental data, see 4), 5) and 6) for example.

The model used in the following analysis assumes that the high velocity gradient at the wall creates a high vorticity in the fluid which in turn generates small spanwise cylindrical eddy structures in the viscous sublayer of the wall. Because of the large local gradient, a small outward perturbation of a portion of the eddy results in a large increase in convection velocity for the perturbed portion. That portion will move ahead of the eddy structure initially creating a hairpin like formation. This faster moving portion will over run other eddy structures and perturb them or absorb them. It may also in turn be overtaken by a larger eddy structure moving at a higher velocity although lower rotational speed.

When the 'eddies' are viewed from a framework moving at their velocity they appear to be rotating about a center. While small, i.e. in the viscous sublayer, the viscous friction with the adjacent fluid forces their vorticity and convection speed to be aligned with the local gradient and velocity.

However, as the 'eddy' grows in size and moves away from the wall to areas of lower gradient, the high rotational momentum generated earlier in the higher gradient but slower speed flow will cause the larger 'eddy' to rotate faster and move slower than implied by the local gradient and velocity. This mismatch will be smaller for fluids with high kinematic viscosity and larger for fluids with low.

The forces generated by this mismatch in speed and rotation will cause the 'eddy', while it convects downstream, to migrate away from the wall until it reaches the area of zero gradient at the center of the pipe, or the edge of the boundary layer for flat plates, where it dissipates its energy. This normal motion of the 'eddies' transfers their mass momentum from the wall to the center of the pipe, slowing the flow in the pipe and greatly increasing the 'effective viscosity' of the fluid.

The volume formerly occupied by these migrating 'eddies' is replaced by faster moving fluid from the pipe center. Hence the effect is to mix the fluid in the pipe.

Each 'eddy', due to its rotary motion, creates a pressure field in its vicinity. The pressure fluctuations caused by these 'eddies' passing along the wall of the pipe provide data from which their speed, their spacing, their distance from the wall and their approximate size may be inferred.

Within the space allotted for this paper it is possible to show in detail only the results mentioned above. Precise information on other aspects of the model will be addressed in the future.

'EDDY' SPACING

Ref. 2) states that the signals from two transducers having a selected longitudinal separation were recorded, then summed with varying time delays. The convection speed of the pressure field is obtained by dividing the separation distance of the transducers by the time delay at which maximum correlation for the selected bandwidth occurs.

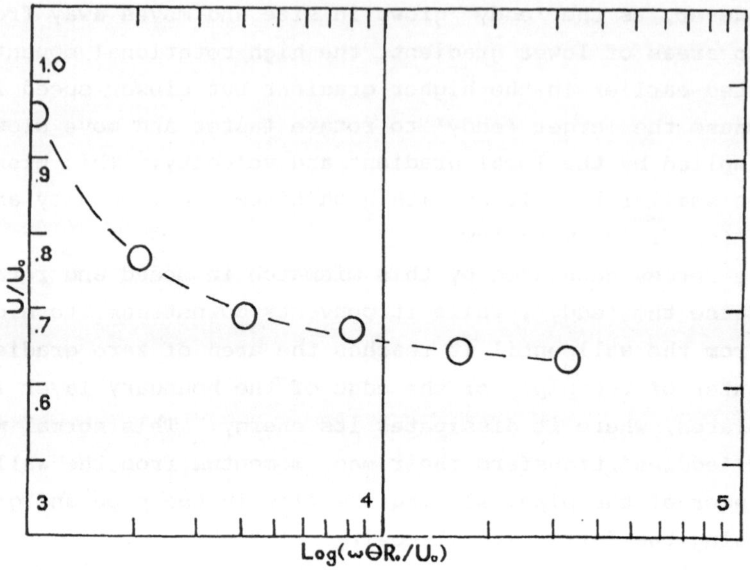

Fig. 1 Convection Velocity vs. Frequency

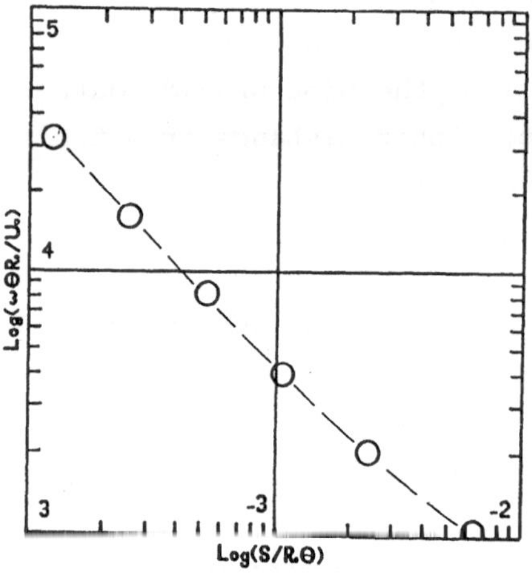

Fig. 2 Freq. vs. Eddy Spacing

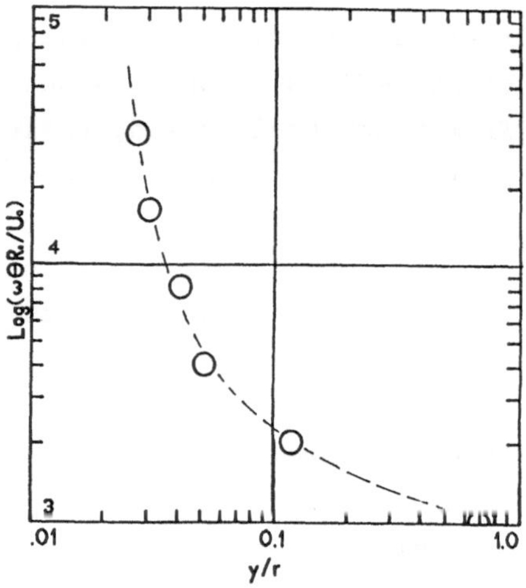

Fig. 3 Freq. vs. Wall Distance

The fluctuating pressure field which generates each frequency band is produced by a train of eddys moving past the transducer, spaced so that their sequential pressure pulses generate the band frequency.

The relation between the spacing, the convection speed and the frequency is: $f = U/S$, where: f = frequency, S = spacing and U = convection speed of the 'eddies'

Hence: $S = U/f$ 1)

Fig. 1) plots data from the experiments which gives the convection speed as a fraction of the centerline velocity versus the non-dimensionalized center of the frequency band, $\omega\theta R/Uo$. where: ω = frequency radians/sec, θ = momentum thickness - .279 cm. and R = reynolds number based on θ and Uo = centerline velocity.

Fig. 2) plots the non-dimensionalized frequency versus the non-dimensionalized 'eddy' spacing as determined from equation 1).

'EDDY' DISTANCE FROM PIPE WALL

Since each 'eddy' train must move at approximately the speed of its surrounding fluid, the convection speed is the average speed of the fluid at the location of the train. Therefore, the velocity profile permits us to infer the distance from the wall at which each frequency is generated. The velocity profile for the fully turbulent pipe flow followed the 1/8 the power law i.e.

$$U/Uo = (y/r)^{1/8}$$

Hence, $y/r = (U/Uo)^8$ 2)

Where U/Uo is the ratio of the velocity at a distance, y, from the wall relative to the velocity at the centerline and where r is the radius of the pipe. Fig. 3 plots the non-dimensionalized frequency versus distance from the wall, obtained from equation 2).

The implication from these plots is that the spacing of the 'eddies' becomes small at high frequencies close to the wall and the frequency high despite the fact that the longitudinal convection velocity is lower.

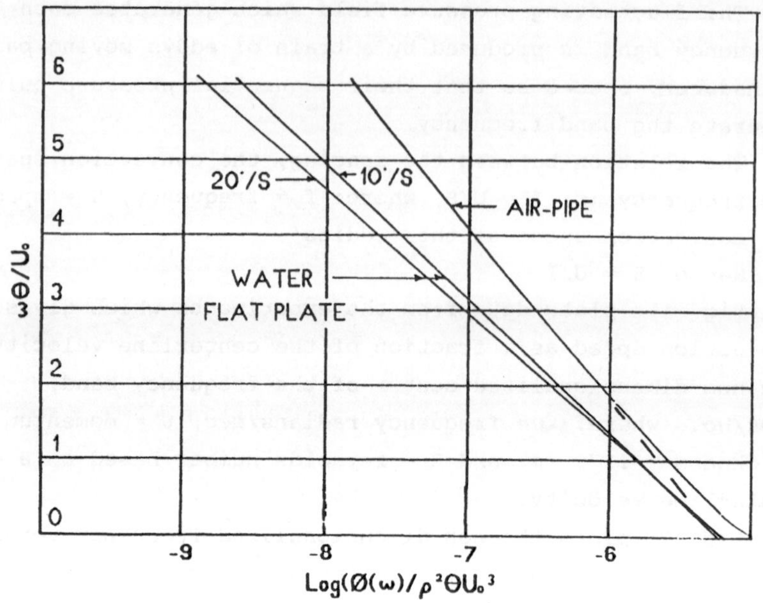

Fig. 4 Frequency Spectra

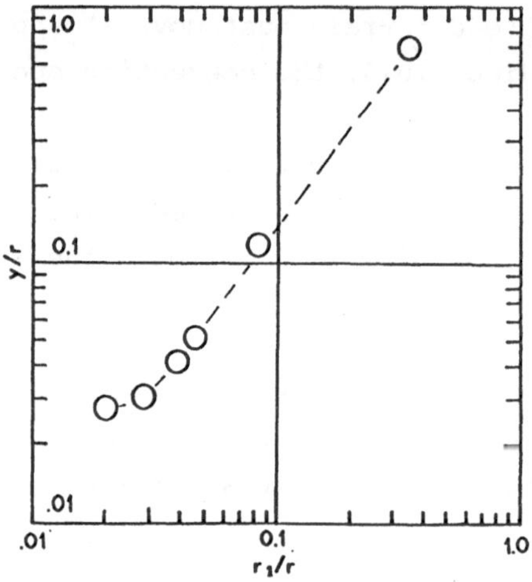

Fig. 5 Wall Distance vs. Eddy Radius

'EDDY' SIZE

Consider a section of a fluid which is rotating such as an 'eddy'. To maintain continuity, it is necessary that the rotating 'eddy' be surrounded with a pressure field greater than the pressure at its center. The pressure differential is given by: $P = s1*\rho*(\omega*r1)^2$
where: P = pressure differential, s1 = non-dimensional size parameter, ρ = mass density, r1 = distance from center of rotation and ω = rotational velocity of eddy
Let: $\phi'=(P^2)/(\rho^2*\theta*Uo^3)/BW=s1^2*(\omega*r1)^4/(\theta*Uo^3)/BW$ 3)
where: BW = bandwidth. Then, the effective radius of the eddy structure is: $r1 = (1/\omega)*\phi'*BW*\theta*Uo^3/s1^2)^{1/4}$ 4)

Since the driving force of the rotation is the gradient, assume that the rotation of the eddy, ω, is proportional to the average velocity gradient at its location in the flow. This will not apply to eddies near the pipe center.

Fig. 4, using data taken from ref. 2), gives the mean frequency spectrum of the pressure fluctuations, $\phi' = \phi(\omega)/(\rho^2*\theta*Uo^3)$, on the pipe wall as a function of the non-dimensional frequency, $\omega\theta/Uo$. Fig. 4 also shows, for comparison, spectra for flat plate water flow for 10 & 20 ft./sec. using data taken from ref. 3). The rise in the low frequency spectrum for pipe flow is probably due to the concentration of energy from the complete circumference on the center of the pipe where the large, low frequency structures accumulate.

When the parameters of the experiment are used in expression, 4), assuming that the pressure of the eddies is proportional to the pressure sensed by the transducers, the size, r1, of the 'eddy' may be estimated.

The pressure data is given as a continuous curve of pressure^2/rad/sec, hence, it is necessary to select a bandwidth to associate with each 'eddy' train in addition to selecting a value for s1 when estimating the size of individual 'eddies'.

Visual depictions of the largest eddies show them to be about one third the size of the boundary layer, and since the use of 1/3 octave band widths with a value of 0.5 for s1 gives vortices of approximately this size, these values were adopted for this discussion.

The non-dimensional radii of the 'eddies', r1/r, for R=10450, as a function of their distance from the wall are plotted in fig. 5.

CONCLUSION

The experimental data when used with the model described above permits an approximate measure of the spacing, distance from the wall, speed and size of the principal structures of the turbulent flow. Since each structure creates a velocity gradient at its own periphery, smaller structures are probably generated there, each again generating smaller 'eddies' and so on to vorticity. The movement of the major 'eddies' may be simulated on a computer to obtain pressure cross-correlations which approximate those obtained experimentally. However, space will not permit discussion of these aspects.

REFERENCES

1. Hunt, J.C.R. Studying Turbulence Using Direct Numerical Simulation: 1987 Center for Turbulence Research NASA Ames/Stanford Summer Program. Journal of Fluid Mechanics 1988 Vol. 190 pp 375-392
2. Bakewell, H.P. Jr., Carey, G.F., Libuha, J.J., Schloemer, H.H. and Von Winkle. Wall Pressure Correlations in Turbulent Pipe Flow: USL Report No. 559, 20 August 1962.
3. Keith, W.L. Measurements of Wall Pressure Fluctuations With and Without Riblet Coatings: Proceedings of the Symposium on Hydrodynamic Performance Enhancement for Marine Applications, Newport, Rhode Island, Oct 31 - Nov 1, 1988 - Chairman R.H. Nadolink NUSC, Newport, R.I. 02891-5047
4. Willmarth, W.W., and B.J. Tu, Structure of Turbulence in the Boundary Layer Near the Wall, Physics of Fluids Supplement, 1967, pp 5134 - 5137
5. Bakewell, H.P., Jr. Comments and Observations on Turbulent Boundary Layer Noise Reduction. A.S.M.E. Pub. Laminar Turbulent Boundary Layers FED - Vol. 11 Feb. 12-16 1984
6. Kline, S.J., Reynolds, W.C., Schraub, F.A., and Runstadler, P.W., "The Structure of Turbulent Boundary Layers,". Journal of Fluid Mechanics Vol. 30, No. 4, 1967, pp. 1880-1889

Behaviour of Separation Bubble with Different Roughness Elements at the Leading Edge of a Flat Plate

By R. S. Azad and B. Doell

Department of Mechanical Engineering
The University of Manitoba
WINNPEG, Manitoba, Canada R3T 2N2

Various grades of roughness surfaces were attached to the leading edge of a flat plate with an elliptical nose to investigate the flow within and behind a two-dimensional laminar separation bubble. To accomplish this, measurements were made of surface pressure, streamwise mean velocity, and turbulence intensity at a Reynolds number, $Re = 2.4 \times 10^4$, based on the plate thickness and freestream velocity. The separation bubble gradually disappeared with the increase of roughness and finally the flow resembled in many respects the flow downstream of a backward-facing step.

1. Introduction

A laminar separation and recirculating vortex flow is often observed near the rounded leading edge of a flat plate, terminated downstream by transition and an abrupt turbulent closure or reattachment onto the surface. The separation bubble can adversely affect the lift and drag coefficients of an airfoil and in turn affect the efficiencies of turbomachines such as turbines and compressors.

An excellent review of leading edge separation bubbles has been written by Tani (1964) who came to the conclusion that the bubble depended on Reynolds number, transition, pressure distribution, surface curvature and free-stream turbulence. Nakamura & Ozono (1987) found that the bubble was shortened by the increase of freestream turbulence; this fact is corroborated by the present study. The pressure generally remains relatively constant after separation until turbulent mixing commences and permits a rapid pressure recovery. Gleyzes et al. (1984) have investigated a separated bubble similar to the present one. Kiya & Sasaki (1983) and Kiya (1986) have studied the separated flow on a two-dimensional flat plate with a rectangular leading edge. Furthermore, Castro & Haque (1987) have measured turbulence quantities through a separation bubble at the leading edge.

In the present study the effects of leading-edge roughness on subsequent separation bubble development and on the flow structure within and downstream of the bubble were investigated. Flow visualization was used to locate the separation bubble and subsequently

A. Gyr (Editor)
Structure of Turbulence and Drag Reduction
IUTAM Symposium Zurich/Switzerland 1989

to estimate its length approximately. Other measurements were made to shed light on the separation bubble.

Section 2 describes the experimental equipment and procedure. Results and discussion are outlined in Section 3. Finally, concluding remarks are made in Section 4.

2. Experimental Equipment and Procedure

The wind tunnel used for the experiment was a closed return type constructed from wood and fibreglass. The air was circulated through the tunnel by a Woods two-stage, counter-rotating tube axial fan driven by a hydraulic motor. For all experiments the velocity was maintained at 15 ms^{-1}.

The test section has a rectangular cross-section of 53 cm x 76 cm and an overall length of 183 cm. The Plexiglas flat plate shown in Figure 1 was positioned horizontally in the test section midway between the upper and lower walls spanning the whole width of the test section.

The surface roughness materials attached at the leading edge of the plate were of 100-, 80-, 60-, 40- and 4-grit. The first four are manufactured by the 3M company under the brand name "Three-M-ite" and consist of abrasive aluminum oxide particles glued to a fabric backing while the 4-grit consists of silicon carbide particles glued to a paper backing. The roughness strips of 25 mm width were attached to the nose of the plate by contact cement on to scotch transparent tape which in turn was attached to the plate.

Pressure probes used for static and total pressure were of United Sensor type and measurement of pressure was made with a Combist micromanometer (± 0.005 mm of water). Alignment of the probes was accomplished with the help of a vernier microscope. The static and Pitot pressures were used to infer the Preston tube pressure to determine the friction velocity, u_*. Calibrations obtained by Kassab (1986) on a Pitot tube of the same diameter as that used in the present study were utilized.

Hot-wire anemometry was used to measure mean velocity $\overline{U}$ and rms fluctuating velocity $\left(\overline{u^2}\right)^{1/2}$ in the streamwise direction. A Dantec constant temperature anemometer (CTA) and associated equipment were used for this purpose. The hot-wire used was the boundary layer type, Dantec 55PO5. Procedures for setting-up and operating the CTA were followed as outlined in the Instruction Manuals of Dantec.

For flow-visualization oil drops were placed on the flat plate surface. The oil used was a mixture of SAE 10W-30, and kerosene with a blue dye (Dayglo Fluorescent Pigment, Horizon blue, A-19) added to make it more visible. The mixture constituents were finally balanced by trial and error so that it was viscous enough not to spread out in a thin layer that would be hard to observe and yet not so viscous that it would not respond to the air flow. The drops of oil moved in the direction of flow near the surface leaving streaks behind them which were then photographed.

3. Results and Discussion

It was found that in the case of the bare nose, 100-, 80- and 60-grit arrangements, a separated flow region was evident; drops of oil placed near the leading edge moved downstream until they seemed to reach a barrier and form into a ridge normal to the flow direction. This was caused by the separation of the flow from the surface. Oil drops immediately downstream of this separation either remained stationary or moved upstream. It was inferred that the backward motion was induced by the backflow next to the surface, commonly found in separation bubbles. The position where a single oil drop flowed upstream and downstream was considered to be in the vicinity of reattachment. Measurements of the length of separation bubble were made from the photographs taken of the oil drop patterns. The oil drops gave fairly consistent results of a separation position at 18 mm from the leading edge in all cases, but the reattachment point was not so unambiguous. For details the reader is directed to consult Doell (1989).

The pressure coefficients presented in Figure 2 are normalized in the manner used by Roshko and Lau (1965) in order to compare with them and with other published results. This reduced pressure coefficient is defined as

$$\tilde{C}_p = (C_p - C_{p\,min}) / (1 - C_{p\,min}) \ , \tag{1}$$

where $C_p = (p - p_r) / (\rho\, U_r^2/2)$, p_r and U_r being reference pressure and velocity respectively, and $C_{p\,min}$ is the minimum C_p in the separation bubble. In the present study, the pressure distribution for the cases having a separation bubble did not show the constant pressure regions evident in the cases of Roshko & Lau (1965) and Castro & Haque (1987). Instead the $C_{p\,min}$ used to calculate $\tilde{C}_p$ was the pressure measured at the station furthest upstream. Figure 2 shows that experimental results collapse onto each other fairly well but differ from those obtained by Roshko & Lau (1965) and Castro & Haque (1987). The difference may be due to different geometry used to produce the flow.

The mean velocity profiles downstream of the bubble in U^+ versus Y^+ form finally fitted the universal log-law for the bare, 100-, 80-, and 60-grit cases, whereas the log-law prevailed over all the stations for the 40- and 4-grits. The displacement thickness, δ^*, momentum thickness, θ, and shape factor, H, were calculated from the mean velocity data for all the roughnesses. Generally, δ^* rose and fell through the separation bubble while θ simply increased. However, both values asymptotically approached a constant after reattachment. The shape factor shown in Figure 3 also comes to a constant value that is indicative of the usual type of turbulent boundary layer. This figure also indicates the presence of the separation bubble for the cases of bare, 100-, 80- and 60-grit and the absence of the separation bubble for the cases of 40- and 4-grit. Gleyzes et al. (1984) have found that after transition to turbulence, the shape factor, H, is 1.6 and in turn they have considered it typical for a turbulent boundary layer. They also concur with the trend in Figure 3 that the high values of H in the bubble decrease rapidly at reattachment. Additionally, it is notable from the figure that all the cases reach the same final value of H = 1.6 and in turn they all eventually become ordinary boundary layers.

The maximum turbulence fluctuating velocity in the streamwise direction $\overline{u^2}_{max}$ was normalized by the freestream reference velocity $U_r = 15\ ms^{-1}$. This normalized intensity, $\overline{u^2}/\overline{U_r^2}$, is plotted against distance from separation normalized by the bubble length in Figure 4. Figure 4 also shows the experimental results of Castro & Haque (1987). Quantitatively, the data from the present study was much lower than that of Castro & Haque (1987). The normalized intensity peaks exist at reattachment for the bare, 100- and 80-grit cases but not for 60-grit case. The normalized intensities finally approach a value of 0.011 as can be seen from Figure 4.

Plots of skin frictin coefficient, C_f, versus the distance downstream from reattachment normalized by bubble length, X*, are shown in Figure 5. The present results of C_f were calculated from Preston tube readings taken downstream from the separation bubble for bare case but results of Bradshaw & Wong (1972), Chandrsuda & Bradshaw (1981), and Adams & Johnston (1988) are from backward-facing steps. All the results show the similar trend of finally approaching an asymptotic value of around 3.

4. Concluding Remarks

By increasing the coarseness of abrasive strips on the leading edge of a flat plate, it was found that the leading edge separation bubble could be slightly shortened and ultimately eliminated. The step-like geometry of the roughness strip could have been the reason for the flow to be like that of a backward-facing step.

The authors are grateful to Dr. Stuart W. Greenwood for reading the original manuscript. The financial support of NSERC of Canada is appreciated.

REFERENCES

Adams, E.W. & Johnston, J.P. 1988 Flow structure in the near-wall zone of a turbulent separated flow. AIAAJ 26, 932-939.

Bradshaw, P. & Wong, F.Y.F. 1972 The reattachment and relaxation of a turbulent shear layer. J. Fluid Mech. 52, 113-135.

Castro, I.P. & Haque, A. 1987 The structure of a turbulent shear layer bounding a separation region. J. Fluid Mech. 179, 439-468.

Chandrsuda, C. & Bradshaw, P. 1981 Turbulence of a reattaching mixing layer. J. Fluid Mech. 110, 171-194.

Doell, B. 1989 Experimental study of effects of roughness on separation bubble. M.Sc. thesis, The University of Manitoba, Winnipeg.

Gleyzes, C., Cousteix, J. & Bonnet, J.L. 1984 Laminar separation bubble with transition-prediction test with local interaction. Rolls-Royce Ltd. Rept. No. PNR-90231. Also translated as NASA N85-18008.

Kassab, S.Z. 1986 Turbulence structure in axisymmetric wall-bounded shear flow. Ph.D. thesis, The University of Manitoba, Winnipeg.

Kiya, M. 1987 Structure of flow in leading-edge separation bubbles. In Proc. IUTAM Symposium on Boundary-layer separation, pp. 57-71. Eds. F.T. Smith & S.N. Brown, Springer-Verlag, New York.

Kiya, M. & Sasaki, K. 1983 Structure of a turbulent separation bubble. J. Fluid Mech. 137, 83-113.

Nakamura, Y. & Ozono, S. 1987 The effects of turbulence on a separated and reattaching flow. J. Fluid Mech. 178, 477-490.

Roshko, A. & Lau, J.C. 1965 Some observations on transition and reattachment of a free shear layer in incompressible flow. In Proc. Heat Trans. and Fluid Mech. Inst. 18, 157-167.

Tani, I. 1964 Low-speed flows involving bubble separations. In Progress in Aeronautical Sciences, vol. 5, pp. 70-103. Eds. D. Küchemann & L.H.G. Sterne, Pergamon Press, Oxford.

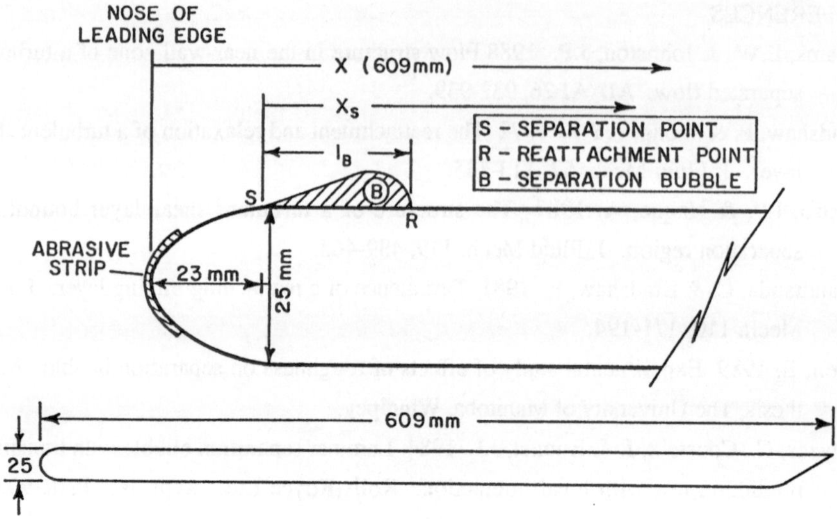

FIG. 1. EXPERIMENTAL PLATE

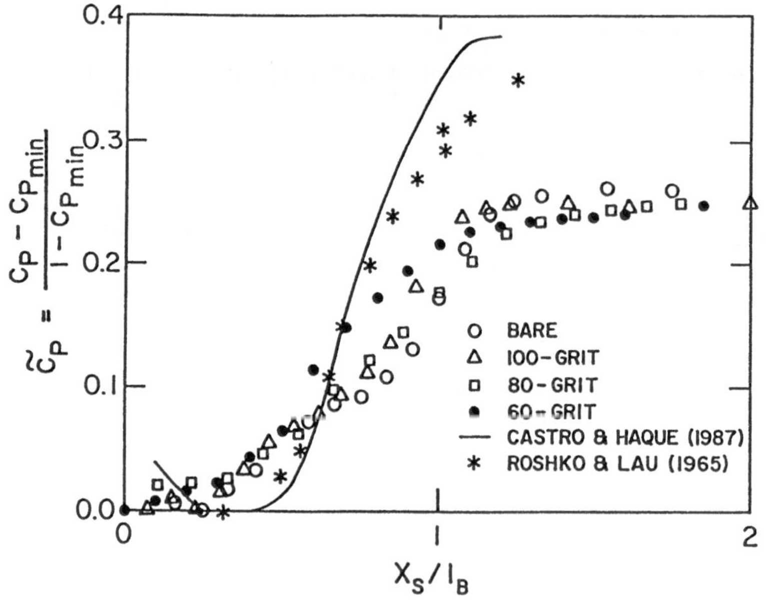

FIG. 2. REDUCED PRESSURE COEFFICIENT DISTRIBUTION

X_S = DISTANCE FROM THE START OF THE BUBBLE

l_B = LENGTH OF BUBBLE

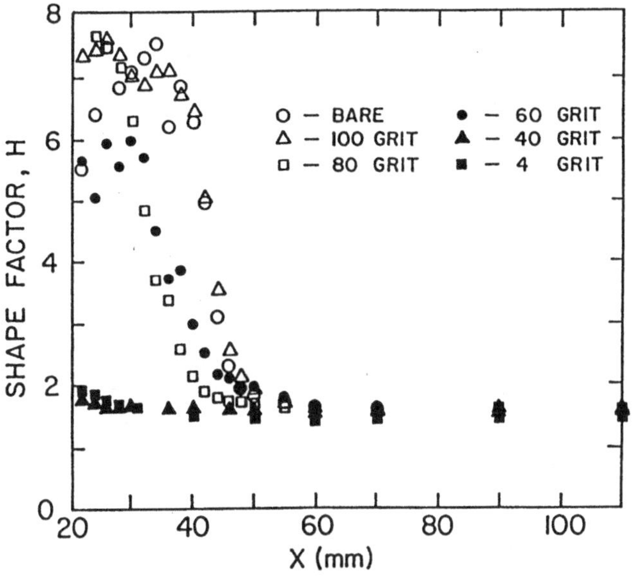

FIG. 3. DEVELOPMENT OF SHAPE FACTOR
X = DISTANCE FROM LEADING EDGE OF THE PLATE

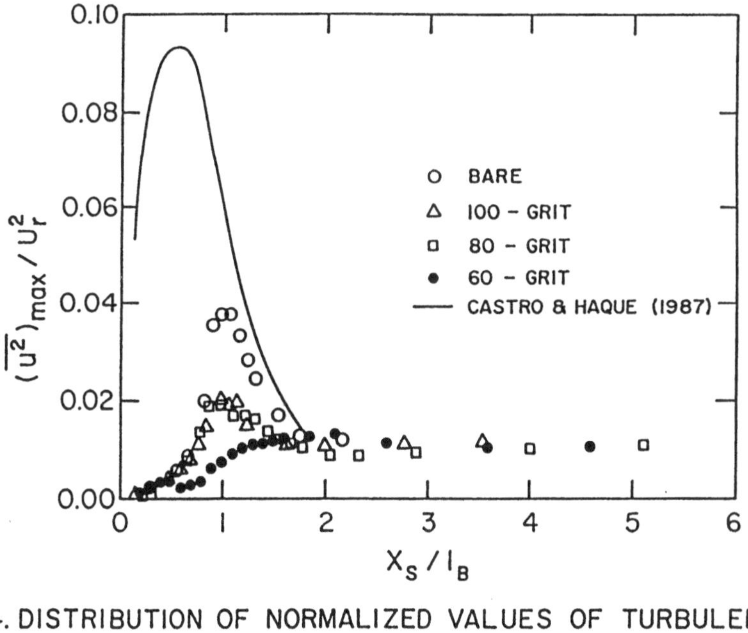

FIG. 4. DISTRIBUTION OF NORMALIZED VALUES OF TURBULENCE

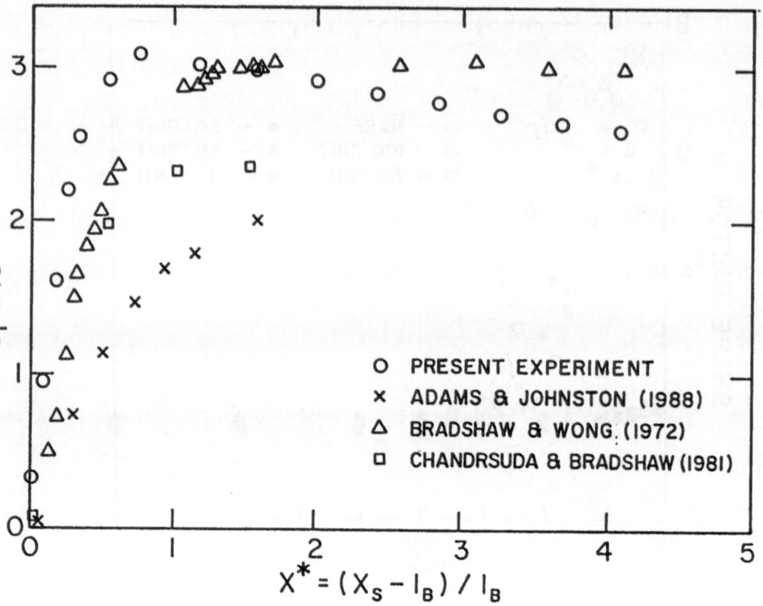

FIG. 5. COMPARISON OF C_f AMONG SEVERAL EXPERIMENTS; WALL STRESS NORMALIZED ON $\rho U_r^2/2$, THE UPSTREAM DYNAMIC PRESSURE

Organized Structures and Transition in the Near Field of a Plane Jet

J. MEYER, A. SEVRAIN, H.C. BOISSON, H. HA MINH
Institut de Mécanique des Fluides de Toulouse
Avenue du Professeur Camille Soula, 31400 Toulouse, France

Abstract

The near field of a plane jet has been investigated by statistical, spectral, correlative and visual analysis, for Reynolds numbers R_d= 1700 - 10000. Cross-correlations between different couples of points in lateral, longitudinal, and diagonal directions have shown several features of the organized structures in the near field of the plane jet. These structures were later confirmed by flow visualization, using laser illumination. These organized structures are shown to be of great importance in the understanding of the physical mechanics involved in the jet development.

1. Introduction

The far field of a turbulent plane jet has been widely investigated and the jet properties in this affinity region are rather well known. Velocities have similar distribution, jet width and the square of the velocity decay have a linear longitudinal evolution (GORTLER [1]). Antisymmetrical organized structures have been reported by OLER & GOLDSCHMIDT [2], based on negative cross-correlation across the mid-plane.

The near field of the plane jet is less documented than its far field. It is generally viewed as a pair of mixing layers, growing independently on both sides of the potential core, which interact at the end of the potential region. Initial dominant frequency is similar to that of single mixing layers (THOMAS & GOLDSCHMIDT [3]). But the frequential evolution, leading to a preferred frequency at the end of the potential region, and the existence and the nature of organized structures in the near field are not clearly understood (HO & HSIAO [4], BROWNE et al. [5]).

In order to detail the mechanisms of the transition in a turbulent plane jet, experiments were conducted in a wind-tunnel, producing a 1cm-wide jet, with laminar boundary layers, for moderate Reynolds numbers, R_d=1700 - 10000. Initial momentum thickness Θ_0 varied from .0195d to .0090d and turbulent intensities were less than 0.6%. Statistical and spectral analysis were conducted up to 50d and 15d, respectively. Cross-correlations were measured between different couples of points in lateral, longitudinal, and diagonal directions. Visualization was obtained by illuminating a thin sheet of the flow with a laser beam, reflected by an oscillating mirror.

A. Gyr (Editor)
Structure of Turbulence and Drag Reduction
IUTAM Symposium Zurich/Switzerland 1989

2. Longitudinal evolution

Several regions have been identified in the longitudinal evolution of the jet parameters, as presented in Fig. 1:

a) A region of extension of confined flow, characterized by exponential growth of the turbulence intensity u'_b, a linear evolution of width b and of momentum thickness Θ, and a tanh-type lateral distribution of mean velocity U/U_o.

b) A region of transition towards unsteady shear layers, characterized by large increase in b and Θ, a reduction of the rate of increase of the turbulence intensity u'_b, and an apparition of continuous frequencies, centered around a dominant frequency f_i, given by $f_i.\Theta_o/U_o = 0.0038$.

c) A region of a pair of unsteady shear layers, characterized by a linear increase of b and Θ and a similar distribution of U/U_m, shown in Fig. 2, typical of turbulent shear layers (with an erf-type dependence, as in GORTLER [1]). On the other hand, the fluctuating intensity u'/U_m is similar only in the outer flow region, and is different from typical u'-distribution for single shear layers, with a maximum in the mid-plane. So the potential region of a plane jet cannot be seen as a pair of independent mixing layers, for moderate Reynolds numbers.

The dominant frequency decreases with a 1/x dependence, as in single mixing layers, Fig. 3. There is a longitudinal shift between frequencies measured at the centerline and the shearlines.

d) A region of interaction between the shear layers, inducing the end of the potential region and the decay of centerline velocity U_m, and a large increase of the widths b and Θ and of the turbulent intensity u'_m/U_m, leading to a maximum for this intensity at the end of this region (Fig. 4).

This maximum value is correlated to the Reynolds number by :

$$(u'^2_m/U_m^2)_{max} \quad \alpha \quad R_d^{-1/2} \quad \alpha \quad \Theta_o/d \ .$$

It is believed that larger initial laminar boundary layers induce higher interaction between the shear layers (FREYMUTH [6]), which is the source of the maximum of u'_m/U_m (ZAMAN & HUSSAIN [7]).

e) A region of plane-jet-type, partial affinity, with linear expansion of the jet, linear decay of the square of the centerline velocity, and similar distribution of U/U_m, with a $sech^2$ law, as in the analysis of GORTLER [1].

f) A region of total affinity, where similar distribution is obtained for u'/U_m also, and asymptotic values for every parameter have been reached, according to known results for a plane jet.

3. Characteristic values

Characteristic length and velocity undergo an evolution as distance from the jet exit increases.

In regions a and b, the characteristic length is the initial momentum thickness Θ_o, while the characteristic velocity is the initial mean velocity U_o. Extension of these regions are related to Θ_o.

In region c, characteristic length becomes the jet initial width d, as the end of this region is not linked anymore with Θ_o, and the dominant frequency at this point is given by $f_c.d/U_o = 0.11$. This indicates the beginning of interaction and transition towards the situation of a jet, and justifies the lack of similarity for the fluctuating velocity u'/U_m.

In the interaction region d, characteristic values become b and U_m, which will be the characteristic values in the affinity regions e and f.

4. Transition

It is shown that transition from the initial 2-D laminar boundary layers to the 3-D turbulent plane jet occurs in several steps, and in a different way for mean and fluctuating velocities.

In region b, there is a first transition from laminar to unsteady shear layers. For mean velocity, there is a transition from 2-D unsteady shear layers to a 3-D turbulent plane jet, in region d. For fluctuating velocity, this transition begins already in region c, and still occurs in region e.

Complementary informations are provided by THOMAS & GOLDSCHMIDT [3], indicating a steep decrease of the 2-D character at the end of the potential region, and by EVERITT & ROBINS [8], indicating the achievement of 3-D flow at the beginning of the total affinity region.

5. Organized structures

In the potential region, symmetrical organized structures are observed (Figs. 5,6) : there are waves, then vortices appear (of dominant frequency f_i), then these vortices undergo a pairing, which is completed at the end of the region b. In the region c, there is no more pairing, but the vortices increase in size.

These features are characteristic of unforced shear layers (WINANT & BROWAND [9], HO & HUANG [10]) and explain the evolution of the jet widths b and Θ, and of the dominant frequency.

Of particular interest are intercorrelations between the shear lines on both sides of the jet (y=+/-b), and temporal signals at these points (Fig.

7): a positive correlation greater than 80% is measured for x= 2-5d. It is shown that those structures on both sides of the jet are symmetrical, and highly correlated between them. There is a spatial correlation, as the presented features occur at the same stations, as well as a temporal one, as both sides react similarly to internal and external perturbations.

Another interesting feature is a longitudinal shift between frequencies measured at the centerline and at the shear lines (see Fig. 3). Cross-correlations and flow visualization show that the vortices are linked vortices, centered at the shear lines, with diagonal convection lines towards the centerline, inducing in the centerline frequencies from upstream positions. Therefore, frequencies in the centerline are traces of the frequencies occurring upstream in the shearlines, as indicated by the spectral analysis, and do not indicate an independent process.

In the region of transition towards the jet, a vortex bursting is observed in the visualizations, explaining the significant reduction in cross-correlation for y=+/-b, and the disappearance of a dominant frequency in the spectra. In this region, a reorganization of the flow occurs towards the situation in the jet-affinity region, where antisymmetrical coherent structures are observed (as in OLER & GOLDSCHMIDT [2]), and indeed, in region f, the cross-correlation becomes negative (Fig. 1).

References

1. GORTLER, H., 1942 Z.angew.Math.Mech., 22, 244-254.
2. OLER, J.W., GOLDSCHMIDT, V.W., 1981 3rd Symp.Turb.Shear Flows, Davies, Cal., U.S.A.
3. THOMAS, F.O., GOLDSCHMIDT, V.W., 1986 J.Fluid Mech., 163, 227-256.
4. HO, C.M., HSIAO, F.B., 1983 in Structure of Complex Turbulent Shear Flow, ed.R.Dumas and L.Fulachier, Springer-Verlag, 121-136.
5. BROWNE, L.W.B., ANTONIA, R.A., CHAMBERS, A.J., 1984 J.Fluid Mech., 149, 355-373.
6. FREYMUTH, P., 1972 in ROCKWELL, D.O., NICCOLLS, W.O., 1972 Trans.ASME, J.Basic Engng, 94, 720-730.
7. ZAMAN, K.B.M.Q., HUSSAIN, A.K.M.F., 1981 J.Fluid Mech., 103, 133-159.
8. EVERITT, K.W., ROBINS, A.G., 1978 J.Fluid Mech., 88, 563-583.
9. WINANT, C.D., BROWAND, F.K., 1974 J.Fluid Mech., 63, 237-255.
10. HO, C.M., HUANG, L.S., 1982 J.Fluid Mech., 119, 443-473.

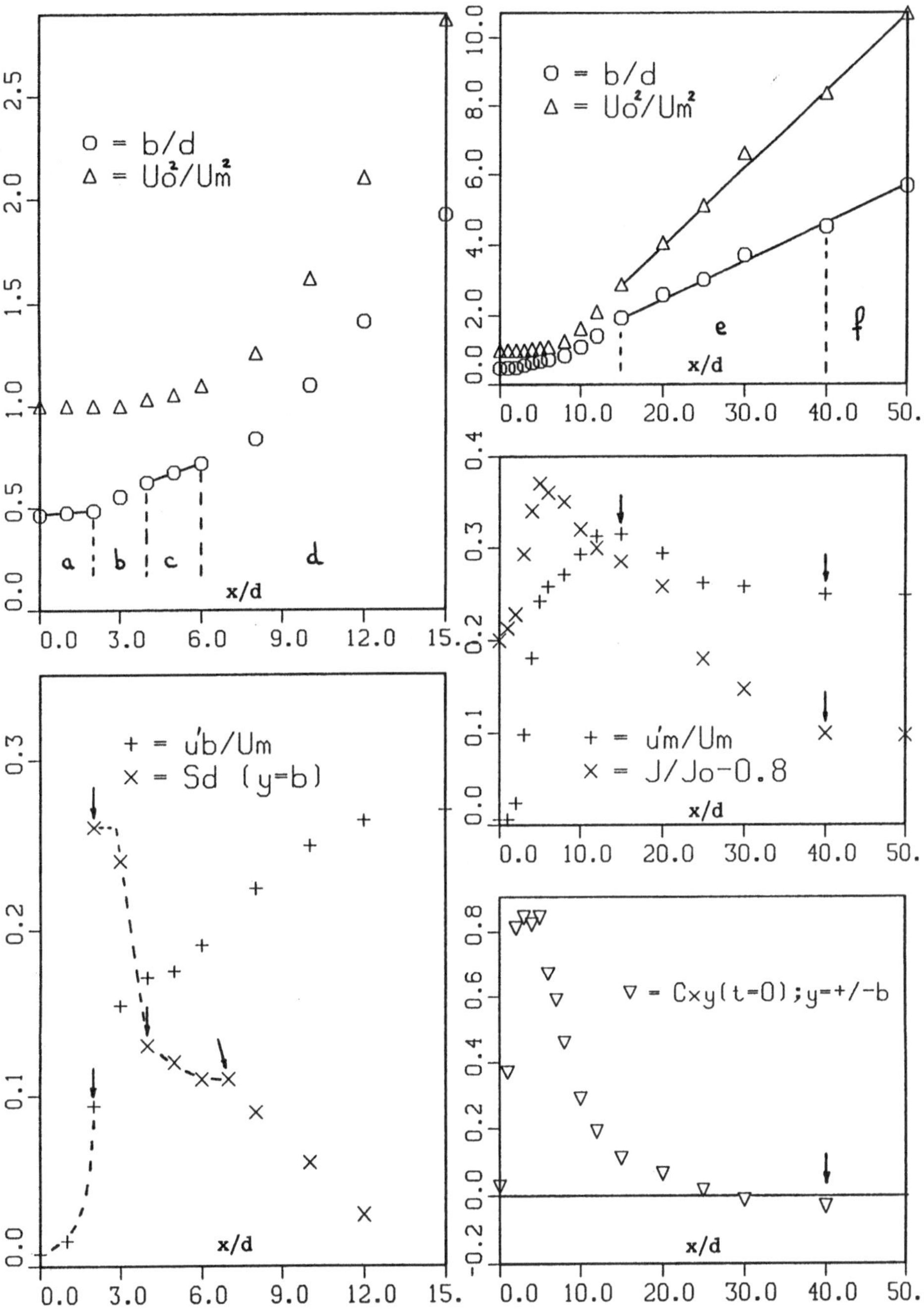

Fig. 1 Longitudinal evolution of jet parameters, for Rd= 3300 :
a) x = 0 - 15d b) x = 0 - 50d

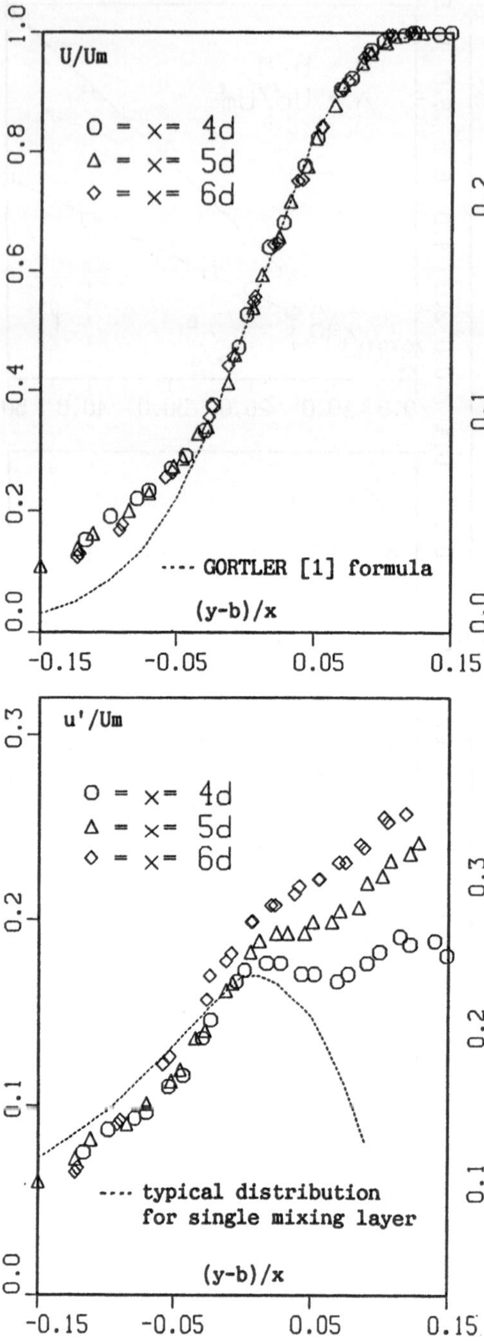

Fig. 2 Lateral distribution of U/Um, u'/Um, in region c, Rd= 3300

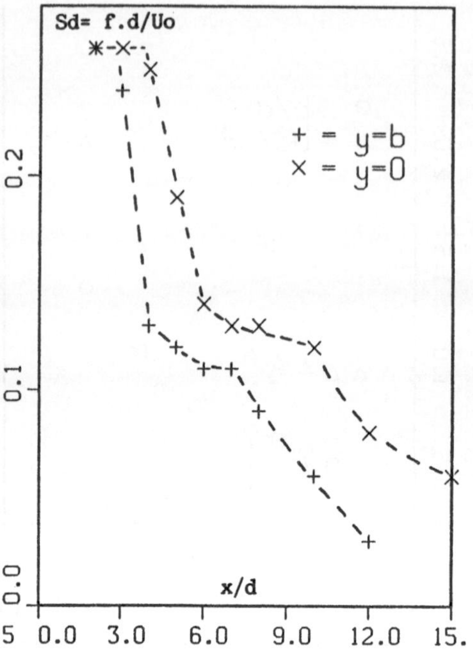

Fig. 3 Longitudinal evolution of dominant frequency, for Rd= 3300, y=0, y=b

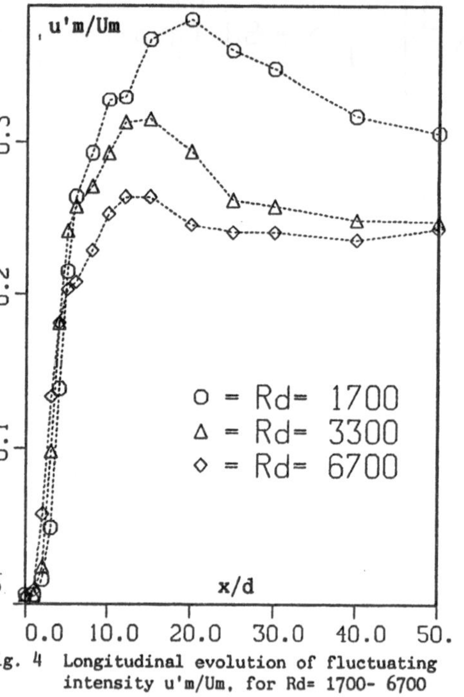

Fig. 4 Longitudinal evolution of fluctuating intensity u'm/Um, for Rd= 1700- 6700

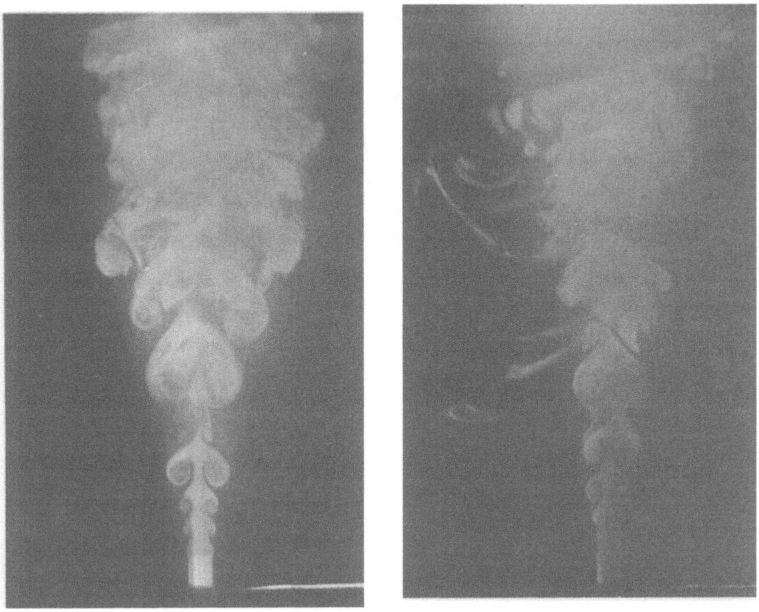

Fig. 5 Flow visualization, Rd= 1700

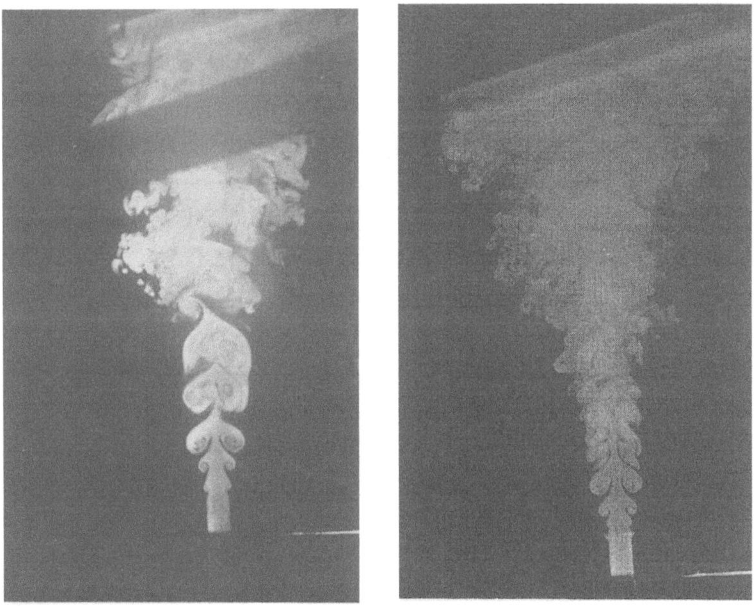

Fig. 6 Flow visualization, Rd= 3300

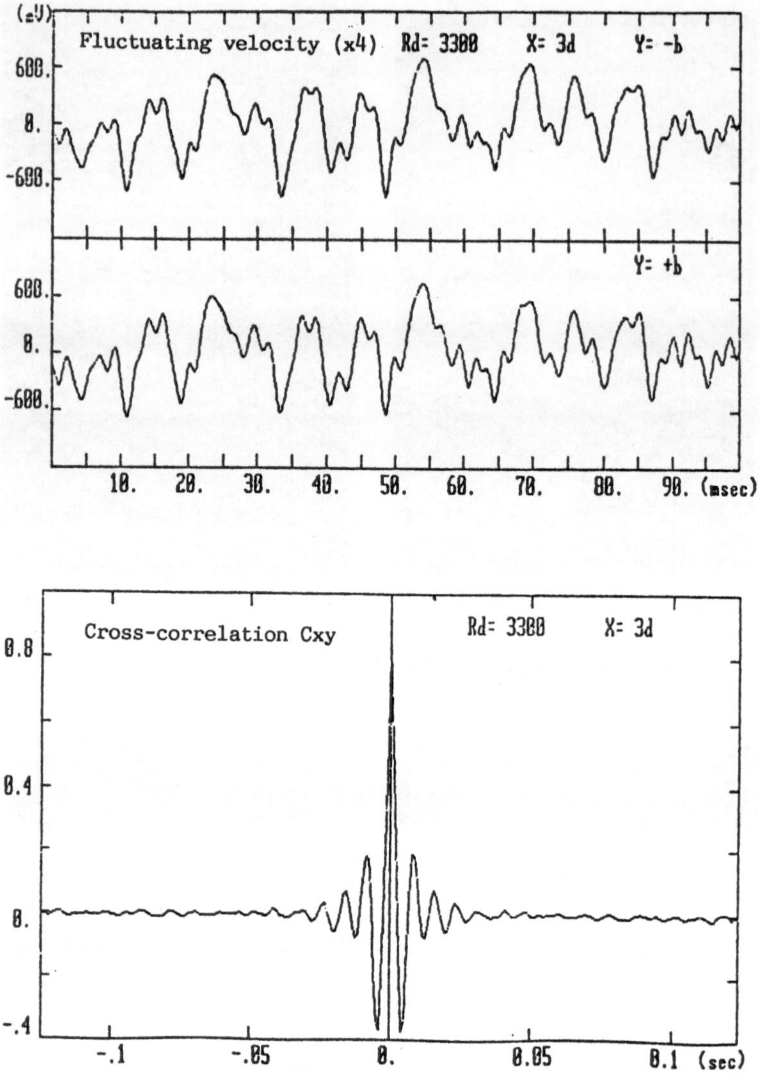

Fig. 7 Temporal signals and cross-correlation between y=-b and y=+b, x=3d, Rd= 3300

Determination of Regular Structures in Turbulent Mixing Processes by Raman Scattering

C. Voßmerbäumer and G. Schweiger
Fachbereich Maschinenbau, Universität Duisburg
Lotharstr. 1, D-4100 Duisburg 1

I. Introduction

The investigation of turbulent mixing processes is of great importance in a wide variety of problems, e.g. combustion processes. The formation of regular structures in the flow field has a significant influence on these mixing processes. Optical measurement techniques have some advantages such as no disturbance of the flow field or the ability of in-situ investigation of concentration fields. The application of laser Raman spectroscopy makes it possible to determine optically the concentration distribution in a multicomponent flow field. Raman scattering is, therefore, at present a highly successful technique for the investigation of the molecular concentration distribution in a variety of flow problems. This technique has been proved to be very important for the analysis of reacting flow field, e.g. where interference between the flow field and the measuring probe is crucial.

In this paper we report on a Raman scattering technique which is tailored to detect and analyse regular structures in the flow. This technique relies on the measurement of the time autocorrelation function of the Raman signal together with the measurement of the amplitude distribution. This makes it possible to recover the fast density variations caused by vortex rolls moving through the measuring volume. In contrast to the work in the past where statistical properties have been analysed [1,2], this method allows the determination of the regular periodic structure of the turbulent vortices. Detecting a light signal which is proportional to the instantaneous concentration of a single gas component this method yields the possibility to evaluate the time dependence of the fast density variations. Whereas visualization methods using dye or laser induced fluorescence give a complete picture of the spa-

A. Gyr (Editor)
Structure of Turbulence and Drag Reduction
IUTAM Symposium Zurich/Switzerland 1989

tial distribution of the flow at one single moment, this method gives the temporal dependence of fluctuations at one single point but measurements at sufficient numbers of different points in the field yield a complete information on the spatial and temporal concentration distributions in the flow.

II. Experimental

This measurement method is applied here to analyse the flow field of free jets emerging from a nozzle with three rectangular parallel chambers with an aspect ratio of 15 and a chamber width of 8 mm. To investigate the influence of the nozzle exit geometry on the formation of regular structures in the flow field, two nozzles with different exit configurations are used. As shown in Fig. 1 the three chambers of one nozzle are separated from each other by blunt walls of 2 mm thickness. The exit geometry of the other nozzle has a slight modification. Here the separating walls between the chambers are knife edge shaped (see detail drawing in Fig. 1). The central chamber of each nozzle was operated by CO_2, the outer two by air. The velocities of the single jets could be varied independently between 0 and 8 m/s to get diffe-

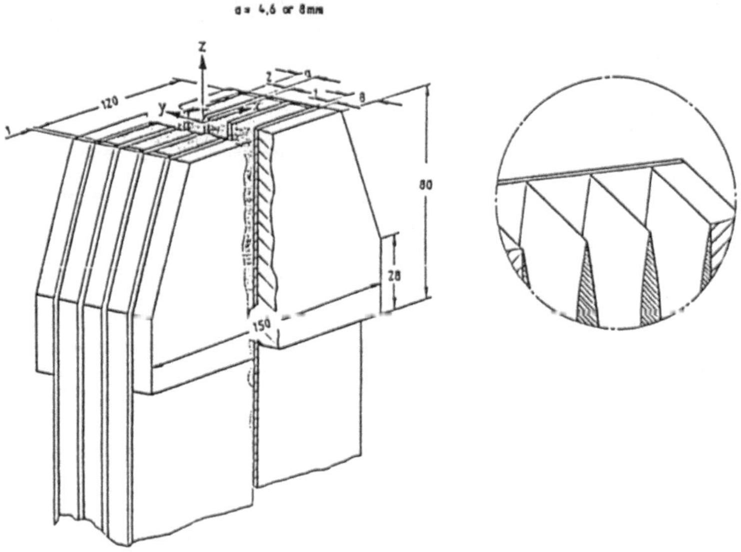

Figure 1: Nozzle exit configuration

rent velocity ratios between the CO_2 and the air jets. CO_2 has a strong vibrational Raman line shifted by 1388 cm^{-1} from the exciting line. The experimental set-up used is sketched in Fig. 2. The beam of an Ar-ion laser was directed through the flow field perpendicular to the long edge of the nozzle and then focussed into the scattering volume. To enhance the incident light intensity in the focus point a multiple pass cell was used. The scattered Raman light was collected under an observation angle of 90 degree and imaged by a lens into the entrance slit of a Spex Tandem Minimate double monochromator tuned to the CO_2 Raman line. A photomultiplier (RCA 31034) cooled to - 22 °C to reduce the dark count rate was used as a detector for the scattered light. The anode pulses were fed to an electronic system which consists of a preamplifier (Ortec model 113), a spectroscopy amplifier (Canberra, model 2010) and a single channel analyser (Ortec model 420A), then counted by a photon counter (PRA model 1770) or analysed by a digital autocorrelator (Malvern, K7025). Data acquisition and experimental control were done by a computer (HP 1000). The laser was operated at a wavelength of 514.5 nm and an output power of 4 Watt. The beam was focussed to a diameter of ≃ 0.1 mm. A pinhole with a diameter of 1.6 mm was located in the entrance plane of the monochromator. The image of this pinhole limited the observed scattering volume to a length of 1.1 mm.

The intensity of the scattered Raman light detected by the photon counting system is proportional to the number of CO_2 molecules in the scattering volume. By relating the actual count rate to the initial count rate in the 100 % CO_2 gas flow region at the exit plane of the nozzle, the CO_2 concentration relative to the initial concentration could be determined. Due to the low scattering cross section and the limited laser power a direct measurement of the time dependence of the fast varying CO_2 amplitudes is impossible. The method used to recover the periodic part of the fast varying concentration with sufficient time resolution was already described previously [3] and only the basic procedute will be given in the following. Both the frequency of reoccurrence and the time autocorrelation function of the Raman signal were determined. The sample time was adjusted to 150 µs, the total measurement

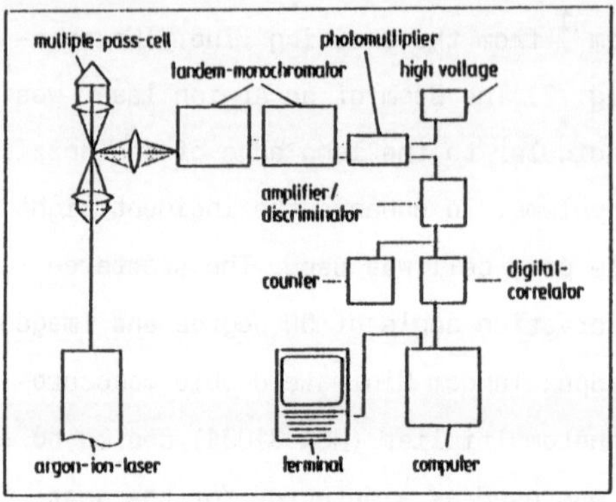

Figure 2: Sketch of the experimental set-up

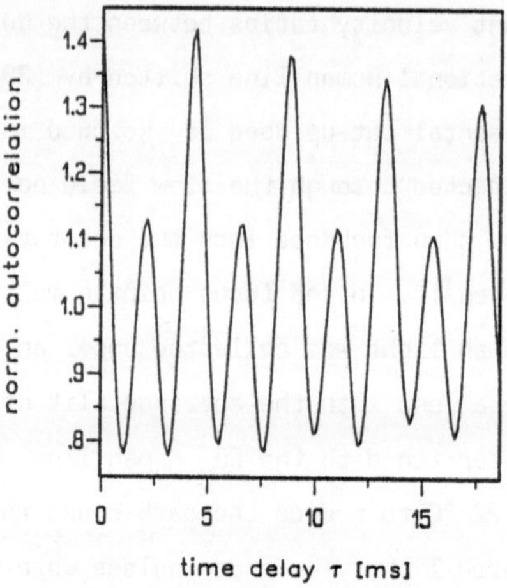

Figure 3: Normalized autocorrelation function

time for each point was set to 1 min. This yields the time average over about 40 000 sampling intervals. The counting rate distribution includes the information which CO_2 concentration amplitudes appear during the measurement time and with which reoccurrence frequency. To get the probability density of these amplitudes the measured distribution has to be Poisson-inverted. Information on the temporal arrangement of the CO_2 amplitudes is contained in the autocorrelation function. As can be seen in Fig. 3, the measured autocorrelation function shows high periodicity and only slight damping. This is due to the high regularity in the formation of vortices streaming through the observation volume. Nevertheless the damping in the autocorrelation function can be used to determine the degree of turbulence in the flow field. To recover the periodic part of the CO_2 fluctuations, one period of the measured autocorrelation function was inverted. Together with the additional information from the probability density a unique solution to the temporal arrangement of the CO_2 amplitudes could be found.

Using this measurement method the time dependence of the periodical CO_2 concentration fluctuations at individual point of the flow field can be determined with a time resolution of 150 µs and a spatial resolution of 1.1 mm limited by the resolution of the present experimental set-up.

III. Determination of Regular Structures

Previously the periodical structures in the flow field of the nozzle with blunt edges were investigated by Schlieren techniques where a pulsed lamp was taken as the light source [4]. These Schlieren pictures gave a general view of the different flow patterns occurring in the flow field. Depending on the velocity ratio between the CO_2 and air jets symmetrical and antisymmetrical structures could be detected. For a quantitative analysis of these flow patterns the measurement method described above was applied to determine the spatial and temporal concentration distributions in the flow. The velocities of the jets were set to $v_{CO_2} = v_{air} = 4.5$ m/s referring to the velocity in the center of each chamber in the exit plane. Both the nozzle with the blunt edges and the one with sharp edges were investigated. To get a y-profile of the concentration distribution for a constant value of z/d, the measurements are made at those points in the flow field which were separated by a distance of 0.5 mm in y-direction. All the measurements are done at constant value of x = 0. For each point a single CO_2 amplitude distribution was evaluated. The autocorrelation function includes no information on the temporal relationship between adjacent distributions, but the Schlieren pictures gave a hint to the shape of the structures and, therefore, to the time relation between adjacent points of observation.

Fig. 4a and 4b show flow patterns at a distance of z/d = 3 from the exit

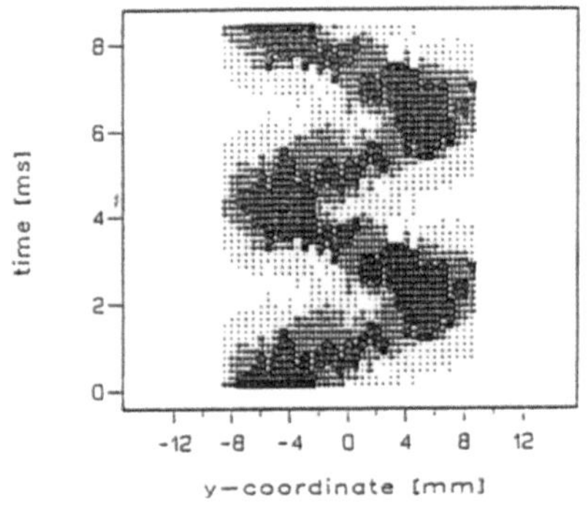

Figure 4a: Flow field of the nozzle at z/d = 3 (blunt edges)

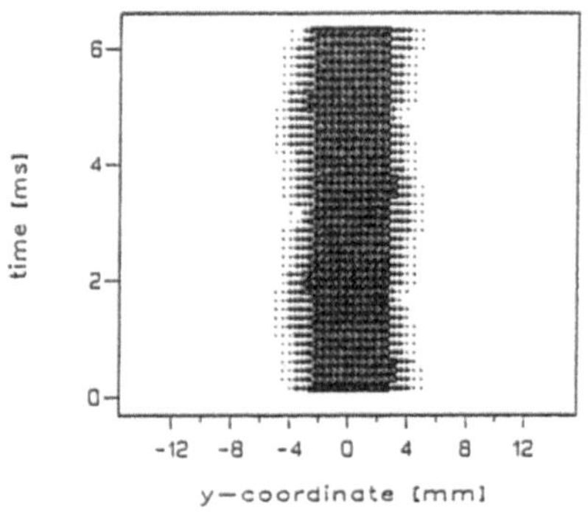

Figure 4b: Flow field of the nozzle at z/d = 3 (sharp edges)

plane for the two nozzles. Two period lengths of the concentration fluctuations streaming through the plane perpendicular to the mean flow direction can be seen. The coordinate y = 0 refers to the center of the mean chamber which is operated by CO_2. For a simplification the range of occurring CO_2 concentration is divided into four intervals each represented by an own symbol. The figures show the influence of the nozzle exit configuration on the formation of structures in the flow. The patterns in the flow field of the nozzle with blunt edges are much more developed than those of the nozzle with sharp edges. Comparing the period length of the periodical structures, it can be seen that the structures of the nozzle with blunt edges show a longer period length than those of the one with sharp edges. These differences in the flow structures might be due to the subpressure region built up by the streaming jets above the blunt end of the separating walls which might promote the formation of regular structures in the flow field.

IV. Investigation of the Mixing Process

Obviously, the formation of regular structures described above has a great influence on the mixing process between the free jets. The information necessary to determine those regions in the flow field where a homogeneous mixture is reached is included in the counting rate distribution, respectively in the probability density of the single gas concentration amplitudes.

Fig. 5 shows the development of the probability density of the CO_2 concentration amplitudes for different distances from the nozzle exit. All these measurements are made at points in the flow field which are located above the center of the nozzle (x = 0, y = 0). The probability density near the nozzle exit includes only those CO_2 amplitudes which correspond to the constant CO_2 concentration in the CO_2 gas flow. The formation of structures in the flow results in the appearance of other than the maximum CO_2 concentrations in the probability density. This is due to the occurrence of pure CO_2 and pure air regions streaming through the observation volume and the begin of mixing between CO_2 and air. Further downstream in the flow field of the nozzle with rectangular edges a homogeneous mixture is reached and CO_2 amplitudes occurring in the probability density corresponds to the mean CO_2 concentration of the mixture. The differences in the mixing process in the flow fields of the two nozzles with different exit configurations can be seen clearly. The flow field of the nozzle with blunt edges reaches the ho-

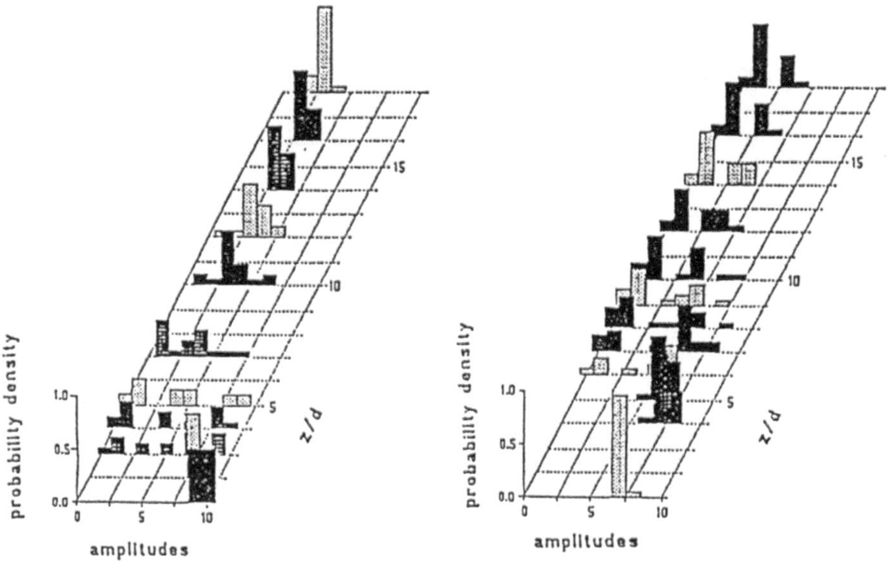

Figure 5: Probability density of CO_2 concentration amplitudes
a) nozzle with blunt edges; b) nozzle with sharp edges

mogeneous mixture at a value of about z/d = 14 from the nozzle exit, whereas the flow field of the nozzle with sharp edges still consists of several different CO_2 amplitudes at this point.

A good measure to determine those points in the flow field where a homogeneous mixture is reached is the normalized factorial moment of the counting rate distribution. It is given by

$$M = \frac{\langle n(n - 1)\rangle}{\langle n\rangle^2}$$

where n refers to the count rate per sample time and the brackets refer to the weighting by the probability of this count rate. Due to the Poisson statistics of the photon counting process the normalized factorial moment equals one for a counting rate distribution which results of a constant light amplitude impinging on the detector. Therefore, a measurement at a point with constant CO_2 concentration gives a counting rate distribution with a moment equal to one. Every broadening of the concentration distribution yields a distribution with a normalized factorial moment greater than one. Corresponding to the CO_2 amplitude distributions shown in Fig. 5 the development of the normalized factorial moment of the same flow conditions is given in Fig.

6a and 6b. Consequently, it seems to be evident that the formation of highly developed structures in a flow field as they can be detected in the flow field of the nozzle with blunt edges promotes the mixing between the jets.

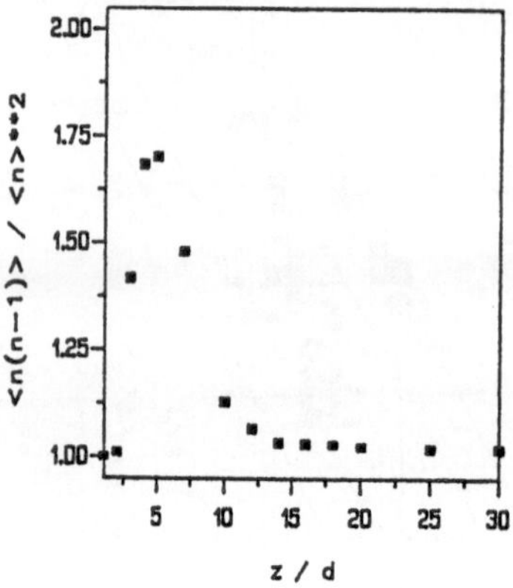

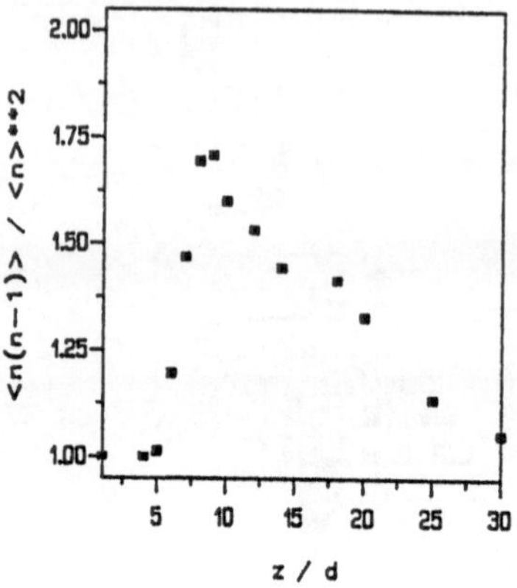

Figure 6a: Normalized factorial moment (nozzle with blunt edges)

Figure 6b: Normalized factorial moment (nozzle with sharp edges)

V. Conclusions

These examples show that the spatial and temporal concentration distribution of a single gas component in the flow field can be determined with a time resolution of 150 µs using linear Raman scattering. The method described here yields the periodic part of the concentration fluctuations. It is, therefore, especially suitable for the determination and analysis of the regular structures in turbulent flows and their role in the mixing process.

References

[1] Birch, A.D., Brown, D.R., Dodson, M.G. and Thomas, J.R., J. Fluid Mech. 88, 431 (1978)
[2] Becker, H.A., Hottel, H.C. and Williams, G.C., J. Fluid Mech. 30, 285 (1967)
[3] Voßmerbäumer, C. and Schweiger, G., 'Determination of the time dependence of periodical optical signals by correlation spectroscopy', in OSA Proceedings on Photon Correlation Techniques and Applications, Vol. 1 (J.B. Abbiss, A.E. Smart, eds.) 1989
[4] Schweiger, G., J. Fluids Eng. 105, 42 (1983)

Wall Layer Eruptions in Turbulent Flows

J. D. A. Walker
Department of Mechanical Engineering and Mechanics
354 Packard Laboratory #19
Lehigh University
Bethlehem, Pennsylvania 18015

ABSTRACT

The near-wall region of a turbulent flow is investigated in the limit of large Reynolds number. When low-speed streaks are present, the governing equations are shown to be of the boundary-layer type. Physical processes leading to local breakdown and a strong interaction with the outer region are considered. It is argued that convected vortices, predominantly of the hairpin type, will provoke eruptions and regenerative interactions with the outer region.

1. Introduction

Most flows occurring in engineering practice are turbulent and at high Reynolds number, and thus the study of the dynamics of turbulent boundary layers is therefore an important area of basic research. However, analysis is extremely difficult due to the complex unsteady environment which contains a rich mixture of scales and three-dimensional vortex structures; such vortices interact with one another and with the viscous flow near solid walls. The interaction with the near-wall flow is the fundamental process of turbulence production whereby the turbulence is sustained. The wall layer is observed to erupt intermittently, at isolated streamwise and spanwise locations, in an event usually referred to as bursting. In this process, new vorticity from the wall is abruptly introduced into the outer region of the boundary layer. In a turbulent boundary layer, the only source of new vorticity is the wall, and thus the important issues of regeneration are necessarily related to the surface interaction. The nature of this interaction and the physical processes involved are the subject of this paper.

The mean turbulent boundary layer is double-structured consisting of: (1) an outer layer where the balance in the mean equations is between convection and Reynolds stress, and (2) an inner wall layer having constant total stress, where viscous and Reynolds stress terms balance and the convective terms are not important to leading order. A significant consequence of the observed[1] coherent structure of the near-wall region (see Walker et al[2] and the references therein) is that the mean normal wall-layer length scale is also appropriate for the time-dependent flow over relatively long periods of time. For a given area of the surface, the wall layer will be observed to be in a relatively quiescent state for a large majority of the total observation time[1,2]. During this quiescent period, the low-speed streaks can be observed, the flow is relatively well-ordered and strong interactions with the outer region do not occur. In this state, the

A. Gyr (Editor)
Structure of Turbulence and Drag Reduction
IUTAM Symposium Zurich/Switzerland 1989

wall layer is a relatively thin viscous region and may be regarded as a developing unsteady flow, driven by the pressure field at the base of the outer region. Indeed, the most plausible cause of the wall-layer streaks is that they are the signature of convected hairpin vortices[2,3,4] near the surface.

The relatively long periods of ordered flow are interrupted on a highly intermittent basis at isolated locations by the bursting phenomenon. The event invariably initiates near a wall-layer streak[2] and is characterized by an abrupt and highly localized eruption of wall-layer fluid. The process may be classified as a strong unsteady viscous-inviscid interaction or equivalently, as a local breakdown of the wall-layer flow. During the relatively short period of breakdown, the two regions of the boundary layer interact strongly and the double structure of the layer is briefly obliterated. In the process, new vorticity from the wall is abruptly introduced into the outer region. A double structure is quickly restored locally with the sweep event, in which high speed outer-region fluid penetrates close to the wall following the burst; the low-speed streaks appear again and locally a new quiescent state begins.

A central issue relates to what dynamical features cause the local breakdown of the wall-layer flow. The most productive theoretical approach is to identify cause and effect relationships in the limit of large Reynolds number for several reasons. First, most practical flow occurs at high Reynolds numbers, and it is desirable to construct an asymptotic description which represents the phenomena over a range of Reynolds numbers. Secondly, the structure of high speed compressible turbulent flows is expected to be similar to the well-studied incompressible case; detailed experimental studies become progressively more difficult with increasing flow speed and a knowledge of the dominant dynamical mechanisms at high Reynolds number provides the basis of an extension to high Mach number flows. The theoretical task is difficult since most of the experimental work is in a limited Reynolds number range; in addition, the complexity of the unsteady environment suggests a multitude of different physical cause and effect relationships. The task in constructing an asymptotic description is to isolate the scales and dominant processes at high Reynolds number. In the next section, the scaled variables describing an evolving quiescent wall-layer flow are developed.

2. The Wall Layer Problem

Consider a nominally steady two-dimensional boundary layer and let $u_\tau(x)$ be the local mean friction velocity. During a typical quiescent period, the wall layer is well-defined and the low-speed streaks are present; all three velocity components are $O(u_\tau)$ and the characteristic length in the spanwise direction is λ, the mean streak spacing. First define a scaled coordinate and velocity in the spanwise direction by

$$Z = u_\tau z/(\lambda^+ \nu) = z/\lambda, \quad W = w/(u_\tau W_1), \tag{1}$$

where λ^+ is observed[5] to have a value of about 100; here ν is the kinematic viscosity and W_1 is a constant, such that $W_1 u_\tau$ is equal to the average magnitude of the spanwise velocity near the outer edge of the wall layer during a typical quiescent state. Measurements of the turbulence intensity $\overline{w'^2}$ for large

$y^+(y^+ \simeq 40 - 100)$ suggest an average value of W_1 of about 2. The scaled normal velocity and distance from the wall are defined by

$$Y = u_\tau y(W_1/\lambda^+)^{1/2}/\nu = y^+(W_1/\lambda^+)^{1/2}, \quad V = v(\lambda^+/W_1)^{1/2}/u_\tau, \tag{2}$$

which follow from balancing: (1) $\partial v/\partial y$ and $\partial w/\partial z$ in the continuity equation and (2) the convective operator $v\partial/\partial y$ with the viscous operator $\nu\partial^2/\partial y^2$ in the momentum equations. A scaled streamwise coordinate and velocity are defined by

$$X = x/L_x, \quad U = u/(U_1 u_\tau). \tag{3}$$

Here L_x is the characteristic length in the streamwise direction, whose value will be discussed subsequently; U_1 is a constant such that $U_1 u_\tau$ is a typical average flow speed in the streamwise direction near the outer edge of the wall layer during the quiescent period. Typical U_1 values of 13~16 can be estimated from the law of the wall for $y^+ = 30{\sim}100$ and consequently U_1 is generally considerably larger than W_1. The time scale follows from a balance between the time derivative and normal viscous diffusion terms and let

$$T = u_\tau^2 t W_1/(\nu\lambda^+) = t^+ W_1/\lambda^+. \tag{4}$$

Finally, the pressure may be written as

$$p = p_\infty(x) + \rho u_\tau^2 U_1^2 P_1 + \rho u_\tau^2 U_1 W_1 P_2/\lambda^+ + \ldots, \tag{5}$$

where $p_\infty(x)$ is the mainstream pressure and ρ is the density. Upon substitution in the continuity equation, it is easily shown that

$$\frac{\gamma}{Re_x}\frac{\partial U}{\partial X} + \frac{\partial V}{\partial Y} + \frac{\partial W}{\partial Z} = 0. \tag{6}$$

Here Re_x, a streamwise Reynolds number, and the parameter γ are defined by

$$Re_x = u_\tau L_x/\nu, \quad \gamma = \lambda^+ U_1/W_1. \tag{7}$$

Substitution in the Navier-Stokes equation yields

$$\frac{\partial U}{\partial T} + \frac{\gamma}{Re_x} U \frac{\partial U}{\partial X} + V \frac{\partial U}{\partial Y} + W \frac{\partial U}{\partial Z} = \frac{\lambda^+}{U_1 W_1} p^+ - \frac{\gamma}{Re_x}\frac{\partial P_1}{\partial X} - \frac{1}{Re_x}\frac{\partial P_2}{\partial X}$$
$$+ \frac{\gamma}{U_1 Re_x^2}\frac{\partial^2 U}{\partial X^2} + \frac{\partial^2 U}{\partial Y^2} + \frac{1}{Re_\lambda}\frac{\partial^2 U}{\partial Z^2}, \tag{8}$$

$$\frac{\partial V}{\partial T} + \frac{\gamma}{Re_x} U \frac{\partial V}{\partial X} + V \frac{\partial V}{\partial Y} + W \frac{\partial V}{\partial Z} = -\gamma U_1 \frac{\partial P_1}{\partial Y} - U_1 \frac{\partial P_2}{\partial Y}$$
$$+ \frac{\gamma}{U_1 Re_x^2}\frac{\partial^2 V}{\partial X^2} + \frac{\partial^2 V}{\partial Y^2} + \frac{1}{Re_\lambda}\frac{\partial^2 V}{\partial Z^2}, \tag{9}$$

$$\frac{\partial W}{\partial T} + \frac{\gamma}{Re_x} U \frac{\partial W}{\partial X} + V \frac{\partial W}{\partial Y} + W \frac{\partial W}{\partial Z} = - \frac{U_1^2}{W_1^2} \frac{\partial P_1}{\partial Z} - \frac{U_1}{Re_\lambda} \frac{\partial P_2}{\partial Z}$$

$$+ \frac{\gamma}{U_1 Re_x^2} \frac{\partial^2 W}{\partial X^2} + \frac{\partial^2 W}{\partial Y^2} + \frac{1}{Re_\lambda} \frac{\partial^2 W}{\partial Z^2}. \qquad (10)$$

Here p^+ is the scaled mainstream pressure gradient defined by

$$p^+ = -\nu(dp_\infty/dx)/(\rho u_\tau^3), \qquad (11)$$

and Re_λ is a spanwise Reynolds number defined by

$$Re_\lambda = W_1\lambda^+ = W_1 u_\tau \lambda/\nu. \qquad (12)$$

This completes the formulation with a minimum number of assumptions. Length and velocity scales consistent with experiment have been introduced to describe a developing wall-layer flow containing low-speed streaks. The objective now is to isolate the type of disturbance which can induce a local wall-layer breakdown and interaction with the outer-layer flow. In the next two sections, several limit problems are considered, all for large streamwise Reynolds number Re_x.

3. The Long Wave Limit

Consider first the limit

$$Re_x \to \infty, \quad (\gamma/Re_x) \to 0, \qquad (13)$$

which implies from equation (7) that $(U_1\lambda)/(W_1L_x) << 1$. Thus motion of relatively long streamwise extent with respect to the streak spacing is of interest. It is easily verified that in this limit equations (6), (9) and (10) involve only the flow components (V, W) in the cross-flow plane. The streamwise vorticity develops independently in the YZ plane and the solution for V and W then feeds into equation (8), thereby influencing the evolution of U. Solutions of the long wave equations have been considered by Walker and Herzog[6] subject to the boundary conditions

$$W = V = 0 \quad \text{at} \quad Y = 0, \quad W \to -\sin(2\pi Z) \quad \text{as} \quad Y \to \infty. \qquad (14)$$

This is the simplest external condition which leads to the representative flow structure depicted in Figure 1. A similar set of equations have been studied by other authors[7,8], with the exception that a periodic time-dependence was assumed in W at large Y; in these studies, the spanwise velocity for large Y reverses direction during the interval under consideration. In contrast, the outer condition (14) is similar to that imposed by a moving hairpin vortex on the flow near the wall[9]; it represents a persistent pumping action at the outer edge of the wall layer and is therefore believed to be a more realistic external condition.

In the limit (13) equations (8), (9), and (10) contain the Reynolds number Re_λ; with average values of $\lambda^+ = 100$, $W_1 = 2$, a typical value of Re_λ is 200.

Walker and Herzog[6] have obtained numerical solutions over a range of Re_λ, using an abrupt imposition of (14) at the start of a typical quiescent state; two routes to breakdown of the flow structure depicted in Figure 1 were identified. The first occurs in situations where the spanwise velocity at the wall-layer edge produces a relatively strong pumping action, so that W_1, and hence Re_λ, is large. In such cases, the adverse pressure gradient due to (14) causes the development of recirculating flow near Z = 0 in the cross-flow plane; the flow evolution is then similar to known examples of vortex-induced separation[10]. Violent updrafts begin to evolve near the recirculating flow, in a region which becomes progressively narrower as the wall layer focusses toward an eruption. This unsteady separation effect will occur whenever a hairpin vortex is close to the wall and/or has a relatively large strength, thus giving rise to elevated levels of W_1. As the strength of the pumping action is decreased (and with it Re_λ), recirculations do appear in the cross-flow plane but the tendency toward a focussed eruption gradually diminishes; at low enough Re_λ, the (V, W) motion evolves toward an apparently steady state. However, the streamwise velocity profiles were observed[6] to develop a strong inflectional character near the center of the recirculation in the cross-flow plane; the configuration is expected to be highly unstable and to break down. This is the second physical process which can lead to the destruction of the assumed flow structure depicted in Figure 1. Finally, at sufficiently low values of Re_λ, cross-flow separation or inflectional streamwise profiles do not develop. The overall implication of these results[6] is that a sufficiently large level of imposed spanwise velocity at the wall-layer edge is required to produce a breakdown of the structure shown in Figure 1.

4. The Full Three-Dimensional Problem

Now consider the limits

$$Re_x \to \infty, \quad (\gamma/Re_x) = O(1), \tag{15}$$

and without loss of generality take $\gamma = Re_x$; this is equivalent to selecting the characteristic streamwise length according to

$$L_x = U_1\lambda/W_1. \tag{16}$$

Using the typical values of U_1 and W_1, equation (16) suggests a characteristic streamwise dimension of 6 to 8 times the spanwise streak spacing, which is consistent with observation[1,2]. It now follows from equation (8) that $\partial P_1/\partial Y = O(Re_x^{-1})$ and therefore $P_1 = P_1(X, Z, T)$. Equations (8), (10), and (6) become, respectively

$$\frac{\partial U}{\partial T} + U\frac{\partial U}{\partial X} + V\frac{\partial U}{\partial Y} + W\frac{\partial U}{\partial Z} = \frac{\lambda^+ p^+}{U_1 W_1} - \frac{\partial P_1}{\partial X} + \frac{\partial^2 U}{\partial Y^2} + \frac{1}{Re_\lambda}\frac{\partial^2 U}{\partial Z^2}, \tag{17}$$

$$\frac{\partial W}{\partial T} + U\frac{\partial W}{\partial X} + V\frac{\partial W}{\partial Y} + W\frac{\partial W}{\partial Z} = -\frac{U_1^2}{W_1^2}\frac{\partial P_1}{\partial Z} + \frac{\partial^2 W}{\partial Y^2} + \frac{1}{Re_\lambda}\frac{\partial^2 W}{\partial Z^2}, \tag{18}$$

$$\frac{\partial U}{\partial X} + \frac{\partial V}{\partial Y} + \frac{\partial W}{\partial Z} = 0. \tag{19}$$

With P_1 assumed known from a specified disturbance flow in the outer region (as well as U and W for large Y), it is evident that equations (17) - (19) are of the boundary-layer type. Once again Re_λ appears as a parameter and solutions can be considered over a range of Re_λ. Note that in the limit $Re_\lambda \to \infty$, equations (17) - (19) give the three-dimensional "laminar boundary-layer equations". An important consequence of this section is that disturbances which induce eruptions in laminar flows will also (when properly scaled) have the same effect on the turbulent wall layer. Recent computational and experimental studies[3,4,11] strongly suggest that the dominant disturbance near the wall in a turbulent boundary layer is the convected hairpin vortex. In the next section the general effects of convected vortex motion are discussed.

5. The Effects of Moving Vortices

In §3 and §4, it has been demonstrated that during the quiescent period the leading-order wall-layer equations are of the boundary-layer type; the pressure, as well as the external spanwise and streamwise velocity distributions drive the motion therein, until ultimately an interaction is induced with the outer flow. The question now is what type of convected disturbance can induce an eruption of the wall layer and what are the relevant dynamics? It is argued elsewhere[9,11] that the principal dynamical feature of turbulent boundary layers is the hairpin vortex. To understand the effects of such a vortex, it is worthwhile to review some aspects of vortex motion.

As a consequence of a number of fundamental studies, it has been possible to identify the general effect of a moving vortex on a viscous flow near a wall. These studies include two-dimensional vortices in an otherwise stagnant flow above a wall[10], vortices convected in a uniform flow[12], counter-rotating vortex pairs[13,14], vortex rings and loops near a surface[15,16] and moving hairpin vortices[3,4,11]. As each vortex convects near a solid surface, a variety of complex unsteady flow patterns are induced in the viscous flow near the wall. However, a common feature is that a vortex always impresses a region of adverse pressure gradient on the near-wall flow, which, if sustained, ultimately provokes an eruption. Consider the configuration shown in Figure 2(a) where a portion of a three-dimensional vortex is in motion above a wall. In Figure 2(b) the local detail in a plane normal to the vortex core is depicted. For the indicated sense of rotation, a region of adverse pressure gradient occurs behind the vortex[12]. In a frame of reference moving with the vortex (at $x = 0$), a sketch of the pressure gradient induced near the wall is shown in Figure 3. The critical aspects are: (1) the distance of the vortex center a from the wall; (2) the vortex strength κ and (3) the vortex Reynolds number $R_v = 2\pi\kappa/\nu$. The size of the vortex core plays no significant role in the nature of the induced flow near the wall. For large vortex Reynolds numbers, the viscous flow near the wall is only able to withstand the action of the adverse pressure gradient for a finite period of time. In most situations[10,15], a secondary recirculating eddy evolves in the near-wall flow with the opposite sense of circulation to the parent vortex. The important consequence is a blocking effect which develops as the near-wall flow is forced to flow up and over the secondary eddy, under the continued action of the external pressure gradient. Strong updrafts then develop on the upstream side of the secondary eddy causing a violent outflow from a zone that is narrow in the streamwise direction. In the latter stages of the process, the near-wall focusses into an explosively growing needle-like region[17,18] containing relatively high levels of vorticity. At this point, the near-wall flow erupts and a strong unsteady viscous-inviscid interaction ensues, culminating in the ejection of fluid from the wall region. The pressure response near the wall becomes quite complex as the interaction develops and a typical behavior in the pressure gradient is sketched as a broken line in Figure 3, near the point of eruption[17,18]. For high Reynolds numbers, it is essentially impossible to resolve this phenomena using a conventional Eulerian description. The location where the eruption will occur is not known a priori; moreover, due to the focussing nature of the phenomenon, it is not possible to resolve the event using a mesh which is fixed in space. Recent progress in the development of Lagrangian algorithms to compute such flows is described by Peridier and Walker[18].

The period required to induce an eruption for a given vortex is $O(\kappa/a)$; thus an eruptive reponse occurs sooner for a stronger vortex and/or the closer the vortex is to the wall. Experiments[15] clearly show an eruptive response persists over a wide range of high Reynolds numbers and at least as low as $R_v = 10^4$; note that it is not a kinematical effect, but an abrupt dynamical event which results in the discrete ejection of near-wall fluid.

Although all vortices carry the pressure signature that can provoke an eruption, it is often argued[9,11] that hairpin vortices are the principal feature of turbulent flows near walls, for a variety of reasons. Hairpin vortices are low-speed streak creators[3,11] and in addition are able to reproduce themselves. A schematic diagram of symmetric and asymmetric hairpin vortices is shown in Figure 4. The symmetric hairpin vortex is the simplest mathematical model, although most hairpins in the turbulence will be asymmetric. There are two features of hairpin vortices which lead to their persistence in the flow. First, calculations show[9] (and experiments confirm[11]) that hairpins interact with a background shear flow to produce new subsidiary vortices to the side of the original hairpin vortex (see Figure 4); in this manner, the hairpin is able to multiply itself in the lateral direction. Secondly, the vortex legs move progressively toward the wall as the vortex convects in a shear flow near the wall, thus hastening an eruptive response. Recent computations and well-controlled experiments[9,11] definitively show that this eruption occurs; in the region of adverse pressure gradient behind the vortex head and between the vortex legs, the creation of secondary hairpin vortices takes place through a viscous-inviscid interaction with the near-wall flow[11]. The process occurs after the vortex has convected over the wall for a period of time and is characterized by a rapidly rising, narrow plume from the wall region; the plume then rolls over into a well-defined secondary vortex[11].

6. Discussion

It has been argued that the equations governing the evolving wall-layer flow are of the boundary-layer type. To explain the production process in the near-wall region, it is necessary to isolate physical phenomena that lead to a strong interaction with the outer flow and which, in the process, destroy the local structure of the wall layer. A convected vortex above the wall layer (which is predominantly of the hairpin type) provides a moving zone of adverse pressure gradient. Note that the outer edge of the wall layer is not at some fixed value of y^+. Rather the boundary between the wall layer and outer region changes continually. In a sweep, the outer region penetrates close to the wall; conversely during a quiescent period the wall layer thickens continuously. When the vortex is strong enough and/or close enough to the wall for a sufficient period of time, an abrupt eruption will be produced; on the other hand, weaker disturbances can provoke an instability in the wall layer[6]. In either case, a local breakdown of the relatively well-ordered near-wall flow occurs.

Finally, it is of interest to obtain an order of magnitude estimate of the period between bursts. For either the long wave problem discussed in §3 or the full three-dimensional problem described in §4, thoeretical investigations[10,12,13] suggest that a moving disturbance, which imposes a O(1) adverse pressure gradient on the wall layer will provoke an eruption for $T = O(1)$; it follows from equation (4) that this occurs for $t = O(\nu\lambda^+/(u_\tau^2 W_1))$ and consequently

$$T_B^+ = O(\lambda^+/W_1). \tag{20}$$

The typical values give $(\lambda^+/W_1) = 50$, which is of the same order as measured[2] values of $T_B^+ \simeq 110$. As with most order of magnitude arguments, the estimate is not precise; however, it may be used to obtain a value for the characteristic length in the streamwise direction. Consider a convected vortex moving with speed $O(U_1 u_\tau)$ at the outer edge of the wall layer; the mean time between bursts is $T_B^+\nu/u_\tau^2$ and may be regarded as the time required for the vortex to provoke an eruption. In this time period, the vortex travels a distance L_x which may be adopted as the characteristic streamwise length and

$$L_x = U_1 u_\tau (T_B^+\nu/u_\tau^+). \tag{21}$$

Using equation (20), it follows that $L_x = O(U_1/W_1)$ in agreement with equation (16).

Acknowledgements

This work was supported by the Air Force of Scientific Research under Contract AFOSR-89-0065.

References

1. Kline, S. J., Reynolds, W. C., Schraub, F. C. and Rundstadler, P. W., "The Structure of Turbulent Boundary Layers", *Journal of Fluid Mechanics*, Vol. 30, p.133, 1967.
2. Walker, J. D. A., Abbott, D. E., Scharnhorst, R. K. and Weigand, G. G., "Wall-Layer Model for the Velocity Profile in Turbulent Flows", *AIAA Journal*, Vol. 72, Part 2, p. 140, 1989.
3. Acarlar, M. S and Smith, C. R., "A Study of Hairpin Vortices in a Laminar Boundary Layer. Part 1. Hairpin Vortices Generated by Hemisphere Protuberances", and "Part 2. Hairpin Vortices Generated by

Fluid Injection", *Journal of Fluid Mechanics*, Vol. 175, p. 1 and p. 43, 1987.

4. Hon, T. L. and Walker, J. D. A., "An Analysis of the Motion of and Effects of Hairpin Vortices", Report FM-11, Department of Mechanical Engineering and Mechanics, Lehigh University, Bethlehem, PA; available NTIS-AD-A187261.
5. Smith, C. R and Metzler, S. P., "The Characteristics of Low-Speed Streaks in the Near-Wall Region of a Turbulent Boundary Layer", *Journal of Fluid Mechanics*, Vol. 129, p. 27, 1983.
6. Walker, J. D. A. and Herzog, S., "Eruption Mechanisms for Turbulent Flow Near Walls", in Transport Phenomena in Turbulent Flows, M. Hirata and N. Kasagi (eds.), Hemisphere, p. 145, 1988.
7. Hatziavramidis, D. T. and Hanratty, T. J., "The Representation of the Viscous Wall Region by a Regular Eddy Pattern", *Journal of Fluid Mechanics*, Vol. 95, p. 655, 1979.
8. Chapman, D. R. and Kuhn, G. D., "The Limiting Behavior of Turbulence Near a Wall". *Journal of Fluid Mechanics*, Vol. 170, p. 265, 1986.
9. Hon, T. L. and Walker, J. D. A., "Evolution of Hairpin Vortices in a Shear Flow", NASA Technical Memoranda 100858, ICOMP-88-9, 1988. Also to appear in *Computers and Fluids.*
10. Peridier, V., Smith, F. T. and Walker, J. D. A., "Methods for the Calculation of Unsteady Separation", AIAA Paper 88-0604, 26th Aerospace Sciences Meeting, Reno, Nevada, 1988.
11. Smith, C. R., Walker, J. D. A., Haidari, A. H. and Taylor, B. K., "Hairpin Vortices in Turbulent Boundary Layers; The Implications for Reducing Surface Drag", this volume, 1990.
12. Doligalski, T. L. and Walker, J. D. A., "The Boundary Layer Induced by a Convected Two-Dimensional Vortex", *Journal of Fluid Mechanics*, Vol. 139, p. 1, 1984.
13. Ersoy, S. and Walker, J. D. A., "Viscous Flow Induced by Counter-Rotating Vortices", *Physics of Fluids*, Vol. 28, p. 2687, 1985.
14. Ersoy, S. and Walker, J. D. A. "Flow Induced at a Wall by a Vortex Pair", *AIAA Journal*, Vol. 24, p. 1597, 1986.
15. Walker, J. D. A., Smith, C. R., Cerra, A. W. and Doligalski, T. L., "The Impact of a Vortex Ring on a Wall", *Journal of Fluid Mechanics*, Vol. 181, p. 99, 1987.
16. Ersoy, S. and Walker, J. D. A., "The Boundary Layer Due to a Three-Dimensional Vortex Loop", *Journal of Fluid Mechanics*, Vol. 185, p. 569, 1987.
17. Chuang, F. J. and Conlisk, A. T., "The Effect of Interaction on a Boundary Layer Induced by a Convected Rectilinear Vortex", *Journal of Fluid Mechanics*, to appear.
18. Peridier, V. and Walker, J. D. A., "Vortex-Induced Boundary-Layer Separation", Report FM-13, Department of Mechanical Engineering and Mechanics, Lehigh University, Bethlehem, PA, October, 1989.

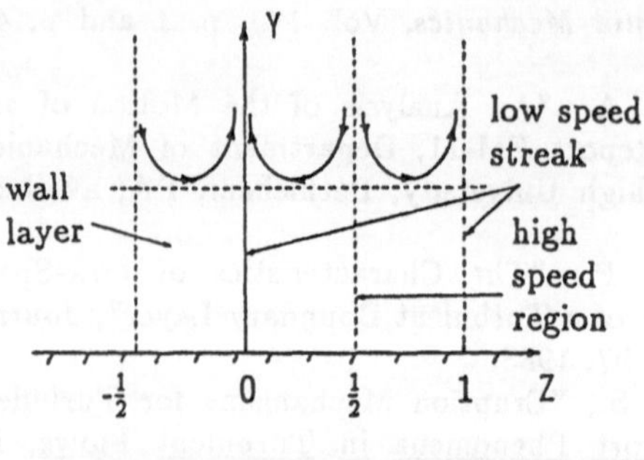

Figure 1. Assumed wall-layer structure during a quiescent period.

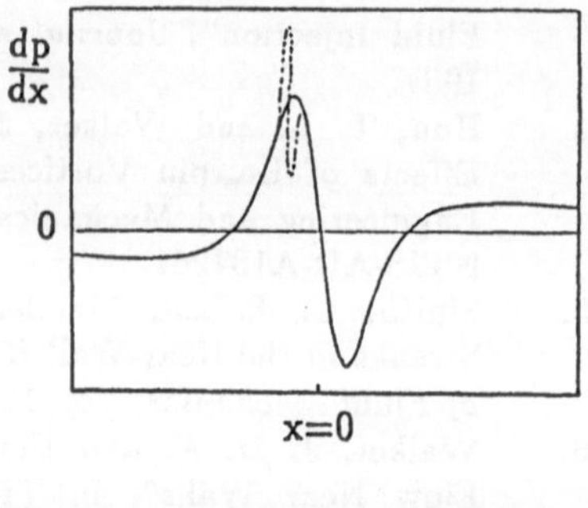

Figure 3. Pressure gradient near the wall due to a moving vortex.

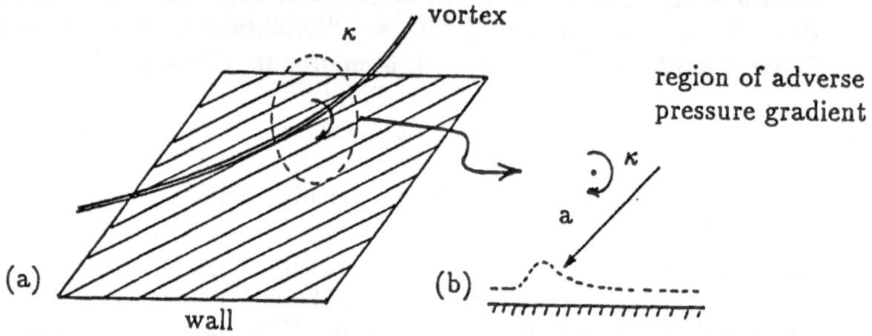

Figure 2. (a) Three dimensional vortex motion above a wall; (b) Details of a slice through the vortex.

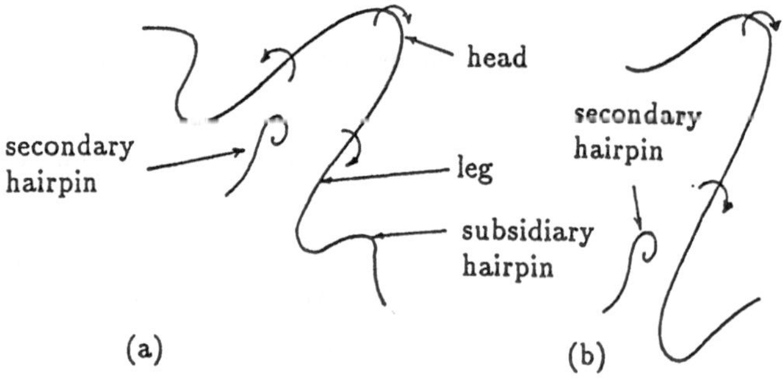

Figure 4. Schematic of (a) symmetric hairpin vortex; (b) asymmetric hairpin.

Non-Linear Coherent Mode Interactions and the Control of Shear Layers

D. E. Nikitopoulos* and J. T. C. Liu**

*Mechanical Engineering Department
Louisiana State University
Baton Rouge, LA 70803, USA

**Division of Engineering
Brown University
Providence, RI 02912, USA

Abstract

A non-linear integral formulation, based on local linear stability considerations, is used to study the collective interactions between discrete wave-modes associated with large-scale structures and the mean flow in a developing shear layer. Aspects of shear layer control are examined in light of the sensitivity of these interactions to the initial frequency parameter, modal energy contents and modal phases. Manipulation of the large-scale structure is argued to be an effective means of controlling the flow, including the small-scale turbulence dominated region far downstream. Cases of fundamental, 1st and 2nd subharmonic forcing are discussed in conjunction with relevant experiments.

Introduction

Following such landmark experiments as those of [1], large-scale coherent structures have been widely accepted as intrinsic features of free shear flows. One of the prevailing trends of thought is to interpret them as wave-like structures, products of hydrodynamic instability in an inflectional mean flow, superimposed upon fine-grained turbulence [2,3]. This idea also has its parallel in turbulent and transitional boundary layers where longitudinal vorticity elements set up local, *three dimensional*, inflectional profiles (see, for instance, [4,5]), leading to high frequency secondary instability modes and turbulence production. According to nonlinear hydrodynamic instability ideas [6], coherent structures have been associated with waves modulated by amplitudes whose variation is much slower than the wave oscillation. The characteristics of these waves are well represented by local linear stability theory, a fact confirmed by experiment [7]. The evolution of the flow is governed by mutual non-linear interactions between the mean flow, the wave-like large-scale structure and the fine-grain turbulence [3]. Observation of "vortex-pairing" phenomena [8] associated with the appearance of subharmonics of the most amplified wave [9], brought forth consideration of the non-linear interactions between different spectral components of the large-scale structure [10,11]. It has been shown, both experimentally [9] and theoretically [12], that wave-wave interactions as well as the downstream development of large-scale

A. Gyr (Editor)
Structure of Turbulence and Drag Reduction
IUTAM Symposium Zurich/Switzerland 1989

structures are very sensitive to upstream "initial" conditions, and particularly to the relative phases of the waves. This sensitivity makes it possible to control the evolution of the flow, *including the small-scale turbulence* [13], by appropriately forcing different spectral components of the large-scale structure. Contrary to the interactions involving small-scale turbulence, the mean flow-wave and wave-wave couplings dominate the early development of perturbed shear flows.

General Formulation

The general problem under consideration is a spatial one involving the simultaneous determination of the streamwise development of the mean flow, the large-scale structure and the fine-grain turbulence. A modified integral method originating from [14] is used in this formulation. Every flow quantity is decomposed into a mean motion component, q^o, a random one, q', representing the small-scale turbulence, and a wave, $\tilde{q}$, associated with the large-scale structure. The wave component is further decomposed into discrete modes:

$$\tilde{q}(\vec{x},t)=\sum_1^k\{q^n(\vec{x};n\beta)e^{-in\beta t}+q^{-n}(\vec{x};n\beta)e^{in\beta t}\}, \quad (1)$$

where q^{-n} is the complex conjugate of q^n, β is the dimensionless frequency, and k is the number of modes that are, either naturally or by means of forcing, present in the flow. In this study we will consider three modes of frequencies β, 2β, and 3β. We shall seek here to investigate the effects of mode interactions on the development of an otherwise laminar viscous shear flow, as the interactions with the fine-grain turbulence have already been studied (e.g. [13]).

Equations of motion for each flow component, including the discrete wave-modes, can be derived from the Navier-Stokes and continuity equations for incompressible flow by straight-forward manipulations and judicious use of time- and conditional-averaging [13]. Kinetic energy equations are obtained from these in order to illustrate the coupling mechanisms between the flow components, and to obtain a solution to the interaction problem [12]. The mean-flow kinetic energy equation is:

$$\frac{\overline{D}}{Dt}\frac{(u_i^o)^2}{2}=\frac{\partial u_i^o}{\partial x_j}\sum_1^3(u_i^k u_j^{-k})+\frac{1}{Re}\left[\frac{\partial^2}{\partial x_j\partial x_j}\frac{(u_i^o)^2}{2}-\frac{\partial u_i^o}{\partial x_j}\frac{\partial u_j^o}{\partial x_i}\right]+\dots \quad (2)$$

The first term on the right is the energy exchange mechanism between the mean flow and the three modes collectively. Depending on its sign, energy can flow to or from the mean flow at the expense or gain of the wave component. The same interpretation holds for the individual modes. The energy equations for the three modes, put in a single equation, are:

$$u_j^0 \frac{\partial}{\partial x_j} |u_i^n|^2/2 = -\frac{\partial u_i^0}{\partial x_j} \Re(u_i^n u_j^{-n}) + \frac{1}{Re}\left[\frac{\partial^2}{\partial x_j \partial x_j}|u_i^n|^2/2 - \left|\frac{\partial u_i^n}{\partial x_j}\right|^2\right] +$$
$$+\delta_{3n}\Re\left[u_i^2 u_j^1 \frac{\partial u_i^{-3}}{\partial x_j} + u_i^1 u_j^2 \frac{\partial u_i^{-3}}{\partial x_j}\right] + \delta_{2n}\Re\left[u_i^{-1} u_j^3 \frac{\partial u_i^{-2}}{\partial x_j} - u_i^2 u_j^1 \frac{\partial u_i^{-3}}{\partial x_j} + u_i^1 u_j^1 \frac{\partial u_i^{-2}}{\partial x_j}\right] +$$
$$+\delta_{1n}\Re\left[-u_i^1 u_j^2 \frac{\partial u_i^{-3}}{\partial x_j} - u_i^{-1} u_j^3 \frac{\partial u_i^{-2}}{\partial x_j} - u_i^1 u_j^1 \frac{\partial u_i^{-2}}{\partial x_j}\right] + \ldots, \forall n \epsilon [1,3], \quad (3)$$

where $\Re$ is the real part function, $|\ |$ denotes magnitude, and δ_{kn} is the Kronecker delta. "Transport" terms have been omitted for the sake of brevity. The first term on the right is the energy exchange with the mean flow. The last three groups of terms provide the inter-mode coupling. Each member of every group represents work done by wave-induced stresses against modal rates of strain. The direction of the energy transfer depends on the sign of these terms, controlled by the relative modal phases. Modes β and 2β can exchange energy directly and can be independent of mode 3β. Two-mode interaction problems have been studied [12,13] assuming that relative phases remain explicitly independent of the streamwise coordinate, and predetermining the direction of intra-modal energy exchanges. To address the problem in a complete manner phase variations must be taken into account explicitly. Governing equations for the modal phases can also be derived from the equations of motion. In single equation form they are:

$$\frac{1}{2} u_j^0 \frac{\partial}{\partial x_j} \ln(u_i^n / u_i^{-n}) = -n\beta |u_i^n|^2 - \frac{\partial u_i^0}{\partial x_j} \Im(u_i^n u_j^{-n}) + \frac{1}{Re}\Im\left[\frac{\partial}{\partial x_j}\left[u_i^{-n}\frac{\partial u_i^n}{\partial x_j}\right]\right] +$$
$$+\delta_{3n}\Im\left[u_i^2 u_j^1 \frac{\partial u_i^{-3}}{\partial x_j} + u_i^1 u_j^2 \frac{\partial u_i^{-3}}{\partial x_j}\right] + \delta_{2n}\Im\left[u_i^{-1} u_j^3 \frac{\partial u_i^{-2}}{\partial x_j} + u_i^2 u_j^1 \frac{\partial u_i^{-3}}{\partial x_j} + u_i^1 u_j^1 \frac{\partial u_i^{-2}}{\partial x_j}\right] +$$
$$+\delta_{1n}\Im\left[u_i^1 u_j^2 \frac{\partial u_i^{-3}}{\partial x_j} - u_i^{-1} u_j^3 \frac{\partial u_i^{-2}}{\partial x_j} + u_i^1 u_j^1 \frac{\partial u_i^{-2}}{\partial x_j}\right] + \ldots, \forall n \epsilon [1,3], \quad (4)$$

$\Im$ denoting the imaginary part. The evolution of the phases is affected by the mean flow (second term on the right) as well as the intra-mode coupling (last three groups of terms).

We shall presently consider the problem posed by the existence of two-dimensional wave-modes in a two-dimensional yet self-similar mean flow. Equations (2–4) are integrated across the shear layer. With the aid of boundary layer-type approximations and use of appropriate shape assumptions for the mean flow and wave-modes [12], we obtain the following set of non-linear, differential equations for the modal energy densities, E_n, modal phases, θ_n, and the growth of the mean flow:

$$\delta\frac{I_n}{E_n}\frac{dE_n}{dx}=I_{rsn}-\frac{I_{dn}}{Re_0\,\delta}+\left[\frac{E_2}{\delta}\right]^{\frac{1}{2}}\left[I^1_{12}+\left[\frac{E_3}{E_1}\right]^{\frac{1}{2}}I^3_{12}\right]\left[\delta_{2n}\frac{E_1}{E_2}-\delta_{1n}\right]+$$
$$+I^2_{13}\left[\frac{E_2E_3}{\delta\;E_1}\right]^{\frac{1}{2}}\left[\delta_{3n}\frac{E_1}{E_3}-\delta_{1n}\right]+I^1_{23}\left[\frac{E_2E_3}{\delta\;E_1}\right]^{\frac{1}{2}}\left[\delta_{3n}\frac{E_1}{E_3}-\delta_{2n}\frac{E_1}{E_2}\right], \qquad (5)$$

$$\delta I_n\frac{d\theta_n}{dx}=n\beta\delta-\frac{\Im(\alpha_n^2)}{Re_0\,\delta}+P^1_{12}\left[\frac{E_2}{\delta}\right]^{\frac{1}{2}}\left[\delta_{2n}\frac{E_1}{E_2}+\delta_{1n}\right]+P^3_{12}\left[\frac{E_2E_3}{\delta\;E_1}\right]^{\frac{1}{2}}\left[\delta_{2n}\frac{E_1}{E_2}-\delta_{1n}\right]+$$
$$+P^2_{13}\left[\frac{E_2E_3}{\delta\;E_1}\right]^{\frac{1}{2}}\left[\delta_{3n}\frac{E_1}{E_3}+\delta_{1n}\right]+P^1_{23}\left[\frac{E_2E_3}{\delta\;E_1}\right]^{\frac{1}{2}}\left[\delta_{3n}\frac{E_1}{E_3}+\delta_{2n}\frac{E_1}{E_2}\right], \qquad (6)$$

$$\delta I_m\frac{d\delta}{dx}=\sum_1^3(E_k\,I_{rsk})-\frac{I_d}{Re_0},\ \forall n\in[1,3], \qquad (7)$$

where δ is the maximum slope thickness of the shear layer. The integral coefficients I_m and I_d are constant, while I_n, I_{rsn}, I_{dn}, I^i_{jk}, and P^i_{jk} depend on $\delta(x)$ through the dependence of the local shape functions of the modes on the local frequency β. In our case the shape functions satisfy Rayleigh's equation for a *tanh* mean-flow profile. The sign and magnitude of I_{rsn} regulate the energy exchange of the modes with the mean flow. The integrals I^i_{jk}, and P^i_{jk} control the intra-mode couplings and are strong functions of the modal phases. Those involving two modes depend on $\theta=(\theta_2-2\theta_1)$ and those involving three modes depend on $\varphi=(\theta_3-\theta_2-\theta_1)$. Equations (5–7) indicate that the development of the wave-modes and the mean flow depend on a multitude of "initial" conditions at the onset of the shear layer. These include initial modal energy densities (E_{no}) and phases (θ_{no}), and the frequency parameter $\beta_0(\delta_0)$. The key feature of the mode interaction is that a "weaker" mode can be considerably amplified or damped by interacting with a "stronger" one, at little expense or gain of the latter.

Discussion and Results

The evolution of the flow is dominated by energy exchanges between individual modes and the mean flow. Regions of fast shear layer growth are due to energy transfer from the mean to the dominant mode and the amplification of the latter. The observed plateaus in the growth of the shear layer are tied to the reversal of the energy transaction with the mean flow leading to saturation and decay of the modal energy content. This is illustrated by Figures 1a and b, where results of a two-mode calculation are shown together with measurements from [9]. The forcing

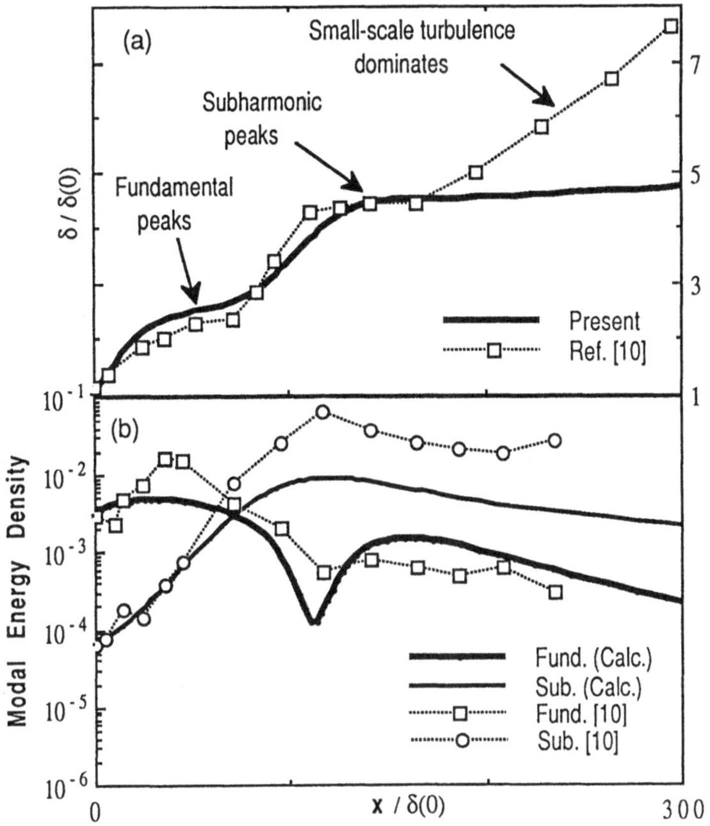

Figure 1: Mean-flow (a) and Modal energy density (b) evolution of a shear layer purturbed by forcing the first subharmonic.

frequency is close to the 1st subharmonic (β_0=0.206) of the most amplified frequency. Such excitation produced two plateaus in the mean flow growth. The first one corresponds to peaking and decay of the fundamental, which saturates first, and the second one corresponds to the subharmonic which has the same fate further downstream. This is shown both by the experimental and the theoretical results which are in excellent qualitative agreement. Flow visualization has indicated [9] that the first roll-over of the shear layer occurs in the neighborhood of the fundamental peak (first plateau). "Vortex pairing" is initiated when the subharmonic and fundamental share the same energy level and is completed near the subharmonic peaking location (second plateau). The linear growth of the mean flow after the second plateau in the experiment is due to extraction of energy from the mean flow by small-scale turbulence, an interaction not accounted for in the calculations.

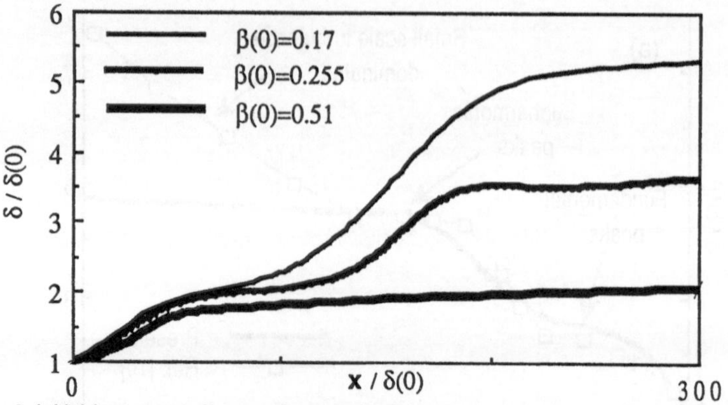

Figure 2: Initial frequency effect on the growth of the shear layer for low amplitude forcing.

The interaction of the large-scale structure with the mean flow is the primary mechanism responsible for the sensitivity of the flow development to the initial forcing frequency parameter. Lower frequencies have longer amplification history, extracting more energy from the mean flow and leading to stronger growth of the latter, as shown by results from a three-mode interaction calculation presented in Figure 2. The fundamental (β_f=0.511) is started in all cases at the same energy level, while the other two modes are approximately two orders of magnitude weaker. The three cases are characterized by different initial frequency parameters $\beta_0 = \beta_f/3$, $\beta_f/2$, and β_f. These conditions correspond to low amplitude forcing of the 2nd subharmonic, the 1st subharmonic and the fundamental respectively. The shear layer roughly doubles and triples

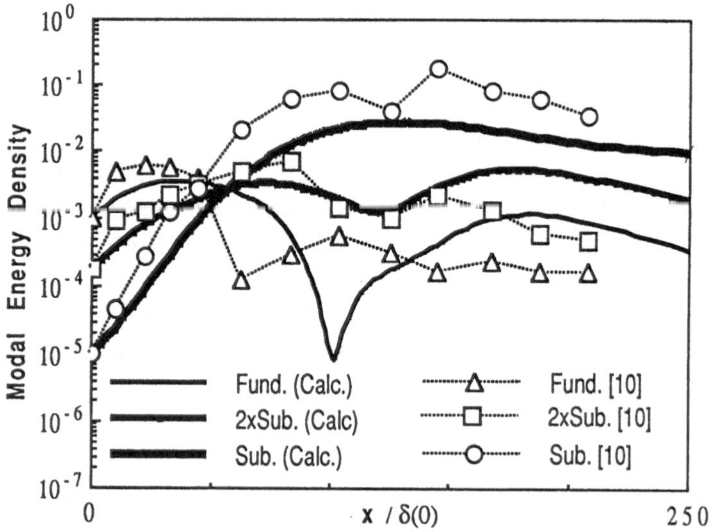

Figure 3: Modal energy density evolution in a shear layer perturbed by forcing the 2nd subharmonic.

when the initial frequency is equal to the 1st and 2nd subharmonic respectively. Modes other than the fundamental and the relevant subharmonic do not cause associated plateaus in the mean flow development because they never become dominant. It is the growth and decay of the occasionally dominating mode that marks the fast growth and saturation of the mean flow. Figure 3 shows the calculated evolution of the modal energy densities for the case of 2nd subharmonic forcing together with measurements from [9]. Mode $2\beta_f/3$ is always overpowered by the fundamental (β_f=0.51) and the 2nd subharmonic ($\beta_f/3$), and is not directly responsible for the fast growth of the shear layer (Figure 2, for $\beta(0)$=0.17). The presence of this mode influences the flow only through interactions with the other modes.

Intra-mode interactions have direct and indirect effects on the evolution of the flow. While the direct effect is weak, the *indirect* one is

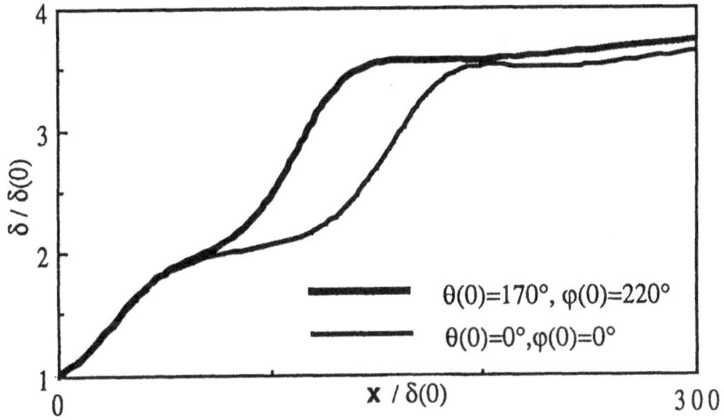

Figure 4: Initial phase effect on the growth of the shear layer (3-mode interaction with 1st subharmonic forcing).

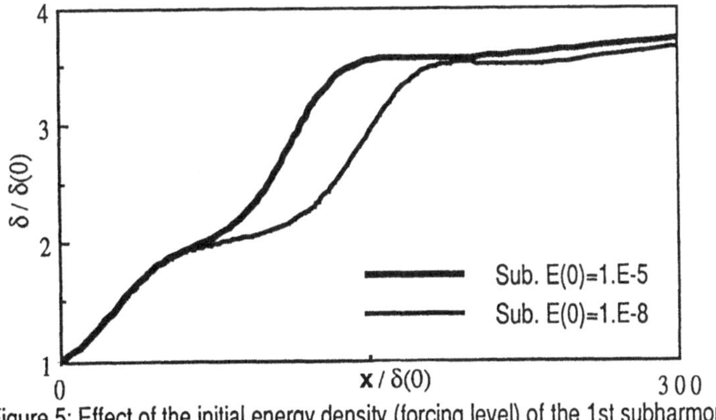

Figure 5: Effect of the initial energy density (forcing level) of the 1st subharmonic on the growth of the shear layer.

of particular importance because it involves a booster mechanism for energy exchanges with the mean flow [12]. This indirect influence causes a significant change in the growth of the shear layer as shown by a case where the intra-mode interactions alone are affected by imposing different initial modal phases (Figure 4). Fast shear layer growth after the first plateau is initiated much earlier in the case where $\theta(0)=170^{\circ}$ and $\varphi(0)=220^{\circ}$ due to extra amplification of the subharmonic induced by energy exchanges with the other two modes. This increased amplification also causes early peaking of the subharmonic and a consequent early saturation of the shear layer spreading rate. The growth of the mean flow in the region of subharmonic dominance can also be affected by use of different subharmonic forcing levels. The calculation of Figure 5 shows that reduction of initial subharmonic strength results in delay of the step-like increase of the shear layer thickness. This is caused by the weakening of the interaction mechanisms (which are proportional to the modal energy density) with the mean flow and the other modes during the infancy of the subharmonic.

The present analysis concentrated on the early stages of shear flow development where interactions of large-scale coherent structure related modes with the mean flow and each other are dominant. We have therefore not taken into consideration the effects of fine-grain turbulence which becomes dominant far downstream when the organized structure is in decline. The development of the small-scale turbulence in an initially turbulent shear layer depends on the history of its interaction with the other scales of motion. These interactions are controlled to a great extent by the initial turbulence energy level and the modal energy contents. It has been shown that the large-scale structure acts as an agent for the transfer of extra energy to the fine-grain turbulence, increasing thus its potential to draw more energy from the mean flow far downstream [13]. This of course enhances the flow-growth in the small-scale dominated region of the flow. It is therefore reasonable to argue that manipulation of the large-scale structure also provides the means for control of small-scale turbulence. It should also be noted that only two-dimensional large-scale structures have been dealt with in this paper. The effect of three dimensional structure on the flow dynamics and its potential role in shear-layer control remains to be investigated.

Acknowledgements

This work is partially supported by the Louisiana State University Council on Research, by NASA/Lewis Research Center through Grant NAG3-1016, and by the National Science Foundation through Grant MSM83-20307.

References

1. Brown, G.L. and Roshko, A., 1974, On density effects and large structure in turbulent mixing layers, *J. Fluid Mech., 64*, 775–816.

2. Liepman, H.W., 1962, Free turbulent flows, In *Mechanique de la turbulance*, CNRS, Paris, 211–27.

3. Liu, J.T.C., 1987, Contributions to the understanding of large-scale coherent structures in developing turbulent shear flows, *Adv. Appl. Mech., 26*.

4. Sabry, A.S. and Liu, J.T.C., 1988, Nonlinear development of Görtler vortices and generation of high shear layers in the boundary layer, I. Symp. to Honor C.C. Lin, Singapore: World Scientific, 175–183.

5. Wilmarth, W.W., 1975, Structure of turbulence in boundary layers, *Adv. Appl. Mech., 15*, 159–254.

6. Stuart, J.T., 1971, Nonlinear stability theory, *Ann. Rev. Fl. Mech., 3*, 347–70.

7. Gaster, M., Kit, E. and Wygnanski, I., 1985, Large-scale structures in a forced turbulent mixing layer, *J. Fluid Mechanics, 150*, 23–39.

8. Winant, C.D. and Browand, F.K., 1974, Vortex pairing, the mechanism of turbulent mixing-layer growth at moderate Reynolds number, *J. Fluid Mech., 63*, 237–55.

9. Ho, C.M. and Huang, L.S., 1982, Subharmonics and vortex merging in mixing layers, *J. Fluid Mech, 119*, 443–73.

10. Nikitopoulos, D.N. and Liu, J.T.C., 1984, Triple-mode interactions in a developing shear layer, *B. Am. Ph. Soc. 29*, 1548.

11. Mankbadi, R.R., 1985, On the interaction between fundamental and subharmonic instability waves in a turbulent round jet", *J. Fluid Mech., 160*, 385–419.

12. Nikitopoulos, D.E., and Liu, J.T.C., 1987, Nonlinear binary-mode interactions in a developing mixing layer, *J. Fluid Mech., 179*, 345–370.

13. Liu, J.T.C., and Kaptanoglu, H.T., 1987, Control of free shear layers, *AIAA* Paper 87-2689.

14. Stuart, J.T., 1958, On the nonlinear mechanics of hydrodynamic stability, *J. Fluid Mech., 4*, 1–21.

Wavenumber-Frequency Spectral Densities of Turbulent Wall Pressure and Wall Shear Fluctuations

Theodore R. Anderson
NUSC, New London, Connecticut, U.S.A.

Abstract

The wavenumber-frequency spectral densities of turbulent wall pressure and wall shear fluctuations are investigated over a rigid flat plate. The spectral densities are determined as an expansion in a complete set of Orr-Sommerfeld eigenfunctions.

It was shown by Landahl (1967) that the lowest discrete Orr-Sommerfeld mode, corresponding to the least-damped Tollmien-Schlichting wave, makes the largest contribution to the inhomogeneous solution. In this research, the lowest order Orr-Sommerfeld mode is used in the expansion of the inhomogeneous solution. The method of Eckhaus (1965) was used to obtain the contribution of discrete Orr-Sommerfeld eigenfunctions to the wavenumber-frequency spectral densities.

However, the discrete spectrum of the Orr-Sommerfeld equation is not a complete set of eigenfunctions in the inhomogeneous solution. It has been shown by Grosch and Salwen (1978) that the Orr-Sommerfeld equation yields a continuous spectrum in addition to the discrete. The combination of the discrete, which are finite in number, and the continuous spectra is a complete set of eigenfunctions for the Orr-Sommerfeld equation. The number of discrete modes increases with Reynolds number. In this research, the method of Eckhaus is extended, and proved to contain the continuous eigenfunctions.

The Fourier-transformed pressure can be expressed in terms of the inhomogeneous solution of the Orr-Sommerfeld equation and its derivatives. The same holds for the Fourier-transformed wall shear fluctuations. Spectral densities can be given for both wall pressure and wall shear fluctuations in terms of the products of the corresponding Fourier transforms.

The nonlinear Reynolds stress terms of the inhomogeneous Orr-Sommerfeld equation are modelled according to the Bark (1975) model of a single burst and bursting statistics utilized by Bark.

Finally, the spectral densities of turbulent wall pressure fluctuations were numerically computed and compared to the experimental results of Martin (1977) and Martin's transformation of Blake's data (1967) fitted to a modified Corcos model. The spectral densities of wall shear fluctuations are in analytical form and have not been numerically computed. The wavenumber-frequency spectral densities for wall pressure fluctuations compared well with Martin's transformation of Blake's data on the convective ridge, but the continuous mode contributions are insignificant there. However, it is shown that the continuous eigenfunction contributions compare well with the low wavenumber, high frequency measurements of Martin. It is expected that the wavenumber-frequency spectral densities for wall shear fluctuations will be dominated by the discrete spectrum on the convective ridge and be about 20 dB lower than the wall pressure spectral densities.

A. Gyr (Editor)
Structure of Turbulence and Drag Reduction
IUTAM Symposium Zurich/Switzerland 1989

1. STATEMENT OF THE PROBLEM

The homogeneous solutions of the Orr-Sommerfeld equation were solved through a turbulent boundary layer. The computer code SUPORT by Scott and Watts (1975) was used for integration and needed orthonormalization. The inhomogeneous fourth-order Orr-Sommerfeld equation, as given by Landahl (1967), is

$$(U - c)(\phi'' - K^2\phi) - U''\phi + \frac{i}{\alpha R}(\phi^{IV} - 2K^2\phi'' + K^4\phi) = \frac{\hat{q}}{\alpha K}\,, \tag{1.1}$$

where ϕ is the inhomogeneous solution, c is the phase speed, $\hat{q}$ is the Fourier Transform of q in the x (streamwise) and z (spanwise) directions, $q = \nabla^2 T_2 - \partial/\partial y(\nabla\cdot T)$, and $T_i = \partial/\partial x_j(\overline{u_j u_i} - u_j u_i)$. The turbulent mean velocity profile was taken from Landahl (1967). Bark modeled the regions of high instantaneous Reynolds stress in a burst as

$$uv = a_1 \delta(t)\delta(x)\delta(z)y^3 \exp(-a_2 y^2)\,. \tag{1.2}$$

Ignoring all other components of $u_i u_j$ (instantaneous Reynolds stress) and $\overline{u_i u_j}$ (average Reynolds stress) except uv, for lack of experimental data, Bark obtained values of a_1 and a_2 from a least-squares fit to the measured data of Kim et al. (1971). Bark indicates that instantaneous values of uv as large as 62 $\overline{uv}$ have been experimentally measured (Willmarth and Lu, 1972) and that the average $\overline{uv}$ term is ignored, leaving only the instantaneous uv component.

Assuming that the time between bursts was distributed according to a Poisson process, Bark was able to use a result (Rice, 1944) which states that the statistics for the response function are given by the response of an individual event. Hence, by Bark's model, the response to the statistics of bursting was reduced to the response to an individual burst. Using the Bark model of the burst, the right-hand side of the inhomogeneous Orr-Sommerfeld equation is

$$Q \equiv \frac{\hat{q}}{\alpha K} = i(2\pi)^{-3/2}\, a_1 y \exp(-a_2 y^2)[Ky^2 + \frac{2}{K}(2a_2^2 y^4 - 7a_2 y^2 + 3)]\,. \tag{1.3}$$

The inhomogeneous solution expanded in discrete eigenfunctions ϕ_n of the Orr-Sommerfeld equation, as given by Eckhaus (1965), for a bounded domain is given here for a semibounded flow over a flat plate:

$$\phi_D = \sum_n A_n \phi_n, \quad \text{where } A_n = \frac{1}{(c_n - c)N(n,n)} \int_0^\infty Q\, \tilde{\phi}_n dy\,, \tag{1.4}$$

where ϕ_D is the inhomogeneous discrete solution with the normalization

$$\int_0^\infty S[\phi_n]\tilde{\phi}_m dy = N(n,m)\delta_{nm} \tag{1.5}$$

and $\tilde{\phi}_n$ are adjoint eigenfunctions satisfying the homogeneous adjoint

Orr-Sommerfeld equation

$$(U - c)(\tilde{\phi}'' - K^2\tilde{\phi}) + 2U'\tilde{\phi}' + \frac{i}{\alpha R}(\tilde{\phi}^{IV} - 2K^2\tilde{\phi}'' + K^4\tilde{\phi}) = 0 \qquad (1.6)$$

$$\tilde{\phi}(0) = \tilde{\phi}'(0) = \tilde{\phi}(0) = \tilde{\phi}'(0) = 0 . \qquad (1.7)$$

$N(n,n)$ is the mathematical norm squared, given as

$$N(n,n) = \int_0^\infty S[\phi_n]\tilde{\phi}_n dy = \int_0^\infty [\tilde{\phi}_n'\phi_n' + K^2\tilde{\phi}_n\phi_n]dy \qquad (1.8)$$

as integration by parts shows.

The method of Eckhaus (1965) was extended to calculate the continuous spectral response to hydrodynamic bursting. The summation over discrete eigenfunctions becomes an integral over continuous eigenfunctions, and the final form of the solution is

$$\phi_C = \int_0^\infty A_k\phi_k dk, \quad \text{where } A_k = \frac{1}{(c_k - c)N(k)}\int_0^\infty Q\tilde{\phi}_k dy \quad , \qquad (1.9)$$

where ϕ_C is the continuous inhomogeneous solution. We get $N(k')$ from the orthogonality property, which states that the norm for the continuous eigenfunctions is defined by a Dirac delta function, unlike a Kronecker delta function for the discrete case (Salwen and Grosch, 1981):

$$\int_0^\infty S[\phi_k]\tilde{\phi}_k' dy = N(k,k')\delta(k - k') \; ; \qquad (1.10)$$

hence,

$$N(k') = \int_0^\infty \int_0^\infty S[\phi_k]\tilde{\phi}_k' \, dydk \; . \qquad (1.11)$$

2. TURBULENT WALL PRESSURE FLUCTUATIONS

The Fourier-transformed pressure $\hat{p}$ used in calculating the wavenumber-frequency spectral density can be expressed in terms of inhomogeneous Orr-Sommerfeld solutions, their derivatives, and Fourier-transformed nonlinear stress terms. Landahl's (1967) derivation evaluated at the wall and using Bark's model of the burst is

$$\hat{p} = \frac{1}{KR}\phi'''(y = 0) \; . \qquad (2.1)$$

Theoretically, the wavenumber-frequency spectral density for turbulent wall pressure fluctuations for a stationary, ergodic, and homogeneous turbulent field in the x and z directions is given as

$$\Phi(K,\omega) = \hat{p}\hat{p}^*\Big|_{y=0} , \quad \text{where } \hat{p} = \hat{p}(K,\omega,y) \qquad (2.2)$$

(see Kinsman, 1967).

In terms of the inhomogeneous solutions of the Orr-Sommerfeld equation, the spectral density becomes

$$\Phi(K,\omega) = \hat{p}\hat{p}^* = (\phi''\ \phi''^*)/K^2R^2 \tag{2.3}$$

where $\phi = \phi_D + \phi_C$, with ϕ_D and ϕ_C being the discrete and continuous inhomogeneous solution to the Orr-Sommerfeld equation. Finally, we can write the discrete, continuous, and cross term contribution to the wavenumber-frequency spectral density as

$$\Phi(K,\omega) = \Phi_D(K,\omega) + \Phi_X(K,\omega) + \Phi_C(K,\omega)\ . \tag{2.4}$$

These terms are numerically evaluated in tables 1 and 2.

3. TURBULENT WALL SHEAR FLUCTUATIONS

The components of the fluctuating stress tensor which contribute to the wall shear fluctuations in cartesian coordinates are σ_{xy} and σ_{zy} evaluated at the surface y=0.

The wavenumber frequency spectral densities for wall shear fluctuations are given by

$$\pi_{xy} = \left.\frac{\hat{\sigma}_{xy}(k,\omega,y)\hat{\sigma}^*_{xy}(k',\omega',y)}{\delta(k-k',\omega-\omega')}\right|_{y=0} \tag{3.1}$$

and

$$\pi_{zy} = \left.\frac{\hat{\sigma}_{zy}(k,\omega,y)\hat{\sigma}^*_{zy}(k',\omega',y)}{\delta(k-k',\omega-\omega')}\right|_{y=0} \tag{3.2}$$

where $\hat{\sigma}_{ik}$ is the Fourier transform of σ_{ik} using kernel $\exp[i(\alpha x + \beta z - \omega t)]$. The complex conjugate of σ_{ik} is σ^*_{ik}.

The Dirac delta function $\delta(k - k', \omega - \omega')$ appears both in the spectral density of the wall shear fluctuations (response) as well as in the wavenumber-frequency spectral density of the hydrodynamic bursting (source). Multiplying the source and response wavenumber-frequency spectral densities by the Dirac delta function and integrating over all numbers and frequencies yields

$$\tilde{\pi}_{xy} = \hat{\sigma}_{xy}(k,\omega,y)\ \hat{\sigma}^*_{xy}(k,\omega,y) \tag{3.3}$$

for the xy-component of the fluctuating wall shear fluctuations wavenumber-frequency spectral densities. Similarly, we have

$$\tilde{\pi}_{zy} = \hat{\sigma}_{zy}(k,\omega,y)\ \hat{\sigma}^{*}_{zy}(k,\omega,y). \tag{3.4}$$

The xy component of the Fourier-transformed shear stress σ_{xy} is given as

$$\hat{\sigma}_{xy} = \frac{\partial\hat{V}_x}{\partial y} + \frac{\partial\hat{V}_y}{\partial x} = \frac{\partial\hat{u}}{\partial y} + \frac{\partial\hat{v}}{\partial x} = \frac{\partial\hat{u}}{\partial y}, \tag{3.5}$$

where $\hat{V}_x = \hat{u}$ and $\hat{V}_y = \hat{v}$.

The y-component of fluctuating velocity $v = v(y)$ is a function of y only and satisfies the inhomogeneous Orr-Sommerfeld equation. To obtain $\hat{u}$, we combine Squire's transformation (Squire, (1933))

$$\alpha\,\tilde{u} = \alpha\,\hat{u} + \beta\hat{w} \text{ and } \alpha\,\tilde{w} = \beta\,\hat{u} - \alpha\,\hat{w} \tag{3.6a,b}$$

with the Fourier-transformed (using kernel exp [i(αx + βz - ωt]) continuity equation

$$i\,\alpha\,\hat{u} + D\,\hat{v} + i\,\beta\,\hat{w} = 0, \text{ where } D = \frac{d}{dy} \tag{3.7}$$

to obtain

$$\hat{u} = \frac{i\alpha}{k^2}\,D\,\hat{v} + \frac{\alpha\beta}{k^2}\,\tilde{w} \text{ and } \hat{w} = \frac{i\beta}{k^2}\,D\,\hat{v} - \frac{\alpha^2}{k^2}\,\tilde{w}\,. \tag{3.8a,b}$$

This technique was utilized by Bark (1975) to obtain u. To calculate velocity spectral densities in the x-direction, Bark (1975) further showed the following:

$$\tilde{w} = L^{-1}\,[(\frac{i\beta}{\alpha})\ (D\ U)\ \hat{v}] + \tilde{w}_H\,, \tag{3.9}$$

where U is the mean velocity profile given by Landahl (1967). The operator L is defined by

$$L\,[\tilde{w}_H] = [(U-c) + \frac{1}{\alpha R}\,(D^2 - k^2)]\,\tilde{w}_H = 0\,, \tag{3.10}$$

When u is obtained from equations (3.8a), (3.9), and (3.11), the wavenumber-frequency spectral density $\tilde{\pi}_{xy}$ can be obtained from (3.3) and (3.5). The y-component of fluctuating velocity $\hat{v}$ satisfies the inhomogeneous Orr-Sommerfeld equation with the solution given as an expansion in discrete and continuous Orr-Sommerfeld eigenfunctions as explained in Section 1. The Orr-Sommerfeld solution ϕ is related $\hat{v}$ by $\hat{v} = -ik\phi(y)$.

The-zy component of shear stress is given by

$$\hat{\sigma}_{zy} = \frac{\partial \hat{W}}{\partial y} + \frac{\partial \hat{V}}{\partial z} = \frac{\partial \hat{W}}{\partial y} , \tag{3.11}$$

where $\hat{V} = \hat{V}(y)$ and $\frac{\partial \hat{V}}{\partial z} = 0$.
The z-component of the fluctuating velocity $\hat{W}$ is obtained from (3.8b). The z-component of fluctuating velocity, as $\hat{u}$ and $\hat{V}$, will be expressed as an expansion in Orr-Sommerfeld discrete and continuous eigenfunctions as explained in Section 1. The wavenumber frequency spectral density $\tilde{\pi}_{zy}$ is now obtained from (3.4) and (3.11).

The final forms of the wavenumber-frequency spectral density will be

$$\tilde{\pi}_{xy} = \tilde{\pi}_{xyD} + \tilde{\pi}_{xyX} + \tilde{\pi}_{xyC} \text{ and } \tilde{\pi}_{zY} = \tilde{\pi}_{zyD} + \tilde{\pi}_{yxY} + \tilde{\pi}_{zyC} , \tag{3.12}$$

where the subscripts D, X, and C designate the contributing of the wavenumber-frequency wall shear fluctuations from discrete Orr-Sommerfeld eigenfunctions, cross products of discrete and continuous Orr-Sommerfeld eigenfunctions, and continuous Orr-Sommerfeld eigenfunctions, respectively.

Although no numerical results are given in this section, the numerical techniques involved are similar to those in Section 2.

CONCLUSIONS

The wavenumber-frequency spectral density predicted from continuous Orr-Sommerfeld eigenfunctions compared well with Martin's low wavenumber, high frequency measurements.

The discrete wavenumber-frequency spectral density ($10 \log \Phi_D(\alpha,0,\omega)$ along the convective ridge (where it dominates) predicts Martin's transformation of Blake's data fitted to a modified Corcos model up to about the -70 dB contour on Martin's figure. Here it starts to deviate with results about 10 dB higher than Martin's results. The continuous component ($10 \log \Phi_C (\alpha,0,\omega)$) of the wavenumber-frequency spectral density agreed well with Martin's low wavenumber measurements. However, the total wavenumber-frequency spectral density ($10 \log \Phi_T (\alpha,0,\omega)$) overpredicted Martin's low wavenumber measurements. This is due to the large contribution of the discrete component of the spectral density into this region. Further work is needed to separate the continuous from the discrete wave contributions to the wavenumber-frequency spectral density.

The wavenumber-frequency spectral densities for turbulent wall shear fluctuations have been given in theoretical form. Numerical results will be forthcoming.

Table 1. Theoretical Values of $\Phi_D(\alpha,\beta,\omega)$ in Decibels and a Comparison With Martin's Transformation of Blake's Data

R = 105,000 (Scaled on boundary layer thickness)

α	ω	Theoretical Value of Φ_D $(\alpha,0,\omega)$ in decibels* (Scaled on displacement thickness)	Martin's Results in decibels
.020	0.135	-6.8	
.370	.2505	-37.9	-40
1.00	.620	-42.9	-50
2.20	1.690	-57.9	-60
6.60	3.790	-59.3	-70
11.49	7.790	-67.88	-80

Table 2. Comparison of the Wavenumber-Frequency Continuous Spectral Density in Decibels With Martin's Low Wavenumber, High Frequency Measurements

R = 50,000 scaled on boundary layer thickness

α	ω	Theoretical Value of Φ_C $(\alpha,0,\omega)$ in decibels (Scaled with displacement thickness)	Martin's Results in decibels
.185	1.58	-99.2	-94
.37	1.75	-99.1	-99
.45	2.21	-101.0	-99
.47	3.4	-106.0	-105
.54	1.8	-98.3	-97
.58	4.82	-109.0	-115

The author is indebted to Professor William Saric for instruction on using the SUPORT code in solving the homogeneous Orr-Sommerfeld equation.

REFERENCES

1. Bark, Fritz, "On the Wave Structure of the Wall Region of a Turbulent Boundary Layer," Journal of Fluid Mechanics, Vol. 70, Part 2, 1975, pp. 229-250.
2. Blake, William, "Turbulent Boundary Layer Wall-Pressure Fluctuations on Smooth and Rough Walls," Journal of Fluid Mechanics, Vol. 44, Part 4, 1967, pp. 637-660.
3. Corcos, G. M., "The Resolution of Turbulent Pressure at the Wall of a Boundary Layer," Journal of Sound and Vibration, Vol. 6, 1967, pp. 59-70.
4. Eckhaus, W., Studies in Non-linear Stability Theory, Berlin, Springer-Verlag, 1965.
5. Grosch, C. E., and Salwen, H., "The Continuous Spectrum of the Orr-Sommerfeld Equation," Part 1, The spectrum and the eigenfunctions, Journal of Fluid Mechanics, Vol. 87, 1978, pp. 33-54.
6. Kim, H. T., Kline, S. J., and Reynolds, W. C., "Journal of Fluid Mechanics, Vol. 50, 1971, pg. 133.
7. Kinsman, B., Wind waves, Prentice-Hall, Inc., Englewood Cliffs, New Jersey, 1967.
8. Landahl, M. T., "A Wave-Guide Model for Turbulent Shear Flow," Journal of Fluid Mechanics, Vol. 29, Part 3, 1967, pp. 441-459.
9. Martin, N. C. and Leehey, P., "Low Wavenumber Wall Pressure Measurements Using a Rectangular Membrane as a Spatial Filter," Journal of Sound and Vibration, Vol. 52(1), 1977, pp. 95-120.
10. Rice, S. O., Bell System Technical Journal, No. 23-24, 1944.
11. Salwen, H., and Grosch, C. E., "The Continuous Spectrum of the Orr-Sommerfeld Equation," Part 2, Eigenfunction expansions, Journal of Fluid Mechanics, Vol. 104, 1981, pp. 445-465.
12. Scott, M. R., and Watts, H. A., "SUPORT -- A Computer Code for Two-Point Boundary-Value Problems Via Orthonormalization," SAND75-0198, Applied Math Divison 2646, Sandia Laboratories, Albuquerque, New Mexico, 1975.
13. Squire, H. B., 1933 Proc. Roy. Soc. A 142, 621.
14. Willmarth, W. W., and Lu, S. S., Journal of Fluid Mechanics, pp. 55, 65, 1972.
15. Willmarth, W. W., and Woolridge, C. E., Journal of Fluid Mechanics, Vol. 14, 1962, pg. 187.

Quadrant Analysis and Instantaneous Momentum Transport – a Critical Review

Andreas Mueller
Institute of Hydromechanics and Water Resources Managment
Swiss Federal Institute of Technology
CH- 8093 Zurich, Switzerland

Summary

When quadrant analysis in the u'-v'-plane is used to study instantaneous momentum transport, it is based on a frame of reference moving with the local mean velocity. In contrast, when the redistribution of momentum by coherent structures is of interest it is proposed to use one frame of reference for an entire structure. In a frame of reference fixed with the flow channel there are only Q1 and Q4 events and they transport axial momentum upward and downward. In a frame of reference moving with a representative convection velocity u_c of the structure there is a gradual shift from Q2 and Q3 event for $\overline{u}(y) < u_c$ to Q1 and Q4 events for $\overline{u}(y) > u_c$.

Introduction

Quadrant analysis of turbulent velocity fluctuations in the u'-v'-plane has become a standard tool in the detection and the analysis of burst events in a boundary layer or channel flow. Its starting point (Wallace et. al., 1972, Willmarth et al., 1972) was the question of how the fluctuations u'(t) and v'(t) produce the average Reynolds stress $\overline{u'v'}$. These authors made an attempt to link Reynolds decomposition with early concepts of coherent structures. Reynolds decomposition is connected with Prandtl's mixing length concept, which relates the fluctuating velocities to the local gradient of the mean flow. The concept of structures, on the other hand, tries to describe motions which are coherent in an entire flow region.

Quadrant analysis allows a classification of events into ejections (Q2) and sweep events (Q4) which transport momentum toward the wall and into wallward (Q3) and outward events (Q1) which transport momentum away from the wall.

A. Gyr (Editor)
Structure of Turbulence and Drag Reduction
IUTAM Symposium Zurich/Switzerland 1989

The quantity $u'(t)\cdot v'(t)$ was later defined as an instantaneous Reynolds stress and statistical properties of the momentum transport by single events were studied (e.g. Nakagawa, Nezu, 1982). In a frame of reference moving with the local mean velocity, $u'v'(t)$ correctly describes the instantaneous momentum transport. The study of the instantaneous momentum transport by the classical quadrant analysis has, therefore, the drawback of a different frame of reference for each level y.

While forces are invariant to transformations into a frame of reference moving at a uniform velocity (Galilei-transformation), the momentum transport is affected. It is, therefore inconvenient to study momentum transport within a coherent structure with frames of reference moving at the local mean velocity.

Momentum transport in a fixed frame of reference

The drawback of having a different frame of reference for each level y can be avoided by describing the local momentum transport within an entire flow region in the same frame of reference.

Conceptually the mean velocity profile $\overline{u}(y)$ can be explained by the superposition of the internal flow of vortical structures which move at a convection velocity u_c (Perry et al. 1982). A vortical structure which has an inclined axis of rotation has regions which transport momentum away from the wall and regions which transport momentum toward the wall. If this redistribution of momentum by vortical structures is of interest then

$$u(t)v(t) = \overline{u}\,\overline{v} + \overline{u}\,v'(t) + \overline{v}\,u'(t) + u'(t)\,v'(t)$$

is the adequate description of the momentum transfer and not $u'(t)v'(t)$ alone. It is inconsistent to retain $u'v'(t)$ as a time function and to neglect the terms $\overline{u}v'(t)$ and $\overline{v}u'(t)$ which are part of the instantaneous momentum transport as well and zero only in the time mean.

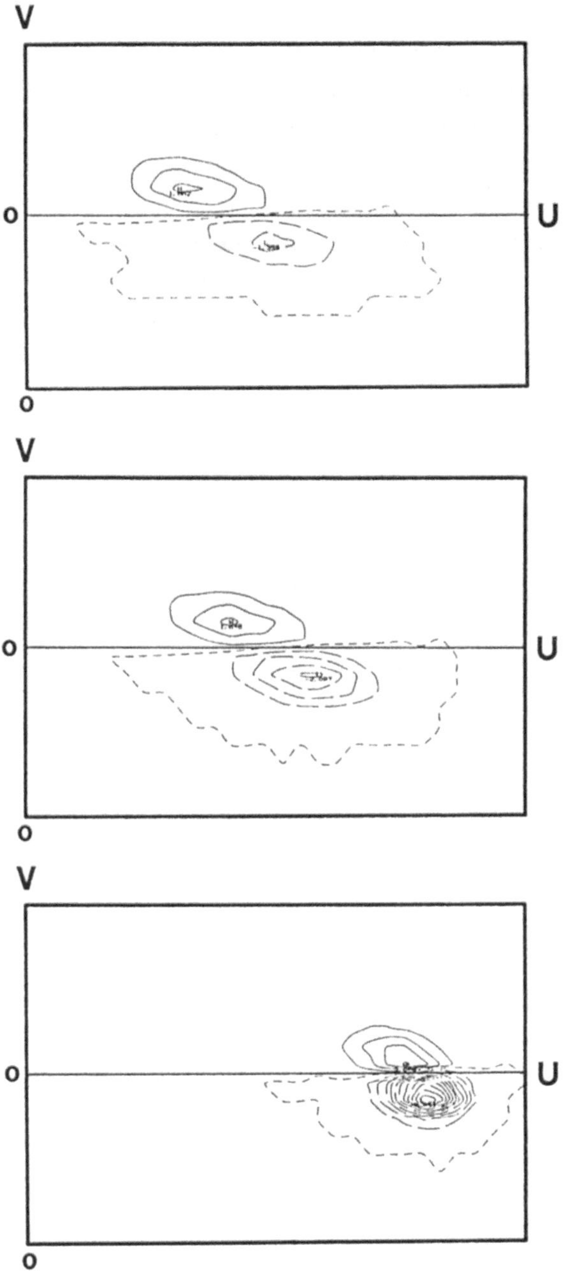

Fig. 1. The redistribution of momentum in an open channel flow represented by the density of the momentum tranport $u \cdot v \cdot p(u,v)/u_*^2$ in the u-v-plane based on a fixed frame of reference.
(1) y/h = 0.037; (2) y/h = =0.062; (3) y/h = 0.45

The joint probability distribution $p(u,v)$ in the u-v-plane is shifted by $\overline{u}$, $\overline{v}$ relative to the usually considered joint probability distribution $p(u',v')$). Since $u(t)$ does not change sign there are only Q1 and Q4 events transporting axial momentum upward or downward, and the complicated arguments regarding the wallward (Q4) and outwards (Q3) events can be avoided. If $\overline{v}$ is not equal to zero due to a secondary flow the term $\overline{u}\,\overline{v}$ is included in the analysis. Figure 1 represents $u \cdot v \cdot p(u,v)/u_*^2$ measured in a channel flow, i.e. the density of the total momentum transport in the u-v-plane (flow conditions see appendix).

Momentum transport in the frame moving with the convection velocity of structures

A frame of reference of special interest is the frame moving with a convection velocity u_c of the structures, which is representative for the flow region of interest. Observations in this frame of reference can provide information on the momentum transport within the vortical structures. Data on the threedimensional velocity field, e.g. obtained from direct simulation (Robinson et al., 1988, Robinson, 1989) or conditionally averaged masurements (Nagib and Guezennec, 1986, Wark and Nagib, 1988) allow an identification of regions with upward and downward transport of momentum.

Measurements of the convection velocity is not straightforward. The moving structures themself are developing, and features which are detected for the determination of u_c are changing. Early correlation measurements in a boundary layer by Favre et al., (1967) showed that u_c is approximately 0.7 U_∞ and at y_o=0.23, (y_o^+= 400), u_c is equal to the local mean velocity $\overline{u}(y_o)$. In an open channel Nakagawa, Nezu (1981) measured u_c=0.93 $\overline{U}$ and the local mean velocity $\overline{u}(y)$ is equal u_c at y_o=0.05h (y_o^+ = 40), where h is the depth and $\overline{U}$ the depth averaged velocity. Other authors (Robinson, 1989) claim that the convection velocity is not constant in a structure and varies with the mean flow profile.

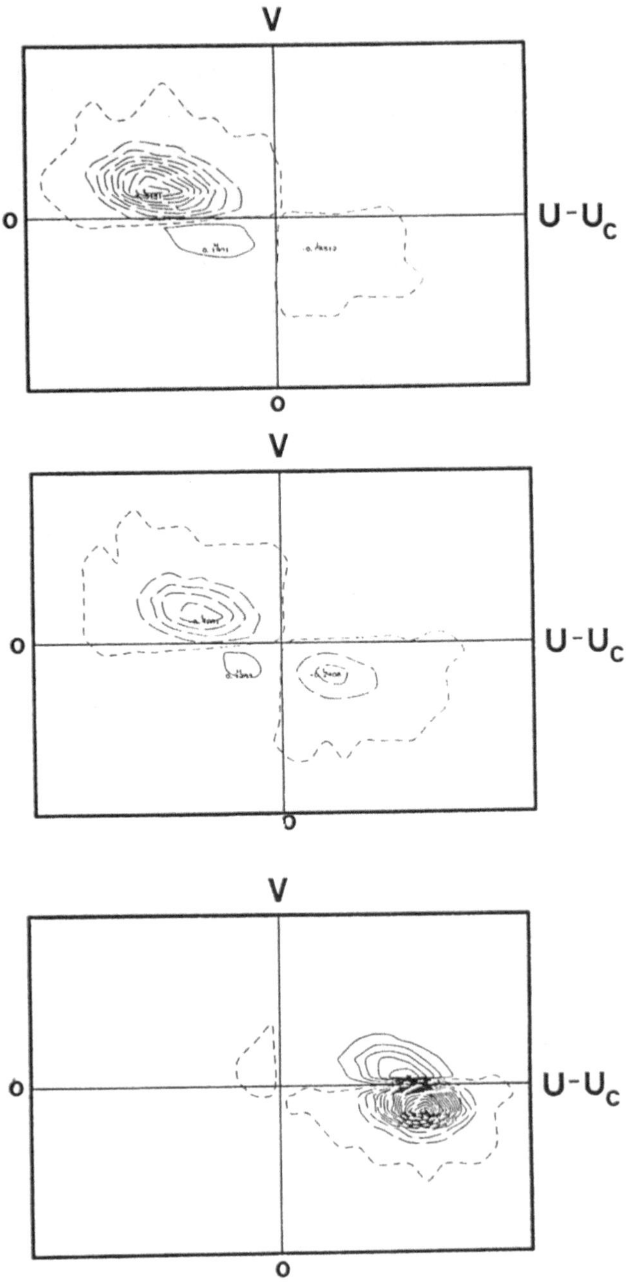

Fig. 2. Momentum exchange in a frame of reference moving with a convection velocity u_c. The density of the momentum transfer $(u-u_c)v\ p(u-u_c,v)/u$ is shifted gradually to the right with increasing $u(y)$.
(1) $y < y_o$, $\bar{u}(y) < u_c$; (2) $y = y_o$, $\bar{u}(y) = u_c$; (3) $y > y_o$, $\bar{u}(y) > u_c$

Quadrant analysis in this representation leads to another interpretation of Q1 - Q4 events. The center of the joint probability distribution $p(u-u_c,v))$ is shifted from negative $\overline{u}(y)-u_c$ values for $y < y_o$ to positive ones for $y > y_o$. In the buffer zone the values $u(t)-u_c$ are always negative. In this zone there is therefore a large transport of momentum toward the wall by Q2 events which is partly compensated by a large upward transport by Q3 events. In the outer layer $\overline{u}(t)-u_c$ is positive, and there is, therefore, a large transport toward the wall by Q4 events which is partly compensated by an outward transport by Q1 events. Figure 2 represents $(u-u_c)v \cdot p(u-u_c,v)/u_*^2$ measured in a channel flow, i.e. the density of the momentum transport in the $(u-u_c,v)$-plane (flow conditions see appendix)

Figure 3 is a conceptual sketch which allows an identfication of regions of a vortical structure which produce a Q1 - Q4 event in a point measurement of the velocity. When a frame of reference moving with the convection velocity of the structure is used, the point of measurements travels across the structure with a velocity $-u_c$. Depending on the orientation and inclination of the structure different sequences of events can be observed in the velocity time history. The Figure shows that for a layer close to

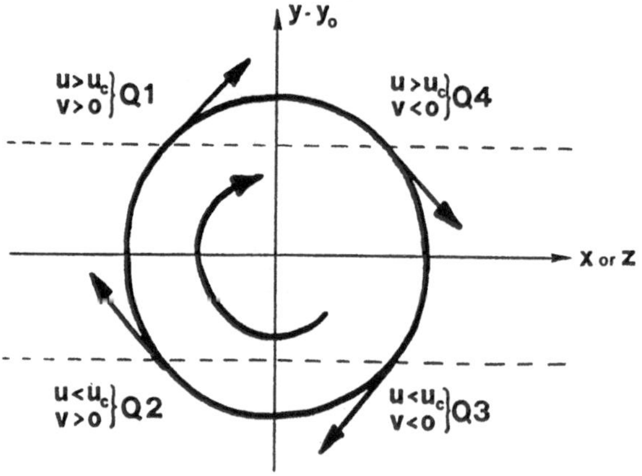

Fig. 3. Conceptual sketch correlating the velocity field of a vortical structure (in a frame of reference moving with the structure) with Q1 - Q4 events as they are observed in the $(u-u_c,v)$-plane of velocity point measurements.

the wall ($y < y_o$) Q2 and Q3 events are dominant, for $y > y_o$ the Q1 and Q4 events are. However, such a model with a circular structures can account only for the redistribution of momentum and not for the net transport $\overline{u'v'}$.

References

Favre, A., Gaviglio, J., Dumas, R., (1967), Structure of velocity space-time correlations in a boundary layer. Physics of fluids supplement, Vol. 10, pp. S138-S145

Mueller, A., Studerus, X., (1979), Secondary flow in an open channel. Proc. 17. IAHR Congress, Cagliari, Vol. 3, pp 19-24

Naguib, H.M., Guezennec, Y.G., (1986), On the structure of turbulent boundary layers. Proc. Tenth Symposium on Turbulence, University of Rolla-Missouri

Nakagawa, H., Nezu, I., (1981), Structure of space-time correlations of bursting phenomena in an open channel flow. J. Fluid Mech., Vol. 104, pp. 1-43

Perry, A.E., Chong, M.S., (1982), On the mechanism of wall turbulence. J. Fluid Mech., Vol 119, pp. 173-217

Robinson, S.K., Kline, S.J., Spalart, P.R., (1988), Quasi-coherent structures in the turbulent boundary layer: Verification and new information from a numerically simulated flat-plate layer. Near Wall Turbulence: Zaric Memorial Conference, Hemisphere, 1989

Robinson, St. K., (1989), A review of vortex structures and associated coherent motions in turbulent boundary layers. Proceedings Second IUTAM Symposium on Structure and Drag Reduction Zurich, Springer-Verlag, Berlin

Wallace, J.M., Eckelmann, H., Brodkey, R.S., (1972), The wall region of turbulent shear flow. J. Fluid Mech. Vol. 54, pp. 39-48

Wark, C.E., Nagib, H.M., (1988), On the character of turbulence-producing events in near-wall turbulence. Near Wall Turbulence: Zaric Memorial Conference, Hemisphere, 1989

Willmarth, W.W., Lu, S.S., (1972), Structure of the Reynolds stress near the wall. J. Fluid Mech., Vol. 55, pp 65-92

Appendix

The concepts presented in the paper are illustrated by means of data of an open channel flow where strong secondary flow cells are induced by longitudinal stripes of roughness (Mueller, Studerus , 1979). They were considered to be a model of stripes of coarse and fine material in an alluvial river. Simultaneously three components of the velocity vector were measured. The secondary flow cells (Figure 4a) transport high momentum fluid into the region above the roughness where the shear is high due to the higher friction. The effect is strong enough to produce a maximum of the mean axial velocity above the roughness (Figure 4b).

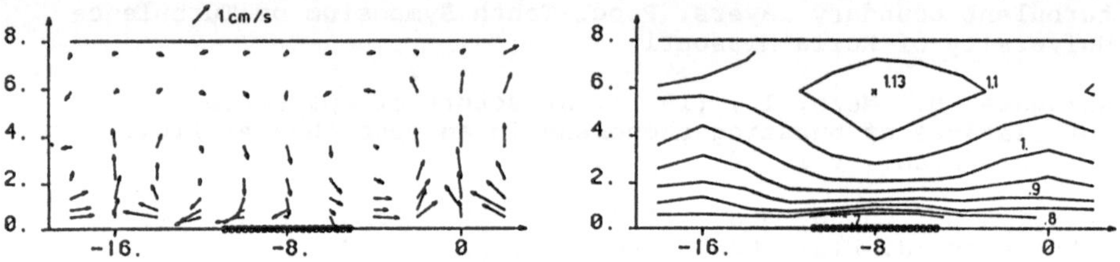

Fig. 4. (a) Secondary flow cells induced by longitudinal stripes of roughness. (b) Isovels $\overline{u}/\overline{U}$ of the axial velocity.

The data shown in Figure 1 and 2 are an evaluation of the velocity time histories of the profile z = -8.0 cm. The convection velocity was not measured and is assumed to be $u_c = 0.7\ U_{max}$, in accordance with the results reported by Nakagawa, Nezu (1981). Figure 5 shows the vertical profiles for mean axial and vertical velocities $\overline{u}(y)$ and $\overline{v}(y)$. The positive value of $\overline{v}(y/h=0.037)$ is a consequence of small rollers which were observed close to the wall.

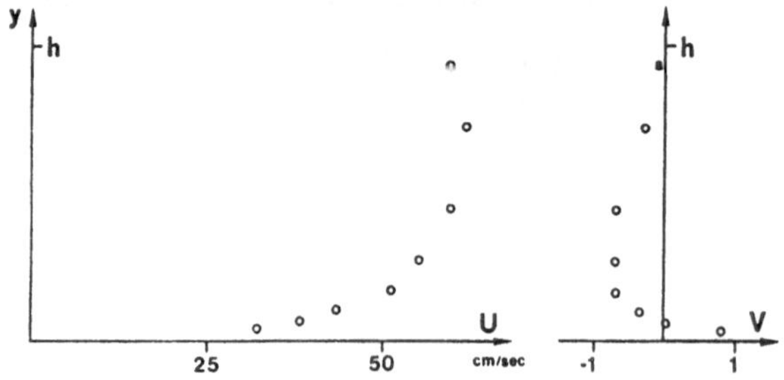

Fig. 5. Profiles $\overline{u}(y)$ and $\overline{v}(y)$ of the horizontal and vertical velocity at z = -8.0 cm

Artificial Control of Turbulent Mixing Layers

Osamu Inoue
Institute of Fluid Science,
Tohoku University
2-1-1 Katahira, Aoba-ku, Sendai 980, Japan

Two-dimensional(2D), spatially growing, turbulent mixing layers are simulated by a vortex method. Special attention is paid to the effect of artificial forcing on the development of a mixing layer. Forcing is applied by imposing velocity disturbances of a sinusoidal form on each discrete vortex that appears at the origin of the mixing layer. Forcing frequencies selected are the predominant frequency of an unforced flow, its subharmonics, high-harmonics, and various combinations of the frequencies. Instantaneous plots of discrete vortices and various statistics up to the second order moment are obtained to see the variation of coherent structures. Results are in good agreement with experiments.

1. Introduction

Forcing is a useful idea of turbulence control. The growth of a mixing layer, for example, can be efficiently manipulated by imposing disturbances on the flow using sound from a loud speaker,[1] using a vibrating flap,[2,3] or by controlling the flow rate.[4] For a comprehensive review, readers are referred to Ho & Huerre[5] and Wygnanski & Petersen.[6]

Computational work on forced mixing layers has been performed extensively.[7-10] Inoue & Leonard[9] simulated 2D forced/unforced mixing layers by a vortex method. They reproduced many of the flow features which have been observed experimentally. For forced mixing layers, computational results showed excellent agreement with experiments. Inoue & Leonard concluded that 2D vortex method is quite effective and useful in simulating two-dimensionally forced mixing layers because the forced mixing layers which are produced experimentally become more 2D than the unforced mixing layers.

This study is an extension to multiple-frequency-forced cases of the work done by Inoue & Leonard for single-frequency-

A. Gyr (Editor)
Structure of Turbulence and Drag Reduction
IUTAM Symposium Zurich/Switzerland 1989

forced mixing layers. One of the goals of this study is to increase our understanding of the effects of multiple-frequency forcing on coherent structures of a mixing layer. For this purpose, we use the same flow model (Model A in Inoue & Leonard) and the same flow parameters (velocity ratio, convection velocity, numerical core radius, etc.) as those used in the study of single-frequency forced mixing layers.

2. Mathematical Formulation and Numerical Procedure

For details of mathematical formulation and numerical procedure, readers are referred to Inoue & Leonard.[9] Here only the new points introduced in this paper are mentioned. As our main interest lies in multiple-frequency-forcing, velocity disturbances which are given to each discrete vortex at the origin (x=0) are assumed to be of the form,

$$v_f = \sum_{k=1}^{K} A_k \sin(2\pi f_k t + \beta_k) \tag{1}$$

where v is the velocity component normal to the flow direction. In this paper we consider forcing up to two frequencies (K=2). The simulation parameters prescribed in this study are as follows.

$$\begin{aligned} &\text{velocity ratio:} \quad r\ (\equiv U_2/U_1) = 0.6, \\ &\text{Forcing amplitudes:} \quad A_1 = A_2 = A = 0.5U_c, \qquad (2) \\ &\text{Forcing frequencies:} \quad f_1,\ f_2 = 0,\ F,\ F/2,\ F/3,\ F/4,\ 2F, \\ &\qquad\qquad \text{and their combinations}, \\ &\text{Phase angles:} \quad \beta_1 = \beta_2 = 0, \end{aligned}$$

where U_1, U_2 ($U_1 > U_2$) are the two freestream velocities, F the predominant frequency of the unforced mixing layer, and the convection velocity is defined as $U_c = (U_1 + U_2)/2$.

3. Results and Discussion

Instantaneous plots of discrete vortices in the case of single-frequency-forced mixing layer are presented in Fig. 1, and the corresponding distributions of momentum thickness in Fig. 5. The growth of a single-frequency-forced mixing layer is characterized by three distinct subregions[6]: two growth regions (regions I and III in ref. 6) separated by one saturation region (region II). With decreasing forcing frequency, the length of region I increases.

A mixing layer forced by two frequencies behaves quite differently. A few examples are presented in Fig. 2 where the forcing frequency is the predominant frequency (F) combined with its first (F/2), second (F/3), and third (F/4) subharmonics, respectively. The corresponding distributions of momentum thickness are shown in Fig. 6. In the figures, for simplicity, the expression $f = f_1 + f_2$ denotes the case that $v_f = A \cdot \sin(2\pi f_1 t) + A \cdot \sin(2\pi f_2 t)$. In every case presented, two saturation regions are observed. Immediately downstream of the origin, the predominant frequency dominates the roll-up process of vorticity. The rolled-up vortices produce the first saturation region where vortex merging is suppressed. Then, vortices tend to merge and the number of merging vortices depends on the subharmonic frequency which is combined with the predominant frequency; that is, every two vortices merge when f=F+F/2 (Fig. 2a), every three vortices when f=F+F/3 (Fig. 2b), and every four vortices when f=F+F/4 (Fig. 2c). For an example of multiple-vortex merging, time development of the forced mixing layer when f=F+F/4 is presented in Fig. 3, where the arrow indicates the merging process of a set of four vortices. New vortices which are produced by multiple-vortex merging lead the second saturation region. Downstream of the second saturation region, the mixing layer recovers its growth. The calculated flow features discussed above are quite similar to those observed experimentally by Ho & Huang.[4]

The effect of combining two frequencies on the growth of a mixing layer is shown in Fig. 4 and Figs. 7 to 9 for the case of f=F+F/2. Figures 4 and 7 suggest that the flow features of a double-frequency forced mixing layer are closer to those of the single-frequency forced flow with the predominant frequency near the first saturation region while closer to those with the subharmonic frequency near the second saturation region. The profiles of r.m.s. u' in Fig. 8 show three peaks at x=140 where two vortices in merging process are nearly laterally aligned, in accordance with the observation in Mode II of Ho & Huang.

The profiles of the Reynolds shear stress show the occurrence of contra-gradient diffusion where the growth of the mixing layer is suppressed.

Reference

1. Zaman,K.B.Q.M. and Hussain,A.K.M.F., "Vortex Pairing in a Circular Jet under Controlled Excitation. Part I. General Jet Response," 1980, J. Fluid Mech., Vol.101, pp. 449-491.

2. Oster, D. and Wygnanski, I., "The Forced Mixing Layer between Parallel Streams," 1982, J. Fluid Mech., Vol.123, pp. 91-130.

3. Mehta,R.D., Inoue,O., King,L.S. and Bell,J.H., "Comparison of Experimental and Computational Techniques for Plane Mixing Layers," 1987, Phys. Fluids, Vol.30, pp.2054-2062.

4. Ho. C. M. and Huang, L.S., "Subharmonics and Vortex Merging in Mixing Layer," 1982, J. Fluid Mech., Vol.119, pp.443-473.

5. Ho. C. M. and Huerre, P., "Perturbed Free Shear Layers," 1984, Ann. Rev. Fluid Mech., Vol.16, pp.365-424.

6. Wygnanski, I. J. and Petersen, R. A., "Coherent Motion in Excited Free Shear Flows," 1987, AIAA J., Vol.25, pp.201-213.

7. Riley, J. J. and Metcalfe, R. W., "Direct Numerical Simulation of a Perturbed Turbulent Mixing Layer," 1980, AIAA Paper 80-0274.

8. Ghoniem, A. F. and Ng, K. K., "Effect of Harmonic Modulation on Rates of Entrainment in a Confined Shear Layer," 1986, AIAA Paper 86-0056.

9. Inoue, O. and Leonard, A., "Vortex Simulation of Forced/Unforced Mixing Layers," 1987, AIAA J., Vol.25, pp.1417-1418. See also AIAA Paper 87-0288.

10. Jacobs, P. A. and Pullin, D. I., "Multiple-Contour-Dynamic Simulation of Eddy Scales in the Plane Shear Layer," 1989, J. Fluid Mech., Vol.199, pp.89-124.

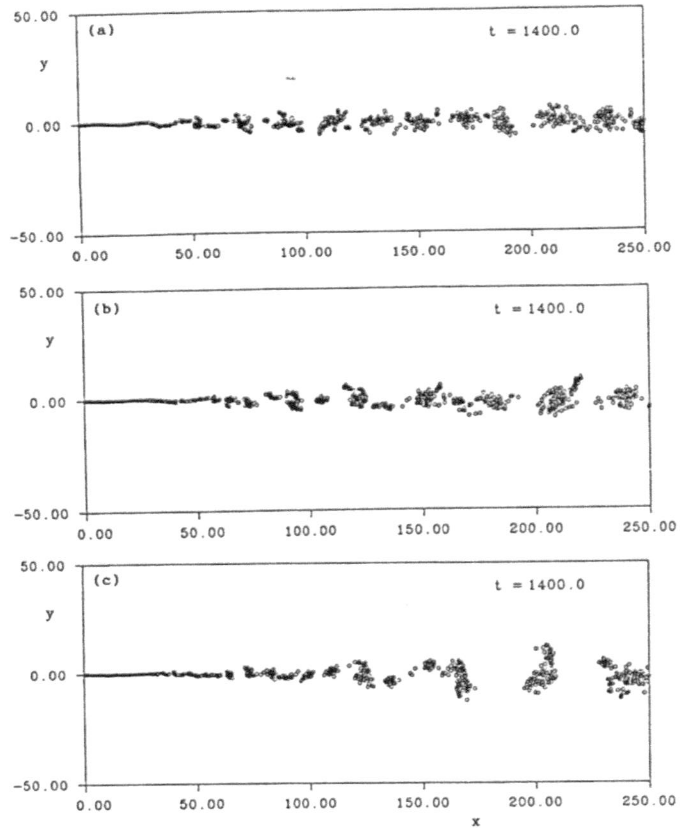

Fig.1. Mixing layer forced by a single frequency. (a) f=F/2, (b) f=F/3, (c) f=F/4.

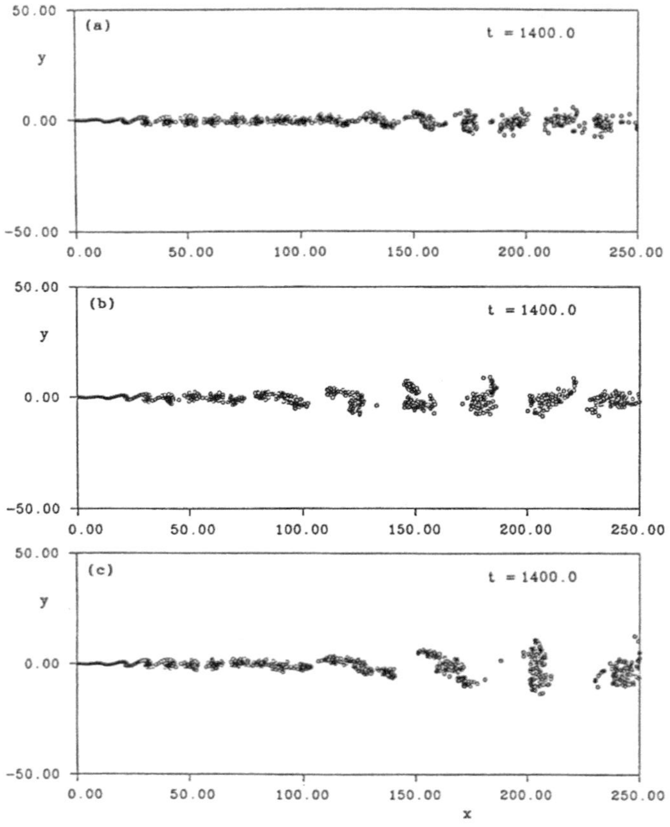

Fig.2. Mixing layer forced by two frequencies. (a) f=F+F/2, (b) f=F+F/3, (c) f=F+F/4.

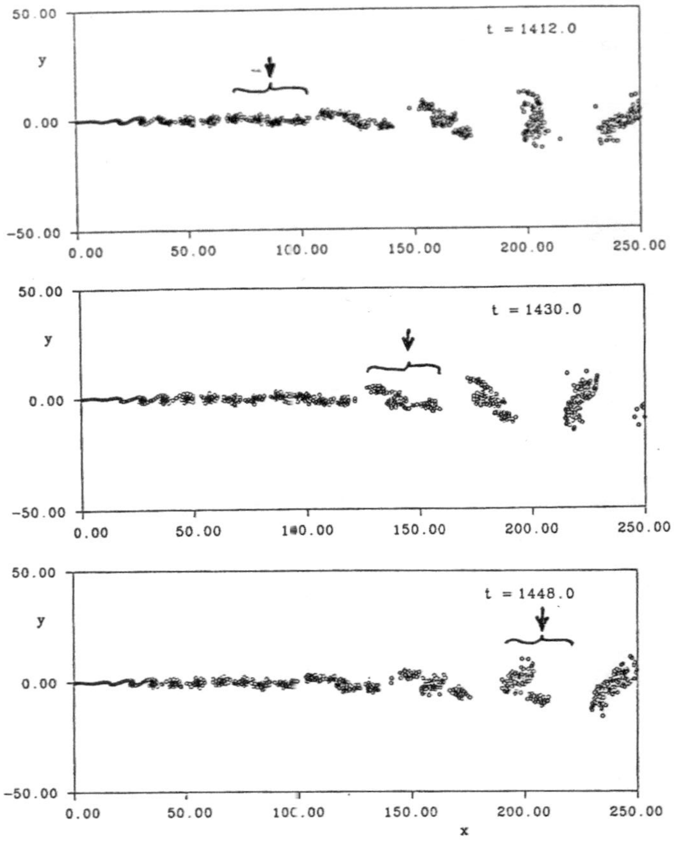

Fig.3. Multiple-vortex merging.
f=F+F/4.

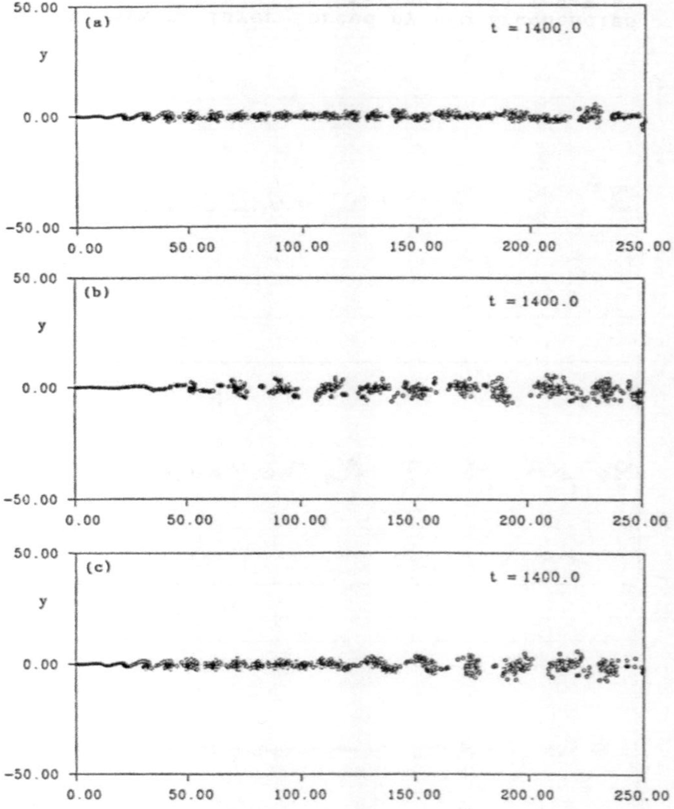

Fig.4. Effect of combining two frequencies.
(a) f=F, (b) f=F/2, (c) f=F+F/2.

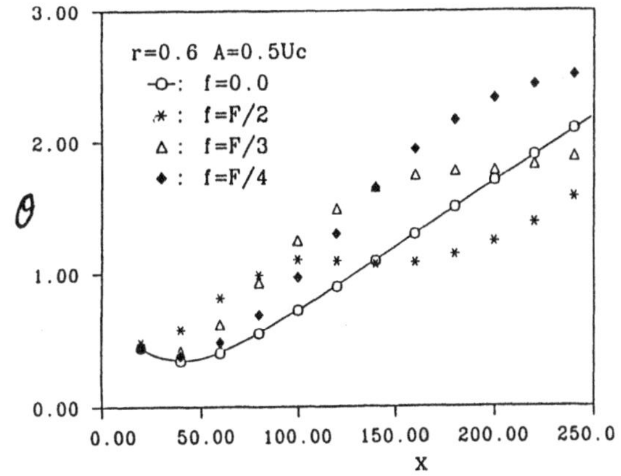

Fig.5. Growth of single-frequency forced mixing layers. (a) f=F/2, (b) f=F/3, (c) f=F/4.

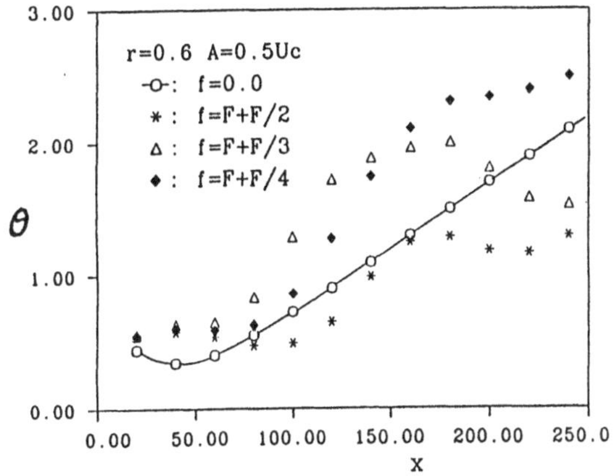

Fig.6. Growth of double-frequency forced mixing layers. (a) f=F+F/2, (b) f=F+F/3, (c) f=F+F/4.

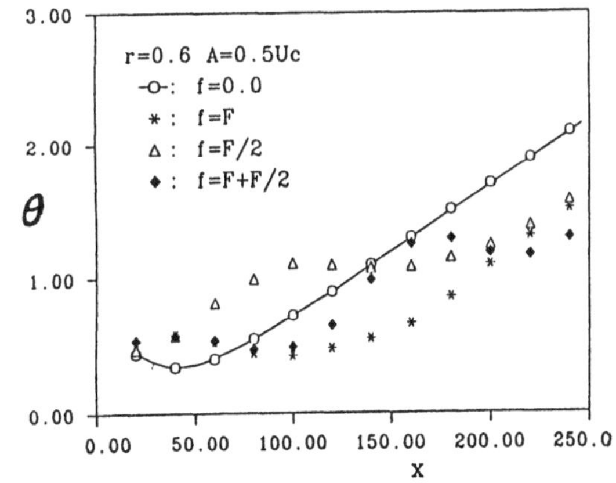

Fig.7. Effect of combining two frequencies on the distributions of momentum thickness. (a) f=F, (b) f=F/2, (c) f=F+F/2.

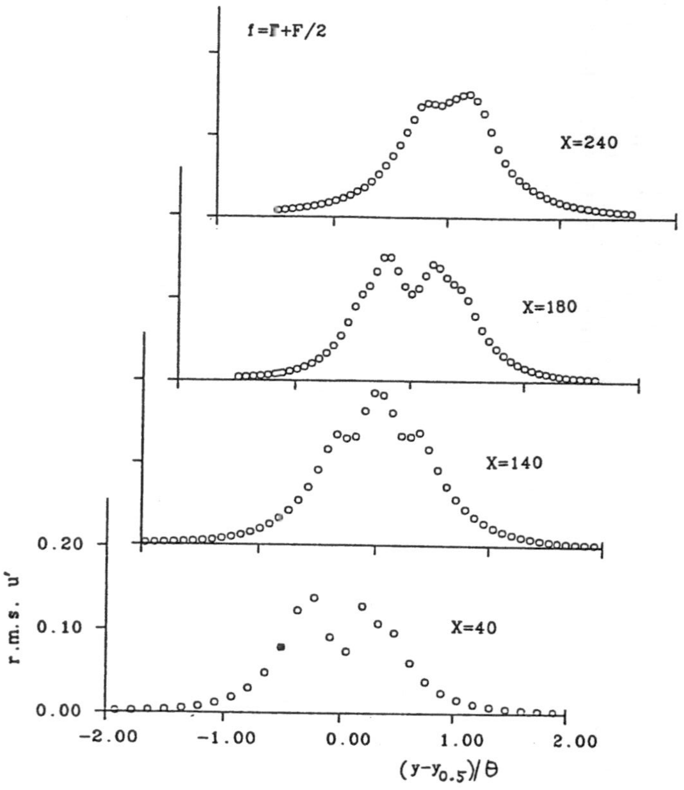

Fig.8. Profiles of r.m.s. u'.
f=F+F/2.

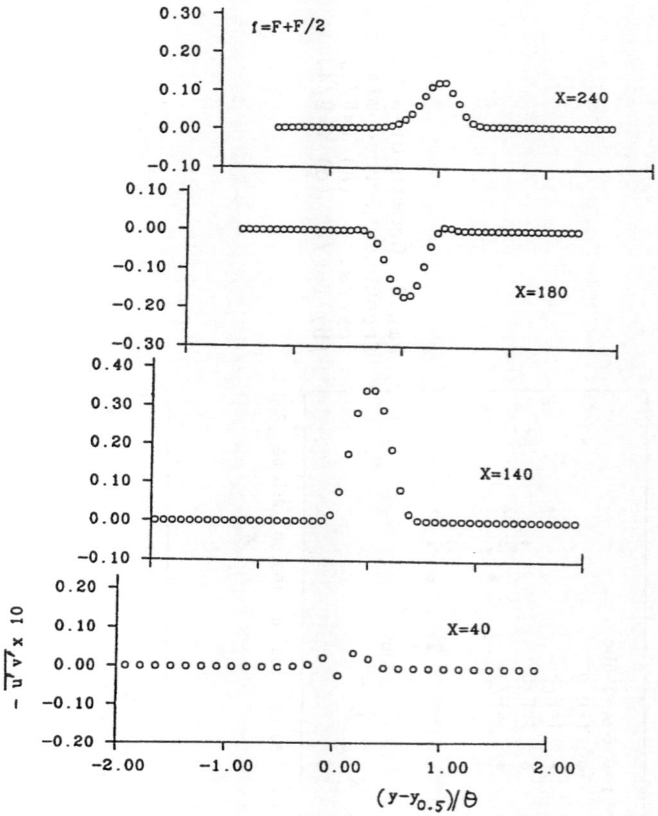

Fig.9. Profiles of $-\overline{u'v'}$.
f=F+F/2.

Part 2
Drag Reduction in Dilute Polymer Solutions & Injections of Concentrated Solutions

Dynamics of Dilute Polymer Solutions

L. Gary Leal
University of California
Department of Chemical and Nuclear Engineering
Santa Barbara, California USA 93106

I. INTRODUCTION

This paper is intended as an overview of the current status of research on the dynamics of dilute polymer solutions, aimed primarily at the "nonspecialist" who is interested in understanding the origins of drag reduction, and other macroscopically observable effects produced by very small quantities of a high molecular weight polymer in a Newtonian fluid. We do this by attempting to address two fundamental questions:

1. What does a polymer molecule do when a dilute solution is subjected to a flow? (and how does it depend on the flow?)
2. What is the effect of the polymer on the flow? (and how is it correlated with any changes of conformation that the polymer may undergo?)

Hence, what do measurements show of changes in the polymer conformation or in the flow? What can we *predict,* and what degree of sophistication is necessary in describing either the polymer, or the flow? What additional research is necessary (if any) to complete this picture?

As far as drag reduction is concerned, we take it for granted that the goal of research is not only to understand the phenomena qualitatively but also to "predict" drag reduction via numerical simulations of turbulence modification in dilute solutions. Such "predictability" would provide the potential for optimization and rational "design" of drag reducing systems for specific applications.

The single, most important, qualitative conclusion from many years of basic research is that dilute polymer solutions have unique properties that distinguish them from almost any other *viscoelastic* fluid. Thus, from a qualitative point of view, it is almost always misleading to apply ideas to dilute solutions that originate with observations or predictions of the behavior of *concentrated* solutions or melts. Especially useless are the classical rheological constitutive models (Oldroyd, Maxwell, etc.) based upon measurements of the rheological properties of *non-dilute* solutions in conventional rheometers that subject the fluid to viscometric (simple-shear) flows. These constitutive models are *only* valid as approximations of the behavior of dilute solutions in the limits of "weak" and/or "slow" flows, and thus do *not* provide a useful basis for theoretical *prediction* of dilute solution behavior for more general flows, including attempts to predict drag reduction.*

To make progress with the development of predictive theories for the motions of dilute polymer solutions, we need constitutive models that are developed expressly for dilute solutions. Furthermore, prior to any attempts to predict turbulent flow behavior, it is essential that these models be studied carefully in well-characterized laminar flows including detailed comparisons between theoretical predictions and experimental data. This latter step is the focus of current research. However, the majority of work to date is of limited value because it focuses only on the conformation changes that occur when a dilute solution undergoes a flow, without simultaneously studying the flow. In *ultradilute* solutions, the influence of the polymer on the flow may be negligible so that the velocity field is unchanged from that for the Newtonian solvent. In general, however, when the polymer conformation is modified by the flow, we should expect that the flow will also be changed significantly, even for dilute solutions (in spite of the fact that the "rheological" properties in the equilibrium state will normally be indistinguishable from the solvent). Thus, it is essential to measure and predict both the polymer conformation and the flow if we are to have a meaningful basis for comparison between experiment and theory. This is the *specific* goal of current research in our laboratory .

In the spirit of providing a basis for discussion of dilute solution behavior in the context of polymer-induced drag-reduction, the present paper is not written in the exhaustive detail of an archival review. Thus, information that is generally known by researchers in the field, and is not perceived as controversial, will often be stated as fact without extensive literature references. Of course, the perception of what is controversial is somewhat in the eye of the beholder, and the reader should be forewarned that the result may necessarily be a somewhat personal point of view. Except where indicated otherwise, we focus our attention on flexible polymers that form a random coil configuration in the rest state. This is the class of materials that is of greatest interest in the context of polymer drag reduction.

II. SOME BASIC "FACTS" ABOUT THE BEHAVIOR OF DILUTE POLYMER SOLUTIONS

It will be convenient for subsequent sections of this paper to begin with a qualitative description of the dynamical behavior of dilute polymer solutions.

In this regard, a useful starting point is the significance of the word "dilute" as it applies to polymer solutions. The basic distinguishing feature of a dilute solution is that the dynamical behavior of a single polymer molecule is solely a consequence of interactions with

* Such models are, perhaps, not useful for the description of the motions of concentrated solutions or melts either, but that is another story.

the solvent, and is *independent* of any interactions with other polymer molecules. The allowable weight percent of polymer in such a solution is not a universal value, but depends strongly on the nature of the polymer, its molecular weight, its conformational state and other factors. The conformational state at equilibrium depends largely on the structure of the molecule, and the properties of the solvent. A discussion of these factors can be found in any textbook on the physical chemistry of macromolecules.[1,2,3]

In the presence of flow, the deformation of the polymer can be a dominant factor, and the polymer can be transformed all the way from a random coil to a highly stretched configuration. Clearly, the magnitude of polymer-polymer interactions will be changed by such changes in conformation. Nevertheless, traditional characterizations of "dilute" have been based upon the solution in its *equilibrium* state. Experimentally, the transition from dilute to non-dilute is identified as the point where some measurable macroscopic property no longer depends linearly on polymer concentration. Theoretically, the usual criteria is based upon an *effective volume fraction,* defined in terms of the volume occupied by spheres which circumscribe the polymers.[4] The concentration corresponding to an effective volume fraction of 0(1) is generally designated as C*, and the theoretical condition for a dilute solution is

$$C \ll C^*$$

Generally speaking, for polymer chains with MW ~ $0(10^6)$ that exist in a random-coil configuration at equilibrium, we may think of "dilute" as corresponding to a concentration of polymer that is less than about 100 ppm by weight. However, this number will decrease sharply with increase of MW, or with deviations from the random-coil configuration due to changes in polymer structure, or solvent conditions. We may note that typical drag-reducing solutions have been mostly in the range from 10-100 ppm.

If we limit our considerations to polymer solutions that are dilute, in the sense described above, then an important distinguishing feature is that the contribution of the polymer to bulk rheological properties is negligible as long as the polymer remains in a near-equilibrium state; *too small* to be measured effectively with any of the classical rheometric techniques. Specifically, in shear or "viscometric" flows, the polymer remains near the equilibrium coil configuration even at relatively large shear rates, and it can be extremely difficult to distinguish a truly *dilute* solution from its solvent. Conversely, if the addition of polymer changes the shear viscosity by more than ~1-2%, this should be recognized as a sign that the solution is *not* dilute. Any implied behavior for dilute solutions based upon measurements in such a non-dilute solution must be viewed with suspicion.

The fact that the polymer contribution to bulk rheological properties is so small in shear flow is because the effective "volume fraction" of polymer in a truly dilute solution is also extremely small when the polymer is in an equilibrium state. Indeed, theories of dilute solutions (as well as suspension, emulsions, etc.) show that the relative magnitude of the polymer contribution to stress is approximately proportional to the *effective volume concentration* based upon circumscribed spheres as described above.[5] The diameter ℓ of these spheres is proportional to the maximum linear dimension of the polymer configuration. Hence, though the polymer contribution to bulk stress is negligible for a dilute solution in an equilibrium configuration (where ℓ is proportional to 2 times the radius of gyration of the polymer coil), an increase of the maximum linear dimension of only $0(10^2)$ – say, via stretching of the polymer – can produce an $0(10^6)$ change in the relative contribution of the polymer to the bulk stress, and it is quite possible that this *enhanced* contribution will be significant even for a dilute solution!

Hence, for dilute solutions at ppm concentrations, the coiled polymer molecule must undergo a major transition away from the equilibrium state to a strongly stretched configuration if there is to be a direct measurable effect of the polymer on the flow. As we shall see, such changes in conformation are sensitive to the details of the flow-field, and tend, in any case, to occur rather locally in laminar flows. Hence, one distinguishing feature of dilute solutions is that the polymer is very "selective" in producing large changes for one type of flow (or in one region of a flow field) but almost no change for other flow fields.

One useful fact is that the maximum extensibility of a polymer relative to its equilibrium radius of gyration is $0(N^{1/2})$, where N is the number of statistical subunits in the chain, each made up of a *minimum* of 15-20 monomer units for a typical "flexible" linear polymer. This is because the radius of gyration is $0(N^{1/2}a)$, and the total contour length of the polymer molecule is 0(Na), where a is the dimension of a statistical subunit. Thus, the maximum extensibility increases with molecular weight, and, for a given weight percent of polymer, the likelihood of a strong flow-effect from the polymer in a stretched state is also increased with molecular weight. This is because the number of chains per unit volume decreases as $\frac{1}{N}$ if molecular weight is increased (but polymer concentration is held constant), while the change in effective "volume" of a single stretched chain goes up as $N^{3/2}$. Hence, all else being equal, the *maximum effective* volume fraction increases as $N^{1/2}$.

A few additional remarks may be useful as a supplement to the basic message of the preceding several paragraphs. First, as is well known, not all polymers exist in a random coil configuration in the rest-state. The main difference between "rod-like" molecules, or polyelectrolytes at low counter-ion concentrations, and the classical random coil is that the molecules already exist in an extended and/or expanded configuration. However, even a relatively weak extension or expansion at equilibrium can make it *much* easier for the flow to produce a stretched state[6,7] – there is less stretching required, and the expanded conformation enhances frictional interactions between the polymer and solvent. Once fully extended, the influence of coils and rigid rods on extensional flows will be *similar* provided the effective volume concentrations are equal, but rigid rods will also produce an effect *of comparable magnitude* on the solution properties in shear flow, and their influence on a turbulent boundary-layer flow may thus be different (certainly, the "onset" behavior for drag reduction is known to be different).[8] We may also note that experimental data (to be discussed later) on steady, planar extensional flows suggests that the "effective" volume fraction for a linear coiled polymer in the extended state must be *extremely* large to produce a measurable influence of the polymer on a flow[9] (i.e. $\sim 0(10^3)$ instead of $0(1)$ as suggested by simple considerations). We shall return to discuss the significance of this observation later.

One important, but parenthetical comment on the relevance of the preceding discussion to drag reduction is that we assume (like most other researchers) that drag reduction actually results from a *strong direct* influence of the polymer on some feature of the turbulent flow. Of course, it is always possible that extremely "weak" local changes in the flow might produce strong global effects – via modified instabilities for example – but we see no persuasive evidence to date to support this suggestion (however, see Reference 10 for an alternative point of view).

III. THE EFFECTS OF FLOW TYPE

The preceding discussion has suggested that the critical requirement for a dilute solution of the random-coil type to produce a strong influence on a flow is that the polymer exists in a highly stretched and oriented state. Thus, it becomes essential to understand when this will occur.

The most important factor is that flows of various types are not all "equal" in their potential for stretching polymer molecules. If we envision the interaction between a polymer molecule and solvent as being similar to a deformable "particle" in a viscous fluid, it will not come as a surprise that the polymer motion and shape will depend strongly on the details of the flow. However, the degree of deformation required of the polymer molecule is much greater than one would normally expect of a deformable particle, and this imposes a greater dependence on the details of the flow. Two simple physical ideas provide the basic principles to understand most of the existing experimental and theoretical observations. First, a polymer molecule cannot stretch faster, or by a larger degree, than a material line element of the fluid. Hence, if the polymer is to be stretched by a factor of $0(10^2)$, this can only occur if material points in the solvent would separate by at least $0(10^2)$ in the same flow. Another way to express the same idea is that the polymer must follow a path through the flow on which the "total strain" experienced by a fluid element is at least $0(10^2)$.

However, this condition on the "total strain" is only a *necessary* condition for the polymer to actually stretch. The second physical idea is that the polymer has a tendency, via random thermal fluctuations of shape ('conformational diffusion'), to return to an equilibrium ("unstretched") configuration in a finite time period. Hence, the *rate* of separation of material points in the solvent must be *fast* enough to overcome this tendency to return to equilibrium, or the polymer will remain in a near-equilibrium configuration regardless of the total strain experienced by a fluid element on the same trajectory.

Together, the two conditions of the preceding paragraph provide a basis for developing a more quantitative description of the *potential* of different *homogeneous* flows for inducing a high degree of stretch in a polymer molecule. This so-called "flow classification" scheme was published several years ago, but the result is sufficiently important that we repeat the basic principles here.[11,12,13]

In particular, let us consider an infinitesimal line element in the solvent that is specified by a vector **R**. Then the rate of change of length and orientation of **R** in a steady, homogeneous (linear) flow is

$$\dot{\mathbf{R}} = G\,\Gamma \cdot \mathbf{R} \tag{1}$$

where $G\,\Gamma$ is the local velocity gradient tensor, $\nabla\mathbf{u}$. The restriction to homogeneous "linear" flow means that

$$\mathbf{u} \equiv G\,\Gamma \cdot \mathbf{x} \tag{2}$$

with $G\,\Gamma$ independent of **x**. Hence, in a steady flow, the velocity gradient experienced by **R** is constant, independent of time. Hence, in this case,

$$R(t) \approx e^{GS^{+}t} \tag{3}$$

where S^{+} is the largest real part of the eigenvalues of Γ. The characteristic equation, which can be used to determine S^{+}, is

$$S^3 - \left(\tfrac{1}{2}\operatorname{tr}\Gamma^2\right) S - \det\Gamma = 0 \tag{4}$$

Hence, all possible linear, homogeneous flows are characterized by the two scalars $\det\Gamma$ and $\operatorname{tr}\Gamma^2$. This allows a pictorial representation of a "flow classification" scheme.

First, make a diagram in the $\det\Gamma$ and $\operatorname{tr}\Gamma^2$ plane that depicts all possible linear flows. As depicted in Figure 1, this turns out to be a finite region, with a shape like a medieval "shield."

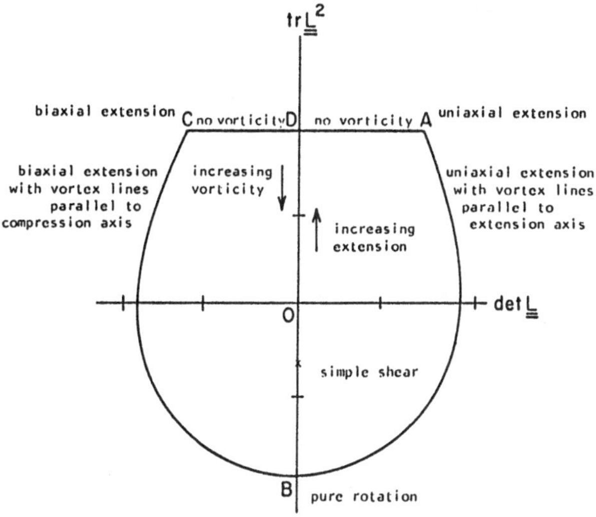

Fig. 1. Schematic diagram identifying particular linear flows with points of the accessible domain (reprinted from Ref. 11).

The complete class of linear, *pure straining* flows turns out to occupy the boundary CDA. All two-dimensional linear flows fall along BOD, with the "hyperbolic" pure straining flow at D, pure rotational flow at B and simple shear flow at O. The outer edges of the shield (i.e. CB or AB) correspond to flows in which the straining part of the flow is either uniaxial (AB) or biaxial (CB) in form, with vorticity of increasing magnitude from top to bottom that is aligned with the principle axis of strain. The common point B, corresponds to a purely rotational, linear flow. Now, for each possible flow (i.e. each point in the (det Γ, tr Γ^2) shield), there is a specific value of S^+ as depicted in Figure 2.

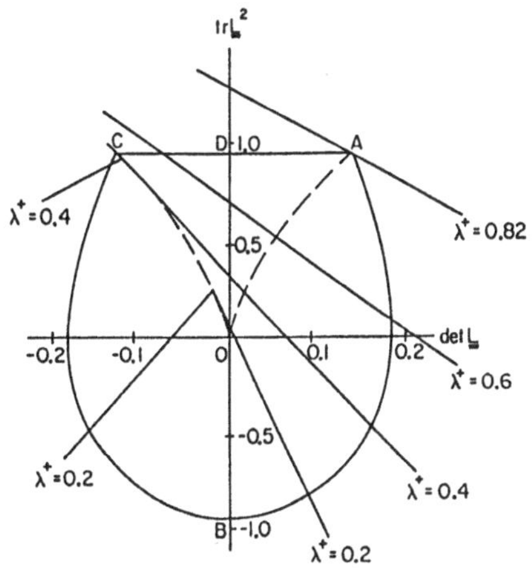

Fig. 2. Representative eigenvalue contours (reprinted from Ref. 11)

The largest value, $S^+ = 0.82$, occurs at the single point A, which corresponds to the pure uniaxial straining flow. However, S^+ is *positive* for all possible linear flow, *except two-dimensional flows on the axis between O and B where the magnitude of the vorticity exceeds the strain-rate.*

We conclude that the *rate* of stretching of a *line-element* in most linear flows is *exponential,* with a characteristic magnitude that depends on G and S^+ (i.e. on the relative strength and orientation of the straining part of the flow and the vorticity). Such flows are commonly known as *strong* flows. For a given G, as we *add vorticity,* the rate of stretching decreases. *Exceptional flows,* with $S^+ = 0$, are two-dimensional with a rate-of-strain that is equal to, or less than the vorticity (*note* – this includes simple shear flow). Such flows are commonly called *weak. All* strong flows have the *potential* to induce an exponential rate-of-stretching in a polymer molecule (or any other deformable body). Whether, the polymer will actually stretch depends on the magnitude of the velocity gradient, G, *relative* to the relaxation timescale for the polymer to return to an equilibrium state. Thus, if "stretching" is a key to the behavior of dilute polymer solutions, we see that the behavior in simple shear flow (the favorite of the rheologist) or any 2D flow in which the magnitude of the vorticity exceeds the strain-rate is neither *important* nor *interesting*. Conversely, "shear-flow" based mechanisms of dilute solution behavior are not likely to be relevant to any macroscopic phenomena, including drag reduction, that is observed for dilute polymer solutions!**

Although the flow classification scheme described above was derived based upon the assumption of a steady and homogeneous (linear) flow, it has frequently been applied to unsteady or heterogeneous flows. In these cases, it is assumed (at least implicitly) that the flow varies slowly enough with respect to either change of position or time, that the velocity gradient tensor Γ can be replaced by a series of "quasi-steady" values each corresponding to the *local,* instantaneous velocity gradient associated with the current position of the material element of interest.

A critical question that has not been addressed is whether the behavior "predicted" by a sequence of "quasi-steady" flows is qualitatively or even quantitatively similar to the actual behavior for an unsteady or heterogeneous flow? From a mathematical point of view, it would not be surprising if there were important differences. The fact is that the change to a velocity gradient that is unsteady from the point of view of a material element transforms the problem of describing the dynamics of a line element from the *autonomous* equation (1) to a *non-autonomous* equation, in which $\Gamma = \Gamma(t)$. The behavior of the solution of a non-

** Assuming, as always, that drag reduction results from a direct O(1) interaction of the polymer and flow, rather than being triggered by some weak and subtle influence of the polymer.

autonomous equation can be much different from the corresponding autonomous equation, even when Γ (t) is "slowly varying." In addition, depending upon the lengthscale of changes in the flow, it may be necessary to generalize the dynamical equation (1) for $\dot{\mathbf{R}}$ to include quadratic or higher order terms.

We are currently exploring the general problem of "flow classification" for flows that are unsteady from the Lagrangian point of view of a material element of fluid.[14] Our indications are that the dynamical behavior of a line element can be qualitatively changed; in particular, we have found examples of flows that yield positive values of S^+ at each point based upon a local homogeneous flow approximation but are *not* capable of inducing a large degree of stretch and vice versa. Clearly, an understanding of conditions in heterogeneous and unsteady flows that produce a high degree of stretch of a line element of the fluid is critical if we are to understand the dynamics of dilute polymer solutions. It is especially noteworthy for drag reduction research that the "strong flow" criteria for exponential stretch of a line element in a steady, homogeneous flow can give unrealistic results if applied directly to prediction of polymer behavior in turbulent flows.

IV. Models for Dilute Solutions in Strong Flows

We have already noted that the interaction between polymer at very dilute concentrations, and flow, is not typical of the behavior of other viscoelastic liquids. One critical difference is that the dilute solution is much more sensitive to the type of flow experienced, in the sense that some flows (where the polymer is only weakly perturbed from equilibrium) may not be affected by the polymer at all, while other flows that produce a coil-stretch transition may be very strongly influenced by the same dilute solution. If we are to model the rheological behavior of a dilute polymer solution, it is evident that we need "constitutive equations" that are obtained specifically for dilute solutions. In general, these start out as molecular models, with bulk-flow properties derived by an averaging process.

The basic premise underlying all models of dilute solutions is that the mean interaction between the polymer molecule and the solvent is analogous to a deformable particle, with bulk properties derived as though the solution were a dilute suspension of these particles. Of course, even if the real polymer dynamics can be captured by the motions of a particle, the actual geometry is *highly* complex, ranging from something resembling a porous near-sphere at equilibrium, to a stretched, wiggly string (possibly with side branches and other complications) under strong flow conditions. Further, even if the solvent can be treated as a continuum for purposes of describing the polymer motion, it is far from certain what boundary conditions should be applied (quite possibly, these should be *slip* conditions) or precisely where they should be applied. The fact is that any attempt at a "faithful" replica of the polymer molecule as a hydrodynamic particle is too complicated, and in any case, the resulting model is probably mathematically intractable when applied to the prediction of nontrivial macroscopic flows.

Instead, a more reasonable and useful objective is to seek a highly simplified model that can capture the complete coil-to-stretch transition and its dependence on the details of the flow, and provide a decent estimate of the polymer contribution to the bulk stress. In my view, the model that holds the greatest promise of satisfying these criteria while still remaining simple enough for fluid mechanics calculations is the so-called elastic dumbbell.[5,15,16] The basic elements of this model are: an *elastic spring* which models the deterministic tendency of random statistical fluctuations in conformation (i.e. conformational diffusion) to drive the polymer toward an equilibrium configuration; a deterministic interaction of the polymer molecule with the solvent that is modelled as a pair of *hydrodynamic point forces* at the ends of the spring (often pictured as "beads" but this is really an unnecessary crutch that often has the unfortunate consequence of being interpreted in some *literal* sense as an attempt to describe the actual polymer "shape"); and, finally, *diffusion,* that accounts for Brownian fluctuations in conformation, not incorporated in the spring, that provide a mechanism to randomly sample all available configurations. One intrinsic limitation in the model, as a consequence of the assumption of an elastic spring, is that the changes in conformation induced by the flow must be slow compared to the timescale for the polymer to randomly sample a large number of configurations. (We shall return to this point later.)

The polymer conformation is described in a statistical sense via the probability density function, $\varphi(\mathbf{r},t)$ in which $\mathbf{r}$ is the end-to-end vector of the dumbbell. This function satisfies the conservation equation

$$\frac{\partial \varphi}{\partial t} + \nabla \cdot (\dot{\mathbf{r}}'\varphi) = 0 \qquad (5)$$

in which the conformation "dynamics" is given by an equation for $\dot{\mathbf{r}}$ that includes the effect of the spring, friction and conformational diffusion as outlined above. The polymer contribution to the bulk stress for the solution is then

$$\tau_p = n \langle \mathbf{r}' \, \mathbf{r}' \, \xi(\mathbf{r}') \rangle \qquad (6)$$

where n is the number density, $\xi(\mathbf{r}')$ is the function that represents the magnitude of the spring coefficient, and $\langle \, \rangle$ signifies an average over possible configurations with $\varphi(\mathbf{r},t)$ as a weighting factor.

The simplest version of the model is the *linear* dumbbell. Here, the "bead" friction coefficient is assumed to be a constant independent of conformation, and the spring is Hookean. The result is an equation for the time-dependent change in the end-to-end vector

$$0 = \underset{\text{(friction)}}{\zeta_o(\dot{\mathbf{r}}' - \Gamma \cdot \mathbf{r}')} + \underset{\text{(spring)}}{\xi_o \mathbf{r}'} + \underset{\text{(diffusion)}}{kT\nabla \ln \varphi} \tag{7}$$

If we nondimensionalize with

$$t_c = \frac{\zeta_o}{\xi_o} \equiv \theta; \quad l_c = Na$$

where N is the number of statistical subunits in the chain and a is the contour length of a subunit, then

$$\dot{\mathbf{r}} = \alpha \, \Gamma \cdot \mathbf{r} - \tfrac{1}{2}\mathbf{r} \tag{8}$$

where α is the dimensionless velocity gradient $G\theta$. In writing (8), we have neglected the *diffusion* term in (7) which tends to be significant only near equilibrium and in "weak" flows where $S^+ = 0$. It should be noted that the model incorporates a *linear* approximation to the flow, based upon the assumption that $|\mathbf{r}'| << L$, where L is the lengthscale for changes in the bulk flow, but does *not* presume that Γ is necessarily constant. The characteristic time-scale $t_c = \theta$ is just the Rouse relaxation time, and is proportional to $(MW)^{3/2}$ as well as η_s.

There are two obvious and well-known flaws with the linear-dumbbell model, both of which result from the fact that the constant friction and Hookean spring assumptions are only valid for small departures from the equilibrium configuration.[17] First, the model exhibits a *singularity* when

$$\alpha S^+ = \tfrac{1}{2} \tag{9}$$

in the sense that $|\mathbf{r}| \to \infty$ (and thus $\tau \to \infty$) as $t \to \infty$. Second, there is only a single relaxation time characterizing the polymer in this model, which is a constant independent of the polymer configuration. This is at odds with experimental observation which generally suggests a spectrum of relaxation times, with a largest value that *increases* with polymer deformation from equilibrium.

These shortcomings of the linear dumbbell model have led to the development of so-called *nonlinear* dumbbell models.[5,16] All such models contain a nonlinear spring-law, in which the spring coefficient increases from its equilibrium value to a very large value as the stretched polymer approaches its full contour length. To date, it has generally been assumed that the details of this spring-law are unimportant, as long as the singularity in $|\mathbf{r}|$ is removed, and thus most investigators have used the nonlinear spring-law of Warner,

$$\xi = \xi_o \left(\frac{1}{1-(r'/Na)^2} \right) \tag{10}$$

which is one of the simplest forms that can avoid the singularity.[5,18,19] Note, however, that I personally do not agree with this point of view. The *form* of the spring-law (i.e. the functional dependence of the spring constant on | **r** |) determines the relationship between conformation (i.e. | **r** |) and the bulk stress. Furthermore, in combination with the friction coefficient, it determines the dependence of the relaxation time on | **r** |. The correct form for the relaxation time will be critical in simulating the behavior of polymer molecules in flows that are strongly time-dependent from the Lagrangian point of view, especially for flows where the polymer is subjected to an alternating sequence of strong and weak conditions. This class of problems is presumably relevant to polymer/model behavior in turbulent flows.

The second modification often introduced into the linear dumbbell model is a hydrodynamic friction law that depends upon conformation. From a polymer dynamics viewpoint, one motivation is simply that the measured (or inferred) relaxation times *increase* with *increasing* deformation from the equilibrium state, whereas the model with constant friction, but non-linear spring, has a relaxation time that *decreases* with increase of | **r** |. From a hydrodynamics point of view, the objective of including conformation dependent friction may be viewed as an attempt to make the mean interaction with the solvent reflect flow-induced changes of the "shape" of the polymer molecule with changes in | **r** |. Two basic approaches have been adopted: in one, the dumbbell is implicitly viewed as representing the actual shape of the polymer, and conformation-dependent friction is then naturally approached as a consequence of a varying degree of bead-bead hydrodynamic interaction as the dumbbell stretches;[15] in the second, the "dumbbell" is viewed *only* as the simplest framework for describing the dynamics of an orientable, stretchable body in a flow, and conformation-dependent friction is directly introduced by making the bead coefficients anisotropic and dependent on | **r** |.[5,16] In principle, the two approaches have much in common. Bead-bead interactions are nothing more than an attempt to model the polymer via a deformable body that adopts a specific sequence of shapes. Thus, the complete dynamics of a *particular* orientable and stretchable body will automatically be captured for this sequence of shapes if the two sphere problem is solved exactly for the full range of configurations from two overlapping spheres in the rest state to two widely separated spheres in the highly stretched state. In its usual implementation as a far-field perturbation from the linear dumbbell limit of two *non-interacting* point forces, however, bead-bead interactions can only account for a slight degree of shape anisotropy. Although this yields somewhat more realistic rheological properties (for example, a slight shear thinning) than predicted by the linear dumbbell, it *cannot* account for the changes of polymer behavior (or bulk properties) that derive from large changes in conformation, such as the coil-stretch transition. Furthermore, the *exact solution* of the two-sphere hydrodynamics problem is highly complex, and it would be difficult to incorporate into a model that is ultimately to be used for bulk flow calculations.

The simpler alternative is an "ad hoc" assignment of a conformation-dependent, anisotropic friction law to the beads, that is designed to yield the correct limiting behavior for isotropic coils and for stretched threads, and to interpolate smoothly in between.[5,16] In this way, the hydrodynamic behavior of the dumbbell can be made to mimic the behavior of a sequence of axisymmetric bodies of increasingly enlongated shape, but without the complexity of obtaining detailed fluid mechanics solutions at the polymer scale. The most important change that occurs for stretched bodies in *strong* flows at low Reynolds number, is that the *magnitude* of the frictional interaction between the body and the flow increases approximately linearly with the longest linear dimension of the body. In addition, of course, the "shape" changes, and the degree of hydrodynamic anisotropy is also modified, but the main axis of the polymer body will tend to be aligned with the *eigenvector* of the velocity gradient tensor in a strong flow, and this change in *anisotropy* will be relatively *unimportant* except in weak flows, or very near to simple shear flow. Thus, for a model that is intended to capture only strong polymer-flow interactions, it is sufficient to assign a conformation-dependent, but *isotropic* friction coefficient to each bead that increases *linearly* with increase of the end-to-end length of the dumbbell. In "weak" flows, such as simple shear, the *rotation* of a body depends critically on its *shape* and it is necessary to adopt an *anisotropic* version of the model if we wish to give a reasonable dynamical description for these flows as well. My personal viewpoint is that the *only* important effects of polymer-on-flow (and vice versa) in a *dilute* solution occur in strong flows where the polymer can undergo a major transition in conformation. For fluid mechanics studies involving dilute solutions, it is nearly irrelevant whether the weak/shear flow behavior is correct! And the use of the *isotropic* friction law greatly simplifies the model, as we shall see.

The full *anisotropic* friction model for an axisymmetric body in a linear flow takes the general form

$$\mathbf{F_f} = \underset{\substack{\uparrow \\ \text{Anisotropic friction} \\ \text{coefficient tensor,} \\ \text{with components} \\ \text{that depend on the} \\ \text{degree of stretch, r'}}}{\zeta}(\mathrm{r'}) \cdot [(\dot{\mathbf{r}}') - (\Omega + \underset{\substack{\uparrow \\ \text{"inefficiency" of rotation} \\ \text{in strain}}}{\alpha(\mathrm{r'})}\mathrm{E}) \cdot \mathbf{r}'] \tag{11}$$

Here, the net frictional force is $\mathbf{F}_f$, and Ω and E are the vorticity and strain-rate tensors, respectively. The necessity for the "inefficiency" factor $\alpha(r')$ is a reflection of the sensitivity of *rotation* in straining flow to shape – note, for example, that a line element rotates with E, but a sphere does not rotate at all in a pure straining flow. For the *linear* dumbbell model

$$\zeta(r') \equiv \zeta_o \mathbf{I} \quad \text{and} \quad \alpha(r') \equiv 1. \tag{12}$$

The most common form of $\zeta(r')$ for application to strong flows is

$$\zeta(r') = \zeta_o \left(\frac{r'}{a\sqrt{N}}\right) \mathbf{I} \tag{13}$$

where ζ_o is the equilibrium friction factor, and the coefficient $r'/a\sqrt{N}$ varies between 1 and $\sqrt{N}$ as r' increases from the equilibrium value $a\sqrt{N}$ for a random coil, to the contour length Na for a stretched chain.

The nonlinear dumbbell model *cannot* exhibit the singularities associated earlier with the linear model. Further, the relaxation timescale for the model is highly dependent on conformation

$$t_c = \frac{\| \zeta(r') \|}{\xi(r')} . \tag{14}$$

In particular, the relaxation time first increases with r' due to the increased friction and this yields extremely slow relaxation from extended states (the ratio of equilibrium to maximum relaxation times is $O(N^{1/2})$). This leads to the possibility, first noted many years ago by Hinch,[5] that an extended polymer can be *maintained* in an extended state by a very *weak* flow. It should be noted, however, that any model with a spring-law that blows up at full extension must eventually suffer a turnaround in the relaxation time t_c as the chain approaches its full contour length. No evidence exists to refute this *predicted* behavior, though substantial evidence *does* exist to support the prediction of a very slow relaxation process from an extended state.

One "cost" of the generalization to a nonlinear dumbbell model is that it is no longer possible to reduce the coupled system (5), (6), (11) and (13) to a standard continuum mechanical constitutive equation.[15] To solve a fluid mechanics problem, one can only use the *coupled* equations, including the equations of motion expressed in terms of the second moment of the end-to-end vector $<\mathbf{r}\,\mathbf{r}>$, and an evolution equation for $<\mathbf{r}\,\mathbf{r}>$ derived from (5), and the generalized form of (7) which incorporates the nonlinear spring and nonlinear friction laws.[20]

V. Model Experiments for Dilute Solutions – Does the Nonlinear Dumbbell Correctly Predict Polymer Conformation?

The obvious question is whether the nonlinear dumbbell model gives correct predictions. To answer this question, we must evaluate the model behavior in strong flows, for which the model is designed. Reasonable behavior in weak/shear flows is a bonus if it occurs, but we should not be too concerned if it does not since the effect of the polymer on the flow and vice versa will be weak in these cases for a *dilute* solution. Thus, we must study solution behavior in extensional (or 3D) flows where $S^+ > 0$.

Of course, the problem is complicated by the coupling between polymer conformation and resulting changes in the flow. As a consequence, both the polymer conformation and the flow must be treated as unknowns. The only exception to this is the case of *ultradilute* solutions. By ultradilute, we mean solutions where changes in polymer conformation, including the full coil-stretch transition, have a negligible effect on the flow. The precise concentration threshold for *ultradilute* behavior depends upon molecular weight – the higher the molecular weight, the lower the allowable polymer concentration. Current experimental estimates of critical conditions for an effect of the polymer on the flow (i.e. conditions when the ultradilute condition is not valid) will be discussed shortly.

In assessing the current state of model evaluation, it is important to recognize that most previous studies have focused entirely upon comparisons between measurements of polymer conformation and model predictions in which the flow is taken to be that of a Newtonian fluid![21,22] With the exception of ultradilute solutions such comparisons are of *qualitative* value, at most.

A. Predictions Confirmed for Dilute Solutions in Steady Flows

In spite of the reservations and difficulties expressed above, there are a number of predictions from the nonlinear dumbbell model that have been confirmed by experimental studies of polymer conformation in two-dimensional flows. In this section, we list four such predictions beginning with those that pertain to the existence of a coil-stretch transition and other features of polymer conformation in *steady* flows.

a. Experimental data confirm the model prediction of an abrupt onset of the transition between a coiled state and an extended and aligned state for solutions of "linear" polymers like polystyrene, polyethylene oxide, etc. Furthermore, the observed critical velocity gradient is found to correlate with the polymer relaxation timescales, and with the flow-type in a manner that is consistent with model predictions (for additional detail, see below).

b. Significant changes in polymer conformation for steady flows are extremely localized in the vicinity of stagnation points, or other regions where the total strain experienced by the polymer is large – again predicted by the models.

c. With the two independent model parameters N and t_c determined, respectively, from the asymptotic birefringence level at high velocity gradients, Δn_∞, and intrinsic viscosity or dynamic light scattering measurements, the nonlinear dumbbell model with a Warner spring and conformation-dependent but isotropic friction gives a quantitative fit to birefringence data for a wide range of velocity gradients and strong flow types, provided account is taken of the actual *molecular weight distribution* (MWD) and the *distribution of total strain* values experienced by the polymer at different points within the birefringent zone.

These conclusions are illustrated below with specific data from a previously published paper. Prior to examining the data, however, there are several subtleties associated with the connection between experimental conditions and the model system that need to be briefly discussed. First, according to either the linear or nonlinear dumbbell models, the transition from a coiled state to a highly stretched configuration should not only occur at a critical, nonzero value of the velocity gradient as stated above, but the transition itself should be *very abrupt,* essentially occurring as a step change from the near-equilibrium state to a fully stretched configuration at a single critical value of the velocity gradient.[23] In the experiment, however, the transition is smooth (i.e. it occurs over a *range* of G), and this has sometimes been used to suggest that the most appropriate version of the model is one with nonlinear spring but *constant* friction since this also produces a smooth transition. However, there is other evidence to suggest that the concept of conformation-dependent friction is correct, as we shall see below, and we believe that the smooth transition in the experimental data is explained by two other factors: First, the molecular weight of the polymer is distributed over a range of values, even for so-called monodisperse samples where the "polydispersity index" is low, $M_w/M_n < 1.1$. Thus, though a collection of identical molecules would undergo a "coil-to-stretch" transition at a single value of G if they were subjected to the same flow, the real polymer sample onsets at a *spectrum* of G values, with the highest molecular weight fraction going first, and the lowest going last. The second factor is that the transition in conformation is usually observed experimentally via the onset of birefringence for an illuminated region of finite lateral extent that is centered at the stagnation point of the flows that we study (and passes through the complete flow field in the third direction including the regions near the top and bottom boundaries). Further, in all laboratory realizations of two-dimensional strong flows, the strong-flow zone occupies only a finite region near the stagnation point.[24] Thus, polymer molecules at different locations within the birefringent region will have been in the stretching part of the flow for different periods, and therefore they would be stretched by different amounts even in an ideal monodisperse system of identical molecules. To understand this, we must recall that it is the *steady-state* ($t \rightarrow \infty$) configuration that is predicted by the model to undergo an abrupt transition at a critical value of G. In the real system, the polymer is subjected to the stretching region of the flow device for a *finite* period, and for most trajectories through the flow system this period is not long enough to achieve a final steady-state configuration. Alternatively, we can note that the total strain on most streamlines is not large enough to allow the polymer to attain a steady degree of stretch. On the stagnation streamline, the residence time in the flow is infinite, and thus, so too is the total strain; but as we move away from the stagnation point, even within the small birefringent zone, there is a sharp decrease in residence time (or total strain).

In order to illustrate the experimental evidence for the first of the three conclusions listed earlier, we present two representative sets of data from a recent publication[9] in Figures 3 and 4.

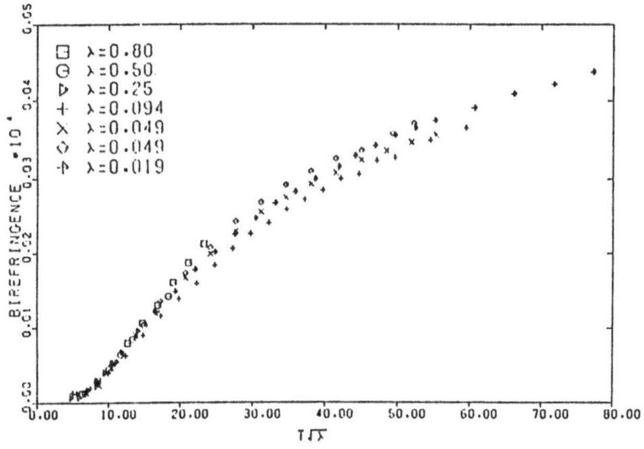

Fig. 3. Flow birefringence at the stagnation point of two- and four-roll mills for a 100 ppm solution of polystyrene (MW = 8.42x10^6, Mw/Mn = 1.17) in Chlorowax LV versus the eigenvalue of the velocity gradient tensor. (reprinted from Reference 9).

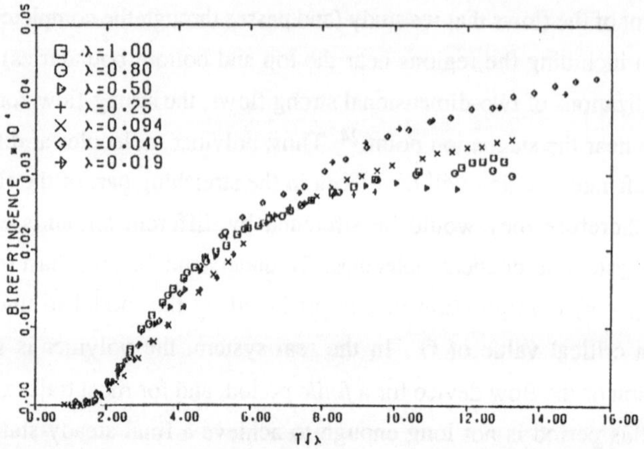

Fig. 4. **Flow birefringence at the stagnation point of two- and four-roll mills for a 100 ppm solution of polystyrene (MW = 2x10^6, Mw/Mn = 1.3) in Chlorowax LV versus the eigenvalue of the velocity gradient tensor. (reprinted from Reference 9).**

These show the measured birefringence levels for polystyrene solutions in a two- and four-roll mill as a function of the magnitude of the principle eigenvalue of the velocity gradient tensor, $G\sqrt{\lambda}$, for a range of different two-dimensional flows that can be represented approximately in the form

$$\mathbf{u} = G \begin{pmatrix} 1+\lambda & 1-\lambda & 0 \\ -1+\lambda & -(1+\lambda) & 0 \\ 0 & 0 & 0 \end{pmatrix} \cdot \mathbf{x} \ .$$

Here, λ is a "flow-type" parameter that ranges in the experiments from approximately 0.02 to 1.0. It may be noted that λ=0 corresponds to linear (simple) shear flow, whereas λ=1.0 is the two-dimensional pure straining flow that is often referred to as hyperbolic flow. The *correlation* of birefringence with $G\sqrt{\lambda}$ for all 2D strong flows is consistent with model predictions. The data in Figures 3 and 4 also illustrate the existence of a *distinct* onset point for birefringence at a *nonzero* value of $G\sqrt{\lambda}$ as predicted by the dumbbell model, with a critical value that depends upon the equilibrium relaxation time for the polymer in solution. Indeed, a compete set of data for the 8x10^6 MW polystyrene sample in a variety of different solvents is shown in Figure 5 to yield a *single* universal curve when expressed in dimensionless terms using independently measured values of the relaxation time for the various solutions.[9]

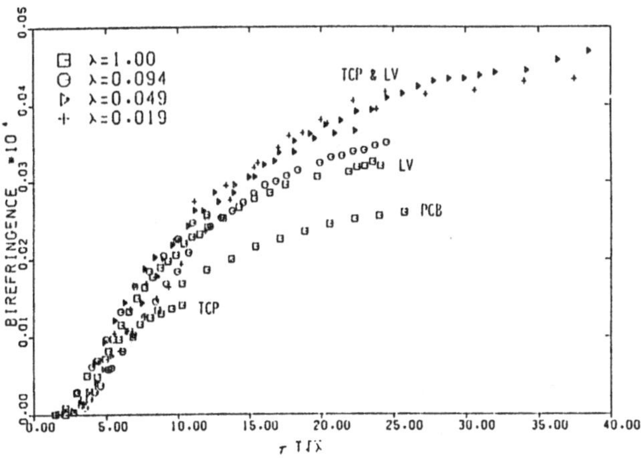

Fig. 5. Flow birefringence versus dimensionless eigenvalue of the velocity gradient tensor, nondimensionalized with the characteristic relaxation time from intrinsic viscosity and light scattering measurements (reprinted from Ref. 9) (TCP = tricresylphosphate, PCB = chlorinated biphenyl (Pyralene 4000), LV = chlorinated paraffin (Chlorowax LV) - Note, for a discussion of the scatter in the data at large velocity gradients, see Ref. 9.

Further, when a calculation of the theoretically expected dependence of birefringence on G is made, including the effects of the measured *polydispersity* of MW, and the *distribution of residence times* for polymer within the birefringent zone, the nonlinear dumbbell with conformation dependent friction agrees *quantitatively* with the measured dependence of birefringence on $G\sqrt{\lambda}$.

We have already noted that an inherent feature of the model is that the polymer will tend to be highly stretched only in the local neighborhood of the stagnation point of the flow. This can be seen experimentally either as detailed measurements of birefringence as a function of spatial position, or more qualitatively by simply taking a photograph to show the region of significant light transmission through a pair of crossed polarizers. We have obtained detailed data,[9] but the interested reader may wish to examine the *photograph* given as Figure 12 in Ref. 22, which shows clearly that the birefringence is limited to a very local region in the vicinity of the stagnation point, and along the outflow axis that emanates from this point. Experimental measurements show that the velocity *gradient* field in the region of view is relatively uniform.[9,25] Thus, the localization of birefringence is clearly a consequence of the distribution of residence times (and total strain) in the strong-flow as explained above. It should be emphasized that the extreme localization of birefringence is characteristic only of *dilute* solutions.

B. Dilute Solutions in Unsteady Flows

Relatively less is known about the behavior of the dumbbell models when applied to strongly time-dependent flows. Of course, even the steady two-roll and four-roll mill flows are unsteady from the Lagrangian view of a polymer molecule, but the transient encountered in these cases is more or less like a step increase in strain-rate as the polymer enters the region between the rollers, and the transient extension of the polymer along any streamline is primarily a reflection of the rate of increase of the total strain, rather than a measure of intrinsic timescales for the polymer. Thus, the steady flow birefringence data provides relatively little insight into the detailed forms of either the spring or friction laws in the model. However, one prediction of the conformation-dependent friction model does appear to be supported by the existing data:

d. The relaxation time increases strongly with the degree of polymer stretch. Thus, the timescale characteristic of relaxation from an extended state is very much longer than indicated by the intrinsic relaxation time for the polymer in the equilibrium or near-equilibrium state.

This is a somewhat controversial feature of the nonlinear dumbbell model with conformation-dependent friction, that has important implications for polymer response to time-dependent flows. Among these is the observation, made some years ago by Hinch and others, that a polymer molecule can be *maintained* in an extended state by a relatively weak flow compared to the strength of the flow that is required to extend it in the first place.

The conclusion (d) is most directly reflected in experiments reported several years ago by Miles and Keller[26] at Bristol. These authors showed that the *decay* of birefringence in a weak flow occurred on a timescale of $0(2\text{x}10^{-2}\text{s})$ for a polymer-solvent system that had a characteristic relaxation timescale at equilibrium of only $0(7.5\text{x}10^{-4}\text{s})$. Additional support for the conclusion of increased relaxation times with increase of polymer deformation is contained in data from our laboratory, which shows that there is a very sharp decrease in the critical velocity gradient for polymer extension when the equilibrium configuration is modified to become more expanded and/or extended.[7] In our experiments, this was accomplished by varying the salt concentration in a dilute polyelectrolyte solution. Closely related to this is the observation from several laboratories that the critical velocity gradient for the coil-stretch transition can be strongly influenced by "pre-shear."[6,27] In particular, if the polymer is subjected to a shearing deformation prior to encountering a region of "strong" flow, then the onset value of G can be sharply reduced compared to its value in the absence of shear. All of these observations can *only* be explained in the context of bead-spring or dumbbell models by a friction coefficient that increases with increase of deformation, either via the assumption of conformation dependent bead-friction or, in principle, via a complete theory of "bead-bead" interactions.

The preceding parts of this subsection have focused on predictions of the nonlinear dumbbell model that have been shown to be either qualitatively or quantitatively correct. It must be emphasized, however, that this section addresses only predictions of polymer conformation in a known velocity field. The more difficult problem of polymer modification of flow is discussed in the next section. Furthermore, there are areas of particular significance for polymer drag reduction where it is *not* yet clear that the existing models are adequate even for predictions of polymer conformation in a *known* flow.

One important problem that remains unresolved is the case of *rapidly* varying flows (in either a Lagrangian or Eulerian sense). If the flow (i.e. $\nabla \mathbf{u}$) varies rapidly or the polymer suddenly encounters a region of large $|\nabla \mathbf{u}|$, then the polymer is forced to try to stretch on a timescale that is fast compared with the timescale for Brownian fluctuations in conformation. Hence, a polymer *cannot* sample many configurations and the statistics of end-to-end dimension associated with the various configurations do not have time to develop – thus, the use of smooth spring-laws is likely to break down. Recent dynamic simulations based upon "many bead" models *may* give some useful insight and point the way to future modification of the dumbbell type models for this class of flows. The qualification inherent in the use of "may" is because the models used for these simulations are extremely simplified and it is not entirely clear whether they represent real physics relevant to polymer molecules, or just irrelevant but complex behavior due to the inclusion of many degrees of freedom. Two such simulations exist, one due to Rallison and Hinch[28] at Cambridge, and the other due to Liu[29] at Madison. Both show that the polymer "stretches," but in the process develops folds that take a "long time" to release. Thus, the polymer will never become fully extended in a flow that tries to induce rapid stretching. This is a manifestation of an extension rate that is too fast to allow diffusion to work.

This suggests the need for inclusion of a rate-dependent resistance to extension in the dumbbell model. One implication of this is that the polymer contribution to the bulk stress may be much higher during a rapid extension process than it is predicted to be on the basis of current linear or nonlinear models. However, the majority of experiments that have been carried out to date do not subject the polymer to sufficiently rapid extension rates to test this point. The ones that do (such as opposing jobs) do not provide data of sufficient detail to differentiate various possible explanations of any discrepancies between theory and experiment. In short, the jury is still out on the application of current models to rapidly varying flows.

A second class of flows that are not yet resolved, even for changes in polymer conformation without changes in the flow, are those involving "complex" strain/strain-rate histories (as seen by the polymer). For example, the polymer may be alternately subjected to periods of strong and weak flow.[30] In this type of situation, which bears some evident analogies with the flow history that one might expect a polymer to encounter in turbulence, the conformational state at any instant will represent an accumulative effect, where accurate simulation of polymer behavior by a model will clearly require that the *details* of the relaxation time versus conformation are correct. In the dumbbell model, this means that it will be important not only that the spring and friction laws be *qualitatively* correct as required by the Eulerian-steady flows studied previously, but also that the *quantitative* dependence on the end-to-end dimension is correct. We have mentioned earlier the prevailing view that it is essential to capture the correct qualitative behavior, but that the detailed forms of the friction and spring laws are more or less irrelevant. This may be true in flows like the steady two- or four-roll mill, because the measurements primarily reflect monotonic extension of the polymer under circumstances where the rate of extension is determined, essentially, by the flow rather than any details of the polymer relaxation timescales. However, if we consider polymer behavior in circumstances where *relaxation* processes are dominant at least part of the time, this will clearly not be true. The polymer extension process in any "strong" flow region will then depend critically on the degree of relaxation of the polymer from the previous extension process. Flows that represent this type of alternation between "strong" and "weak" regions are very easy to produce in the laboratory – any laminar flow that is "chaotic" will exhibit precisely this type of behavior. But, as yet, there have not been any detailed experimental studies, even for the ultradilute limit.

In summary of this section, then, we have seen that the current nonlinear dumbbell models of dilute solutions give *reasonable* predictions of polymer conformation in steady, strong flows, including the prediction of a rapid increase in the relaxation timescale with polymer extension (at least over the region of conformations where the nonlinear spring-law does not dominate the relaxation behavior). However, the ability to model polymer response to flows that vary more rapidly, or exhibit more complex time-histories from the Lagrangian viewpoint of the polymer, is currently unknown. Of course, one need hardly emphasize with this audience that it is likely to be precisely the behavior in these latter problems that is *most* relevant to understanding and modeling of the polymers' role in turbulent drag-reduction!

VI. MODEL EXPERIMENTS FOR DILUTE SOLUTIONS – DOES THE NONLINEAR DUMBBELL MODEL CORRECTLY PREDICT POLYMER EFFECTS ON FLUID MECHANICS?

The most essential question is not whether current models give correct predictions of polymer conformation in known flows (i.e. in flows that are unchanged from Newtonian fluid behavior), but whether the models can correctly predict the motion of a dilute polymer solution in circumstances *where the flow is changed* because of the polymer! Surprisingly, perhaps, this is still largely an *open question*. Since the expected "rheological" behavior of a dilute solution is unique, relative to normal viscoelastic fluids, the many studies of the rheological behavior of more concentrated polymer solutions, or polymer melts, provide little guidance. Nor is there much, if any, insight to be anticipated from recent attempts to compare the flow behavior of normal viscoelastic fluids with numerical predictions obtained using conventional (Maxwell/Oldroyd or other) constitutive equations.

The key feature of the "rheology" of the nonlinear dumbbell is that the largest resistance to motion is predicted to occur for extensional (or "strong") flows, when the polymer is in an extended state. The predicted effect of the polymer *during* the stretching process is relatively weak, because the rapid increase in spring-tension only kicks in (with current models) when the polymer is already significantly extended. The rheological predictions for simple shear flow include a range of shear rates where the "viscosity" is shear-thickening. However, the *magnitude* of polymer effects in simple shear (or other weak flows) is expected to be extremely small (see Section III). Thus, though the dumbbell model with conformation-dependent but isotropic friction is *known* to be an oversimplification for such flows, this is an "intentional" cost of developing the simplest model that can be expected to give reasonable predictions in *strong* flows, which is the *only* class of motions where the polymer has the *potential* to have a strong effect on the motion. The transition from a coiled to a stretched state requires: (a) a critical magnitude of $G > 0$ (which depends on the flow-type, including whether the flow is homogeneous or not), and (b) large total *strain* along the trajectory of a polymer molecule (including the possibility of an accumulative effect associated with a sequence of stretching regions with insufficient time for polymer relaxation in between). The development of a meaningful test of dilute solution models thus requires an experimentally realizable flow that meets these conditions, but still remains simple enough to allow analytical or numerical solutions of the fluid mechanics problem.

A. "Steady" Laminar Flows in 2- and 4-Roll Mill

One major problem in assessing the prediction capabilities of current models, even assuming that we could obtain analytical or numerical solutions of the dilute solution model equations for nontrivial flows, is the lack of experimental data that include systematic measurements of both the polymer behavior, and the velocity field, in a nontrivial flow geometry. The only case where this has been done, to date, is our own studies of steady flow in two- and four-roll mills. However even there, the data base is still relatively limited.[9] Furthermore, this type of problem is not an optimal choice for the necessary fluid mechanics studies.[24] This is largely a consequence of the fact that the interactions between the polymer and the flow are *extremely localized,* as noted earlier. This leads to a number of difficulties in carrying out the desired comparisons between experimental data and theoretical predictions:

1. *The problems are difficult to solve numerically –*
   They involve extremely localized effects, with large stress and conformation gradients that occur in the "middle" of the flow domain.
2. *It is difficult to make meaningful measurements –*
   If we attempt *local* measurements of polymer conformation, and especially velocity or velocity gradient fields, there is a major problem with spatial resolution. If we attempt *"global"* measurements, we get only the consequences of a big *local* effect averaged over a large region, and there is a serious loss of sensitivity (for example, this problem clearly influences attempts to measure extensional viscosities of dilute solutions via *pressure* changes in opposing jet rheometers).
3. *Many of the flows that can be studied most easily are "homogeneous" in the region of interest –* Thus the changes in polymer conformation or stress may be much larger than the dynamically significant effects on $\nabla \cdot \tau$. As a result, changes in the flow may be relatively subtle.

In addition, the available data suggests (but does not prove in view of the second point above) that a measurable effect of the polymer on the flow (determined via light scattering measurements of the velocity gradient) only appears when the polymer becomes sufficiently extended that there is a *very high degree of polymer-polymer interaction.* In particular, the data of Dunlap and Leal[9] leads to the conclusion that the effective volume "fraction" based upon the volume occupied by spheres that circumscribe the extended polymer, must exceed

$$\phi_{effective} > 0(10^3)$$

before we find any measurable effect of the polymer on the flow. From a dynamics point of view, the solution must undergo a transition from "dilute" – where polymer molecules do not interact – to a locally "concentrated" solution where they interact extensively. Since the existing dilute solution *models* are derived strictly *excluding* polymer-polymer interaction effects, this might seem a fatal flaw in the theoretical approach. Certainly, one cannot take it for granted that predictions of conformation and stress from the dumbbell (or other dilute solution) models will give reasonable results when pushed to ranges of concentration and polymer extension where polymer interactions must become important. However, a related study of the rheology of a suspension of aligned rods does give some hope that the model *could* be better than one might expect. In this case, Batchelor[31] showed that the effective viscosity for a non-dilute suspension of interacting, aligned rods differs from predictions for a dilute suspension (evaluated at the same concentration) only by a numerical factor that is generally $O(1)$. Thus, it may be that the predictions of dilute solution models for strong flows, where the polymer is highly aligned, may not be too bad even though polymer interactions appear to become important. Of course, a final conclusion must await detailed comparisons between experiment and theory.

At the present moment, however, no such comparisons exist. Experimental data exists for polymer conformation and the velocity gradient field in two- and four-roll mills. Numerical solutions of the nonlinear dumbbell model equations also exist (for the simplified model with nonlinear spring and constant friction) but only for streaming flow past a solid sphere.[20] We are currently in the process of solving the full nonlinear model equations in the two-roll mill geometry. When complete, these solutions will provide the first real indication of the usefulness of the dumbbell for fluid mechanics predictions of the motions of dilute polymer solutions.

B. Other Steady, Laminar Flows

Although definitive model comparisons are not available, it may be worthwhile to briefly summarize other experimental studies of flow modification in dilute polymer solutions.

One area of current investigation is the behavior in flows with *very* large strain-rates. Such flows are easiest to achieve in flow through contractions, and a number of studies (including some proposals for extensional rheometers) have utilized this approach.[32] The fact is that it is possible to achieve extremely large strain-rates by simply increasing the volumetric flow rates. However, such flows contain no stagnation point and thus the total strain experienced by the polymer in a single pass is always limited. Furthermore, though few investigators have addressed the issue, it is *always questionable* at high flow rates whether such flows are actually *stable* and *laminar*. Nevertheless, the evidence of flow modification is interesting. Experiments involving flow into slits and circular orifices suggest strongly that a measurable influence of the polymer on the flow (i.e., changes in the ratio pressure drop by flow rate) only appears for extension rates that are between 40 and 400 times the inverse *relaxation times* for the polymer in its equilibrium configuration![28] In contrast, the effects of polymer on flow in the two- and four-roll mill (i.e., of changes in $\nabla\mathbf{u}$) appear at velocity gradients that are very close to the inverse relaxation time (as shown earlier). This suggests that it may be the *inhibited* extension associated with *rapidly* extending flows (identified via dynamic simulations of many-bead models – see above) that is responsible for the effectiveness of dilute polymer solutions in modifying the class of motions where the total *strain* is limited but the strain-rate is very high. However, the only detailed comparisons between predictions and observations for this type of problem used a dilute solution model that does not seem to be consistent with the more recent dynamical simulations,[33] and it is unclear whether any version of inhibited extension in dumbbell models will be adequate.

The flow of dilute solutions in turbulent drag reduction is one example of a problem where there is a measurable overall ("global") characteristic of the motion that changes due to the presence of polymer in solution. Another, that has received extensive investigation is the motion of dilute solutions in *porous media,* where there is a very pronounced decrease in the ratio of volumetric flowrate to pressure drop at some critical flowrate.[34,35] This has been widely interpreted as a global manifestation of the onset of a coil-to-stretch transition in the polymer, and this may well be true. However, the porous media problem does not offer much opportunity for detailed investigations. First of all, the geometry is exceedingly complex, and in most instances it is not possible to "see" into the interior to make detailed measurements. In addition, the polymer used for most of the porous media studies has been ill-characterized, with very broad molecular weight distributions, and this greatly complicates comparison between experiment and theory.

For purposes of both experimental and theoretical investigation, it would be advantageous to study laminar flows that show very strong *global* effects. However, we do not know of any such problem where the flow is *steady,* and the geometry is simple enough to allow detailed studies.

C. Chaotic Laminar Flows - Future Directions

It is evident that an understanding of dilute polymer solutions that is adequate to assess polymer behavior in drag reduction, requires investigation of complex time-dependent flows that are closer to those encountered by the polymer in a turbulent velocity field. Of course, all real laminar flows (except simple shear) are nonhomogeneous (i.e., $\nabla \mathbf{u}$ depends on spatial position), and thus time-dependent from the viewpoint of the polymer. However, they are not generally complex enough to provide a meaningful framework for understanding polymer dynamics when the timescales of extension by the flow and of diffusion-induced relaxation of the polymer both play a significant role. Further, in laminar flows that do not involve *complex* geometry (like porous-media), the requirement of large total strain to induce a coil-stretch transition generally translates into extremely *localized* effects along streamlines that pass very close to the stagnation points of the flow.

It is our view that an ideal class of problems for future investigation is time-dependent, laminar flows that are "chaotic" over large parts of the flow domain. Many of the limitations and difficulties cited above disappear for this class of problems, as explained in detail below. One example of a chaotic laminar flow that we have begun to study in our own laboratory is the blinking or time-modulated co-rotating two-roll mill.[36] In the *blinking* mode, the two cylinders are *alternately* rotated in the same-direction at the same speeds. In the time-modulated mode, the *relative* rates of rotation are oscillated periodically.

A chaotic flow is "globally" strong – in the sense that material points separate exponentially "on average" in the limit of large times, $t \gg 1$. In particular, in the chaotic region, the flow at any material point is alternately "strong" and "weak," but the long time average corresponds to exponential extension of a line element of the fluid, at a rate that depends upon the magnitude of the so-called Lyapunov exponent. The polymer behavior will depend critically on the relaxation timescales of the polymer compared with the extension and rotation timescales of the flow, since there will be regions of stretching and re-orientation interspersed with relaxation toward equilibrium where the flow is "weak." The ability to predict the behavior of a polymer molecule in such a flow constitutes a critical test of the *details* of the model – i.e., the specific spring and friction laws – that has not been possible for steady strong flows, but is clearly a pre-requisite to any believable *predictions* of polymer behavior in *turbulent* flows.

In addition to this general objective, there are several *key* advantages of chaotic, laminar flows that deserve mention. First, since the flow can be chaotic over large regions of the flow domain, we may also expect that polymer-flow interactions can be strong in large regions of space – i.e., instead of the very localized effects that occur in the steady two-roll mill, we expect major overall effects of the polymer on the flow. Second, there are *many* fundamental questions to be answered that are sufficiently *generic* as to be directly relevant to understanding and/or modelling the polymer behavior in drag reduction applications. For example, fluid elements separate exponentially, as a long-time average, at a rate that is determined by the Lyapunov exponent of the flow; thus, what is the relationship (if any) between the Lyapunov exponent and the onset conditions for polymer stretch, or between Lyapunov exponents and the *rate* of polymer extension? There are two fundamental components inherent in any answer to such questions: the first is the dynamics of a polymer molecule in any non-homogeneous or time-dependent flow (and an understanding of the transition from *autonomous* to *non-autonomous* dynamical equations); the second is the direct effect of the chaotic dynamics in the flow. A third general feature of many laminar, chaotic flows for a Newtonian fluid is that they can be described *exactly* via analytic and/or numerical solutions.[37] If the flow is *not* effected by the polymer (ultra-dilute systems), this provides a basis for exact predictions of polymer conformation using the polymer model of choice. The existence of numerical procedures for solution of the steady two-roll mill problem for a Newtonian fluid, and the current development of such solutions for dilute solutions, should also provide a basis to study the complete polymer-flow interaction problem for this particular time-dependent, chaotic flow. Thus, since *time-dependent measurements* of the velocity gradient and polymer conformation are also possible for this flow in our laboratory,[38] we have the potential to evaluate polymer models in complex, time-dependent flows, where polymer-flow interactions should be strong and *not* local.

Of course, all of the material in this subsection represents *current* research rather than accomplished results. Thus an evaluation of these prognostications will have to await some future conference on drag reduction!

VII. CONCLUSIONS

The goal of this paper has been to provide an overview of our current understanding of the dynamics of dilute polymer solutions, with a particular focus on mathematical descriptions of polymer behavior that could be combined with direct simulations of turbulent flows to provide a basis for predicting and evaluating turbulent drag reduction applications.

The conclusion that I want to emphasize most strongly is that we are *not* currently in a position to make such predictions. Conventional constitutive models for viscoelastic, polymer solutions or melts do not provide a meaningful basis to predict the behavior of dilute solutions, but the "molecular" models that are being developed specifically for dilute polymer solutions are not currently at a state where predictions in complex, time-dependent flows can be trusted. One critical feature of both models and actual dilute solution behavior, is that it is *extremely* sensitive to the details of the flow. Sensitivity to the details of the model is also expected in a complicated time-dependent flow. Thus, to get the "right answer" for the "right reasons" in any attempt to explain or predict polymer drag reduction, it is *essential* to have *the correct polymer model* (down at least to the details of the spring and friction laws for a dumbbell model) and to have *a very faithful simulation of the details of the turbulent flow*, especially at the smallest length scales where the polymer is directly active. Glossed-over details may *seem* to give *correct predictions* for global parameters of the flow, but will *likely* give these results for the wrong reasons.

Current research on the dynamics of dilute polymer solutions is aimed at assessing dumbbell models for prediction of steady laminar flows, as well as understanding and modelling polymer behavior in more complex, time-dependent flows. It is my personal opinion that the class of laminar, chaotic flows provides the most useful basis for the development of polymer models that can ultimately yield believable results in numerical simulations of turbulent drag reduction, but that is a prediction that awaits future confirmation.

References

1 Flory, P.J. *Principles of Polymer Chemistry*, Cornell Univ. Press, Ithaca (1953).

2 Tanford, C. *Physical Chemistry of Macromolecules*, Wiley, New York (1961).

3 Meares, P. *Polymers: Structure and Bulk Properties*, Van Nostrand, London (1965).

4 Doi, M. and Edwards, S.F. *The Theory of Polymer Dynamics*, Clarendon Press, Oxford (1986).

5 Hinch, E.J. "Mechanical models of dilute polymer solutions in strong flows," *Phys. Fluids* **20**, 522 (1977).

6 James, D.F. and Saringer, J.H. "Flow of dilute polymer solution through converging channels," *J. non-Newt. Fl. Mech.*, **11**, 317 (1982).

7 Dunlap, P.N., Wang, C.-H. and Leal, L.G. "An experimental study of dilute polyelectrolyte solutions in strong flows," *J. Poly. Sci. Part B: Polymer Physics*, **25**, 2211 (1987).

8 Virk, P.S. "Drag reduction by collapsed and extended polyelectrolytes," *Nature*, **253**, 109 (1975).

9 Dunlap, P.N. and Leal, L.G. "Dilute polystyrene solutions in extensional flows: birefringence and flow modification," *J. non-Newt. Fl. Mech.*, **23**, 5 (1987).

10 Keyes, D.E. and Abernathy, F.H. "A model for the dynamics of polymers in laminar shear flows," *J. Fluid Mech.* **185**, 503 (1987).

11 Olbricht, W.L., Rallison, J.M. and Leal, L.G. "Strong flow criteria based on microstructure deformation," *J. non-Newt. Fl. Mech.* **10**, 291 (1982).

12 Tanner, R.I. and Huilgol, R.R. "On a classification scheme for flow fields," *Rheol. Acta*, **14**, 959 (1975).

13 Tanner, R.I. "A test particle approach to flow classification for viscoelastic fluids," *AIChE J.* **22**, 910 (1976).

14 Szeri, A.J., Wiggins, S. and Leal, L.G. "On the dynamics of suspended microstructure in heterogeneous fluid flows," to be submitted *J. Fluid Mech*, (1990).

15 Bird, R.B., Hassager, O., Armstrong, R.C. and Curtiss, C.F. *Dynamics of polymeric Liquids, Vol. 2: Kinetic Theory* 2nd Ed. Wiley, New York, N.Y., (1987).

16 Phan-Thien, N., Manero, O. and Leal, L.G. "A study of conformation-dependent friction in a dumb-bell model for dilute solutions," *Rheol. Acta*, **23**, 151 (1984).

17 Leal, L.G., Fuller, G.G. and Olbricht, W.L. "Studies of the flow-induced stretching of a macromolecule in a dilute solution," in *Viscous Drag Reduction*, Vol. 72, Prog. in Astronautics and Aeronautics, 351 (1980).

18 Warner, H.R. "Kinetic theory and rheology of dilute suspensions of finitely extendable dumbbells," *I & EC Fundamentals* **11**, 379 (1972).

19 Armstrong, R.C. "Kinetic theory and rheology of dilute solution of flexible macromolecules: I. Steady State " *J. Chem. Phys.* **60**, 724, (1974); "II. Linear," *J. Chem. Phys.* **60**, 729, (1974).

20 Chilcott, M.D. and Rallison, J.M. "Creeping flow of dilute polymer solutions past cylinders and spheres" *J. non-Newt. Fl. Mech.* **29**, 381 (1988).

21 Keller, A. and Odell, J.A. "The extensibility of macromolecules in solution: a new focus for macromolecular science," *Colloid and Polymer Science*, **263**, 181 (1985).

22 Fuller, G.G. and Leal, L.G. "Flow birefringence of dilute polymer solutions in two-dimensional flows," *Rheol. Acta* **19**, 580 (1980).

23 Fuller, G.G. and Leal, L.G. "The effects of conformation-dependent friction and internal viscosity on the dynamics of the onlinear dumbbell model for a dilute polymer solution," *J. non-Newt. Fl. Mech.* **8**, 271 (1981).

24 Leal, L.G. "Studies of flow-induced conformation changes in dilute polymer solutions," AIP Conf. Proceedings, **137**, 5 (1985).

25 Fuller, G.G., Rallison, J.M., Schmidt, R.L. and Leal, L.G. "The measurement of velocity gradients in laminar flow by homodyne light-scattering spectroscopy," *J. Fluid Mech.* **100**, 555 (1980).

26 Miles, M.J. and Keller, A. "Conformational relaxation time in polymer solutions by elongational flow experiments: 2. Preliminaries of further developments: chain retraction; identification of molecular weight fractions in a mixture" *Polymer* **21**, 1295 (1980).

27 Bewersdorff, H.W., private communication (1989).

28 Rallison, J.M. and Hinch, E.J. "Do we understand the physics in the constitutive equation?" *J. non-Newt. Fl. Mech.* **29**, 37 (1988).

29 Liu, T.W. "Flexible polymer chain dynamics and rheological properties in steady flows," *J. Chem. Phys.* **90**, 5826 (1989).

30 Nollert, M.J. and Olbricht, W.L. "Macromolecular deformation in periodic extensional flows," *Rheol. Acta* **24**, 3 (1985).

31 Batchelor, G.K. "The stress generated in a non-dilute suspension of elongated particles by pure straining motion," *J. Fluid. Mech.* **46**, 813 (1971).

32 James, D.F., Chandler, G.M. and Armour, S.J. "A converging channel rheometer for the measurement of extensional viscosity," preprint (1989).

33 Ryskin, G. "Calculation of the effect of polymer additive in a converging flow," *J. Fluid Mech.* **178**, 423 (1987).

34 Durst, F. and Haas, R., "Dehnströmungen mit verdünnten polymerlösungen: Ein theoretisches modell und seive experimentelle verifikation" *Rheol. Acta* **20**, 179 (1981).

35 Haas, R. and Durst, F., "Die charakterisierung viskoelastisches fluide mit hilfe ihrer strömungseigenschaften in kugelschüttungen" *Rheol. Acta* **21**, 150 (1982).

36 Ng, R. C.-Y., James, D.F. and Leal, L.G. "Chaotic mixing and transport in a two-dimensional time periodic Stokes flow--The blinking two-roll mill (BTRM): I. Newtonian Fluids II. Dilute polymer solutions" to appear.

37 Kaper, T., Shapira, M., Ascoli, E. and Leal, L.G. "An analytic and numerical study of flow in a two-roll mill" to appear.

38 Geffroy, Enrique, PhD Thesis, Calif. Institute of Technology (1989).

The Effect of Dilute Polymer Solutions on Viscous Drag and Turbulence Structure

WILLIAM G. TIEDERMAN

School of Mechanical Engineering
Purdue University, W. Lafayette, IN 47907 USA

Summary

All recent results confirm that the major differences between dilute polymer and Newtonian flows occur in an expanded buffer region near the wall. The wall-normal velocity fluctuations decrease and the principal axes for the Reynolds shear stress rotate toward the streamwise and wall-normal directions. Lower threshold uv events in quadrants 2 and 4 are damped while higher threshold uv events are not damped. The mean period of the burst cycle increases as drag reduction increases in the same way that streak spacing increases. While the wall-layer structures are modified in dilute polymer flows, they still contain all of the features of the Newtonian wall-layer structures.

Introduction

Based on the personal accounts [1,2] of their work, the discovery that high molecular weight additives could lower the friction factor for flow of a polymer solution below that of the Newtonian solvent was totally unexpected. Perhaps the non-intuitive nature of this discovery is the reason that the published literature which began in 1949 [3,4,5] grew rather slowly during the first decade. However, Savins [6] noted that almost 300 journal articles, reports, theses and patents appeared in 1969 alone. Review articles also began to appear in 1969 when Lumley discussed the major phenomenological features of drag-reducing flows obtained from pressure drop, flow rate data and from mean velocity profiles [7]. Hoyt's Freeman Scholar Lecture which was published in 1972 [8] and Virk's review in 1975 [9] are comprehensive accounts of the literature that show there were only a few experiments prior to 1976 which reported either structural features or higher-order velocity statistics for drag-reduced flows.

There were several reasons for the relatively slow beginning of "structural" studies in drag-reducing flows. Instrumentation is an issue in dilute polymer flows because measurement errors have been noted for both Pitot tubes and cylindrical hot-film probes [7,8]. As a result the laser velocimeter, whose development began in

A. Gyr (Editor)
Structure of Turbulence and Drag Reduction
IUTAM Symposium Zurich/Switzerland 1989

the 1960s, is the instrument of choice for aqueous polymer solutions. The lack of consensus among turbulence "experts" as to both the meaning and relevance of structural studies in Newtonian wall turbulence that developed during the 1970's must have also inhibited the extension of structural studies to dilute polymer flows. Finally, one only has to refer to the review articles [8,9] to see the diversion caused by the large number of additive-solvent systems and the wide variety of phenomena that can be studied with simple instrumentation.

When the first IUTAM Symposium on Structure of Turbulence and Drag Reduction was held in 1976, only a few studies had been performed with laser velocimeter and the debates over the significance of various turbulent structures that had been identified were intense. Since that time, advances in computer technology and the Kline-Robinson study [10] have had a large impact on these two topics. Full numerical solutions of turbulent flow in a two-dimensional channel [11] and in a zero pressure gradient boundary layer [12] have provided spatial information about turbulent structures. When coupled with the careful survey of the experimental database, this information has resolved many of the issues of previous debates. Laboratory computers have transformed the laser velocimeter from a novelty to a fully useful tool for dilute polymer flows. As a result, the quality of our experimental information as well as our ability to interpret dilute polymer experiments is much improved. This will be evident in this review of fully developed, homogeneous dilute polymer flows in pipes and in large aspect ratio channel flows. However, these enhanced capabilities have not yet led to a full understanding of dilute polymer flows.

Non-dimensional Parameters

One of the most frustrating aspects of dilute polymer flows is the absence of a set of non-dimensional parameters that are sufficient to correlate the friction factor. In fully developed Newtonian flows the friction factor, f, is a unique function of the Reynolds Number, Re, and the relative roughness, e/D. That is

$$f = f(Re, e/D) \tag{1}$$

All of the parameters can be measured and it is a straightforward matter to correlate data and design piping systems. It is well known that for a given concentration of a specified additive and solvent, Equation 1 will not correlate the pressure drop and flow rate for dilute polymer solutions in pipes of different sizes [13]. This means that at least one additional parameter is needed. Frequently the proposed parameter is the ratio of some characteristic time associated with the solution and a characteristic time associated with the wall layer [14,15]. This ratio is called the Deborah number. Usually, the characteristic time for the solution is calculated based on a model for the solution [16,17]. The result depends upon molecular properties of the solution which are often poorly controlled (distribution of molecular weights, degradation) and usually not measured directly. An exception is the method of Darby and Chang [18] which uses apparent viscosity data to obtain the required rheological information. While the Deborah Number approach is reasonable and popular, the results vary [13,18] and there is not yet a universally accepted method for correlating the friction factor for dilute polymer flows.

Mean Velocity Profiles

Most features of interest in the mean velocity profiles of dilute polymer flows have been discussed in previous reviews [9,19,20]. The common approach is to compare drag-reducing profiles with a Newtonian profile at equal wall shear stress using inner scaling of kinematic viscosity, ν, and shear velocity, u_τ, to normalize the mean streamwise velocity, $\overline{U}$, and the distance from the wall, y. (The nondimensional variables are designated by the superscript +.) Semi-log co-ordinates are used so that the inner region representation of the overlap region

$$U^+ = \frac{1}{\kappa} \ln y^+ + B \tag{2}$$

appears as a straight line. Since the wall shear is the same, both Newtonian and dilute polymer profiles are the same very near the wall where $U^+ = y^+$. Since there is a higher flow rate for the dilute polymer, drag-reducing flows, these profiles then fall above Equation 2 for $y^+ \gtrsim 30$. In general the linear extent of the sublayer is the same for Newtonian and drag-reducing flows. The buffer region is extended in drag-reducing flows yielding a thicker wall layer (sum of the linear sublayer and buffer region). For many additive-solvent systems, there is a logarithmic region for lower values of drag reduction given by

$$U^+ = \frac{1}{\kappa} \ln y^+ + B + \Delta B \tag{3}$$

Most investigators report that only the intercept in Equation 2 changes. However, recent measurements in polyacrylimide solutions show changes in both the intercept and slope of the logarithmic region at levels of drag reduction between 10 and 40% [21,22]. Clearly, the possibility that the slope can change is a potential explanation why scale-up procedures based up ΔB [23] do not always work.

Caution should be used when mean velocity profiles normalized with u_τ are compared for flows of unequal shear velocity. Unlike Newtonian flows, U^+ is not a unique function of y^+ for dilute polymer flows with the possible exception of flows at maximum drag reduction [19,24] where the wall region extends to the center-line. In all other cases, as with the friction factor, at least one additional parameter is required to produce a universal correlation and this universal correlation (if it exists at all) has not yet been determined.

When drag-reducing and Newtonian velocity profiles are compared at equal flow rates or equal Reynolds number a different view is obtained (see Figure 1). The wall slope of the Newtonian profile is larger. Then, as y increases, the slope of the drag-reducing profile becomes larger than the slope of the Newtonian profile. For both flows in Figure 1, the Reynolds number is 17,800, U_m is the average velocity over the cross section, a is the channel half-height and the polymer flow has a drag reduction of 22% [25]. In this comparison the drag-reducing slope becomes larger for $y^+ \gtrsim 15$ and remains larger until $y^+ \sim 100$ where the drag-reducing profile begins to be described by Equation 2. The slope of the mean velocity profile for a drag-reducing flow with wall shear equal to the Newtonian flow is larger over approximately the same region [25].

Turbulence Intensity Profiles

Laser velocimeter measurements of the root-mean-square of fluctuations about the mean of the streamwise velocity component, u', have been relatively common [21,22,25,26,27,28,29,30]. Since the most interesting changes occur in the wall region, inner scaling is typically used. All of the data show that the peak level for u' occurs at larger values of y^+ than in the Newtonian case and that the peak is broader. Maximum values of u'/u_τ vary a great deal from study to study. Some of the variation is due to changes in u_τ. In two-dimensional channels [22,25,26] the peak value of u'/u_τ increases from the Newtonian value of 2.5 to 4.0 as drag reduction increases for flows at equal wall shear stress. This increase must be due in part to the higher mean velocities in the outer region of these drag-reducing flow. Comparison of u'/U_o profiles at equal Reynolds number (U_o is the center-line velocity) shows the broader peak and the movement of the peak away from the wall, but the peak values are essentially the same [25]. In circular pipes the peak values of u'/u_τ can be similar [27], smaller [21,28] or larger [29] than the Newtonian value.

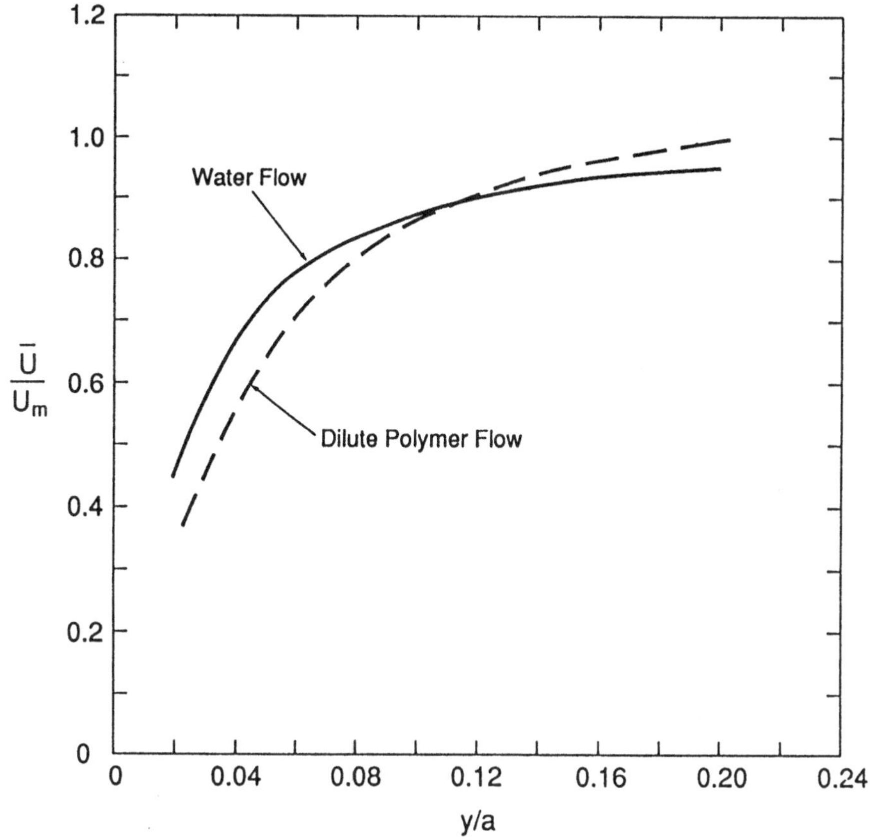

Fig.1 Comparison of the mean streamwise velocity profiles in the buffer region for flows at equal Reynolds number.

Measurements of the root-mean-square of the velocity fluctuations normal to the wall, v' are relatively rare [21,22,25,30]. However, the results consistently show reduced levels of v' in the extended buffer region of dilute polymer channel and pipe flows.

Reynolds Stress

Measurements of the fluctuating velocity product $\overline{uv}$ in dilute polymer channel flows also show that changes occur in an extended buffer region [21,22,25,30]. In some cases [21,22,25], these changes are simply somewhat lower levels of $\overline{uv}$ with the sum of the viscous stress ($\mu\frac{d\overline{U}}{dy}$) and the Reynolds stress ($-\rho\overline{uv}$) adding to yield the expected linear variation in total stress. That is

$$\tau = \mu\frac{d\overline{U}}{dy} - \rho\overline{uv} = \tau_w(1-y/a) \tag{4}$$

However, in one homogeneous flow of a polyacrylimide solution [22] and in one flow of a polyethylene oxide solution [30], the Reynolds stresses were decreased dramatically and Equation 4 did not yield the total shear stress. A similar result was obtained with the central injection of a concentrated thread that remained intact [21]. Willmarth et al. [30] conjectured that their solution may not have been well mixed and the injected polymer solution was not seeded. However, Harder's [22] channel flows were all both well mixed and homogeneously seeded. The non-Newtonian effect occurred for the flow with both the highest strain rate (4000 s^{-1}) and the highest concentration (5 ppm) that Harder tested. If non-Newtonian rheology can effect the shear stress, it is conceivable that the effects would occur under these conditions. For example, Bewersdorff and Berman [31] suggested that the Reynolds stress deficit can be explained by using a local viscosity which includes an elongational component.

Quadrant analysis of the uv products occurs when the data are placed in fluctuating velocity or u-v space. For example a fluid particle moving slower than the streamwise mean velocity ($u<0$) and away from the wall ($v>0$) is a quadrant 2 motion while fluid particles moving faster than the mean streamwise velocity ($u>0$) and toward the wall ($v<0$) is a quadrant 4 motion. Luchik and Tiederman [25] showed that the lower Reynolds stresses they measured occurred because there were fewer lower threshold uv products in quadrants 2 and 4 while the higher threshold uv products occurred at the same or a higher rate. This result is consistent with the observation of Schmid [32] and Walker and Tiederman [33] that the probability density distributions of uv in the buffer region of drag-reducing channel flows rotate such that the principle axes are more aligned with the laboratory axes. The results of this rotation are higher u', lower v', lower $\overline{uv}$ and a lower correlation coefficient for $\overline{uv}$.

Turbulence Production

Recently [22,33] large enough ensembles of two-component velocity data have been taken to make accurate estimates of the production term, $(\overline{uv})d\overline{U}/dy$, in the

Reynolds stress equation for $\overline{u^2}$. In a Newtonian flow this production term is balanced primarily by destruction from viscous dissipation and by redistribution to the other normal stresses through the pressure-strain correlation. The production term for $\overline{u^2}$ is important because there is no equivalent term in fully developed, two-dimensional channel flow for the other two normal stress terms. In particular, the major source term for $\overline{v^2}$ is the pressure-strain correlation which increases $\overline{v^2}$ while decreasing $\overline{u^2}$ [34].

In homogeneous channel flows, the production of $\overline{u^2}$ is lower in dilute polymer flows than for water flows at equal Reynolds number [22]. This is surprising since the level of $\overline{u^2}$ itself is higher while $\overline{v^2}$ is lower in the dilute polymer flows. Since the scales at which viscous dissipation occurs are isotropic, it seems likely that viscous dissipation for both $\overline{u^2}$ and $\overline{v^2}$ would be similar. Consequently it appears that the polymer solution inhibits the transfer of energy from $\overline{u^2}$ to $\overline{v^2}$ and this could occur through inhibiting the Newtonian transfer mechanism provided by the pressure-strain correlation.

Streak Spacing

The persistent theme is that the turbulence is altered in the buffer region. Knowledge that the dilute polymer solution must be in the wall region to be effective [35] preceded the velocity measurements described earlier. In this same time period, wall-layer studies at Stanford [36,37] and Ohio State [38] were documenting the active role of the wall region in producing turbulence through a quasi-periodic event called a wall-layer burst. Associated with the burst structure are longitudinal streaks that contribute low momentum fluid to the ejections (quadrant 2 momentum transport) within the burst events. Three early studies [39,40,41] showed that the spanwise spacing of the streaks increased when drag reduction occurred. By carefully controlling the way in which streaks were marked, Oldaker and Tiederman [42] offered an explanation for previous differences in the streak spacing results. An essential element in this explanation is the concept that the low momentum streaks are formed by eddy structures above the linear sublayer. As the wall is approached, the essentially passive sublayer is better able to resist these motions in dilute polymer flows. Therefore streak spacing measurements must be made in the region $y^+ \geq 2$ in order to count all of the streaks. When this is done, then the spanwise spacing increase is correlated well by

$$\lambda^+ = 1.9\ D_R + 99.7 \tag{5}$$

Here D_R is the percent drag reduction, and λ is the spanwise spacing of the streaks. Notice that when there is no drag reduction, Equation 5 yields the Newtonian spacing for wall-layer streaks.

Wall-layer Bursts

In a parallel study [43], A.J. Smith used a fluroescent dye and side-lighting to demonstrate that the flow visualization procedures for marking and counting *all* of the ejections were more complex than previously imagined. The streamwise locations downstream of the marker injection which give the correct count depend upon the amount of marker placed in the sublayer and upon the amount of drag reduction. All of the issues raised by Smith's study were not resolved until a series of papers from Purdue presented accurate methods for deducing the average period between bursts from wall-slot flow visualization [44] and from velocity sensors [45,46,47]. These new techniques show that for homogeneous, dilute polymer flows, the increase in the average time between bursts for drag-reducing flows was equal to the increase in the spanwise streak spacing when compared to water flows at equal wall shear [25]. A larger increase in the average time between bursts was obtained near an injector where the polymer solution was still mixing with the channel flow of water [48].

Other features of the burst characteristics have been measured in low concentration flows of polyacrylimide solutions [25]. Perhaps the most significant result is that all features of the Newtonian events occur in the dilute polymer flows. Some are modified (damping of low threshold quadrant 2 and quadrant 4 motions) but none are eliminated. This is illustrated in Figures 2 and 3 which compare conditionally averaged signals for the streamwise velocity fluctuation, u, the wall normal velocity fluctuation, v, and the uv product for flows at Reynolds number. Note that in both flows there are more distinct peaks of $\langle v \rangle$ and $\langle uv \rangle$ associated with the leading edge of the burst than with the trailing edge. Also evident in Figure 3 is the characteristic VITA signal associated with the sharp internal shear layer at the trailing edge of the burst [49].

Discussion

The mechanism which leads to the changes described in previous sections has not yet been determined in a conclusive way. It is important to note that dilute polymer flows are *not* laminar. The turbulence structure is modified in various ways

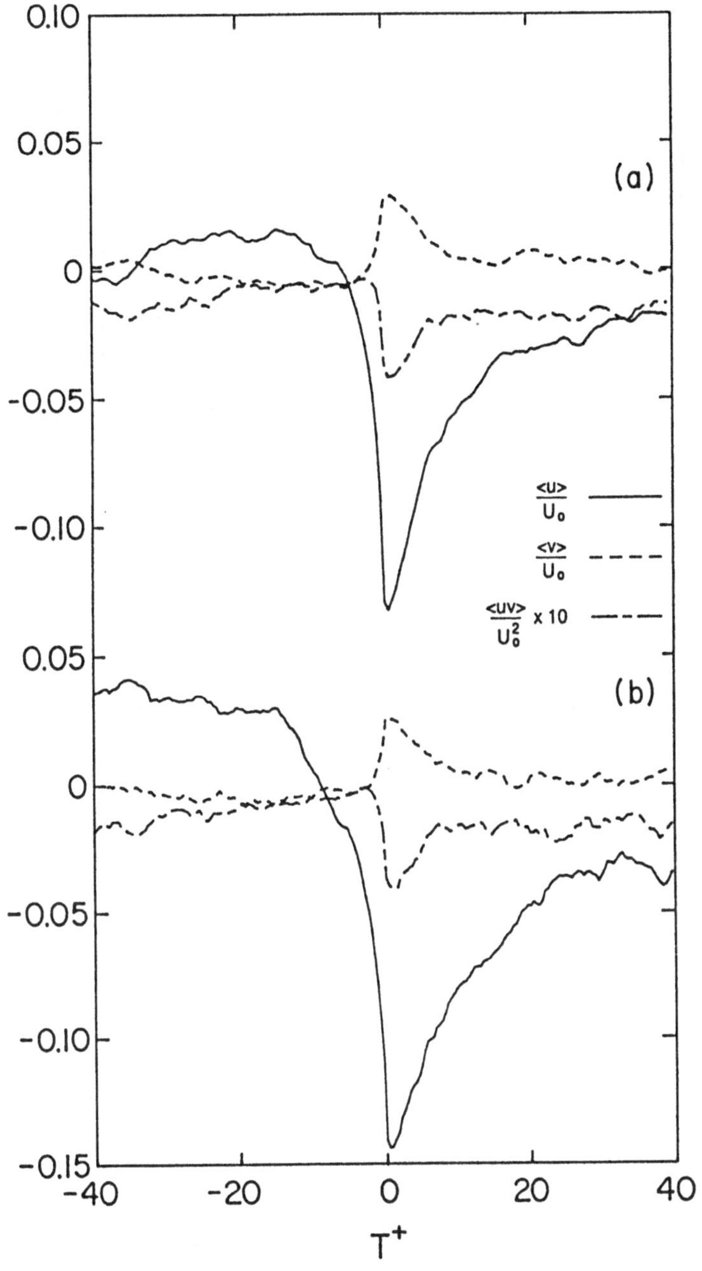

Fig.2 Conditionally averaged velocity signals centered on the leading edge of a burst; a) water flow; b) 22% drag reducing flow.

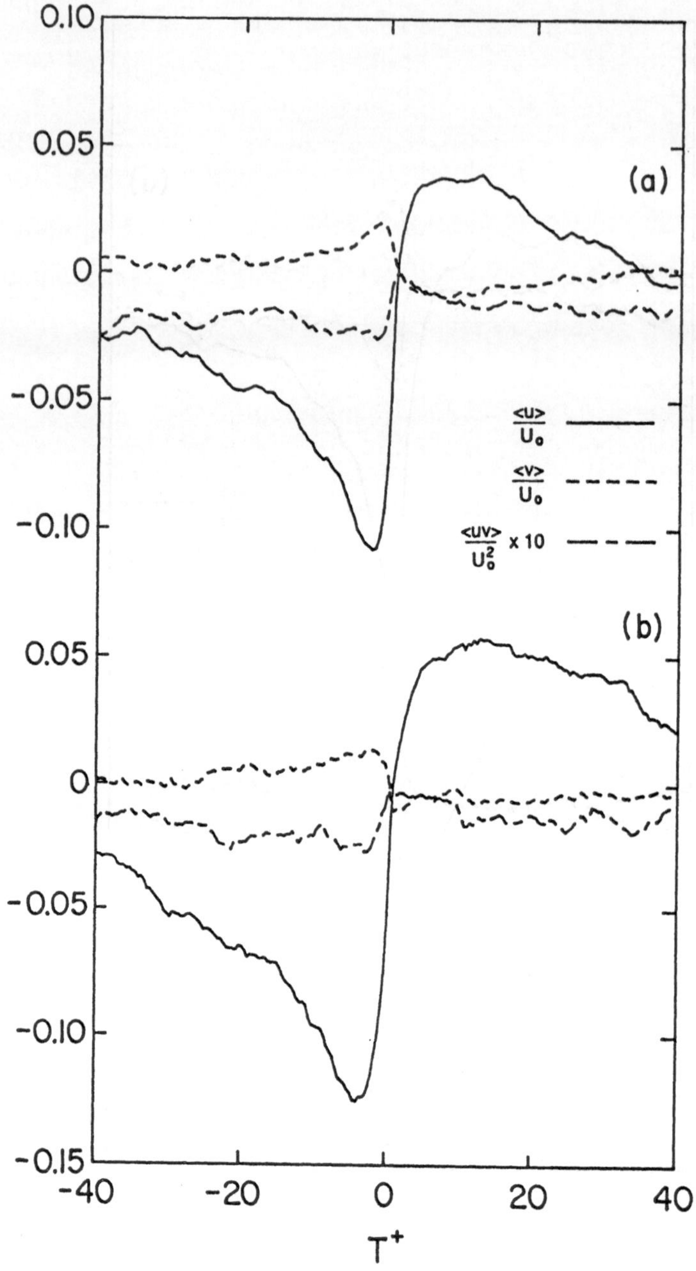

Fig.3 Conditionally averaged velocity signals centered on the trailing edge of a burst; a) water flow; b) 22% drag-reducing flow.

but all features of the Newtonian wall structure remain. This led Luchik and Tiederman [25] to postulate that the turbulence and the polymer molecules in dilute polymer flows reach a statistical equilibrium where the remaining turbulent stretching keeps the molecule extended so that lower threshold motions can be damped by the viscoelastic properties of the dilute polymer solution. Gyr [50] was more specific and attributed the required molecular extension to a wall-layer structure of Λ vortices. The difficulty, of course, is that rheological measurements have not been made in the dynamical situation of wall turbulence. The experiments of James et al. [51] are an interesting model of turbulent flow and more activity of this type is needed to delineate the mechanism.

References

1. Mysels, K.J.: Early Experiences with Viscous Drag Reduction. Chemical Engineering Progress Symposium Series 67, 111 (1971) 45-49.
2. Toms, B.A.: On the Early Experiments on Drag Reduction by Polymers. Phys. Fluids 20, 10, Pt. II (1977) S53-S55.
3. Oldroyd, J.G.: A Suggested Method of Detecting Wall Effects in Turbulent Flow Through Pipes. Proc. Intl. Congr. on Rheol., II, North Holland, Amsterdam (1949) 130-134.
4. Mysels, K.J.: Flow of Thickened Fluids. U.S. Patent 2, 492, 173 (1949).
5. Toms, B.A.: Some Observations on the Flow of Linear Polymer Solutions Through Straight Tubes at Large Reynolds Numbers Proc. Intl. Congr. on Rheol., II, North Holland, Amsterdam (1949) 135-141.
6. Savins, J.G.: Preface for Drag Reduction Chemical Engineering Progress Symposium Series 67, 111 (1971) v-vi.
7. Lumley, J.L.: Drag Reduction by Additives. Ann. Rev. Fluid Mech. 1 (1969) 367-384.
8. Hoyt, J.W.: The Effect of Additives on Fluid Friction. ASME J. Basic Engrg. 94D (1972) 258-285.
9. Virk, P.S.: Drag Reduction Fundamentals. AIChE J. 21 (1975) 625-656.
10. Kline, S.J.: Quasi-Coherent Structures in the Turbulent Boundary Layer: Part I. Status Report on a Community-Wide Summary of the Data. In Near Wall Turbulence ed. by S.J. Kline and N.H. Afgan, Hemisphere, New York, 1989 (in press).
11. Kim, J., Moin P. and Moser, R.D.: Turbulence statistics in fully developed channel flow at low Reynolds number. J. Fluid Mech. 177 (1987) 133-166.
12. Spalart, P.R.: Direct numerical simulation of a turbulent boundary layer up to $Re_\theta = 1410$. J. Fluid Mech. 187 (1988) 62-98.

13. Sellin, R.H.J. and Ollis, M.: Effect of Pipe Diameter on Polymer Drag Reduction. Ind. Eng. Chem. Prod. Res. Dev. 22 (1983) 445-452.

14. Hershey, H.C. and Zakin, J.L.: A Molecular Approach to Predicting the Onset of Drag Reduction in the Turbulent Flow of Dilute Polymer Solutions. Chem. Engrg. Sci. 22 (1967) 1847-1857.

15. Seyer, F.A. and Metzner, A.B.: Turbulent Flow Properties of Viscoelastic Fluids. Can. J. Chem. Engrg. 45 (1967) 121-126.

16. Zimm, B.H.: Dynamics of Polymer Molecules in Dilute Solution: Viscoelasticity, Flow Birefringence and Dielectric Loss. J. Chem. Phys. 24 (1956) 269-278.

17. Bird, R.G., Hassager, O., Armstrong, R.C. and Curtis, C.F.: Dynamics of Polymeric Liquids 2. New York: J. Wiley & Sons 1977.

18. Darby, R. and Chang, H.D.: Prediction of Turbulent Drag Reduction in Polymer Solutions from Rheological Properties. In Drag Reduction ed. R.H.J. Sellin and R.T. Moses, p.A.4-1-A.4-4 University of Bristol, UK, 1984.

19. Berman, N.S.: Drag Reduction by Polymers. Ann. Rev. Fluid Mech. 10 (1978) 47-64.

20. Lumley, J.L. and Kubo, I.: Turbulent Drag Reduction by Polymer Additives. In the Influence of Polymer Additives on Velocity and Temperature Fields ed. B. Gampert, p.3-21, Springer-Verlag, Berlin, 1985.

21. Bewersdorff, H.-W.: Heterogene Widerstandsverminderung bei turbulenten Rohstromungen. Rheol Acta, 23 (1984) 522-534.

22. Harder, K.J.: Effect of Wall Strain Rate, Polymer Concentration and Channel Height Upon Drag Reduction and Turbulent Structure. MSME thesis, Purdue University, 1989.

23. Granville, P.S.: Scaling-up of Pipe Flow Frictional Data for Drag Reducing Polymer Solutions. Proceeding, 2nd Int'l Conf. on Drag Reduction, British Hydromechanics Research Association, Cambridge, UK, 1978.

24. Virk, P.S.: An Elastic Sublayer Model for Drag Reduction by Dilute Solutions of Linear Macromolecules. J. Fluid Mech., 45 (1971) 417-440.

25. Luchik, T.S. and Tiederman, W.G.: Turbulent structure in low-concentration drag-reducing channel flows. J. Fluid Mech. 190 (1988) 241-263.

26. Reischman, M.M. and Tiederman, W.G.: Laser-Doppler anemometer measurements in drag-reducing channel flows. J. Fluid Mech. 70 (1975) 369-392.

27. Berner, C. and Scrivener, O.: Drag Reduction and Structure of Turbulence in Dilute Polymer Solutions. In Viscous Flow Drag Reduction ed. by G.R. Hough, p.290-299, American Institute of Aeronautics and Astronautics, New York, NY 1980.

28. Mizushina, T. and Usui, H.: Reduction of eddy diffusion for momentum and heat in viscoelastic flow in a circular tube. Phys. Fluids 20, 10, Pt. II (1977) S100-S108.

29. McComb, W.D. and Rabie, L.H.: Part II: Laser-Doppler Measurements of Turbulent Structure. AIChE J. 28 (1982) 558-565.

30. Willmarth, W.W., Wei, T. and Lee, C.O.: Laser anemometer measurements of Reynolds stress in a turbulent channel flow with drag reducing polymer additives. Phys. Fluids, 30 (1987) 933-935.

31. Bewersdorff, H.-W. and Berman, N.S.: The influence of flow-induced non-Newtonian fluid properties on turbulent drag reduction. Rheol Acta 27 (1988) 130-136.

32. Schmid, A.: Experimental Investigation of the Influence of Drag Reducing Polymers on a Turbulent Channel Flow. In Drag Reduction ed. R.H.J. Sellin and R.T. Moses, p.B.12-1 - B.12-7, University of Bristol, UK, 1984.

33. Walker, D.T. and Tiederman, W.G.: Turbulent structure in a channel flow with polymer injection at the wall. Submitted to J. Fluid Mech. 1989.

34. Mansour, N.N., Kim, J. and Moin, P.: Reynolds-stress and dissipation-rate budgets in a turbulent channel flow. J. Fluid Mech. 194 (1988) 15-44.

35. Wells, C.S. and Spangler, J.G.: Injection of a drag-reducing fluid into turbulent pipe flow of a Newtonian fluid. Phys. Fluids 10 (1967) 1980-1984.

36. Kline, S.J., Reynolds, W.C., Schraub, F.A. and Runstadler, P.W.: The structure of turbulent boundary layers. J. Fluid Mech. 30 (1967) 741-773.

37. Kim, H.T., Kline, S.J. and Reynolds, W.C.: The production of turbulence near a smooth wall in a turbulent boundary layer. J. Fluid Mech. 50 (1971) 133-160.

38. Corino E.R. and Brodkey, R.S.: A visual study of turbulent shear flow. J. Fluid Mech. 37 (1969) 1-30.

39. Donohue, G.L., Tiederman, W.G. and Reischmann, M.M.: Flow visualization of the near-wall region in a drag-reducing flow. J. Fluid Mech. 56 (1972) 559-575.

40. Eckelman, L.D., Fortuna, G. and Hanratty, T.J.: Drag reduction and the wavelength of flow-oriented wall eddies. Nature 236 (1972) 94-96.

41. Achia, B.U. and Thompson, D.W.: Structure of the turbulent boundary layer in drag reducing pipe flow. J. Fluid Mech. 81 (1977) 439-464.

42. Oldaker, D.K. and Tiederman, W.G.: Spatial structure of the viscous sublayer in drag-reducing channel flows. Phys. Fluids 20, No.10, Pt.II (1977) S133-S144.

43. Tiederman, W.G., Smith, A.J. and Oldaker, D.K.: Structure of the viscous sublayer in drag-reducing channel flows. In Turbulence in Liquids, 1975 ed. by J.L. Zakin and G.K. Patterson, p.312-322 University of Missouri-Rolla, USA, 1977.

44. Bogard, D.G. and Tiederman, W.G.: Investigation of flow visualization techniques for detecting turbulent bursts. In Symposium on Turbulence, 1981 ed by X.B. Reed, G.K. Patterson and J.L. Zakin, p.289-302, University of Missouri-Rolla, USA, 1983.

45. Bogard, D.G. and Tiederman, W.G.: Burst detection with single-point velocity measurements. J. Fluid Mech. 162 (1986) 389-413.

46. Luchik, T.S. and Tiederman, W.G.: Timescale and structure of ejections and bursts in turbulent channel flows. J. Fluid Mech. 174 (1987) 529-552.

47. Tiederman, W.G.: Eulerian Detection of Turbulent Bursts. In Near Wall Turbulence ed. by S.J. Kline and N.H. Afgan, Hemisphere, New York, 1989 (in press).

48. Tiederman, W.G., Luchik, T.S. and Bogard, D.G.: Wall-layer structure and drag reduction. J. Fluid Mech. 156 (1985) 419-437.

49. Bogard, D.G. and Tiederman, W.G.: Characteristics of ejections in turbulent channel flow. J. Fluid Mech. 179 (1987) 1-19.

50. Gyr, A.: Direct Evidence that Drag Reduction is an Effect of the Elongation of the Polymer Molecules. In Drag Reduction ed. by R.H.J. Sellin and R.T. Moses, p.B.10-1-B.10-7, University of Bristol, UK, 1984.

51. James, D.F., McLean, B.D. and Saringer, J.H.: Presheared Extensional Flow of Dilute Polymer Solutions. J. Rheology 31 (1987) 453-481.

Aspects of Mechanisms in Type B Drag Reduction

P.S. Virk and D.L. Wagger
Department of Chemical Engineering, Massachusetts Institute of Technology, Cambridge, MA 02139, USA

ABSTRACT.

Drag reduction by macromolecular additives is briefly reviewed and shown to comprize two extreme forms of gross flow behaviour, Types A and B, that are respectively associated with random-coiling and extended additive conformations. The mechanism of drag reduction is discussed within the framework of an "Additive-Burst Matrix", devised such that its rows represent macromolecular states while its columns connote turbulent burst events. Experiments are presented wherein Type A and Type B behaviour were both induced by polyelectrolyte additives of the identical backbone structure and molecular weight. Results for Type B drag reduction revealed that, at fixed $Re\sqrt{f}$, flow enhancement increased almost linearly with additive concentration; also, a typical gross flow trajectory commenced on the maximum drag reduction asymptote and then exhibited "retro-onset" into the polymeric regime of less than maximum drag reduction. Comparisons between selected Type A and B trajectories also delineated "additive equivalence" at "iso-slip" points, where different polymer solutions achieved the same flow enhancement. Interpretation of these experimental results suggests that Type B drag reduction is likely more fundamental to the mechanism of additive-turbulence interaction than the more widely studied Type A.

INTRODUCTION.

The object of this work is to explore the role of the additive in turbulent drag reduction, particularly the effects of macromolecular coil extension.

The drag reduction phenomenon has by now been extensively studied and much reviewed[1,2,3,4,5]. It has been shown to comprize two asymptotic, additive-insensitive flow regimes, of zero and maximum drag reduction, respectively. These envelop a third "polymeric" regime wherein the properties of the additive exert a decisive influence. Within the polymeric regime, two extreme types of gross flow behaviour, termed Types A and B, are observed[6]. In type A drag reduction, a family of additive solutions yields friction factor segments fanning outward from a common "onset" point on the Prandtl-Karman line, their slopes increasing with increasing concentration and drag reduction increasing with increasing $Re\sqrt{f}$. This is characteristic of random-coiling macromolecules. In type B drag reduction, a family of additive solutions yields segments that are roughly parallel to, but displaced upwards from, the P-K line, with drag

A. Gyr (Editor)
Structure of Turbulence and Drag Reduction
IUTAM Symposium Zurich/Switzerland 1989

reduction essentially independent of $Re\sqrt{f}$ but increasing with increasing concentration. This behaviour is exhibited by a variety of additives, including extended polyelectrolyes, fibers, soaps, and clays. Despite its ubiquity, Type B drag reduction has not yet been studied as thoroughly as Type A.

Turning to the drag reduction mechanism, three lines of evidence suggest that a seminal interaction occurs between an additive and a turbulent burst. First, at the onset of drag reduction, ratios of characteristic macromolecular to wall turbulence length and time scales are invariant, implying a linkage. Second, the onset of drag reduction occurs at the same wall shear stress in hydrodynamically smooth and fully rough pipes, implying that the additive-turbulence interaction commences outside the viscous sublayer. Third, mean velocity (and turbulent intensity) profiles during drag reduction show that a characteristic new region, the "elastic sublayer", arises in the vicinity of $y^+ \simeq 15$, where the turbulent energy production and dissipation associated with bursts is at a maximum. Based on the foregoing, we have devised an "Additive-Burst Matrix", Figure 1, to frame a discussion of mechanism. In this A-B matrix, rows represent macromolecular states, namely (1) random-coiled, (2) coil $\leftrightarrow$ stretch transitional, and (3) extended, while columns connote turbulent burst events, namely (1) lift-up, (2) growth, and (3) breakdown. The object of the mechanistic search is to identify the matrix element(s) most intimately associated with drag reduction. Regarding the polymeric additive, with

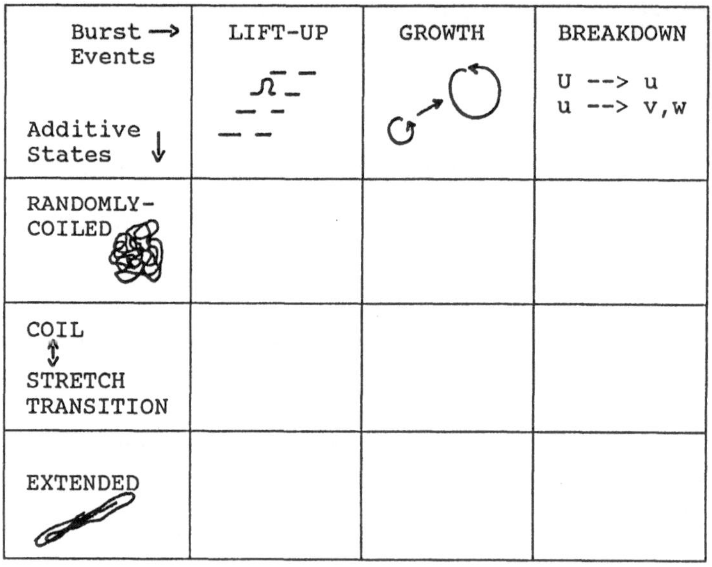

Burst Events →
Additive States ↓
LIFT-UP
GROWTH
BREAKDOWN
U --> u
u --> v,w
RANDOMLY-COILED
COIL ↕ STRETCH TRANSITION
EXTENDED

FIGURE 1. An Additive-Burst Matrix for Drag Reduction by Macromolecules.

which we are concerned here, the equilibrium random-coiled conformation, Row 1, is unlikely to be very important, because drag reduction by spheres is invariably rather small. Row 2 of the A-B matrix is likely related to Type A drag reduction, observed with solutions of initially randomly-coiled macromolecules. Thus onset, at a characteristic flow strength $Re\sqrt{f}^*$, might represent incipient coil → stretch transition, while the subsequent increase in drag reduction with $Re\sqrt{f}$ possibly reflects the greater proportion of macromolecules populating stretched states. Row 3, representing the extended conformation, is almost certainly relevant to drag reduction, being associated with Type B behaviour. Two questions of particular interest are: Does the act of macromolecular elongation (Row 2) contribute to drag reduction or does this latter depend solely upon the final amount of elongated macromolecules (Row 3)? Also, what skeletal attributes govern the effectiveness of an extended additive? The present work addressed these questions experimentally, using polyelectrolyte additives in saline solutions. This permitted variation of the initial macromolecular conformation over the entire range from collapsed random-coil, at high salinity, to greatly extended, at vanishing salinity, while maintaining the identical backbone structure and chain links in the additive.

EXPERIMENTAL.

A stainless steel test pipe, 14.6 mm id x 3033 mm long, was used, installed in a gravity-drained single-pass flow system that, with an entrance trigger, provided fully turbulent flow for solvent Reynolds numbers from 3000 to 30000. All experiments were run at $T = 20 \pm 1$ C. The additives employed were commercially available flocculants[7] we call C832 and C836. Both were partially-hydrolysed polyacrylamides, abbr PAMH, of molecular weight $M_w \simeq 15 \times 10^6$, with C832 having a higher degree of backbone hydrolysis than C836. Additive concentrations were varied from 1 to 30 wppm in each of five saline solutions containing from 0.00001 to 0.10 m NaCl in distilled water. The intrinsic viscosities of C832 in 0.10 and 0.00001 m NaCl, respectively $[\eta] \simeq 7300$ and 110000 cm3/g, offer an indication of the great differences in hydrodynamic coil size between its collapsed and extended conformations. Special pains were taken to minimize polymer degradation during solution makeup and to maintain identical polymer concentrations in solutions of different salinities.

RESULTS.

The profound effect of additive conformation upon drag reduction is illustrated in Figure 2, a gross flow diagram, with solvent-based Prandtl-Karman coordinates, that shows data for 20 wppm solutions of C836 in 0.10, 0.01, 0.001, and 0.0003 m NaCl. The three solid lines, labelled L, N, and

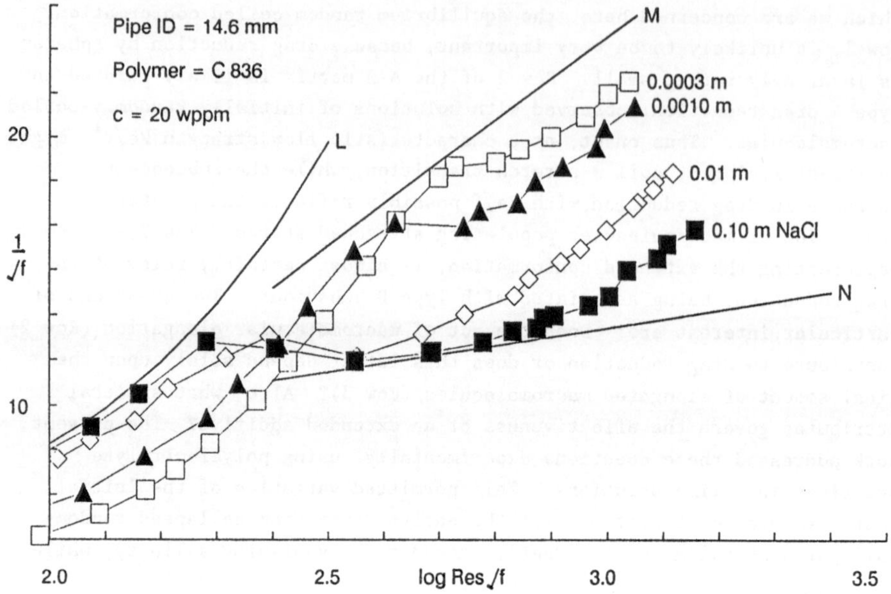

FIGURE 2. The Effect of Additive Conformation on Drag Reduction.

M, respectively represent Poiseuille's law for laminar flow, the Prandtl-Karman law for Newtonian turbulent flow, and the maximum drag reduction asymptote:

(L) $1/\sqrt{f} = Re\sqrt{f}/16$

(N) $1/\sqrt{f} = 4.0 \log Re\sqrt{f} - 0.4$

(M) $1/\sqrt{f} = 19.0 \log Re\sqrt{f} - 32.4$

Data for the most collapsed conformation, in 0.10 m NaCl, adheres closely to Poiseuille's law L for $2.0 < \log Re_s\sqrt{f} < 2.3$, then undergoes laminar-to-turbulent transition from $2.3 < \log Re_s\sqrt{f} < 2.55$, adheres closely to the P-K law N for $2.55 < \log Re_s\sqrt{f} < 2.85$, and exhibits an onset of drag reduction at $\log Re_s\sqrt{f}^* \simeq 2.85$. Beyond onset, the polymer solution exhibits drag reduction, with $1/\sqrt{f_p}$ exceeding the solvent $1/\sqrt{f_n}$ at constant $Re_s\sqrt{f}$. At the highest $\log Re_s\sqrt{f} = 3.15$, for example, $1/\sqrt{f_p} = 16$ versus the solvent $1/\sqrt{f_n} = 12$; this "apparent slip" $S' = 4$ units of $1/\sqrt{f}$ corresponds to a fractional flow enhancement of 0.33 relative to solvent. As salinity is systematically reduced from 0.10 m to 0.0003 m, inducing polyelectrolyte coil expansion, the gross flow trajectories exhibit large systematic changes. For the most extended conformation, at the lowest salinity of 0.0003 m, the trajectory lies roughly parallel to, but about 5 units of $1/\sqrt{f}$ below, L for $2.0 < \log Re_s\sqrt{f} < 2.6$; then at $\log Re_s\sqrt{f} = 2.7$, $1/\sqrt{f} = 18$, it almost touches the maximum drag reduction asymptote M, and for 2.8 <

log $Re_s\sqrt{f} < 3.0$ exhibits $19 < 1/\sqrt{f_p} < 23$. At the highest log $Re_s\sqrt{f}$, the $S' = 11$ corresponds to a fractional flow enhancement of 0.91 relative to solvent. Evidently, extending the polyelectrolyte conformation greatly increases drag in laminar flow while greatly reducing drag in turbulent flow. The laminar drag increase, due to the greater hydrodynamic volume of the extended coils, is well understood. The greater turbulent drag reduction by the more extended conformations has been previously observed[6], but its mechanistic implication has not hitherto been fully appreciated. Namely, that the progressively greater drag reduction efficacies of the more extended conformations implies that they must ever more closely resemble the additive state that ultimately interacts with the turbulence. Since the limit of conformational extension is a macromolecule stretched to its contour length L_c, an additive state akin to this likely participates in the elementary steps of the drag reduction mechanism.

Types A and B drag reduction, as respectively exhibited by 0.10 m and 0.00001 m NaCl solutions of C832, are illustrated in Figures 3 and 4.

In Figure 3, solutions of C832 in 0.10 m NaCl (collapsed conformation) exhibit the "fan" pattern characteristic of Type A behaviour. At the lowest log $Re_s\sqrt{f}$ all solutions lie close to the solvent line N; the onset of drag reduction occurs at log $Re_s\sqrt{f^*} \simeq 2.75$; and for log $Re_s\sqrt{f} >$ log $Re_s\sqrt{f^*}$, polymer solution trajectories fan outward from the onset point, their slopes exceeding the solvent slope by increments δ that increase with increasing polymer concentration. Both the onset wall shear stress, $\tau_w^* \simeq 6.5$ dyn/cm2, as well as the intrinsic slope increment, $(\delta/\sqrt{c}) \simeq 3$, derived from these data accord with previous studies[4] of Type A drag reduction by randomly-coiled PAMH additives.

In Figure 4, solutions of C832 in 0.00001 m NaCl (extended conformation) exhibit a "ladder" pattern characteristic of type B drag reduction. These solutions reduce drag at the lowest turbulent $Res\sqrt{f}$, so that no onset is visible. Indeed, for the case c = 10 wppm the trajectory attains the maximum drag reduction asymptote M, and then exhibits a "retro-onset", at log $Re_s\sqrt{f_*} \simeq 2.75$, into the envelope of lesser drag reduction lying between M and N. For c = 20 wppm, the data strive towards M, implying that retro-onset lay beyond our highest $Res\sqrt{f}$. Within the drag reduction envelope, all trajectories exhibit relatively shallow slopes, so, for fixed c, flow enhancement increases only slowly with increasing log $Re_s\sqrt{f}$. However, at all log $Re_s\sqrt{f}$, flow enhancement increases strongly with increasing c, giving rise to the distinctive ladder pattern. At fixed log $Re_s\sqrt{f} = 3.0$, solutions of c = (1, 2, 5, 10, 20) wppm respectively yield apparent slip S' = (0.6, 1.2, 4.5, 9, >14), suggesting that S' is approximately proportional to c. The ratio [S'/c], called "intrinsic slip", measures the slip induced per additive molecule, and is thus characteristic of additive efficacy in Type B drag reduction. For C832 in 0.00001 m NaCl the present data yield [S'/c] = 0.74 ± 0.17; by way of

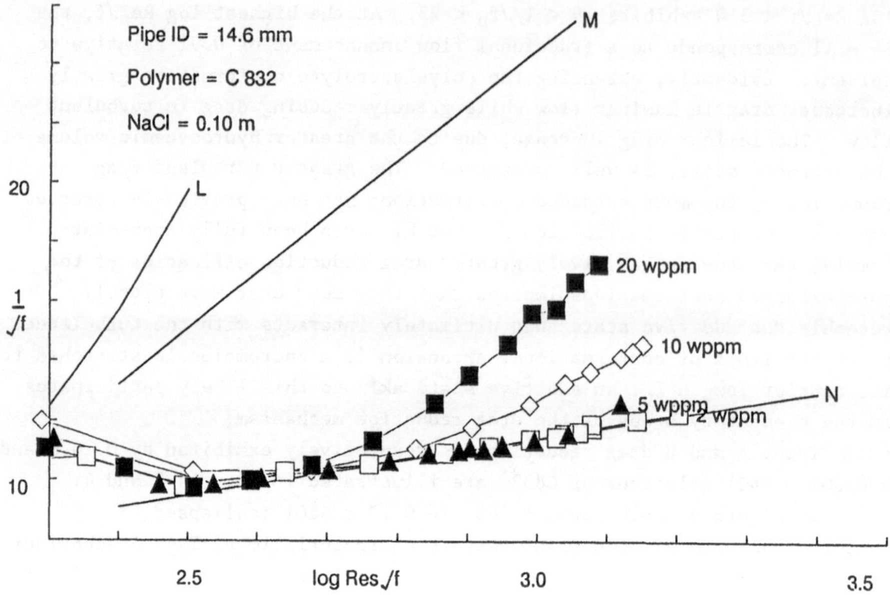

FIGURE 3. Type A Drag Reduction "Fan" for Collapsed Conformations.

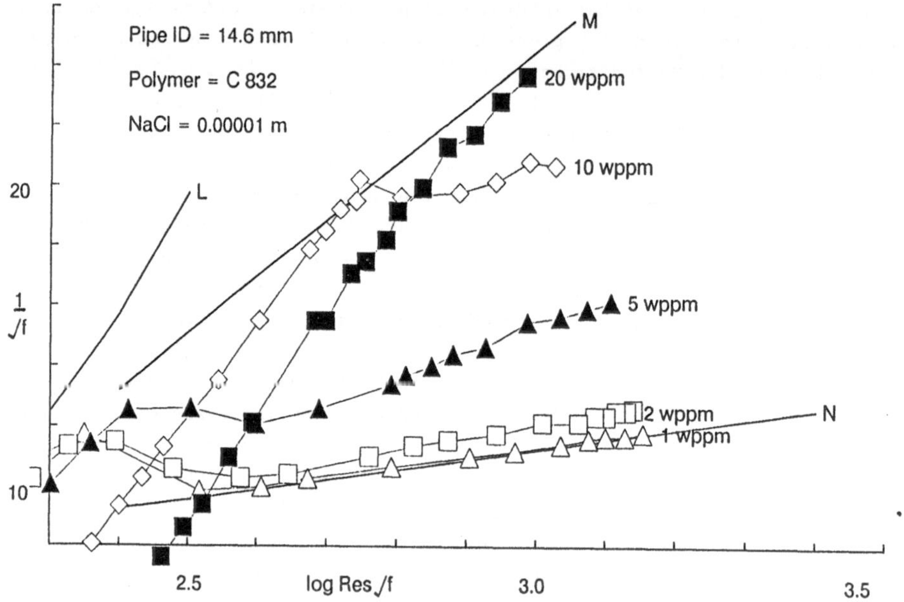

FIGURE 4. Type B Drag Reduction "Ladder" for Extended Conformations.

comparison, analysis of previous work[6] for extended conformations of a similar PAMH additive B1110 yielded [S'/c] = 0.72.

DISCUSSION.

We consider first a preliminary analysis of Type B drag reduction results reported in the literature, and then a comparison between the present Type A and Type B results obtained with the C832 PAMH additive.

Figure 5 summarizes apparent slip results obtained with a variety of additives in Type B drag reduction. These include the polyelectroyte macromolecules PAMH[this work,6,8] and xanthan[9], and fibers of asbestos[10,11], paper[12], and nylon[13]. Some aspects of the works comprizing Figure 5 are briefly noted. Oliver & Bakhtiyarov[8] used a PAMH of Mw ≃ 23x10[6] in "Birmingham tap water" and reported percent drag reduction over a narrow range of Re ≃ 15000; in the absence of actual trajectories, we presumed that their PAMH additive was likely to adopt an extended conformation and exhibit Type B behaviour. Rochefort & Middleman[9] studied 10 to 110 wppm solutions of a Xanthan gum, Mw ≃ 2x10[6], in a mixed glycerol-water solvent; their data showed clear Type B trajectories. Both Sharma, Seshadri & Malhotra[10] and McComb & Chan[11] used the same nominal asbestos additive, with fibril diameter and length (d_f, l_f) ≃ (0.00004, 1.5) mm; gross flow results of the former investigators were suggestive of Type B behaviour, so we presumed the latter was also of this type. The data of Lee & Duffy[12]

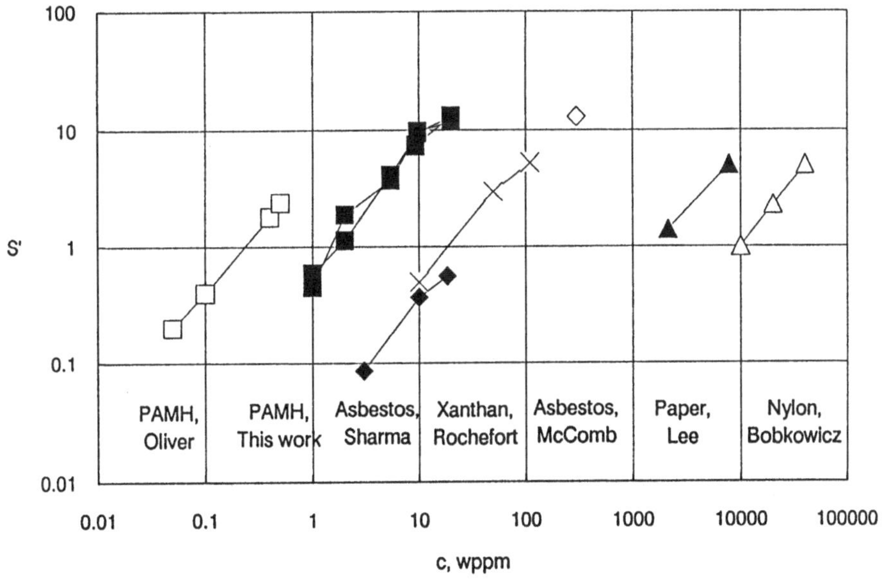

FIGURE 5. The Intrinsic Slip of Additives in Type B Drag Reduction.

for papermaking fibers, $(d_f, l_f) = (0.03, 2.7)$ mm, and of Bobkowicz & Gauvin[13] for nylon fibers, $(d_f, l_f) = (0.02, 1.03)$ mm, were both of Type B. Figure 5 reveals that additives capable of inducing Type B drag reduction can differ widely in their efficiencies, with intrinsic slips spanning five orders of magnitude from $[S'/c] \simeq 4$ for a PAMH to $[S'/c] \simeq 0.00011$ for a nylon fiber. For the two PAMH entries, $[S'/c]$ decreases from $\simeq 4$ for Oliver's Mw $\simeq 23\times10^6$ to $\simeq 0.74$ for the present Mw $\simeq 15\times10^6$, implying a 2 to 3 power dependence of intrinsic slip upon backbone chain links for fixed skeletal structure. Comparing the three fiber entries, the asbestos $[S'/c] \simeq 0.025 \gg$ than both the paper $[S'/c] \simeq 0.00067$ and nylon $[S'/c] \simeq 0.00011$, despite all fiber lengths being approximately equal, $l_f = 2 \pm 1$ mm; in these cases the aspect ratios (l_f/d_f), respectively asbestos 40000 >> paper 90 and nylon 50, seem to mirror additive efficacy.

To physically interpret the information in Figure 5, let the amount of additive required to induce unit flow enhancement be measured by a characteristic concentration $C = 1/[S'/c]$ wppm, and, separately, let L be the length scale of an additive particle, which is presumed to solely occupy a sphere of that diameter. The additive-pervaded volume fraction X_v at unit flow enhancement is then:

Macromolecules: $X_v = (\pi L_c^3/6)(CN\rho/\mathrm{Mw}10^9)$

Fibers: $X_v = (\pi l_f^3/6)(C(\rho/\rho_f)(4/\pi d_f^2 l_f 10^6))$

where L_c is macromolecule contour length, N Avogadro's number 6.02×10^{23}, ρ solvent density, l_f fibril length, d_f fibril diameter, ρ_f fiber density, and appropriate conversion factors have been applied for lengths in mm, densities in g/cm^3, and concentrations in wppm. As an example, for the authors' PAMH, Mw $= 15\times10^6$, $L_c = 0.053$ mm, $\rho = 1$ g/cm^3, and the observed intrinsic slip $[S'/c] = 0.74$, so $C = 1.3$ wppm and $X_v = 4300$. Thus, if each molecule of our PAMH were fully stretched to its contour length, and solely occupied a solvent sphere of that diameter, then the total additive-pervaded volume fraction at unit flow enhancement would be $X_v = 3100$. Since X_v cannot physically exceed unity, either all the molecules were not fully stretched out to L_c, or, if they were, then they aligned with inter-particle spacing $< L_c$, causing multiple occupancy of the solvent volume associated with L_c. Similar calculations for Oliver's PAMH yield $X_v \simeq 1800$, of the same order of magnitude as ours; also, for the xanthan gum (of uncertain molecular characterization), $X_v \simeq 40$. Among the fibers, asbestos $X_v \simeq 8500$, paper $X_v \simeq 5$, and nylon $X_v \simeq 20$; in all cases, for unit flow enhancement, the calculated $X_v \gg 1$, implying additive alignment. It is also curious that the PAMH macromolecules, with numerous backbone chain links $\simeq 500000$, and the asbestos fibers, with high aspect ratio $\simeq 40000$, seem to exhibit one magnitude of $X_v \simeq 4000$, while the xanthan gum, with relatively few backbone links $\simeq 4300$, and the paper and nylon fibers, with relatively low aspect ratios $\simeq 70$, exhibit a much lower magnitude of $X_v \simeq 15$. Thus X_v, together with an additive parameter akin to backbone chain links and aspect ratio, might eventually offer a universal basis for correlating Type B drag reduction by diverse classes of additives.

Figures 6 and 7 effect some quantitative comparisons between drag reduction by the collapsed and extended conformations of the same additive, C832 PAMH. In Figure 6, results for 20 wppm in 0.10 m NaCl, depicted by the solid symbols and heavy line, are superimposed upon the data for several concentrations in 0.00001 m NaCl, shown by hollow symbols and light lines. The trajectory for 20 wppm in 0.10 m NaCl, abbr 20A, intersects the trajectory for 5 wppm in 0.00001 m NaCl, abbr 5B, at coordinates log $Re_s\sqrt{f}$ = 3.0, $1/\sqrt{f}$ = 15.5; this is termed an "iso-slip" point, since both solutions exhibit the identical flow enhancement, S' = 4.2. At this iso-slip point, 20 wppm of the collapsed conformation and 5 wppm of the extended conformation are evidently equivalent in their ability to reduce drag. We use the term "additive equivalence" to describe such a case, where two different additive solutions induce the same drag reduction. (Strictly, the foregoing is only a case of gross flow equivalence, with identical friction factors; we envision extensions, in ascending echelons of stringency, to mean flow and turbulence structure. The more stringent equivalences imply the less stringent, but not v.v.; thus mean flow equivalence implies gross flow equivalence, because friction factors are merely integrated velocity profiles, but there is no rigorous way to infer mean flow equivalence from gross flow equivalence.) Returning to our example, at the iso-slip log $Re_s\sqrt{f_i}$ = 3.0, the experimental trajectory for 20 wppm in 0.00001 NaCl exhibits $S' \simeq 14$, having reached the maximum drag reduction asymptote M. However, since the lesser concentrations were earlier seen to provide an intrinsic slip $[S'/c] \simeq 0.74$ at log $Re_s\sqrt{f}$ = 3.0, one can reasonably infer that 20B would have exhibited slip $S' \simeq 15$, had this value not caused it to exceed M. We can thus generate, at log $Re_s\sqrt{f_i}$ = 3.0, a slip triad S'(20B, 20A, 5B) = (15, 4.2, 4.2), which reveals that the collapsed to extended slip ratio at constant concentration, R_{sc} = S'(20A)/S'(20B) = 0.28, is roughly the same as the ratio of extended to collapsed concentrations at constant slip, R_{cs} = c(5B)/c(20A) = 0.25. This result is sufficiently striking to warrant displaying another example.

Figure 7 further elaborates additive equivalence between collapsed and extended conformations of C832 PAMH, showing data for 10 wppm in 0.10 m NaCl (10A) and both 2 and 10 wppm in 0.0003 m NaCl (2B and 10B). In the turbulent regime, the intersection of trajectories 10A and 2B defines an iso-slip point, coordinates log $Re_s\sqrt{f}$ = 3.05, $1/\sqrt{f}$ = 13.4, where both solutions exhibit the same S' = 1.9. Too, at the iso-slip log $Re_s\sqrt{f_i}$ = 3.05, the entirely experimental slip triad S'(10B, 10A, 2B) = (9.4, 1.9, 1.9), shows the collapsed to extended slip ratio at constant concentration, R_{sc} = S'(10A)/S'(10B) = 0.20, about the same as the ratio of extended to collapsed concentrations at constant slip, R_{cs} = c(2B)/c(10A) = 0.20. This reinforces the analogous observation in Figure 6. The foregoing examples of additive equivalence offer a noteworthy physical insight. Namely, that

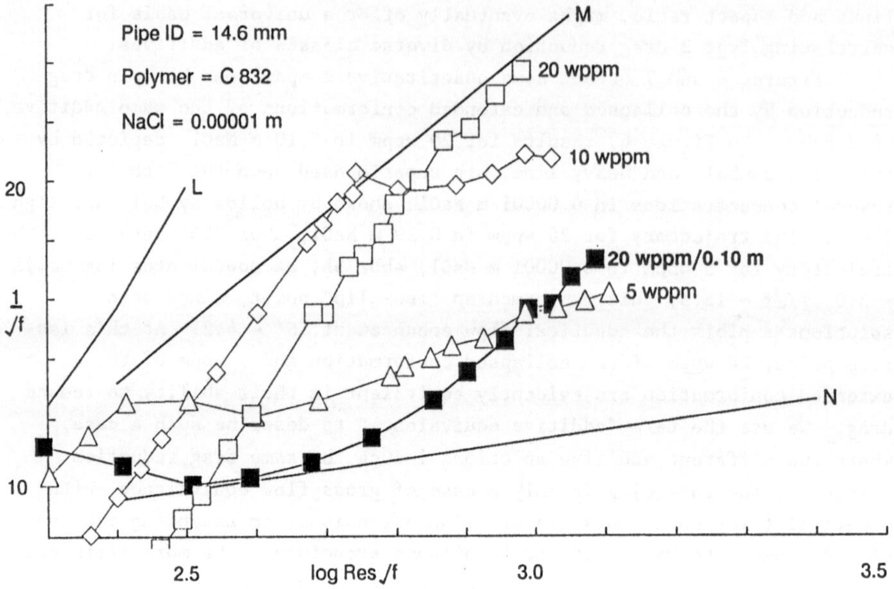

FIGURE 6. "Iso-Slip" Points and Gross Flow "Additive Equivalence".

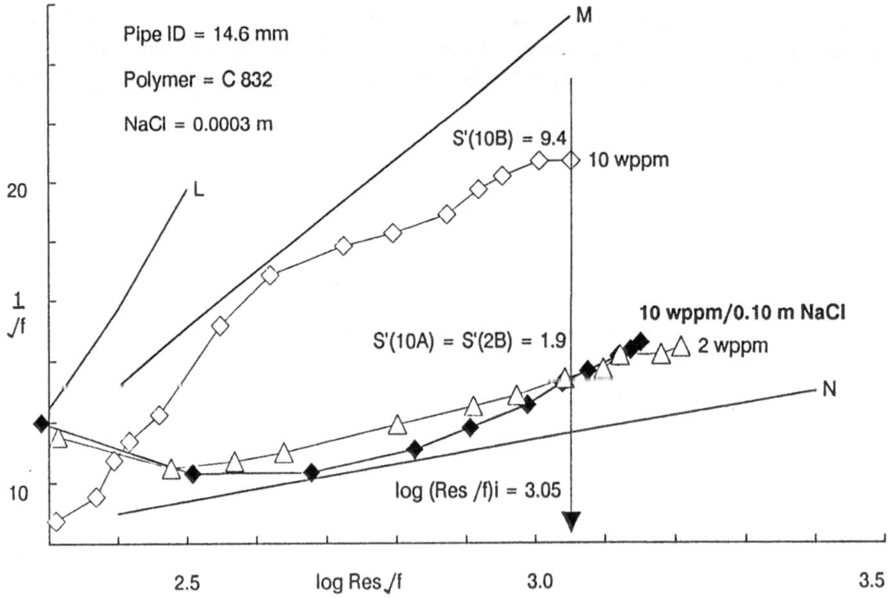

FIGURE 7. Additive Equivalence Among Extended and Collapsed Conformations.

for fixed total additive concentration, the collapsed to extended slip ratio R_{sc} observed at any $Re_s\sqrt{f}$ must simply represent the fraction of originally collapsed macromolecules that have become extended *in situ*, with only the extended macromolecules being effective in drag reduction. The validity of this inference may be directly verified from the respective equalities of R_{sc} and R_{cs} obtained at the iso-slip points in Figures 6 and 7; too, the observations made at the experimental $Re_s\sqrt{f_i}$ must apply to all (turbulent) $Re_s\sqrt{f}$, because the coordinates of the observed iso-slip points are arbitrary consequences of our experimental conditions. If further work verifies the preceding inference that the extended additive states are the active species responsible for drag reduction, then Type B behaviour, and the associated row 3 of our A-B Matrix, should emerge as a fundamental feature of the phenomenon, with the better known Type A behaviour perhaps seen as a special case brought about by the extension, hence activation, of initially randomly-coiled macromolecular additives by the turbulent flow field.

Additive equivalence has not previously been reported in the drag reduction literature, and there appear to be no direct measurements of macromolecular extension during turbulent drag reduction, against which the present inferences might be compared. Interestingly, though, features of our Type A "fan", including onset, can be seen in certain birefringence vs. strain rate curves reported for the deformation of randomly-coiled polystyrene in laminar extensional flows[14,15]; the possible analogy between drag reduction and these rheological observations merits study.

SUMMARY.

1. An Additive-Burst Matrix was formulated to frame theoretical discussion of the drag reduction mechanism. A-B matrix rows represent additive states, from coiled to extended, while columns connote turbulent burst events, from lift-up to breakdown.
2. Types A and B drag reduction were induced experimentally using two polyelectrolytic PAMH additives, Mw $\simeq 15\times10^6$, in saline solutions containing from 0.10 m to 0.00001 m NaCl.
3. Progressive extension of the initial additive conformation increased drag in laminar flow, while greatly reducing drag in turbulent flow. For 20 wppm of PAMH C836, a reduction of salinity from 0.10 m to 0.0003 m caused the laminar drag to increase almost two-fold, while at the highest turbulent log $Re_s\sqrt{f} = 3.15$ the fractional flow enhancement increased from 0.33 to 0.91 relative to solvent. For fixed polyelectrolyte backbone chain and concentration, turbulent flow enhancement by extended conformations always exceeded that from collapsed conformations.
4. Features of Type B drag reduction were delineated. These included: A characteristic "ladder" structure, with flow trajectories roughly

parallel to the Newtonian Prandtl-Karman line. "Retro-onset", where trajectories commence on the maximum drag reduction asymptote at low $Re_s\sqrt{f}$ and then depart into the polymeric region of lower drag reduction. An almost linear increase in flow enhancement with increasing additive concentration; PAMH C832 yielded an intrinsic slip $[S'/c] = 0.74$ at $\log Re_s\sqrt{f} = 3.0$.

5. A preliminary analysis of the Type B drag reduction literature was presented, including both polyelectrolyte and fiber additives. These exhibited intrinsic slips from 4 to 0.000011. For all additives, the characteristic concentrations required to induce unit flow enhancement correspond to additive-pervaded volume fractions $X_v >> 1$, which suggests that the additives are aligned, with inter-particle separations < their extended lengths, during drag reduction.
6. Gross flow "additive equivalence" was detected at "iso-slip" points, where different polymer solutions induced equal flow enhancements. Remarkable equalities were found among certain ratios of flow enhancements and the concentrations of collapsed and extended conformations at iso-slip points. These led us to conclude that, for fixed total additive concentration, the ratio of collapsed to extended flow enhancements at any $Re_s\sqrt{f}$ likely represents the fraction of originally collapsed macromolecules that have been extended by the flow.
7. If extended additive states are the active species that induce drag reduction, as appears likely, then Type B behaviour may well prove more fundamental to the mechanism of additive-turbulence interaction than the hitherto more widely studied Type A.

Acknowledgement: The authors gratefully acknowledge the support of Dr. R.D. Cooper, ONR Fluid Dynamics Branch, for initiating the present work, and the Hertz Foundation, for awarding a fellowship to one of us (D.L.W.).

REFERENCES

1. Hoyt, J.W.: *ASME J. Basic Eng.*, 94, 258, (1972).

2. Lumley, J.L.: *Macromol. Rev.*, 7, 263, (1973).

3. Hoyt, J.W.: p A1 in "Proc. 2nd Intern. Conf. on Drag Reduction," B.H.R.A., Cambridge, UK (1977).

4. Virk, P.S.: *A.I.Ch.E.Jl.*, 21, 625, (1975).

5. Virk, P.S.: p 149 in "Biotechnology of Marine Polysaccharides", R.R. Colwell, Ed., Hemisphere Publ. Corp., N.Y. (1985).

6. Virk, P.S.: *Nature*, 253, 109, (1975).

7. American Cyanamid Co., Ind. Prod. Div., Wayne, NJ 07470, USA (1987).

8. Oliver, D.R.; Bakhtiyarov, S.I.: *J. Non-Newt. Fl. Mech.*, 12, 113, (1983).

9. Rochefort, S.; Middleman, S.: p 117 in "Polymer-Flow Interaction", Y. Rabin, Ed., AIP Conf. Proc. 137, New York (1985).

10. Sharma, R.S.; Seshadri, V.; Malhotra, R.C.: *J. Rheol.*, 22, 643, (1978).

11. McComb, W.D.; Chan, K.T.J.: *J. Fl. Mech.*, 152, 455 (1985).

12. Lee, P.F.W.; Duffy, G.G.: *A.I.Ch.E.Jl.*, 22, 750, (1976).

13. Bobkowicz, A.J.; Gauvin, W.H.: *Can. J. Chem. Eng.*, 43, 87, (1965).

14. Pope, D.P.; Keller, A.: *Coll. & Polym. Sci.*, 256, 751, (1978).

15. Leal, L.G.; Fuller, G.G.; Olbricht, W.L.: p 351 in "Viscous Flow Drag Reduction", G.R. Hough, Ed., Prog. Astro. Aero., 72, New York (1980).

Change of Structures Close to the Wall of a Turbulent Flow in Drag Reducing Fluids

Albert Gyr and Hans-Werner Bewersdorff

Institute of Hydromechanics and Water Resources Management, ETH Zürich, Switzerland

Abstract

In flows of drag reducing polymer solutions a Reynolds stress deficit occurs in the momentum balance. When this deficit is interpreted as an increase of the local effective viscosity, it produces a viscous lid at the outer edge of the large counterrotating longitudinal near-wall vortices. The longitudinal vortices in the near-wall region are calculated by solving the Navier-Stokes-equations under the slender body assumption, and by using the concept of an effective viscosity which increases with the wall distance. The results show an increase in size and a decrease of the vorticity of these vortices.

Detailed measurements of the turbulence structure in order to understand the basic mechanism responsible for drag reduction in dilute polymer solutions, revealed the following picture:

1. In the time-averaged velocity profile drag reduction is accompanied by an alteration of the buffer zone between the viscous sublayer to an elastic sublayer and the core region, whereas the core region exhibits a parallel shift. The size of the elastic sublayer increases with the drag reduction until in case of Virk's maximum drag reduction [1] the turbulent core vanishes and the elastic sublayer expands to the centre of the pipe or channel.

2. In pipe or channel flows of drag reducing polymer solutions the absolute values of the turbulence intensities are generally reduced in comparison with those of the Newtonian solvent. If normalized by the friction velocity u_τ the maximum of the axial turbulence intensity exhibits no general tendency, i.e. it can decrease or increase [2]. With increasing drag reduction the maximum is shifted towards the centre of the pipe or channel. The radial turbulence intensity in pipe flows or the transverse turbulence intensity in channel flows is strongly reduced in drag reducing fluids, even when normalized by u_τ [3, 4]. An important property for characterizing the turbulent fluctuating motion is the ratio of both turbulence

A. Gyr (Editor)
Structure of Turbulence and Drag Reduction
IUTAM Symposium Zurich/Switzerland 1989

intensities. In flows of drag reducing polymer solutions the anisotropy of the fluctuating motion becomes more pronounced. The anisotropies of the fluctuating motion are shown in Fig.1 where the joint probability density distribution of the axial and transverse velocity fluctuations of a drag reducing polymer solution and the Newtonian solvent are compared at nearly the same dimensionless wall distances. Significant differences occur between the buffer zone of the Newtonian solvent and the elastic sublayer of the polymer solution.

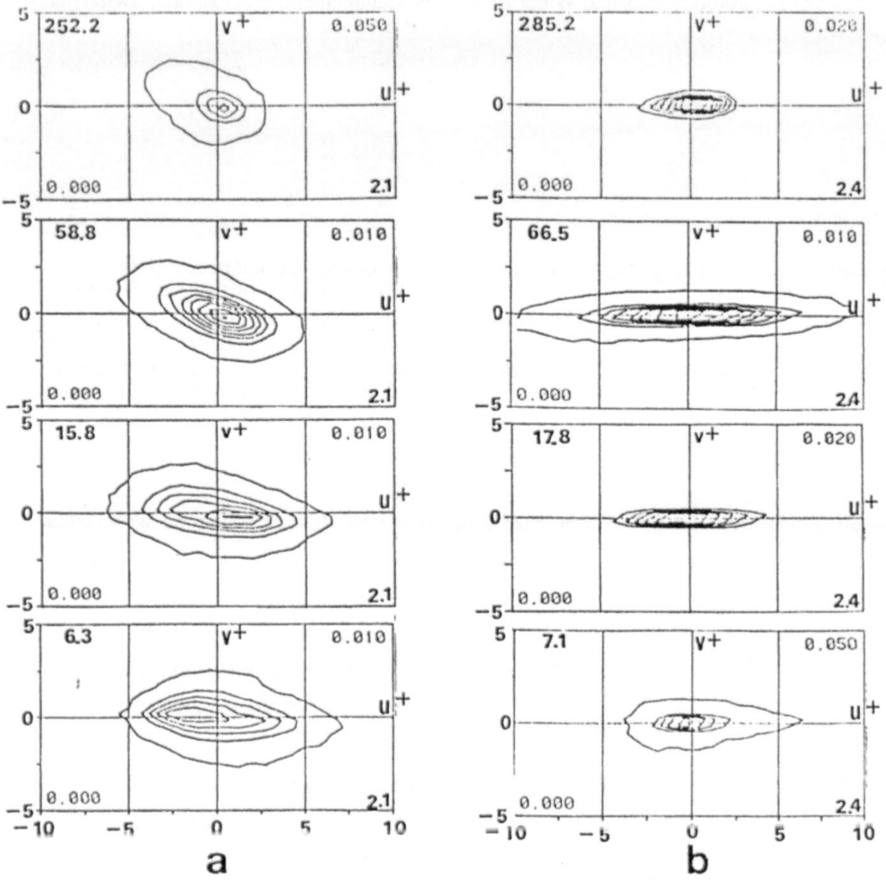

Fig. 1: Joint probability density distribution of the u'- and v'-velocity fluctuations normalized by u_τ in an open channel flow. (a) solvent, (b) polymer solution, polyacrylamide Separan XD 30207. The numbers in the top left corner of each plot designate y^+, the one in the top right corner the increment of the probability curves, and the one to the right of the bottom u_τ

Dilute drag reducing polymer solutions exhibit the following rheological properties: They are viscoelastic fluids. Due to the low concentrations used for drag reduction the deviations in the shear viscosity in comparison with that of the solvent can be neglected, whereas the elongational viscosity, which is in uniaxial elongational flows three times the shear viscosity, can be increased in polymer solutions. However, there exist drag reducing polymer solutions which exhibit only an increased elongational viscosity in comparison with that of the solvent when pre-sheared [5, 6]. In these pre-shearing experiments the fluids are first subjected to a laminar Couette or channel flow before they experience the additional elongational field produced by an orifice mounted in the wall. The experimental results described above demonstrate the difficulties in describing the rheological behaviour of dilute drag reducing polymer solutions. The most spectacular behaviour in the flow behaviour of dilute polymer solutions in comparison with the solvent therefore is the drag reduction itself.

Dilute polymer solutions exhibit an onset behaviour, i.e. below a certain wall shear gradient their friction behaviour does not differ from that of the solvent. When a certain wall shear gradient is exceeded the friction factor is lowered, i.e. drag reduction occurs. This onset behaviour finds its analogy in the rheological pre-shearing experiments described above. However, in a turbulent flow due to the eddy motion a spectrum of shear and strain rates occur. Therefore, and due to the fact that the flow behaviour of viscoelastic fluids depends on the deformation history, it is difficult to introduce the correct rheological behaviour for modelling the turbulence structure in turbulent flows of drag reducing polymer solutions. Measurements of the Reynolds shear stresses in pipe or channel flows of drag reducing fluids demonstrate that in the Reynolds stress balance a deficit problem exists [3, 7, 8]:

$$- \frac{\overline{u'v'}}{u_\tau^2} = 1 - \frac{y}{R} - \frac{dU^+}{dy^+} - G \qquad (1)$$

In eq.(1) G stands for the Reynolds stress deficit or the "elastic stresses" as it was called in [3]. This deficit occurs in flows of drag reducing polymer solutions whenever the shear viscosity is used for calculating y^+. These experimental results give an indication that the shear and elongational processes alter the viscosity in the buffer zone. Figure 2 shows the Reynolds stress deficit in a pipe flow at maximum drag reduction conditions [3].

The elastic stresses show up as a damping of the turbulent shear stresses. However, eq.(1) which is a momentum balance provides no information whether the production or the dissipation is influenced.

An alternative attempt to explain the drag reduction phenomeneon is based on the hypothesis that certain classes of anisotropic flow fields are less energy consuming. Tsinober [9] showed that some properties in MHD-flows exhibit a drag reducing behaviour. These flows have a predominant helicity or are nearly two-dimensional.

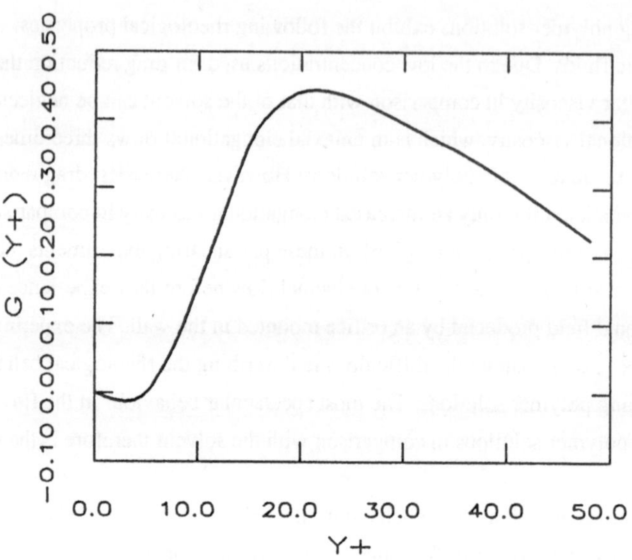

Fig. 2: Reynolds stress deficit G in a pipe flow of a polymer solution at maximum drag reduction conditions from [3]

It is therefore interesting to know in which way rheological properties of dilute polymer solutions interact with the structures of a turbulent flow field and to look whether they influence the anisotropies in the near-wall region or not. This can be done in two different ways: First, by considering structures of turbulence with fields of high strain which influence the rheological properties of the drag reducing fluid as done in [10] or, secondly, by considering structures of turbulence with low strain which do not influence the rheological properties of the polymer solution. In elongational fields, mainly in the so-called ejections of the near-wall burst process, the polymer molecules will be stretched which results in an increased elongational viscosity. Events with high internal strain fields therefore can not be modelled by the shear viscosity behaviour of dilute polymer solutions. However, it is possible to consider flow structures with low strain by means of an "effective" viscosity as introduced in [7, 8].

Well-known coherent structures with low strain fields are the counterrotating vortices in the near-wall region [11]. These vortices are oriented in flow direction and behave quasi periodic in the spanwise direction. They are responsible for the low speed streaks close to the wall. It is assumed that they are the essential structure responsible for the burst mechanism due to two instability processes:

(1) A second instability of the vortices themselves, a "flapping" of them in crossflow direction, occurs and results in a vortex breakdown.

(2) They produce an inflexional profile in the regions where low speed fluid is lifted up as a result of the vortex motion.

In the present work rheological properties of polymer solutions will be related to this class of coherent structures of the near-wall turbulence.

When the Reynolds stress deficit G in eq.(1) is attributed to an increased local effective viscosity

$$\eta = \frac{\nu_{eff}}{\nu_o} \tag{2}$$

as suggested in [8], eq.(1) becomes [9]

$$-\frac{\overline{u'v'}}{u_\tau^2} = 1 - \frac{y}{R} - \eta \frac{dU^+}{dy^+} \tag{3}$$

A plot of the Reynolds stress deficit of the data of Fig.2 in terms of the effective viscosity is shown in Fig.3. The effective viscosity increases with the wall distance in the elastic sublayer. It acts therefore like a "viscous lid". The new effective viscosity η exhibits its maximum at a dimensional wall distance y_m^+ which is of the order of the outer edge of the longitudinal vortices. The viscous lid therefore should stabilize both instabilities mentioned above.

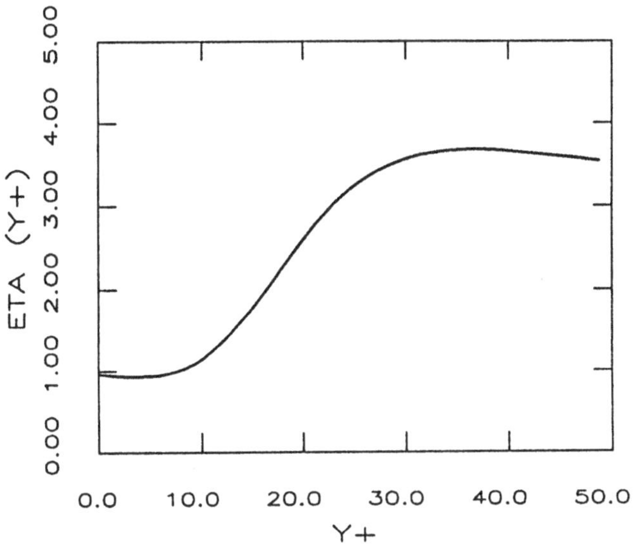

Fig. 3: Effective viscosity η as a function of the dimensionless wall distance for a pipe flow at maximum drag reduction conditions

Based on the effective viscosity data shown in Fig.3 the question arises in which way the counterrotating vortices as detected in Newtonian fluids are altered by the changed rheological properties of the drag reducing polymer solution. Due to the complexity of this problem, and remembering that the flow behaviour of viscoelastic fluids depends on the deformation history, it cannot be expected that the alteration of the flow field can be investigated in a single step. The problem is therefore treated in a very simplified way. However, at least the relevant properties have to be preserved and the problem is simplified by introducing the following assumptions:

1. The flow is considered to be steady. This implies that the lifetime of the vortices are long enough to investigate the relevant flow structures by means of the steady form of the momentum equations.
2. The problem is treated under the slender body assumption. Since the measured longitudinal vortex structures remain two-dimensional, at least in the short time approximation, it becomes obvious to separate the spanwise from the streamwise flow field.

With these approximations the spanwise flow field can be described by the stream function ψ and the streamwise vorticity ω_x

$$\omega_x = \frac{\partial w}{\partial y} - \frac{\partial v}{\partial z} \tag{4}$$

with

$$v = -\frac{\partial \psi}{\partial z} \tag{5}$$

and

$$w = \frac{\partial \psi}{\partial y} \tag{6}$$

Therefore ψ and ω_x are related by the Poisson equation

$$\omega_x = \frac{\partial^2 \psi}{\partial y^2} + \frac{\partial^2 \psi}{\partial z^2} \tag{7}$$

and the vorticity equation

$$\frac{\partial}{\partial z}\left(\frac{\partial \psi}{\partial y}\omega_x\right) - \frac{\partial}{\partial y}\left(\frac{\partial \psi}{\partial z}\omega_x\right) = \eta\left(\frac{\partial^2 \omega_x}{\partial z^2} + \frac{\partial^2 \omega_x}{\partial y^2}\right) + 2\,\frac{\partial \eta}{\partial y}\,\frac{\partial \omega_x}{\partial y} + \\ + \frac{\partial^2 \eta}{\partial y^2}\left(\frac{\partial^2 \psi}{\partial y^2} - \frac{\partial^2 \psi}{\partial z^2}\right) \tag{8}$$

which contains the changed rheological behaviour.

These two equations are identical to those of Hanratty and coworkers [12-14] for a Newtonian fluid ($\eta = 1$) and they were used to explain the mechanism of drag reduction. Hanratty et al. showed that an increase of the diameter of the longitudinal vortices results in drag reduction. Since such an increase is observed in flows of polymer solutions they argued that this would explain the drag reduction mechanism. Following their arguments an alteration in the rheology therefore should result in an increase of the vortex size.

To study the influence of the changed rheological properties on the longitudinal vortices the equations (7) and (8) were iteratively integrated by using a difference scheme with the non-slip condition at the wall as a first boundary condition. The boundary conditions were completed by two additional conditions:

1. The flow field is treated periodically in the spanwise direction with a periodicity of $\lambda_z{}^+ = 100$, the magic number of the near-wall vortices.
2. Instead of an upper boundary condition an energy preserving argument was used. In the whole flow field which is limited by $\lambda_z{}^+ \cdot y_m{}^+$ the kinetic energy $v^2 + w^2$ was kept constant.

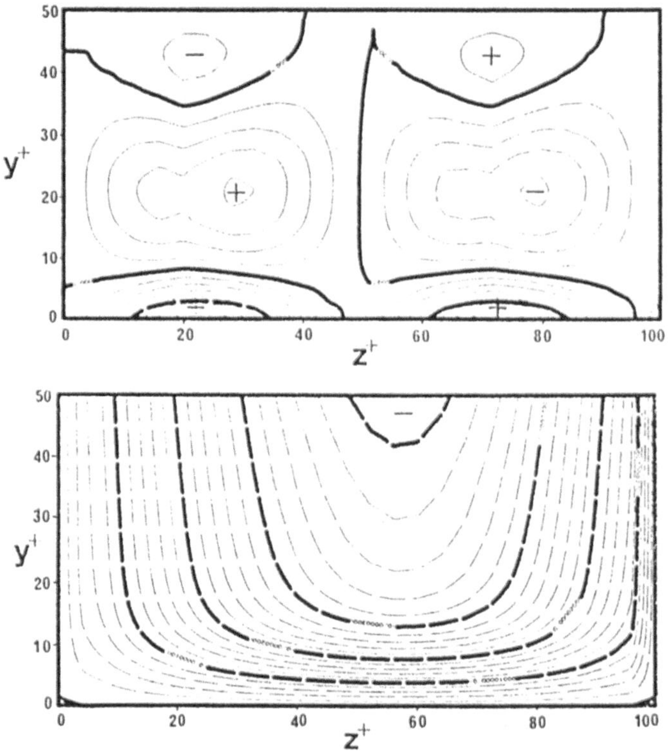

Fig. 4: The near-wall longitudinal vortices in a Newtonian fluid (a) and in a drag reducing fluid (b) using the concept of an increased local effective viscosity with the values given in Fig.3. The solid lines are positive and the dashed negative iso-vorticity lines. In Fig. (a) $\Delta\omega_x^+ = 0.0025$, whereas in Fig. (b) $\Delta\omega_x^+ = 0.000002$

The results of the calculations are shown in Fig.4. In Figure 4 (a) the well-known counter-rotating longitudinal vortices with their centre located at about $y^+ = 20$ can be recognized. Due to the non-slip condition at the wall a strong vorticity with the opposite sign in comparison to that of the longitudinal vortices situated above exists in the viscous sublayer. At about $z^+ = 50$ the centre of a low speed streak appears in the viscous sublayer.
Figure 4 (b) exhibits the shape of the vorticity when the new rheology is introduced. Instead of two counterrotating vortices only one vortex is established in the perodicity area of the Newtonian fluid. It is increased in size, its centre is shifted to a larger dimensionless wall distance, and its vorticity is decreased by two orders of magnitude. Furthermore, the modulation of the viscous sublayer changes. The calculations are not sensitive enough to resolve the structures close to the boundaries where high shear layers must exist.
The aim of the calculations was not a simulation of the complete turbulent flow field in the near-wall region. The calculations were done to elucidate in which way a changed rheology can influence a well-known structure of turbulence. The experimental results of the joint density probability distribution of the velocity fluctuations shown in Fig. 1 can be interpretated by the development of a higher helical flow.

Acknowledgement

The second author thanks the Deutsche Forschungsgemeinschaft (DFG) for supporting the work by research grant Be 1056/2.

References

[1] Virk PS (1975) AIChE J 21:625
[2] Berman NS (1978) Ann Rev Fluid Mech 10:47
[3] Schümmer P, Thielen W (1981) Chem Eng Commun 4:593
[4] Willmarth WW, Wei T, Lee CO (1987) Phys Fluids 30:933
[5] Vissmann K in: Giesekus H, Hibberd MF (eds) Proc 2nd Conf European Rheologists, Prague (1986) Steinkopff, Darmstadt, 252
[6] James DF, McLean B, Saringer JH (1987) J Rheol 31:453
[7] Giesekus H (1981) Von Kármán Institute for Fluid Dynamics, Lecture Series 1981-86, Non-Newtonian Flows
[8] Bewersdorff HW, Berman NS (1988) Rheol Acta 27:130
[9] Tsinober A (1989) in: Gyr A (ed) Structure of Turbulence and Drag Reduction, Springer, Berlin
[10] Gyr A (2-5 July 1984) 3rd Intern Conf Drag Reduction, Bristol, England, paper B10
[11] Blackwelder RF, Eckelmann H (1979) J Fluid Mech 94:577
[12] Hanratty TJ (1989) in: Gyr A (ed) Structure of Turbulence and Drag Reduction, Springer, Berlin
[13] Nikolaides C (1984) Ph.D. thesis, University of Illinois, Urbana
[14] Lyons SL, Nikolaides C, Hanratty TJ (1988) AIChE J 34:938

The Influence of Polymer Additives on the Coherent Structure of Turbulent Channel Flow

B. Gampert, C. K. Yong

Universität-GH-Essen, Strömungsmechanik
Schützenbahn 70, 4300 Essen 1,
West Germany

Abstract

The intension of this paper is to identify changes in the ejection and sweep events due to the addition of the ionic polyacrylamide Praestol 2360 (Fa. Stockhausen) to deionized water. Laser-Doppler measurements have been carried out in fully developed turbulent square duct flow of polymer solutions of comparatively low Re numbers between Re = 4800 and 16000. Drag reductions were close to the maximum values as given by Virk. Mean velocity profiles, Reynolds stress values and rms-values of the velocity fluctuations are presented. Joint probability density functions are given for (u',v') and ($\partial u'/\partial t$, $\partial v'/\partial t$). Time fractions, contributions to the Reynolds stress and to ($\partial u'/\partial t$, $\partial v'/\partial t$) in the four quadrants of the u'-v'-plane have been measured.

Introduction

The influence of polymers on the wall near structures in turbulent flow has been investigated first in papers by Fortuna & Hanratty [1] and Donohue, Tiederman & Reischman [2]. In recent years papers on this subject by Tiederman, Luchik & Bogard [3], Schmid [4], Gyr [5], and Luchik & Tiederman [6],[7] have appeared. The main intension of the present paper is to identify changes of the ejection and sweep events due to the addition of polymers to fully developed turbulent square duct flow of water. Such changes are of relevance for the qualitative and quantitative understanding of the mechanism of drag reduction by polymer additives.

Experimental method

The experimental configuration is similar to the arrangement shown in the paper of Gampert & Delgado [12]. The test chamber which is part of a recirculating system is 3.3 m in length. The cross section is quadratic with a width of 19 mm. The test section itself is situated 2.7 m from the channel inlet; this corresponds to approximately 160 hydraulic diameters.
A modular one-colour LDV-system from Dantec was used. It was

A. Gyr (Editor)
Structure of Turbulence and Drag Reduction
IUTAM Symposium Zurich/Switzerland 1989

operated as a two component configuration whereby the seperation of the individual velocity components was achieved by using the reference beam method in the forward scattering mode.
Velocity measurements were restricted to the symmetry plane of the channel. The light source was a He-Ne-laser with a power output of 15 mW. Two photomultipliers constituted the light gathering detectors. The Doppler signals were processed by two frequency trackers which in turn produced two analog signals corresponding to the velocities being measured. Both analog signals from the trackers were digitalized by the A-D-converter of a computer (Digital PDP-11-23). When larger amounts of data were required a dual-channel digital recorder was used; the data were then transmitted over a IBM-AT personal computer to the university mainframe computer for the data evaluation and analysis. The polymer used was the commercially available polyacrylamide Praestol 2360 (Fa. Stockhausen) with a weight-average molecular mass of $\overline{M}$ = $3\text{-}4 \cdot 10^6$ g/mol solved in deionized water.

Results and discussion

In figure 1 mean velocity profiles for pure water and for polymer solved in deionized water are presented in the non-dimensional wall-layer coordinates $u^+=\overline{u}/u_\tau$ and $y^+= y_\tau/\nu$. The viscosity ν in the case of polymer solutions has been obtained for each turbulent flow from the shear stress and the shear rate at the wall. The percentage drag reduction DR was calculated by comparing the friction coefficient in a solvent and a drag reducing flow at equal flow rate. Drag reductions DR are close to the maximum values DR_{max} as given by Virk.
In the viscous sublayer the data from all volocity measurements collapse onto one curve. For y^+>8-10 the data depart from the curve $u^+=y^+$ and follow for y^+>30 rather closely Virks ulitmate profile. In figure 2 the rms-values of the streamwise velocity fluctuation in drag reducing and solvent flows are plotted. For drag reducing flow u'_{rms}/u_τ strongly depends on the Re number and increases with increasing Re number (Re = $\overline{u}_m$ d/ν, ν= viscosity of the solution). In the comparatively low Re number regime investigated the values of u'_{rms}/u_τ have been strongly reduced

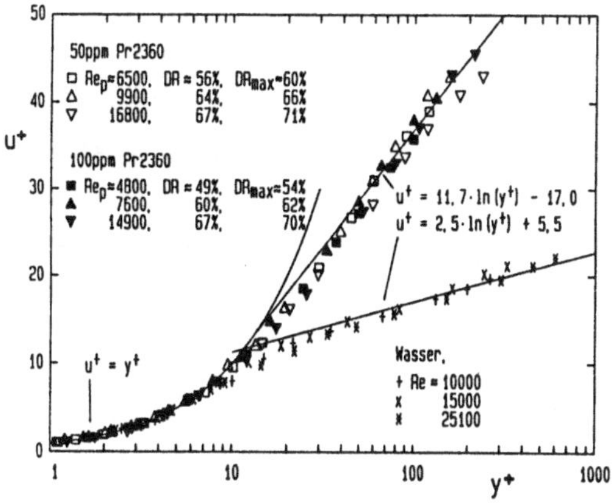

Figure 1: Mean streamwise velocity profile

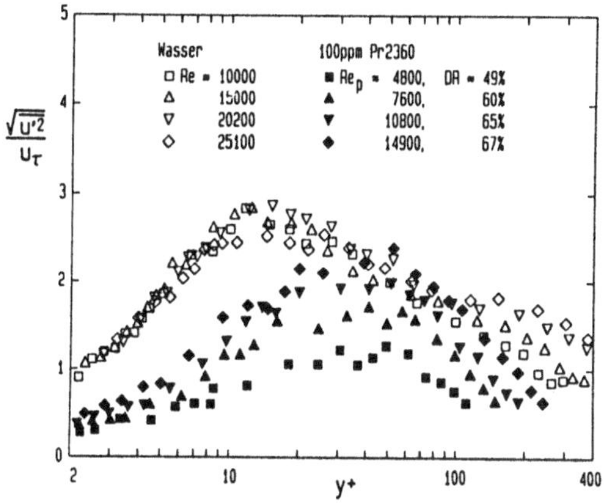

Figure 2: Forward velocity fluctuations normalized by u_τ

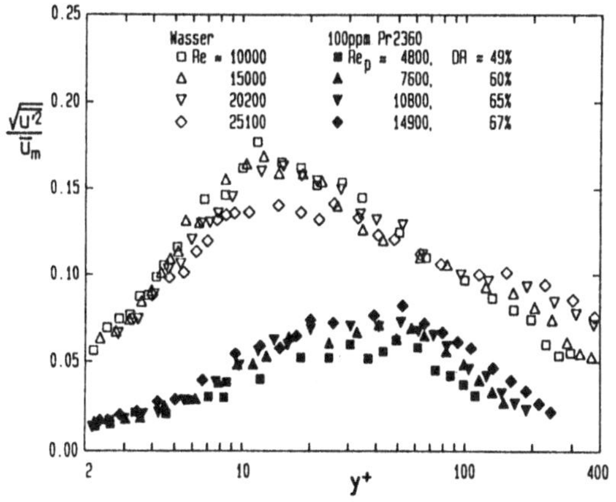

Figure 3: Forward velocity fluctuations normalized by $\bar{u}_m$

by the addition of polymers while the maxima have moved away from the wall to $y^+_{max} = 30\text{-}60$.

As pairs of Re numbers correspond to the same flow rate the absolute fluctuation values for water and polymer solutions can be compared directly when u'_{rms} is non-dimensionalized b $\bar{u}_m$. Figure 3 shows that u'_{rms} is drastically reduced by polymers when $\bar{u}_m$ is held constant. In general the data of the figures 1-3 do not confirm the square duct laser measurements of Rudd [9] but are in good accordance with the results of Reischman & Tiederman [10] and of Gampert & Delgado [8] which were obtained in two-dimensional flow.

For Re numbers larger as presented in figures 1-3 we found that u'_{rms}/u_τ is slightly increased by polymers. But the drastic increase shown by Rudd did not appear.

We cannot exclude the possibility that the influence of secondary flows will be of greater importance for Re numbers as large as investigated by Rudd

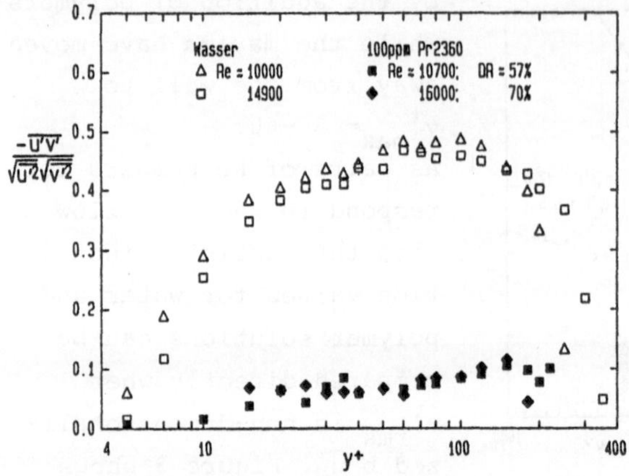

Figure 4: Reynolds stress profiles

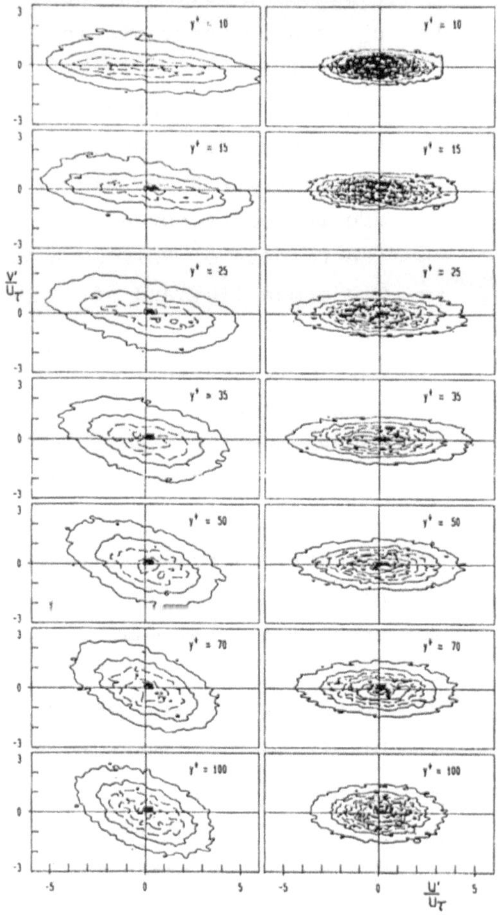

(Re=50000). But our results show no tendency from which we could conclude that this is very likely.

In case that the Reynolds stress is diminished only by reduced fluctuation values the expression $-\overline{u'v'}/(u'_{rms} v'_{rms})$ remains unchanged. In figure 4 the water values quickly increase with larger wall distances and obtain values between 0.4 and 0.5 in a large region of the profile. The polymer solution data show that the drag reduction by polymers not only results from diminished fluctuations but in addition from a decoupling of the u'- and v'-fluctuations.

The instantaneous signals u' and v' are recorded in figure 5 where the joint probability density function $JPDF^+(u',v')$ of the forward and the normal velocity fluctuation is presented.

Figure 5: $JPDF^+$ of water (left) and 100 ppm Pr 2360 (right) Re=15000. Min $JPDF^+$ =0.01, Δ=0.025

Along the elliptic contour lines $JPDF^+$ is constant. First we regard the results of the flow of pure water (Re ≈ 15000). In the region of maximum energy production ($y^+ \approx 10-15$) the ellipses are very slender and orientated almost in the x-direction. High fluctuations u' can appear with fluctuations v' being small simultaneously when the coherent structures in y-direction are comparatively homogenous. The turning of the ellipses which becomes stronger with increasing wall distance corresponds to an enhanced negative correlation and enlarged Reynolds stresses $-\rho \overline{u'v'}$. Simultaneously the ellipses become smaller. Positive values of the Reynolds stress result from the second quadrant (u'<0,v'>0) and the fourth quadrant (u'<0,v'>0) underlining the great importance of the events belonging to these parts of the u'v'-plane. The contribution from the first and the third quadrant to the Reynolds stress is negative and comparatively small.

Regarding the data for the polymer solution flow (Re ≈ 15000, DR= 67%, DR_{max}=70%) one realizes that the turbulence field is strongly modified in the productive, or buffer, region. The ellipses are very slender and indicate that the flow field is strongly non-isotropic.

The instantaneous fluctuation velocities behave very similar to the rms-values. The fluctuations are reduced and maxima are moved away from the wall by polymer additives. Another important change is the extremely reduced correlation of the u' and v' fluctuations which is indicated by the almost horizontal orientation of the ellipses. If the value of $JPDF^+(-u',v')$ is the same as that of $JPDF^+(u',v')$ this corresponds to a decoupling of the phase relationship between u' and v' and $\overline{u'v'} = 0$ results.

The joint probability density function of the time derivates $\partial u'/\partial t$, $\partial v'/\partial t$ gives an impression of the intensity of the events and is presented in figure 6 where $\partial u'/\partial t$ and $\partial v'/\partial t$ have been non-dimensionalized by inner parameters corresponding to a multiplication by (ν/u_τ^3). One finds that the fluctuations $\partial u'/\partial t$ decrease towards the centre of the channel while fluctuations $\partial v'/\partial t$ become more intensive. Substructures appear in the first and third quadrant of the $\partial u'/\partial t$ - $\partial v'/\partial t$-plane.

In the flow of polymer solutions the long and slender geometry

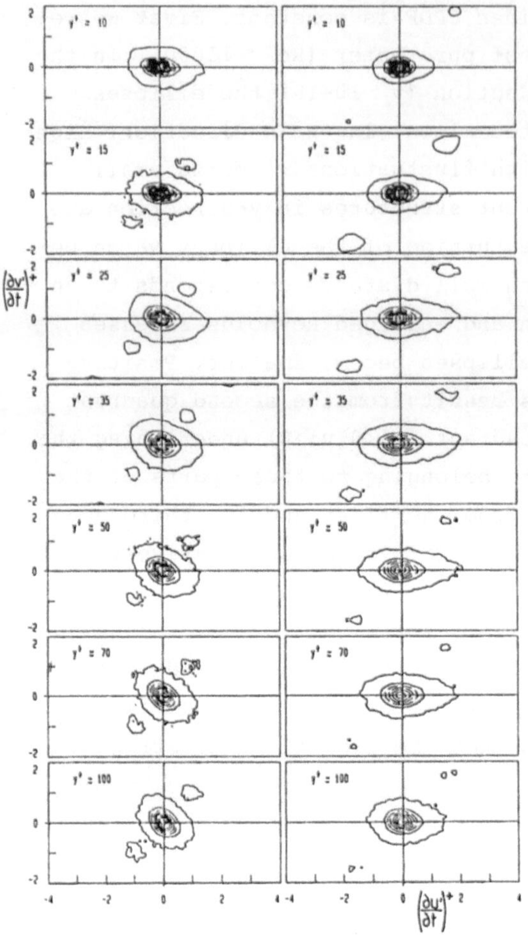

Figure 6: $JPDF^+(\partial u'/\partial t, \partial v'/\partial t)$ of water (left) and 100 ppm Pr 2360 (right), Re= 15000, Min. $JPDF^+$= 0.01, Δ=0.025

of the ellipses indicating that the flow field is non-isotropic is present as well. But a decoupling of $\partial u'/\partial t$ and $\partial v'/\partial t$ appears.

In addition remarkable changes of the ellipses with increasing y^+ can be observed. The largest extension appears at y^+=35, at the point of maximum turbulence production for polymer solutions. The substructures in the first and third quadrant can be found for polymers as well.

Contributions to $-\rho\overline{u'v'}$ in the four quadrants of the u'v'-plane have been computed in order to obtain a more detailed picture of the turbulence structure. The four quadrants can be connected with special kinds of motion discovered by flow visualization techniques. With the first quadrant are connected outward interactions (u'>0, v'>0), with the second quadrant ejections (u'<0,v'>0), with the third quadrant wallward interactions (u'<0,v'<0) and with the fourth quadrant sweeps (u'>0,v'<0). Not each motion of the second or fourth quadrant necessarily corresponds to a sweep or an ejection as originally discovered as wall near, coherent structures. In order to have names available in what follows the fluid motions of the second and fourth quadrant will be called ejections and sweeps instead of the difficulties just mentioned.

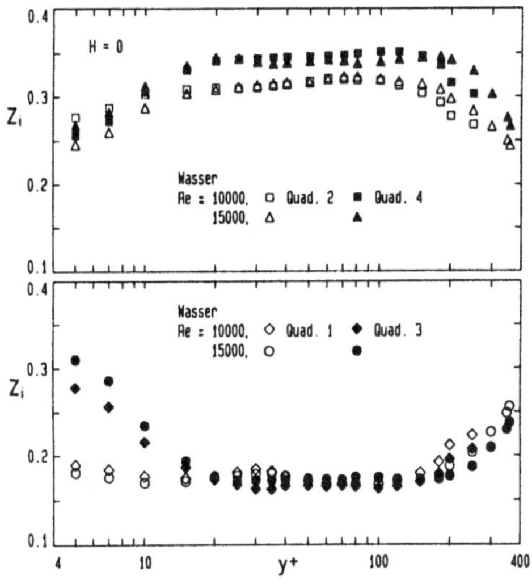

Figure 7: Time fractions Z(i) of pure water flow in the four quadrants (H=0)

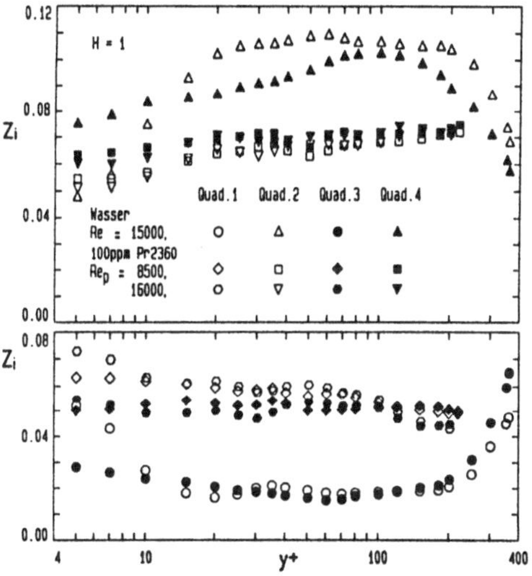

Figure 8: Time fractrions Z(i) of water and polymer solutions in the four quadrants (H=1)

The first step consists in characterizing the four quadrants according to their time fractions. We define Z(i) = T(i)/T, where T(i) is the time during which the u'v'-signal is present in the i-th quadrant and T is the total time. The relative time fractions Z(i) of the four quadrants are shown in figure 7 for pure water flow. In the buffer layer all time fractions are almost constant but different in value. Sweep events have a larger time fraction than ejections. These two classes of events dominate the two other classes. Willmarth & Lu [11] have published time fractions Z(i) at $y^+=30$ for turbulent boundary layer flow. Our results are in good agreement with those published in [11]. The behaviour of those events which posess high instantaneous u'v'-values can be studied by defining a threshold. The criterion used to determine when the product u'v' is sampled is that $|u'v'|$ should have attained a certain value $|u'v'|>H \cdot u'_{rms} \cdot v'_{rms}$. The threshold value H=1.0 is chosen. Time fractions calculated for this condition are presented in figure 8. In figure 7 Z(i=4)>Z(i=2) but when the hole criterion is applied one has Z(i=2)>Z(i=4). Only in the immediate vincinity of the wall ($y^+>10$) the time fraction for sweeps is larger than for ejections, see figure 8.

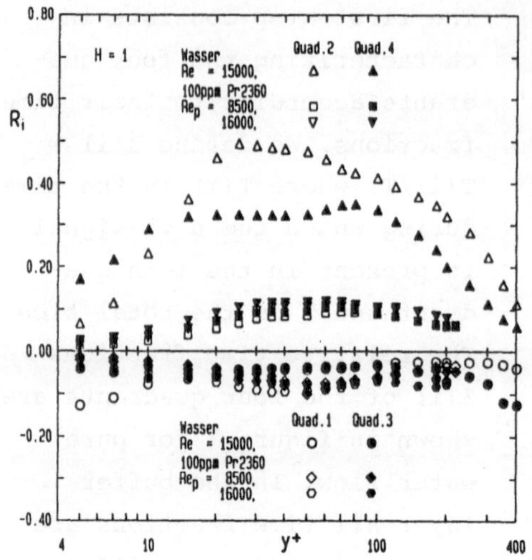

Figure 9: Reynolds stress contributions in the four quadrants for water and polymer solutions (H=1)

As compared with water sweeps and ejections are definetly reduced for $y^+ > 20$ by polymer additives. The interaction events are increased and their time fractions are about the same as those for sweeps and ejections. The effect of polymer additives mainly consists of an equalization of the time fractions of the four classes of events. Remarkable is the influence on the ejections as their time fraction no longer is larger than the time fraction of the sweeps.

Figure 8 shows that for a large fraction of the time |u'v'| is very small relative to shorter intervals of intense activity. Roughly speaking the intermittency factor for u'v' (at $y^+=30$) is 0.75. In pure water flow and in polymer solution flow the signal u'v' is in the hole 75% of the time.

Of special importance is the question how much each of the four quadrants contributes to the Reynolds stress. The contributions of the correlation $\overline{u'v'}$ in each quadrant were non-dimensionalized by u^2_τ. The relative contributions R(i) which were obtained for H=1 are plotted in figure 9.

In pure water flow the largest contribution of the Reynolds stress occurs in the quadrant(u'<0,v'>0) which represents ejection events. There is, however, also a large contribution to the Reynolds stress in the fourth quadrant, where u'>0 and v'<0. Sweep events even dominate for $y^+ \leq 10$-15. The interaction events of the first and third quadrant give only small negative contributions. The results of our four quadrants analysis in water are in substantial agreement with results presented in the literature.

In polymer solution flow the Reynolds stress is reduced drastically. This results from a supression of the sweep and ejection events. For all y^+-values the contributions of sweeps and ejections are of similar strength. Interaction events are increased by polymers. The contributions to the Reynolds stress of all four quadrants are approximately equal.

Finally the joint probability density function of the time derivates $JPDF^{+}(\partial u'/\partial t, \partial v'/\partial t)$ has been distributed to the four quadrants of the u'v'-plane, see figures 10,11.

In the case of polymer solutions substructures appear only at $y^{+}=15$. One interpretation is that substructures preceed interaction events. Their high acceleration values mean that incidents which lead to interaction processes proceed abruptly and intensively.

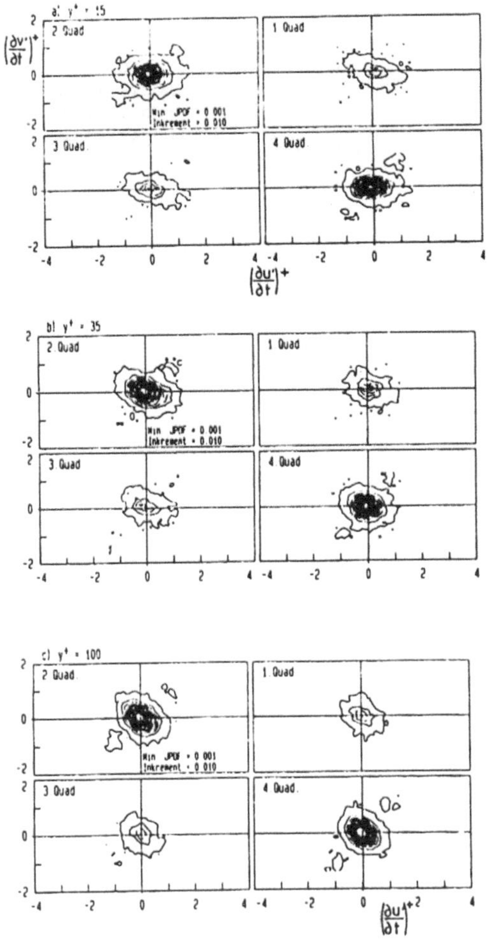

Figure 10: Contributions of the four quadrants to $JPDF^{+}(\partial u'/\partial t, \partial v'/\partial t)$ for water (Re=15000), H=1.

This research was supported by the "Ministerium für Wissenschaft und Forschung des Landes Nordrhein-Westfalen" and by the Universität Essen (Forschungspool). The authors gratefully acknowledge Dr. A.Gyr (ETH Zürich) for his helpful comments and suggestions. Futher acknowledgement goes to Fa. Stockhausen for their generous donation of the drag-reducing additives.

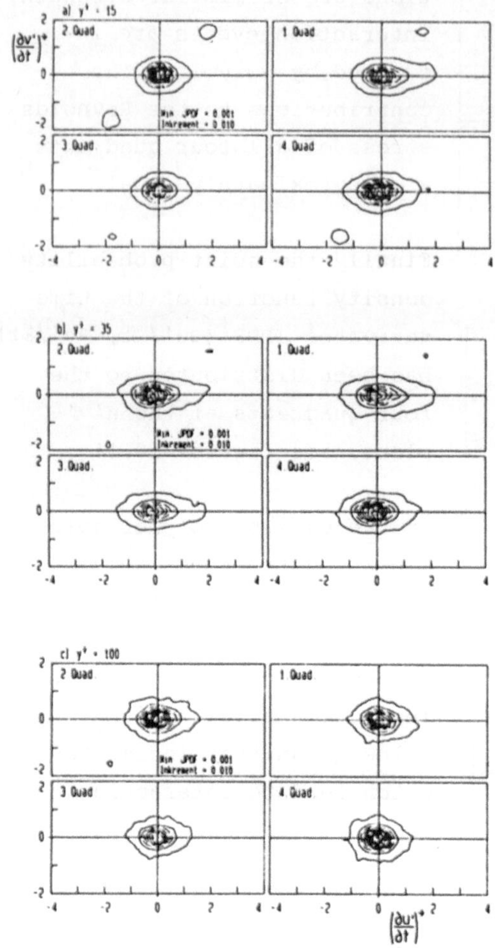

Figure 11: Contributions of the four quadrants to JPDF$^+$($\partial u'/\partial t$, $\partial v'/\partial t$) for polymer solution flow (100 ppm Praestol 2360, Re=15000), H=1

1. Fortuna,G.;Hanratty,T.J.: The influence of drag-reducing polymers on turbulence in the viscous sublayer. J.Fluid.Mech. 53 (1972) 575-586.
2. Donohue,G.L.;Tiederman,W.G.;Reischman, M.M.: Flow visualization of the near-wall region in drag-reducing channel flow. J.Fluid.Mech. 56 (1972) 559-575.
3. Tiederman,W.G.;Luchik,T.S.;Bogard,D.G.: Wall-layer structure and drag reduction. J.Fluid.Mech. 156 (1985) 419-437.
4. Schmid,A: Experimental investigation of the influence of drag reducing polymers on a turbulent channel flow. In (ed. R.H.J.Sellin):"Third Int. Conf. on Drag Reduction", (1984).
5. Gyr,A.: The vorticity diffusion of Λ-vortices in drag reducing solutions. In: (ed. B.Gampert): "The influence of polymer additives on velocity and temperature fields", Springer-Verlag (1985) 233-247.
6. Luchik,T.S.;Tiederman,W.G.: Timescale and structure of ejections and bursts in turbulent channel flows. J.Fluid.Mech. 174 (1987) 529-552.
7. Luchik,T.S.;Tiederman,W.G.: Turbulent structure in low-concentration drag-reducing channel flows. J.Fluid.Mech. 190 (1988) 241-263.
8. Gampert,B.;Delgado,A.: The Reynolds stress tensor in two-dimensional drag reducing channel flows. Turbulent shear Flows 5 (1985) 9.47-9.52.
9. Rudd,M.J.: Velocity measurements made with a laser dopplermeter on the turbulent pipe flow of a dilute polymer solution. J. Fluid.Mech. 51 (1972)673-685.
10. Reischman,M.M.;Tiederman,W.G.: Laser-Doppler anemometer measurements in drag-reducing channel flows. J.Fluid.Mech. 70 (1975) 369-392.

Interaction of Molecules and Turbulent Flow in Dilute Polystyrene Solutions

Richard H. Nadolink
Naval Underwater Systems Center
Newport, Rhode Island U.S.A.

ABSTRACT

Friction reduction by addition of trace quantities of certain polymers to the turbulent flow of liquids have fascinated interdisciplinary scientists and engineers for 35 years. The obvious advantages of reduced energy production and dissipation in turbulence have inspired numerous aquatic applications, although a precise explanation of the mechanism has remained unexplained. One of the shortcomings of previous investigations which has been overcome in this work is the relatively uncharacterized nature of the high molecular weight polymers, and the violation of dilute solution rules.

The onset, turbulent flow, and characterization data collected was obtained using narrow fractions of polystyrene ranging in molecular weight from 600 to approximately 14×10^6. (Most samples were produced by a novel experimental process by the author, but that will not be discussed). The length, time and energy scale theories were then tested using experimental results and a most precise measure of the resulting polymer solution principal relaxation time, i.e.,

$$\tau = \overline{M}\, \eta_s\, [\eta]/\lambda \mathbb{R} T$$

A large range of relaxation times was obtained by systematic variation of mean molecular weight, $\overline{M}$, solution viscosity, η_s, intrinsic viscosity, $[\eta]$, and also the absolute temperature, T. The characteristic flow time was taken to be the strain rate, $\dot{\gamma}$ (sec^{-1}) at the wall of cylindrical tube flow. Further variation of the polymer solution parameters was obtained by utilizing solvents which ranged from excellent to poor (theta conditions). The experimental variables, then, afforded a test of polymer molecule interaction with the structure of fully developed turbulent flow processes.

The experiments also included anomalous samples of polystyrene whose molecular weight distribution was broad and some that were bi-modal with high molecular weight "tails" of very low overall content. The results with these samples illustrate that the distribution function is an essential parameter. Average molecular weight values will easily lead to large errors in correlating experimental results which rely on comparing length, time and energy scales.

As a final experimental validation, the exact polymer solutions which were tested under turbulent flow conditions were also characterized rheologically in a capillary tube viscometer to obtain extrapolations to zero shear and intrinsic viscosity for each solution. In this way, a single Mark-Houwink relation was found to correlate each polymer/solvent pair for the narrow distributions.

1. INTRODUCTION AND PURPOSE

Parts-per-million quantities of high molecular weight polymers have been shown to greatly reduce the frictional resistance of ordinary Newtonian solvents under turbulent flow conditions. The effect, most commonly known as the Toms phenomenon, after the discovery

A. Gyr (Editor)
Structure of Turbulence and Drag Reduction
IUTAM Symposium Zurich/Switzerland 1989

by Toms (1949) and confirmation by Mysels (1949), has received continuous attention through the previous thirty-five years due to the strong desire for applications in the areas of irrigation pumping, fire fighting and marine vehicles, to mention a few.

Major reviews have appeared periodically, Zakin and co-workers (1971), Hoyt (1972), Lumley (1969, 1973), Landhahl (1973), Berman (1978) and Virk (1975) and in each case emphasis was placed on the author's research specialty. The latter review by Virk stands as the most relevant background material to the work reported here. Recent studies of water soluble polyacrylamides by Gampert and Wagner (1985), and Kulicke and Klein (1985) on the influence of molecular parameters have not yet been covered by reviews and stand as the most similar and relevant work in relation to the work reported here.

The purpose of the experiments conducted here was to clarify some results, interpretations, and theories presented previously which were confusing due to three practical considerations:

a. The polymers in question have been characterized only by type and average molecular weight specified by the manufacturer.

b. The same "popular" water soluble polymers have been brought to solution by many methods, most of which alter their final solution properties; therefore, the "same" polymer-solvent systems yield dramatically different results under the same gross flow conditions. There is also very poor batch to batch consistency in water soluble polymers sold in bulk under the same number series.

c. Very few studies have been conducted in true dilute solution, i.e., where individual polymer molecules are non-interacting in the sense of macromolecular entanglement parameters.

Because of these effects, which obviously require and will receive further explanation, the majority of past work have introduced mechanisms which in themselves probably contribute to the anomaly (in some positive or negative way).

2. EXPERIMENTAL APPROACH

The approach was to concentrate on obtaining unique and quality friction reduction and viscometric/solution property data utilizing a polymer which yields two important characteristics absent in previous work.

a. Polymers which have been well characterized in regard to molecular weight distribution and intrinsic viscosity over a wide range of solvent and flow variables such as solvent type, solvent power, high and low viscosity, good and theta solvents; polymers which are nearly monodisperse in a practical sense.

b. By varying the molecular weight, solvent type, and temperature, produce a number of polymer-solvent combinations which provide a wide range of molecular size, relaxation time, and viscosity in each case conforming to the definition of dilute solution.

The principal objective was to use calibrated polystyrene polymer candidates to test the validity of current length, time, and energy theories in a broad experimental domain.

The results, then, will serve to verify or refute current ideas as well as contribute to a framework for theory expansion or new insights.

A number of polystyrene samples were obtained by the present author (1975), and are described in Table 1.

A complete characterization was performed on each sample, an example of which is shown in Figure 1 (Gel Permeation Chromatography was also completed on most samples.)

A very versatile, yet small turbulent pipe flow apparatus was built and exercised in order to obtain friction coefficients and onset points as a function of Reynolds number, polymer/solvent and thermodynamic condition. A sketch is shown in Figure 2, which is self-explanatory.

The size of the device was dictated by the small quantities of polystyrene which were available, i.e., approximately 5 gms each, while still permitting Reynolds numbers in the 10^2-10^5 range.

As an example of the friction reduction potential of the polystyrenes produced, Figure 3 shows performance of two samples in toluene over the 10 to 1000 ppmw concentration range. While the 2 x 10^6-14b sample reaches 50 percent potential at 1000 ppmw, the 7 x 10^6,-7D sample reaches 50% at 10 ppmw which represents a factor of 100 improvement. Note the early saturation effect with sample 7D which is fully defined by Virk's maximum drag reduction asymptote for all turbulent flow conditions where concentration is above 50 ppmw.

Reduced, and intrinsic viscosity measurements were made with each polymer/solvent/temperature/concentration series using a number of Ubbelohde type devices, and in each case the exact viscometrically measured solution sample was used in the friction pipe apparatus within a 24 hour period. A list of experimental combinations is shown in Table 2. The upper limit for "diluteness" was taken to be $c[\eta]\sim 1.08$ as described by Zakin to be the value where overlap of polymer chains becomes important.

3. RESULTS AND CONCLUSIONS

The full set of parametric results for all polymer/solvent/thermodynamic conditions cannot, obviously, be presented here. Figure 4 shows sample results from the friction factor experiments in the classical Prandtl-Karman plot style for a mid-range polystyrene-toluene combination.

Figure 5 shows drag reduction performance compared to the maximum asymptote on a Deborah number scale (actually, a partial set of the data of Figure 4 replotted).

The data supports a strong dependence on the relaxation time of molecules in dilute solution effecting and regulating all aspects of the friction reduction phenomena. One test of this theory is the correlation of onset Deborah number with relaxation time. All of the Deborah number data was plotted in Figure 6 versus calculated relaxation time. The correlation, while not fitting the ideal case of $D_o = 1$, does show that over two decades of polymer relaxation time, the mean value of onset Deborah number is 4.54 with a standard deviation of 2.53.

From the total data set accumulated by Nadolink (1986), the following major conclusions can be drawn:

a. The polymer solution relaxation time(s) control the onset, degree of effectiveness, and saturation, when narrow distribution high molecular weight polystyrenes were used in true dilute solution. The relaxation time is a function of several variables:

$$\tau = \tau\,(\bar{M},[\eta]\,,\eta_s,T,C,\text{Solvent Power})$$

with $$\tau = M\,\eta_s\,[\eta]/\lambda RT$$

was found to regulate each phase of the drag reduction process.

1. Increasing mean molecular weight increases overall effectiveness and reduces the onset strain rate.

2. Increasing solution viscosity has exactly the same effect, independent of how it is increased, either by solvent or adding non-friction-reducing polymer of the same species.

3. Intrinsic viscosity has the same effect as molecular weight (with narrow distribution) and this fact is redundant because of the functional relationship between the two variables given by the Mark-Houwink relation.

4. Effectiveness increases and onset strain rate decreases with increasing concentration (as well as slight increases in relative viscosity) up to a saturation point, beyond which adding polymer can only produce stamina of the solution in regard to shear degradation. Stamina is used here in the sense of resisting shear wear by multiple passes.

b. The distribution of molecular weight was found to be a controlling variable, both as an experimental challenge to control during polymerization and as an essential parameter to measure.

c. Combined with the above is the finding that very subtle changes in the molecular weight distribution, not reflected in the mean viscometrically determined properties, alter all aspects of friction reduction.

The highest molecular weight material dominates and onset is correlated with relaxation times indicative of these higher fractions, rather than the mean value.

d. Intrinsic viscosity as an indicator of mean molecular weight and, therefore, as a performance parameter, is useful only when the distribution of molecular weight is narrow or fairly normally distributed.

e. The range of Deborah numbers from onset to saturation was found to be from approximately 1 at onset to 100 at saturation.

f. No single previous theory is unanimously supported by the results collected. The time scale, Deborah number hypothesis is clearly the most attractive from the experimental viewpoint.

g. The data for onset and maximum asymptote friction reduction suggest that an added elongational viscosity is responsible for interaction of the small scale unsteady flow and natural modes of the extending molecular coils.

h. Energy storage within the extensional modes of individual macro-molecular coils is experienced during the production phase of turbulence generated by a steep velocity gradient. Production is inhibited by absorption at coincident time scales, and then taken up by the contraction of molecules in areas of small gradients and/or mismatched time scales.

i. The same argument as above can be made for dissipation of turbulent energy, and to the extent that production and dissipation are balanced, and occur in the same region of tube flow, they probably are equivalent.

4. REFERENCES:

a. Berman, N.S.: Annual Reviews of Fluid Mechanics, Vol. 10, Sears, W.R. (ed), Annual Reviews, Inc., Palo Alto, California (1978)

b. Gampert, B., Wagner, P.: Proceedings of Symposium. The Influence of Polymer Additives on Velocity and Temperature Fields, Essen, Gampert, B. (ed), Springer-Verlag, p. 71 (1985)

c. Hoyt, J.W.: American Society Mechanical Engineering, Journal of Basic Engineering, Vol. 94, p. 248 (1972)

d. Kulicke, W. -M., Klein, J. (same as b) p. 44 (1985)

e. Landahl, M.T.: Proceedings of the 13th International Congress on Theoretical and Applied Mechanics, Moscow (USSR), Becker, E. and Mikhailov, G.K. (eds), Springer-Verlag, Berlin, p. 177 (1973)

f. Lumley, J.L.: Annual Reviews of Fluid Mechanics, Vol. 1, Sears, W.R. (ed), Annual Reviews, Inc., Palo Alto, California (1969)

g. Mysels, K.J.: United States Patent 2,492,173, December 27, (1949)

h. Nadolink, R.H.: Naval Underwater Systems Center Technical Report No. 4422, Part 1 (1975)

i. Nadolink, R.H.: Doctoral Thesis, University of California, San Diego (1986)

j. Toms, B.A.: Proceedings of the First International Congress on Rheology, Vol. II pp. 135-141, North Holland, Amsterdam (1948)

k. Virk, P.S.: American Institute of Chemical Engineering Journal, Vol. 21(4), p. 625 (1975)

l. Zakin, J.L., Liaw, G.G., and Patterson, G.K.: American Institute Chemical Engineering Journal, Vol. 17, p. 391 (1971)

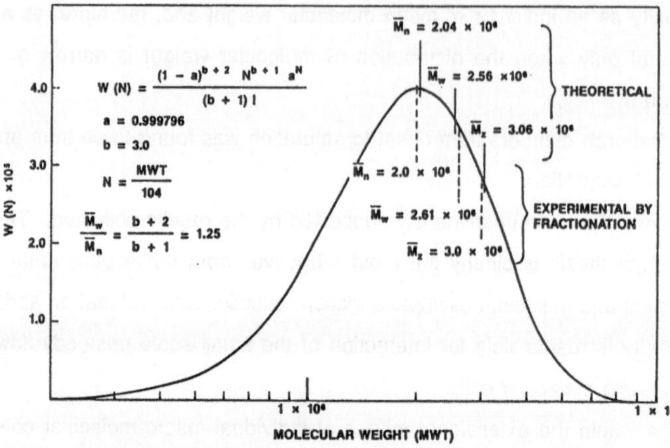

Figure 1. Comparison of mean molecular weight distribution properties of polystyrene sample 14b by experimental and theoretical techniques.

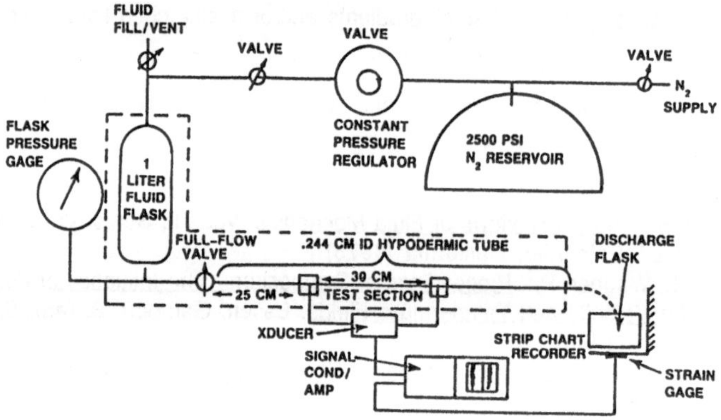

Figure 2. Variable Reynolds number-temperature turbulent pipe flow apparatus.

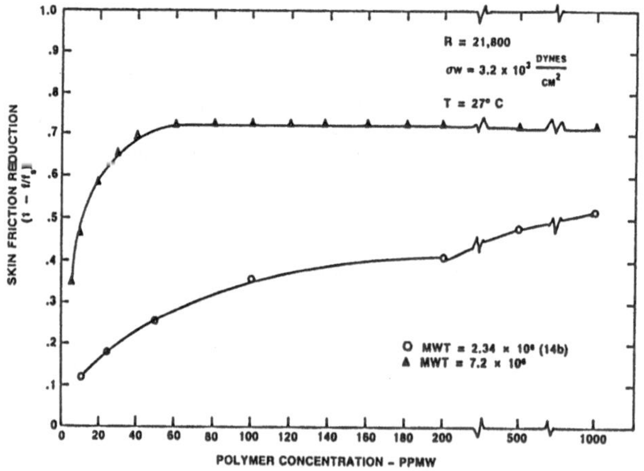

Figure 3. Friction reduction versus concentration for polystyrene-toluene.

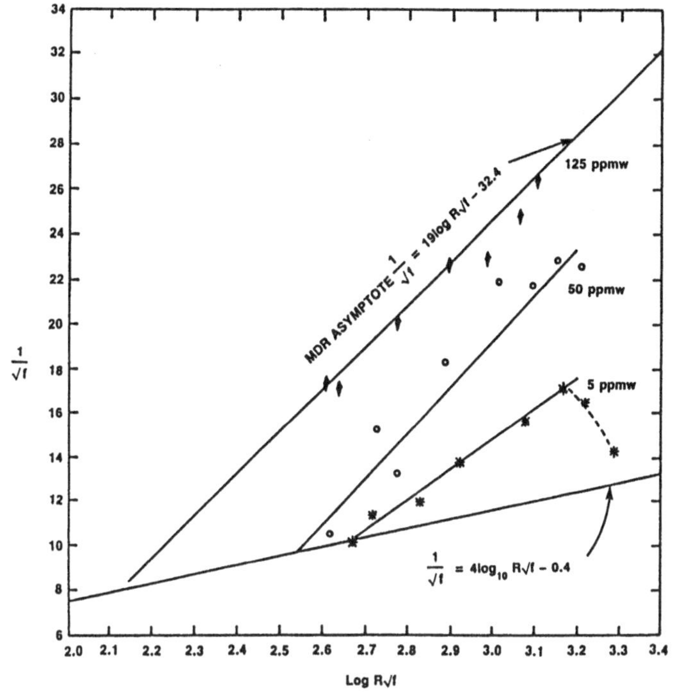

Figure 4. Saturation effect due to high molecular weignt and good solvent. Prandtl-Karman plot of PS-7D-toluene as a function of concentration [$\bar{M}_w$ = 7.25x 10^6 ; [η] = 7.27 dl/gm; T - 22°C; d = 0.244 cm].

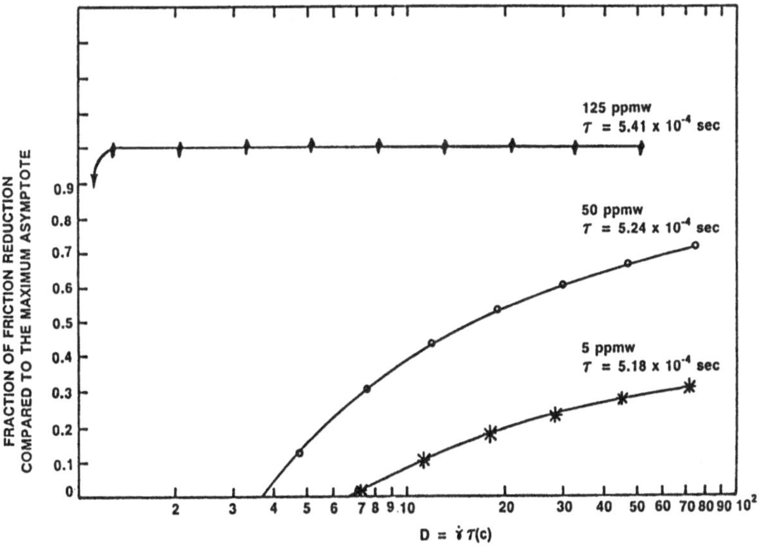

Figure 5. Saturation effect. Performance compared to the maximum asymptote versus Deborah number for PS-7D toluene at various concentrations.

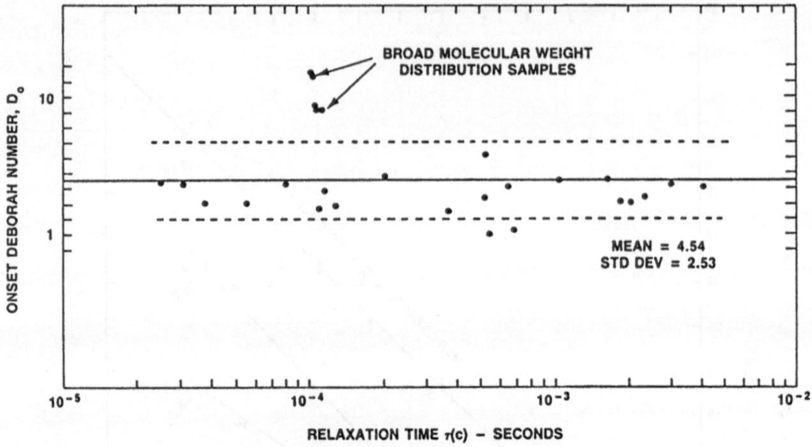

Figure 6. Onset Deborah number versus polymer relaxation time for all polystyrene friction reduction data sets.

Polystyrene Sample No.	Viscosity Average Molecular Weight, $\bar{M}_v$	Polydispersity M_w/M_n
PS-1	600	1.10
PS-2	2100	1.10
PS-3	4000	1.10
PS-4	10^4	1.06
PS-5	5×10^4	1.06
PS-6	9.72×10^4	1.06
PS-1C	2×10^5	1.06
PS-0.5	5×10^5	1.20
PS-0.7	6.7×10^5	1.15
PS-09	8.6×10^5	1.15
PS-14b	2.34×10^6	1.25
PS-7	2.3×10^6	1.25 (High WMT Tail)
PS-9A	6.5×10^6	>5 (broad)
PS-7D	7.25×10^6	1.20
PS-25	14.34×10^6	?

Table 1. Inventory of Polystyrene samples with mean properties.

Solvent	Temperature(s) - °C	TYPE, comment
Toluene	0, 11, 22-25	Very good, $\alpha = 0.71$
Cyclohexane	$34.5 = T_\theta$	Poor, Theta
2-nitropropane	$25.0 = T_\theta$	Poor, Theta
Toluene/PS 0.5	22	Very good, moderate η
di-octylphthalate	$22 = T_\theta$	Poor, very high η

Table 2. List of Solvents, temperatures to be used with various polystyrene samples for viscometric and friction reduction tests.

Λ-Shaped Vortices in Dilute Polymer Solutions

H. MIZUNUMA, H. KATO and T. KURITA
Faculty of Engineering, Tokyo Metropolitan University, Tokyo, Japan

The effect of polymer additives on plane Poiseuille flow was investigated in a turbulent flow and a transitional flow. In a turbulent flow, frictional drag was measured. In a transitional flow, we maintained a laminar flow at a Reynolds number higher than the minimum critical Reynolds number and visualized the growth of Λ-shaped vortices artificially created in the flow. The results showed that polymer additives reduced the frictional drag in a turbulent flow and increased the spanwise wave length of Λ-shaped vortices in a transitional flow. It is known that polymer additives increase the spanwise spacing of bursts in a turbulent flow. Therefore polymer additives increase the spanwise spacing of the coherent structures in both a turbulent flow and a transitional flow.

1. Introduction

A trip wire or a vibrating ribbon introduces a two-dimensional wave into a transitional flow, and generates Λ -shaped vortices as shown in Fig.1. This early stage of three-dimensional development can be described by the mechanism of secondary instability[1]. This secondary instability shows that low-amplitude three-dimensional waves are rapidly amplified in the presence of a finite-amplitude two-dimensional wave. In experiments the three-dimensional components of the background turbulence are rapidly amplified and develop to Λ -shaped vortices. The horseshoe vortices appear

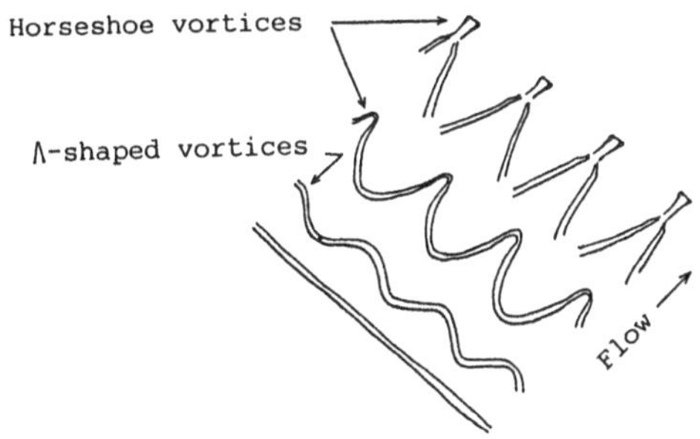

Fig.1 Λ-shaped vortices in a transitional flow.

A. Gyr (Editor)
Structure of Turbulence and Drag Reduction
IUTAM Symposium Zurich/Switzerland 1989

at the head of the growing Λ-shaped vortices and are broken up in high frequency fluctuation. This scenario in the turbulent transition is similar to that in turbulent bursts[2]. In the bursting process, horseshoe vortices are also formed and broken up. It is known that polymer additives increase the spanwise spacing of turbulent bursts with an increase of the drag reduction. Therefore it is predicted that polymer additives affect the scale of the Λ-shaped vortices in a transitional flow. In §3.2 we prove experimentally that this prediction is valid.

The structure of background turbulence affects the scale of the Λ-shaped vortices, because the Λ-shaped vortices result from the linkage of a two-dimensional wave and background turbulence. We produced artificial three-dimensional disturbances to change the structure of background turbulence. These disturbances link a two-dimensional wave generated by a vibrating wire and develop into Λ-shaped vortices. It is investigated in §3.3 whether the structural change of background turbulence has the same effect on the Λ-shaped vortices as polymer additives do.

2. Experimental Equipment and Procedure

Experiments were conducted on plane Poiseuille flow. The rectangular channel is 3980mm in length, 15mm in width and 300mm in height. The test fluids are water and polyacrylamide (PAA) solutions with a concentration of 200wppm. We added this high concentration of polymer to cancel the mechanical degradation caused by a pump. The rate of degradation approached zero, after PAA solutins were pumped for about 30 minutes. Since PAA solutions have weak non-Newtonian viscosity, we used the Reynolds number for a power-law fluid. The Reynolds number for plane Poiseuille flow is

$$Re=\left(\frac{n}{2n+1}\right)^n \frac{6\rho V^{2-n}h^n}{K},$$

where V is mean velocity, h is channel half width, and n and K are material constants in a power-law model,

$$\tau = K\dot{\gamma}^n.$$

Here, τ is shear stress and $\dot{\gamma}$ is the shear rate. The frictional coefficient f (= $\tau_w/\frac{1}{2}\rho V^2$, τ_w:wall shear stress, ρ:density of fluid) is

$$f=\frac{12}{Re},$$

in a laminar flow.

In Newtonian fluids, the minimum critical Reynolds number is 1300 and the critical Reynolds number from a liner stability theory is 7700.

Although PAA solutions show the Toms effect, the minimum critical Reynolds number of PAA solutions is 1300[3], which is the same as that of Newtonian fluids. By reducing the inflow turbulence by an inlet settling chamber, a laminar flow was maintained for the flow of above a Reynolds number of 1300. Our experiments were conducted in a range of the Reynolds number from 1300 to 3000. When we measured the frictional drag in a turbulent flow, we inserted a tube (2.6mm O.D.) across the channel inlet as shown in Fig.2 and forced the flow to change into a turbulent flow. When we investigated the Λ -shaped vortices in a transitional flow, we removed the tube and maintained a laminar flow. In a laminar flow the frictional coefficient agrees with the formula $f=12/Re$.

The lambda-shaped vortices generally appear downstream from a two-dimensional step, a trip wire or a vibrating ribbon. We attached a trip wire on the wall to generate the Λ-shaped vortices. This trip wire was a diameter of 1.2mm and capable of vibrating in an amplitude of 0.1 to 0.2mm. When we vibrated the trip wire, we chose the frequency about equal to that of the two-dimensional wave detected for a fixed trip wire.

The lambda-shaped vortices have a correlation with the three-dimensional structure of background turbulence. To investigate this correlation, we used the capillary tubes as shown in Fig.3 and changed the three-dimensional structure of background turbulence. The capillary tubes were located at regular intervals (p in Fig.3) in the spanwise direction. The streamwise distance from the capillary tubes to the trip wire was 653mm (87h). We threaded a stainless steel wire (0.1mm O.D.) through the capillary tubes and held the wire straight in the channel. The disturbances by the capillary tubes have a spanwise wave length equal to their distance p and have an intensity proportional to their diameter d and length l. The

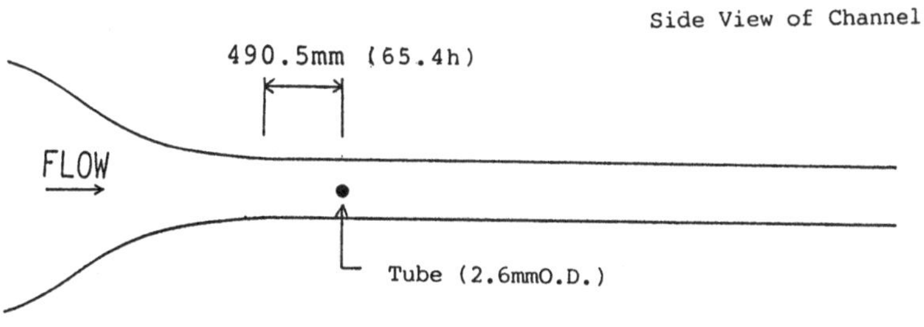

Fig.2 Experimental equipment for a turbulent flow.

distance p was 20 to 50mm. The sizes of the capillary tubes were 0.3 to 0.6mm for d and 2 to 14mm for l.

The growth of the Λ-shaped vortices was visualized by dye streak lines or dye sheets. The streak lines were fed through fourty-nine wall holes each separated by 5mm in the spanwise direction.

3. Experimental Results and Considerations

3.1 Turbulent drag reduction

We inserted a tube across the channel inlet as shown in Fig.2 and forced the flow to change into a turbulent flow. Figure 4 gives the frictional

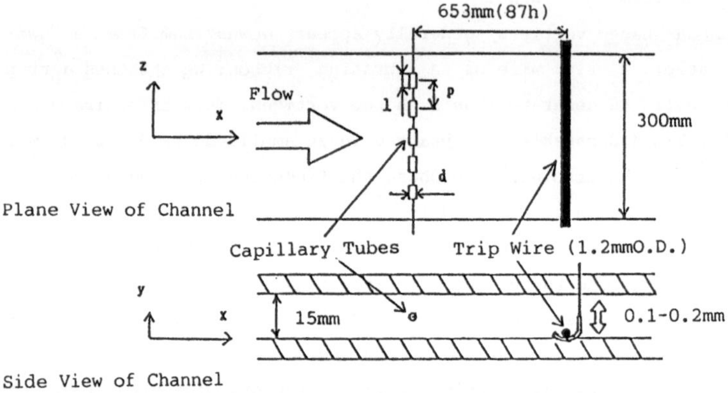

Fig.3 Experimental equipment for a transitional flow.

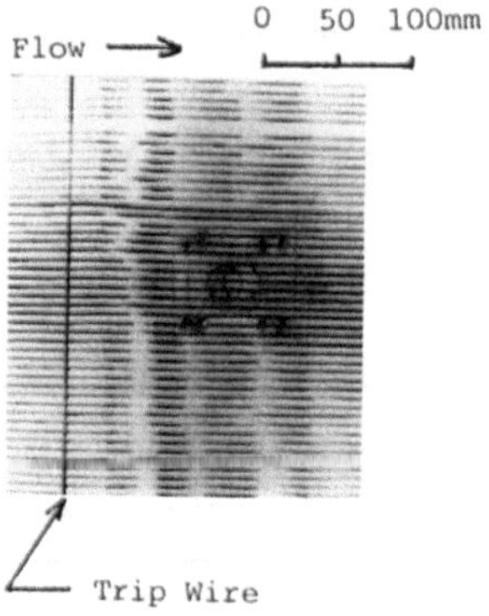

Fig.5 Two dimensional disturbances by a trip wire; water, Re=1400.

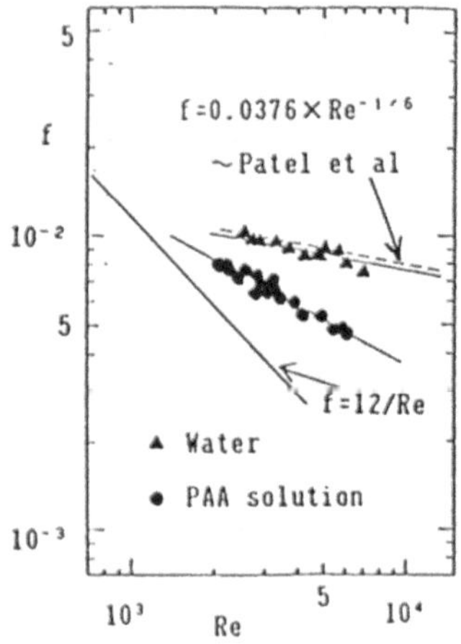

Fig.4 Turbulent frictional coefficient.

coefficients in a turbulent flow. The results for water were about equal to those by Patel & Head[4]. Polymer additives reduced the frictional drag by 20 to 30% over the range of 2000<Re<3000.

3.2 Influence of polymer additives on Λ-shaped vortices

Experiments were conducted with a trip wire (1.2mmO.D.) on the wall surface. At about the minimum critical Reynolds number, the disturbance by a trip wire was nearly a two-dimensional wave as shown in Fig.5. With increasing Reynolds number, the disturbances developed to the Λ-shaped vortices as shown in Fig.6. For water in Fig.6(a), the spanwise wave length was 20 to 25mm. This result is the same as the one by Nishioka, Asai & Iida[5] and Kozlov & Ramazanov[6] in the higher Reynolds numbers range of 5000 to 6500. Both the channels used by them were 15mm in width, and of the same width as our channel. Therefore the non-dimensional wave number σ ($= 2\pi h/\lambda$, λ:spanwise wave length of Λ-shaped vortices) is about 2 in all experiments.

For PAA solution, the spanwise wave length of the Λ-shaped vortices was 40 to 50mm, as shown in Fig.6(b). This length is about twice the one in Newtonian fluids. It is also known that polymer additives increase the spanwise spacing of bursts in a turbulent flow. These similar changes of coherent structures suggest the structural similarity between a transitional flow and a turbulent flow.

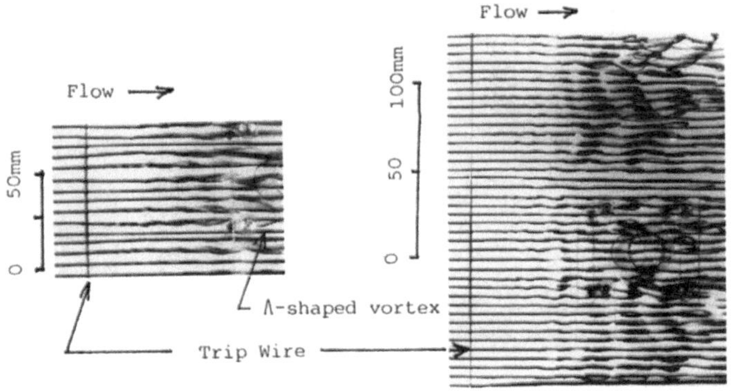

(a) Water (b) PAA solution

Fig.6 Λ-shaped vortices at Re=3000.

The growth of Λ-shaped vortices is activated by vibrating the trip wire as shown in Fig.7. However, the spanwise wave length of the Λ-shaped vortices remains unchanged by vibrating the trip wire for both water and PAA solutions.

3.3 Influence of background turbulence on Λ-shaped vortices

If we regulate the three-dimensional structure of background turbulence by means of the capillary tubes in Fig.3, the Λ-shaped vortices appear more uniformly at equal spanwise distances as shown in Fig.8. These results show that the Λ-shaped vortices are affected by the three-dimensional structure of background turbulence. Therefore two possible causes of the effect described in §3.2 are considered as follows: 1) Polymer additives change the structure of background turbulence, and so the disturbances with greater spanwise wave length become dominant. 2) Polymer additives change the mechanism of the secondary instability, and so the disturbances with greater spanwise wave length are selectively amplified from the background turbulence. If the first conjecture is right, the structural change of background turbulence will have the same effect on the Λ-shaped vortices as polymer additives do. To change the spanwise wave length of background

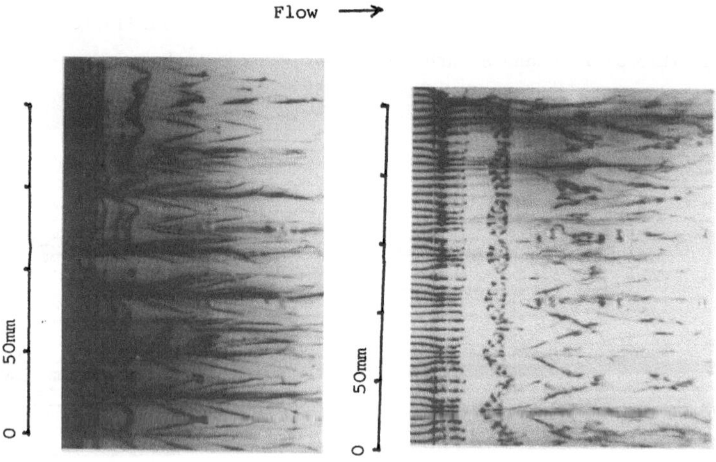

(a) Water, Re=2640.
Frequency of vibrating wire=4Hz.

(b) PAA solution, Re=3240.
Frequency of vibrating wire=6Hz.

Fig.7 Λ-shaped vortices without capillary tubes upstream.

turbulence, we changed the distance of the capillary tubes, p, up to 50mm for water. Corresponding to the distance of the capillary tubes, the spanwise wave length of the Λ-shaped vortices changed. However, the maximum wave length was about 40mm. The Λ -shaped vortices with a spanwise wave length of greater than 30-40mm did not appear with reproducibility for water. Figure 9(a) shows the results with the capillary tubes of p=50mm. The spanwise width of the Λ-shaped vortices is about 35mm in Fig.9(a) and less than the distance of the capillary tubes, p. Furthermore, the Λ-shaped vortices with such spanwise width often split into two Λ-shaped vortices with a spanwise width of about 20mm. These results show that the structural change of the Λ-shaped vortices in PAA solutions is not due the structural change of the background turbulence. The effect by polymer additives is supposedly due to a change of instability mechanism.

Figure.9 shows the growth of the Λ-shaped vortices for water and PAA solutions. The distance of the capillary tubes, p, is 50mm for both fluids. For water, horseshoe vortices appear at the head of the growing Λ-shaped vortices and are broken up. For PAA solutions, the horseshoe vortices appear similarly as for water. The influence of polymer additives on the horseshoe vortices is the subject of a future study.

4. Conclusions

Polymer additives reduce the frictional drag by 20 to 30% in a turbulent flow and approximately double the spanwise wave length of the Λ-shaped vortices in a transitional flow. This effect on a transitional flow is supposedly caused by a change of the flow instability mechanism which determine the scale of the three-dimensional disturbances selectively amplified from the background turbulence. It is known that polymer additives increase the spanwise spacing of bursts in a turbulent flow. Therefore polymer additives increase the spanwise spacing of the coherent structures in both turbulent flow and transitional flow.

We wish to express our thanks to Mr.S.Komiya for assisting with some of the experiments. This work was supported in part by the Saneyoshi Fundation.

References

1. Herbert,T.: Ann. Rev. Fluid Mech. 20 (1988) 487-526.
2. Nishioka,M., Asai,M. & Iida,S.: Proc. Symposium on Transition and Turbulence. ed. Meyer,R.E. (Academic Press, 1981) 113-126.
3. Mizunuma,H. Kato,H: JSME Int. J. Ser.II 31-2 (1988) 209-217.
4. Patel,V.C. & Head,M.R.: J.Fluid Mech. 38-1 (1969) 181.

5. Nishioka,M.,Asai,M. & Iida,S.: Proc.IUTAM Symposium on Laminar-Turbulent Transition. eds. Eppler,R. & Fasel,H. (Springer-Verlag, 1980) 37-46.
6. Kozlov,V.V. & Ramazanov,M.P.: J.Fluid Mech. 147 (1984) 149-157.

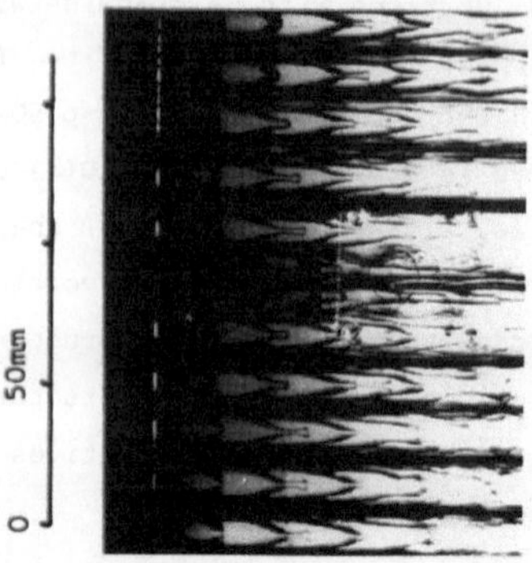

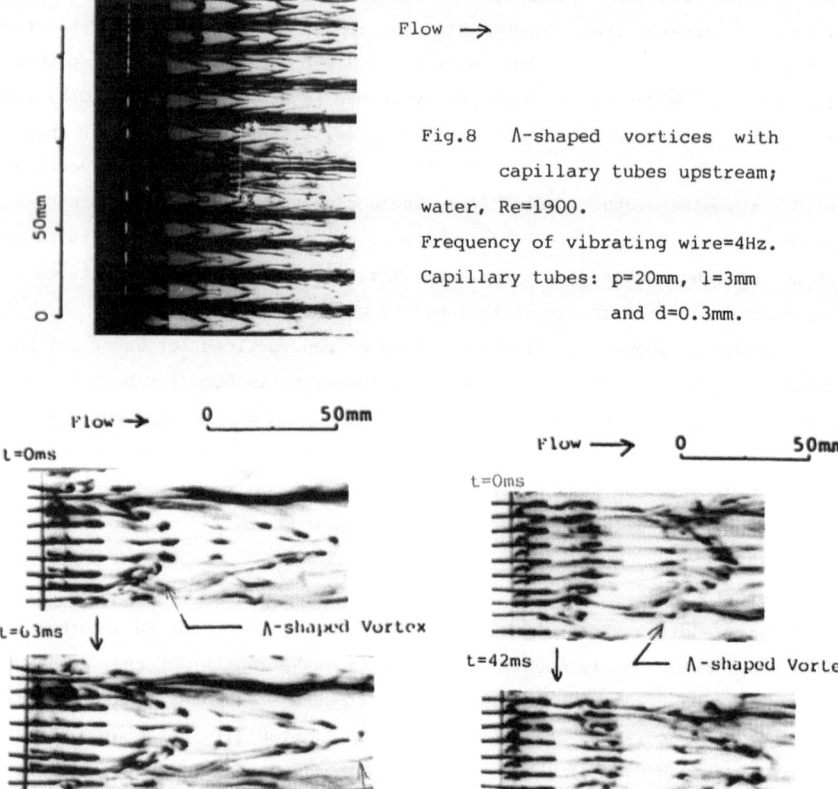

Fig.8 Λ-shaped vortices with capillary tubes upstream; water, Re=1900. Frequency of vibrating wire=4Hz. Capillary tubes: p=20mm, l=3mm and d=0.3mm.

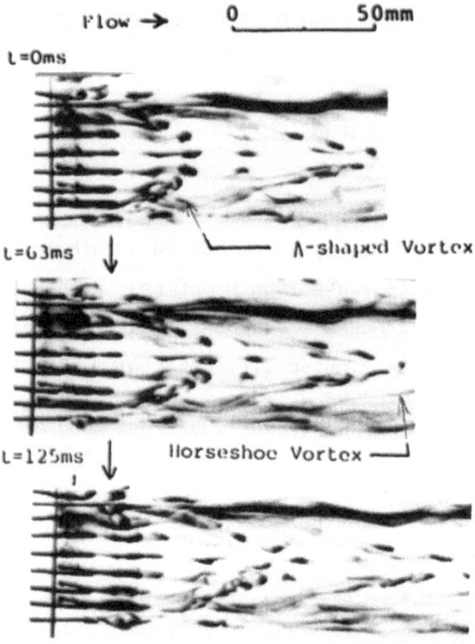

(a) Water. Frequency of vibrating wire=4Hz. Capillary tubes: p=50 mm, l=2mm and d=0.4mm. t=0ms t=42ms t=83ms

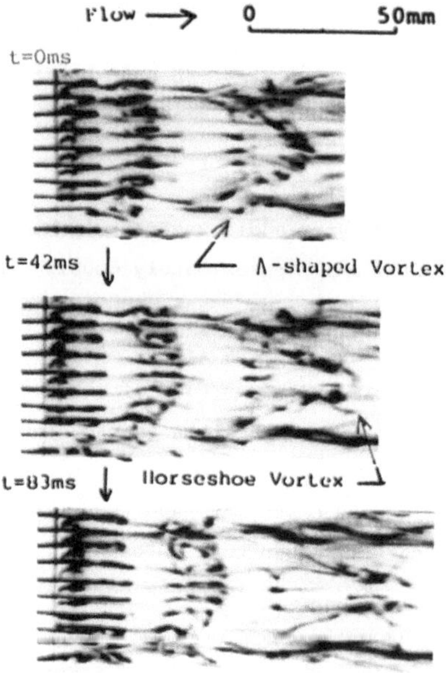

(b) PAA solution. Frequency of vibrating wire=6Hz. Capillary tubes: p=50mm, l=14mm and d=0.6mm.

Fig.9 Growth of Λ-shaped vortices at Re=2300.

Wake Flows in Dilute Polymer Solutions

AN . A. BORISOV; B. P. MIRONOV; B. G. NOVIKOV; V. D. FEDOSENKO

Institute of Thermophysics,
Siberian Branch of the USSR Academy of Sciences,
630090 Novosibirsk, USSR

The known results of investigation of the effect of dilute polymer solutions on free turbulent flows (wakes /1/, jets /2 -5 /, flows behind the grid /6 - 9/) are rather contradictory. The possible reason for these contradictions can be due to the fact that measurements were performed mainly at relatively small distances from the turbulence source. Therefore different experimental conditions including the methods of polymer solution supply into the flow could affect the measuring results. In the presented investigations an attempt is made to exclude, if possible, the effect of small experimental details on the measuring results.

The measurements were performed in the system of unmoved liquid. Rotation ellipsoid and "cup-body" falling freely along vertical guide string served as a turbulence source. The ellipsoid diameter was 40 mm, the ellipsoid length l = 560 mm. The internal diameter d of the "cup-body" was 25 mm and the length l = 37.5 mm. Both bodies had an orifice along the axis for the string guide. In this connection pressure at the "cup-body" bottom could not be everywhere equal to velocity head. The coefficient of the "cup-body" drag in water was 0.83, i.e. considerably less than the "cup-body" drag without an axial orifice. The unmoved medium filled in the vertical tube with a height of 7 m and a diameter of 0.5 m. A test chamber with optical glasses for taking pictures of tracing particles suspended in a liquid was installed in the centre of the tube.

The chosen experimental conditions allowed to decrease essentially the effect of polymer destruction on the measuring results. The concentrated solution was prepared 24 hours before the experiment. The required solution concentration was prepared 1 hour before the experiment. A free body falling excluded the necessity to use towing devices perturbing the flow. All measurements are performed at relatively large times of the flow development which correspond to large distances between the body and the measuring point.

The flow under study is statistically nonstationary in the coordinate system of the unmoved liquid. In this case averaging its characteristics over time is eliminated. So averaging over the set of instantaneous values of the characteristic under study in the volume of the test chamber was used. To this end, stereo pictures of the motion of suspended polysterene spheres were taken. Exposure was performed by a series of 5 pulse flashes. Knowing three coordinates of the tracing particles at the given moment t one could calculate the field of instantaneous velocity of each tracing particle by varying t.

The instantaneous profiles of three average velocity components in a cylindrical coordinate system were calculated using the methods of regressive analysis /10/. The difference between instantaneous and average velocities at the corresponding point of the space was used

A. Gyr (Editor)
Structure of Turbulence and Drag Reduction
IUTAM Symposium Zurich/Switzerland 1989

as an instantaneous value of velocity fluctuations. Besides, the instantaneous profiles of three components of RMS velocity pulsations and Reynolds stresses were calculated.

In the analysis of the measuring results the instantaneous value of the falling body velocity U_∞ in the thickness of the pulse introduced into the flow was used as a length scale. The spread in the measured instantaneous values of average velocity $U_{mo}(t)/U_\infty$ and halfwidth of the profile of average velocity $r_o(t)/\delta^{**}$ for different realisations of the flow did not exceed ± 10%. The analogous spread in maximum values of three velocity fluctuation components and halfwidth of corresponding profiles did not exceed 20%.

By comparison, measurements are performed in water and dilute polyethyleneoxide and polyacrylamide solutions. The ellipsoid was tested in water with and without turbulizer. At the Reynolds number $Re = U_\infty l/\nu \approx 4\ 10^6$ without turbulizer the length of the laminar boundary layer depends essentially on the external turbulence intensity, model vibration, occasional surface roughness and a number of other factors. Owing to this the coefficient of ellipsoid drag in water without turbulizer in different realizations varied in the range of

$$C_x = X/\frac{\rho U_\infty^2}{2}\cdot\frac{\pi d^2}{4} \approx 0.07 \text{ to } C_x \approx 0.17,$$

in the presence of a turbulizer C_x is about 0.16-0.19. In polyethyleneoxide and polyacrylamide solutions the coefficient of the ellipsoid drag without turbulizer in all realizations is $C_x \approx 0.06$-0.08. All instantaneous profiles of average velocity in water and polyethyleneoxide and polyacrylamide solutions are concentrated near the curve

$$U_o(r,t) = U_{mo}(t)e^{-0,693(r/r_o(t))^2}.$$

The maximum intensity of all three velocity fluctuation components $U_{mi}(t) = \max \overline{U_i'(r,t))^{1/2}}$ in polyethyleneoxide and polyacrylamide solutions is 2 times smaller than the corresponding intensity in water. In water the normalized profiles of velocity pulsations are also approximated by the error distribution curve $U_i(r,t)/U_{mi}(t) = e^{-0,693(r/r_i(t))^2}$.

In polymer solutions the shape of instantaneous profiles of velocity pulsations diverge concentrating near the curves $\exp[0.693(r/r_i(t))^2]$ and $\exp[0.693(r/r_i(t))^4]$ with $i = 1, 2, 3$.

In the flows generated by the "cup-body" the maximum values of the average velocity $U_{mo}(t)$, and all three velocity pulsation components in polyacrylamide solution are roughly by 30% smaller than the corresponding values in water. The shape of average velocity profile both in water and polyacrylamide solution is approximated by the curve $U_o(r,t) = U_{mo}(t)\exp\{-0,693[r/r_o(t)]^4\}$.

All the profiles of velocity pulsations have their maximum at $r = 0,5r_i(t)$, but $U_{mi}(0,t)$ is about $0,9U_{mi}(t)$.

In all investigated cases, both in water and solutions, the current scales of velocity $U_{mi}(t)$ and width $r_i(t)$, where (i = 1, 2, 3), are approximated by the formulas

$$U_{mi}(t)/U_\infty = 1/2\{[S_i U_\infty (t-t_{oi})/\delta^{**}]\}^{-2/3};$$

$$r_i(t)/\delta^{**} = 1/2\{[S_i U_\infty (t-t_{oi})/\delta^{**}]\}^{1/3};$$

$$S_i = r_i(t)/U_{mi}(t)(t-t_{oi}) = \text{const.}$$

In this case the equality $t_{00} = t_{01} = t_{02} = t_{03}$ is consistent at the time t_0 when the body passed the test section.

Wake flows are a good object for visualizations made with dilute rhodamine solution. In the coordinate system of unmoved liquid the velocities of all motions are small. The observer can see the development of individual vortex formations from the stage of their origin to destruction.

In water and dilute polymer solutions the characteristic form of organized structures are the annular vortices. All the observed annular vortices were generated beyond the flow axis. The axes of annular vortices were always inclined to the flow axis. A part of the annular vortices left the turbulent region and moved from it for large distances attaining frequently the tank walls.

Another frequently observed form of ordered motion are formations of the "roll" type with almost straight axis and distinct end boundary between dyed and nondyed liquids. The difference between the development of ordered structures observed in water and in solutions became noticeable only at the stage of their destruction. At this stage wash-out of the boundary between dyed and nondyed liquid occurs considerably more intensively than in the solution.

Figs.1-3 present the photographs of the flow visualized by rhodamine behind the falling sphere in water (Fig.1), and polyethyleneoxide (Fig.2) and polyacrylamide (Fig.3) solutions. For flow visualization several drops of concentrated rhodamine solution were applied to the sphere surface. To evaluate the effect of nonstationary flow in the trailing part of the sphere on generation of organized structures, visualization of the flow generated by a rapid motion of dyed filament along its axis is performed. Fig.4 presents an example of such a flow.

References

1 Pokryvailo N.A. et al (1973). On flow of polymer solutions in a wake of poorly streamlined bodies // Inzh. - Fiz. Zh. Vol. 25, N 6, pp. 993 - 998

2 Jackley D.N. (1967). Drag-reducing fluids in a free turbulent jet // Int. Shipbuild Progr. Vol. 14, N 152, pp. 159 - 165

3 Shul'man V.P. et al (1973). On measuring the structure of turbulent flow of polymer solution submerged jets // Inzh. - Fiz. Zh. Vol. 25, N 6, pp. 977 - 986

4 Thorne P.F. (1974). Drag reduction in fire-fighting: Proc. Intern. Conf. on Drag reduction, Cambridge 1974, P.H. 40

5 Hoyt J.W. (1974): Proc. Intern. Conf. on Drag reduction, Cambridge, P.H.1 - 1

6 Friehe C.A., Schwarz W.H. (1970). Grid-generated turbulence in dilute polymer solutions // J. Fluid Mech. Vol. 44, N 1, pp.173 - 193

7 Created C.A. (1969). Effect of polymer additive on grid turbulence // J. Nature, Vol. 224, pp. 1196 - 1197

8 Mc Comb W.D., Allan J., Created C.A. (1977). Effect of polymer additives on the small-scale structure of grid-generated turbulence // Phys. Fluids, Vol. 20, N 6, pp. 873 - 879

9 Barnard B.J.S., Sellin R.H.J. (1969). Grid turbulence in dilute polymer solutions // J. Nature, Vol. 222, pp. 1160 - 1162

10 Dreiner, Norman, Smith (1973). Applied Regressive Analysis. M. Statistics, p. 83.

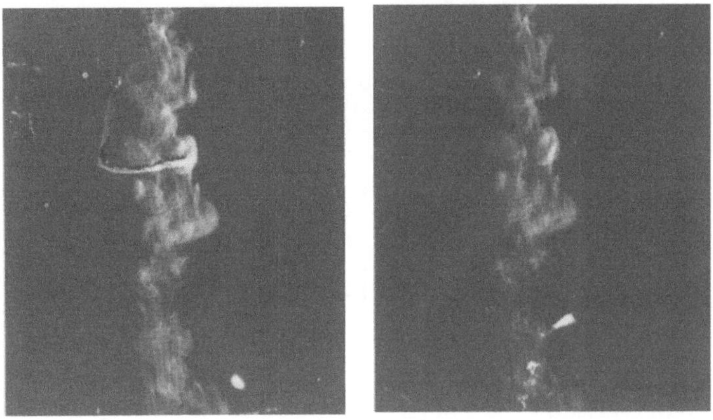

Figure 1. Stereo pair of a flow behind a sphere in a water.

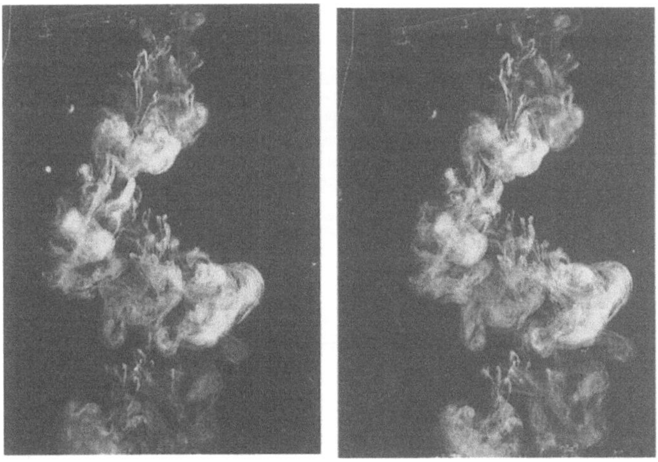

Figure 2a. Stereo pair of a flow behind a sphere in PEO-solution (C = 1 ppm).

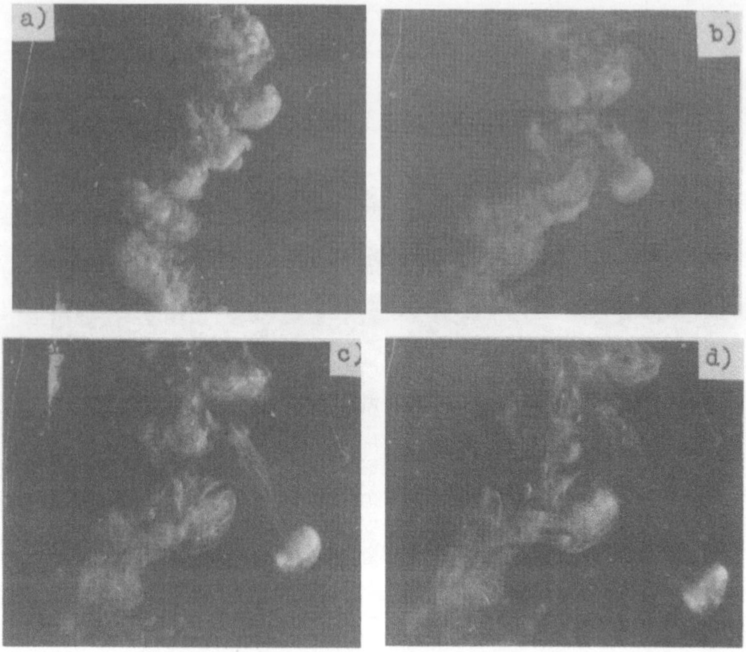

Figure 2b. Development of a flow behind a sphere in PEO - solution (C = 1 ppm).

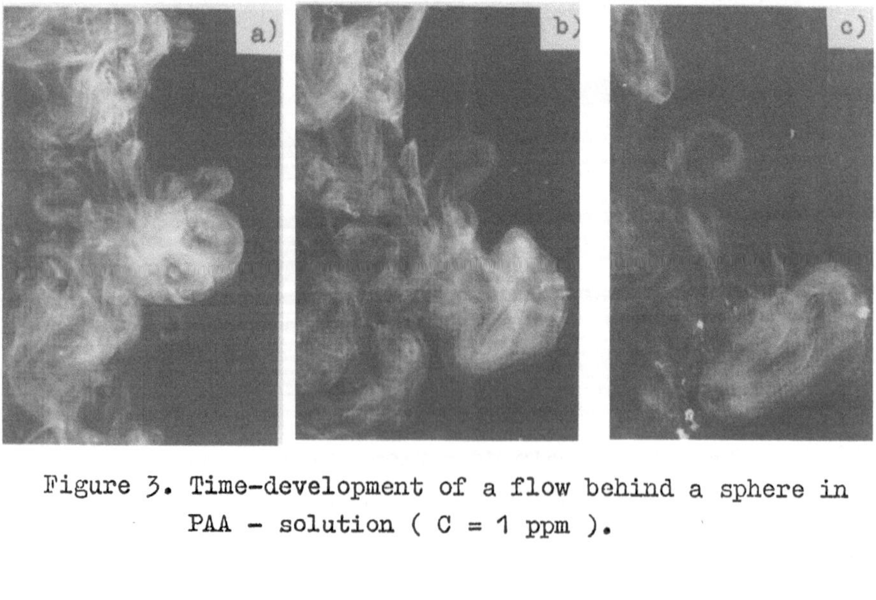

Figure 3. Time-development of a flow behind a sphere in PAA - solution (C = 1 ppm).

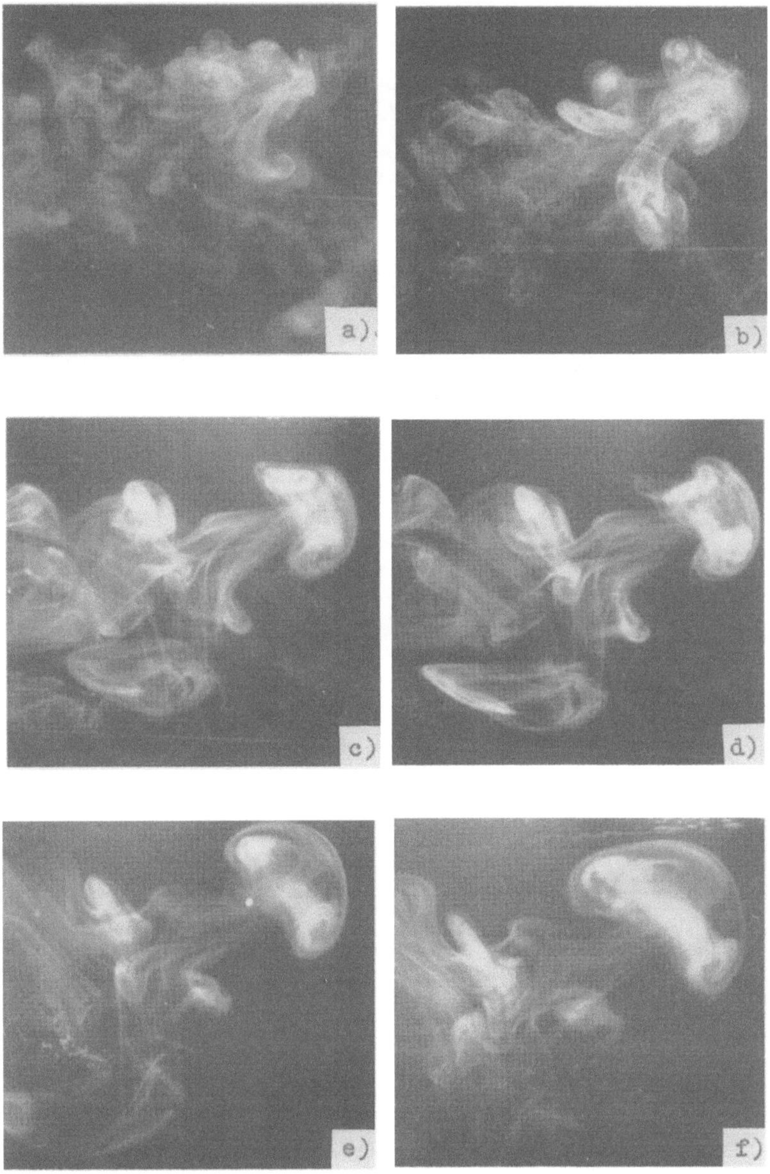

Figure 4. Time-development of a flow generated by a filament.

Drag Reduction Caused by the Injection of a Polymer Solution into a Pipe Flow

HIROMOTO USUI

Department of Chemical Engineering
Yamaguchi University
Tokiwadai, Ube 755, Japan

Summary

Drag reduction caused by the injection of concentrated polymer solutions into a turbulent pipe flow is reviewed. The experimental results obtained by means of a tracer particle method are examined in detail, and they are compared with the previous LDV measurements. It is suggested that the wall turbulence structure might be controlled by suppressing the large eddy motion in the turbulent core region.

1. Drag reduction by polymer additives

Turbulent drag-reducing phenomenon by use of a polymer additive is well known as an effective turbulence controlling technique. However, the detailed mechanism of the interaction between turbulent eddies and polymer additives is not fully understood. The addition of polymers in well premixed conditions has been investigated for thirty years. Experimental results such as the increase in the scale of turbulent eddies (1,2), the damping of the turbulent fluctuations (3), the increase in the bursting periods (4), etc., have been revealed through these investigations. Also the anomalously high value of the extensional viscosity in dilute polymer solutions has been reported by many investigators (5,6). Combining these

A. Gyr (Editor)
Structure of Turbulence and Drag Reduction
IUTAM Symposium Zurich/Switzerland 1989

experimental results, a hypothesis (7) has been proposed in which the drag reduction is brought about by the interaction of the extensional secondary flow in wall turbulence with the molecular extension of the polymer additives. This hypothesis appears to be the most plausible. On the other hand, many anomalous phenomena caused by the polymer additive have been reported. For example, the decrease in the rate of heat transfer and the rate of vortex shedding in cross flow around a circular cylinder (8,9,10), the change in the scale of turbulence in the grid generated decaying turbulent field (11,12) or in a round free jet (13,14) and the increased drag through a porous media (15). Some of these may be attributed to the molecular stretching of linear flexible molecules. However, the interaction between turbulent eddies and polymer additives is still open to question.

2. Drag reduction by inhomogenous injection of concentrated polymer solutions

Recently, the drag reduction caused by the injection of concentrated polymer solutions into a pipe flow has been extensively investigated, with many different aspects being reported (16-21). For example, the injected polymer is not dispersed completely as is observed in premixed polymer solutions. Instead, the injected polymer solution exists as a polymer thread, and the flow condition becomes a heterogeneous drag-reducing system. Many investigators using laser Doppler anemometors found similar results such as a thickening of the elastic sublayer (17,20), an enlargment of the integral scale of the auto-correlation of the fluctuating velocity (17), a decrease of the high-frequency energy spectra (18) and an increase in the bursting time (18). However, a precise comparison of previous experimental results for heterogeneous drag-reducing systems has revealed that there exists an essential difference in the interpretation of the polymer-turbulent eddy interaction among the previous investigations. One hypothesis (17) is that the interaction of the injected polymer thread with the turbulent eddies near the wall is indispensable in producing the drag reduction in a heterogeneous flow. This hypothesis is based on the experimental data obtained for the multi-fine polymer threads (17). The other hypothesis (16,18) based on the drag-reducing flow for a single thread is as follows. The injected polymer thread can interact with the large scale turbulent eddies which exist in the central part of a pipe flow. The change in the large scale turbulence structure caused by the interaction with the injected polymer thread affects the structure of the wall turbulence. Recent experimental results appear that the second hypothesis is most plausible.

The author has carried out a flow visualization experiment for a heterogeneous drag-reducing flow. The injection of a colored polymer thread into a pipe flow showed that the flow pattern of the polymer threads can be divided into three categories when the Reynolds number and the pipe diameter are fixed at $Re = 10^4 \sim 3\times10^4$ and $D=51.3$ mm, respectively. When polyethylene oxide ($M \simeq 4\times10^6$) is used as the polymer additive, the injected polymer having a concentration of less than 500ppm, disperses rapidly in the turbulent pipe flow and the observed rate of drag reduction is small. When the concentration of the polymer solution is increased to 2,000~4,000ppm, i.e. the middle concentration, the injected solution divides into fine threads, and these fine threads are observed to flow with the large eddy motion in the turbulent pipe flow. The photograph of the heterogeneous flow is shown in Fig. 1-A. Here Re, D and x are Reynolds number, the pipe diameter and the longitudinal distance from the injection point, respectively. Drag reduction rate, DR, is defined as

$$DR = 100\ (f_N - f_P)/f_P \quad (\%) \qquad (1)$$

where f_N and f_P are the friction factors in a Newtonian fluid flow and a polymer induced drag-reducing flow, respectively. When the polymer concentration is further increased to as high as 8,000ppm, i.e. the highly concentrated condition, the injected polymer thread is

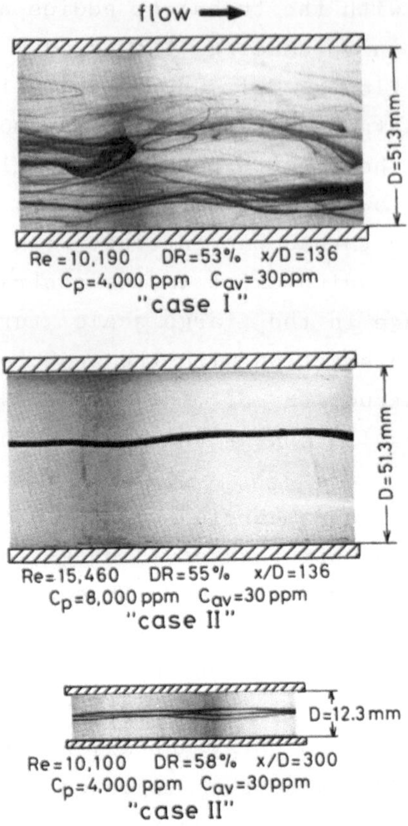

Fig. 1. Flow visualization results in the centerline injected drag-reducing flow.

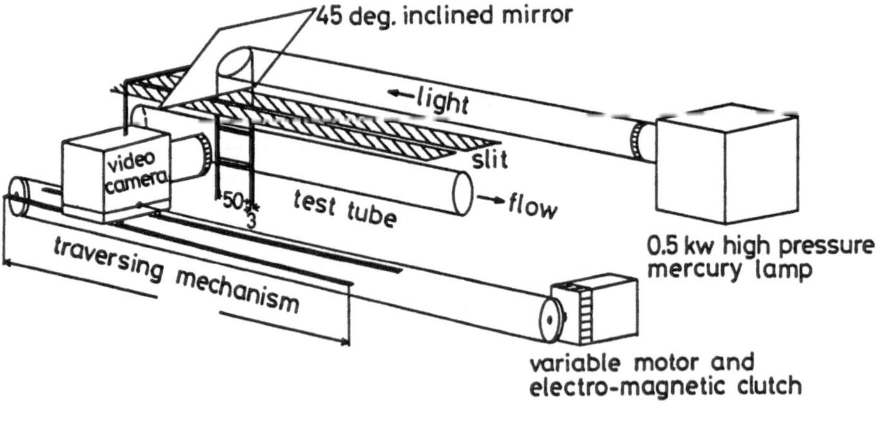

Fig. 2. Experimental setup for the measurement of tracer particle trajectories.

stable and flows mainly in the center portion of the pipe flow in a wave like manner as shown in Fig. 1-B. Similar drag reduction rates were observed for the last two cases. When the pipe diameter is reduced to D=12.3 mm, the same flow condition as for the case of 8,000ppm is observed even at the middle polymer concentration (4,000ppm) as shown in Fig. 1-C. The polymer threads shown in Fig. 1-B and C are similar to the result obtained by Bewersdorff (16). We define the heterogeneous drag reduction caused by the fine threads as "case I", and the drag reduction caused by a relatively stable thread flowing in the center portion of the pipe in a wave like manner as "case II".

3. Experimental apparatus and procedure for the particle tracer measurements with polymer injection

Although the flow visualization of the injected polymer thread is easily accomplished, quantitative analysis of the polymer thread movement is rather difficult. Therefore, the present author has employed a tracer particle method which can separate the velocity information of the polymer threads from that obtained for the water phase in the heterogeneous drag-reducing flow. For this purpose, the tracer particles were introduced into the main flow at the beginning of the experiments, i.e. in the water phase only. After measurements in the water phase were completed, the tracer particles were added only in the polymer solution. It was visually determined that the tracer particles that passed through the injection nozzle remained only in the flowing polymer threads. Measurements of the time averaged velocity distributions, the turbulent intensity and the Reynolds stress distributions were carried out for two of the previously mentioned polymer concentrations, i.e. the middle and high concentrations of injected polymer solutions, as well as for a premixed flow and for a Newtonian (water) flow. As the experimental apparatus is the same as reported previously by Usui et al.(21,22), only the brief description is given in this paper.

The test section was an 8 m long × 51.3 mm i.d. horizontal acrylic pipe which formed a part of the once through flow system. Polyethylene oxide (grade Alcox E-180 supplied by Meisei Chemical Corp.) was used as the polymer additive. The aqueous solution of the polymer additive containing C_p=4,000 ppm or C_p=8,000 ppm Alcox E-180 was pressurized to be injected into the water flow through the injection nozzle. The centerline injection mode was accomplished by a single nozzle (4 mm i.d. nozzle for 4,000 ppm and 3 mm i.d. nozzle for 8,000 ppm) located at the center of the test tube. The injection flow rate was adjusted so that the bulk averaged polymer concentration, C_{av}, becomes 30 ppm. The Reynolds number Re was based on the viscosity of water.

A schematic diagram of the experimental setup for the measurement of the tracer particle trajectories is shown in Fig. 2. The light from a high pressure mercury lamp was set parallel to the test tube. A 45 degree inclined mirror and the slit above the test tube created the illuminated sampling volume (51.3 mm height × 50 mm longitudinal width × 3 mm depth). The polyvinyl chloride particles having 0.1 mm dia. were dispersed as tracer. The density of the tracer particles was adjusted to the density of the test fluid by coloring with a fluorescent paint because the particle density was sensitive to the amount of paint spread on the particle surface. The particle images in the sampling volume were recorded by a videotape recorder which was moved in the downstream direction at the same velocity as the averaged velocity in the test tube. At the beginning of the experiment, the tracer particles were added into the main flow, i.e., into the water phase. After the measurements in the water phase were completed, the tracer particles were added only into the polymer solution stored in a pressurized vessel and to be injected. It was visually determined that the tracer particles injected through the injection nozzle existed only in the flowing polymer threads.

The data acquisition was performed over the range of x/D which varied from 110 to 150 where the flow was fully

Table I Experimental conditions

	C_P (ppm)	C_{AU} (ppm)	$\nu \times 10^6$ (m^2/s)	Re (-)	f (-)	$u_\tau \times 10^2$ (m/s)	DR (%)
water	—	—	1.06	10,140	0.0079	1.31	—
injection;							
"case I"	4,000	30	1.06	10,190	0.0037	0.91	53
"case II"	8,000	30	1.27	15,460	0.0032	1.54	55

developed even for the case of polymer injection (Usui et al. (22)). The displacement of each tracer particle image was analyzed, and both the longitudinal and radial velocities were calculated by taking into account the movement of the video camera. The radial position was subdivided into ten blocks, i.e, y/R=0.0-0.1, 0.1-0.2, ····, 0.9-1.0, where y and R are the distance from the wall and the pipe radius, respectively. The data included in the same block were averaged to obtain the time averaged velocity and other turbulent properties.

4. Experimental results and discussion

4.1 Velocity distribution

The time averaged velocity distributions are shown in Fig. 3, where the longitudinal velocity U and the distance from the wall y are normalized by wall shear velocity u_τ and kinematic viscosity ν. The results of mean velocity profiles for the water phase of the polymer injection system are in good agreement with velocity profiles in a heterogeneous drag-reducing pipe flow measured by a Laser Doppler Velocimeter (17,18,20). As the volumetric flow rate of the injected polymer thread was 0.75 % ("case I") and 0.375 % (case II") of the volumetric flow rate of water phase, it is reasonable that the good agreement between the previous LDV measurements and the present velocity profiles is obtained for the water phase. The local velocity of fine threads ("case I") is higher than that of the water phase

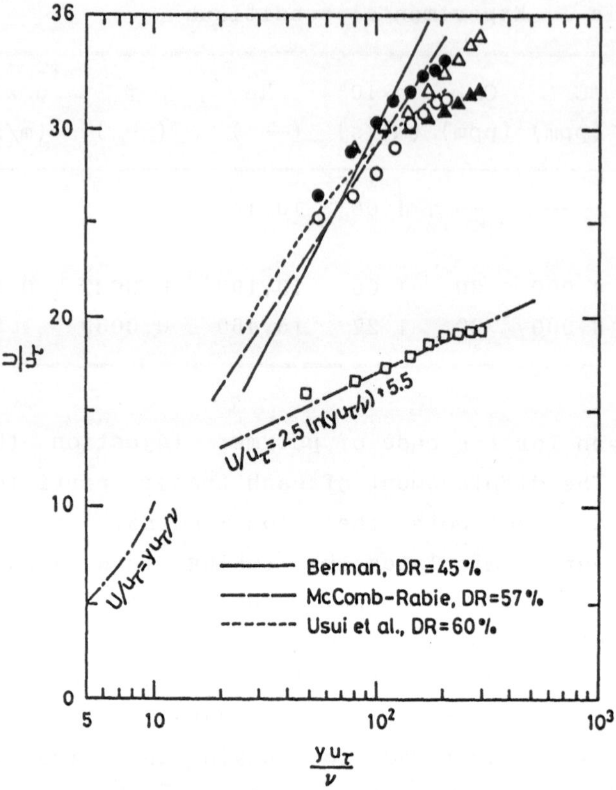

Fig. 3. Mean longtudinal velocity profiles during the centerline injection of polymer solutions.
□ ;Newtonian (water) data
○ ;data for the water phase flow in "case I"
● ;data for the polymer thread in "case I"
△ ;data for the water phase flow in "case II"
▲ ;data for the polymer thread in "case II"

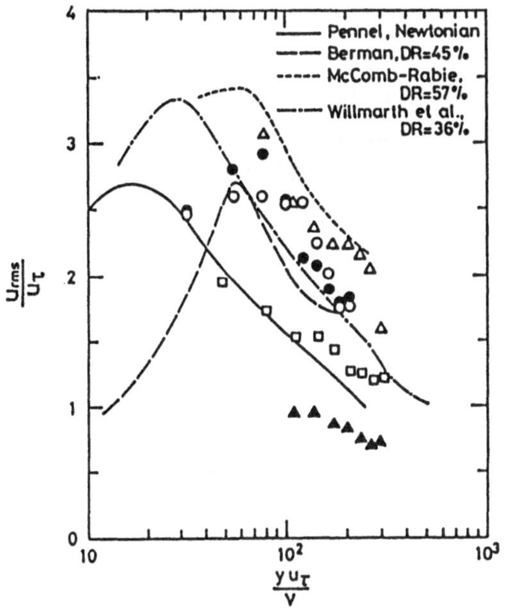

Fig. 4. Profiles of turbulent intensity in the longitudinal direction.
□ ;Newtonian (water) data
○ ;data for the water phase flow in "case I"
● ;data for the polymer thread in "case I"
△ ;data for the water phase flow in "case II"
▲ ;data for the polymer thread in "case II"

as indicated by a solid circle symbol. The existence of the velocity difference between the polymer thread and the water phase was previously reported for "case I" by the present author (21). The most plausible interpretation of this phenomenon was given as follows: (1) the centerline injected polymer threads exist only in the high speed large eddies in a pipe flow; and (2) the low speed eddies ejected from the wall layer scarcely contain the polymer threads in them. It is noteworthy that the fine threads follow the large eddy motion, in other words, the fine threads of "case I" are included inside of the large eddies.

On the other hand, the local velocity of a relatively stable thread moving in a wave like manner ("case II") is lower than that of the water phase as indicated by a solid triangular symbol. The fluid turbulent motion acts to deform the flowing thread. However, the strong viscoelasticity of an injected polymer solution prevents from the large deformation of thread. The polymer thread in "case II" appears to flow at an uniform velocity which does not depend on the radial position. In fact, the velocity variation of thread was narrow (from 46.6 cm/s to 49.3 cm/s) as shown in Fig. 3. Thus, the velocity of the injected thread in "case II" is lower than that of the water phase. The injected thread was scarcely observed at $yu_\tau/\nu < 100$.

4.2 Turbulent intensities

Turbulent intensity profiles in the longitudinal direction are shown in Fig. 4. The present results for a water flow show reasonable agreement with tho a Newtonian fluid data of Pennel at al.(23) The present experimental results for the water phase and the LDV data in the heterogeneous drag-reducing system of Berman (18), McComb-Rabie (17) and Willmarth et al.(19) show that the peak location of the turbulent intensity profiles is located at yu_τ/ν = 30~80. The upward shift of turbulent intensity for drag-reducing flows does not maen an increase in the exact turbulence level. The turbulence intensities at the pipe centerline normalized by the maximum velocity, u_{rms}/U_{max}, were 0.064, 0.054 and 0.053 for the water flow and the water phase data of the injection experiments in "case I" and "case II", respectively. Thus, only a little suppression of longitudinal turbulent fluctuation is observed in the core region of a polymer injected drag-reducing flow.

The turbulent intensity profiles of injected threads are indicated by solid symbols in Fig. 4. These data were obtained by analyzing the movement of the tracer particles contained in the injected polymer threads. The clear difference between the water phase data and the polymer threads in "case I" is not observed. This observation supports indirectly that the injected fine threads in "case I" follow the large eddy motion. On the other hand, a significant suppression of turbulence level of the thread in "case II" is observed. As the injected thread in "case II" has strong viscoelasticity and it stretches and relaxes in the longitudinal direction, the local fluctuation of the thread may be suppressed.

The radial turbulent intensity profiles are shown in Fig. 5. The previous experimental results of radial turbulent intensity in a heterogeneous drag-reducing system given by Willmarth et al.(19) are compared in this diagram. It is evident that the radial fluctuation in the water phase in both "case I" and "case II" is significantly damped. The discrepancy between the present experimental results and the data of Willmarth et al. may be attributed to the difference of the drag reduction rate, DR. The radial turbulence intensities at the pipe centerline normalized by the maximum velocity, v_{rms}/U_{max}, were 0.065, 0.027 and 0.017 for the water flow and the water phase data of the injection experiments in "case I" and "case II", respectively. The radial fluctuation of polymer threads is more anomalously suppressed than the case of the water phase flow. Many investigators (24,25) have reported that a significant damping of radial turbulent intensity was observed for the case of large-eddy breakup devices (LEBU) experiments. The suppression of the radial eddy motion appears to be essential in the drag-reducing flow.

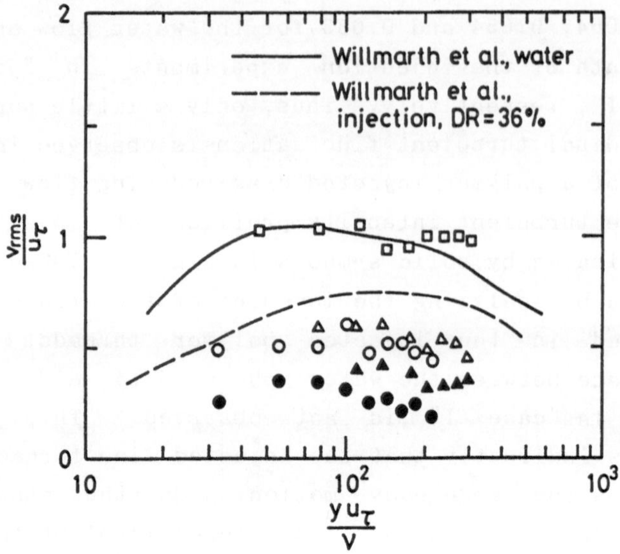

Fig. 5. Profiles of turbulent intensity in the radial direction.
□;Newtonian (water) data
○;data for the water phase flow in "case I"
●;data for the polymer thread in "case I"
△;data for the water phase flow in "case II"
▲;data for the polymer thread in "case II"

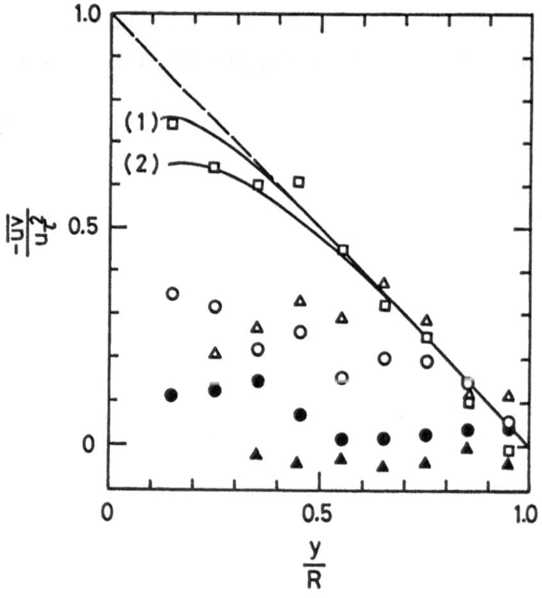

Fig. 6. Reynolds stress distributions,
□;Newtonian (water) data
○;data for the water phase flow in "case I"
●;data for the polymer thread in "case I"
△;data for the water phase flow in "case II"
▲;data for the polymer thread in "case II"

4.3 Reynolds shear stresses

The experimental results of Reynolds stress measurements are shown in Fig. 6. Measurements for water are in good agreement with the Reynolds stress level [curved line (1)] calculated from the following equation:

$$\frac{-\overline{uv}}{u_\tau^2} = 1 - \frac{y}{R} - \frac{\nu dU/dy}{u_\tau^2} \quad (2)$$

The injection of polymer threads results in a drastic damping of the Reynolds stress. In particular, the polymer thread data of "case I" in the core region and all the data of "case II" show almost zero Reynolds stress level. This experimental evidence means that not only the injected threads have no contribution to the Reynolds stress transport, but also they would obstruct the turbulent motion of fluid.

The measurements for the water phase of the polymer injection experiments show that the Reynolds stress is less than the calculated value [curved line (2)] from Eq.(2). Bewersdorff (16) reported that the Reynolds stress was strongly reduced near the wall, but the Reynolds stress was unaffected for distances from the wall greater than 0.4 times the pipe radius. The results for the "case II" are in good agreement with the observation by Bewersdorff. As already mentioned in the previous section of this paper, the configuration of an injected thread in "case II" is similar to that observed by Bewersdorff. Recent measurements by Willmarth et al.(19) show that the Reynolds stress is less than the calculated value from Eq.(2) all across the channel. The present results for the "case I" indicated by an open circle symbol show the same tendency as reported by Willmarth et al. It is not clear that the configuration of the thread in Willmarth's experiments corresponds to the "case I" with fine threads, because they did not give a flow visualization result. However, Willmarth et al. injected 1,000 ppm polymer solution at four points in the settling chamber. It is reasonable to suppose that the experimental results of Willmarth et al. corresponds to the "case I". Thus, it may be concluded that the profile of the Reynolds stress in the turbulent core region depends on the flow regime of the injected thread, i.e. "case I" or "case II". The difference between the calculated value of the Reynolds stress distribution and the measurements for the water phase flow in the heterogeneous drag-reducing flow is not clearly explained at the present stage. The experimental evidence suggests that we should consider a new momentum transport mechanism in addition to the usual momentum transport, i.e. the viscous transport and the Reynolds stress transport. The discussion on this point is an important subject in the future research on the heterogeneous drag reduction.

5. Concluding remarks

The above mentioned discussion may give the following concluding remarks on the interaction between the injected threads and turbulent eddies.

1. For the case of the middle concentration ("case I"), the fine polymer threads were contained mainly in the high speed large eddies of the pipe flow. Both the radial fluctuation and the Reynolds stress for the injected polymer threads were significantly suppressed. The suppression of Reynolds stress is significant even at the center portion of the pipe flow.

2. For the case when the high concentration was injected ("case II"), the polymer threads were relatively stable in the pipe flow, and did not divide into fine threads. The higly viscoelastic polymer thread flowing mainly at the center portion of the pipe flow has non-turbulent but a wave like motion. The injected polymer threads flowed at rather uniform velocity, and the velocity was lower than the velocity of the water phase flow at the centerline of a pipe flow. A significant suppression of the radial fluctuation was observed. The Renolds stress was strongly reduced near the wall, but the Reynolds stress was unaffected for distances from the wall greater than 0.6 times the pipe radius.

The experimental results obtained thus far indicate that the configuration of the polymer threads and the manner of interaction with the turbulent eddies are quite dependent on the polymer concentration and the Reynolds number. However, it is evident that the interaction between the injected polymer threads and the turbulent eddies is quite different from that observed in a premixed drag-reducing flow. Based on the knowledge that the injected polymer thread is scarcely observed very near the wall and the turbulent characteristics in the core region of the heterogeneous drag-reducing flow are anomalously changed, it is suggested that the main interaction occurs in the central portion of the pipe flow and the structure of the wall turbulence may be controlled by suppressing the large scale turbulent motion in the turbulent core region.

References

1. Oldkar, D. K. and W. G. Tiederman,"Spatial Structure of the Viscous Sublayer in Drag-Reducing Channel Flows", Phys. Fluids, 20, (1977) s133-s144

2. Berner, C. and O. Scrivener,"Drag Reduction and Structure of Turbulence in Dilute Polymer Solutions", Prog. Astronaut. Aeronaut., 72, (1980) 290-299

3. Reishman, M. M. and W. G. Tiederman,"Laser-Doppler Anemometer Measurements in Drag-Reducing Channel Flows" , J. Fluid Mech., 70, (1975) 369-392

4. Achia, B. U. and D. W. Thompson,"Strauctures of Turbulenct Boundary in Drag-Reducing Pipe Flow", J. Fluid Mech., 81, (1977) 439-464

5. Baid, K. M. and A. B. Metzner," Rheological Properties of Dilute Polymer Solutions Determined in Extensional and in Shearing Experiments", Trans. Soc. Rheol., 21, (1977) 237-260

6. James, D. F. and J. H. Saringer,"Extensional Flow of Dilute Polymer Solutions", J. Fluid Mech., 97, (1980) 655-671

7. Lumley, J. L.,"Two-Phase and Non-Newtonian Flows", in Topics in Applied Physics, Vol.12, ed. by P. Bradshaw, Springer-Verlag, Berlin (1976)

8. James, D. F. and A. J. Acosta,"The Laminar Flow of Dilute Polymer Solutions Around Circular Cylinders", J. Fluid Mech., 42, (1970) 269-288

9. Piau, J. M.," A Characteristic Length for Dilute Drag Reducing Polymer Solutions", Rheol. Acta., 19, (1980) 724-730

10. Usui,H., T. Shibata and Y. Sano,"Karman Voltex Behind a Circular Cylinder in Dilute Polymer Solutions", J. Chem. Eng. Jap., 13, (1980) 77-79

11, Bernard, B. S. J. and R. H. J. Sellin,"Grid Turbulence in Dilute Polymer Solutions", Nature, 222, (1969) 1160-1162

12. McComb, W. D., J. Allan and C. Greated,"Effect of Polymer Additives on the Small=Scale Structure of Grid-Generated Turbulence", Phys. Fluids, 20, (1977) 873-879

13. Berman, N. S., J. C. Berman, M. Parkino, W. Lindquist and J. J. Hwang,"Initial Development of Submerged Circular Jet of Dilute Polymer Solutions", Proc. 8th Symp. on Turbulence, Missouli-Rolla, (1983) 62-69

14. Usui, H. and Y. Sano,"Turbulence Structure of Submerged Jet of Dilute Polymer Solutions", J. Chem. Eng. Jap., 13, (1980) 401-404

15. Haas, R. and F. Durst,"Viscoelastic Flows of Dilute Polymer Solutions in Regarding Packed Bed", Rheol. Acta. , 21, (1982) 566-571

16. Bewersdorff, H. W.,"Effect of a Centrally Injected Polymer Thread on Drag in Pipe Flow", Rheol. Acta., 21, (1982) 587-589 or Proc. of Third Int. Conf. on Drag Reduction, ed. by R. H. L. Sellin and R. T. Moses, Univ. of Bristol,(1984) B.4, 1-8

17. McComb, W. D. and L. H. Rabie,"Local Drag Reduction due to Injection of Polymer Solutions into Turbulent Flow in a Pipe, Part I: Dependence on Local Polymer Concentration, Part II: Laser-Doppler Measurements of Turbulence Structure", AIChE. J., 28, (1982) 547-565

18. Berman, N. S.,"Velocity Fluctuations in Non-Homogeneous Drag Reduction", Chem. Eng. Commun., 42, (1986) 37-51

19. Willmarth W. W., T. Wei and C. O. Lee,"Laser Anemometer Measurements of Reynolds Stress in a Turbulent Channel Flow with Drag Reducing Polymer Additives", Phys. Fluids , 30,, (1987) 933-935

20. Usui, H., M. Kodama and Y. Sano,"Laser-Doppler Measurements of Turbulence Structure in a Drag-Reducing Pipe Flow with Polymer Injection", J. Chem. Eng. Jap., 21, (1988) 134-140

21. Usui, H., K. Maeguchi and Y. Sano,"Drag Reduction Caused by the Injection of Polymer Thread into a Turblent Pipe Flow", Phys. Fluids, 31, (1988) 2518-2523

22. Usui, H., M. Kodama and Y. Sano,"Drag Reduction with Heterogeneous Polymer Injection into a Turbulent Pipe Flow", AIP Document No. PAPS PFLDA-31-2518-14, (1988)

23. Pennel W. T., E. M. Sparrow and E. R. G. Eckert, "Turbulence Intensity and Time-mean Velocity Distributions in low Reynolds Number Turbulent Pipe Flow", Int. J. Heat Mass Transfer, 15, (1972) 1067-1074

24. Guezennec, Y. G. and H. M. Nagib,"Documentation of Mechanism Leading to Net Drag Reduction in Manipulated Turbulent Boundary Layers", AIAA Paper No.88-0519 (1985)

25. Veuve, M., T. V. Troung, and L. L. Ryhming,"Detailed Measurements Downstream of a Tandem Manipulator in Pressure Gradients", Proc. of the Three Day Int. Conf. on Turbulent Drag Reduction by Passive Means, published by Royal Aeronaut. Soc. (1987)

Large Eddies and Polymer Strings

Neil S. Berman
Department of Chemical, Bio and Materials Engineering
Arizona State University
Tempe, Arizona 85287-6006, USA

Summary

The Doppler shift of forward scattered light from particles and the fluorescence intensity at 90 degrees from the incident light forming the same probe volume were measured simultaneously to study the interaction of injected polymer strings with the flow. In drag reducing turbulence when clusters of strings are observed, the velocity fluctuations are damped, but isolated strings appear to have no effect on the flow.

Introduction

In non-homogeneous drag reduction, relatively high concentration polymer solutions by homogeneous drag reduction standards are injected into a turbulent pipe flow through small tubes placed inside the pipe. The injected polymer threads are broken into much finer strings by the turbulent flow, but the strings do not dissolve in the solvent phase. Drag reduction produced by these strings does not have the same onset behavior, pipe diameter dependence, or velocity profile as drag reduction by homogeneous premixed solutions. Some differences only arise when the amount of drag reduction is low, and highly drag reducing cases for both types are similar. Previous studies of non-homogeneous drag reduction by Bewersdorff [1] and Berman [2] have suggested that the interaction of the polymer strings with large eddies is an important factor in the mechanism of non-homogeneous drag reduction. In this study laser Doppler velocity (LDV) measurements were combined with fluorescence detection of the marked threads to examine the influence of threads on the large eddies.

Although the turbulent transfer mechanisms are different in submerged jets and pipe flows, polymer solution jets can show the same formation of fine threads [3]. When an experiment to determine the correlation time by monitoring particle arrivals in two probe volumes was performed [4], the apparent integral time scale for the polymer solution jet was less than the integral time scale for a Newtonian fluid jet. However, the integral time scale estimated from power spectra of the velocity fluctuations was higher for the polymer solutions compared to a Newtonian fluid. Since the injection of polymer into a pipe flow is

A. Gyr (Editor)
Structure of Turbulence and Drag Reduction
IUTAM Symposium Zurich/Switzerland 1989

similar to the submerged jet but stretched out along the pipe axis, the pipe study and the jet study are related.

Experimental

The one-dimensional set-up for the pipe flow and the two- dimensional set-up for the jet flow have been described previously [2,3]. All of the experiments reported herein were run with a 12.7 mm diameter smooth tube at a Reynolds number of 9500. The probe volume was located 150 pipe diameters from the 2 mm diameter center-line injector. A new window was added to the plastic block which housed the 0.025 mm thick mylar film used as the pipe wall at the measurement location. This window was at 90 degrees to the line of the laser beam so that the LDV beam intersection could be imaged on a photodetector when light scattered from particles had a minimum in intensity.

Velocity measurements were made with two focused beams in a plane parallel to the pipe axis to give axial velocity, v_x , and with four beams focused in pairs at 45 degree angles to the pipe axis to give two components of the velocity. In the later case the v_x and v_y components were measured when the long axes of the beam intersections were perpendicular to a radius, and the v_x and v_φ components were measured when the long axis was along a radius. The $1/e^2$ intensity dimensions of the probe volume were 0.14 by 1.4 mm. For most of the reported experiments at low drag reduction a single probe volume was six by 60 in y_+ units. Therefore, when measuring the single v_x component the wall could be approached to within approximately three y_+ units. The limits of approach to the wall for the four beam experiments were about 15 y_x units for the v_y orientation and 30 for the v_φ orientation.

The injected fluid contained 4000, 6000 or 8000 ppm Polyox WSR 301 poly(ethylene oxide), one ppm sodium fluorescein, one drop of a 20% by volume mixture of 0.45 micrometer polystyrene spheres, and 0.01 molar sodium borax in one liter of solution. The water phase contained the same concentrations of particles and borax. The borax was added to keep the solutions at constant pH so that the fluorescence was constant. Normal tap water had a pH in the range where the fluorescence was highly sensitive to pH. An orange optical filter was placed in front of the fluorescence phototube detector to remove any stray blue light from the laser. Samples of the effluent from the tube containing injected polymer solution and water phase were allowed to mix and then the fluorescein concentration was determined in a separate flow cell. The experiments reported here all had final well mixed concentrations of 20 to 25 ppm of polymer. The pressure drop along a length of the pipe upstream of the laser beam probe volume was measured with a differential pressure gage.

In order to obtain long records to analyze, the fluorescence phototube voltage and the voltages from one or two TSI trackers were recorded on three channels at 15 inches per second with an Ampex FM tape recorder. The results were played back a factor of eight slower. Some data was digitized at 280 or 140 Hertz to form 8000 points per channel records. These records were separated into polymer phase and water phase parts and analyzed by Hasenauer [5]. Other records were digitized at 70 to 280 Hertz with an Applescope analog to digital converter and printed out as strip chart records for visual analysis. Only two channels could be printed at a time so the thread channel served as a reference to match the other components.

Results and Discussion

The drag reduction was 15% for the 8000 ppm injection, but well mixed polymer at the same final concentration showed no drag reduction. When a polymer thread passed through the probe volume, the phototube voltage changed sharply and then returned to the baseline. For 6000 ppm and 4000 ppm the drag reduction was about 40%, but there was a baseline shift in addition to the sharp peaks. Therefore, we only analyze in detail the 8000 ppm case where the polymer and water phases could be clearly separated.

The time averaged velocity profiles for the polymer threads and the water phase are shown in Figure 1. The polymer phase velocities are larger than the water phase. The turbulent intensity profiles are shown in Figure 2. The axial velocity fluctuations are less for the polymer threads than for the surrounding fluid except for some points for y_+ between 15 and 30. There is a lot of scatter in the velocity profile data in this range of y_+. Very few threads were seen for y_+ less than 15 and the points for the polymer phase may not be significant.

Visual examination of the velocity records shows that v_x and v_y fluctuations are damped when several threads are seen in the record within a short time, but no damping is seen when an isolated thread passes through the probe volume. A typical example of the record is shown in Figure 3. The single thread passes through the two beam intersections as it moves toward the wall at a higher axial velocity than the average for that location. Still photographs of a laser sheet illumination of the flow were taken with a 35 mm camera at 1/500 second using Kodak T Max 3200 film developed at a film speed in excess of 20,000. The threads were seen to be oriented at a slight angle to the tube axis. Hasenauer measured 5 ms. for the average time for the thread to move through a probe volume so the thread is at least 10 mm. long. The transit time was the same for all radial locations when y_+ was greater than 30. If the threads go through the thinnest part of the beam in the radial direction the average thread velocity is then 3 cm/s which is about the same as the root mean square radial fluctuation velocity. For the multiple thread episode shown the axial velocity remains slightly higher than the average, the radial velocity shows that most of the threads are moving away from the wall, and there is a strong reduction in the fluctuations. A more detailed analysis of thread motions at eight different y_+ locations is given in Table 1.

Table 1. Analysis of v_x and v_y fluctuations when threads are present. Quadrant one (Q-1) is when v_x is higher than average and v_y is away from the wall, etc. The numbers for the quadrants are in percent. The last column is the ratio of threads that are moving toward the wall to those moving away from the wall.

y_+	*Threads*	*Time*	*Q-1*	*2*	*3*	*4*	v_-/v_+
265	142	3.6	14	29	15	42	1.33
158	126	3.6	5	27	16	53	2.15
102	106	7.2	17	30	15	38	1.12
69	302	21.6	20	23	35	22	1.34
55	119	14.4	21	15	24	40	1.77
38	332	43.2	17	16	29	39	2.06
30	411	43.2	27	26	26	20	0.87
18	194	57.6	14	23	23	40	1.70

The threads in quadrant 2 for the locations close to the wall were checked for those that had low axial velocities which could be associated with bursts. At $y_+ = 18$ there were none out of 44 threads, at $y_+ = 30$ there were 22 out of 108 and at $y_+ = 38$ there were 2 out of 52 threads. Again there is a different behavior at $y_+ = 30$. A possible explanation is that the thread passage time through the probe volume (average of 4-5 ms.) is the same as the turbulence integral scale ky/u_τ, where k^{-1} is the slope of the velocity profile on a log scale. For probe locations farther from the wall, the integral scale is larger than the average thread passage time, and for locations closer to the wall, the integral scale is smaller than the thread passage time. Thus we expect fluctuations in the data at $y_+ = 30$ when the thread is expected to undergo changes during the time it takes to go through the probe volume. It appears that on the average the threads move toward the wall except possibly at $y_+ = 30$.

Since the optical set-up must be realigned to move the beam intersections to measure the angular component of velocity, only one location could be measured for this study. A typical result of the velocity record is shown in Figure 4. The velocity amplification is 2.5 times that in Figure 3. Visual inspection of the axial-angular records indicates that threads are present when the velocities are well correlated and never present when the records are not correlated. The records are correlated much of the time when threads are not present.

Conclusions

The significant observations are the damping of fluctuations when clusters of threads are present perhaps giving a region large enough to interact with large eddies and the rapid movement of the threads through the probe volume. The latter movement of threads independent of the turbulence near the center of the pipe is the reason the two point correlation experiments gave an integral scale unrelated to the turbulent scale. The threads appear to have a large influence in the buffer layer where they may turn back toward the center of the pipe and inhibit the movement of slow moving fluid coming from the wall.

References

1. Bewersdorff, H.W.: Heterogene Widerstandsverminderung bei Turbulenten Rohrstromungen. Rheol. Acta 23 (1984) 522-543.
2. Berman, N.S.: Velocity Fluctuations in Non-Homogeneous Drag Reduction. Chem. Engineering Comm. 42 (1986) 37-51.
3. Berman, N.S.; Tan, H.: Two Component Laser Doppler Velocimeter Studies of Submerged Jets of Dilute Polymer Solutions. AIChE Journal 31 (1985) 208-215.
4. Berman, N.S.: Lagrangian Dispersion in Turbulent Flow from Laser Transit Anemometry. AIChE Journal 32 (1986) 782-788.
5. Hasenauer, L.: The Correlation Between Concentration and Velocity Fluctuations for Polymer Solutions in Turbulent Mixing Processes. Diplomarbeit University of Dortmund.

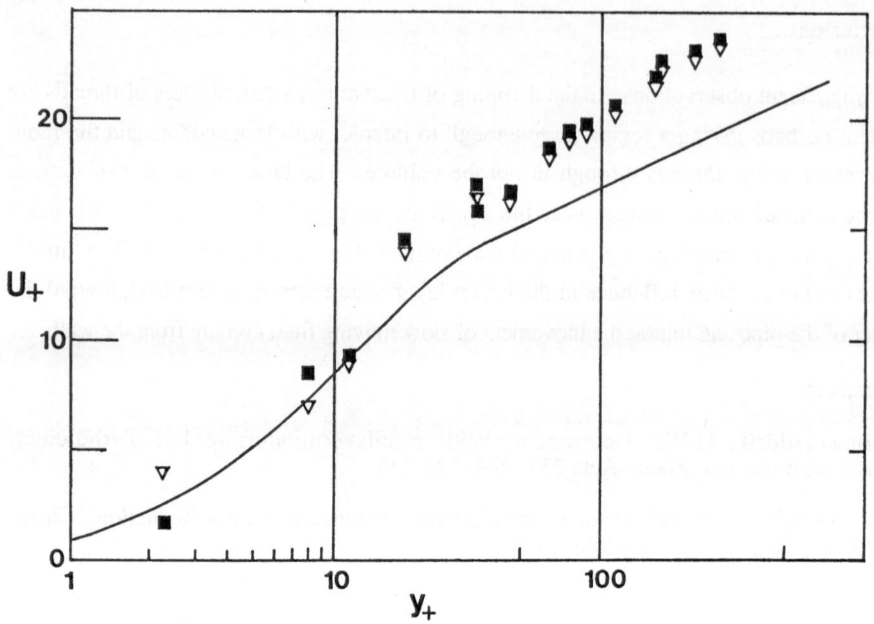

Figure 1. Velocity profiles for 8000 ppm injection. The solid line is the Newtonian fluid curve. The filled in squares are for polymer threads and the open inverted triangles area for the water phase.

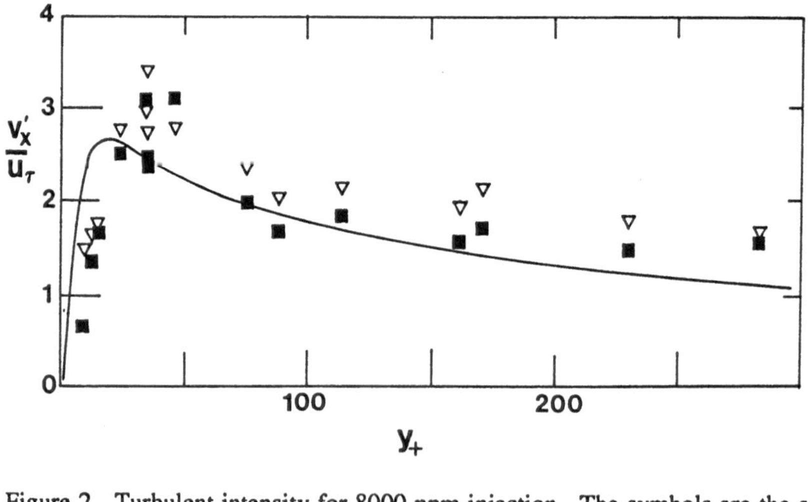

Figure 2. Turbulent intensity for 8000 ppm injection. The symbols are the same as in Figure 1.

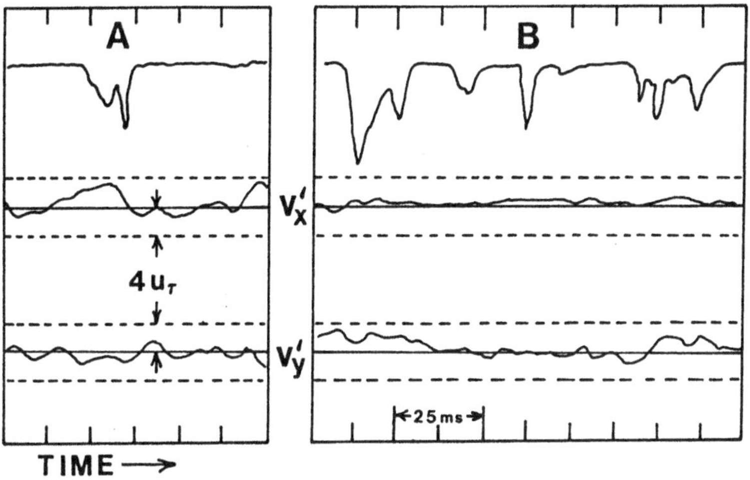

Figure 3. Trace of fluorescence, axial velocity and radial velocity for A, an isolated thread, and B, a cluster of threads, at $y_+ = 100$.

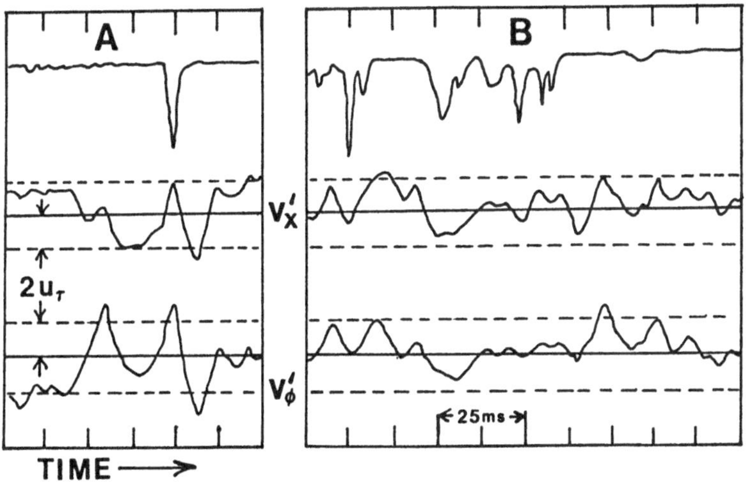

Figure 4. Trace of fluorescence, axial velocity and angular velocity for A, an isolated thread, and B, a cluster of threads, at $y_+ = 120$.

Friction Drag Reduction by Injection of Polyethyleneoxide Solution in a Turbulent Boundary Layer Through Slot and Perforated Section

V. N. MAMONOV; B. P. MIRONOV; PANOV S. V.

Institute of Thermophysics,
Siberian Branch of USSR Academy of Sciences,
Novosibirsk, USSR

It is known that injection of high molecular polymer solutions in a turbulent boundary layer is an effective means for turbulent friction drag reduction /1,2/. In this paper results of an experimental study on friciton drag reduction with injection of a polyethyleneoxide solution ($M = 4.10^6$ g/mol) in a boundary layer are presented.

The experiments were carried out on a hydrodynamic stand (Fig.1) incorporating a horizontal operating section 4 (length = 3600 mm, height = 100 mm and width = 30 mm). The measuring section (length = 1200 mm) was placed immediately behind contraction 3. On the wide side wall of the section at X = 160 mm from the channel inlet there was either a slot (width = 0.7 mm, length = 80 mm, slope = 7°), or a perforated section (70 x 80 x 10 mm with 1460 orifices, 1mm in diameter, pitch = 2 mm) or a porous section (70 x 80 x 10 mm). On the same wall at X = 310, 540, 760 and 1040 mm, friction gauges, working on the principle of a "floating" element, with a 24 mm diameter of sensitive area, were mounted.

The stand was equiped with an apparatus for the polymer concentration profile measurements in a boundary layer by method of holographic interferometry /3/ and a system of stroboscopic visualization for the velocity profile measurements.

The experimental conditions are:

- Velocity in operating section U_o = 7m/s
- Flow rate of the injected polymer solution Q = 0.08, 0.03, 0.05, 0.10, 0.20 l/s
- Concentrations of the injected polymer solution: C_o = 100, 500 and 1000 ppm.

All measurements were made with the help of a specially developed information - measurement system (IMS) connected to a minicomputer. The experimental results are processed in the form of $R = f(q_x)$,

A. Gyr (Editor)
Structure of Turbulence and Drag Reduction
IUTAM Symposium Zurich/Switzerland 1989

where

$$R = [(\tau_s - \tau_p)/\tau_s]\ 100\%. \qquad (1)$$

Here τ_s and τ_p are tangential stress on the wall without injection; respectively with injection of a polymer solution. The formula for q_x will be given below.

In Fig.2 dependences of $R = f(x)$ are given. One can see that for injection through the slot and through the perforation at C_o = 100 ppm, the value of R along the plate length decreases for all values of Q. This corresponds to a decrease of the wall polymer concentration in downstream direction caused by turbulent diffusion. However, by injecting a solution C_o = 1000 ppm at a significant distance downstream behind the location of injection, an increase of the value of R is observed. It is evident that in this case the following occurs: All at once, behind the location of injection the polymer solution concentration exceeds the saturation concentration. The viscosity of the solution at such a concentration significantly exceeds the viscosity of water, which naturally increases the wall friction. Downstream the injection, both the polymer concentration and the local viscosity near the wall decrease due to turbulent diffusion. So when the polymer concentration exceeds the saturation concentration, the value of R increases downstream. The natural fall of the effect is observed by a further decrease of the concentration below the saturation value. The authors of /4/ show a possibility of the described effect.

As it is known, for the generalization of integral characteristics of friction of any body or plate the reduced specific consumption

$$q = QC_o/U_oS \qquad (2)$$

(where S is a square of the wetted surface) a parameter is successful. In the case of generalization of the local characteristics of the wall friction as S it is logical to choose some thickness of the boundary layer multiplied by a unit or by a characteristic scale of the channel width. However, for this it is necessary to know the velocity in each section at a given X. Various methods of polymer supply into a boundary layer (through a porous wall, a perforation, a slot with various rates of supply and various initial concentrations and various polymers and degrees of their dissolution), provide a possibility, within defined limits, to change the conditions of polymer distribution and flow velocity on a given height of the boundary layer. For example, in Fig.3 concentration profiles at identical conditions of injection of PEO solution ($M = 4\cdot10^6$q/mol) and NaCl are shown. In Fig.4 velocity profiles in a boundary layer according to the data of this paper and paper /5/ are given. Even when the polymer consumption varies by a factor 2 to 5, the velocity profiles can be the same. Such features can arise due to unsteady-state processes, especially near the location of injection.

Thus a great variety of possible situations can be observed, a detailed study of which involves great difficulties. Since understanding the set of different variants for the determination of the parameter q_x can be very large (it would logically begin with the variant $S_x = \delta_c^* \cdot l$, where δ_c^* is the displacement thickness of the diffusion boundary layer, l is the channel width), the authors decided to act very simply: as a first approximation the thickness of a dynamic boundary layer without any polymer injection is taken.

Considering the results for various sections at a given X with this approach, the authors intended a whole set of circumstances and a display of some effects. First, for each section X with such an approach some mean concentration at different relative wall distances is more precise evaluated (because the specific parameter q_x is calculated by a square of the cross-section of the boundary layer S_x). In the second place, if assuming any features of the polymer supply (slot or perforation), they will be approximately identical in all the studied four sections. Thirdly, if the supply rate of polymer solution is less, the action of supply conditions is less significant. Fourthly, if X is increased, distinction of intitial conditions must be less obviuos etc.

In Figs. 5 to 8 results of all investigations in the form

$$R = f(q_x), \text{ where } q_x = QC_o/U_o\delta l \qquad (3)$$

are given (values δ are accounted for according to technique /5/). From Figs. 5 to 8 follows:

1. At injection through both a slot or a perforation at small and moderate values of Q (0.008 - 0.05 l/s) the dependences of $R = f(q_x)$ for various cross-sections of the boundary layer have a tendency to generalize in spite of C_o changing from 100 ppm to 1000 ppm.

2. The same is observed over the range of Q's, but in a narrower variation of q_x-changes (from small to moderate ones). For the experimental conditions given, this region corresponds to small and moderated values of C.

3. At large values of Q and q_x a separation of the curves of $R = f(q_x)$ at the sections X is observed. Here the values of R are smaller for the sections situated closer to the location of injection. Such a separation is more apparent in perforation injection.

4. Dependences of $R = f(q_x)$ with a slot injection (Fig.7) are shifted to the right along the q_x-axis with increasing Q. The same applies for perforation injection.

5. Within a comparatively small spread of experimental data at a small and moderate injection, the dependences of $R = f(q_x)$ for various X's are not only close to each other for the slot and perforation injection (for each case in particular), but are close enough between themselves, i.e. differences between the slot and perforation injection effectiveness at small and moderate consumptions of Q are not so essential.

6. The fact that, for a very small injection (Q = 0.008 l/s) the dependence $R = f(q_x)$ for perforation injection is significantly shifted to the left, relative to other curves $R = f(q_x)$ obtained at a more intensive injection through perforation, is very interesting.

Some features of different curves $R = f(q_x)$ can be explained by determine the q_x-parameter more precisely. For example, by introducing the value $\delta_c \cdot 1$ instead of $\delta \cdot 1$ into (3) (where δ_c is thickness of diffusion boundary layer) the differences in $R = f(q_x)$, observed in Figs. 1 - 6 will decrease. Then, analysing the dynamics of the change of $R = f(q_x)$ versus the parameters Q, x and the methods of injection (Figs. 5 to 8) by neglecting the effects of mechanical degradation, one can estimate the characteristic changes of the diffusion boundary layer. These estimations lead to the following considerations:

a) Slot injecton (Fig.5). If Q increases the injection velocity V_w increasing to. Thus, a more favourable conditions exists for the wash-out of near-wall jet of polymer. (Already at Q = 0.10 l/s the ratio between the velocity on the boundary of the viscous sublayer and the velocity in a jet $U_{vs}/V_w \approx 1$). With the increase of Q the thickness δ_c increases. Introducing δ_c into the parameter q_x the curves $R = f(q_x)$ in Figs. 5 and 7 would fall together.

b) Perforation injection (Fig.6). Since the injected flow is perpendicular to the main flow, perturbations are strong enough at moderate values of Q. This is because the value of δ_c at these regimes must be higher than for the slot injection. A correlation for δ_c between slot and perforation injection would allow the curves $R = f(q_x)$ to fall together for various methods of supply (Figs. 6 and 7).

c) Very small injection (Q = 0.008 l/s). $R = f(q_x))$ coincides for slot and perforation injection for all values of X because, due to the small perturbations, a development of δ_c in these cases is almost the same. $R = f(q_x)$ does not depend on the way the polymer is supplied into the boundary layer.

d) Slot and perforation injection. $R = f(q_x)$ at large values of X (Figs.5 and 6) for equal Q are close enough because, far enough downstream of the injection site, the characteristics of the diffusion boundary layer become more comparative.

e) Noticed in 6. This feature is explained by small perturbations of the near-wall flow due to injection. This is because the thickness δ_c must be notably less than for other values of Q. Comparisons with the data for porous injection (Fig.7), at which perturbations of the near-wall zone are too small, are revealing.

f) Mentioned in 3. The separation of $R = f(q_x)$ is connected with effects of the polymer concentration in a flow when it is higher than the saturation concentration.

The considerations made in a) to f) are confirmed by quantitative observations of the thickness of the diffusion boundary layer. At the present, qualitative measurements are being carried out.

In Fig.8 the data of Figs.5 - 6 are given in integral form, where $R_\Sigma = [(F_s-F_p)/F_s]\ 100\%$. Here F_s and F_p mean the integral frictional force on the washed-out surface S without a polymer injection and accordingly with an injection $q_\Sigma = QC_o/V_oS$.

References

1 Pilipenko V.N. (1980). Vliyanie dobavok na pristennie turbu lentnie techenia // Mekhanika zhidkosti i gaza. M., T 15, pp. 156 - 257 (Igoti nauki i tekhniki. VINITI AN SSSR)

2 Hoyt J.W. (1972). The Effect of Additives on Fluid Friction // Trans. ASME. V 94, ser.D., N 2, pp. 1 - 31

3 Eskin V.E. (1979). Rasseyanie sveta rastvorami polimerov. L. Nauka

4 Aleskin V.A., Mihailu A.G. & Pilipenko V.N. (1983). Vliyanie raspredelennoi podachi polimernogo rastvora na kharakteristiki turbulenthogo pogranichnogo sloya // Izv. AN SSSR. Mekhanika zhidkosti i gaza. N 5, pp. 58 - 64

5 Vanin Yu.P. & Khodaev A.M. (1978). Issledovanie kharakteristic pristennoi turbulentnosti v potoke s peremennoi koncentraciei polimernukh dobavok i uprugosti poverhnosti na pristennuyu turbulentnost. Novosibirsk. pp. 16 - 23

6 Fedyaevski K.K., Ginevski A.S. & Kolesnikov A.V. (1973). Raschet turbulentnogo pogranichnogo sloya neszhimaemoi zhidkosti. L., Sudostroenie

7 Makarenkov A.P., Vinogradnyi G.P., Skripachev V.V. & Kanarski M.V. (1973). Vliyanie polimernych dobavok na pulsacii davleniya v pogranichnom sloe // Inzh. Fiz. zhurn. N 6, pp. 1006 - 1009.

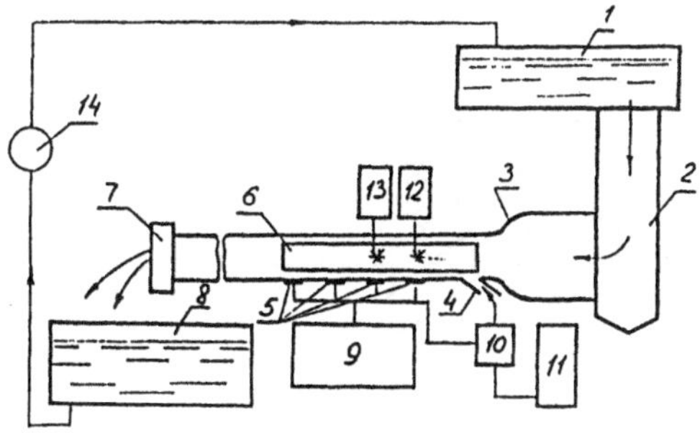

Figure 1

1 - pressure tank

2 - pressure channel

3 - contraction

4 - slot

5 - friction gauges

6 - optical windows

7 - tap

8 - overflow tank 9 - IMS

10 - flowmeter

11 - tank with polymer solution

12 - system for measurement the polymer concentration profiles

13 - system of stroboscopic visualisation

14 - pump

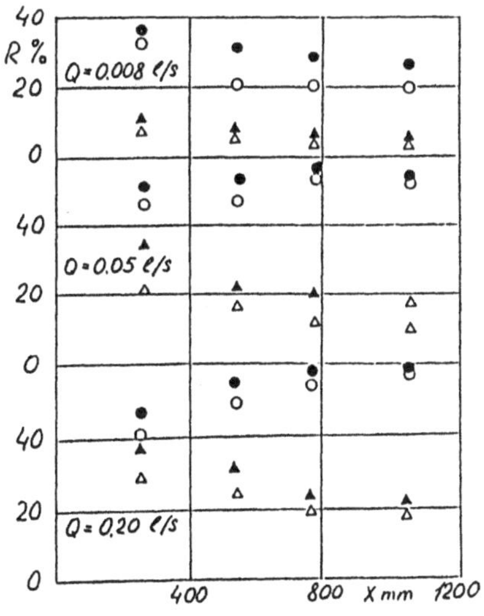

Figure 2

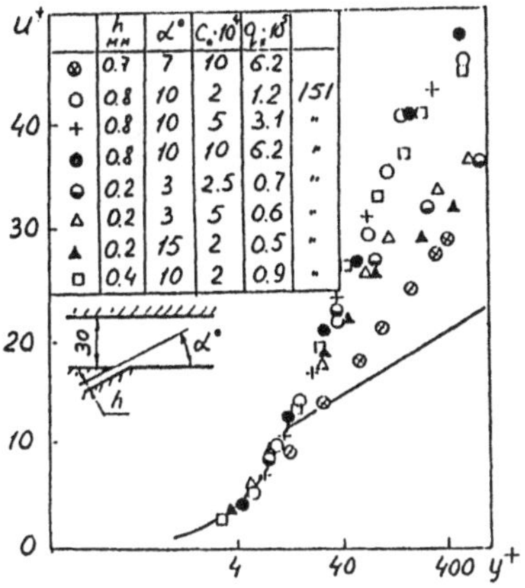

Figure 3

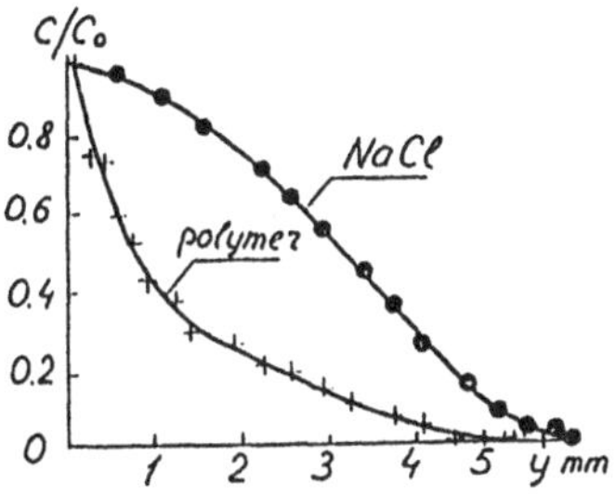

Figure 4

● ▲ - slot

o Δ - perforation

● o - C_o = 1 10^{-3} 1000 ppm

▲ Δ - C_o = 1 10^{-4} 100 ppm

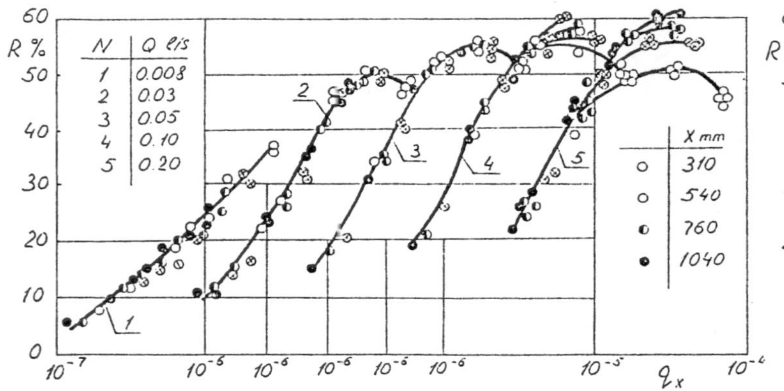

Figure 5. (A slot).

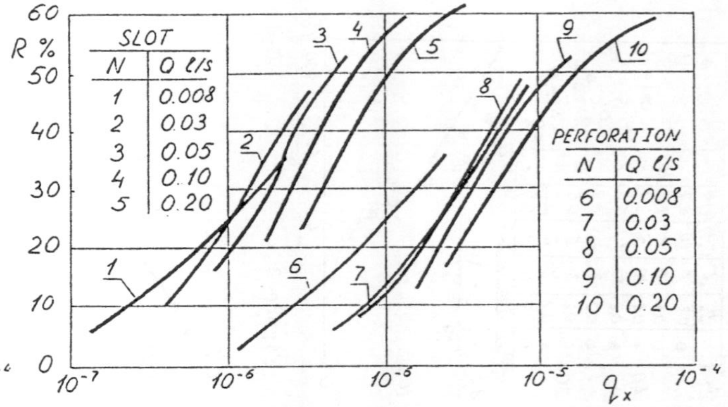

Figure 7.

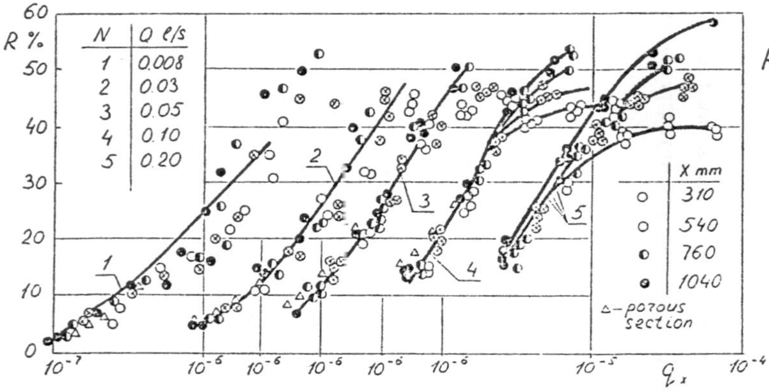

Figure 6. (Perforation).

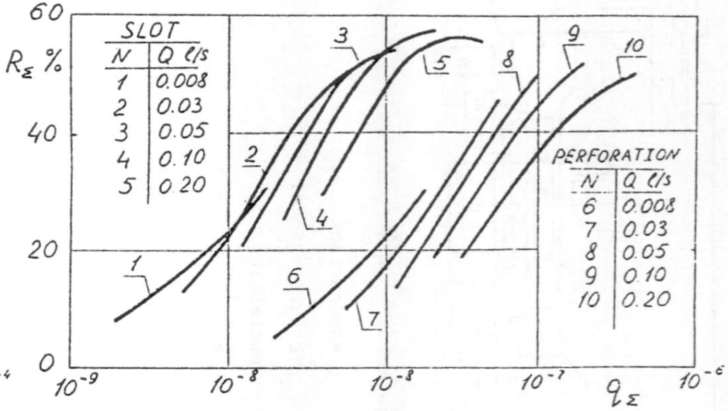

Figure 8.

Part 3
Drag Reduction by Other Means

Drag Reduction in Surfactant Solutions

Hans-Werner Bewersdorff, Department of Chemical Engineering, University of Dortmund, F.R. Germany

Present address: Institute of Hydromechanics and Water Resources Management, ETH Zürich, Switzerland

Abstract

Drag reducing surfactant solutions are characterized by the presence of rod-like micelles which are formed by single surfactant molecules above a characteristic concentration. This critical micelle concentration strongly depends on the temperature and the electrolyte concentration. Shear viscosity measurements of drag reducing surfactant solutions show that at shear rates above a critical value the viscosity suddenly increases due to the formation of a shear-induced state (SIS) in which the micelles coalescence to larger structures and are completely aligned in flow direction. The turbulent friction behaviour of these surfactant solutions is characterized by a critical wall-shear stress. The observed loss of drag reduction beyond this critical wall-shear stress is reversible.
Small-angle neutron scattering (SANS) experiments demonstrate that the alignment of the rod-like micelles in flow direction correlates with drag reduction. At low Reynolds numbers in the turbulent flow regime the dimensionless velocity profiles are very similar to those observed in dilute polymer solutions, whereas at maximum drag reduction conditions the shape of the profiles is similar to a laminar profile. The axial turbulence intensity is increased whereas the transverse turbulence intensity and the Reynolds shear stresses are strongly damped. An attempt is made to explain theses changes by an increased effective viscosity.

Introduction

Althogh one of the first publications about drag reduction by additives deals with surfactant solutions [1] this type of additives received less attention than others, especially polymers. The advantage of using drag reducing surfactants instead of polymers in practical applications is that no degradation occurs over long time periods [2-10]. This feature is also a definite advantage for studying the structure of turbulence under drag reducing conditions in closed circuits, because the solution properties and therefore the drag reduction conditions do not change with time.

A. Gyr (Editor)
Structure of Turbulence and Drag Reduction
IUTAM Symposium Zurich/Switzerland 1989

"Surfactant" is an abbreviation of "surface-active agent". Surfactant molecules consist of a hydrophilic and a hydrophobic part. In water these molecules tend to assemble to aggregates, called micelles, when the concentration exceeds a certain critical value, the critical micelle concentration, CMC. For a given surfactant system the CMC strongly depends on the temperature and the electrolyte concentration. Above the CMC globular micelles are usually formed. However, there are systems where disc-shaped or rod-like micelles are built-up. The CMC's of various surfactant systems are tabulated in [11]. Drag reducing surfactant solutions are characterized by the presence of rod-like micelles. Among the surfactants used for drag reduction one has to distinguish between anionic, cationic and nonionic surfactants. In the concentration range for drag reduction, typically 200-2000 ppm by weight, the length of the rods is about 25-200 nm, whereas its diameter, which is twice of the length of the extended hydrocarbon chain, is about 2-5 nm. It then follows that, at rest, the rods can rotate without interfering each other in these surfactant solutions [3]. However, drag reducing surfactant solutions exhibit a viscoelastic flow behaviour [6,12, 13].

It is the aim of the present study to combine the experimental results of the rheological and friction behaviour, the turbulence structure, and small angle neutron scattering in order to achieve a deeper insight into the interaction between the micelles and the turbulence which leads to drag reduction.

Rheological behaviour

In laminar pipe, channel and Couette flows dilute drag reducing surfactant solutions behave like normal Newtonian fluids at low shear rates. However, when a critical shear rate is exceeded, the shear viscosity suddenly increases due to the formation of a shear-induced state (SIS) [3,9,10,12, 13]. The critical shear rate depends on the surfactant and additional electrolyte concentration, the temperature, and, surprisingly, also on the geometry of the viscometer, eg. on the slit width of a Couette viscometer. It increases with temperature and the electrolyte concentration and decreases with the surfactant concentration and the slit width of the viscometer [3]. The magnitude of the increase of shear viscosity due to the SIS increases with the surfactant concentration and the slit width of the viscometer, and it decreases with temperature and electrolyte concentration. In Figure 1 the shear viscosity behaviour of a 700 ppm solution of tetradecyltrimethyl ammonium salicylate (C_{14}TASal) for different temperatures is shown as an example. In addition to the surfactant this solution contained an equimolar amount of the salt NaBr because it was prepared by dissolving equimolar mixtures of the salts tetradecyltrimethyl ammonium bromide and sodium salicylate. In the SIS the rods become completely aligned in flow direction as shown by flow birefringence [14,15] and small angle neutron scattering (SANS) experiments [16]. The increase in shear viscosity can only be explained by the formation of larger superordered structures. The size of

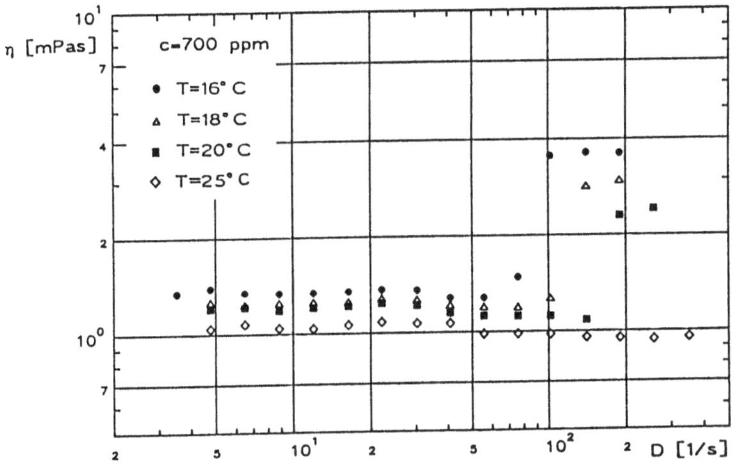

Fig. 1: Shear viscosity of a C_{14}TASal-solution in deionized water for c = 700 ppm and different temperatures

the superordered structures appears to reach the order of the gap width of a Couette viscometer, because in such a viscometer with a flexible suspension of the inner cylinder, this cylinder moved elliptically within the outer one and no steady state value of the torque could be obtained [10]. Two models for the spatial arrangement of the rod-like micelles in the SIS were discussed, the formation of pseudo-lattices from individual rods with lattice planes sliding past each other caused by cooperative electrostatic interaction energies or the coalescence of the rod-like micelles to very flexible rods, that act like high molecular polymers [3,14,15].

In the SIS the surfactant solutions exhibit viscoelastic properties and the elongational viscosity is larger than that of the solvent [17,18]. In these experiments the solutions were first subjected to a laminar shear gradient in a channel [17] or in a Couette [18] flow before they entered an elongational field produced by an orifice mounted in a wall. The elongational viscosity of viscoelastic surfactant solutions decreases with the elongational rate and also at high shear rates. Therefore it seems that the superordered structures, which are built-up in the SIS, can be destroyed by elongation and also by high shear forces.

Friction behaviour

The turbulent friction behaviour of drag reducing surfactant solutions depends on the size, number and surface charge of the micelles [3]. Since these properties change with temperature, and the concentrations of the surfactant and the additional electrolytes, the friction behaviour of surfactant solutions is influenced in a similar way by these properties. It is characteristic for the friction behaviour of surfactant solutions that the drag reduction disappears when a critical wall shear stress is exceeded. In contrast to the friction behaviour of drag reducing polymer solutions, where similar effects due to degradation of the polymer molecules occur, this loss of drag reduction is reversible, i.e. the drag reduction recovers when the wall shear stress is lowered again [3,6,7,9,10]. The critical wall shear stress is independent of the pipe diameter and depends on the properties mentioned above. SANS-experiments in a turbulent channel flow of a drag reducing surfactant solution [19] demonstrated that the orientation of the micelles and their superordered structures vanish above the critical wall shear stress, but the micelles are not destroyed. These results are in agreement with the findings of [3] where it was shown that the time constants for the formation of rod-like micelles of the surfactants used in [19] are of the order of 10^3-10^4 s. If the micelles would be destroyed by exceeding the critical wall shear stress the drag reduction could not recover when the wall shear stress is lowered again without a large time delay.

Figure 2 shows the friction behaviour of a C_{14}TASal-solution in a 15 mm pipe for different temperatures. Due to the limit in maximum flow-rate of the pump in the flow-circuit the break-down of drag reduction could not be reached. In this plot the Reynolds number is based on the solvent viscosity. This is a convenient choice in plots of the friction behaviour of dilute polymer solutions because, according to Virk [20], the shear thinning behaviour of dilute polymer solutions has only a small effect on the turbulent friction behaviour. However, due to the different rheological

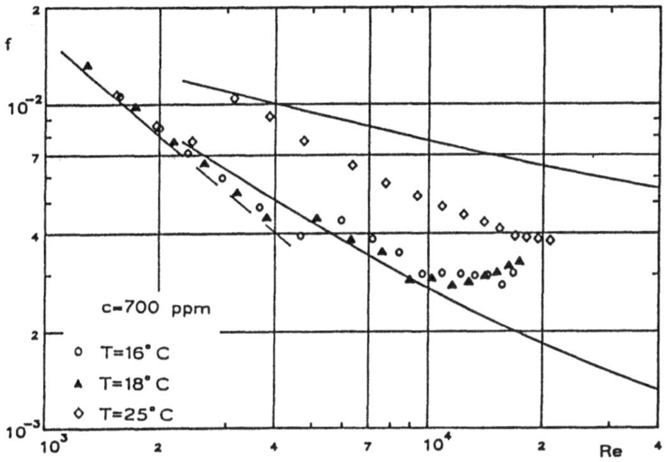

Figure 2: Friction behaviour of a C_{14}TASal-solution in de-ionized water for different temperatures in a 15 mm pipe at c = 700 ppm

behaviour of surfactant solutions the question remains whether this choice is useful. Using this choice provides the advantage that whenever the SIS influences the friction behaviour a "bump" in the flow curve should occur at a certain Reynolds number.

The drag reduction regime is limited by the Prandtl-Kármán law for a Newtonian fluid

$$\frac{1}{\sqrt{f}} = 4 \log(Re \sqrt{f}) - 0.4 \tag{1}$$

and by the empirical law of maximum drag reduction [20] of polymer solutions

$$\frac{1}{\sqrt{f}} = 19 \log(Re \sqrt{f}) - 32.4 \ . \tag{2}$$

This relation also appears to be valid for surfactant solutions. At low Reynolds numbers the friction data in Figure 2 follow the equation $f = 16/Re$ which results from Hagen-Poiseuille's law of a laminar pipe flow. For the lower temperatures the friction data follow the laminar curve up to Re = 4000-5000 before deviating to Virk's maximum drag reduction asymptote.

Therefore, the transition laminar-turbulent is shifted to higher Reynolds numbers. This observed shift of the transition is in agreement with stability calculations for stiff rods [21]. At T = 25°C the surfactant solution exhibits an onset phenomenon, i.e. deviations from the Newtonian turbulent friction behaviour only occur when a critical Reynolds number is exceeded. The onset phenomenon, which is well-known for dilute polymer solutions, can be used to check the influence of the SIS on drag reduction.

Figure 3 shows the friction behaviour of C_{14}TASal-solutions at different temperatures in a 15 mm pipe. To produce an onset behaviour in the turbulent flow regime 2 mmol/l NaBr were added as an additional electrolyte. By assuming that at the onset-point the turbulent velocity profiles of a surfactant solution and a Newtonian fluid do not differ, the turbulent velocity profile of a Newtonian fluid [22] can be used to calculate the dimensionless wall distance y^+ where the critical shear gradient for the SIS occurs. At onset conditions this critical shear gradient is found in the buffer zone (y^+ = 4.7-6.4). This means that drag reduction in surfactant solutions occurs when shear-induced structures can be built-up in the

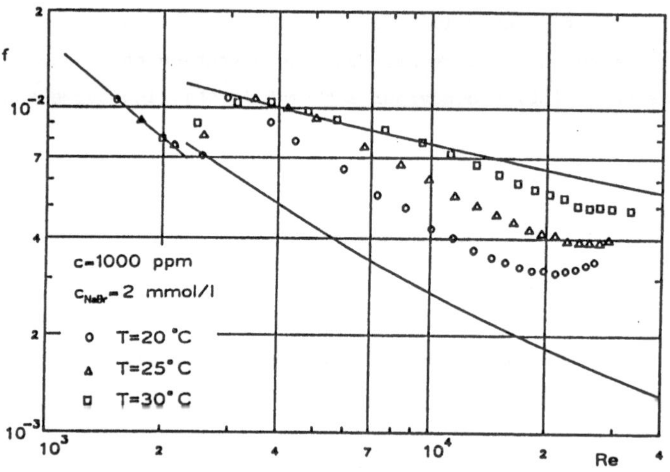

Figure 3: Friction behaviour of a C_{14}TASal-solution in de-ionized water for different temperatures in a 15 mm pipe at c = 1000 ppm with an additional electrolyte concentration of c_{NaBr} = 2 mmol/l

buffer zone. SANS-experiments [16] showed that no anisotropies in the scattering pattern existed below the onset of drag reduction, i.e. the micelles were oriented statistically in the flow. After the onset of drag reduction the anisotropy of the scattering pattern is correlated with the drag reduction, and when the friction factor approached Virk's maximum drag reduction asymptote a nearly complete alignment of the rod-like micelles in the flow direction was detected from the scattering pattern. The SANS-experiments demonstrate that the micelles are oriented parallel to the flow direction in laminar Couette as well as in turbulent drag reduced pipe flows when a critical shear gradient is exceeded.

Turbulence structure

Turbulent velocity profiles are usually presented in terms of wall variables where the dimensionless velocity

$$U^+ = \frac{U}{u_\tau} \tag{3}$$

is plotted versus the logarithm of the dimensionless wall distance

$$y^+ = \frac{y\ u_\tau}{\nu} \quad . \tag{4}$$

In these two equations U is the time-averaged local velocity, y the wall distance, ν the kinematic viscosity, and u_τ the friction velocity defined by

$$u_\tau = \sqrt{\frac{\tau_w}{\rho}} \tag{5}$$

with τ_w being the wall shear stress and ρ the density of the fluid. In drag reducing dilute polymer solutions the viscous sublayer $U^+ = y^+$ near the wall appears to be unchanged in comparison to a Newtonian fluid, whereas the turbulent core region exhibits a parallel shift ΔB

$$U^+ = 2.5 \ln y^+ + 5.5 + \Delta B. \tag{6}$$

Between these two regions a buffer zone with

$$U^+ = 11.7 \ln y^+ - 19 \tag{7}$$

is found. This relation is often called Virk's ultimate profile. The buffer zone expands with increasing drag reduction. In analogy to the friction behaviour in eq.(4) the solvent viscosity is generally used. This may cause problems, when drag reducing surfactant solutions are used which exhibit a strongly shear-dependent viscosity in the SIS. Results of turbulent velocity profile measurements of drag reducing surfactant solutions are reported in both channel [23-26] and pipe flows [9,10]. The viscous sublayer generally seems to remain unchanged, whereas in the other parts of the velocity profile the shape of the profile exhibits no general tendency. The shape of the velocity profiles in the buffer zone and in the core region depends on the flow regime in the friction behaviour [10]. The limits of these flow regimes occur at different Reynolds numbers in pipes of different diameters; the limits of the flow regimes are shifted to higher Reynolds numbers with increasing diameter [3,10]. Therefore it is neccessary to discuss in detail the relationship between the SIS, i.e. the rheological behaviour, the friction behaviour, and the turbulence structure. In the following, the shear viscosity behaviour, the friction behaviour, and turbulent velocity profiles of a C_{14}TASal-solution obtained in a 15 mm pipe by Laser-Doppler anemometry will be presented.

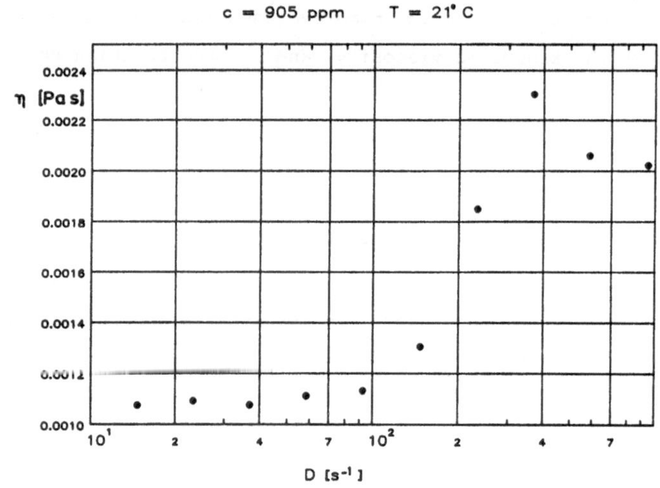

Figure 4: Shear viscosity of a C_{14}TASal-solution in de-ionized water at c = 905 ppm and T = 21°C

Figure 4 shows the shear viscosity of a C_{14}TASal-solution. By exceeding a shear gradient of $D = 100\ s^{-1}$ the shear viscosity increases due to the formation of the shear-induced state. The shear viscosity in the SIS is about twice as large as at the lower shear gradients, where the solution behaves Newtonian.

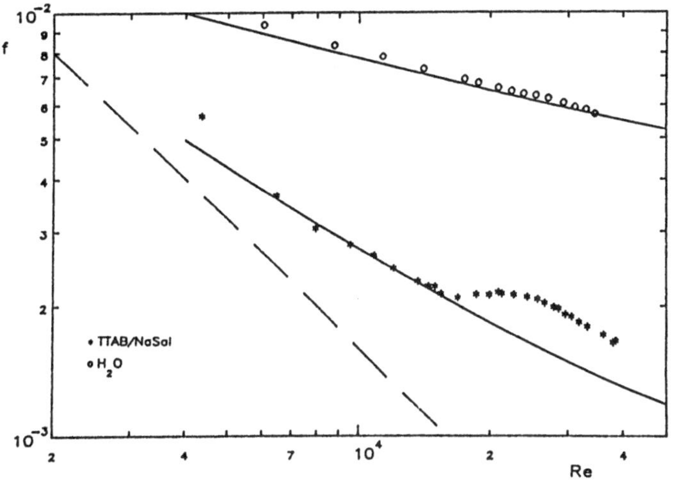

Figure 5: Friction behaviour of a C_{14}TASal-solution in de-ionized water at $c = 905$ ppm and $T = 21^{o}C$ in a 15 mm pipe

In the friction behaviour of the C_{14}TASal-solution in Figure 5 different flow regimes can be detected. At low Reynolds numbers in the turbulent flow regime the friction factor approaches Virk's maximum drag reduction asymptote. At Re = 19000 a flow regime occurs in which the friction factor is nearly constant. A further increase in the Reynolds number results in a further decrease of the friction factor. The same flow regimes were also found for salt free solutions of C_{14}TASal/C_{16}TASal in de-ionized water for different temperatures [3] and in pipes of different diameters [10].
At Re = 4200 where the friction factor lies above Virk's maximum drag reduction asymptote, eq.(2), the velocity data in the buffer zone follow Virk's ultimate profile (Fig.6) and exhibit a core region starting at $y^+ = 40$. The slope in this core region is larger than for the Newtonian solvent which is in contrast to observed velocity profiles of drag reducing dilute polymer solutions where a parallel shift of the velocity profile, eq.(6), occurs. The increased slope is an indication for a change of the structure

of turbulence in the core region. Increased slopes in the core region are also reported for the so-called heterogeneous or non-homogeneous drag reduction by polymer threads [27,28], where the turbulent structure in the core region is affected as well. This flow regime in the friction behaviour

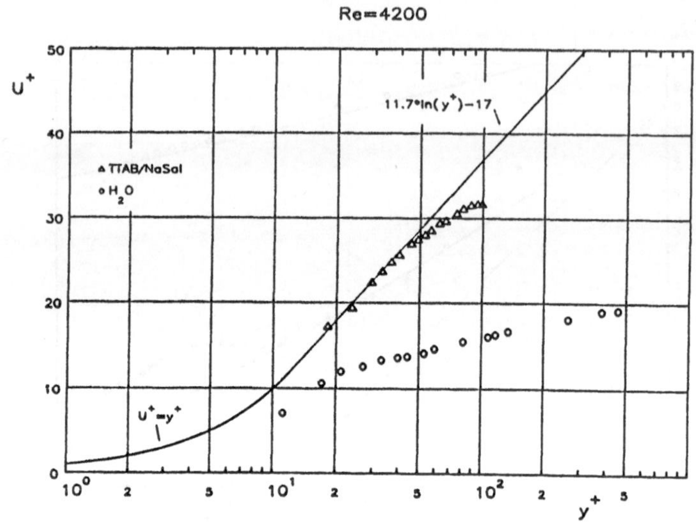

Figure 6: Dimensionless velocity profile of a C_{14}TASal-solution at Re = 4200, c = 905 ppm and T = 21°C obtained in a 15 mm pipe

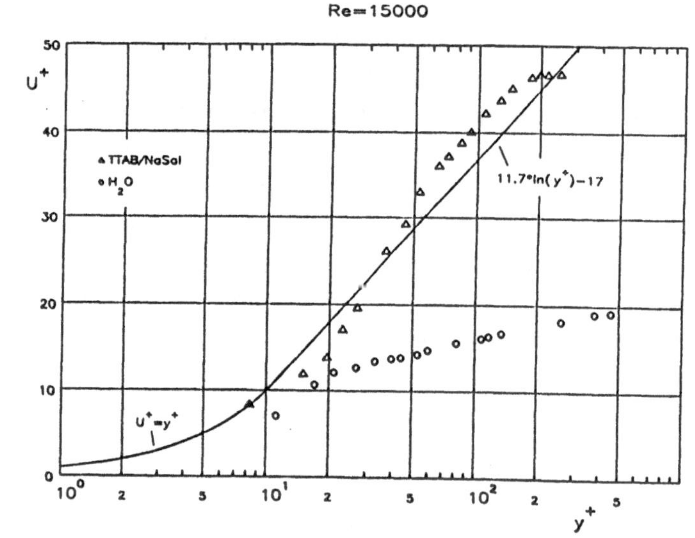

Figure 7: Dimensionless velocity profile of a C_{14}TASal-solution at Re = 15000, c = 905 ppm and T = 21°C obtained in a 15 mm pipe

(Fig.5) only occurs when the critical shear gradient for the SIS is not exceeded in the laminar flow regime.

When the friction factor becomes close to Virk's maximum drag reduction asymptote deviations from Virk's ultimate profile, eq.(7), occur. The velocity profiles in the buffer zone (Fig.7) then become S-shaped, i.e. meaning they deviate from Virk's ultimate profile to lower dimensionless velocity values at smaller wall distances, to higher ones in the medium range, and again towards lower ones at larger wall distances [10, 24]. Therefore the shape of the dimensionless velocity profiles at maximum drag reduction conditions is more similar to a laminar than to a logarithmic turbulent profile.

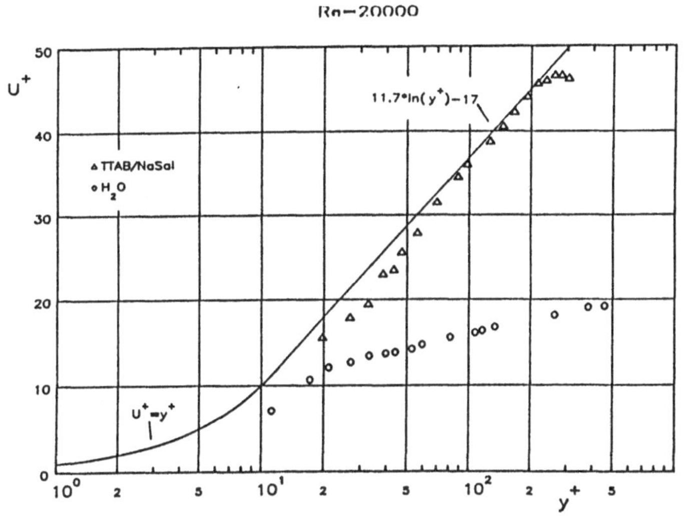

Figure 8: Dimensionless velocity profile of a C_{14}TASal-solution at Re = 20000, c = 905 ppm and T = 21°C obtained in a 15 mm pipe

In the flow regime where the friction factor is constant (see Fig.5) a core region starts forming (Fig.8). The velocity data in the buffer zone at low wall distances are still lower than those of eq.(7), as it was found at maximum drag reduction conditions (Fig.7). However, in the region $y^+ = 100$-220 they follow Virk's ultimate profile.

In the flow regime at high Reynolds numbers of Fig.5, in which the friction factor again decreases with the Reynolds number, the core region of the velocity profile (Fig.9) expands and exhibits an increased slope in comparison with the solvent. In the buffer zone Virk's ultimate profile is a good approximation of the velocity data.

The S-shaped velocity data at maximum drag reduction conditions were also

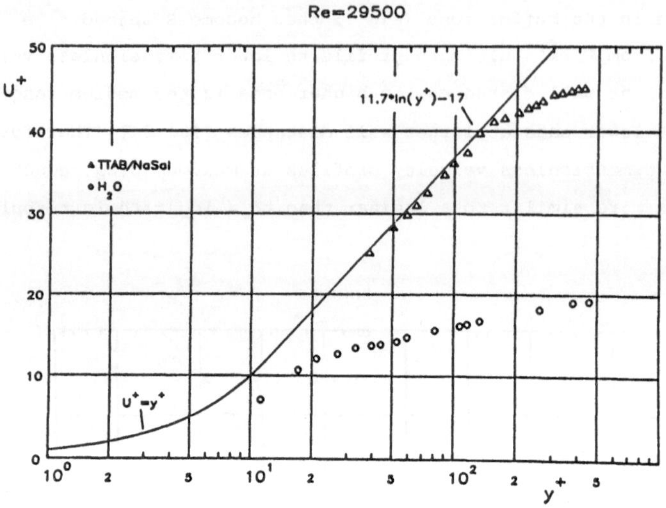

Figure 9: Dimensionless velocity profile of a C_{14}TASal-solution at Re = 29500, c = 905 ppm and T = 21°C obtained in a 15 mm pipe

observed in [10] and [24-26] where a solution of the surfactant "Metaupon" was studied in a channel flow. Depending on the pressure drop and surfactant concentration velocity profiles were found which look quite similar to those of the present work.

Apart from the velocity profile measurements, additional information on the turbulence characteristics is needed in order to understand the mechanism of interaction of these additives with turbulence. Turbulence intensity profiles in drag reducing surfactant solutions are reported in [10,23-25]. The maximum of the axial turbulence intensity which is found in Newtonian fluids at y^+ = 12-13 is shifted with increasing drag reduction to higher y^+-values in surfactant solutions. Depending on the flow regime both decreases and increases of the axial turbulence intensity normalized by the friction velocity u_τ occurred [24]. This behaviour of the axial turbulence intensity is similar to that found in drag reducing dilute polymer solutions [29]. The transverse turbulence intensity, even normalized by u_τ, is damped drastically in surfactant solutions in comparison with a Newtonian fluid. It is reduced over the whole cross-section of the channel or pipe.

An important property for characterizing the turbulent fluctuating motion is the ratio of both turbulence intensities. As shown in flows of polymer

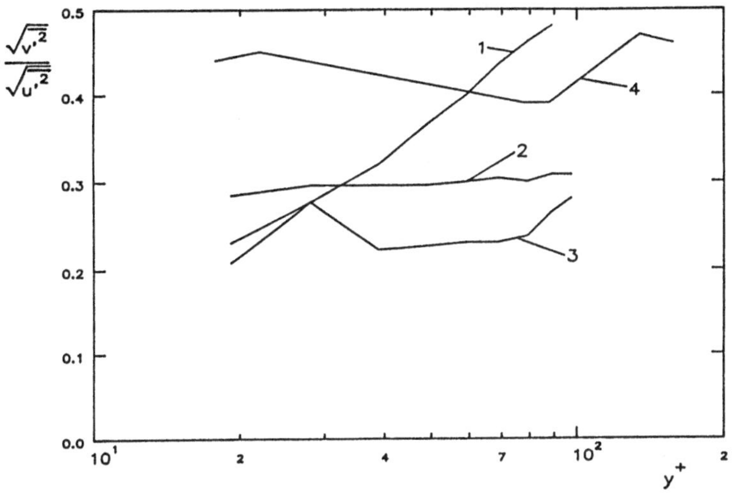

Figure 10: Anisotropy of velocity fluctuations in drag reducing surfactant solutions. Curve 1-3 are channel flow data from [23], 1 - water, 2 - C_{16}TAchloride-α-naphtol-solution (DR = 66%), 3 - Potassium oleate-solution (DR = 76.2%), 4 - pipe flow of a C_{14}TASal-solution, Re = 35000 (DR = 74,8%)

solutions the anisotropy of the fluctuating motion also becomes more pronounced in flows of surfactant solutions (s. Fig.10). The strong anisotropies which decrease in Newtonian turbulence with the wall distance seem to be nearly independent of the wall distance for the surfactants. Measurements of the higher moments of turbulence, the skewness and the flatness, which were done in [23] in a channel flow of a surfactant solution, demonstrated that the zero crossing of the skewness and the minimum of the flatness were shifted to higher dimensionless wall distances in comparison to the Newtonian solvent. Therefore the zone of maximum turbulent energy generation also occurs at larger wall distances. Furthermore, it was found in [23] that in the zone of maximum energy generation a large percentage of the fluctuations are weak and the number of the strong fluctuations is smaller than in Newtonian turbulence. Integral scales and the microscales of turbulence were calculated in pipe flows of drag reducing surfactant solutions in [10] from the autocorrelation function of the axial velocity fluctuations. The microscale of the surfactant solution was found to be increased over the whole cross-section of the pipe in comparison with the Newtonian solvent. The largest increase of the microscale occurred in the near-wall region. These results can be interpreted as an increase of the size of the eddies in the dissipation range which could be caused by an increased local effective viscosity. The integral scale was also increased in the surfactant solution over the whole cross-section of the pipe. This means that the large-scale structure is affected as well.

The turbulent momentum transfer is characterized by the Reynolds shear stresses. The distribution of the Reynolds shear stresses $-\rho\,\overline{u'v'}$ can be calculated in pipe flows of Newtonian fluids by

$$-\frac{\overline{u'v'}}{u_\tau^2} = 1 - \frac{y}{R} - \frac{dU^+}{dy^+} \tag{8}$$

where R is the radius of the pipe. Measurements of the Reynolds shear stresses were done in pipe [9] and channel flows [23]. The results of these measurements demonstrate that when the solvent viscosity is used for calculating the viscous shear stresses a Reynolds stress deficit G occurs

$$-\frac{\overline{u'v'}}{u_\tau^2} = 1 - \frac{y}{R} - \frac{dU^+}{dy^+} - G\ . \tag{9}$$

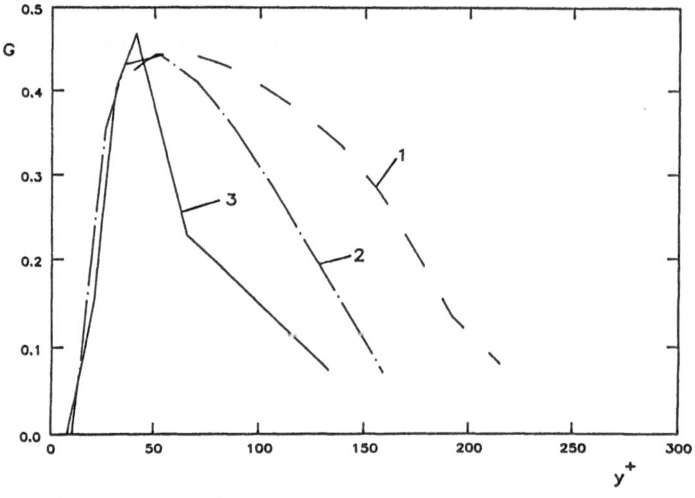

Figure 11: Literature data of the elastic stresses G as a function of the dimensionless wall distance. 1 - pipe flow of a poly (acryl amide) solution at maximum drag reduction conditions from [30], 2 - channel flow of a surfactant solution (DR = 66%) from [23], 3 - pipe flow of a C_{14}TASal-solution (DR = 48.5%) from [9]

G stands for the Reynolds stress deficit or for the "elastic stresses" as it was called by the first authors [30] recognizing this deficit in pipe flows of drag reducing polymer solutions.

When the elastic stresses are related to an increase of the local viscosity as suggested in [31] eq.(9) becomes

$$-\frac{\overline{u'v'}}{u_\tau^2} = 1 - \frac{y}{R} - \frac{\nu_{eff}}{\nu_o}\frac{dU^+}{dy^+} . \qquad (10)$$

In Figure 12 the relative effective viscosity

$$\eta_{eff} = \frac{\nu_{eff}}{\nu_o} \qquad (11)$$

is plotted versus the dimensionless wall distance. The relative viscosity increase exhibits maximum values between 3.8 and 6.3. In all regions of the velocity profile except for the viscous sublayer the effective viscosity is increased. The occurence of the increased effective viscosity in surfactant solutions could be due to the increase of the shear viscosity in the SIS.

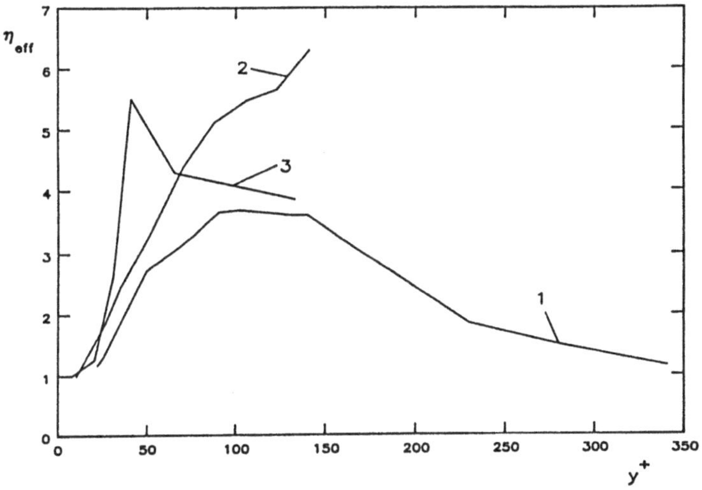

Figure 12: Literature data of the effective viscosity η_{eff} as a function of the dimensionless wall distance calculated from the data in Fig.11 using eq.(10)

Taking into account the results of the SANS-experiments [16] that the micelles are aligned in flow direction in turbulent drag reduced flows as well as in the SIS, it appears that large shear-induced micelle superstructures exhibiting an increased shear viscosity cause the Reynolds stress deficit.

The reason for the occurrence of the Reynolds stress deficit in flows of drag reducing polymer solutions was discussed in [32], where a model for calculating the maximum effective viscosity increase was presented. This model for viscoelastic fluids interpretes the velocity profiles at maximum drag reduction conditions as "pseudo-laminar", and predicts an effective viscosity increasing with the wall distance. The model provides values of the effective viscosity which are similar to those obtained from the Reynolds stress measurements.

By assuming a pseudo-laminar profile as in [32] the effective viscosity was then calculated for the velocity profiles presented in Figures 6-9 (and others at different Reynolds numbers of the same series of experiments), using the equation

$$y^+ - \frac{1}{\sqrt{2\,Re_o}} \int_0^{y^+} \sqrt{\eta_{eff}}\; y^+ \, dy^+ = A \ln(\eta_{eff}\, y^+) - B \qquad (12)$$

where the constants A and B were estimated for each velocity profile from the experimental data. Figure 13 shows the results of the calculations. The

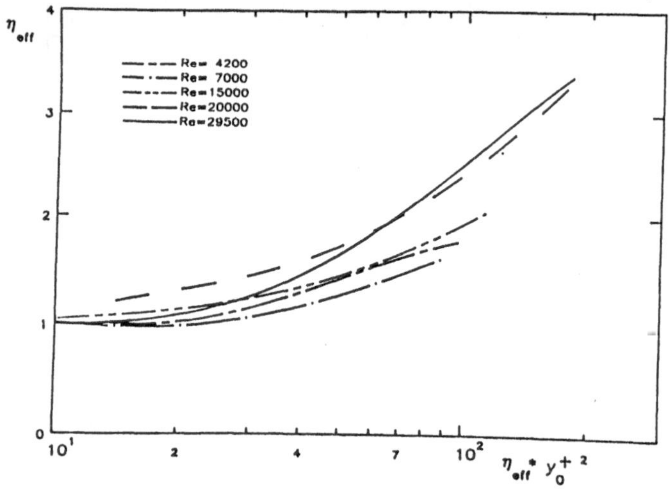

Figure 13: Relative viscosity increase

maximum relative effective viscosity increase is about 3 near the centre of the pipe at the higher Reynolds numbers, whereas the viscometric measurements (s. Fig.4) gave an increase of the shear viscosity in the SIS by a factor of 2. Taking into account that the shear viscosity in the SIS increases with the gap width it seems quite reasonable that the measured velocity profiles are pseudo-laminar profiles of a viscoelastic fluid.

Conclusions

Drag reduction in surfactant solutions requires the presence of rod-like micelles [3,10,23]. Rod-like micelles are aggregates of surfactant molecules which are formed in some surfactant systems when a critical surfactant concentration is exceeded. This critical micelle concentration of a surfactant depends on the temperature and on the concentration of additional electrolytes. In the concentration range used for drag reduction the micelles can rotate freely without hindering each other in shear flows, because the rotational volumes do not overlap. However, in simple shear flows surfactant solutions do exhibit a shear-induced state when a critical shear gradient is exceeded and become viscoelastic. In the SIS the micelles are completely aligned in flow direction [14-16] and the shear viscosity is increased. Since the shear viscosity should decrease and not increase as observed when the rods are aligned in flow direction larger superordered structures must therefore be present in the SIS. In the SIS surfactant solutions also exhibit an increased elongational viscosity [17,18].
The turbulent friction behaviour of drag reducing surfactant solutions depends on the size, number, and surface charge of the micelles [3] which can be influenced by changing the temperature, and the concentrations of the surfactant and the electrolytes (see Figs.2 and 3). In each surfactant system exhibiting drag reduction there exists a reversible loss of drag reduction when a critical wall shear stress is exceeded. This wall shear stress is independent of the pipe diameter [3,6,7,9,10]. Drag reduction occurs when the shear gradients in the buffer zone exceed the critical value for the SIS. This can be concluded from the rheological and friction behaviour of surfactant solutions exhibiting an onset behaviour. In pipes of small diameters this critical shear gradient is normally exceeded in the laminar flow regime. Therefore in theses pipes no onset-behaviour can be detected and drag reduction occurs as soon as the flow becomes turbulent. This behaviour in small pipes is also found for high molecular weight polymer solutions. Sometimes the transition laminar-turbulent in surfactant solutions is shifted to higher Reynolds numbers which can be explained by

the presence of stiff rods [21]. The micelles become more and more aligned in flow direction with increasing drag reduction, until they become completely aligned in flow direction when Virk's maximum drag reduction asymptote is approached [16]. When the critical wall shear stress is exceeded, the micelle superstructure is destroyed and the micelles loose their orientation in the flow but remain intact [19]. This is in agreement with the rheometric measurements which show that the SIS vanishes at high shear gradients. In addition, the elongational viscosity, which decreases with the elongational rate, also approaches a Newtonian behaviour at high shear rates [18].

Dimensionless turbulent velocity profiles at low drag reduction look quite similar to those found for drag reducing dilute polymer solutions [10,24]. Virk's three layer model [20] seems to be a good description for the velocity profiles in the viscous sublayer and in the buffer zone. Deviations occur only in the core region. Surfactant solutions exhibit an increased slope in the core region in comparison to the Newtonian solvent. This phenomenon which is also found in heterogeneous drag reduction [27,28] indicates a changed turbulence structure in this region.

However, large deviations from Virk's ultimate profile, eq.(7), occur when the friction factor approaches maximum drag reduction conditions. The dimensionless velocity profiles in the buffer zone become S-shaped, meaning that they deviate from eq.(7) to lower dimensionless values at smaller wall distances, to higher ones in the medium range, and again to lower ones near the centre of the pipe or channel. Therefore their shape is more similar to a laminar than to a turbulent logarithmic velocity profile. The maximum of the axial turbulence intensity is shifted towards the centre of the pipe or channel whereas the transverse turbulence intensity is strongly damped. Therefore the anisotropy of the fluctuating motion becomes more pronounced than in Newtonian turbulence and nearly independent of the wall distance. The strong damping of the transverse component becomes understandable because the micelles respectively their superordered structures are completely aligned in flow direction as it was found by SANS [16].

A Reynolds stress deficit has been found in flows of drag reducing polymer solutions [30,32]. In drag reducing surfactant solutions it may be caused by the increased shear viscosity in the SIS. An increased shear viscosity due to the SIS in the buffer zone would cause a strong damping of the Reynolds shear stresses and an increase of the microscale of turbulence. It would also imply an enlargement of the size of the longitudinal counterro-

tating vortices in the near wall region supposed the size of these vortices remains constant in y^+-units. The integral scale of turbulence in the buffer zone should therefore increase as well.
In conclusion, all the available measurements of the turbulence structure in connection with SANS-measurements point in the same direction: Drag reducing surfactant solutions appear to exhibit an increased shear viscosity in the buffer zone of the velocity profile. By considering the measured velocity profiles as pseudo-laminar, relative viscosity increases are obtained, which are compatible with the results of the analysis of the measured Reynolds shear stresses and the shear viscosity measurements in the SIS.

References

[1] Mysels KJ (1949) US Patent 2 492 173

[2] Zakin JL, Lui HL (1983) Chem Eng Commun 23:77

[3] Ohlendorf D, Interthal W, Hoffmann H (1986) Rheol Acta 26:468

[4] Myska J, Slanec K (1980) Acta Polytechnica 11:57

[5] Ohlendorf D, Interthal W, Hoffmann H (1984) IXth Intern Congr Rheology, Mexico, 2:41, Elsevier Sci Publ, Amsterdam

[6] Savins JG (1967) Rheol Acta 6:323

[7] Elson TP, Garside J (1983) J Non-Newtonian Fluid Mech 12:121

[8] Wells CS (1969) in: White A (ed) Viscous Drag Reduction, 297, Plenum Press, New York

[9] Bewersdorff HW, Ohlendorf D (1985) Proc 5th Symp on Turbulent Shear Flows, Ithaca, 9:41

[10] Bewersdorff HW, Ohlendorf D (1988) Colloid & Polymer Sci 266:941

[11] Mukerjee P, Mysels KJ (1971) Critical micelle concentration of aqueous solutions, NRRDS-BBS, No. 36, USA

[12] Gravsholt S (1976) J Colloid Interface Sci 57:575

[13] Hoffmann H, Platz G, Rehage H, Schorr W, Ulbricht W (1980) Ber Bunsenges Phys Chem 85:2255

[14] Rehage H, Wunderlich I, Hoffmann H (1986) Progr Colloid Polym Sci 72:51

[15] Löbl M, Thurn H, Hoffmann H (1984) Ber Bunsenges Phys Chem 88:1102

[16] Bewersdorff HW, Dohmann J, Langowski J, Lindner P, Maack A, Oberthür R, Thiel H (1989) Physica B 156&157:508

[17] Wunderlich AM, James DF (1987) Rheol Acta 26:522

[18] Vissmann K, Bewersdorff HW (1989) 4th Intern Conf Drag Reduction, Davos, Switzerland

[19] Bewersdorff HW, Frings B, Lindner P, Oberthür RC (1986) Rheol Acta 25:642

[20] Virk PS (1975) AIChE J 21:625

[21] Bark FH, Tinoco H (1978) J Fluid Mech 87:321

[22] Schlichting K (1960) Boundary Layer Theory, McGraw-Hill, New York

[23] Povkh IL, Stupin AV, Aslanov PV (1988) Fluid Mech-Sov Res 17:65

[24] Povkh IL, Stupin AB, Maksjutenko SN, Aslanov PV, Simonenko AP (1980) in: Mekhanika turbulentnykh potokov, Moscow, pp. 44-69

[25] Aslanov PV, Maksjutenko SN, Povkh IL, Simonenko AP, Stupin AB (1980) Izv Akad Nauk SSR, Mekh Zhidk Gaza, pp. 36-43

[26] Povkh IL, Stupin AB, Maksjutenko SN, Aslanov PV, Roshchin EA, Tur AN, (1975) Inzh Fiz ZH 29:522

[27] Bewersdorff HW (1984) Rheol Acta 23:522

[28] Berman NS (1986) Chem Eng Commun 42:37

[29] Berman NS (1978) Ann Rev Fluid Mech 10:47

[30] Schümmer P, Thielen W (1981) Chem Eng Commun 4:593

[31] Giesekus H (1981) Structure of turbulence in drag reducing fluids, Lecture Series 1981-86, Von Kármán-Institute for Fluid Dynamics, Rhode-Saint-Genèse, Belgium

[32] Bewersdorff HW, Berman NS (1988) Rheol Acta 27:130

Turbulent Drag Reduction Versus Structure of Turbulence

Arkady TSINOBER
Department of Fluid Mechanics and Heat Transfer
Faculty of Engineering, Tel Aviv University
Tel Aviv 69978, Israel

"...most flows of practical interest involve turbulent boundary layers and an alternate approach is to permit the flow to remain turbulent but somehow to reduce the turbulent shear forces."

Gary R Hough 1972, Viscous drag reduction,
Progress in Astronautics and Aeronautics, 72, p. XV.

1. Introduction

Turbulent drag reduction by whatever methods is generally not associated with reduction of turbulence intensity. It may not be necessary to suppress the turbulence or even its production but to let the flow remain turbulent *somehow* altering its structure in such a way as to reduce turbulent dissipation of energy and thereby the turbulent drag. In other words the question is not whether we stop or not the turbulence production but, rather whether we can influence this process in such a way that the resulting turbulence will be low dissipative or "disabled" (Narasimha and Sreenivasan[1]). Is it possible and if so *how* it can be done? Before answering this very difficult question, it is appropriate to ask another somewhat *simpler question whether low dissipative turbulent structures do exist and what they are.* An attempt is made in this paper to shed some light on this question. First a review of the few known examples of low dissipative turbulent flows is given. These include an important case of turbulent MHD-flow in the presence of an azimuthal magnetic field, in which the body-force is applied directly to the turbulence only and does not interact with the mean flow. This case is of special interest since it exhibits essentially laminar drag (i.e. approximately vanishing Reynolds stresses) but possesses high level of turbulence. Similar behaviour has been observed in other drag reducing situations, like dilute and heterogeneous polymer drag reduction and some

A. Gyr (Editor)
Structure of Turbulence and Drag Reduction
IUTAM Symposium Zurich/Switzerland 1989

others. Of particular importance are also situations with so called "negative (eddy) viscosity" in which Reynolds stresses are of opposite sign to the "normal" one. These kind of flows are known in astro- and geophysical contexts and are quite a rarity in laboratory observations.

The simplest idealized (but in some sense turbulent) flows where no turbulent dissipation is expected are the two-dimensional flows and flows with maximum helicity (Beltrami flows). In both the enstrophy generation process (vortex stretching) does not exist and therefore the turbulent dissipation is essentially zero. The important common feature of these flows is their anisotropy. The existing experimental evidence and simple theoretical examples (both rather scarce) indicate quite clearly that the flow anisotropy is the most important property necessary (but not sufficient) to maintain small turbulent dissipation, its production and/or even a negative eddy viscosity with concomitant reverse energy cascade leading in some cases to energy transfer from turbulence to the mean flow.

2. Magnetohydrodynamic drag reduction

Mean flow characteristics: The discovery of drag reducing effects of an external magnetic field on turbulent liquid metal flow is due Hartmann and Lazarus[2].

In their classical work, all the main features of the influence of a transverse magnetic field on the drag for flow in pipes of rectangular cross section with nonconducting walls were established. They showed, as did many subsequent investigators (for further references see Tsinober[3]), that as the magnetic field is increased, the drag decreases, reaches a minimum, and then increases and becomes very close to the value corresponding to laminar flow. This sort of behaviour is typical for a wide range of Reynolds numbers and for all investigated shapes of channel cross section. An example of results similar to that of Harmann and Lazarus results in shown in Fig. 1. It is seen that the drag reduction effect in this specific configuration can be very large (5 times). Corresponding changes happen in the mean velocity structure. As can be seen from Fig. 1 simultaneously with the drag reduction the velocity profile becomes very close to laminar. The Harmann number is defined as $Ha = Bd\{\sigma/(\rho\nu)\}^{1/2}$, where B - magnetic field induction, d - lenth scale. σ - electrical conductivity, ρ - density, ν - kinematic viscosity.

The change in mean flow properties (drag, etc.) is usually explained in terms of suppression of turbulence and the so-called Harmann effect. The first one is associated with the "direct" action of the magnetic field on turbulence, e.g. turbulent shear, while the second one is associated with the changes of the mean velocity profile via formation of thin layers ($\sim Ha^{-1}$) at insulating boundaries normal to the magnetic field and other similar phenomena. Among the situations where the "indirect" action of the magnetic field on the flow turbulence is strongly reduced to a minimum is the configuration with the so-called azimuthal magnetic field, i.e. where the magnetic field is perpendicular to the short side of the cross section. In this case, one expects much stronger turbulent drag reduction effects as was confirmed experimentally. The example shown in Fig. 1 belongs to this kind of configuration. Ideally, one would like to get rid of all the walls normal to the magnetic field, since, in this case, the mean flow does not interact with the magnetic field entirely and, therefore, the drag after reaching its laminar value (at $Ha/Re \sim 10^{-2}$) will not change anymore.

Turbulence structure: From the general point of view, it is beyond any doubt that any appreciable drag reduction effect should be a consequence of an essential modification of the structure of turbulent flow. An extreme example is total suppression of turbulence but, as we shall see in the following, it is not the only case when an essential drag reduction effect is present.

As mentioned above, studies on the integral characteristics of turbulent MHD flows show that turbulent transfer of momentum and energy is strongly inhibited by a magnetic field. It was expected on this basis that at supercritical values of Ha/Re (according to the drag) the intensity of the velocity fluctuations should be close to zero for coincidence of the drags of the laminar and laminarized flows. However, by the first measurements of the fluctuations for flows in insulated channels and pipes in a transverse magnetic field have shown that this expectation was wrong. In a number of papers (for further references see Tsinober[3]), it was found that a high level of turbulence (30–50% of the value for B=0) was maintained in the flow for supercritical values of Ha/Re over the entire length of the experimental section.

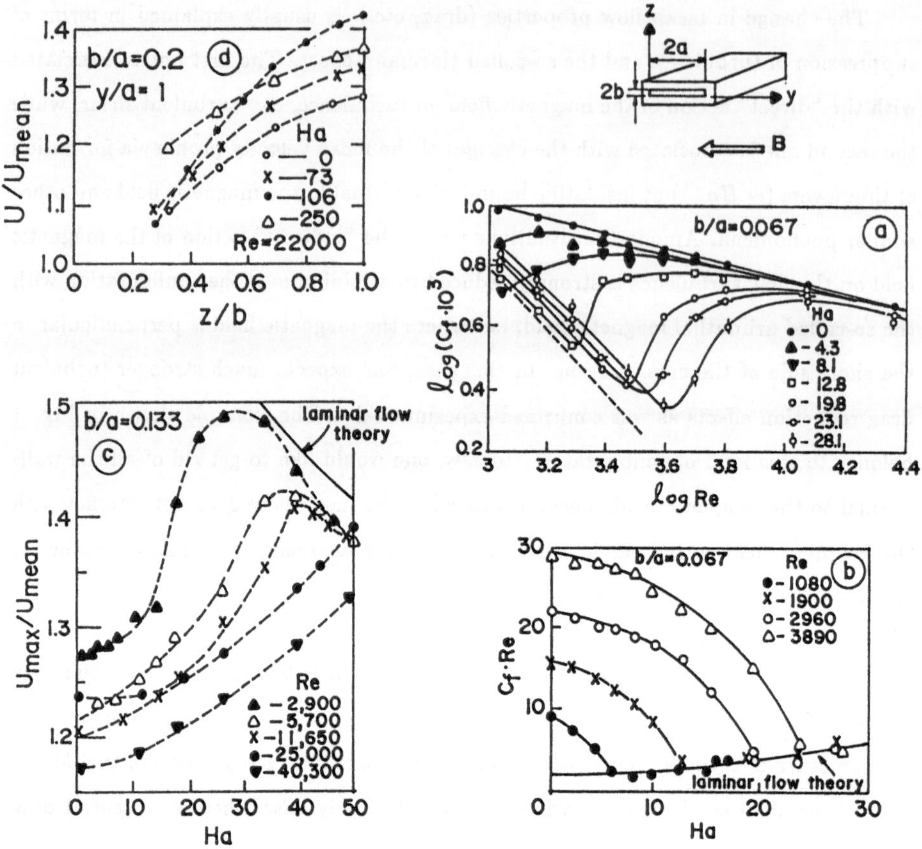

Fig. 1. Drag reduction effect of an azimuthal magnetic field on liquid metal flow[4]. a, b) results for b/a = 0.067 in different representations. The full line in b) corresponds to the exact laminar solutions[5].

Influence of magnetic filed on mean velocity.[6,7] c) experimental data for velocity at the center of the channel (the full line corresponds to the exact solution[5], d) mean velocity profiles in the center plane perpendicular to the magnetic field.

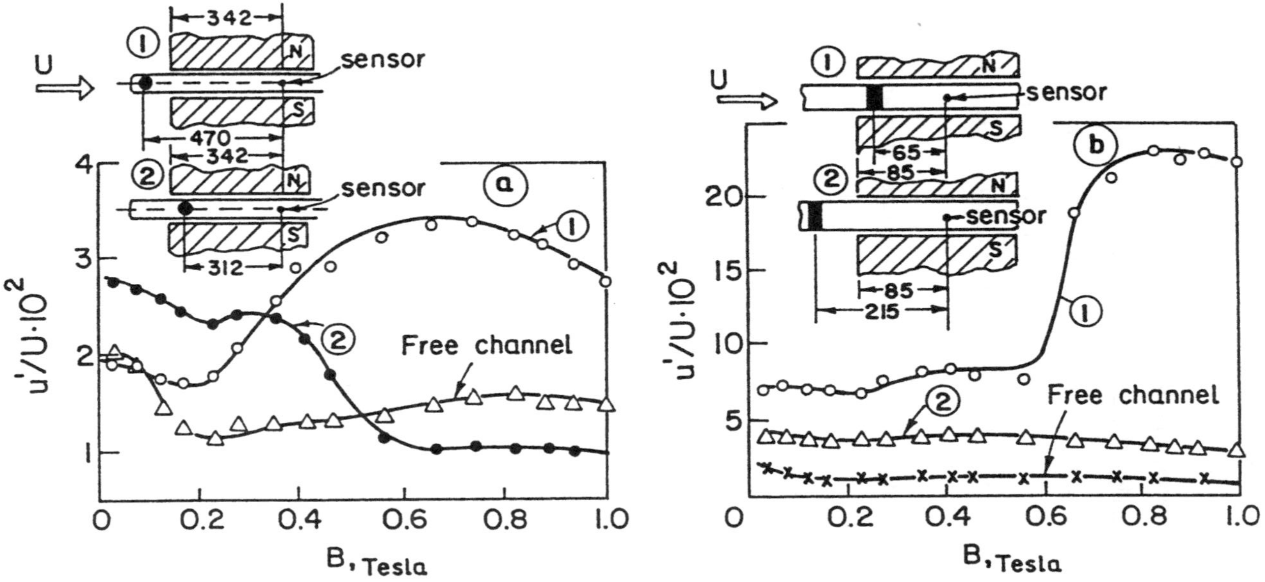

Fig. 2. The effect of magnetic field on the perturbations on the wake behind a cylinder[8].
a) cylinder axis perpendicular to the magnetic field, b) cylinder axis parallel to the magnetic field.

Further analysis has shown that the disturbances that are retained do not abstract energy from the mean flow (to within the 3-5% accuracy attained in the experiment). Furthermore, in certain special cases, especially when the boundary geometry permits flows that are close to two-dimensional in a plane perpendicular to the magnetic field, the turbulence intensity is 2–3 times greater than in a zero magnetic field as can be seen from an example shown in Fig. 2. Similar results were obtained for other configurations like gratings with their cylinders parallel to the magnetic field etc. (see Tsinober[3] and Fig. 4).

On the basis of the facts mentioned above, it was suggested (Kit and Tsinober[9]) that the structure of the turbulence becomes strongly anisotropic with increasing magnetic field and approximates to two-dimensional. Moreover, the absence of any effect of a magnetic field on a arbitrary two-dimensional flow in a plane perpendicular to the field (Gotoh[10]) and the stabilizing effect of magnetic field on the three-dimensional disturbances in this configuration (Wooler[11]) led to the suggestion that a flow of an electrically conducting fluid in sufficiently strong magnetic fields can be used to realize experimentally ordinary(nonmagnetohydrodynamic) two-dimensional nonlinear flows and, in particular, two-dimensional turbulence. Subsequent qualitatively different experiments confirmed unequivocally the above conjectures. Some examples are shown in Fig. 3.

In the first experiment (Fig. 3a), a thermoanemometer was used to measure the spatial correlation of longitudinal velocity fluctuations; the results showed that the correlation coefficient in the direction of the magnetic field approaches unity with increasing field, remaining much less that unity in other directions.

In the second experiment (Fig. 3b), a study was made of the mass transfer in a flow in which disturbances were artificially created by means of a grid made up out of cylinders with axes parallel to the magnetic field.. The results showed that in a strong field mass transfer occurs primarily in a plane perpendicular to the magnetic field, whereas in a zero field mass transfer is much the same in all directions. It is significant that the rate of (two-dimensional) mass transfer intensifies in a magnetic field.

An attempt was made to detect in the energy spectra of the fluctuations intervals characteristic for two-dimensional turbulence. The simplest flows were utilized for this purpose: flow past a grid made up out of cylinders with axis parallel to the field and flow in an insulated channel of rectangular cross section. It was found that the spectra indeed contained the expected intervals, being described by a k^{-3} law in the high-frequency part of the spectrum and by a $k^{-5/3}$law at lower frequencies. A comparison of the spectra at different distances from the cylinder showed that there was no energy transfer upwards through the spectrum (i.e., from large scales to small) and that at least qualitatively, a reverse energy cascade occurred (see Fig. 3c).

Reed[14] and Reed and Lykoudis[15] performed direct measurements of turbulent stress (along with RMS of longitudinal and transverse turbulent fluctuations) in a turbulent MHD - flow in a high aspect ratio channel (~ 6) with the long side of the cross section perpendicular to the magnetic field. One of the most essential results of this work is the existence of large regions in the flow where at high enough Ha/Re the turbulent Reynolds stress is essentially vanishing while the turbulent energy is still high enough and is ~ 40% of that without the magnetic field.

It should be emphasized that in these regions the turbulent intensity of velocity fluctuations parallel to the magnetic field is far from being zero though it is suppressed much stronger than the longitudinal component. *This seems to be the first direct experimental indication that highly anisotropic turbulence (but three-dimensional) may have strongly reduced turbulent drag and dissipation.*

It was suggested by this author that more reliable observations of the reverse energy cascade as well as the so-called negative eddy viscosity effect (i.e. momentum and energy transfer from the fluctuations to the mean flow) can be performed in laboratory experiments using the above techniques along with an external source of energy input into the disturbances directly, e.g., by electromagnetically driven vortices by a set of electrodes via interaction of electrical currents with magnetic field. The suggestion was used later by Sommeria et al[16] (see also references therein) to reproduce much more convincing evidence for the existence of a reverse energy cascade, e.g. a $k^{-5/3}$ spectrum in low wavenumber range. Other qualitative results related to the negative eddy viscosity effects are reviewed in Tsinober[3]. In particular it seems to be possible to realize genuine negative eddy viscosity effect by an appropriate application of the above method in a shear flow.

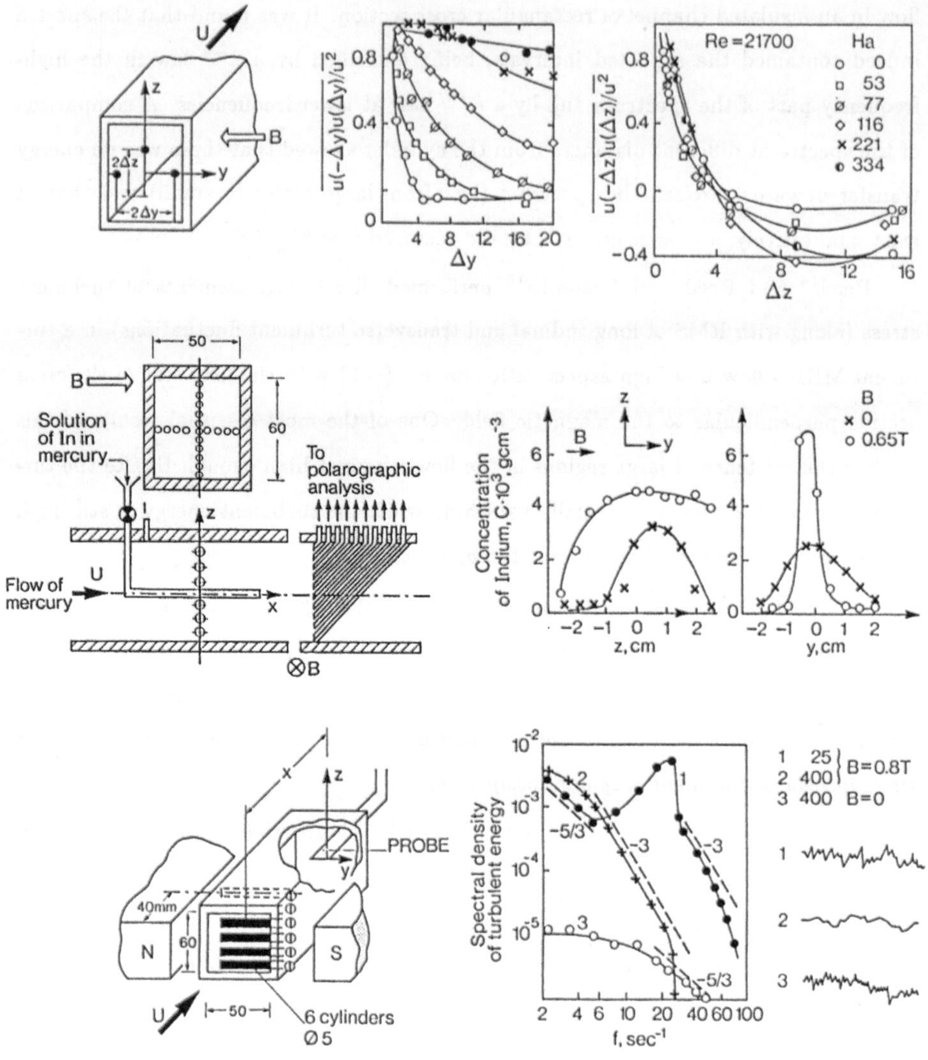

Fig. 3. Experimental results confirming the two-dimensional structure of turbulence in strong magnetic field. a) spacial correlations[12], b) mass transfer and c) fluctuations spectra in the flow past a two-dimensional grid.[13]

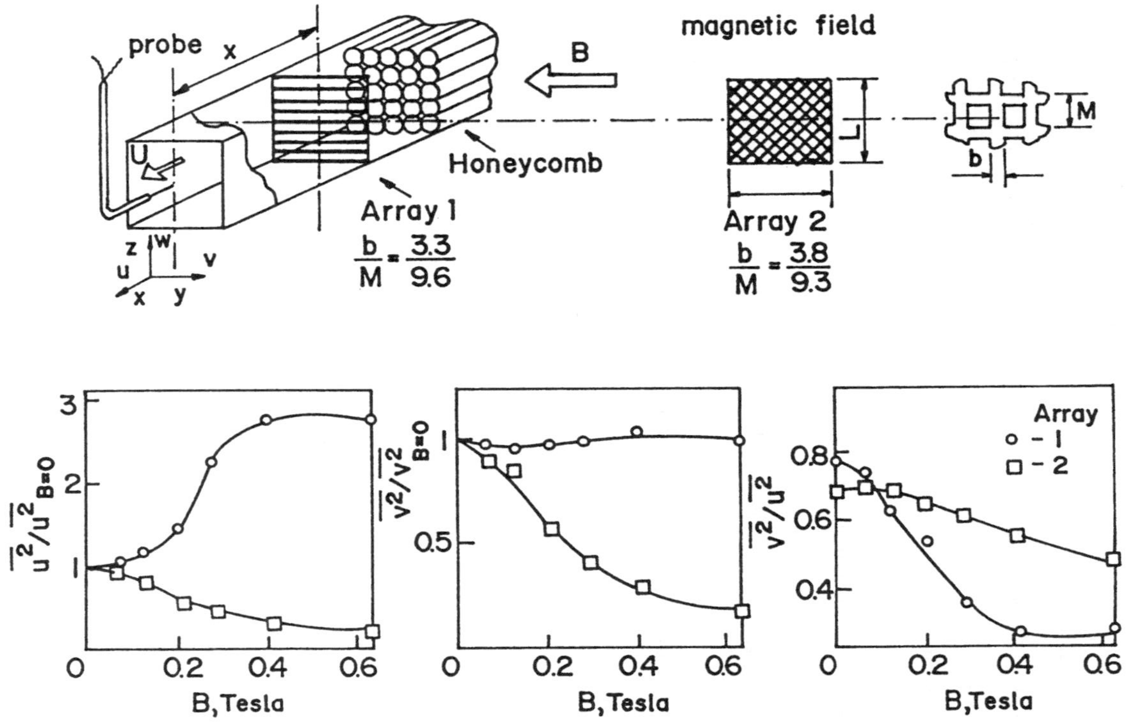

Fig. 4. Turbulence past grids of different geometry in a transverse magnetic field[18] It is seen that the grid in the form of a grating with cylinders parallel to the magnetic field (i.e. favouring 2-D disturbances with their axis parallel to magnetic field) produces turbulence of much higher intensity than without magnetic field.

All that has been said above relates to the case when the boundaries of the flow perpendicular to the magnetic field are electrically nonconducting and the electric field induced in the flow gives rise to currents that flow only over shunting Hartmann layers of relatively high electrical resistance. If, however, the walls of the channel perpendicular to the field have a good electrical conductivity, then all disturbances (including two-dimensional) are suppressed in a sufficiently strong field (Platnieks and Freibergs[17]). In other words, the flow confinement by insulating walls normal to the magnetic field is essential to suppress most of three-dimensional disturbances and to retain the two-dimnensional ones, i.e., it is necessary that the boundary conditions should favour the existence of two-dimensional disturbances. The same is true also in respect with the flow geometry and, in particular, *geometry influencing the initial nature of the disturbances* (see Fig. 4).

We note that the glows described above, which approximate to two-dimensional in a sufficiently strong field, cannot be regarded as being "purely" two-dimensional due to the presence of a wall perpendicular to the field, even in the case when the dimension of this wall is small. Purely two-dimensional flows can be realized in an annular tube in an azimuthal magnetic field of the form $B_\theta \sim r$.

3. Polymer drag reduction

The very early results have shown that generally the polymer drag reduction is not associated with reduction in turbulent intensity again indicating that change in turbulence structure and not its suppression is responsible for drag reduction. In fact, the turbulence intensity may be even increased (see the review by White and Hemmings[19]). Later experiments confirmed this observation (e.g. Bartels et al[20], Berman[21], Berman and Tan[22], Maksimovich[23] and some other papers in Gampert[24], Bewersdorff and Olendorf[25], Luchick and Tiederman[26] and references therein).

There is some evidence that in low-concentration drag reducing flows and heterogeneous drag reduction, the turbulence structure becomes more anisotropic (Bartels et al[20], Berman and Tan[22], Maksimovich[23], Bewersdorf[27,28], Usui et al[29]), e.g. the "decrease in the wall shear stress in low concentration polymer flows is associated with damping of the velocity fluctuations normal to the wall"[26].

The mechanisms of drag reduction are usually associated with various mechanisms of rheological nature (e.g. Berman[21], Lumley and Kubo[30]), which operate at small scales. In this connection, it is noteworthy that in dilute polymer solutions the *large scale structures* can be influenced significantly. An example of such observation was given by Riediger[31] (see also references therein) in a mixing layer in which "at 320mm downstream of the splitter plate, the structures in water were nearly destroyed whereas in the polymer solution, the flow field shows the characteristic shape for large vortices of the plane mixing layer. This is caused by a stronger dissipation of energy in the water", i.e., with polymer additives the *turbulent* dissipation is reduced and most probably the flow structure is more anisotropic (cf. Berman and Tan[22]). This is quite analogous to what happens in the wake of a cylinder in a MHD flow with magnetic field parallel to its axis (see Fig. 2) and flow past gratings in similar geometry (Fig. 4).

Perhaps the most extraordinary phenomenon of drag reduction ever observed is the so-called heterogeneous drag reduction which occurs for example in centrally injected concentrated polymer threads (Bewersdorff[27,28], Berman[21], Hoyt and Sellin[32], Usui et al[29] and references therein). In these flows "the thread" seems to persist through the length of the pipe and little, if any, diffusion of polymer ... is apparent"[32] and significant drag reduction effect occurs only when the "thread" is "wandering" throughout the pipe cross-section (Berman[21], Bewersdorff (1988, private communication), Hoyt and Sellin[32]). Since *the thread* occupies a *very small fraction* of the fluid volume and all the *rest of it is essentially a Newtonian fluid*, it seems that the drag reduction *is not due to rheological effects.* A possible explanation is that the "thread wandering" produces a highly helical low dissipative anisotropic flow structure (Bewersdorff and Gyr (1988), private communication). Hence, the importance of this phenomenon from the general basic point of view.

A similar change of flow structure was proposed in dilute polymer drag reducing (Gyr and Bewersdorff (1989), this issue). It should be emphasized that there seems to be a *qualitative* difference between the dilute polymer drag reduction and the heterageneous one in that the former primarily acts and modifies directly the small scale structure of the flow, while the latter - mostly the large scale one[21].

4. Negative Eddy Viscosity Phenomena and some related processes.

This term is usually used to denote flow situations in which the turbulent transport of momentum occurs against the mean velocity gradient, i.e. the Reynolds stress and the mean velocity gradient are of *opposite sign.* These kinds of processes are known to take place in large scale flows in geo- and astrophysics (for review and references see McWilliams[33], Monin and Ozmidov[34], Rhines[35], Starr[36]), the most prominent of which is believed to be generation of mean flows by large scale fluctuations, e.g. jet flows in the atmosphere and ocean, zonal circulation in the outer layers at the sun, etc. (see the above references and also Busse[37], Monin and Seidov[38], Monin[39]).

An explanation (at least qualitative) of this "anomalous" phenomenon is usually given via properties and by analogy with two-dimensional turbulence (for description of these see Lesieur[40], Kraichnan and Montgomery[41] and Numero special of the Journal de Mecanique Theorique et Applique[42]) which, under certain conditions, exhibit negative eddy viscosity and other "anomalous" properties. The first attempt of this kind was made by Lorenz[43]. The analogy, however, is qualitative only and geophysical and other flows with negative eddy viscosity (and energy production reversal) at best can be considered as coexistence of quasi-two-dimensional structures (mostly in large scales) with more three-dimensional smaller scales. What seems to be sure is that all negative eddy viscosity flow configurations are considerably anisotropic (more than "normal" flows) due to some external influences (geometrical constraints, rotation, density stratification, magnetic field, etc.).

Laboratory experiments: The popular example is the jet in which $-\langle u_1 u_2\rangle dU/dx_2 < 0$ in the region between the points where $dU/dX_2 = 0$ and $-< u_1 u_2 >= 0$ (e.g. Schwartz and Cosart[44], Eskinazi and Erian[45], for other references see Sreenivasan et al[46]). This region is rather narrow and the mean velocity gradient there is zero or very small. It was therefore argued (among other arguments of limited importance of this phenomenon) that though "locally the mean velocity may gain energy, however, when integrated over the entire cross section, the effect will be a decrease of the total kinetic energy of the mean motion" (Hinze[47], see also Wilson[48]). Another example is the boundary layer at a convex wall. The effect of curvature in the outer part of it, in some cases, leads to a reversal of the turbulent shear (So and Mellor[49,50], for other references see Gibson[51].

In this connection of special importance are experiments in highly *coherently* forced mixing layer performed by Oster[52], Oster and Wygnansky[53], Weibrot[54] and Weisbort and Wygnansky[55]. They found a *whole region* in the flow field with turbulent shear opposing the mean velocity gradient over the *entire flow cross section* ($0.55 < x < 0.84m$, Fig. 5). It is noteworthy that, unlike other flows, this phenomenon is especially strong close to the middle of the mixing layer cross section where the *mean velocity gradient is maximal.* Browand and Ho[56] have given a simple *kinematic* explanation of the process in terms of orientation changes (tilting) of elliptically shaped vortices similar to one proposed by Starr[36] in a geophysical context (see also Busse[37]). Though it is tempting to explain the above phenomenon in simple kinematic (two-dimensional) terms or more generally in terms of (dynamical) properties of *two-dimensional* turbulence, it should be emphasized that the above phenomenon seems to be essentially *three-dimensional.* A clear indication of this is found in measurements by Oster[52] and Oster and Wygnansky[53]. It follows from their results that in the zone of opposing turbulent shear ($x \approx 1.2 \div 1.4m$)$w'/u' \approx 0.4 \div 0.5$($v'/u' \approx 2$), while for a regular (nonforced) mixing layer $w'/u' \approx 0.9 \div 1.0$($v'/u' \sim 1$), i.e. the forced mixing layer becomes more anisotropic but still remains far from being two-dimensional.

Recent experiments by Hilberg and Fiedler[57] deserve special mentioning. They performed experiments on mixing layer with lateral extent an order of magnitude larger that its spanwise width and observed that the coherence and energy contents of large scale structures were highly above the values found in the normal shear layers flow. One can expect that due to much higher anisotropy a negative eddy viscosity phenomenon may exist in this kind of flow. Similar phenomena were observed in a shallow jet by Giger and Giger et al[58].

Theoretical examples. As mentioned above, the first attempt to treat theoretically the phenomenon of negative eddy viscosity was made by Lorenz[43]. He considered a two-dimensional (turbulent) flow consisting of mean and fluctuative components, *both unsteady,* and the interaction between the two. In case when the scales of the smaller motions of the mean flow and of the larger ones of the fluctuations overlap the energy of the fluctuations can be transferred to the mean flow, i.e. the eddy viscosity becomes

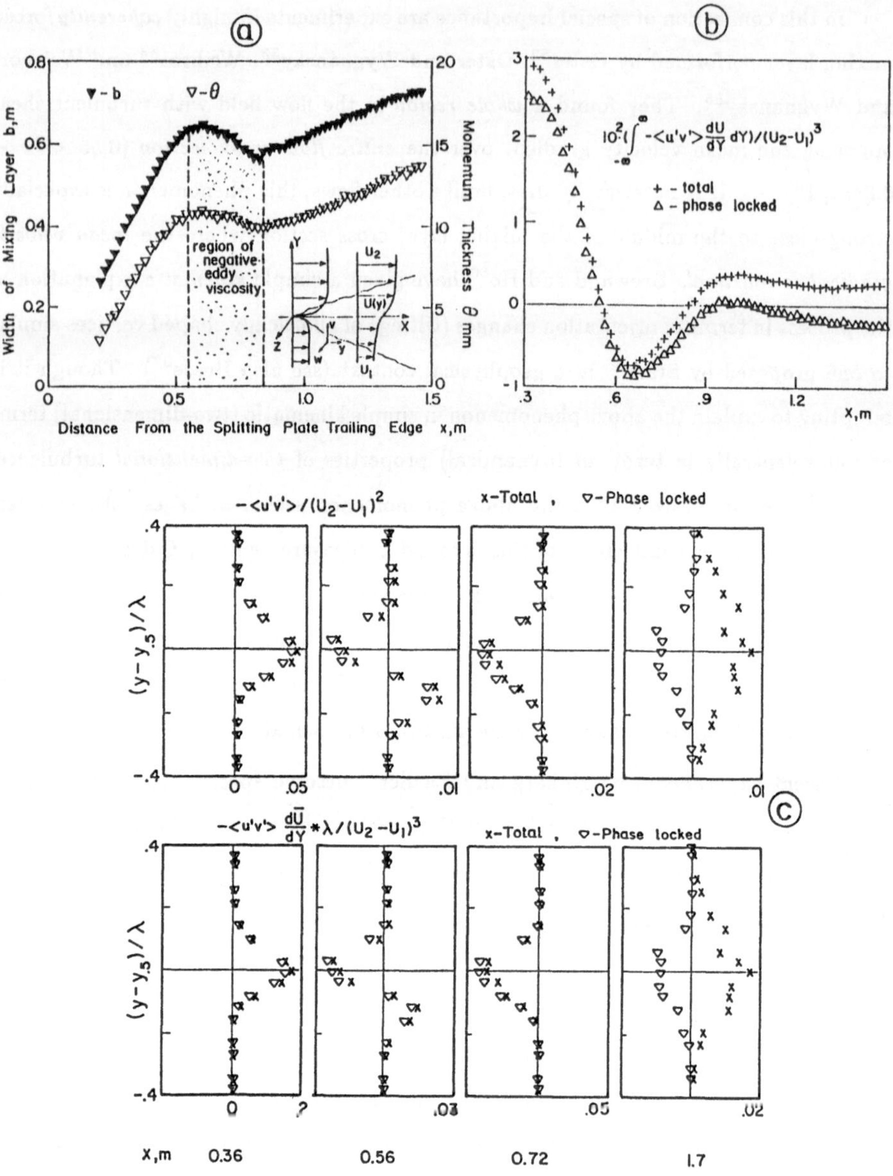

Fig.5. Negative eddy viscosity phenomenon in forced mixing layer[54]. a) width and momentum thickness, b) integrated turbulent energy production of Reynolds stress and turbulent energy productions at four cross-sections. Note the region $0.55 < X < 0.84$m where the mixing layer becomes narrower and the Reynolds stress and turbulent energy production reverse their signs. c) Profiles of Reynolds stresses and turbulent energy production at several longitudinal locations.

negative. Krause and Rudiger[59] using methods of mean field hydrodynamics have shown that in the case of homogeneous and isotropic two-dimensional turbulence the eddy viscosity (defined in the strict sense for isotropic turbulence) has no definite sign (while in 3-D case it is definitely positive), is zero when the molecular viscosity is zero and is negative in some special cases. Negative eddy viscosity understood as a long-wave instability (i.e. collective or cooperative phenomenon) of a system of forced small scale eddies was considered by Sivashinsky[60,61], and Sivashinsky and Yakhot[62] in simple periodic two-dimensional configurations.

Recently, a number of three-dimensional flows of simple geometry were considered which in some sense exhibit similar behaviour (Bayly and Yakhot[63], Frisch et al[64], Shtilman and Sivashinsky[65], Sulem et al[66], Yakhot and Pelz[67], Yakhot and Sivashinsky[68]). All of them are concerned with the large scale instability and consequent generation of large scale structure by small scale flows. The common feature of these examples is that *the background small scale* flow should be in some sense *anisotropic* to be able to develop a large scale instability. Frisch et al[64] and Sulem et al[66] considered the so called AKA effect (anisotropic kinetic alpha effect): anisotropic small scale flow lacking parity-invariance (i.e. lacking any center of symmetry) is able to generate strongly helical structures at large scales. It is noteworthy that the mean field hydrodynamics approach used in[64,66] was formulated much earlier in[59]. The main feature of their approach is that the average Reynolds stress tensor due to the small scale flow is supposed to depend on the mean (i.e. large scale) velocity vector (linearly when the large scale flow is weak) and not on its gradients. An interesting aspect is that the AKA-effect saturates in a highly helical flow, i.e. with parallel velocity and vorticity fields. At present, it is not clear whether any of the above approaches enable to explain the experimental results on negative eddy viscosity phenomena. In fact, no theoretical framework is available to describe the kinematical and dynamical features of these flows. Further extension to 3-D flows of the approaches suggested by Lesieur et al[69] and Liu[70,71] seem to be promising.

5. Simple aspects of low dissipative structure of turbulent flows.

General considerations. Obviously a low dissipative turbulent flow should possess special structure in some sense. As we have seen from above examples the existing evidence shows that this structure should be considerably anisotropic. On the other hand the flow structure should be such as to suppress the process of enstrophy (ω^2) generation - *the only mechanism sustaining high level of turbulent dissipation.* The common explicitly known (idealized) cases are:

i - two dimensional turbulent flows, i.e. flows the *instantaneous* structure of which depends on two Cartesran coordinates only.

ii - the Beltrami flows, i.e. flows in which $\boldsymbol{\omega} = \text{curl } \mathbf{u}$ is parallel to u.

In both the generation process of mean enstrophy does not exist, i.e. $< \omega_i \omega_k S_{ik} >= 0$, $S_{ik} = \frac{1}{2}(\partial u_i/\partial x_k + \partial u_k/\partial x_i)$ and therefore the turbulent dissipation $\epsilon =< \omega^2 >$ can be expected to be very small.

The above two examples are special cases of a more general situation when the enstrophy generation vanishes. Since $< \omega_i \omega_k S_{ik} >=< \boldsymbol{\omega} \text{ curl } (\boldsymbol{\omega} \times \mathbf{u}) >=$
$< (\boldsymbol{\omega} \times \mathbf{u}) \text{ curl } \boldsymbol{\omega} >=< (\boldsymbol{\omega} \times \text{curl } \boldsymbol{\omega})\mathbf{u} >=< (\mathbf{u} \times \text{curl } \boldsymbol{\omega})\boldsymbol{\omega} >$, the general condition is that one of the expressions in $< \cdots >$ should be a divergence, i.e. $\boldsymbol{\omega} \text{ curl } (\boldsymbol{\omega} \times \mathbf{u}) = \text{div } \mathbf{A}$, etc. The general structure of fields satisfying these conditions is not known. However, rather broad classes can be found as a consequence of solenoidality of vectors u, $\boldsymbol{\omega}$, e.g.

$$\boldsymbol{\omega} \times \mathbf{u} = \text{grad } \beta, \; \boldsymbol{\omega} \times \text{curl } \boldsymbol{\omega} = \text{grad } \beta, \; \mathbf{u} \times \text{curl } \boldsymbol{\omega} = \text{grad } \gamma \qquad (1)$$

where α, β, γ are some scalar fields. These three relations can be interpreted as constraints imposed respectively on the interaction between energy containing eddies (ECE) and eddies of the order of Taylor microscale (ETM), ETM and eddies of the order of Kolmogoroff scale (EKS) and ECE and EKS. If possible each of these constraints can be effective in drag reduction.

The Beltrami flows are the special case of the first configuration in which the Lamb vector $\boldsymbol{\lambda} = \boldsymbol{\omega} \times \mathbf{u}$ is a potential vector. This configuration is of special interest for the following reason. It corresponds to fixed points of the Euler equations and it has been suggested (Moffatt[72–74]) that turbulent flow may spend a large proportion of its time in a neighbourhood of such fixed points. This implies that in turbulent flow the potential part of the Lamb vector can be expected to be larger than its solenoidal part. Indeed, direct numerical computations performed by Kraichnan and Panda[75] showed that the quantity $< (\boldsymbol{\omega} \times \mathbf{u} + \nabla h)^2 >$ is only 0.57 of the corresponding value for a Gaussian field with the same energy spectrum. Shtilman and Polifke[76] corroborated their results and showed that the Lamb vector contains a substantial potential part by calculating the probability density function of the angle between the Fourier image of the Lamb vector $\boldsymbol{\lambda}(\mathbf{k})$ and the wave vector $\mathbf{k}$. They found that there was a high probability that $\boldsymbol{\lambda}(\mathbf{k})$ and $\mathbf{k}$ were aligned, which implies that in physical space the Lamb vector should have a large potential part.* They found also, that this property is characteristic for an initial Gaussian field and that the flow dynamics is such that it enhances this tendency. Moreover it was shown analytically by Tsinober[77] that the Lamb vector $\boldsymbol{\lambda} = \boldsymbol{\omega} \times \mathbf{u}$, $\boldsymbol{\omega} = \nabla \times \mathbf{u}$ for a random Gaussian solenoidal vector field $\mathbf{u}$ is 70% potential in the sense that 70% of the mean square $< \lambda^2 >$ is contributed by its potential part. The first implication of this result is that there is a strong reduction of nonlinearity in turbulent flows due to purely kinematic constraints in agreement with computations by Shtilman and Polifke[78]. This reduction can be as large as 70%. As we already mentioned, the dynamics of the flow is such that this tendency is amplified further[75,76]. Another aspect is that, again for purely kinematic reasons, the turbulent flow is quite close (in the above sense) to the set of fixed points of the Eulerian equations, which imposes certain constraints on the topology of the flow field, i.e. vectors $\boldsymbol{\omega}$ and $\mathbf{u}$ are not "far" from being tangent to α-surfaces, where α is the potential of the potential part of the Lamb vector $\boldsymbol{\lambda}$.

It is remarkable that the remaining 30% (possibly even less in real turbulent flows), are able to produce such strong effects as they do, e.g. vortex stretching, enstrophy generation and high level of turbulent dissipation. It is precisely this solenoidal part

* this can be checked directly by carrying out a Helmholz decomposition of the Lamb vector obtained from 3-D computations and comparing, say, the rms values of the Lamb vector and its potential and solenoidal parts, pdf distributions of the cosine of the angles between them etc.

of Lamb vector which should be suppressed in order to achieve sensible turbulent drag reduction effect, since only this part contributes to the enstrophy generation, since $<\omega_i\omega_j S_{ij}>= -<\text{curl}\,\boldsymbol{\beta}\cdot\text{curl}\,\boldsymbol{\omega}>$, where curl $\boldsymbol{\beta}$ is the solenoidal part of the Lamb vector. Note that two-dimensional flows satisfy the second condition (1), since in 2-D flows $\boldsymbol{\omega}\times\text{curl}\boldsymbol{\omega}=\text{grad}(\frac{1}{2}\omega^2)$.

Role of initial and environmental conditions. It is well known that many turbulent flows are highly sensitive to initial and environmental conditions (for comprehensive and updated reviews see Liu[70,71]). Hence their importance in control problems like drag reduction. The significance of these conditions is likely to increase in cases when at some initial moment the enstrophy generation is suppressed below its viscous dissipation, i.e.

$$\frac{d\langle\frac{1}{2}\omega^2\rangle}{dt}=\langle\omega_i\omega_j S_{ij}\rangle-\nu\langle(\text{curl}\,\boldsymbol{\omega})^2\rangle<0, \tag{2}$$

If the turbulent dissipation (i.e. mean enstropy) in this state is low enough, e.g. configurations satisfying one of the relations (1), it will survive for much longer time then otherwise. It is likely that such states are maintained in large scale coherent structures, e.g. large scale persisting atmosperic and oceanic vortices, mixing layers, etc. Since low dissipative states are expected to exhibit quasi-Eulerian behaviour it is possible that existence of inviscid invariants in addition to kinetic energy, i.e. enstropy in 2-D and helicity in 3-D, supports the persistence of two-dimensional and highly helical three-dimensional long living structures.

We know very little as to how to create these states in a controlled way. There are, however, a number of indications that this is possible as seen from previous sections. As pointed up above the common feature of flows of this kind is their distinct *anisotropy.* Here we bring additional simple examples of initially low dissipative flows.

Strongly helical flows. Andre and Lesieur[78] (see also Lesieur[70]) have shown, using an elaborated two point closure computation that homogeneous turbulent flow with large enough mean helicity has a much slower rate of decay comparing to that with zero mean helicity. This is due to the depression of the overall turbulent transfer in the presence of large helicity as was inferred by Kraichnan[79]. These results were conformed by Polifke and Shtilman[80] via direct numerical computations. A related suggestion was made by

Moffatt[72] and by this author (see Pelz et al[81]) that regions of high helicity should be low dissipative. The opposite is generally not true, e.g. two-dimensional flows or flows with $\omega \times \mathbf{u} = \text{grad}\ \alpha, \alpha \not\equiv 0$, etc. (see (1)). In other words the class of low dissipative flows is much broader than that of highly helical ones. This along with confusing between local and global aspects of helicity has lead to controversial results about the relation between dissipation and helicity density (see Rogers and Moin[82] and references therein). In any case the above suggestion finds support by the results of direct numerical simulations by Metcalfe (see Hussain[83]), Melander and Hussain[84] and Hussain et al[85], which show distinct separation between the regions of large helicity density and regions with high dissipation.

Turbulence subjected to rotation. Similar behaviour was observed in a related problem of turbulence decay subjected to rotation both theoretically and numerically, and experimentally (Cambon and Jacquin[86], Jacquin et al[87,88], Teissedre and Dang[89]). Moreover an angular spectral energy flux was observed corresponding to energy transfer from three-dimensional to two-dimensional flow modes in the plane perpendicular to rotation axis (an analogous process was conjectured by Kit and Tsinober[9] for MHD-turbulence). This effect along with inhibition of energy transfer to small scales results in decrease of rate of decay and enhanced anisotropy. For reviews of earlier work see Hopfinger[90], Tritton[91,92]. Turbulence in a stably stratified fluid without vertical shear can be expected to behave in a similar way (e.g. Maxworthy et al[93], Herring et al[94], for reviews see Hopfinger[95], Thorpe[96]).

Decay of anisotropic grid turbulence. Such a cigar-like turbulence was created past a honeycomb installed past a conventional grid (Hidenaru et al[97]). The longitudinal velocity component was considerably larger than the two other components. The rate of decay of this turbulence was observed to be substantially slower than that for quasi-isotropic turbulence.

A final remark we would like to emphasize that *all known structures which build up as a result of some inverse cascade (e.g. two dimensional or highly helical) are low dissipative.*

6. Concluding Remarks

The main points we wanted to put forth are as follows.

Turbulent drag reduction mostly *is not associated with suppression of turbulence production*, both in active drag reducing situations as described above and in passive drag reduction, e.g. "riblet drag reduction is probably not due primarily to alteration of turbulence production" (Wilkinson et al[98]), see also Djenidi et al[99,100]. The most prominent (counter) example is (quasi-) two-dimensional turbulent shear flow. The existing evidence indicates that *what is suppressed in such flows is the turbulent dissipation and its production (i.e. production of enstrophy).*

Turbulent drag reduction and related phenomena are normally characterised by an *increased degree of anisotropy* of some kind and vice versa the indication seems to be that a wide class of highly anisotropic flows should exhibit strongly reduced turbulent drag or, in other words, turbulent dissipation, i.e. highly anisotropic turbulence can be low dissipative. The most familiar examples are some quasi-two-dimensional turbulent flows and highly helical flows, e.g. rapidly rotating storms.

It follows that along with looking for methods of suppressing turbulence production it is necessary to look for methods of reducing turbulent dissipation and its production. The answer seems to be in identifying and studying the properties of highly anisotropic turbulent flows which are low dissipative and the underlying mechanisms sustaining such flows. Of particular interest are the following questions:

How increased anisotropy is related to suppression of turbulent dissipation and its production (i.e. enstrophy production) and which kinds of anistropy are good for this and which are not?

Can some of the so called coherent structures (claimed to be observed in various turbulent flows) be characterised as highly anisotropic entities with reduced dissipation and enstrophy production, e.g. quasi-two-dimensional, highly helical or more generally approximately satisfying one of the conditions (1), and thereby surviving for a longer time that just their turnover time? Are they created and sustained by one or several mechanisms which maintain low turbulent dissipation flows, negative eddy viscosity phenomena, etc.?

What is the role of inviscid invariants (enstropy in 2-D and helicity in 3-D) in the build up and maintaining of low dissipative turbulent flows? It is beyond any doubt that helicity invariant plays an essential role in a much broader context (see page 123 of the founding paper by Moffatt[101]).

As a starting point it would be instructive to perform a direct numerical simulation (DNS) of a two-dimensional turbulent flow like Poiseulle flow at supercritical Reynolds number (i.e. Re > 7000) - the expectation is rather high level of turbulence and almost laminar drag. The next step would be a DNS of a highly anisotropic flow maintained by some external factor which does not interact directly with the mean flow, e.g. azimuthal magnetic field, rotation with $\boldsymbol{\Omega} \parallel U_{mean}$ etc. Experimentally along with MHD and rotating flows the polymer drag reducing flows (especially heterogeneous) seem to be among the most appropriate providing effective means to study in a controlled way the relations between the degree of anisotropy, turbulent dissipation (and in particular turbulent drag) etc. in various turbulent flows. Methods involving direct influence on turbulence structure without direct interaction with the mean flow (like MHD, rotation, polymer especially heterogeneous) seem to be also very promising inputs into the black box of turbulence (Clauser[102]) which will produce effective outputs towards its better understanding.

And a final remark. It is likely that careful studying of the properties of the already known low dissipative turbulent flows, negative eddy viscosity phenomena etc. may be particularly useful as a starting in looking for drag reduction methods. This seems to be quite appropriate in view of rather low drag reduction by passive means like riblets and especially LEBU's (Prabhu et al[103], Savill and Mumford[104], Sahlin et al[105], for a recent review see Bushnell and McGinley[106]).

The mechanisms maintaining these spectacular phenomena are very purely understood (if at all). One can hope that progress in such an understanding will promote not only the advance of the turbulent drag reduction problem but will comprise a further step towards the disclosure of the enigma of turbulence.

Acknowledgements

The author is grateful to Drs H.W. Bewersdorff, A. Gyr and E. Kit for useful discussions.

My special thanks goes to Dr. A. Gyr for his continuous encouragement to write this paper.

The final stage of preparation of this work was supported by SERC during the author's summer stay at DAMTP, University of Cambridge, U.K.

POSTSYMPOSIUM NOTES ADDED IN PROOF. The common notion was that large scales are more effective in transferring momentum, and LEBUs, for example, were suggested on the basis of this view. As has been demonstrated in this paper, large scales may be ineffective in transporting momentum if the flow is anisotropic enough, e.g. quasi-two-dimensional turbulent flows. Promoting this kind of a flow behind a LEBU device can probably be done by forcing in a coherent way the trailing edge of the LEBU device in a manner similar to that in mixing layer experiments[52–5] and creating a long-living large scale quasi-two-dimensional flow structure (i.e. with reduced spanwise variations), which will hopefully dominate the flow over a long distance (as in mixing layer experiments) past the LEBU device. In this sense it may not be necessary to "switch off" the "coherent structures", but to produce the low dissipative ones, which may appear much easier. The apparent failure of LEBUs and the rather limited effect of riblets clearly indicate that influencing a narrow range of scales only (large or small) is too simple to expect significant drag reduction for such a complicated phenomenon as turbulence . This can be expected when the *interaction of scales is altered* (see eq. (1) above), which happens at least in some flows with enhanced anisotropy, as was reported in a number of papers at this Symposium too. We would like to point out that flow possessing the properties (1) should be anisotropic as indicated in[77] .

It is noteworthy that in drag reducing flows the Reynolds analogy is violated and while the momentum transport is inhibited the transfer of a passive scalar can be either enhanced (as in some MHD flows[3]) or inhibited much stronger than the drag (as in polymer drag reducing flows). The former case is related to anisotropy as in quasi-two-dimensional flows (e.g. u_1 , u_2 >> u_3) while the latter is related to anisotropy as in some highly helical flows (e.g. u_1 >> u_2 , u_3). Very little is known about the anisotropy of small scales, i.e. of the field of velocity derivatives.

REFERENCES

1. Narasimha, R. and Sreenivasan, K.R., 1979. "Relaminarization of fluid flows", *Adv. Appl. Mech.*, **19**, 221–309.
2. Hartmann, J. and Lazarus, F., 1937. Hg-Dynamics II, *Det. kgl. Dansk. Vidensk. Selsk., Mat.-fys. Medd.*, **15**, No. 7.
3. Tsinober, A., 1989. "MHD - flow drag reduction", to be published in *"Viscous drag reduction"* volume, eds. Bushell, D.M. and Hefner, D.M., *AIAA Prog. Astron. Aeron.* series.
4. Branover, G.G. and Gelfgat, Yu. M. and Tsinober, A., 1966. "Turbulent magnetohydrodynamic flows in prismatic and cylindrical ducts", *Magnetohydrodynamics*, **2**, 3–21.
5. Shercliff, J.A., 1953. "Steady motion of conducting fluids in pipes under transverse magnetic fields", *Proc. Cambr. Phil. Soc.*, **49**, 136–144.
6. Branover, G.G., Vasil'ev, A.S. Tsinober and Shkerstene, A. Ya., 1968. "Distribution of flow velocities in a rectangular pipe situated with the long side of its cross section in the direction of a magnetic field", *Magnetohydrodynamics*, **4**, 42–43.
7. Branover, G.G. Slyusarev, N.M. and Shcherbinin, E.V., 1970. "Measurements of velocity profiles and fluctuations in a two-dimensional channel with the long side of cross-section parallel to the transverse magnetic field", *Magnetohydrodynamics*, **6**, 492–496.
8. Kit, E., Turuntaev, S.V. and Tsinober, A., 1970. "Investigation with conduction anemometer of the effect of magnetic field on disturbances in the wake of a cylinder", *Magnetohydrodynamics*, **5**, 331–335.
9. Kit, E. and Tsinober, A., 1971. "Possibility of creating and investigating two-dimensional turbulence in a strong magnetic field", *Magnetohydrodynamics*, **7**, 312–318.
10. Goto, K., 1960. "Stokes flow of an electrically conducting fluid in a uniform magnetic field", *J. Phys. Soc. Jap.*, **15**, 696–705.
11. Wooler, P.T., 1961. "Instability of flow between parallel planes with coplanar magnetic field", *Phys. Fluids*, **4**, 24–27.
12. Platnieks, I.A., 1972. "Correlation study of the transformation of a field of turbulence velocity perturbation in a MHD duct", *7th Riga Conf. MHD*,**1**, p. 31 (in Russian).
13. Kolesnikov, Yu. and Tsinober, A., 1972. "An experimental study of two-dimensional turbulence behind a grid", *Fluid Dynamics*, **9**, 621–624.
14. Reed, C.B. and Lykoudis, P.S., 1978. "The effect of transverse magnetic field on shear turbulence", *J. Fluid Mech.*, **89**, 147–171.
15. Reed, C.B., 1976. "An investigation of shear turbulence in the presence of magnetic field", *Ph.D. Thesis*, Purdue University.
16. Sommeria, J., N'guyen Duc, J.M. and Caperan, P., 1988. "Two-dimensional MHD turbulence", *IUTAM Symposium on Liquid Metal MHD*, May 16–20, 1988, Riga, Latv. SSR. See also: N'guyen Duc, J.M., Caperan, P. and Sommeria, J., 1988. "Experimental investigation of the two-dimensional inverse energy cascade", *Prog. Astr. Aeron.*, **112**, 78–86.
17. Platnieks, I. and Freibergs, Ya. Zh., 1972. "Turbulence and some stability problems in flows with M-shaped velocity profiles", *Magnetohydrodynamics*, **10**, 164–168.

18. Selyuto, S.F., 1984. "Effect of magnetic field on formation of turbulence structure behing arrays of different configurations", *Magnetohydrodynamics*, **20**, 268–273.
19. White, A. and Hemmings, J.A.G., 1976. "Drag reduction by additives", *Review and bibliography.* BHRA, Fluid engineering.
20. Bartels, P.V., Markus, H. and Smith, J.M., 1985. "The turbulent mixing of viscoelastic fluids in pipe", in: *The influence of polymer additives on velocity and temperature fields*, ed. B. Gampert, Springer, pp. 279–289.
21. Berman, N.S., 1985. "A qualitative understanding of drag reduction by polymers", in: *The influence of polymer additives on velocity and temperature fields*, ed. B. Gampert, Springer, pp. 294–310.
22. Berman, N.S. and Tan, H., 1985. "Two-component laser Doppler velocimeter studies of submerged jets of dilute polymer solutions", *AIChE*, **31**, 208–215.
23. Maksimovich, C., 1985. "Turbulence structure of a developing duct flow with near wall injection of drag reducing polymers", in: *The influence of polymer additives on velocity and temperature fields* ed. Gampert B., Springer, pp. 359–368.
24. Gampert, B. (ed.) 1985. "Influence of polymer additives on velocity and temperature field", Springer.
25. Bewersdorff, H.W. and Olendorf, D., 1988. "The behaviour of drag reducing cationic surfactant solutions", *Colloid Polym. Sci.*, **266**, 941–953.
26. Luchik, T.S. amd Tiederman, W.G., 1988. "Turbulence structure in low- concentration drag-reducting channel flows", *J. Fluid Mech.*, **190**, 241–263.
27. Bewersdorff, H.W., 1984. "Effect of centrally injected polymer thread on turbulent properties in pipe flow", in: *3rd Int. Conf. Drag Reduction*, Bristol, ed. R.H.J. Sellin and R.H. Moses, B.4-1, see also *Rheol. Acta*, **23** 522–543.
28. Bewersdorff, H.W., 1985. "Heterogeneous drag reduction in turbulent pipe flow", in: *The influence of polymer additives on velocity and temperature fields*, ed. B. Gampert, Springer, pp. 337–348.
29. Usui, H., Maeguchi, K. and Sano, Y. 1988. "Drag reduction caused by injection of polymer thread into a turbulent pipe flow", *Phys, Fluids*, **31**, pp. 2518–2523.
30. Lumley, J.L. and Kubo, I., 1985. "Turbulent drag reduction by polymer additives: A survey" in: *The influence of polymer additives on velocity and temperature fields*, ed. Gampert, Springer, 3–21.
31. Riediger, S., 1988. "The influence of drag reducing additives on the coherent structures in a face shear layer", *2nd Europ. Turb. Conf.*, Berlin, Aug. 30–Sept. 2, 1988; see also *Proc. 6th Symp. Turb. Shear Flows*, Toulouse 1987, pp. 14.5.1–14.5.2.
32. Hoyt, J.W. and Sellin, R.H., 1988. "Drag reduction by centrally injected polymer threads", *Rheol. Acta*, **27**, 518–522.
33. McWilliams, J.C., 1983. "On the relevance of two-dimensional turbulence ot geophysical fluid motions", *J. Mech. Theor. Appl.*, Numero Special, 83–97.
34. Monin, A.S. and Ozmidov, R.V., 1985. "Turbulence in the ocean", Reidel.
35. Rhines, P., 1979. "Geostropic turbulence", *Ann. Rev. Fluid Mech.*, **11**, 401–41.
36. Starr, V., 1968. "Physics of negative viscosity phenomena", McGraw Hill.
37. Busse, F.H., 1983. "Generation of mean flows by thermal convection", *Physica*, **9D**, 287–299.
38. Monin, A.S. and Seidov, D.G., 1983. " On the generation of jet flow by negative viscosity", *Dokl. AN SSSR*, **268**, 454–457.

39. Monin, A.S., 1987. "On the negative viscosity in global circulations", *Dokl. AN SSSR.*,**293**, 70–73.
40. Lesieur, M., 1987. "Turbulence in Fluids", Martinus Nijhoff.
41. Kraichnan, R.H. and Montgomery, D., 1980. "Two-dimensional turbulence", *Rep. Prog. Phys.*, **43**, 547–619.
42. Two-dimensional turbulence, 1983. *J. de Mec. Theor. et Appl.*, Numero Special.
43. Lorenz, E.N., 1953. "The interaction between mean flow and random disturbances", *Tellus*, **5**, 238–250.
44. Schwartz, W.H. and Cosart, W.P., 1961. "The two-dimensional turbulent wall-jet", *J. Fluid Mech.*, **10**, 481–495.
45. Eskinazi, S. and Erian, F.F., 1969. "Energy reversal in turbulent flows", *Phys. Fluids*, **12**, 1988–1998.
46. Sreenivasan, K.R., Tavoularis, S. and Corrsin, S., 1982. "A test of gradient transport and its generlization", in: *Turbulent shear flows*, **3**, (ed. Bradbury L.O.S. et al), Springer, pp. 96–112.
47. Hinze, J.O., 1970. "Turbulent flow regions with shear stress and mean velocity gradient of opposite sign", *Appl. Sci. Res.*, **22**, 163–175.
48. Wilson, D.J., 1974. "Turbulent transport of mean kinetic energy in countergradient shear stress regions", *Phys. Fluids*, **17**, 674–675.
49. So, R.M. and Mellor, G.L., 1982. "An experimental investigation of turbulent boundary layers along curved surfaces", NASA CR-1940.
50. So, R.M. and Mellor, G.L., 1973. "Experiment on convex curvature effects in turbulent boundary layers", *J. Fluid Mech.*, **60**, 43–62.
51. Gibson, M.M., 1985. "Effects of streamwise curvature on turbulence", in: *Frontiers of Fluid Dynamics*, eds. Davis S.H. and Lumley, J.L., Springer, 184–198.
52. Oster, D., 1980. "The effect of an active disturbance on the development of the two-dimensional turbulent mixing layer", Ph.D. Thesis, Tel Aviv University.
53. Oster, D. and Wygnanski, I., 1982. "The forced mixing layer between parallel streams", *J. Fluid Mech.*, **123**, 91–130.
54. Weisbrot, I., 1984. "A highly excited turbulent mixing layer", Ph.D. Thesis, Tel Aviv University.
55. Weisbrot, I. and Wygnanski, I., 1988. "On coherent structures in a highly excited mixing layer", *J. Fluid Mech.*, **195**, 137–159.
56. Browand, F.K. and Ho, C.M., 1983. "The mixing layer: An example of quasi-two-dimensional turbulence", *J. Mec. Theor. Appl.*, Numero special, 99–120.
57. Hilberg, D. and Fiedler, H.E., 1988. "The turbulent shear layer in a narrow slit", *2nd Europ. Turb. Conf.*, Berlin, Aug. 30–Sept. 2, 1988.
58. Giger, M., Dracos, T. and Jirka, G., 1989. "Plane turbulent jets in bounded fluid layer", submitted to *J. Fluid Mech.* ; Giger, M. 1987. "Der obene Freistrahl in flachem Wasser", *Dissertation*, ETH, Zurich.
59. Krause, F. and Rüdinger, G., 1974. "On the Reynolds stresses in mean field hydrodynamics. I. Incompressible homogeneous isotropic turbulence. II. Two-dimensional turbulence and the problem of negative eddy viscosity", *Astron. Nachr.*, **295**, 93–99 and 185–193.
60. Sivashinsky, G.I., 1983. "Negative eddy viscosity effect in large-scale turbulence. Long-wave instability of a periodic system of eddies", *Phys. Lett.*, **95A**, 152–154.

61. Sivashinsky, G.I., 1985. "Weak turbulence in periodic flows", *Physica*, **17D**, 243–255.
62. Sivashinsky, G.I. and Yakhot, V., 1985. "Negative viscosisty effect in large-scale flows", *Phys. Fluids*, **28**, 1040–1042.
63. Bayly, B.J. and Yakhot, V., 1986. "Positive and negative effective viscosity phenomena in isotropic and anisotropic Beltrami flows", *Phys. Rev.*. **A34**, 381–391.
64. Frisch, U., She, Z.S. and Sulem, P.L., 1987. "Large scale flow driven by the anisotropic kinetic alpha effect", *Physica*, **28D**, 382–392.
65. Shtilman, L. and Sivashinsky, G., 1986. "Negative viscosity effect in three- dimensional flows", *J. Physique*, **47**, 1137–1140.
66. Sulem, P.L., She, Z.E., Scholl, H. and Frisch, U., 1988. "Generation of large-scale structures in three-dimensional flows lacking parity invariance", *J. Fluid Mech.* **205**, 341–358.
67. Yakhot, V. and Pelz, R., 1987. "Large-scale structure generation by anisotropic small-scale flows", *Phys. Fluids*, **30**, 1272–1277.
68. Yakhot, V. and Sivashinsky, G., 1986. "Negative viscosity phenomena in three-dimensional flows", *Phys. Rev.*, **A35**, 815–820.
69. Lesieur, M., Staquet, C., Le Roy, P. and Comte, P., 1988. "The mixing layer and its coherence from the point of view of two-dimensional turbulence", *J. Fluid Mech.*, bf 192, 511-534.
70. Liu, J.T.C., 1988. "Contributions to the understanding of large scale coherent structures in developing free turbulent shear flows", *Adv. Appl. Mech.*, **26**, 184–309.
71. Liu, J.T.C., 1989. "Coherent structures in transitional and turbulent free shear flows", *Ann. Rev. Fluid Mech.*, **21**, 285–315.
72. Moffatt, HJ.K., 1985, 1986a. "Magnetostatic equilibria and analogous Euler flows of arbitrary complex topology". Part 1. Fundamentals, *J. Fluid Mech.*, **159**, 359–378; Part 2. Stability considerations, *J. Fluid Mech.*, **166**, 359–378;
73. Moffatt, H.K., 1986b. "On the existence of localized rotational disturbances which propagate without change of structure in an inviscid fluid." *J. Fluid Mech.*, **173**, 289–302.
74. Moffatt, H.K., 1989. "Fixed points of turbulent dynamical systems and suppression of nonlinearity," *Whither turbulence* workshop, Cornell University, March 22–24, 1989.
75. Kraichnan, R.H. and Panda, R. 1988. "Depression of nonlinearity in decaying isotropic turbulence", *Phys. Fluids*, **31**, 2395–2397.
76. Shtilman, L. and Polifke, W. 1989. "On the mechanim of the reduction of nonlinearity in incompressible turbulent flows", *Phys. Fluids*, **A1**, 778–780.
77. Tsinober, A., 1989. "On one property of Lamb vector in isotropic turbulent flow," to be published.
78. Andre, J.C. and Lesieur, M. 1977. "Influence of helicity on the evolution of isotropic turbulence at high Reynolds number". *J. Fluid Mech.* **81**, 187–297.
79. Kraichnan, R. 1973. "Helical turbulence and absolute equilibrium", *J. Fluid Mech.*, **59**, 745–752.
80. Polifke, W. and Shtilman, L. 1989. "On initial conditions and time stepping schemes in numerical simulations of decaying turbulence," *Phys. Fluids* (in press).

81. Pelz, R.B., Yakhot, V., Orstag, S.A., Shtilman, L. and Levich, E., 1985. "Velocity-vorticity patterns in turbulent flows", *Phys. Rev. Lett.*, **54**, 2505–8.
82. Rogers, M. and Moin, P. 1987. "Helicity fluctuations in incompressible turbulent flows", *Phys. Fluids*, **30**, 2662–2671.
83. Hussain, A.K.M.F. 1986. "Coherent structures and turbulence", *J. Fluid Mech.*, **173**, 303–356.
84. Melander, M.V. and Hussain F. 1988. "Cut and-connect of two parallel vortex tubes, *Proc. Summer Progr*: Studying turbulence using numerical simulation databases II. Rep. **CTR-S88**, pp. 257–286.
85. Hussain, F., Moser, R., Colonis, T., Moin, P. and Rogers, M.M. 1988. "Dynamics of coherent structures in a plane mixing layer", *Proc. Summer Progr.*: Studying Turbulence using numerical simulation databases II. Rep.-**CTR-S88**, pp. 49–55.
86. Cambon, C. and Jacquin, L. 1989. "Spectral approach to non-isotropic turbulence subjected to rotation", *J. Fluid. Mech.*, **202**, 295–317.
87. Jacquin, L., Leuchter, O. and Geffroy, P., 1987. "Experimental study of homogeneous turbulence in the presence of rotation", in: *Proc. Sixth Symposium on Turbulent Shear Flows*, Toulouse, 1987, pp. 3.5.1–3.5.6;
88. Jacquin, L., Geffroy, P. and Leuchter, O. 1988. "Experimental study of rotation effects on grid generated turbulence with different mesh sizes", *Second European Turbulence Conf.*, Berlin, Aug. 30–Sept 2, 1988.
89. Teissedre, C. and Dang, K., 1987. "Anisotropic behaviour of rotating homogeneous turbulence by numerical simulation", AIAA paper 87–1250, *AIAA Fluid and Plasma Dynamics Conference*, Honolulu, Hawaii.
90. Hopfinger, E.J., 1989. "Turbulence and vortices in rotating fluids", in: *Theoretical and Applied Mechanics*, Proc. XVII Int. Congr. IUTAM, eds. Germain, P. et al, pp. 117–138.
91. Tritton, D.J., 1978. "Turbulence in rotating fluids", in: *Rotating Fluids in Geophysics*, eds. Roberts, P.H. and Soward, A.M., pp. 105–138.
92. Tritton, D.J., 1985. "Experiments on turbulence in geophysical fluid dynamics. I. Turbulence in rotating fluids", in: *Turbulence and Predictability in Geophysical Fluid Dynamics and Climate Dynamics*, Ed. Ghil, A., R. Benzi and Parisi, G., pp. 172–192, North-Holland.
93. Maxworthy, T. Caperan, Ph. and Spedding, G.R., 1987. "Two-dimensional turbulence and vortex dynamics in a stratified fluid", *Proc. 2nd Int. Symp. Strat. Fluids*, **2**, Pasadena, Feb. 3–5.
94. Herring, J.R., McWilliams, J.C., Metais, O. and Gamage, N., 1987. "Vortical turbulence in a stratified fluid", *Proc. 2nd Int. Symp. Strat. Fluids*, **2**, Pasadena, Feb. 3–5.
95. Hopfinger, E.J., 1987. "Turbulence in stratified fluids: A review", *J. Geophys. Res.*, **92C**, 5287–5303.
96. Thorpe, S.A., 1987. "Transitional phenomena and the development of turbulence in stratified fluids: A review", *J. Geophys. Res.*, **92C**, 5231–5248.
97. Hidenaru, M., Takao, I. and Akiyoshi, I., 1988. "An experimental study of axisymmetric turbulence". Ist report: "The generation and decay process of a cigar-shaped axisymmetric turbulence field", *Trans. Jap. Soc. Mech. Engn.* B. **54**, (No. 505), 2408–2415.

98. Wilkinson, S.P., Anders, J.B., Lazos, B.S. and Bushnell, D.M., 1987. In: Turbulent drag reduction by passive means", *Proc. Int. Conf.*, **1**, *Roy. Aeronaut. Soc.*, pp. 1–32.
99. Djenidi, L., Anselmet, F. and Fulachier, L., 1987. "Influence of a riblet wall on boundary layers, in: Turbulent drag reduction by passive means", *Proc. Int. Conf.*, **II**, *Roy. Aeronaut. Soc.*, 310–329.
100. Djenidi, L., Liandrat, J., Anselmet, F. and Fulachier, L., 1988. "About the mechanism involved in a turbulent boundary layer over riblets", *2nd Europ. Turb. Conf.*, Berline, Aug. 30–Sept 2, 1988.
101. Moffatt, H.K., 1969. "The degree of knottedness of tangled vortex lines", *J. Fluid Mech.*, **35**, pp. 117–129.
102. Clauser, F.H., 1956. "The turbulent boundary layer", *Adv. Appl. Mech.*, **4**, 1–51.
103. Prabhu, A., Vasudevah, B., Kailasnath, P., Kulkarni, R.S. and Narasimha, R., 1987. "Blade manipulators in channel flow", in: *Turbulence Management and Relaminarization*, ed. Liepmann, H.W. and Narasimha, R., Springer, 97–103.
104. Savill, A.M. and Mumford, J.C., 1988. "Munipulation of turbulent boundary layers by outer layer devices: Skin friction and flow visualization results", *J. Fluid Mech.*, **91**, 389–418.
105. Sahlin, A., Johanson, A.V. and Alfredson, P.H., 1988. "The possibility of drag reduction by outer layer manipulators in turbulent boundary layers", *Phys. Fluids*, **31**, 2814–2820.
106. Bushell, D.M. and McGinley, C.B., 1989. "Turbulence control in wall flows", *Ann. Rev. Fluid Mech.*, **21**, 1–20.

Drag Reduction by Sand Grain Roughness

K. Abe, A. Matsumoto, H. Munakata

Department of Aerospace Engineering
Nihon University, Funabashi, 274 Japan

and I. Tani

National Aerospace Laboratory
Jindaiji, Chofu, Tokyo, 182 Japan

Summary

On the premise of Coles law of the wall and law of the wake, an evaluation is made of the local skin friction coefficient c_f, the roughness shift F, and the wake parameter Π from experimentally measured mean velocity profiles for the turbulent boundary layer on a flat plate roughened with sifted sand grains. It is found that F is negative for a limited range of the roughness Reynolds number $u_\tau h/\nu$, but becomes positive with increasing $u_\tau h/\nu$, where u_τ is the friction velocity and h is the scale of the sand grain roughness. It is in the region of negative F that a reduction in c_f is achieved when combined with the increase in Π, namely, when Π reaches a value higher than that for a smooth plate. With reference to the case of drag-reducing riblet, the sand grain roughness behaves somewhat similarly, although the drag reduction is smaller.

Introduction

Of the various methods tested so far for reducing turbulent skin friction, one which has been well established is the use of riblets, or small streamwise grooves, originated with Michael Walsh and his associates [1][2][3] at NASA Langley Research Center. A reduction of much as 8 percent in skin friction may be achieved with v-groove riblet with a height h and a spacing s of about 15 wall units, namely, $h = s = 15\nu/u_\tau$, where ν is the kinematic viscosity and u_τ is the friction velocity. This amount of reduction is really surprising in view of a sizeable increase in wetted area. A closer observation by Hooshmand, Youngs and Wallace [4] revealed that the velocity gradient at the wall is enhanced near the peak of the riblet, but substantially reduced deep within the valley, and a net drag reduction ensues.

A. Gyr (Editor)
Structure of Turbulence and Drag Reduction
IUTAM Symposium Zurich/Switzerland 1989

Strictly speaking, therefore, the boundary layer velocity distribution over the peak is different from that over the valley, the difference being noticeable only in the immediate neighborhood of the wall. Since the dimensions of the groove are small compared to the boundary layer thickness, it may safely be assumed that the boundary layer is two-dimensional and has a friction velocity u_τ corresponding to its average over peak and valley. The assumption of a two-dimensional boundary layer is expected to be equally applicable to the case of wall roughened with sand grains, which was first investigated by Nikuradse [5]. For a two-dimensional turbulent boundary layer, the velocity profile is expressed with sufficient accuracy by the Coles [6] law of the wall and law of the wake of the form

$$\frac{U}{u_\tau} = \frac{1}{\kappa}\ln\frac{u_\tau y}{\nu} + C - F + \frac{\Pi}{\kappa}w \qquad (1)$$

outside the viscous sublayer ($y > 50\nu/u_\tau$), where U is the mean velocity in the x-direction, x and y are coordinates parallel and normal to the reference smooth wall, respectively, κ is the Karman constant (= 0.41), C is the smooth-wall constant (= 5.0), F is the constant shift due to wall modification such as by riblet or roughness, Π is the wake parameter, and w is the wake function of y/δ, δ being the boundary layer thickness. As pioneered by Coles [7][8] for smooth-wall turbulent boundary layers (F = 0), the analytical expression (1) may be used for evaluating u_τ and Π by fitting it to experimentally measured velocity profiles.

An integral version of the evaluation has been put forward by Tani and Motohashi [9] with a view to dispensing with the elaborate procedure of fitting the velocity profiles. An improved approximation to the modified wake function (Lewkowicz [10]),

$$w = 2\left(\frac{y}{\delta}\right)^2\left(3 - 2\frac{y}{\delta}\right) - \frac{1}{\Pi}\left(\frac{y}{\delta}\right)^2\left(1 - \frac{y}{\delta}\right)\left(1 - 2\frac{y}{\delta}\right), \qquad (2)$$

has been used in the analysis. To evaluate u_τ, Π and F from rough wall boundary layer measurements, additional use has to be made of the boundary layer momentum integral (Tani [11]). After setting $U = U_0$ and $y = \delta$ in equation (1) and introducing the conventional local skin friction coefficient $c_f = 2(u_\tau/U_0)^2$, where U_0

is the free-stream velocity, we arrive at the law of friction

$$\sqrt{\frac{2}{c_f}}\exp\kappa\sqrt{\frac{2}{c_f}} = \frac{U_o\delta}{\nu}\exp(2\Pi + \kappa C - \kappa F). \tag{3}$$

Differentiation for a constant value of $U_o\delta/\nu$ gives

$$\frac{dc_f}{c_f} = \frac{2\kappa dF - 4d\Pi}{1 + \kappa\sqrt{(2/c_f)}}, \tag{4}$$

which means that skin friction reduction ($dc_f < 0$) is achieved when F is reduced ($dF < 0$) and Π is increased ($d\Pi > 0$).

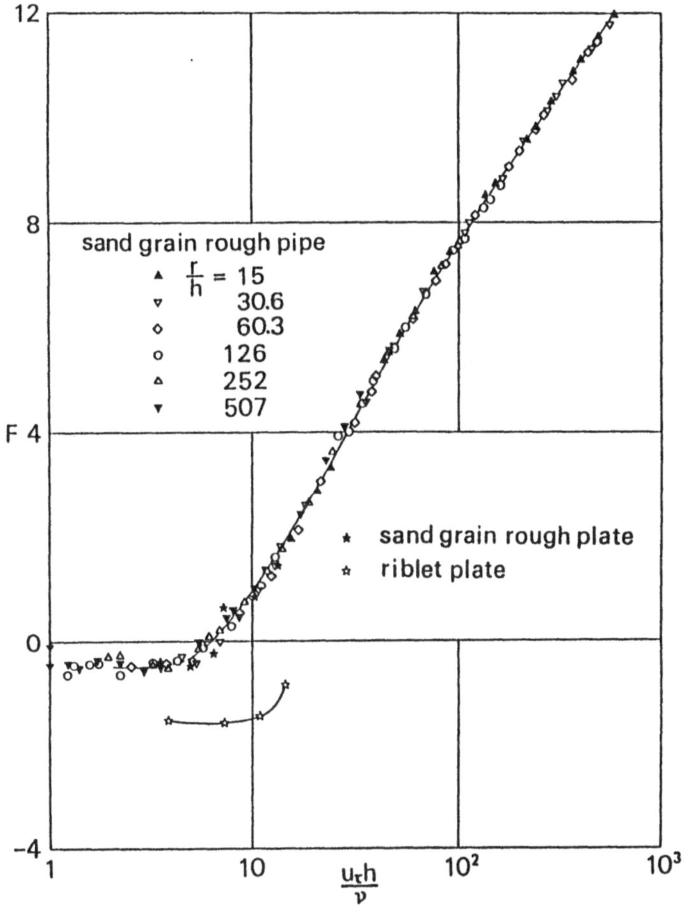

Fig. 1. Roughness shift plotted against roughness Reynolds number. Turbulent flow in sand grain rough pipe, and turbulent boundary layer on a sand grain rough plate and a riblet plate.

The method of evaluation was first applied to Nikuradse's experimental data for sand grain rough pipes (Tani [11][12]). It was found among others that F is negative for a limited range of the roughness Reynolds number $u_\tau h/\nu$, but becomes positive with increase in $u_\tau h/\nu$ (Fig. 1). In the region where F is negative, dF is negative since F = 0 for smooth pipe, but $d\Pi$ is also negative since Π is found to be smaller than that for smooth pipe, resulting in a nearly zero reduction in c_f, as surmised from equation (4). Nevertheless, the reduction of F is a matter of uncommon occurrence, while an increase of Π may commonly be found in rough-wall turbulent boundary layers. Motivated by this interpretation, an experimental investigation has been proceeding on turbulent boundary layers over a flat plate roughened with sand grains.

Experimental arrangement

The experiment was made in the Nihon University general-purpose wind tunnel, of the closed-circuit type and at a speed up to 50 m/s. A flat plate of 5 m length and 2 m span was mounted in the working section with a cross section of 2 m by 2 m in a near zero pressure gradient. The sand grain roughness was generated by attaching sifted steel shots to the flat plate, beginning at 0.4 m downstream from the leading edge. The sand grains were packed not most closely, in accordance with Nikuradse's expectation that the existence of interspace is prerequisite for the size of grain to amount to the hydrodynamically effective scale of roughness.

Measurements of the mean velocity profile were made with a hot-wire probe at several stations along the centerline of the plate for two sizes of sand grain, h = 0.2 and 0.4 mm, and also for three values of the unit Reynolds number, U_0/ν = 444, 649 and 842 mm^{-1}. The boundary layer was tripped by sparsely distributed sand grains near the leading edge to provide a fully turbulent, but not excessively perturbed flow at the beginning of the test surface. In addition to the main experiment on the sand grain rough plate, similar measurements were made also on a reference smooth plate and a plate coated with riblet film, which was manufactured at 3M Decorative Products Division, having a nominal scale of h =

0.006 inch. The small scale of the riblet necessitated operation at such high wind speeds that most of the measurements were made with the pitot tube.

Results and discussion

In view of the equations (3) and (4), it is convenient to present results of evaluation from measured velocity profiles as shown in Fig. 2, where $U_0 x/\nu$, Π and c_f are plotted as functions of $U_0\delta/\nu$.

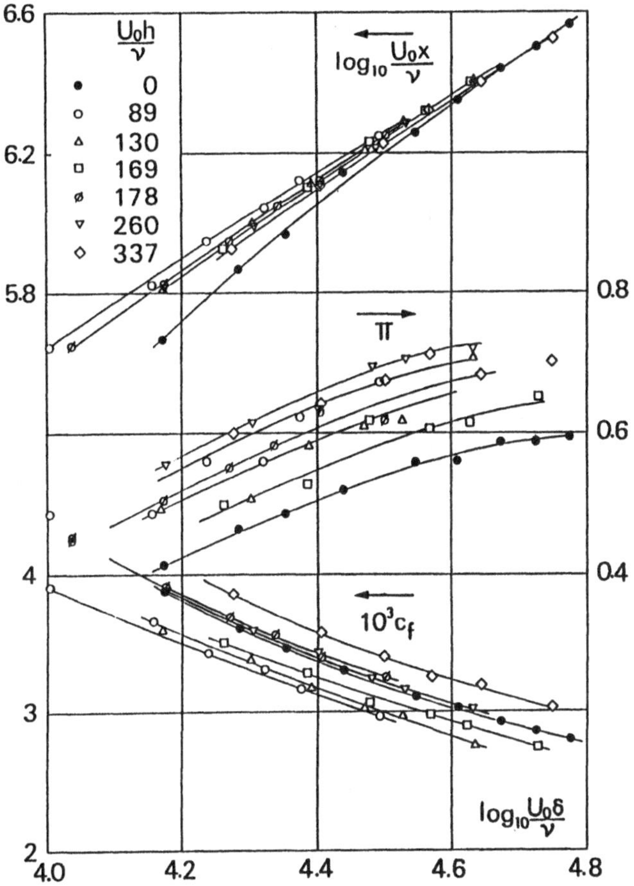

Fig. 2. Streamwise distance Reynolds number, wake parameter and local skin friction coefficient plotted against thickness Reynolds number. Turbulent boundary layer on a flat plate roughened with sand grains.

Turbulent boundary layer on sand grain rough plate

h mm	U_o/ν mm^{-1}	$U_o h/\nu$	$u_\tau h/\nu$	F	$d\Pi$	dc_f/c_f
0.2	444	89	3.5	-0.38	0.13	-0.07
0.2	649	130	5.0	-0.50	0.08	-0.06
0.2	842	169	6.4	-0.26	0.05	-0.03
0.4	444	178	7.3	0.69	0.10	0.02
0.4	649	260	10.3	0.89	0.15	0.01
0.4	842	337	13.4	1.50	0.11	0.07

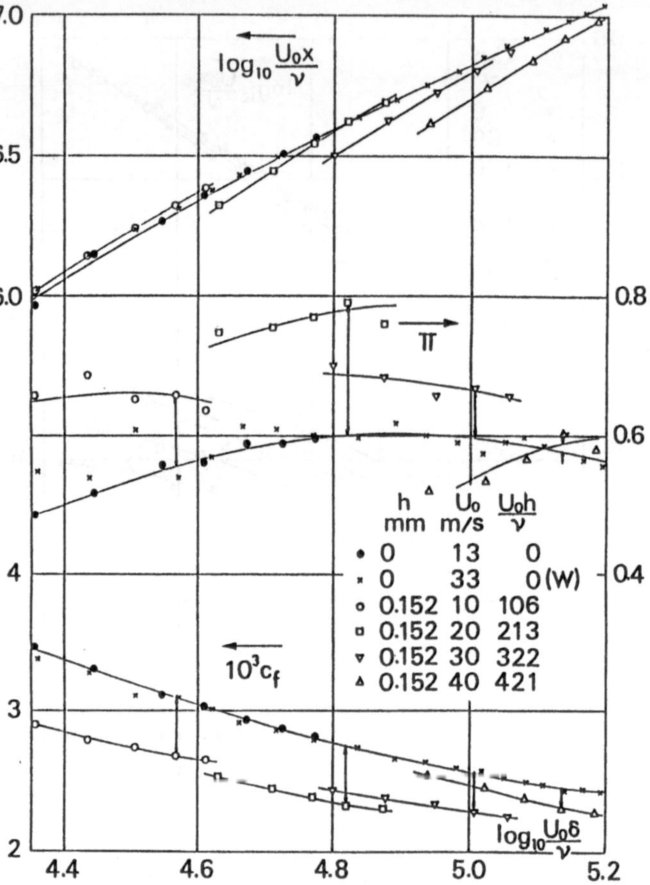

Fig. 3. Streamwise distance Reynolds number, wake parameter and local skin friction coefficient plotted against thickness Reynolds number. Turbulent boundary layer on a flat plate coated with riblet film. (W) refers to Wieghardt's data on smooth plate [8].

Here, x is the distance from the leading edge and δ is the thickness of the boundary layer defined as $\kappa\delta^*\sqrt{(2/c_f)}/(59/60 + \Pi)$, δ^* being the directly measurable displacement thickness. Some numerical data are also tabulated on the preceding page. It is seen from these results that for a limited range of $u_\tau h/\nu$ the sand grain roughness produces favorable effect in reducing δ, F and c_f and in increasing Π, when compared to those for the smooth plate at the same value of $U_0\delta/\nu$. When $u_\tau h/\nu$ exceeds a value of about 7, however, F becomes positive, making the performance to change from drag-reducing to drag-increasing.

Values of F determined for the sand grain rough plate are also shown in Fig. 1, indicating a good agreement with those previously evaluated from rough pipe data. The difference between the two evaluations is perceived in the behavior of Π, which is increasing ($d\Pi > 0$) for rough plates, but decreasing ($d\Pi < 0$) for rough pipes, as already mentioned.

Results of evaluation for riblet plate are shown similarly in Fig. 3. Values of F are also included in Fig. 1. With reference to these results it may be said that the sand grain roughness behaves somewhat similarly to the riblet, although the amount of drag reduction is smaller. The mechanism of drag reduction remains still unknown, but is likely to originate in the nearly quiescent flow of fluid, deep within the valley or interspace.

Finally it is to be noted that the evaluation of c_f, Π and F is not direct, but dependent on the validity of the Coles law (1), including the choice of the constants κ and C. We have used the standard values $\kappa = 0.41$ and $C = 5.0$, which appear, however, to overestimate the drag reduction. It was found that the maximum reduction of the evaluated skin friction amounts to about 15 per cent for the riblet plate (Fig. 3), almost twice as much as the reduction reported by Walsh. The disagreement may be reduced by the choice of κ and C different from the standard set of values, for example, by keeping κ at 0.41 but slightly increasing C.

References

1. Walsh, M. J.; Weinstein, L. M.: Drag and heat transfer on surfaces with small longitudinal fins. AIAA Paper 78-1161 (1978).

2. Walsh, M. J.: Turbulent boundary layer drag reduction using riblets. AIAA Paper 82-0169 (1982).

3. Walsh, M. J.; Lindemann, A. M.: Optimization and application of riblets for turbulent drag reduction. AIAA Paper 84-0347 (1984).

4. Hooshmand, D.; Youngs, R.; Wallace, J. M.: An experimental study of changes in the structure of a turbulent boundary layer due to surface geometry changes. AIAA Paper 83-0230 (1983).

5. Nikuradse, J.: Strömungsgesetze in rauhen Rohren. VDI Forschungsheft Nr. 361 (1933).

6. Coles, D.: The law of the wake in the turbulent boundary layer. J. Fluid Mech. 1 (1956) 191-226.

7. Coles, D.: The turbulent boundary layer in a compressible fluid. Rand Corp. Rep. R-403-PR (1962).

8. Coles, D.: The young person's guide to data. Proc. 1968 AFOSR-IFP-Stanford Conference on Computation of Turbulent Boundary Layers, 2 (1969) 1-45.

9. Tani, I.; Motohashi, T.: Non-equilibrium behavior of turbulent boundary layer flows. Proc. Japan Acad. B 61 (1985) 333-340.

10. Lewkowicz, A. K.: An improved universal wake function for turbulent boundary layers and some of its consequences. Z. Flugwiss. Weltraumforsch. 6 (1982) 261-266.

11. Tani, I.: Turbulent boundary layer development over rough surfaces. Perspectives in Turbulence Studies (eds. H. U. Meier and P. Bradshaw), Springer-Verlag 1987, 223-249.

12. Tani, I.: Drag reduction by riblet viewed as roughness problem. Proc. Japan Acad. B 64 (1988) 21-24.

Effect of Wall Suction on the Transport of a Scalar by Coherent Structures in a Turbulent Boundary Layer

ANSELMET, F.[+], ANTONIA, R.A.[*], BENABID, T.[+], FULACHIER, L.[+].
+ Inst. Méc. Stat. Turb., Univ. Aix-Marseille II, 13003 Marseille, France
* Dept. of Mech. Eng., Univ. of Newcastle, N.S.W.2308, Australia.

ABSTRACT

Two kinds of measurements are described. In the vicinity of the wall the longitudinal heat flux is studied through the correlation coefficient and the joint pobability density function; particular attention is paid to temperature measurements. Conditioned measurements based on the detection of temperature fronts are performed using a five cold-hot-wire arrangement. The influence of wall suction on the importance of coherent motions is investigated.

1. INTRODUCTION

The main aim of the present paper is to study the organised motion in a slightly heated boundary layer which is subjected to a step change in wall velocity by surface suction. It is focused on the heat transport and it is a continuation of two recent papers [1] [2].

The idea of perturbing a turbulent boundary layer at the wall and examining its response can provide very useful knowledge about the overall flow behaviour. It is noteworthy that, with suction, the inertial terms do not vanish at the wall, and so, this flow differs fundamentally from that on an impervious surface.

The main strategy is to introduce temperature as a passive marker of the flow. In particular, the temperature fronts, characterized by a spatially coherent rapid decrease in temperature with time (coolings), give a plausible dynamic link between the bursting phenomenon and the large scale structures. Coolings can be identified with the back of bulges in the outer region of the boundary layer and they can be associated with vortex loops or hairpin vortices [1]. Heatings, characterized by a spatially coherent increase in temperature, are not considered herein: from the point of view of contributions to Reynolds stresses, for instance, heatings are less important than coolings [2]. Two kinds of results are presented hereafter. The first ones concern the very near wall region where it is practically impossible to measure all the velocity components: the longitudinal heat flux is investigated up to the viscous sublayer. The other set of results is related to measurements conditioned on the occurrence of the coherent structures associated with

A. Gyr (Editor)
Structure of Turbulence and Drag Reduction
IUTAM Symposium Zurich/Switzerland 1989

coolings: velocity vectors, spanwise vorticity, strain rate, temperature variances and heat fluxes.

2. EXPERIMENTAL CONDITIONS AND PROCEDURES

2.1. Facility

Experiments are carried out in a fully developed turbulent boundary layer; the first section of the wall (Fig. 1) is a smooth metal surface, whereas the second one is a porous surface (1.8 m long). They are both heated to a constant temperature T_w relative to the ambient temperature T_1 of 10 K. The measuring station is located at a distance of 640 mm from the start of the porous surface. Measurements are made at a nominal free-stream velocity U_1 of 12.6 m/s and the longitudinal pressure gradient is practically equal to zero. The boundary layer thickness is δ = 62 mm, the friction velocity U* = 0.48 m/s and the friction temperature T* = 0.42 K. The effect of suction can be quantified by either the suction rate $A = -\rho_w V_w/\rho_1 U_1$ or the parameter $A+ = -V_w/U*$ where ρ is the density, V_w is the wall-suction velocity and the subscripts w and 1 correspond to wall and free stream conditions respectively. Only one suction rate is used herein A = 0.003 (i.e. A^+ = 0.005) for which δ = 57 mm, U* = 0.67 m/s and T* = 0.55 K.

2.2. Measurements close to the wall

Particular attention is paid to cold wire temperature measurements mainly in the wall vicinity. Indeed end effects are crucial for cold wires [3] and they are even more important when the velocity is small, as it is the case in the viscous sublayer, because of the development of thermal boundary layers along the prongs. A systematic study of the influence of the wire length l, as well as of the type of cold wire with or without "Wollaston" casing, is performed. Figure 2 presents two examples of results obtained with these two kinds of wires; the temperature r.m.s. θ' normalized by T*, i.e. θ'^+, is given as a

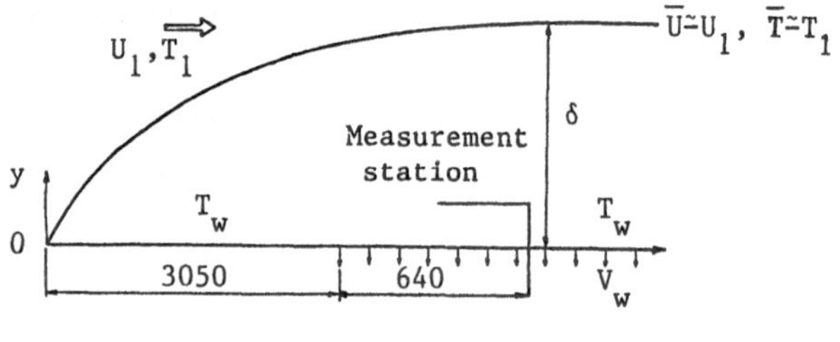

Fig.1. Schematic experimental configuration.

function of y^+ (= y U*/ν, with ν kinematic viscosity). Temperature sensors are made of 0.63 μm diameter Pt-10% Rh wire and the aspect ratio l/d is about 1200 for both of them. These cold wires are operated with constant current circuits built in-house, with a current of 0.15 mA small enough to avoid velocity contamination. The attenuation on the temperature r.m.s. θ' obtained with the fully etched wire is important : these measurements are similar to those of Elena [4] who had also used a fully etched wire (l/d = 800). Hishida and Nagano [5], with Wollaston and l/d = 400, Bremhorst and Bullock [6], with Wollaston and l/d = 560, and Fulachier [7], without Wollaston and with l/d = 400, obtain a more important attenuation. On the contrary, Krishnamoorthy [8], with Wollaston and l/d = 1600, reports a trend similar to present measurements with no attenuation. Moreover, non attenuated measurements are in good agreement with the direct numerical simulation of a turbulent channel flow performed by Kim and Moin [9].

The longitudinal velocity fluctuation u and the temperature fluctuation θ are measured with a pair of parallel wires aligned in the spanwise z direction and at the same distance from the wall. Both wires are etched from Wollaston, the upstream cold wire having an active length l of 1.2 mm and the dowstream hot wire having a diameter d' of 5 μm and a length l' of 1 mm. The streamwise separation is about 0.4 mm, i.e. about 630 d, large enough to avoid perturbation of the hot wire by the cold wire wake [10]. The hot wire is operated with a constant temperature circuit with a resistance ratio of 1.6: the possibility of this wire affecting the cold wire by heat transfer has been examined as in [10].

2.3. Conditioned measurements

The detection of coolings is obtained from a three cold wire arrangement, namely, two of them are at a fixed distance from the wall (y_f = 5 mm, y_f/δ = 0.081 for A = 0) and the third one is moving along the y direction. The spanwise separation between the fixed cold wires is about 6 mm ($\simeq 0.1\delta$): the correlation between the signals from these two wires should be sufficiently large for the purpose of detection since at this y_f position the spanwise lengthscale is about 0.4δ. The moving probe is made of the third cold wire located 0.9 mm upstream of an X wire, in order to obtain simultaneous measurements of temperature T and longitudinal and normal to the wall velocity components U and V respectively.

The detections of temperature fronts are based on a procedure called WAG (window average gradient) which is applied to the three cold wires: it is

described in [2]. In order to estimate the contribution that the organized motion makes to the conventional second order moments ($\overline{\theta^2}$, $\overline{\theta u}$ ...), use is made of the triple decomposition: $F = \overline{F} + \langle f \rangle + f_s$ ($= \overline{F} + f = \langle F \rangle + f_s$) where F stands for the instantaneous quantities U, V or T; $\overline{F}$ is the usual mean, $\langle f \rangle$ is due to the large-scale motion and f_s is the small-scale contribution. For products which involve two fluctuations f and g we obtain:
$\langle fg \rangle = \langle f \rangle \langle g \rangle + \langle f_s\, g_s \rangle$ with the assumption that $\langle f \rangle$ (or $\langle g \rangle$) and f_s (or g_s) are uncorrelated.

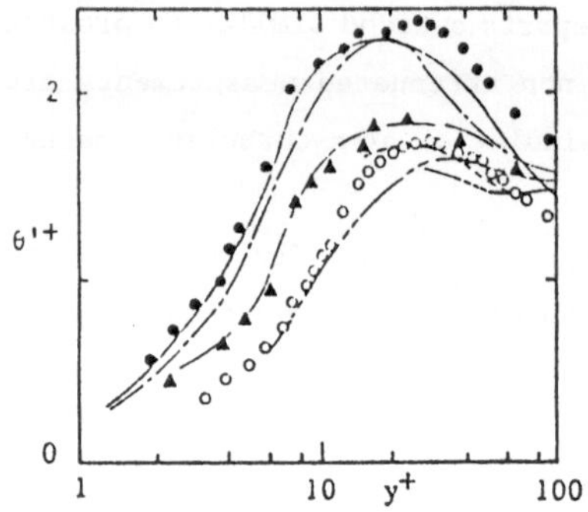

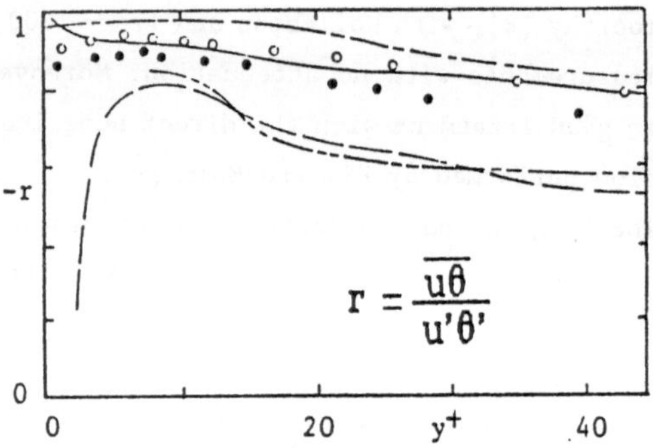

Fig.2. Temperature r.m.s. Fig.3. Correlation coefficient between θ and u. Present measurements : A = 0, • Wollaston wire, ▲ fully etched wire; A = 0.003, o ; —— —— : [4]; —— – – —— : [5]; —— – – – —— : [6]; – – – – : [7]; ———— : [8]; —— – —— : [9].

3. RESULTS

3.1. Some properties of the inner region

As was established in previous studies (see [1]) the r.m.s. intensities u'^+, v'^+, w'^+, normalized by U*, and θ'^+ are reduced by suction. The reduction in u'^+ is of the same order of magnitude as that in w'^+: suction increases the stability of the viscous sublayer by decreasing the amplitude of both longitudinal and spanwise oscillations which is consistent with flow visualisations. The r.m.s. of the temperature fluctuations, which provide a global picture of the turbulent field, are given on Figure 2. Near the wall, the gradient $d\theta'^+/dy^+$ with suction is equal to 0.42 times that without suction, whereas the values of du'^+/dy^+ and dw'^+/dy^+ for A = 0 are about 0.55 times those for A = 0.003. It can be shown that $B = (\overline{q^2}/\overline{\theta^2})^{1/2}\,(d\overline{T}/dy)/(d\overline{U}/dy)$ introduced by [7] is practically equal to 1 throughout the boundary layer for A= 0 whereas it is equal to about 1.3 for A= 0.003 (where $\overline{q^2} = \overline{u^2} + \overline{v^2} + \overline{w^2}$).

Figure 3 presents the results relative to the correlation coefficient r between θ and u. Without suction measurements are close to those of [10] and to computations of [9]. They strongly differ from the results of [4]: this would be due to the end effects. Nevertheless, it is noteworthy that present measurements are obtained with two amplifiers of different type (constant current for θ and constant temperature for u), and it could be argued that possible phaseshift might increase -r. However, this criticism is not to be considered since such a phaseshift would have to be strongly changing over the region investigated herein where the cospectra between θ and u are not significantly modified [10]. In addition the trend of experimental points is very similar to that of computational results. Suction does not have a strong influence on r even very close to the wall. More insight into the relative behaviour of u and θ within the sublayer can be inferred from the joint probability function p(u, θ). The plots of figure 4 highlight the narrowness of iso-value contours: such stretched ellipses, with their major axis aligned with the $u/u' = -\theta/\theta'$ equation line, are characteristic of the very strong negative correlation coefficient at this distance from the wall. Suction tends to make this relation even stronger, and the occupancy times from quadrants 2 ($u < 0$, $\theta > 0$) and 4 ($u > 0$, $\theta < 0$) are increased whereas those from 1 and 3 are reduced. It also appears that quadrant 2 is more sensitive to the boundary condition ($U = 0$, $T = T_w$) without suction than with suction, which may be due to the fact that since U* is in-creased with suction (owing to the wall deflected momentum) the position $y^+ = 3$ with suction is quite further from the wall than the position $y^+ = 2.5$ without suction. The agreement with contours inferred from direct simulations for A = 0 [9] is very good, in particular, the position of the peak contour (located within quadrant 2) is very similar. In [10], for A = 0, additional contours, parallel and on either side of the main contours, were detected. These contours were associated with very small magnitude local peaks of p (u,θ) and a trimodal form for p was suggested; they are not found here. At distances considered here ($y^+ \simeq 2.5$) quadrant 2 can be associated mainly with low speed streaks. The signals u and θ are found in this quadrant for 47% of the total record duration, with and without suction; the rest of the time is especially accounted for in quadrant 4 (about 40%). The contribution to $\overline{u\theta}$ is only about 42% from quadrant 2 and 57% from quadrant 4, with and without suction.

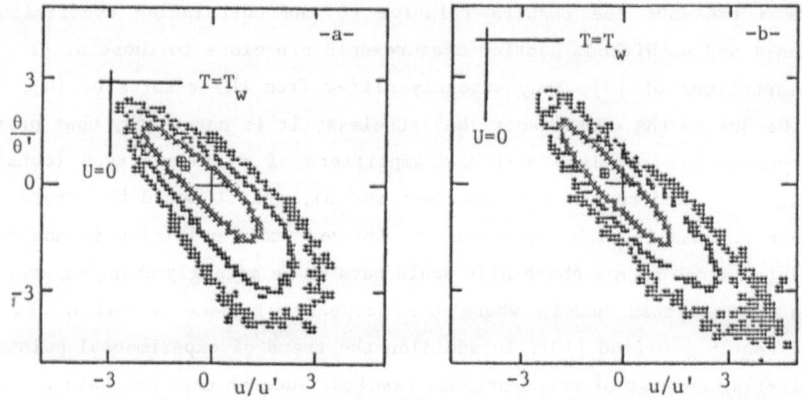

Fig. 4. Joint probability density function (outer to inner contours: 0.001; 0.01; 0.10; 0.38 (a), 0.55 (b)). -a- $A = 0$, $y^+ = 2.5$; -b- $A = 0.003$, $y^+ = 3$.

3.2. Some results on the organized motion

Among the various conditioned quantities, Figure 5 presents the velocity vector field associated with coolings in the (x,y)-plane, as well as the contours of spanwise vorticity and temperature variance due to the large-scale motion. The reference location used is y_r; at other y distances, the detection instant is delayed by the maximum τ_m of the distribution of the relative time delays between detections at y and at y_r. Taylor's hypothesis is used to convert time delays to streamwise separations $\Delta x = -U_c \tau$ taking the convection velocity U_c as the characteristic velocity [2]. U_c is assumed constant over the investigated region, equal to 0.67 U_1 for $A = 0$ and to 0.79 U_1 for $A = 0.003$. These results are presented in a reference frame which translates in the flow direction (right to left on Figure 5) with a velocity U_c . Salient features in the velocity vectors (Fig. 5a) are the lines defined by the coolings and the rotational patterns which the vectors delineate on either side of these lines. The coolings are directed along the diverging separatrices through the saddle points. Foci points are less clearly defined: they are better identified when heatings are detected. With suction the flow organization is enhanced and longitudinal and lateral scales are greater than for $A = 0$. Contours of the conditional spanwise vorticity $<\Omega>$ (= $\partial<V>/\partial x$ - $\partial<U>/\partial y$) (fig. 5b) and temperature variance $<\theta^2>$ occur near the wall and the largest concentration resides along the cooling. With suction, the contours are stretched away from the wall and the largest $<\Omega>$ and $<\theta^2>$ contours are flattened near the wall. The conditional strain rate (not shown here) has a behaviour very similar to $<\Omega>$. This suggests that coolings captured by the present scheme may be associated with hairpin vortices. When the contribution to second order moments from the organized motion associated with coolings is quantified (not shown here), it is found that this motion transports heat more effectively than momentum, and that suction increases the coherent transport. For instance, the cooling contribution to the temperature variance is maximum at $y/\delta \simeq 0.1$ and is equal to about 0.33 for A = 0 and 0.43 for A = 0.003.

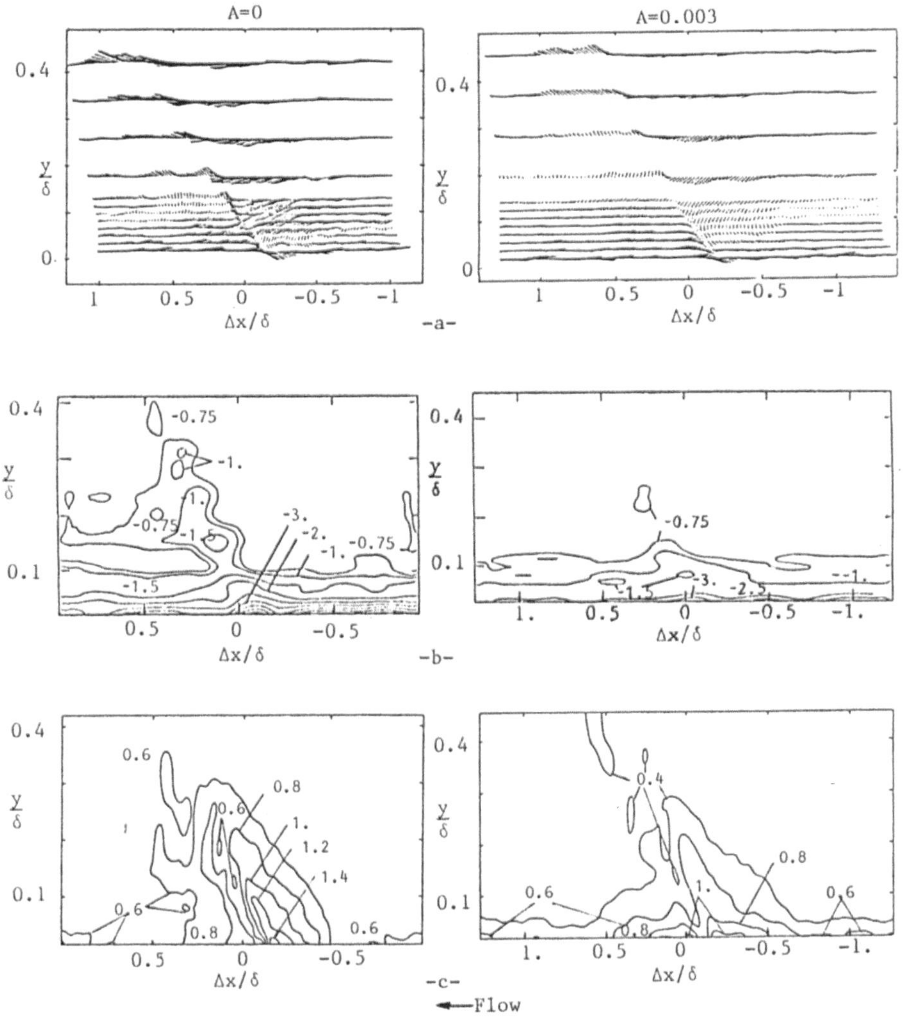

Fig. 5. Localisation in the (x,y) plane of quantities associated with coolings - a - velocity vectors ($<U> - U_c$, $<V>$); - b - spanwise vorticity $<\Omega>\ \delta/U_1$; - c - temperature variance $100\ <\theta^2>/(T_w - T_1)^2$.

4. CONCLUSION

Close to the wall, with or without suction, the u and θ fluctuations are very similar and the low-speed or high-temperature streaks are important in terms of space occupancy but coolings are more effective in the longitudinal heat transport. Suction enhances the organization of the flow and then increases the stabilization of the near-wall flow.

REFERENCES

1. Antonia, R.A, Fulachier, L, Krishnammorthy, L.V, Benabid, T. and Anselmet, F., 1988, *Influence of wall suction on the organized motion in a turbulent boundary layer*, J. Fluid Mech., 190, pp. 217-240.

2. Antonia, R.A. and Fulachier, L., 1989, *Topology of a turbulent boundary layer with and without wall suction*, J. Fluid Mech., 198, pp. 429-451.

3. Paranthoen, P., Petit, C. and Lecordier, J.C., 1982, *The effect of the thermal prong-wire interaction on the response of a cold wire in gaseous flows (air, argon and helium)*, J. Fluid Mech, 124, pp. 457-473.

4. Elena, M., 1975, *Etude des champs dynamique et thermique d'un écoulement turbulent en conduite avec aspiration à la paroi*, thèse de Doctorat ès Sciences, IMST, Univ. Aix-Marseille II. Also Rapport C.E.A. R-4843, 1977.

5. Hishida, M. and Nagano, Y., 1979, *Structure of turbulent velocity and temperature fluctuations in fully developed pipe flow*, J. Heat Transfer, 101, pp. 15-22.

6. Bremhorst, K. and Bullock, K.J., 1970, *Spectral measurements of temperature and longitudinal velocity fluctuations in fully developed pipe flow*, Int. J. Heat Mass Transfer, 13, pp. 1313-1329.

7. Fulachier, L., 1972, *Contribution à l'étude des analogies des champs dynamique et thermique dans une couche limite turbulente: effet de l'aspiration*, thèse de Doctorat ès Sciences; IMST, Univ. de Provence.

8. Krishnamoorthy, L.V. 1987, *Measurements in the near-wall region of a turbulent boundary layer*, P.H.D. Thesis, Univ. of Newcastle.

9. Kim, J. and Moin, P., 1987, *Transport of passive scalars in a turbulent channel flow*, 6th Symp. on Turb. Shear Flows, pp. 5.2.1 - 5.2.6.

10. Antonia, R.A., Krishnamoorthy, L.V. and Fulachier, L., 1988, *Correlation between the longitudinal velocity fluctuation and temperature fluctuation in the near-wall region of a turbulent boundary layer*, Int. J. Heat Mass Transfer, 31, pp. 723 - 730.

Flow Control by Suction

Mohamed Gad-el-Hak
Department of Aerospace and Mechanical Engineering
University of Notre Dame
Notre Dame, IN 46556
U.S.A.

1. INTRODUCTION

A viscous fluid which is initially irrotational will acquire vorticity when an obstacle is passed through the fluid. This vorticity controls the nature and structure of the boundary layer flow. For an incompressible, two-dimensional, wall-bounded flow, the flux of spanwise vorticity at the wall, and hence whether the surface is a sink or a source of vorticity, is affected by the wall motion (e.g. in the case of a compliant coating), transpiration (suction or injection), pressure gradient in the ambient flow, wall curvature, and viscosity gradient near the wall (e.g., heating/cooling of the wall or introducing a shear-thinning additive into the boundary layer). These alterations separately or collectively control the shape of the mean velocity profile which in turn determines the skin friction at the wall, the boundary layer ability to resist transition and separation, and the intensity of turbulence and its structure. In the case of a turbulent flow, the situation is complicated in that modulations such as pressure gradient or suction can dramatically change the production of Reynolds stress in the wall region and hence the momentum transport in the boundary layer.

Several active and passive techniques are available to reduce either the pressure drag or the skin-friction on land, air and sea vehicles. In addition to drag reduction, other control goals include lift enhancement, relaminarization, and turbulence augmentation. The present paper will employ the example of wall suction as a generic control tool used to delay transition or separation, achieve an asymptotic turbulent boundary layer, relaminarize an already turbulent flow, and, perhaps, achieve a turbulent skin-friction reduction. Effects of transpiration on the low-speed streaks, the bursting events, and the production of Reynolds stress in the wall region will be elaborated.

2. GOVERNING EQUATIONS

For a steady, incompressible flow around a two-dimensional or axisymmetric surface of small curvature, the continuity and streamwise momentum equations can be integrated in the normal direction to yield the von Karman integral equation:

$$C_f = 2\,\frac{d\delta_\theta}{dx} + 2\,\delta_\theta \left[\left(2 + \frac{\delta^*}{\delta_\theta}\right) \frac{1}{U_\infty}\,\frac{dU_\infty}{dx} + \frac{1}{R}\,\frac{dR}{dx} \right] - 2\,\frac{v_0}{U_\infty}\,, \qquad (1)$$

where C_f is the local skin-friction coefficient, δ^* and δ_θ are the displacement and momentum thicknesses, respectively, U_∞ is the freestream velocity, R is the radius of curvature of the wall, and v_0 is the normal velocity of fluid injected through the surface

A. Gyr (Editor)
Structure of Turbulence and Drag Reduction
IUTAM Symposium Zurich/Switzerland 1989

(positive for injection and negative for suction). Equation (1) is valid for both laminar and turbulent boundary layers.

A second useful equation is obtained from the time-averaged streamwise momentum equation as the wall is approached ($y \rightarrow 0$). Assuming a wall which is being heated or cooled, a viscosity gradient is generated due to the existence of a temperature gradient. At the fixed but porous wall, the equation becomes after rearranging:

$$\rho v_0 \left[\frac{\partial \overline{U}}{\partial y} \right]_0 + \frac{dP_0}{dx} - \frac{d\mu}{d\overline{T}} \left[\frac{\partial \overline{T}}{\partial y} \, \frac{\partial \overline{U}}{\partial y} \right]_0 + \rho \left[\frac{\partial \overline{uv}}{\partial y} \right]_0 = \left[\mu \, \frac{\partial^2 \overline{U}}{\partial y^2} \right]_0 , \tag{2}$$

where $\overline{T}$ is the mean temperature field, and the subscript $[\ ]_0$ indicates flow quantities computed at the wall. The right-hand side of (2) is the flux of mean spanwise vorticity, $\Omega_z = - \partial \overline{U} / \partial y$, at the surface. In the absence of suction/injection, pressure gradient, and surface heating/cooling, the first three terms on the left-hand side of (2) vanish.

3. SUCTION

Wall transpiration provides a good example of a single control technique that is used to achieve a variety of control goals. Suction can be employed to delay laminar-to-turbulent transition, postpone separation, achieve an asymptotic turbulent boundary layer, relaminarize an already turbulent flow, and, perhaps, achieve a net turbulent skin-friction reduction.

As seen from equation (2), small amounts of fluid withdrawn from the near-wall region of the boundary layer change the curvature of the velocity profile at the wall and can dramatically alter the stability characteristics of the boundary layer. Additionally, suction inhibits the growth of the boundary layer, so that the critical Reynolds number based on thickness may never be reached. Although laminar flow can be maintained to extremely high Reynolds number provided that enough fluid is sucked away, the goal is to accomplish transition delay with the minimum suction flow rate. Not only will this reduce the power necessary to drive the suction pump, but also the momentum loss due to suction, and hence the skin friction, is minimized.

The case of a uniform suction from a flat plate at zero incidence is an exact solution of the Navier-Stokes equation. The asymptotic velocity profile in the viscous region is exponential and has a negative curvature at the wall. The displacement thickness has the constant value $\delta^* = \nu / |v_0|$, where ν is the kinematic viscosity and $|v_0|$ is the absolute value of the normal velocity at the wall. In this case, (1) reads: $C_f = 2\, C_q$. Bussmann and Münz (1942) computed the critical Reynolds number for the asymptotic suction profile to be $R_{\delta^*} \equiv U_\infty \delta^* / \nu = 70{,}000$. From the value of δ^* given above, the flow is stable to all small disturbances if $C_q \equiv |v_0| / U_\infty > 1.4 \times 10^{-5}$. The amplification rate of unstable disturbances for the asymptotic profile is an order of magnitude less than that for the Blasius boundary layer (Pretsch, 1942). This treatment ignores the development distance from the leading edge needed to reach the asymptotic state. When this is included into the computation, a higher C_q (1.18×10^{-4}) is required to ensure stability (Iglisch, 1944; Ulrich, 1944).

Suction can be also used to postpone separation. Prandtl (1904) applied suction through a spanwise slit on one side of a circular cylinder. His flow visualization photographs convincingly showed that the boundary layer adhered to the suction side of the cylinder over a considerably larger portion of its surface. By removing the decelerated fluid particles in the near-wall region, the velocity gradient at the wall is increased, the curvature of the velocity profile near the surface becomes more negative, and separation is avoided. Prandtl (1935) used the momentum integral equation, (1), to make a simple estimate of the required suction coefficient to prevent laminar separation from the cylinder's surface: $C_q = 4.36\ Re^{-0.5}$, where Re is the Reynolds number based on the cylinder diameter and the freestream velocity.

Several researchers have used similar approximate methods to calculate the laminar boundary layer on a body of arbitrary shape with arbitrary suction distribution (see, e.g., Chang, 1970; Schlichting, 1979). A particularly simple calculation is due to Truckenbrodt (1956). He reduces the problem to solving a first-order ordinary differential equation. As an example, for a symmetrical Zhukovskii airfoil with uniform suction, Truckenbrodt predicts a suction coefficient just sufficient to prevent separation of: $C_q = 1.12\ Re^{-0.5}$,where Re is the Reynolds number based on the airfoil chord and the freestream velocity.

For turbulent boundary-layers, semi-empirical methods of calculation are inevitably used due to the closure problem. Suction coefficients in the range of $C_q = 0.002 - 0.004$ are sufficient to prevent separation on a typical airfoil (Schlichting, 1979). Optimally, the suction should be concentrated on the low-pressure side of the airfoil just a short distance behind the nose where, at large angles of attack, the largest local adverse pressure gradient occurs.

The suction rate necessary for establishing an asymptotic turbulent boundary layer independent of streamwise coordinate ($d\delta_\theta / dx = 0$) is much lower than the rate required for relaminarization ($C_q \simeq 0.01$), but still not low enough to yield net drag reduction. For $Re = 0\,[10^6]$, Favre et al. (1966), Rotta (1970) and Verollet et al. (1972), among others, report an asymptotic suction coefficient of $C_q \simeq 0.003$. From equation (1) rewritten for a zero-pressure-gradient boundary layer on a flat plate, the corresponding skin-friction coefficient is $C_f = 2\ C_q = 0.006$, indicating higher skin friction than if no suction was applied. To achieve a net skin-friction reduction with suction, the process must be further optimized. The results of Eléna (1975; 1984) and more recently of Antonia et al. (1988) indicate that suction causes an appreciable stabilization of the low-speed streaks in the near-wall region. The maximum turbulence level at $y^+ \simeq 13$ drops from 15 to 12% as C_q varies from 0 to 0.003. More dramatically, the tangential Reynolds stress near the wall drops by a factor of 2 for the same variation of C_q. The dissipation length scale near the wall increases by 40% and the integral length scale by 25% with the suction.

Gad-el-Hak and Blackwelder (1987; 1989) suggest that one possible means of optimizing the suction rate is to be able to identify where a low-speed streak is presently located and apply a small amount of suction under it. Assuming that the production of turbulent energy is due to the instability of an inflectional $U(y)$ velocity profile, one needs to remove only enough fluid so that the inflectional nature of the profile is alleviated. An alternative technique that could conceivably reduce the Reynolds stress is to inject fluid selectively under the high-speed regions. The immediate effect would be to decrease the viscous shear at the wall resulting in less drag. In addition, the velocity profiles in the spanwise direction, $U(z)$, would have a smaller shear, $\partial U/\partial z$, because the injection would create a more uniform flow. Since Swearingen and Blackwelder (1984) have found that inflectional $U(z)$ profiles occur as often as inflection points are observed in $U(y)$ profiles, injection under the high-speed regions would decrease this shear and hence the resulting instability. In all cases, the shear associated with the inflection points has been reduced. Since the inflectional profiles are all inviscidly unstable with growth rates proportional to the shear, the resulting instabilities would be weakened by the suction/injection process.

The feasibility of the selective suction as a drag-reducing concept has been demonstrated by Gad-el-Hak and Blackwelder (1989). Low-speed streaks were artificially generated using the method of Gad-el-Hak and Hussain (1986), and a hot-film probe was used to record their near-wall signature. An equivalent suction coefficient of $C_q = 0.0006$ was sufficient to eliminate the artificial events and prevent bursting. This rate is five times smaller than the asymptotic suction coefficient for a corresponding turbulent boundary layer. If this result is sustained in a naturally developing turbulent boundary layer, a skin friction reduction of close to 60% would be attained. Gad-el-Hak and Blackwelder propose to combine suction with non-planar surface modifications. Minute longitudinal roughness elements if properly spaced in the spanwise direction greatly reduce the spatial randomness of the low-speed streaks (Johansen and Smith, 1986). By withdrawing the streaks forming near the peaks of the roughness elements, less suction should be required to achieve an asymptotic boundary layer. Recent experiments by Wilkinson and Lazos (1987) and Wilkinson (1988) combine suction/blowing with thin-element riblets. Although no net drag reduction is yet attained in these experiments, their results indicate some advantage of combining suction with riblets as proposed by Gad-el-Hak and Blackwelder (1987; 1989).

REFERENCES

Antonia, R.A., Fulachier, L., Krishnamoorthy, L.V., Benabid, T., and Anselmet, F. (1988) J. Fluid Mech. 190, pp. 217-240.

Bussmann, K., and Münz, H. (1942) Jahrb. Dtsch. Luftfahrtforschung 1, pp. 36-39.

Chang, P.K. (1970) Separation of Flow, Pergamon Press, Oxford, England.

Eléna, M. (1975) Thèse de Doctorat ès Sciences, Université d' Aix-Marseille, Marseille, France.

Eléna, M. (1984) Phys. Fluids 27, pp. 861-86.

Favre, A., Dumas, R., Verollet, E., and Coantic, M. (1966) J. Mecanique 5, pp. 3-28.

Gad-el-Hak, M., and Blackwelder, R.F. (1987) AIAA Paper No. 87-0358.

Gad-el-Hak, M., and Blackwelder, R.F. (1989) AIAA J. 27, pp. 308-314.

Gad-el-Hak, M., and Hussain, A. K. M. F. (1986) Phys. Fluids 29, pp. 2124-2139.

Iglisch, R. (1944) Schr. Dtsh. Akad. Luftfahrtforschung 8 B, pp. 1-51.

Johansen, J.B., and Smith, C.R. (1986) AIAA J. 24, pp. 1081-1087.

Prandtl, L. (1904) Proc. Third Int. Math. Congr, pp. 484-491, Heidelberg, Germany.

Pretsch, J. (1942) Jahrb. Dtsch. Luftfahrtforschung 1, pp. 54-71.

Rotta, J.C. (1970) Seventh Congress of the International Council of the Aeronautical Sciences," ICAS Paper No. 70-10, Rome, Italy.

Schlichting, H. (1979) Boundary-Layer Theory, Seventh Edition, McGraw-Hill, New York.

Swearingen, J.D., and Blackwelder R.F. (1984) Bull. Am. Phys. Soc. 29, p. 1528.

Truckenbrodt, E. (1956) Forschg. Ing.-Wes. 22, pp. 147-157.

Ulrich, A. (1944) Schriften Dtsch. Akad.Luftfahrtforschung 8 B, p. 53.

Verollet, E., Fulachier, L., Dumas, R., and Favre, A. (1972) in Heat and Mass Transfer in Boundary Layers, Vol. 1, eds. N. Afgan et al., pp. 157-168, Pergamon Press, Oxford.

Wilkinson, S.P. (1988) AIAA Paper No. 88-3670-CP.

Wilkinson, S.P., and Lazos, B.S. (1987) in Turbulence Management and Relaminarization, eds. H.W. Liepmann and R. Narasimha, pp. 121-131, Springer-Verlag, New York.

Inertial Interaction of Spherical and Fibre-like Solid Particles with Turbulent Flowing Liquids

M. KROL
Universität Kaiserslautern
6750 - Kaiserslautern, FRG

1. Technical Background: The flow of concentrated suspensions of slender particles is encountered in a variety of practical applications such as: pneumatic and hydraulic transport, particle deposition and reentrainment, paper making, or reinforcement of plastics or metals to name a few.

2. Problems: The efficiency of the processes in question depends on the volume fractions of the disperse material in the respective flow configuration. Typical or desired values of the particle volume fractions are in the range of 15-50%. Crucial to the problem of drag reduction is the prediction and control of the bulk stresses in dense suspensions of irregular particles or ,in other words, of determining the "suspension viscosity". Densities of the particles usually exceed those of the fluid by more than 100% giving rise to significant inertial effects. Fundamental design parameters depend in case of fibre-like particles in a very special way on the orientations of the particles in the flow [1]. Verification of theoretical predictions requires simultaneous measurement of particle velocities and local particle orientations at practically relevant particle volume fractions (C_v > 1%).

2.1 Concentration Effects: Analysis of earlier experimental studies [2] have indicated that important motion parameters of suspension components , closely related to the suspension viscosity, may drastically be changed under presence of even small amounts of particles in the flow.
Considered is the pipe flow of a suspension of nearly mono-disperse glass spheres in water. Reynolds number is ca. 40000.
The mean diameter of the particles is 53 μm. We ask how fast would

A. Gyr (Editor)
Structure of Turbulence and Drag Reduction
IUTAM Symposium Zurich/Switzerland 1989

a cloud of fluid points change its longitudinal dimensions in:

a) particle free flow

b) flow at particle volume fraction of ca.4% ?.

Turbulent diffusion coefficients D_t as defined by G.I. Taylor [3] represent adequate flow parameters to characterize the situation in question:

$$D_t(T) = \frac{1}{2} \frac{d\,\overline{X^2}(T)}{dT} \quad (1)$$

It is convenient here to remind of the close relationship between the coefficient of molecular diffusion D_m derived in the kinetic gas theory and the dynamic gas viscosity μ:

$$D_m = \mu/\varrho \quad (2)$$

Fig. 1 compares the longitudinal turbulent diffusion coefficients of the fluid points in particle free flow (upper shaded region) with corresponding values in the suspension (lower shaded region).

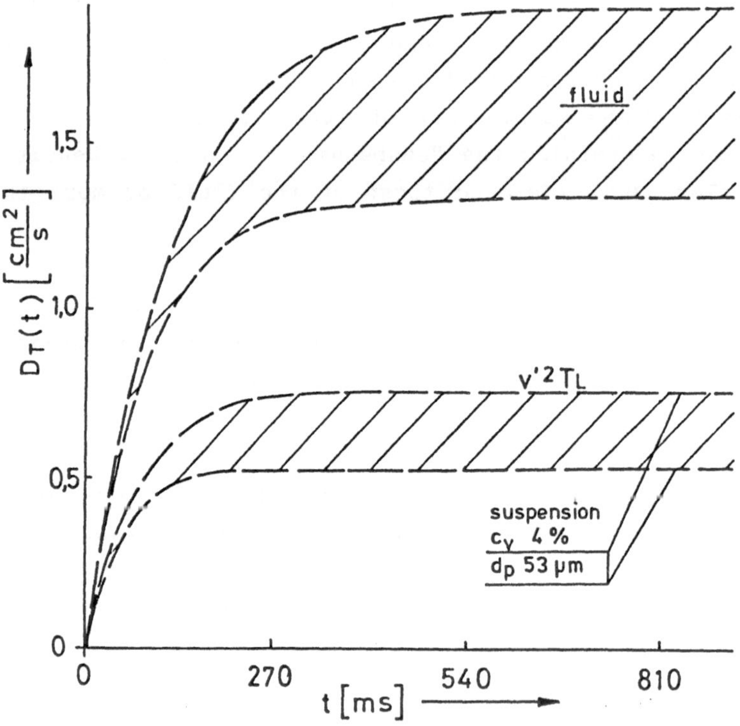

Fig.1. Turbulent diffusion coefficients: fluid vs. suspension.

The intensities of the fluctuating velocities of the fluid points in the core region of the pipe were found to exceed those of the particles by ca. 6% only.

The reduction of the longitudinal fluid diffusivities by a factor of ca. 0.4 taking place at particle volume fraction of merely 4% could be estimated using the following relation between the parameters of the particle free (index 0) and particle loaded (index C_v) flow [2]:

$$\lim_{T \to \infty} \frac{D_{l,0}(T)}{D_{l,cv}(T)} \approx \frac{v'^4_0}{v'^4_{cv}} \left(1 + C_v\, \rho_P / \rho_F\right)^{3/2} \tag{3}$$

Remarkable is here the high order of dependence of the quotient of diffusion coefficients on the fluctuating velocities of the particle free flow and of the particle loaded flow.

At the same time the longitudinal diffusivities of the particles were found to exceed considerably those of the fluid.

2.2 Inertial Effects: The above shown results were interpreted in terms of a model of inertial interaction of the particles contained in fluid elements performing curvilinear motion within the turbulent shear flow [4]. Considered is the case of particle densities larger than fluid densities. To be determined is the range of magnitudes of the fluctuating velocities of the fluid elements and of therein suspended particles for the mean direction of fluid motion.

According to the model velocity fluctuations collinear with the average motion of the fluid continuum contribute to the fluctuating velocities of the particles smaller than those of the fluid:

$$v'_{F,l} > v'_{P,l} \tag{4}$$

Velocity fluctuations transversal to the average fluid motion contribute to the fluctuating velocities of the particles generally larger than those of the fluid:

$$v'_{F,1} < v'_{P,1} < v'_{F,1} + 2\,|v_{rel,1}| \qquad (5)$$

Thus the resultant fluctuating velocities of the particles in the considered turbulent boundary layer flow are the result of the interplay of collinear and transversal velocity fluctuations in respect to the mean velocity vector. The magnitude of the relative particle-fluid velocity controls here the resultant longitudinal, fluctuating particle velocity. Since the particle-fluid force coupling is bidirectional there must be at least qualitative correspondence between the resultant fluctuating velocities of the particles and of the fluid elements. Weighting coefficient of the interaction is thereby determined by the local particle volume fraction and the quotient of particle and fluid densities.

Components of the relative velocities of spherical particles are invariant respectively to the particle orientations. On the contrary the relative particle-fluid motion of non-spherical particles is determined by both the inertial and significant orientational effects as well.

2.3 Orientation Effects: Without going into the details of the calculation based on the theory developed by Happel and Brenner for creeping flows [5] one may show that rod-like particles exhibit strong dependence of all velocity components on the particle orientation ϕ relatively to the vector of local acceleration $\underline{a}$ of the fluid element. Fig. 2, curve 3, shows exemplarily the velocity component collinear with the local acceleration vector plotted over the corresponding angular displacement of the major particle axis. Influence of the particle motion on the velocity field of the fluid is neglected.
Calculation of the velocity fields influenced by a multitude of non-spherical particles represents still a challenging theoretical and numerical problem [6,7]. The known exact solutions of the problem of particle-fluid and , consequently , particle-particle

interaction are limited to systems comprising very few particles only.

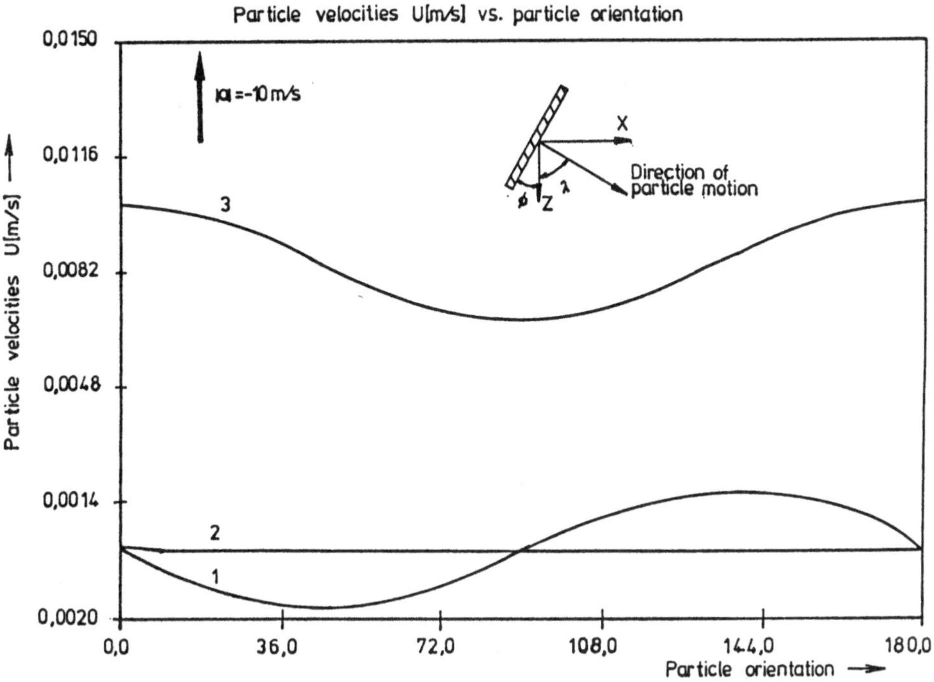

Fig. 2. Velocity components of a rod-like particle relative to an accelerated fluid element. Curve 3: particle velocity component U_z collinear with the acceleration vector a.

Crucial to the solution of practical problems is the correct prediction of the probability density functions of particle orientations in the flow. Gallily et al. performed such calculations for laminar and turbulent flows [1]. However, theoretical predictions pertaining to turbulent flow depend always on the quality of the used turbulence model and require further experimental verification.

<u>2.4 Simultaneous Measurement of Particle Velocities and local Particle Orientations:</u> This objective could be fullfilled using optical probes shown in Fig. 3. Two circular sensors aligned with

the mean direction of flow (Fig. 3) were used to determine the velocities of the particles.

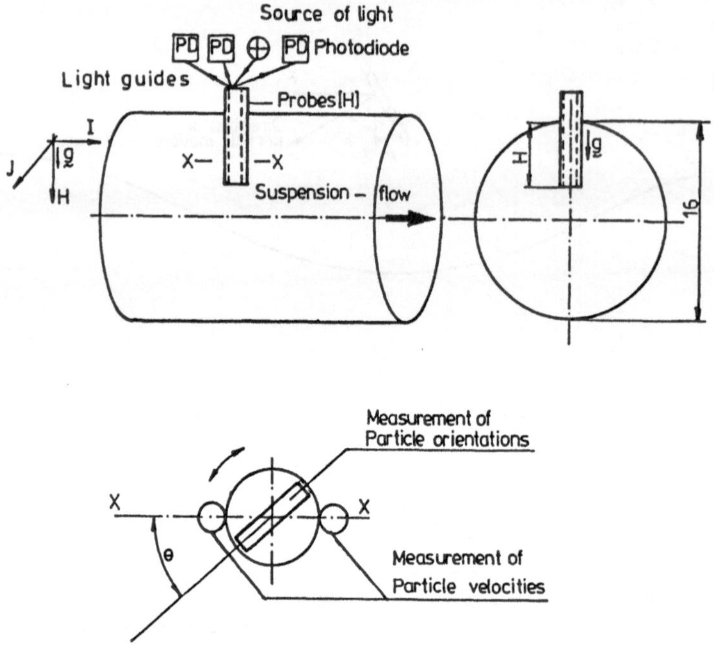

Fig. 3. Measurement principle, sensor arrangement.

Correlative methods tested and described earlier [2,8] were used to extract the information about particle velocities from time histories of light scattered by the particles and registered at sensor location. Elongated, rectangular sensor was used to determine the preferential orientations of the cylindrical, nearly monodisperse, glass particles with mean particle lengths of ca. 830 µm and diameters of 40 µm. Thereby the intensity of light received by the sensor reaches its maximum when the longitudinal axis of the sensor is aligned with the longitudinal axis of the particle.

Fig. 4 shows the results of test measurements conducted using a regular arrangement of the fibres together with measurements conducted in turbulent pipe flow at varied particle volume fractions.

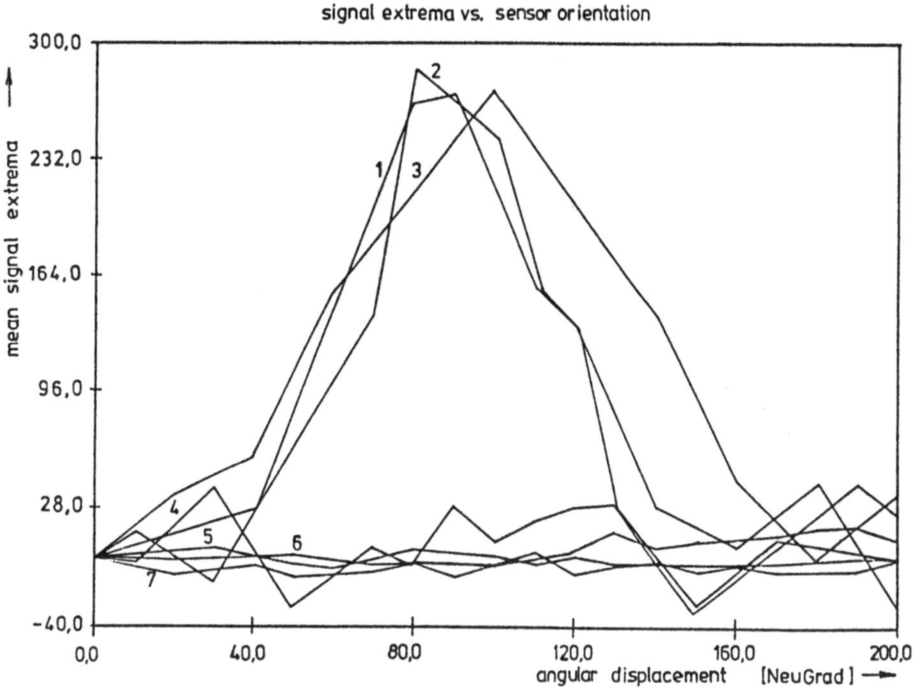

Fig. 4. Results of test measurements for (curves 1,2,3) regular fibre arrangements and (curves 4,5,6,7) for fibres in turbulent pipe flow Re ≈ ca.16000.

In case of the regular fibre arrangement the existence of a preferential fibre orientation may easily be recognized. It corresponds to the locations of the signal extrema. On the other hand measurements carried out in the turbulent pipe flow at Re of ca. 16000 exhibit no deterministic features exceeding the noise level.

3. Conclusions: Small changes in particle volume fraction may cause significant changes of the relevant flow parameters. These phenomena may qualitatively be explained in terms of the associated inertial effects. The developed measurement system

allows simultaneous determination of the local mean and fluctuating particle velocities and of the local preferential particle orientations at medium and high particle volume fractions. Preliminary results obtained till now seem to confirm the theoretical predictions of Gallily et. al.[1] for turbulent flows: orientations of macroscopic particles in turbulent flows appear to be random.

ACKNOWLEDGMENTS
This paper was prepared in grateful memory of professor Isaiah Gallily of the Hebrew University of Jerusalem.

REFERENCES

[1] Gallily, I.;Krushkal, E.M.; On the orientation distribution function of nonspherical aerosol particles in a general shear flow I. The laminar case; J. Colloid Int. Sci., 99,No.1(1984) II. The turbulent case; J. Aerosol Sci.,19,No.2,197-211,1988

[2] Krol,M.; Ebert, F.; Particle diffusivities and particle-fluid interaction in turbulent suspension flows at high particle concentrations; PCH PhysicoChemical Hydrodynamics; 6(1985)

[3] Taylor, G.I. Diffusion by Continuous Movements, Proc. London Math. Soc., Ser. 2,20,196(1921)

[4] Krol, M. Prediction of relative particle-fluid displacements in flowing suspensions at medium particle volume fractions by means of a simple stochastic model; Particulate Phenomena and Multiphase Transport; Hemisphere Publishing Corp.; Vol.4(1988)

[5] Happel, J.; Brenner, H.; Low Reynolds number hydrodynamics, Noordhoff Int. Publishing (1973)

[6] Kim, S.; Singularity Solutions for Ellipsoids in low-Reynolds-Number Flows; Int. J. Multiphase Flow, 12(1986)

[7] Durlofsky, L.J.; Brady, J.F.; Dynamic simulation of bounded suspensions of hydrodynamically interacting particles; J. Fluid Mech.; 200(1989)

[8] Krol, M.; Experimental Determination of Velocities and Spatial Orientations of Monodisperse, Fibre-Like Solid Particles in Turbulent Suspension Flows; 4th European Symp. Particle Characterization, PARTEC 89, Nürnberg 1989.

Part 4
Stability and Computations

Hydrodynamic Instability and Coherent Structures in Turbulence

Marten T. Landahl
Massachusetts Institute of Technology

Abstract

The role of hydrodynamic instability in the creation of coherent structures in wall-bounded turbulence is reviewed. Because the mean velocity profile for a flat-plate boundary layer or channel flow is highly stable to small wave-like disturbances, the classical Tollmien-Schlichting instability mechanism is not operative for the excitation of the fluctuation field. Instead, this must involve other instability mechanisms and/or nonlinearity. For such stable flows it is found that the transfer of energy from the mean to the fluctuations takes place primarily through algebraic instability [1]. This gives rise to structures of long-lived alternating low and high-speed regions in the near-wall region, i.e., streaks, which give contributions to the Reynolds shear stress proportional to their lifetime. By restraining the spanwise fluctuating velocity component so as both to reduce the spanwise vortex stretching, thereby inhibiting the formation of internal shear layers, and the spanwise asymmetry of the structures, their contribution to the Reynolds stress, and hence the total turbulent stress and the skin friction, may be reduced.

Introduction

One of the original aims of the study of hydrodynamic instability was to arrive at an understanding of the mechanisms that control fully developed turbulent flows. Despite the massive research efforts devoted to this field for more than a century the role of hydrodynamic instability in fully developed turbulent flows still remains incompletely understood. The turbulent flow over a flat plate, and in a channel, has a mean flow that is highly stable to small wave-like disturbances [2]. The excitation of the fluctuations in flows that are hydrodynamically stable in the classical sense must therefore involve finite amplitudes and/or other instability mechanisms. By regarding the mean flow as a linear system driven by the fluctuating turbulent stresses modelled as concentrated sources Landahl [2] constructed a wave-guide model for the fluctuating field, which brought out the dominating role of the damped travelling wave modes in the wave number-frequency spectrum (see also Bark [3]). However, this model does not give an answer to how the turbulence is created and maintained. For turbulent flows which are unstable in the mean, such as free shear layers and boundary layers in adverse pressure gradients,

A. Gyr (Editor)
Structure of Turbulence and Drag Reduction
IUTAM Symposium Zurich/Switzerland 1989

however, linear instability may provide the dominating mechanism for creation of the large scales, see the recent work by Gaster et al.[4].

The problem of generation and maintainance of turbulence in a boundary layer was adressed in Reference 5. By considering the interaction between disturbances of widely different scales, it was shown that, in a flow of small viscosity, the large-scale disturbance initiated by a localized region of small-scale instability, which may arise, for example, in a region of inflectional velocity distribution, could produce an almost permanent 'scar' of velocity defect or excess. (Additional discussions of secondary instability breakdown were presented in the previous IUTAM Symposium on the subject [6]). In a later paper [1] adressing the question of how small disturbances may evolve in an inviscid parallel shear flow it was shown that not only do all such flows when disturbed leave behind such a nondecaying remnant of a three-dimensional disturbance, whether or not they are unstable in the Rayleigh sense, but for a certain class of initial disturbances the streamwise dimension of the disturbed region, and hence also the total disturbance energy, will continue to grow forever at a constant rate proportional to the local shear. In a viscous flow this algebraic growth will eventually be quenched by viscous diffusion, but in a fluid with small viscosity the total growth may be quite large before viscosity takes over.

The characteristics of algebraic instability are highly suggestive of the most outstanding phenomenon observed in the near-wall region of a turbulent flat-plate boundary layer, namely the formation of streamwise streaks of alternating low- and high speed flows. Their significance was first clearly brought out by the flow visualization experiments of Kline et al. [7]. The streaky structure of the turbulent eddies near a wall have since then been extensively studied by many researchers, most recently with the use of conditional sampling of numerically simulated wall flows (Alfredsson et al., [8]).

Most of the efforts to date to explain the streaks from theory have been concerned with their average spanwise spacing and have met with some limited success. Barks' [3] calculation of the wave number-frequency spectrum of the streamwise velocity fluctuations based on the wave-guide model [2] gave a peak of the streamwise fluctuation spectrum at a spanwise wave number corresponding to a streak spacing of about 150 in viscous wall units, which is to be compared to the observed value of about 100. Jang et al. [9] proposed that the streak formation is a result of a resonance between the vertical vorticity mode (the "Squire mode") and the Orr-Sommerfeld mode. They found such a resonance at a spanwise wave number corresponding to a streak spacing of 90 in viscous units.

Hatziavramidis and Hanratty [10] applied a '2 1/2-dimensional' model to the streaks in which the cross-flow velocity (v,w-) field was approximated as a two-dimensional one in the cross-flow (y,z-)plane. The near-wall flow was considered driven by the fluctuating pressure originating from the action of the flow outside and, accordingly, they represented the outside spanwise flow as a periodic one with a spanwise length scale of 100 viscous units. From this model, they found good agreement with measured statistical proprties, particularly for $y^+ < 15$.

In this paper the evolution of a coherent structure in the near-wall region will be studied on the basis of linear theory. It will be demonstrated that, according to this theory, a three-dimensional inviscid disturbance grows into a streamwise elongated streaky structure in a manner quite similar to what has been observed in experiments and in numerical simulations. The instability mechanism involved is the algebraic one [1]. Nonlinearity acts to initiate the structures through secondary instability of internal shear layers produced by the shearing of the convected eddies. It is found that spanwise asymmetry is necessary for the structures to evolve into long streaks. The consequences of these findings for the possibilities of achieving turbulent drag reduction through active or passive means are discussed.

Formulation of a model for the streaky structures

As an idealized model we consider unsteady disturbances of velocity components $u_i(x_i,t)$ and pressure $p(x_i,t)$ in an incompressible and parallel mean shear flow, $\bar{U}(x_2)\delta_{i1}$. The coordinate system used is the usual one in boundary layer theory with x_1=x in the mean streamwise direction, x_2=y normal to the wall (located at y=0), and x_3=z in the spanwise direction. From the Navier-Stokes equations the following equations for the disturbed flow field are obtained:

$$U_i(x_j,t) = \bar{U}(y)\,\delta_{i1} + u_i(x_j,t) \tag{1}$$

$$\frac{\bar{D}u_i}{Dt} + \bar{U}'(y)u_2\,\delta_{1i} = -\frac{1}{\rho}\left[\frac{\partial p}{\partial x_i} + \frac{\partial}{\partial x_j}\left(\mu\frac{\partial u_i}{\partial x_j} + \tau_{ij}\right)\right] \tag{2}$$

$$\frac{\partial u_i}{\partial x_i} = 0. \tag{3}$$

where

$$\frac{\bar{D}}{Dt} = \frac{\partial}{\partial t} + \bar{U}\frac{\partial}{\partial x} \tag{4}$$

$$\tau_{ij} = -\rho\left(u_i u_j - \overline{u_i u_j}\right) \tag{5}$$

By elimination of the pressure one can combine these into a single equation of the form of a nonhomogeneous Orr-Sommerfeld equation with all the nonlinear terms grouped on the right-hand side [2],

$$\frac{\bar{D}\nabla^2 v}{Dt} - v_x \bar{U}'' - \nu\nabla^4 v = q \tag{6}$$

where

$$q = \nabla^2 T_2 - \frac{\partial^2 T_i}{\partial x_i \partial x_2} \approx \frac{\partial^2}{\partial y^2}\left[\frac{\partial}{\partial x}(uv) + \frac{\partial}{\partial z}(vw)\right] + \ldots \tag{7}$$

$$T_i = \frac{1}{\rho}\frac{\partial \tau_{ij}}{\partial x_j}, \tag{8}$$

The fluctuation velocity field near the wall is found to have a boundary layer character in that it is highly flattened. Therefore one may anticipate that the nonlinear terms singled out in (7), which contain the highest y-derivatives, would be the dominating ones in this region.

The numerical simulations [11] have shown that the vertical component (but not the horizontal ones) is highly intermittent in the near-wall region as reflected in the extremely large values of the flatness for the distribution function for this component. A meaningful idealized model for the near-wall fluctuating field may therefore be one in which the vertical component is subjected to instantaneous accelerations by the nonlinear interactions at distinct bursting instants t_n. Thus one sets

$$\frac{\bar{D}\phi}{Dt} - \bar{U}''\frac{\partial v}{\partial x} = Q_n(x,y,z)\delta(t-t_n) \tag{9}$$

where Q_n gives the spatial distribution of the instantaneous pulse in the Laplacian of v. Integration over time from $t=t_n - 0$ to $t=t_n + 0$ gives:

$$\phi \equiv \nabla^2 v = Q_n(\xi_n,y,z) + \bar{U}''\frac{\partial l}{\partial x} \tag{10}$$

$$\xi_n = x - \bar{U}(y)(t-t_n) \tag{11}$$

and where

$$l = \int_{t_n}^{t} v(x_1,y,z,t_1)dt_1 \quad , x_1 = x - \bar{U}(y)(t-t_1) \tag{12}$$

Here l is the fluid element liftup in linear approximation, c. f.Prandtl [12]. For $t > t_n$ one may formulate the problem as an initial value one with

$$\frac{\bar{D}\nabla^2 v}{Dt} - \bar{U}''\frac{\partial v}{\partial x} = 0 \tag{13}$$

and, on the assumption that $l=0$ for $t=t_n$,

$$\phi=\phi_n=\nabla^2 v_n = Q_n(x,y,z) \quad \text{for } t = t_n + 0 \tag{14}$$

Inversion of the Laplacian gives

$$v = -\frac{1}{4\pi}\int_{-\infty}^{\infty} dx_1 \int_{-\infty}^{\infty} dz_1 \int_{0}^{\infty} \phi(x_1,y_1,z_1,t)(1/R_1 - 1/\bar{R}_1)dy_1 \tag{15}$$

where

$$R_1 = [(x-x_1)^2 + (y-y_1)^2 + (z-z_1)^2]^{1/2} \tag{16}$$

$$\bar{R}_1 = [(x-x_1)^2 + (y+y_1)^2 + (z-z_1)^2]^{1/2} \tag{17}$$

and where $\phi(x,y,z;t)$ is given by (10), (11). The fluid element liftup, l, may be obtained from (12) by direct integration. After a reversal of the orders of integration one finds the following result:

$$l = -\frac{1}{4\pi}\int_{-\infty}^{\infty} dx_1 \int_{-\infty}^{\infty} dz_1 \int_{0}^{\infty} \frac{P - P^*}{\bar{U}-\bar{U}_1}(1/R_1 - 1/\bar{R}_1)dy_1 \tag{18}$$

where

$$P(x_1,y_1,z_1;t) = \int_{-\infty}^{\xi_1} Q_n(x_2,y_1,z_1)dx_2 + \bar{U}''(y_1)l(x_1,y_1,z_1;t) \tag{19}$$

with

$$\xi_1 = x_1 - \bar{U}(y_1)(t-t_n)$$

and where

$$P^*(x_1,y_1,y,z_1;t) = \int_{-\infty}^{\xi^*} Q_n(x_2,y_1,z_1)dx_2 + \bar{U}''(y_1)l^*(x_1,y,y_1,z_1;t) \tag{20}$$

$$l^* = \int_{t_n}^{t} v(\xi^{**},y_1,z_1,t_1)dt_1 \tag{21}$$

with

$$\xi^{*}= x_1 - \bar{U}(y)(t-t_n)\ ,\ \ \xi^{**}= x_1 - \bar{U}(y)(t-t_1)$$

For the streamwise fluctuation component one has

$$\frac{\bar{D}u}{Dt} = -v\,\bar{U}'(y) - \frac{1}{\rho}\frac{\partial p}{\partial x}\ (+\text{ nonlinear terms}) \tag{22}$$

It can be shown that, for structures that have large streamwise dimensions compared to their spanwise ones, the dominating contribution to the streamwise perturbation velocity comes from the first term the contribution from the pressure gradient term being of the order $(l_3/l_1)^2$ times that from the first term. Thus, neglecting this term and the nonlinear terms, one finds

$$u \approx u_n(\xi_n, y, z) - \bar{U}'(y) l \tag{23}$$

(c.f. Prandtl [11]). The first term represents a purely convected part of the disturbance with v=0, which may be treated separately. It will not contribute directly to the formation of the streaky structures, or to the Reynolds stress (since the associated vertical component is zero), and will therefore not be included in the numerical examples presented below.

The expression (18) for the fluid element liftup contains on the right-hand side both the liftup itself, and v, inside the integral. Only for $\bar{U}''=0$ will their contribution vanish making it possible to calculate l directly from the initial values for arbitrary times. The results will then be exactly equivalent to those found from rapid distortion theory. For a nonzero $\bar{U}''$, the solution must be found by a step-by-step integration in time. For this purpose an explicit method having second order accuracy in the size of the time step was developed.

Long-time behavior - the role of algebraic instability

The long-time behavior of the disturbance may be directly inferred from the solution (18) - (21). Consider an initial local disturbance of streamwise length 2L confined between x > -L and x<L. Also, set

$$\int_{-\infty}^{\infty} Q_n(x_1, y, z)\,dx_1 = P_{n\infty}(y,z) \tag{24}$$

where $P_{n\infty}$ is assumed to be nonzero. Then, following (18), for large t, there will be a streamwise region of length $\Delta x_1 = |\bar{U}-\bar{U}_1|t-2L$ for which the contribution to the integrand from the initial condition will be independent of x_1. Also, since for a stable flow $v \equiv Dl/Dt \rightarrow 0$ for $t \rightarrow \infty$, the term

under the integral proportional to $\overline{U}''(y_1)$ will give a comparatively small contribution. Furthermore, also since its streamwise integral is zero. Thus, for large t, l will have large extended regions with properties independent of x. Hence, for large times, the flow will tend to become two-dimensional in the cross-flow (y,z-) plane over a streamwise length growing linearly with time.

This phenomenon is related to the phenomenon of algebraic instability explored by Landahl [1]. He showed through streamwise averaging of the perturbation equations that for an inviscid shear flow without inflection, and having $u = u_0$, $v = v_0$, for $t = 0+$,

$$\int_{-\infty}^{\infty} v(x,y,z,t)dx = \int_{-\infty}^{\infty} v_0(x,y,z)dx \tag{25}$$

$$\int_{-\infty}^{\infty} u(x,y,z,t)dx = -t\overline{U}'(y) \int_{-\infty}^{\infty} v_0(x,y,z)dx + \int_{-\infty}^{\infty} u_0(x,y,z)dx \tag{26}$$

It follows from this that, for an inviscid shear flow that is stable in the Rayleigh sense, the total streamwise momentum, and the perturbation energy, will grow linearly in time for all times after the initiation of the disturbance. As it is clear from the above analysis, the amplitude of the disturbance does not grow for all times, instead the streamwise length of the disturbance continues to grow. Of course, in a real viscous fluid this growth will eventually cease. Nevertheless, for a fluid of small viscosity the flow will admit a substantial algebraic growth before the slow viscous decay takes over, in particular if the mean shear is large (the long-time effects of viscosity will be adressed later). Therefore, three-dimensional disturbances initially localized in the spanwise direction will evolve into higly elongated ones, i. e., will produce a streaky structure incorporating eddies having a large ratio of streamwise to spanwise scales. For the calculation of the long-time behavior of such eddies it is appropriate to employ the "slender-body" approximation familiar in aerodynamics (see, e.g., Ashley & Landahl, [13]). It reduces the triple integral in (18) to a two-dimensional one in which the streamwise variable only enters parametrically, i.e.,

$$v = \frac{1}{2\pi} \int_0^{\infty} dy_1 \int_{\infty}^{\infty} f(\xi_n, y_1, z_1; t) \ln(r_1/\overline{r_1}) dz_1 \tag{27}$$

with

$$r_1 = [(y-y_1)^2 + (z-z_1)^2]^{1/2} \qquad (28)$$

$$\bar{r}_1 = [(y+y_1)^2 + (z-z_1)^2]^{1/2} \qquad (29)$$

Very close to the wall the solution may be simplified even further so as to be expressible through the following line integrals:

$$v \approx \int_0^y \phi(\xi_n, y_1, z)(y-y_1)dy_1 - y\int_0^\infty \phi(\xi_n, y_1, z)dy_1 \qquad (30)$$

The numerical results presented below are all based on the near-wall approximation (30).

Reynolds stresses

The contribution from the streaks to the mean Reynolds shear stress.will be determined on the basis of linear inviscid theory. The gradient of the turbulent shear stress is obtained from the average of

$$-\frac{\partial}{\partial y}(uv) = -uv_y - u_y v \qquad (31)$$

We want to determine the contribution from each streak by integration over time and horizontal space. It is convenient to express this in terms of the liftup variable l. By introducing (19) and neglecting the terms involving streamwise derivatives one finds (taking $u_n=0$ in (23) after integration with respect to time

$$-\int_0^t \frac{\partial}{\partial y}(uv)dt = \frac{1}{2}\frac{\partial}{\partial y}[l^2\bar{U}'(y)] \qquad (32)$$

or

$$-\int_{-\infty}^{\infty}dx\int_{-\infty}^{\infty}dz\int_0^t(uv)dt = \frac{1}{2}\bar{U}'(y)\int_{-\infty}^{\infty}dx\int_{-\infty}^{\infty}l^2dz \qquad (33)$$

which differs from Prantdl's [12] famous formula

$$-\langle uv\rangle = l_m^2(\bar{U}'(y))^2 \qquad (34)$$

l_m being the root-mean square of l, which Prandtl obtained under the assumption that $v \approx -u$. For a two-dimensional flow one finds in a similar manner that

$$-\int_0^t \frac{\partial}{\partial y}(uv)dt = \frac{1}{2}l^2\bar{U}''(y) \tag{35}$$

which is equivalent to Taylor's [14] result obtained from the vorticity transport equations. On taking the streamwise average and noting from (18) that for large t the integrand in (33) may be taken to be independent of x and equal to $P_\infty = \int_{-\infty}^{\infty} Q_n dx$ over a distance equal to $|\bar{U}-\bar{U}_1|t$ one finds that

$$-\int_{-\infty}^{\infty}dx \int_{-\infty}^{\infty}dz \int_0^t (uv)dt =$$

$$\frac{1}{2}t\,\bar{U}'(y)\int_0^{\infty} H(y,y_1)P_{n\infty}(y_1;z)dy_1 \int_0^{\infty} H(y,y_2)P_{n\infty}(y_2;z)dy_2 \tag{36}$$

where, in the near-wall approximation,

$$H(y,y_1) = 0.5(|y-y_1|-y-y_1) \tag{37}$$

In this inviscid flow model the shear stress contributed by a single streak thus continues to grow linearly in time for all times, in consistency with the prediction of algebraic instability theory. That the linear inviscid theory gives a Reynolds stress that grows linearly with time was found also by Moffatt [15] in an analysis based on rapid distortion theory. In a real fluid visçous diffusion will eventually cause the streak to decay, but for a fluid of low viscosity this is a very slow process. More likely, the streak will be broken up by local instabilities after a finite life time. The long-term effects of viscosity and secondary instability will be discussed further below. For the total average Reynolds stress contributed by the near-wall streaks one thus has

$$-\langle uv\rangle = \frac{t_b}{2T_b}U'(y)\int_0^{\infty} dy_1 \int_0^{\infty} H(y,y_1)H(y,y_2)R(y_1,y_2)dy_2 \tag{38}$$

where t_b and T_b are, respectively, the the average lifetime of the streak and the time between bursts, and where one may approximate R with

$$R(y_1,y_2) = \langle P_{n\infty}(y_1,z)P_{n\infty}(y_1,z)\rangle \tag{39}$$

<> denoting ensemble average. It follows from (38) (39) that the Reynolds stress varies like y^2 for small y. This supports

the model proposed recently by Haritonidis [16]. For large y, $R(y_1,y)$ will become a function of y_1-y_2, only, and the average

Reynolds stress will then become proportional to y $\overline{U}'$. In the constant stress region this will lead to a logarithmic velocity profile.

Application of conditional sampling

For the study of coherent structures it is necessary to apply a suitable conditional sampling method in order to bring out the characteristic structural features of interest. Since the model evolution equations developed are linear, the averaging over the sample leads to a simple linear superposition of the individual events selected by the sampling criterion applied giving

$$<v> = -\frac{1}{4\pi}\int_{-\infty}^{\infty}dx_1\int_{-\infty}^{\infty}dz_1\int_{0}^{\infty}<\phi(\xi_1,y_1,z_1)>(1/R_1 - 1/\overline{R}_1)dy_1 \quad (40)$$

$$<\phi(\xi,y,z)> = \frac{1}{N}\sum_{i=1}^{N}\phi_n(\xi_n,y,z)\ , \quad N\rightarrow\infty \quad (41)$$

$$\xi = x - \overline{U}(y)(t-t_{event})\ ;\ \xi_1 = x - \overline{U}(y_1)(t-t_{event}) \quad (42)$$

from which the sampled fluid element liftup $<l>$ and hence the sampled streamwise velocity

$$<u> = <u_n(\xi_n,y,z)> - \overline{U}'(y)<l> \quad (43)$$

may be obtained.

A sampling method that has been used with some success in the literature is the variable interval time averaging (VITA) sampling criterion (Blackwelder & Kaplan, [17]), which sorts out events for which the local variance, evalulated for a selected time interval, T, exceeds a preset threshhold factor, k, times the mean square fluctuation level of the streamwise velocity component, i. e.,

$$\mathrm{var}(u,T) \equiv \frac{1}{T}\int_{t_e-T/2}^{t_e-T/2}u^2dt - [\frac{1}{T}\int_{t_e-T/2}^{t_e-T/2}udt]^2 > k\,u^2_{rms} \quad (44)$$

As pointed out by Johansson and Alfredsson [18] the integration time acts primarily as a filter so that small T tends to sort out events that are short-lived, whereas large T single out long-lived events. Since the measurements are carried out in a fixed laboratory frame of reference, a short-lived event has a corresponding short streamwise scale, and vice versa for the long-lived ones. The application of such a criterion supplies the instants of the events around which the sampling is done.

One has found that the sampled signal normalized with the root-mean-square fluctuation amplitude and with the square root of the threshhold k is approximately independent of the threshhold level. Landahl [19] has shown that this is consistent with a linear model.

Convected eddy

For an inviscid shear flow without inflection a linear disturbance will decay at a rate faster than t^{-1}. Hence, for long times, $p \rightarrow 0$ and the equation for u becomes simply

$$\frac{\bar{D}u}{Dt} = 0 \tag{45}$$

with the solution

$$u = u_\infty(\xi, y, z)\ ,\quad \xi = x - \bar{U}(y)t \tag{46}$$

where u_∞ is the final distribution of the u-perturbation velocity after the initial formation period with nonzero v has subsided. The result (46) is nothing but a statement of Taylor's "frozen field" hypothesis. It cannot hold forever in a flow with viscosity because

$$\frac{\partial u}{\partial y} = -t\bar{U}'(y)\ u_{\infty\xi} + u_{\infty y} \tag{47}$$

i.,e., shear intensifies linearly with time. The interaction with the mean shear will hence lead to the formation of an internal shear layer in regions where $u_{\infty\xi}<0$ which may be susceptible to local inflectional instability, as noticed by Gill [20]. However, for large times the viscous shearing stress will become important and the disturbance will be governed by the diffusion equation

$$\frac{\bar{D}u}{Dt} - \nu\nabla^2 u = 0 \tag{48}$$

It is possible to include the effects of a small viscosity in a simple approximate manner by setting

$$u = F(\xi, y, z, t) \tag{49}$$

which for large times gives

$$\nabla^2 u = t^2(\bar{U}')^2 F_{\xi\xi} + O(t) + O(1) \tag{50}$$

which, after application of the transformation

$$T = \nu t^2(\bar{U}')^2 t^3/3 \tag{51}$$

and neglect of the unimportant terms at large times, gives

$$F_T - F_{\xi\xi} = 0 \ , \ F(\xi,y,z,0) = u_\infty \tag{52}$$

This may be solved with the standard methods for the heat equation. For a disturbance in the form of a Gaussian 'hat',

$$u_\infty \sim \exp[\,-(\frac{\xi}{l_1})^2] \tag{53}$$

one finds

$$u \sim (l_1/L) \ \exp[-(\xi/L)^2] \tag{54}$$

where

$$L = [\, l_1{}^2 + 4\nu(\bar{U}')^2 t^3/3]^{1/2} \tag{55}$$

This result shows that the viscosity simply tends to make the disturbance grow in the streamwise direction and decay at the same rate so as to keep the total streamwise momentum defect constant. It follows from (55), (56) that the maximum shear occurs after a finite time given by

$$t = (3 \ l_1{}^2/2 \ \nu\bar{U}'^2)^{1/3} \tag{57}$$

For the linearly growing portion of the streak in which u eventually becomes independent of x this analysis will not hold and the variation of u with time will have to be determined from the full diffusion equation (48) with the term u_{yy} included.

Model for the nonlinear excitation

It follows from the above analysis of a convected eddy that, because of the spanwise stretching of the mean shear, local regions of high shear will always arise in a disturbed three-dimensional flow of low viscosity. Such regions can be expected to suffer from local instability of the inflectional type. Because the time scale of such an instability is inversely proportional to the shear, the instability will develop very quickly and thereby provide a local strong mixing that could act as the intermittent driving term in the analysis given above. Although there also could be other mechanisms giving locally a strong nonlinear activity, we will assume the dominating one to be local inflectional instability.

Provided the scale of the unstable motion is much smaller than the background large-scale motion which is causing it, one can use the classical hydrodynamic instability theory in conjunction of kinematic wave theory to study the exponential growth of a wave train, or wave group. One thus analyzes an instability of the form

$$u = \hat{u}(y)\, E \ , \ v = \hat{v}(y) E, \ \ w = \hat{w}(y)\, E \tag{58}$$

where

$$E = \exp[i\alpha(x-ct) + i\beta z]\ ,\ c = c_r + ic_i \tag{59}$$

(real part is implied) The complex amplitudes $\hat{u}$, $\hat{v}$, $\hat{w}$ could also be allowed to vary slowly in x,z-space and time.The disturbances develop on a parallel flow field of the form

$$U(y) = (U(y),W(y))\ \ U = U + \tilde{u},\ W(y) = \tilde{w} \tag{60}$$

the tilde symbol referring to the large-scale coherent motion which is assumed to vary slowly in x,z-space and time. Manipulation of he inviscid linearized disturbance equations in a manner similar to what was employed above in the calculation of the contribution to the Reynolds stress from a local disturbance structure gives:

$$-\frac{d}{dy}(\hat{u}\hat{v}^*) = A \sin \Theta_w + B \cos \Theta_w \tag{61}$$

$$-\frac{d}{dy}(\hat{w}\hat{v}^*) = -A \cos\Theta_w + B \sin \Theta_w \tag{62}$$

where

$$A = \frac{1}{2}\tan\,(\Theta_w - \Theta_u)\frac{d}{dy}[\tilde{U}U'(y)\frac{d}{dt}|\hat{l}|^2] \tag{63}$$

$$B = \frac{1}{2}\tilde{U}''\frac{d}{dt}|\hat{l}|^2 = \frac{\tilde{U}''|\hat{v}^2|c_i}{(\tilde{U}-c_r)^2 + c_i^2}\ ,\ \alpha\tilde{U} = \alpha U + \beta W \tag{64}$$

Here $\hat{l}(y)$ is the complex amplitude of the fluid element liftup. Highest growth rate is found for waves with fronts normal to the local flow direction. Hence, for the most unstable waves:

$$\Theta_w \approx \Theta_u\ ,\ A \approx 0 \tag{65}$$

or, for the total momentum transfer due to the instability

$$-\int^{t_n}\frac{d}{dy}(\hat{u}\hat{v}^*)Dt \approx \hat{B}\cos \Theta_w \tag{66}$$

$$-\int^{t_n}\frac{d}{dy}(\hat{v}\hat{w}^*)Dt \approx \hat{B}\sin \Theta_w \tag{67}$$

where

$$\hat{B} = U''\frac{1}{2}|\hat{l}|^2 \tag{68}$$

Accordingly, we choose for the conditionally sampled driving term <Q>:

$$\langle Q\rangle \equiv \frac{\partial^2}{\partial y^2}\left(\frac{\partial}{\partial x}\langle uv\rangle + \frac{\partial}{\partial z}\langle vw\rangle\right) = \frac{\partial^2}{\partial y^2}\left(\frac{\partial \hat{B}}{\partial x}\cos\Theta_w + \frac{\partial \hat{B}}{\partial z}\sin\Theta_w\right) \tag{69}$$

and for $\hat{B}$ a modified Gaussian "hat",

$$\hat{B} = \hat{C}\, y^3\left[1 - 2\left(\frac{z}{l_3}\right)^2\right] \exp\left[-\left(\frac{x}{l_1}\right)^2 - \left(\frac{y}{l_2}\right)^2 - \left(\frac{z}{l_3}\right)^2\right] \tag{70}$$

$\hat{C}$ being an arbitrary constant. The factor $1 - 2\left(\frac{z}{l_3}\right)^2$ is included to make $\int_{-\infty}^{\infty}\hat{B}dz = 0$, which is necessary in order for the spanwise velocity perturbation to vanish at $z=\pm\infty$. The resulting structure may be subdivided into a symmetrical one, denoted by subscript s, multiplying $\cos\Theta_w$ in (69), and an antisymmetrical one, denoted by subscript a, multiplying $\sin\Theta_w$. Both of these may be determined from a single base solution, denoted by subscript b, with initial conditions determined by of an intial stream function Ψ_0 for the initial cross-flow velocity field of

$$\Psi_0 = -\frac{1}{2}\hat{C}\, l_3^2 y^3 \exp\left[-\left(\frac{x}{l_1}\right)^2 - \left(\frac{y}{l_2}\right)^2 - \left(\frac{z}{l_3}\right)^2\right] \tag{71}$$

such that

$$(v_b)_{t=0} = -\frac{\partial \Psi_0}{\partial z}, \quad (w_b)_{t=0} = \frac{\partial \Psi_0}{\partial y} \tag{72}$$

The symmetrical and antisymmetrical portions are then found from

$$v_s = \frac{\partial v_b}{\partial x}\cos\Theta_w \quad v_a = \frac{\partial v_b}{\partial z}\sin\Theta_w \tag{73}$$

and similarly for the other components. Since, following (73),

$$\int_{-\infty}^{\infty} v_s(x_1, y, z, 0)\, dx_1 = 0,$$

only the antisymmetrical portion will be responsible for the streak formation for large times.

Results for the model structures and comparisons with numerical simulations

For the calculation of the evolution of the model streaky structures the mean velocity and its first and second derivatives are needed, for this the expression proposed by Reichardt [21] was used. The initial streamwise velocity component for the disturbance was taken as zero on the assumption that the local instability on the average creates just enough Reynolds stress to remove the disturbance producing the instability [6]. The scale factors l_1, l_2, and l_3, were chosen such as to correspond approximately to the conditions in the investigation of Alfredsson et al. [8]. In their conditional sampling of the numerically generated data for the streamwise fluctuation component they used the VISA ('Variable Intregral Space Averaging') with an integration length of $L^+=$ 200, which was found to corresponds approximately to $T^+=10$ in the VITA studies of Johansson &Alfredsson [18]. Accordingly, in the theoretical model the streamwise scale l_1 was selected such as to make the variance maximum, for a given value of the mean square of the streamwise disturbance velocity component, for that value of the integration time (see [19]). This gave a value for l_1 of approximately 100 in viscous wall units, which was hence adopted. The vertical scale l_2^+ in the model (98) was chosen as 15 , which was found to give a y-distribution of the Reynolds stress in reasonable agreement with the measured Reynolds stresses during bursting found by Kim et al. [22]. Finally, $l_3^+ = 30$ was selected so as to give a spanwise scale of the u-signature close to what was obtained in the conditional sampling results of Alfredsson et al. [8].

The numerical quadrature of the appearing integrals was carried out with the aid of a cubic spline fit routine. Satisfactory accuracy was achieved by representing the integrands in 21 equidistant points between $x^+=-200$ and $x^+= 200$ and 21 points between $y^+=0$ and $y^+=50$. All the results presented were obtained from the near-wall approximation (56). In that z appears solely as a parameter so that only one x,y-plane for each of the symmetric and antisymmetrical distributions needed to be evaluated, the variation with z determined simply through multiplication with the appropriate functions of z. The step-by-step integration in time was carried out with a step size of $\Delta T^+=0.5$. Some test calculations with half this step size produced insignificant differences showing that the second order explicit method developed for the time integration gave satisfactory convergence for this step size.

In Figures 1 - 6 are presented results for the evolution of both symmetrical structures , and for structures having a small amount of asymmetry. Figure 1 illustrates the early development, at $t^+=5$, of the symmetrical structure. The vertical component, Figure 1a) has a region of positive values downstream from one of negative ones. This is im consistency with the sign of the spamwise moment of momentum component produced by a positive Reynolds stress patch, as shown by the

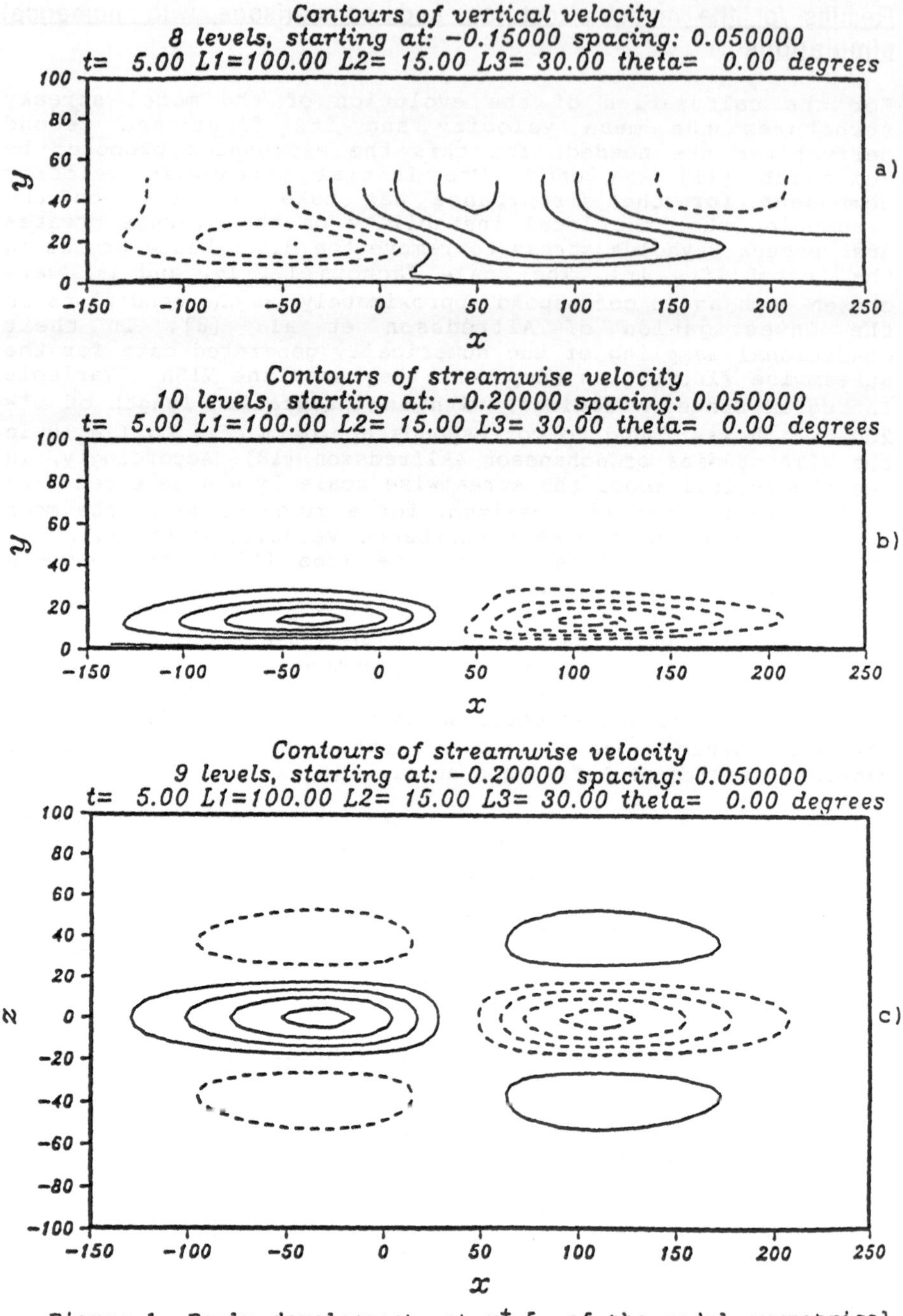

Figure 1. Early development, at $y^+=5$, of the model symmetrical structure: a) plane $z^+=0$, vertical perturbation velocity component, v ; b) streamwise component, u ; c) streamwise component in the plane $y^+=15$.

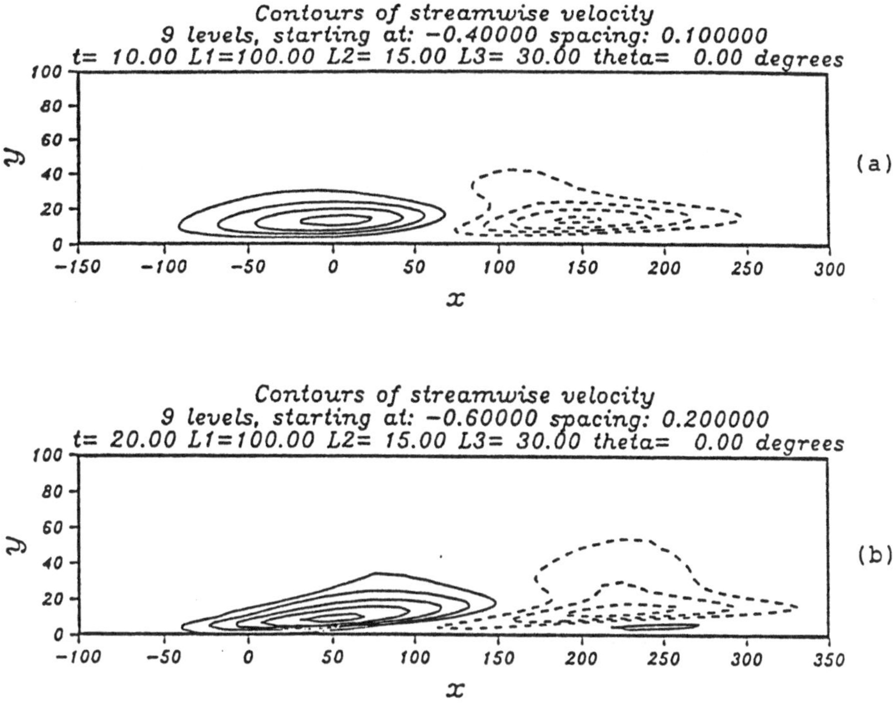

Figure 2. Contours of constant u in the plane $y^+=15$;

a) $t^+= 10$; b) $t^+= 20$.

analysis of [4]. The streamwise component shows coorespondingly a low-speed region downstream of a high-speed one. The low-speed region results from the lift-up of fluid elements ('ejection') and the high-speed region from inflow ('sweep') towards the wall. Figure 2 illustrtes the development of a shear layer. As time increase, the shear layer intensifies as seen by the increasing closeness of the lines of constant u near the interface between the positive and negative regions. Note that for this plane only the symmetrical part of the solution gives a contribution since the antisymmetrical one is zero for $z^+= 0$.

The long-time evolution for $t^+=20$ - 80 of the symmetrical and antisymmetrical structures is illustrated in Figures 3-6 giving the u-contours at the plane $y^+=15$ for the two kinds of disturbances. The elongation of the asymmetrical structure for $t^+>20$ is clearly evident. Also, for very large times (Figures 5 and 6) the asymmetrical structures have developed an irregular wavy appearance very reminiscent of the long structures found

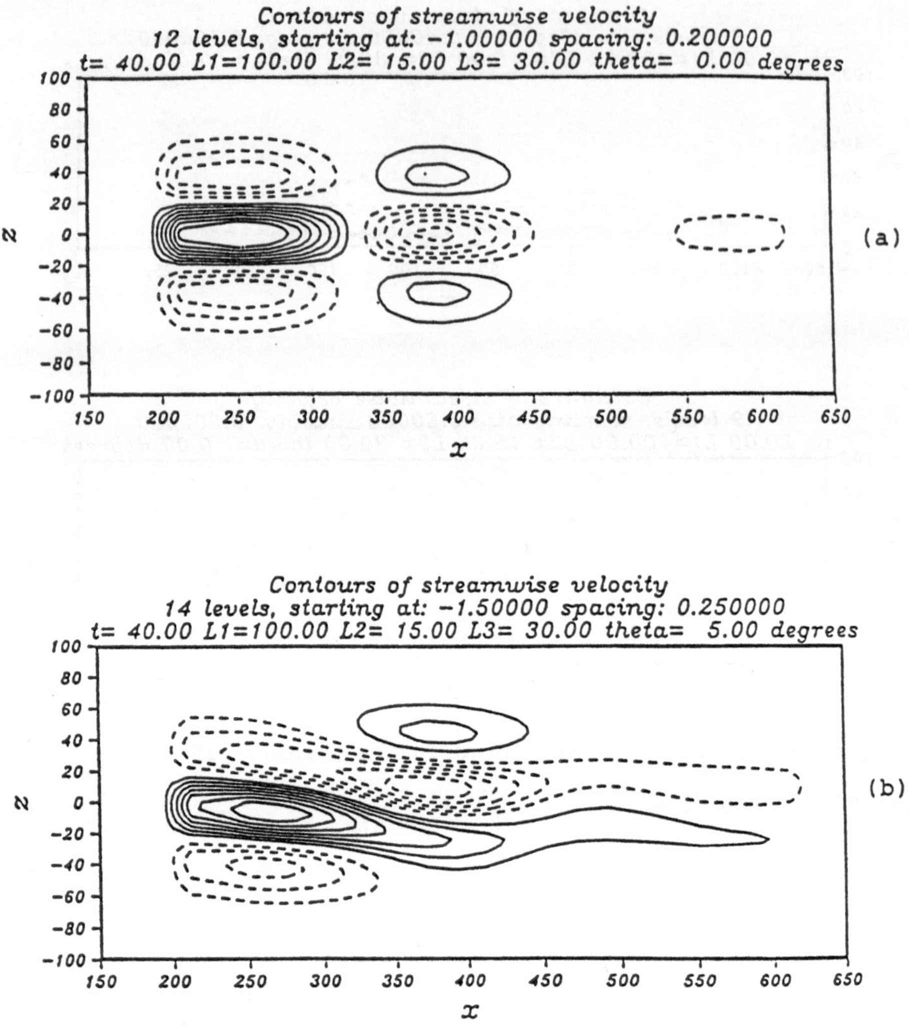

Figure 3. Contours of constant u in the plane $y^+=15$, $t^+= 40$: a) symmetrical structure; b) asymmetrical structure, $\theta=5^\circ$.

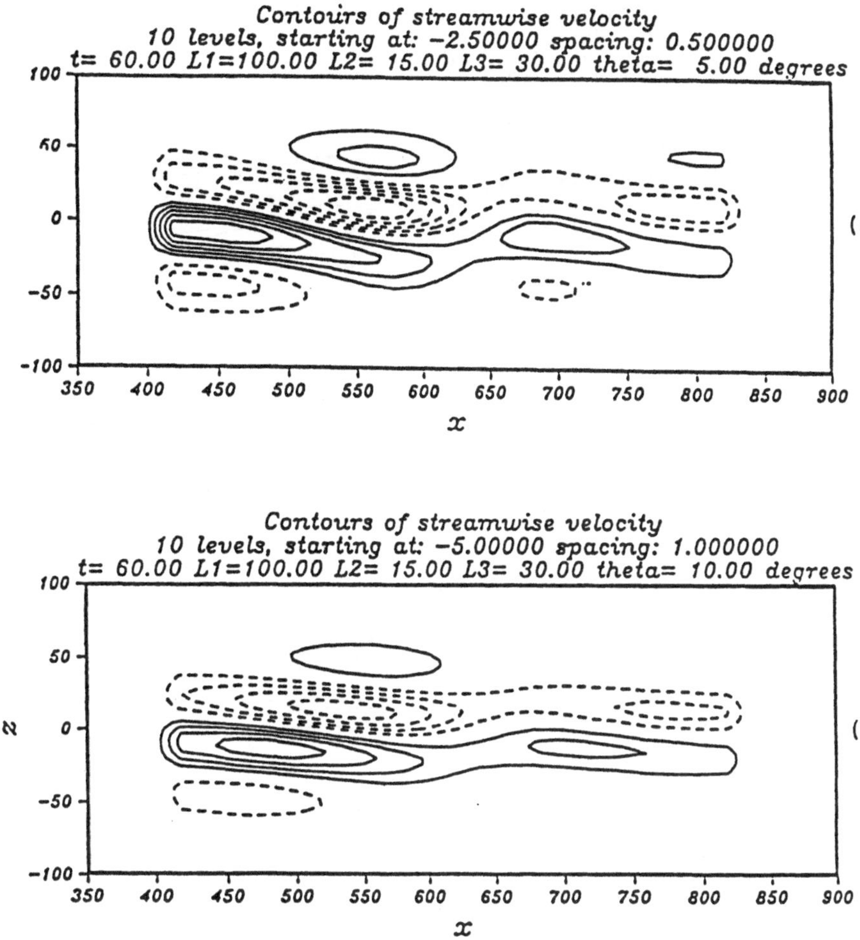

(a)

(b)

Figure 4. Contours of constant u in the plane $y^+=15$, $t^+=60$: a) asymmetrical structure, $\theta=5^o$; b) asymmetrical structure, $\theta=10^o$.

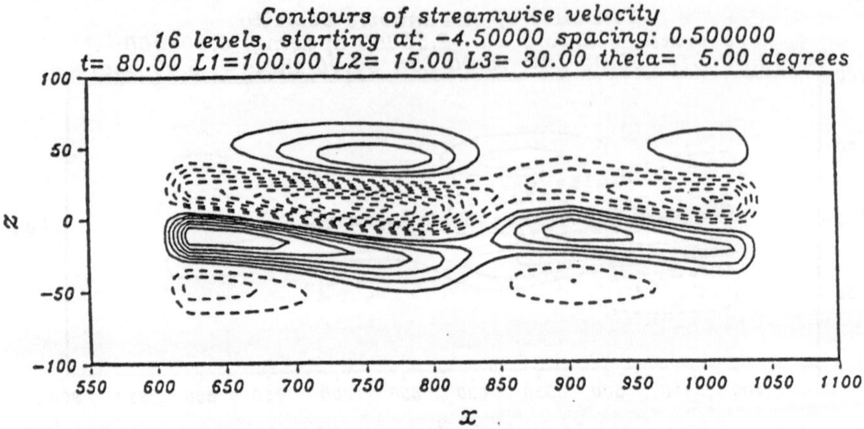

(a)

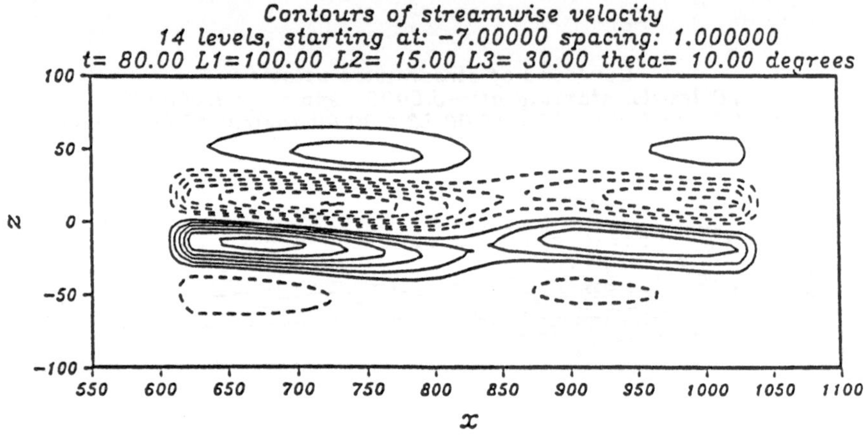

(b)

Figure 5. Contours of constant u in the plane $y^+=15$, $t^+=80$: a) asymmetrical structure, $\theta=5^o$; b) asymmetrical structure, $\theta=10^o$.

near the wall in the numerical simulations of Moin and Kim [11]. Since these structures are advected downstream with the mean velocity, in a laboratory frame of reference they will appear as oscillations. This result is also in accordance with the experiments of Kline et al.[7] in which it was found that the streaks began to oscillate before they broke up. It is primarily the symmetrical portion of the solution that contributes to the wavy appearance as is clear from the comparison of the results for two values of obliquity, $\theta=5^o$ and $\theta=10^o$. It is seen that the results for the higher obliquity, $\theta=10^o$, show less waviness.

The symmetrical structures show little, if any, streamwise growth; in contrast, for large times they show a breakup into smaller cells. This behavior is evident by comparing the results for $t^+=20$ and $t^+=40$, (see Figures 5 and 6) and has the character of a breakup of the structure into cells of about half the original streamwise scale with the cells weakening with increasing distance downstream.

The results presented may be compared with those obtained from numerical simulation of turbulent wall-bounded flows. In Figure 6 is reproduced the instantaneous streamwise fluctuation velocity contours for the plane $y+=6.14$ obtained by Moin and Kim [11] in their Navier-Stokes simulation of turbulent channel flow. The width of the spanwise section shown is approximately 600 in viscous wall units. The spanwise alternation of high- and low-speed elongated regions is clearly evident. Particularly interesting is the wavy appearance of the longer streaks, also exhibited by the model results for the older structures. This behavior may be a consequence of the excitation of the damped waves and points to a possible explanation for the streak oscillations seen in the visualization experiments discussed above.

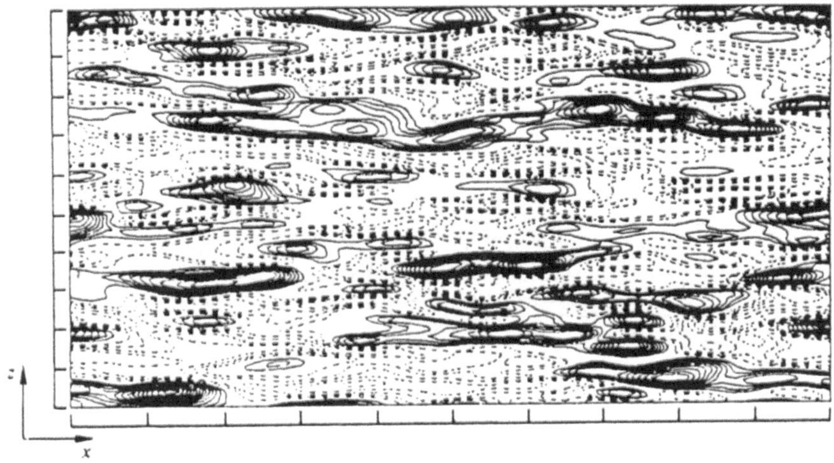

Figure 6. Contours of fluctuating streamwise velocities in the plane $y^+=6.14$ obtained by numerical simulation for a channel flow. From Moin and Kim [11]

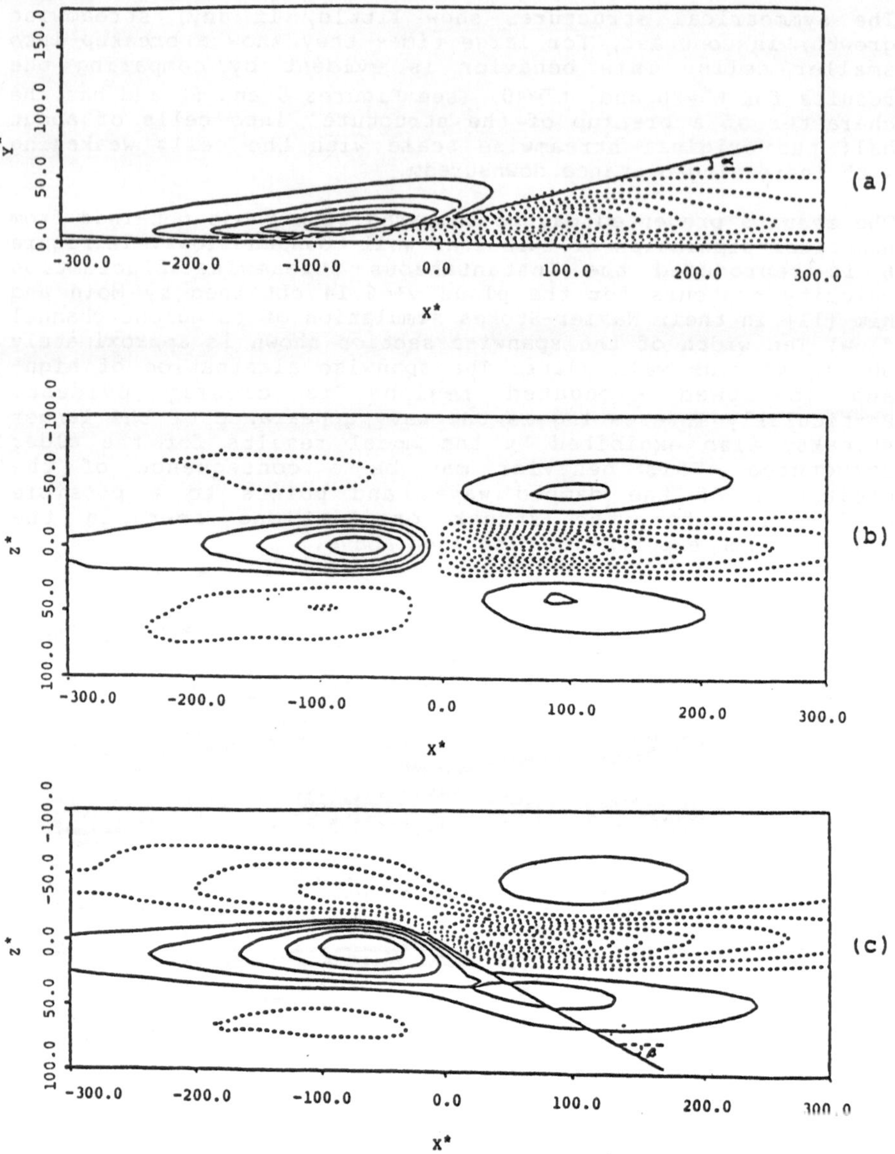

Figure 7. Contours of constant u for VISA-educed structures obtained by Johansson and Alfredsson [23] from numerically simulated turbulent channel flow using the NASA Ames laboratory data bases: a) detection x,y-plane, symmetrical structure; b) plane $y^+=15$, symmetrical structure; c) $y^+=15$, asymmetrical structure.

The NASA Ames numerical data bases were also used in the recent investigations by Alfredsson et al.[8] and Alfredsson and Johansson [23] to study the near-wall turbulence structure with the use of the VISA conditional sampling technique (the spatial counterpart of the VITA). The straight-forward application of any one-point detection scheme, such as the VITA, will sort out only structures with spanwise symmetry because of the statistical homogeneity in z of the fluctuating field. However, simple theoretical arguments [4] indicate that asymmetrical structures may be important. In order to study asymmetrical structures, Alfredsson et al. [8] devised a special scheme, in which the individual structures were switched with respect to the x,y-mid-plane according to the sign of the spanwise derivative of u at the detection point. The resulting u-pattern in the y+=15 plane (taken from [23]) is shown in Figure 7c) at the time of detection of the structure. The corresponding symmetrical structure is illustrated in Figure 7 a),b). The point x+=0 in their diagram is located at the point of detection, which is approximately at the point where $\partial u/\partial x$ is maximum. The contours of u from the model for an asymmetric structure with $\theta = 5^{o}$ show a remarkable similarity with those obtained from the conditionally sampled numerical data. The model symmetrical structure ($\theta=0^{o}$) results are also in good qualitative agreement with these data.

The model structures do show some differences in details from those exhibited by the numerical data, however. The model results show a much more abrupt onset of the disturbed region than what the numerically generated ones do. This may possibly be a consequence of the extreme temporal intermittency assumed in the idealized model, with the Dirac delta function behavior of the nonlinear source term. Also, the sample selected for the construction of the average structure include disturbances of all different obliquities, which would produce a smearing effect on the conditionally averaged structure. Such a smearing will not effect the model structures which are calculated for one particular oblique angle.

Conclusions

The major conclusion that can be drawn from the simple flow model considered here is that instability mechanisms do play an essential role for the creation and maintenance of the primary stress-carrying coherent structures in the wall region, namely the streaks, and that the one controlling the flow is the algebraic one. Practically all of the emphasis in the literature on hydrodynamic instability has in the past been on Tollmien-Sclichting type waves and their nonlinear interaction. This consequently misses the algebraic instability, which corresponds to the spectral component with a zero streamwise wave number. In the present model the fluctuating flow field is assumed to be governed primarily by the linear interaction with the mean shear, the nonlinearity acting only during a vanishingly short time to supply the initial conditions for the subsequent linear evolution. The justification for such a radical idealization is the observed high intermittency of the v-fluctuations, as reflected in the very high values of the flatness found for this component in the near-wall region, and which is also seen in the bursting phenomenon. It is significant that only the v-component shows this intermittency. In contrast, the u-component evolves over a protracted time producing continuously growing streaky regions of low and high speed flow.

The formation of the streaky structure can be shown to be a direct consequence of algebraic instability [1] which would occur in any strong shear flow. Streaks will grow from any initial local and three-dimensional disturbances having a nonzero net vertical momentum along lines of constant y,z. As suggested in an earlier paper [4], such a disturbance may be initiated by a local inflectional instability region with spanwise asymmetry, as would be the case for an oblique wave-like disturbance, perhaps resulting from a cross-flow instability induced by large-scale spanwise motion.

The Reynolds stress contributed by the streaks may be expressed in a simple manner in terms of the mean square of the fluid element liftup, as in Prandtl's mixing-length model. However, the present model gives a different multiplication factor, $(1/2T_b)(\partial U/\partial y)$, T_b being the average time between bursts, than the factor $(\partial U/\partial y)^2$ proposed by Prandtl [12]. The model also suggests that the streaks account for the major part of the Reynolds stresses, and hence for the turbulence production, in the near-wall region.

The structures obtained from the model show a remarkable similarity with those seen in visualization experiments, and with those found from numerical simulations. Thus, the continuous linear growth predicted by the present linear inviscid model is reflected in the longevity of the streaks , and their large eventual streamwise growth, found both in laboratory and numerical experiments. Also, the irregular wavy appearance of the older streaky structures seen in the results from the theoretical model, as well as in the numerical simulations, is strongly suggestive of the oscillations that Kline et al. [7] observed to occur just before the streak breakup.

The present model gives certain interesting hints as to how the skin friction reduction methods that have been tried with some success may work. The two key instability mechanisms singled out in the model are the local inflectional instability of the internal shear layers formed as the near-wall coherent structure is convected downstream and the algebraic instability mechanism leading to the subsequent creation and evolution of a streaky structure. With the use of simplified constitutive models for the fluid modelled as a suspension of aligned rigid rods it was found that drag reducing polymer additives do have an effect on inflectional instability; they reduce the growth rate of unswept instability waves [25] but may in fact destabilize swept three-dimensional waves (Tinoco & Bark,[26]). However, a more likely stabilizing effect of the drag reducing additives is that they restrain, through the hightened extensional viscosity they produce, the stretching of spanwise vorticity and hence reduce the rate at which internal shear layers form [24], [27]. Riblets may act in a similar manner.

A particularly interesting new finding is that spanwise asymmetry is necessary for the streaks to form. The importance of asymmetry in the excitation of fluid element liftup was noticed already in [4], and the strong and persistent production of Reynolds stresses in such structures was seen in the study by Alfredsson et al. [8]. One may conclude from these and the present findings that alignment of the structures in the streamwise direction through, e.g., the use of riblets of the correct dimensions, is likely to help to make the emerging structures have less spanwise asymmetry and hence reduce the algebraic instability and the production of Reynolds stress.

Obviously, the model presented has a bearing also on the transition to turbulence. For sufficiently strong initial three-dimensional disturbances, the interaction with the mean shear flow will produce internal shear layers that may intensify and break down due to inflectional instability long before the Tollmien-Schlichting waves will have had time to grow to any substantial amplitude. This nonlinear bypass mechanism has been explored in a recent Ph.D. thesis by Breuer [28].

Acknowledgements

Partial support for this research from the Aeronautical Research Institute of Sweden and from NASA Lewis Research Center under Grant No NCC3-177 is gratefully acknowleged.

References

1. Landahl, M.T. "A note on algebraic instability of inviscid parallel shear flows." J. Fluid Mech. 98, p. 243, 1980.

2. Landahl, M.T. "A wave-guide model for turbulent shear flow." J. Fluid Mech 27, p.443, 1967.

3. Bark, F.H. "On the wave structure of the wall region of a turbulent boundary layer." J. Fluid Mech. 70, p.229, 1975.

4. Gaster, M., Kit, E., and Wygnanski, I. "Large-scale structures in a forced turbulent mixing layer." J. Fluid Mech. 150, p.23, 1983.

5. Landahl, M.T. "Wave breakdown and turbulence." SIAM J. Applied Math. 28, p.775, 1975.

6. Landahl, M.T. "Dynamics of Boundary Layer Turbulence and the Mechanism of Drag Reduction." Phys. Fluids 20(10), Part II, p.S55, 1977.

7. Kline, S.J., Reynolds, W.C., Schraub, F.A. & Runstadler, P.W. "The structure of turbulent boundary layers." J. Fluid Mech. 50, p.133, 1967.

8 Alfredsson, P.H., Johansson, A.V. & Kim, J. "Turbulence production near walls, the role of flow structures with spanwise asymmetry." Proceedings, 2d Summer Program of the Center for Turbulence Research. p.131, NASA Ames/Stanford University, 1988.

9. Jang, P.S., Benney, D.J. & Gran, R.L. "On the origin of streamwise vortices in a turbulent boundary layer." J. Fluid Mech. 169, p.109, 1986.

10. Hatziavramidis, D.T. and Hanratty, T.J. The representation of the viscous wall region by a regular eddy pattern. J. Fluid Mech. 95, p.655, 1979.

11. Moin, P. & Kim, J. "Numerical investigation of turbulent channel flow." J. Fluid Mech. 118, p.341, 1982.

12. Prandtl, L. "Bericht uber Untersuchungen zur ausgebildeten Turbulenz." Z. angew. Math. Mech. 5, p.136, 1925.

13. Ashley, H. & Landahl, M.T. Aerodynamics of Wings and Bodies. Addison-Wesley, 1965.

14. Taylor, G.I. "Eddy motion in the atmosphere." Phils. Trans. R. Soc. London, Ser. A 215, p.1, 1915.

15. Moffatt, K. "The interaction of turbulence with strong wind shear." In Proceedings of the URSI-IUGG International Colloquium on Atmospheric Turbulence and Radio Wave Propagation. eds. Yaglom A.M. & Tatarsky, V.I., Nauka, Moscow, p.139, 1965.

16.Haritonidis, J. "A model for near-wall turbulence." Phys. Fluids A,1, p.302, 1989.

17. Blackwelder, R.F. & Kaplan, R.E. "On the wall structure of the turbulent boundary layer." J. Fluid Mech. 76, p.89, 1967.

18. Johansson, A.J & Alfredsson, P.H. "On the structure of turbulent channel flow." J. Fluid Mech. 122 , p.295, 1982.

19. Landahl, M.T. "On the dynamics of large eddies in the wall region of a turbulent boundary layer." in Turbulence and Chaotic Phenomena in Fluids (ed. T. Tatsumi), Elsevier, p.467, 1984.

20. Gill, A.E. "A mechanism for instability of plane Couette flow and of Poiseuille flow in a pipe." J. Fluid Mech. 21, p.503, 1965.

21. Reichardt, H. "Vollständige Darstellung der turbulenten Geschwindigkeitsverteilung in glatten Leitungen" Z.A.M.M., 31, p.208, 1951.

22. Kim, H.T., Kline, S.J. & Reynolds, W.C. "The production of turbulence near a smooth wall in a turbulent boundary layer." J. Fluid Mech. 50, 133, 1971.

23. Alfredsson, P.H. & Johansson, A.V., "Turbulence experiments - Instrumentation and processing of data." in Proceedings of the 2d European Turbulence Conference, Sept. 1988. eds. H. Fiedler & H. Fernholz, Springer, 1988.

24. Landahl, M.T. "Effects of Additives on Turbulent Bursting Dynamics." in Proceedings: Progress in Aeronautics and Astronautics 72, American Institute of Aeronautics and Astronautics, p. 300, 1980.

25. Landahl, M.T. "Drag reduction by polymer addition." Proceeding, XIII IUTAM Congress, eds. E. Becker & G.K. Mikhailov. Springer Verlag, p.177, 1973.

26. Tinoco, H. and Bark, F.H. "Inflectional instability of some parallel flows of a dilute suspension of fibres." KTH-TRITA-MEK 80-9, The Royal Institute of Technology, Stockholm, 1980.

27. Landahl, M.T. & Henningson, D. S. "The Effects of Drag Reduction Measures on Boundary Layer Turbulence Structure - Implications of an Inviscid Model." AIAA-85-0560, 1985.

28. Breuer, K.S. "The development of a localized disturbance in a boundary layer." Mass. Inst. Tech. FDRL Report 88-1, 1988.

Effect of Three-dimensional Surface Elements on Boundary Layer Flow

by L. Håkan Gustavsson and Stefan Wallin*
Division of Fluid Mechanics
Luleå University of Technology
S-95187 Luleå, Sweden

ABSTRACT

The perturbation velocity field induced by a three-dimensional surface distorsion in a boundary layer flow is considered. For small amplitudes, the kinetic energy is shown to be composed of two factors; one associated with the surface structure and the other with the velocity profile. Level curves of the profile factor, in the (α,β) wave-number plane, are ridge-like and approach the β-axis as the Reynolds number increases. Thus, in the inviscid limit, the kinetic energy is confined to structures infinitely extended in the streamwise direction. For a certain class of surface structures, also the level curves for the kinetic energy have been determined. It is shown how a spanwise modulation and an aspect ratio of the surface distorsion change the position of the level curves and the amplitudes.

1. INTRODUCTION

When predicting the drag reducing capacity of a given surface structure only limited guidance is provided by boundary layer theory. This is due to limits in the basic understanding of the mechanism(s) responsible for the creation and maintenance of turbulent flow. The relative success of riblets for drag reduction (Walsh, 1983) indicates a coupling to sublayer streaks and therefore the calculations of Bartenwerfer & Bechert (1987) is an interesting approach to account for the effects on the flow by this type of surface structures. Since natural surfaces, such as shark skin, may exhibit intriguing three-dimensionality on a micro-level (Raschi & Musick, 1986), and only in a cooperative effect are riblet-like, it seems worthwhile also to investigate the effect on the flow by 3-D surface structures. The experiments by Acarlar & Smith (1987) have shown how hairpin vortices can form behind surface elements and it is possible that 3-D structures may affect the hairpin vortices observed in the breakup of the sublayer streaks.

From a theoretical viewpoint it is not obvious, however, what aspects of the interaction between the surface and the flow should be considered. Increased stability, and thereby delayed transition, influence on streak generation and instability, are but a few possible points for study, in addition to the obvious problem of vorticity generation.

In the present paper the question of how single surface elements change a boundary layer flow will be addressed and, in particular, the kinetic energy of the perturbation field will be considered. The approach is rather basic since only small amplitude effects are considered. The results should therefore be viewed as indications on the sensitivity of the flow rather than accurate predicitions of the flow behaviour.

---

* Present address: Volvo Flygmotor AB, Trollhättan, Sweden

A. Gyr (Editor)
Structure of Turbulence and Drag Reduction
IUTAM Symposium Zurich/Switzerland 1989

2. ANALYSIS

2.1 Assumptions and basic equations

Consider the steady laminar boundary layer flow over a surface distorsion defined by $y=y_B(x,z)$, where x and z are in the streamwise and the spanwise directions, respectively. The velocity profile for the corresponding flat surface is given by U(y). The surface structure is assumed to have a small amplitude so that $y_B(x,z) = \varepsilon A(x,z)$, where $\varepsilon \ll 1$ and $A = O(1)$.
The basic equations governing the flow field are scaled with the free stream velocity and the boundary layer thickness and become

$$(\mathbf{u}\cdot\nabla)\,\mathbf{u} = -\nabla p + \frac{1}{R}\nabla^2\mathbf{u} \tag{1}$$

for momentum, and

$$\nabla\cdot\mathbf{u} = 0 \tag{2}$$

for continuity. R is the Reynolds number.

Since the amplitude of the surface structure is small, the induced velocity perturbations are also assumed to be small. Therefore, expansions for the velocity components and the pressure field are sought of the form

$$u = U + \varepsilon u_1 + \dots\ ;\ v = \varepsilon v_1 + \dots\ ;\ w = \varepsilon w_1 + \dots\ ;\ p = P + \varepsilon p_1 + \dots\ , \tag{3a-d}$$

where capital letters refer to the flat surface case. Substitution of (3a-d) into (1) and (2), subtraction of the equations for the unperturbed flow, followed by linearization and Fourier-transformation (^) in x and z, lead to the following equations for the Fourier-transforms of v_1 and the vertical vorticity perturbation ($\omega_1 = \frac{\partial u_1}{\partial z} - \frac{\partial w_1}{\partial x}$):

$$(D^2-k^2)^2\,\hat{v}_1 - i\alpha R[\,U(D^2-k^2)\hat{v}_1 - U''\hat{v}_1] = 0\ , \tag{4}$$

and

$$(D^2-k^2)\,\hat{\omega}_1 - i\alpha R U\,\hat{\omega}_1 = i\beta R U'\hat{v}_1 \tag{5}$$

where α and β are the transform variables, $k^2=\alpha^2+\beta^2$, $D=d/dy$ and $U'=dU/dy$.

2.2 Boundary conditions

On the surface the perturbation velocities are zero and, since y_B is small, a Taylor expansion around y=0 gives

$$\hat{v}_1(0) = 0\quad ,\quad D\hat{v}_1(0) = i\alpha U'(0)\hat{A}(\alpha,\beta)\quad ,\quad \hat{\omega}_1(0) = -i\beta U'(0)\hat{A}(\alpha,\beta) \tag{6a-c}$$

Far from the surface the perturbation field is assumed to decay so that

$$\hat{v}_1 \to 0 \;,\; D\hat{v}_1 \to 0 \;,\; \hat{\omega}_1 \to 0 \qquad \text{as} \qquad y \to \infty \tag{7a-c}$$

2.3 Formal solution for $\hat{v}_1$

The solution for $\hat{v}_1$ can be expressed in terms of the four linearly independent solutions to (4). These are denoted by $\phi_1, \phi_2, \phi_3, \phi_4$ and thus satisfy

$$\phi^{(4)} - (k^2 + Q)\phi'' + (k^2Q + Q'')\phi = 0 \quad , \tag{8}$$

where

$$Q = k^2 + i\alpha RU. \tag{9}$$

In the free stream, (8) can be solved exactly but only two functions are decaying as $y \to \infty$. These are ordered such that $\phi_1 \sim e^{-ky}$ and $\phi_2 \sim e^{-\mu y}$ as $y \to \infty$, where $\mu^2 = k^2 + i\alpha R$. Then the solution for $\hat{v}_1$ becomes

$$\hat{v}_1 = i\alpha U'(0)\hat{A}(\alpha,\beta)\frac{\phi_1\phi_{20} - \phi_2\phi_{10}}{\phi'_{10}\phi_{20} - \phi_{10}\phi'_{20}} \tag{10}$$

where the second index in ϕ denotes values at y=0.

2.4 Formal solution for $\hat{\omega}_1$

Knowing $\hat{v}_1$, a formal solution for $\hat{\omega}_1$ can be obtained by solving (5). The solution is constructed from the linearly independent solutions of

$$\psi'' - Q\psi = 0 \tag{11}$$

which are labeled such that $\psi_1 \sim e^{-\mu y}$ and $\psi_2 \sim e^{\mu y}$ in the free stream. Using standard techniques, the solution for $\hat{\omega}_1$ becomes

$$\hat{\omega}_1 = \frac{i\beta R}{2\mu}\left[\psi_1\frac{\psi_{20}}{\psi_{10}}\int_0^\infty U'\hat{v}_1\psi_1 d\eta - \psi_1\int_0^y U'\hat{v}_1\psi_2 d\eta - \psi_2\int_y^\infty U'\hat{v}_1\psi_1 d\eta\right] - \frac{i\beta\hat{A}U'(0)}{\psi_{10}} \tag{12}$$

from which follows that $\hat{\omega}_1$ is non-zero only if the surface structure has a spanwise variation.

2.5 Solution for $U(y)=1-e^{-y}$

The evaluation of $\hat{v}_1$ and $\hat{\omega}_1$ must in general be done numerically. However, it is realized that, for any form of U(y), equations (8) and (11) can be combined to give

$$Q''\psi\phi = (\psi'\phi'' - \psi\phi''')' + k^2(\psi\phi' - \psi'\phi)'. \tag{13}$$

Using the definition of Q in (13), it follows that, if U' is proportional to U", the integrals appearing in (12) can be evaluated analytically. This is the case for $U(y)=1-e^{-y}$ and the expression for $\hat{\omega}_1$ then becomes

$$\hat{\omega}_1 = i\beta\hat{A}(\alpha,\beta)\left[\frac{(\phi_1''-k^2\phi_1)\,\phi_{20}-(\phi_2''-k^2\phi_2)\,\phi_{10}}{\phi_{10}'\phi_{20}-\phi_{10}\,\phi_{20}'} - \frac{\psi_1}{\psi_{10}}\left(1+\frac{\phi_{10}''\phi_{20}-\phi_{10}\,\phi_{20}''}{\phi_{10}'\phi_{20}-\phi_{10}\phi_{20}'}\right)\right] \tag{14}$$

2.6 The kinetic energy of the perturbation field

The expressions for $\hat{v}_1$ and $\hat{\omega}_1$ will be used to determine the kinetic energy of the perturbation field which, to $O(\varepsilon^2)$, is

$$T=\frac{1}{2}\int(u_1{}^2+v_1{}^2+w_1{}^2)dV\,, \tag{15}$$

where the integral is taken over the region in (x,y,z)-space occupied by the disturbance. T may as well be evaluated in the (α,y,β) space, by the use of Parsevals theorem. Substituting u_1 and w_1 via continuity and the definition of ω_1, leads to

$$T=\frac{1}{2}\int_\alpha\int_y\int_\beta\frac{1}{k^2}\left(|\hat{\omega}_1|^2+k^2\,|\hat{v}_1|^2+|D\hat{v}_1|^2\right)d\alpha\,dy\,d\beta \tag{16}$$

which, after integration in the y-direction, using the expressions for $\hat{v}_1$ and $\hat{\omega}_1$ derived above, can be written as

$$T=\frac{1}{2}\int_\alpha\int_\beta|\hat{A}(\alpha,\beta)|^2\;F(\alpha,\beta,R)\,d\alpha\,d\beta \tag{17}$$

where

$$F(\alpha,\beta,R) = (\alpha^2 I_v + \beta^2 I_\omega)/k^2. \tag{18}$$

The factors I_v and I_ω are the integrals

$$I_v=\int_0^\infty\frac{k^2|\phi_1\phi_{20}-\phi_2\phi_{10}|^2+|\phi_1'\phi_{20}-\phi_2'\phi_{10}|^2}{|\phi_{10}'\phi_{20}-\phi_{10}\,\phi_{20}'|^2}dy \tag{19}$$

and

$$I_\omega = \int_0^\infty \left| \frac{(\phi_1'' - k^2\phi_1)\,\phi_{20} - (\phi_2'' - k^2\phi_2)\,\phi_{10}}{\phi_{10}'\phi_{20} - \phi_{10}\phi_{20}'} - \frac{\psi_1}{\psi_{10}}\left(1 + \frac{\phi_{10}''\phi_{20} - \phi_{10}\phi_{20}''}{\phi_{10}'\phi_{20} - \phi_{10}\phi_{20}'}\right) \right|^2 dy \tag{20}$$

T is thus seen to consist of two factors; one due to the shape of the surface element and the other due to the velocity profile. The profile factor is seen to consist of a two-dimensional part, associated with v_1, and an oblique part, associated with ω_1.

3. NUMERICAL TECHNIQUES

The special choice of U(y) allows series expansions to be used in the calculations of ϕ and ψ. Introducing $q=e^{-y}$, series solutions of the type $q^\sigma \Sigma a_n q^n$ were sought where the coefficients a_n follow simple recursion relationships. These were used to estimate the number of terms in the series required to obtain accurate results. The series were then summed using Horner's scheme, and the integrals evaluated with standard quadrature techniques.

4. RESULTS

The contour curves F= constant were first determined at two Reynolds numbers, R=200 and R=1000, and the results are shown in Figures 1a,b.

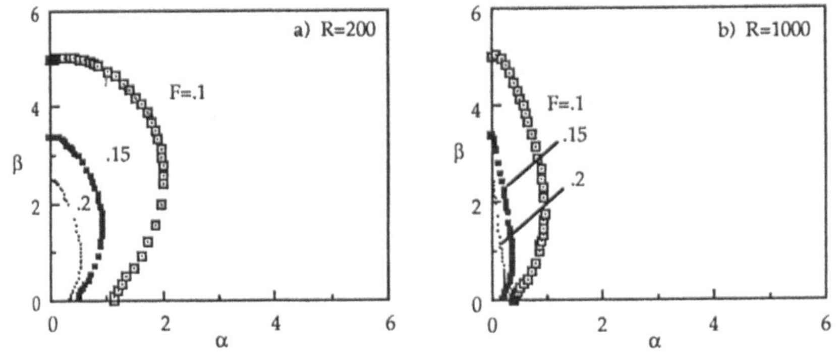

Figure 1: Level curves of F in wave-number plane: a) R=200 , b) R=1000

The starting point on the β-axis is seen to be independent of R but the extent in α decreases as R increases. Therefore, as R→∞, the contour curves approach the β-axis and the energy will thus be confined to structures which are increasingly elongated in the streamwise direction.

Next, the product $|\hat{A}|^2$ F was studied. The form of the surface distorsion was chosen such that the volume under it is zero, since it is possible to show that an O(ε) term appears in the energy otherwise. Also, it was desirable to be able to evaluate the Fourier-transform analytically. The following specific form of A was chosen:

$$A(x,z)=(1+a_1x^2)e^{-b_1x^2}\cos(c_1x)(1+a_2z^2)e^{-b_2z^2}\cos(c_2z) \quad , \qquad (21)$$

where the zero-volume condition gives $a_1=4b_1^2/(c_1^2-2b_1)$, and similarly for a_2. In (21), b_1 and b_2 determine the global amplitude decay in either direction and their ratio can be viewed as an aspect ratio for the surface structure. The parameters c_1 and c_2 enable modulations of the amplitude to be applied. For the combination of (b_1,c_1,b_2,c_2)=(1,0,1,0), A is shown graphically in Figure 2.

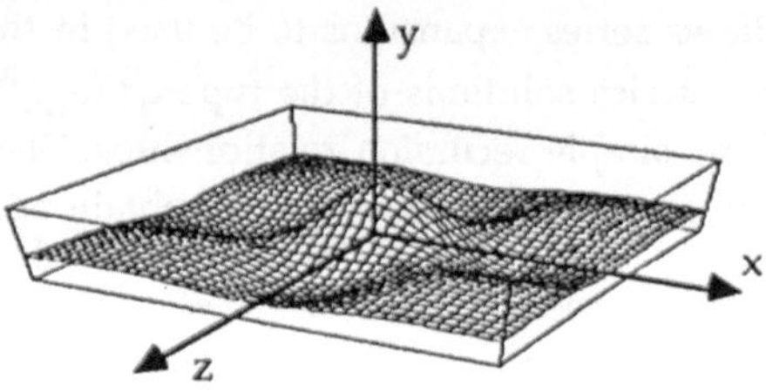

Figure 2 : Graphic display of A(x,z) for (b_1,c_1,b_2,c_2)=(1,0,1,0)

Three cases were studied in some detail, all at R=200. The first was the above combination of (1,0,1,0), corresponding to the same spatial behaviour in x and z. The level curves of $|\hat{A}|^2$ F are shown in Figure 3a. The second case was the combination (1,0,1,8) where a spanwise modulation has been applied. The result is shown in Figure 3b, where the shift in β is seen to match the modulation frequency.

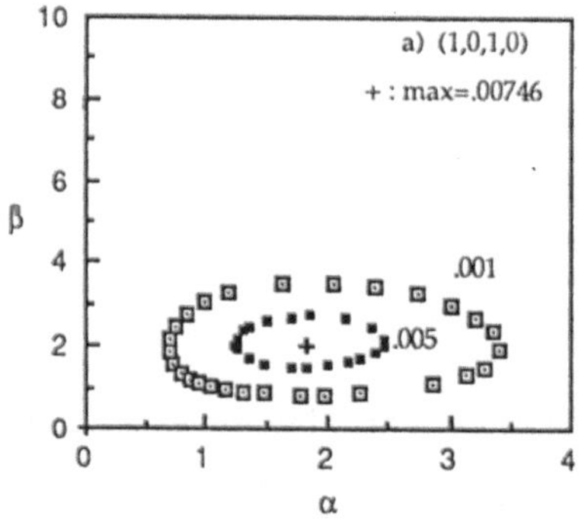

Figure 3a)

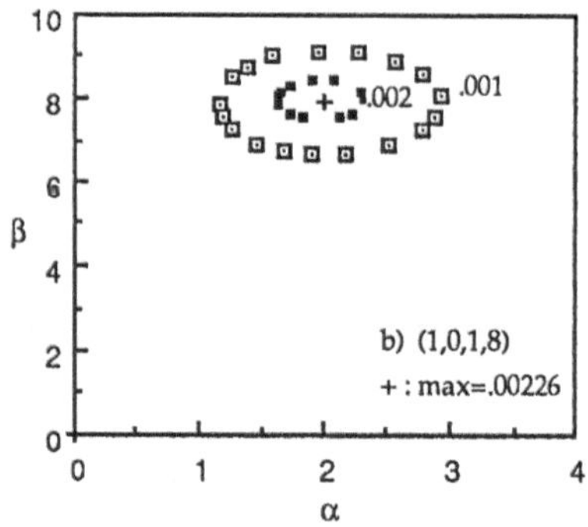

Figure 3b)

Comparing with the first case, the energy is confined to a narrower region and the maximum is also decreased. This implies that a spanwise modulation acts to decrease the kinetic energy of the perturbation field.
The last case, (1,0,9,0), indicates that the spanwise scale is decreased thus providing a surface elongated in the streamwise direction. Figure 3c shows a drastic decrease of the maximum value but also a larger extent of the level curves.

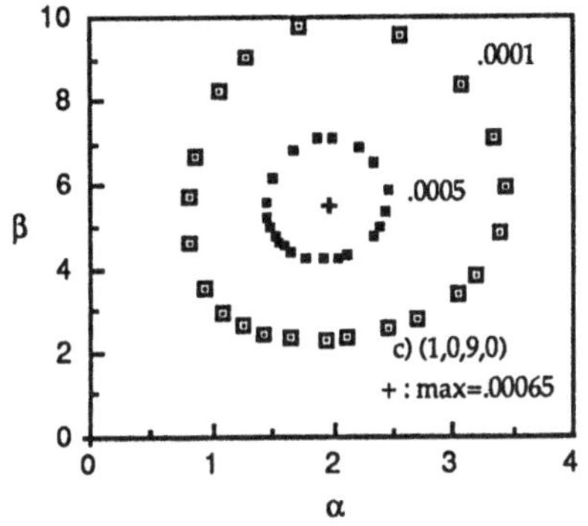

Figure 3c)

Figure 3: Level curves of $|\hat{A}|^2 F$ at R=200 for the combinations of (b_1, c_1, b_2, c_2): a) (1,0,1,0); b) (1,0,1,8); c) (1,0,9,0)

5. DISCUSSION

The velocity profile chosen is an exact solution to the Navier-Stokes equations if a uniform suction is applied at the surface. For that flow,the perturbation analysis differs from the one used here and the system of equations changes (cf . Drazin & Reid, pp. 227). In the present work, the suction profile has been used only to provide analytic simplicity and the results should be viewed accordingly. For other velocity profiles, the structure of (17) still holds, with F depending on the actual shape of the profile. However, during the calculations it was found that the main contribution to F, when integrating in y, is confined to the near-wall region. Therefore, it may be expected that the results for a mean velocity such as the Blasius profile will resemble those obtained here. It is a natural continuation of the present work to investigate the character of the deviation. Another point of interest is how higher approximations in the perturbation series will change the results. In that case, higher derivatives of U will appear in the boundary conditions and the corrections will be increasingly sensitive to the actual shape of the profile.
Using the present results in a drag reducing context requires that the kinetic energy in some sense is coupled to drag mechanisms. Independent of their character, it seems likely that if much energy is generated at scales where the instability mechanism(s) of the flow operate, the flow would easily become unstable. Since these scales have so far not been identified, the results of the present work calls for a parallel effort to further investigate the nature of the transition and other instability processes in boundary layers.

REFERENCES

Acarlar, M.S. & Smith, C.R. 1987, A study of hairpin vortices in a laminar boundary layer. Part 1. Hairpin vortices generated by a hemisphere proturberance. J. Fluid Mech. 175, pp 1-41.

Bartenwerfer, M. &Bechert, D.W. 1987, Die viskose Strömung über Oberflächen mit Längsrippen. DFVLR-FB87-21.

Drazin, P. &Reid, W. 1981, "Hydrodynamic stability", pp. 227, Cambridge University Press.

Raschi, W.G. & Musick, J.A. 1986, Hydrodynamic aspects of shark scales, NASA Contract Report 3963

Walsh, M.J. 1983, Riblets as a viscous drag reduction technique. AIAA Journal 21, 485.

Interpretation of Polymer Drag Reduction in Terms of Turbulence Producing Eddies Close to the Wall

Thomas J. Hanratty and S. L. Lyons, Univ. of Illinois, Urbana, IL, 61801, USA;
John McLaughlin, Clarkson Univ., Potsdam, NY, 13676, USA

Abstract

Results on the eddy structure and the scaling of the viscous wall layer, which are relevant to understanding the production of turbulence and drag-reduction, are presented. It is suggested that drag-reduction can be associated with an increase in the dimensionless dissipation of turbulence energy.

Introduction

A limiting factor in developing a theoretical understanding of drag-reduction has been the lack of information about the eddy structure in the viscous wall region ($y < 30$-40)*, and its relation to turbulence production. This paper uses experimental and computational studies, that have been carried out at the University of Illinois over the past 25 years, to present a conceptual model which could provide a useful basis for interpreting drag-reduction results.

Measurements of wall turbulence in the 1950's showed that both the production of turbulent energy and the dissipation of turbulent energy are quite large, and of the same magnitude, in the viscous wall region. However, the difference of these two large numbers is such that there is a net positive production which supplies energy to the outer region where the local dissipation is the same (in the "log-layer") or greater (in the core) than local production.

Injection of dye at the wall and various instrumental measurements have revealed elongated, organized structures in the viscous wall layer which are associated with long streaks of low velocity fluid having a spanwise spacing $\lambda = 100$. An important relationship between the characteristics of the streaky structure and drag-reduction was established by the finding that an increase in the streak spacing, λ, is associated with a

*Unless otherwise stated, all quantities in this paper are made dimensionless with the friction velocity, u^*, and the viscous length, ν/u^*.

A. Gyr (Editor)
Structure of Turbulence and Drag Reduction
IUTAM Symposium Zurich/Switzerland 1989

decrease in drag (1) (2). Consequently, one of the main focuses of this paper is to provide an understanding of what governs the magnitude of λ.

2. Calculations Based on a 2 1/2 D Model

Laboratory studies of the turbulent flow field in the viscous wall layer are summarized in a recent review (3). Measurements from arrays of electrochemical wall shear stress probes show that the spanwise variation of the streamwise, S_x, and spanwise, S_z, velocity components are characterized by a wavelength of 100 wall units, as indicated in figure 1. The large magnitudes of S_x and S_z, their relative phases, and their spatial scales in the flow direction indicate that elongated eddy patterns of the type shown on the top of figure 1 are intermittently encountered.

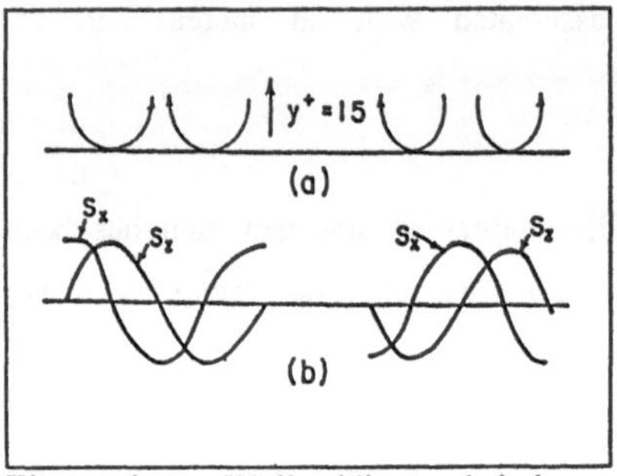

Figure 1 Wall eddies and their effect on two components of the wall velocity gradient.

These eddies are pictured to create streamwise velocity fluctuations, u, (and low velocity streaks) by bringing high momentum fluid to the wall, exchanging momentum with the wall, and carrying low momentum fluid away from the wall. Since the high momentum fluid is transported by negative normal velocities, v, and low momentum fluid by positive normal velocities, these flows are associated with large values of -uv. They are, thus, important contributors to the Reynolds stress, $-\overline{uv}$, and the production of turbulence, $-\overline{uv}\,\frac{d\bar{U}}{dy}$.

At the outer part of the viscous wall region spatial correlation measurements have indicated that both the w and u- velocity components are dominated by larger scale motions in the z-direction, but that the v-velocity has a spanwise scale of about 100^+ (wall units). Combined measurements with multiple wall probes and probes in the fluid have indicated that the $\lambda = 100$ eddies, depicted in figure 1, extend from the wall to the edge of the viscous wall region, y_o. They also seem to suggest that the $\lambda = 100$ eddies are closed during periods when the production of Reynolds stress is large.

These type results motivated the development of a simple computer representation of the time varying flow in the viscous wall region (4) (5). The observation that flow at the wall is elongated prompted the assumption of slender body turbulence whereby all terms involving x-derivatives are neglected. The resulting equations are a 2 1/2 D model in that they solve for the three velocity components in the y-z plane.

An important aspect of this model is that it allowed the exploration of non-physical situations (not possible in laboratory experiments) to obtain an understanding of the physics of the viscous wall layer. Of particular interest to the presentation in this paper are the calculations with different λ (5).

Calculated mean velocity profiles show that U at the edge of the viscous wall region and, therefore, drag-reduction increase with increasing λ. In the viscous wall region, the total dimensionless dissipation decreases with increasing λ. The difference of the total production and the total dissipation in the viscous wall region was calculated as a function of λ. For small λ there is a net dissipation and for large λ there is a net production. At $\lambda \cong 93$ the calculated net production is just sufficient to provide the net energy dissipated in the outer flow. This is close to the measured spacing of the streaky wall structure and suggests that the Reynolds stress producing eddies adjust their size so that a balance of energy is maintained. It is noted that the net dissipation in the outer flow is quite small compared to the total production or the total dissipation in the viscous wall layer. This means that, for a Newtonian fluid, the viscous wall layer is close to equilibrium, in that the total production of energy is roughly equal to the total dissipation.

3. Computer Simulation of Channel Flow

The recent successful computer calculation, by Kim, Moin & Moser, of the time-varying velocity and pressure fields for fully-developed turbulent flow in a channel has prompted the implementation at the University of Illinois of a similar code, by John McLaughlin and Stephen Lyons (6). The three-dimensional, time dependent Navier-Stokes equations are solved pseudo-spectrally with a fractional-step algorithm developed by Orszag and Kells and a modification proposed by Marcus. Fully developed turbulent channel flow at a Reynolds number of 2260, based on the half channel height and the bulk average velocity, is considered.

Figure 2 gives a picture of a portion of the calculated instantaneous flow field in the y-z plane at a fixed x. The orientation of the instantaneous vectors gives the direction of flow in this plane. The length gives the speed, in accordance with the scale indicated in the figure. The highlighted portions of the figure are regions where the instantaneous value of -uv are greater than unity. Since turbulence production equals

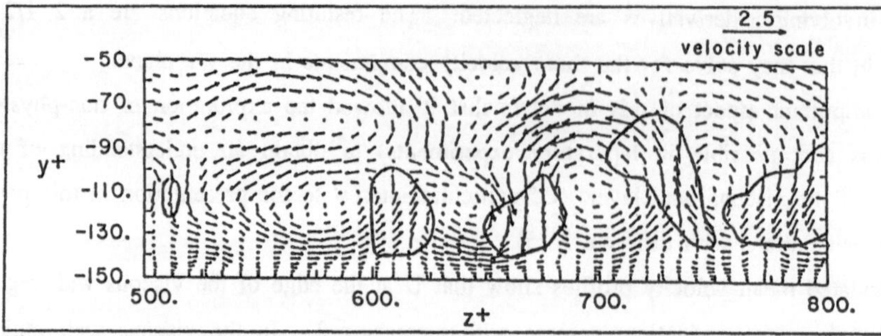

Figure 2 Instantaneous velocity vectors in the y-z plane. Regions where -uv > 1 are highlighted.

$-\overline{uv}\ \frac{d\overline{U}}{dy}$, they also represent regions in which turbulence is being produced. An examination of figures such as this reveals that in the viscous wall region turbulence production occurs intermittently at strong updrafts and downdrafts, often associated with closed eddies that are attached to the wall and are, usually, paired asymmetrically.

A main focus of the work of Lyons (6) was to determine the average characteristic dimension of the turbulence producing eddies in the viscous wall region. Consequently it was desired to identify the mean characteristics of the eddies which are large contributors to the Reynolds stress at y = 11.4, where the turbulence production is a maximum. The location of all of the events for outflows (quadrant 2) and for inflows (quadrant 4) revealed a streaky structure. The streaks have a spanwise spacing of 100 and lengths as large as 1000. Lyons (6) gives conditionally averaged vector patterns in the y/z plane for outflows at y = 11.4 with -uv > 1.6. This corresponds to 50 per cent of the total contribution of quadrant 2 events to the Reynolds stress. The flow pattern that resulted was a pair of closed eddies having spanwise dimensions of about 50 and having their centers located at a distance of about 22 from the wall. This conditionally averaged pattern was the same for quadrant 4 events and for different triggering levels of -uv.

More recently we have explored the effect of changing the y at which the triggering is done. Figure 3 shows an example in which the pattern was conditionally averaged on outflows at y = 25.3 with -uv > 1.4. Again the threshold Reynolds stress was selected to capture 50 per cent of the contribution of quadrant 2 events. This shows a spanwise dimension of about 50, but the center is slightly farther from the wall, at y = 30. Conditional averaging for quadrant 2 events at y = 38.9 produced eddies with their centers 40 units from the wall and separated by a distance of 50. Eddies conditionally

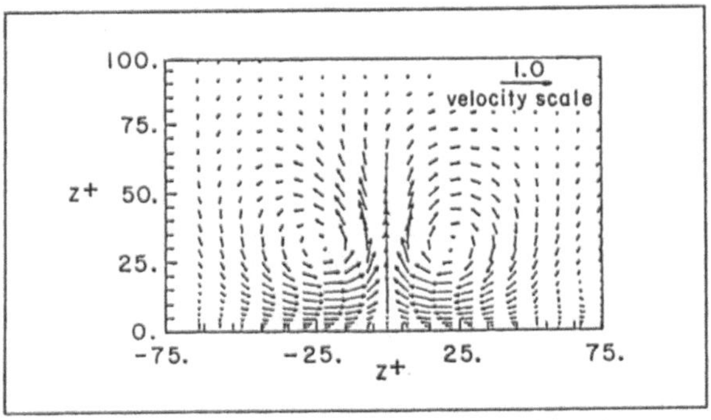

Figure 3 Eddies in the y/z plane conditionally averaged for -uv > 1.4 at y = 25.3.

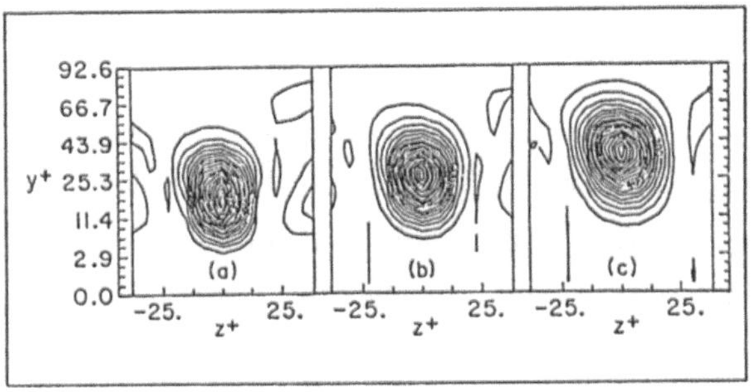

Figure 4 Reynolds stress contours conditionally averaged for a quadrant 2 event with -uv > 1.4: (a) at y = 11.4, normalized with -uv = 3.43 (b) at y = 25.3, normalized with -uv = 3.58 (c) at y = 38.9, normalized with -uv = 3.71.

averaged for quadrant 4 events at y = 25.3 and y = 38.9 showed slightly larger spanwise dimensions than when conditionally averaged at y = 11.4, respectively being 60 and 66.

The spatial characteristics of the Reynolds stress producing eddies is displayed in figures 4 and 5, which show Reynolds stress contours in the y-z and x-y planes conditionally sampled for quadrant 2 events. These clearly indicate a spanwise extent of $\lambda \cong 50$ and a streamwise extent of 300 - 400.

The stress producing eddies can be identified at the edge of the viscous wall region by the distance between zero crossings of v in the z-direction. Define a microscale, λ_g, as $\overline{(\partial v/\partial z)^2} = 2\ \overline{v^2}/\lambda_g^2$. For a Gaussian signal, that is homogeneous in the z-direction, the distance between zero crossings is $\pi\ \lambda_g/\sqrt{2}$. The spanwise length of the stress producing eddies then becomes $\lambda = \pi\ \lambda_g/\sqrt{2}$ (or $\lambda = \sqrt{2}\ \pi\ \lambda_g$). Calculations of Lyons (6) give λ = 36, 40 at y = 30, 40, in approximate agreement with the conditionally averaged value of 50.

4. Scaling

Work outlined in the previous sections indicates that net turbulence production in the viscous wall layer is controlled by eddies with a spanwise length of 50 and a length of 300-400. Scaling relations developed by Finnicum & Hanratty (4) are useful in understanding how these results are related to surface drag.

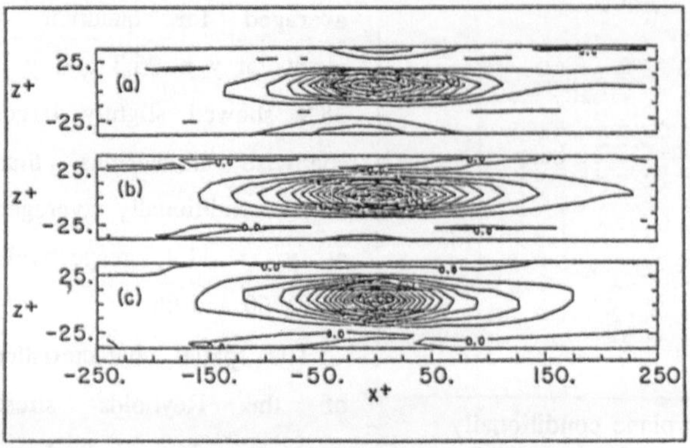

Figure 5 Reynolds stress contours conditionally averaged for a quadrant 2 event with $-uv > 1.4$ (a) $y = 11.4$ normalized with $-uv = 3.43$ (b) $y = 25.3$ normalized with $-uv = 3.58$ (c) $y = 38.9$ normalized with $-uv = 3.71$.

If the flow is isotropic at y_o, the dimensionless dissipation of turbulent energy is given as

$$\varepsilon = 15\, q^2/\lambda_s^2$$

This relation is often used in non-isotropic flows with $q^2 = (\overline{u^2} + \overline{v^2} + \overline{w^2})/3$. At $y_o = 40$, $q^2 = 1.6$ and $\varepsilon = 0.05$. A value of $\lambda_s = 22$, $\bar{\lambda} = 49$ is then obtained from the isotropic relation, in agreement with the experimental value of 50.

The outer edge of the viscous wall regions, y_o, is located at the beginning of the log-layer where $d\overline{U}/dy = \kappa\sqrt{\tau/\rho}/y$. Here $\overline{U}$ is the dimensional mean velocity, τ, the local shear stress, κ, the von Kármán constant and y, the dimensional distance from the wall. The location y_o may be defined so that the viscous stress, $\mu\, d\overline{U}/dy$, is a small fixed fraction of the total stress, τ. As Finnicum & Hanratty have pointed out $y_o \sqrt{\tau/\rho}/\nu = 40$ is the location at which the viscous stress is 6 per cent of the total.

If the isotropic relation is used to relate $\bar{\lambda}$ to the energy dissipation and if it is assumed that q scales with $\sqrt{\tau/\rho}$ then the dimensional energy dissipation is given as

$$\varepsilon = a_1\, \nu \left(\frac{\tau}{\rho}\right)/\bar{\lambda}^2$$

At the edge of the viscous wall region the dimensional production is

$$P = \left(\frac{\tau}{\rho}\right)^{3/2} \kappa/y_o$$

Since $\varepsilon = P$ at y_o the following relation is derived between y_o and $\bar{\lambda}$:

$$y_o = \bar{\lambda}^2\, (\tau/\rho)^{1/2}\, a_2/\nu$$

5. <u>Drag Reduction Due to Polymers</u>

The 2 1/2 D analysis outlined in Section 2 suggests that the value of λ (required so that the total production of turbulence is approximately equal to the total dissipation in the viscous wall region) will increase if the presence of polymers increases the dimensionless dissipation of turbulent energy. Consequently it is of interest to inquire regarding possible effects of polymer addition on dissipation in order to understand the observed increase of λ. Two have received considerable attention: the influence of viscoelasticity on oscillating shear layers in the turbulent field and the occurrence of large elongational viscosity because of the unraveling of the polymer molecule.

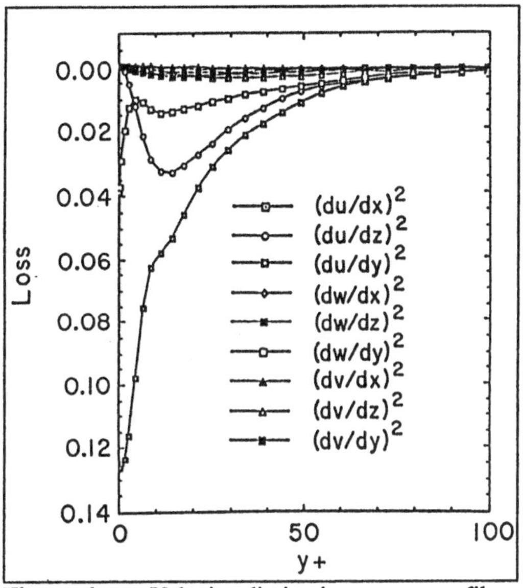

Figure 6 Velocity dissipation tensor profiles.

Figure 6 plots the different terms contributing to viscous dissipation calculated by Lyons (6). It is noted that the dominant term is $\overline{(\partial u/\partial y)^2}$. The importance of the oscillating shear layer at the wall is, thus, particularly evident. The median circular frequency of oscillation (3) has been determined to be $\omega = 0.009 \times 2\pi\ v^{*2}/v$. One would expect the effective viscosity experienced by the oscillations to be changed when $1/\omega$ is of the order of the relaxation time of the polymer molecules, θ. This requires $\theta\ \overline{S} = 18$, where $\overline{S}$ is the mean velocity gradient at the wall. An examination of figure 6 reveals that the contributions of elongational flows, such as $\overline{(\partial v/\partial y)^2}$, to dissipation in the viscous wall region are quite small. This suggests that the effective elongational viscosity would have to be increased by at least an order of magnitude for the presence of polymer molecules to start having an effect on viscous dissipation.

The scaling relations, discussed earlier, defined the edge of the viscous wall region as $y\ \sqrt{\tau/\rho}/v_{PS} \cong 40$. Here v_{PS} is the effective kinematic viscosity of the time-averaged shear at y_o. The dimensionless thickness of the viscous wall region defined in terms of the kinematic viscosity at the wall is $40\ v_{PS}/v$; it will therefore increase if v_{PS}/v is greater than unity. The scaling relating the characteristic dimension of the stress producing eddies, λ, to the thickness of the viscous wall region gives

$$\frac{y_o\ (\tau/\rho)^{1/2}}{v_{PS}} = \frac{a_2\ \lambda^2\ (\tau/\rho)}{v^2}\ (\frac{v}{v_{PS}})\ (\frac{v}{v_{PD}})$$

Here ν_{PD} is the effective viscosity of the polymer solution for viscous dissipation of turbulence. The left side equals 40, by definition, so the above relation indicates that dimensionless λ (or dimensionless λ) will be larger than for a Newtonian fluid if $\nu_{PS}\,\nu_{PD}/\nu^2$ is greater than unity.

These scaling results are consistent with the discussion of the energy producing eddies. They show that both y_o and λ can increase if the effective viscosity of the polymer solution is greater in the turbulent flow outside the viscous sub-layer, y < 5.

6. An Energy Dissipation Principle for Drag-Reduction

The motion that an increase in λ can be associated with an increase in the dimensionless rate of dissipation of turbulent energy implies that drag-reduction will result if turbulence structures with a fixed velocity and length parameter become more dissipative. This could be a general principle underlying a number of observations, so it deserves additional discussion.

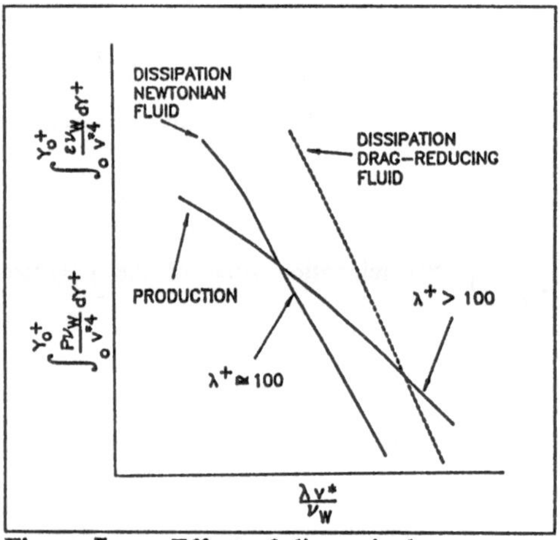

Figure 7 Effect of dimensionless turbulent energy dissipation on λ.

The idea is illustrated for flow in a pipe in figure 7. In the viscous wall layer $\varepsilon/(v^*)^3(v^*/\nu_w)$ and $P/(v^*)^3(v^*/\nu_w)$ are functions of yv^*/ν_w, where v^* and ν_w/v^* scale the velocity and size of turbulent structures. In the core of the pipe the proper length parameter is the pipe radius, a, so that $a\varepsilon/(v^*)^3$ and $Pa/(v^*)^3$ are functions of y/a. In the log-layer which overlaps these two regions $P = \varepsilon$ and the proper length is the distance from the wall y. The self-consistency of this scaling can be seen by evaluating $\int_0^{Y_o}(P - \varepsilon)\,dy$. This shows that the flow of turbulent energy out of the viscous wall region over a length Δx varies as $\rho v^{*3} a \Delta x$. The evaluation of the net dissipation outside the viscous wall region can be broken into two parts $\int_{Y_o}^{0.15a} + \int_{0.15a}^{a}$. The first of these will be zero since $P = \varepsilon$ in the log-layer. The use of core scaling to evaluate the second integral gives a net rate of dissipation of energy outside the viscous wall region which scales as $\rho v^{*3} a \Delta x$.

The two solid curves in figure 7 sketch the influence of λ on the production and on the dissipation of turbulent energy in the viscous wall region suggested by the calculations of Lyons, et al (5). Flows with small λ are impossible since dissipation is greater than production. For $\lambda \cong 100$ the net production of energy by a Newtonian fluid in the viscous wall region is just sufficient to supply the small net dissipation in the core. If the role of a drag-reducing agent is to increase $\varepsilon v_w/v^{*3}$ in the viscous wall region, then dissipation would be represented by the dashed curve in figure 7. A larger dimensionless λ would be required for the viscous wall layer to supply the energy needed in the core.

These considerations suggest that drag-reduction would also result from an increase in the dimensionless dissipation in the core, $\varepsilon a/v^{*3}$. Such an increase would require a larger net production in the viscous wall region and, from figure 7, a larger λ.

Acknowledgment

This work was supported by the Office of Naval Research under Grant N00014-82K0324 and by the National Science Foundation under Grant NSF CBT 88-00980.

References

1. Fortuna, G. and T. J. Hanratty, "The Influence of Drag-Reducing Polymers on Turbulence in the Viscous Sublayer", *J. Fluid Mech.*, **53**, 575-586 (1972).
2. Eckelman, L. D., G. Fortuna and T. J. Hanratty, "Drag Reduction and the Wavelength of Flow-Oriented Wall Eddies", *Nature Physical Science*, **236**, 94-96 (1972).
3. Hanratty, T. J., "A Conceptual Model of the Viscous Wall Region", published in "Near Wall Turbulence", Hemisphere Publishing Corporation, Washington, D.C. (1989).
4. Finnicum, D. S. and T. J. Hanratty, "Effect of Favorable Pressure Gradients on Turbulent Boundary Layers", *A.I.Ch.E. Journal*, **34**, 529-540 (1988).
5. Lyons, S. L., C. Nikolaides and T. J. Hanratty, "The Size of Turbulent Eddies Close to a Wall", *A.I.Ch.E. Journal*, **34**, 938-945 (1988).
6. Lyons, S. L., "A Direct Numerical Simulation of Fully Developed Turbulent Channel Flow with Passive Heat Transfer", Ph.D. thesis, University of Illinois, Urbana (1989).

Active Turbulence Control in Wallbounded Flow Using Direct Numerical Simulations

J. Kim, P. Moin and H. Choi

Center for Turbulence Research
NASA Ames Research Center, Moffett Field, California 94035
and
Stanford University, Stanford, California 94305

An exploratory study of concepts for active control of turbulent boundary layers using the direct numerical simulation technique was performed. Significant drag reduction was achieved when the surface boundary condition was modified such that it could suppress the large-scale structures present in the wall region. This was achieved by prescribing the normal component of velocity at the wall to be 180° out of phase with the normal velocity slightly above the wall at each instant. The drag reduction was accompanied with significant reduction in the intensity of the wall-layer structures and reductions in the magnitude of Reynolds stresses throughout the flow. Suitability of wall-pressure and shear-stress fluctuations for detection of flow structures above the wall was examined. A preliminary result obtained by applying the present control strategy to a transitional flow is also briefly described, from which one can infer a possible linkage between the control strategy and flow stability.

1. INTRODUCTION

The potential benefits of managing and controlling turbulent flows that occur in various engineering applications are known to be significant. For example, it is estimated that a 20% reduction in the fuselage skin-friction drag of commercial airplanes alone would save approximately one billion dollars per year in fuel consumption.[1] It is recognized that organized structures in turbulent flows play an important role in turbulence transport. Therefore, attempts for control of turbulent flows for engineering applications have focused on manipulation of coherent structures. Most turbulence control strategies for wall-bounded turbulent flows to date have used passive approaches. For example, a flow device, such as riblet or LEBU (Large-Eddy-Break-Up device), is placed in the boundary layer in an attempt to suppress the formation (or interaction) of organized flow structures. Such a device plays a passive role in the sense that there exists no feedback loop as the flow structures are modified. The present study is aimed at active control of dynamically significant coherent structures to achieve skin-friction reduction. The control strategy will respond instantly through a feedback loop as instantaneous flow structures are being modified.

The widely observed coherent structures in the wall layer are the streaks: elongated regions of low- and high-speed fluid alternating in the spanwise direction. From flow visualization data, Kline et al.[2] pointed out that the production of turbulence in boundary layers is largely due to the bursting event which consists of lift-up, oscillation, and violent breakup of the streaks. The sweep event which is described as the inrush of high speed fluid towards the wall[3] is also believed to be a major contributor to turbulence production. Johansson et al.[4] also reported that much of turbulence production takes place within the so-called internal shear-layer structure associated with the bursting process. Some of these event-oriented descriptions of important phenomenon in turbulent boundary layers, however, is beginning to change, largely from examination of

A. Gyr (Editor)
Structure of Turbulence and Drag Reduction
IUTAM Symposium Zurich/Switzerland 1989

direct numerical simulation databases. For example, the bursting event may not be anything more than a consequence of the convection of a single streamwise vortex past a fluid marker.[5] The passage of the vortex lifts the marker, and as the marker wraps around the vortex, it appears as oscillations in a side view. In this paper we use the terms sweep and ejection at a point to simply denote the flow direction towards or away from the wall respectively.

We seek physical algorithms that can suppress the bursting process with subsequent net skin-friction reduction. We use the direct numerical simulation technique to achieve this goal. Owing to availability of all flow variables at many spatial locations, and the ability to readily alter flow boundary conditions, numerical simulation technique provides a unique laboratory for testing and design of turbulence control concepts. Although some of the concepts may not turn out to be feasible for implementation, the simulations can provide data on what may be possible to achieve just from fluid dynamical considerations.

A brief description of the numerical procedures used in this study is given in section 2. In section 3, results from a numerical experiment for active control of a turbulent channel flow are presented. Prospects for detection of important structures near the wall by wall-mounted sensors are examined in section 4. This is important for practical considerations. A brief summary is given in section 5.

2. NUMERICAL PROCEDURES

The numerical technique used in this study is identical to that of Kim et al.,[6] to which the reader is referred for detail descriptions. The time advancement for the convective terms was made with a third-order Runge-Kutta method instead of the original Adams-Bashforth advancement. The boundary condition for the normal component of the velocity is modified according to the particular control strategy, which resulted in a pentadiagonal system compared to a tridiagonal system for the no-slip boundary condition of the unperturbed channel.

The base flow is the fully-developed channel flow. Preliminary experiments were performed using $32 \times 65 \times 32$ spectral modes (streamwise, normal to the wall, and spanwise, respectively) at Reynolds number, $Re = 1800$ based on the centerline velocity and the channel half-width. Starting with the same initial field several different boundary conditions were tried to achieve an optimum result. The final computation was made using $128 \times 129 \times 128$ spectral modes at $Re = 3300$, from which most of results presented in this paper were obtained. The particular Reynolds number was chosen to compare with the results of the unperturbed channel.[6]

In this paper x, y, z denote the streamwise, normal to the wall, and the spanwise directions, respectively. The velocities are u, v, and w in x, y, and z directions respectively which are used interchangeably with subscripted variables u_1, u_2, and u_3. The subscript w indicates the value at the wall, and the superscript $+$ indicates a non-dimensional quantity scaled by the wall variables: e.g. $y^+ = yu_\tau/\nu$, where ν is the kinematic viscosity and $u_\tau = (\tau_w/\rho)^{\frac{1}{2}}$ is the wall-shear velocity.

3. RESULTS FROM AN ACTIVE CONTROL EXPERIMENT

Our aim was to examine whether we can reduce the wall-skin friction by suppressing the sweep and ejection events. We applied suction or injection on the channel walls exactly opposite to the normal component of velocity at a given y-location. At each instant the boundary condition for v at $(x, y = 0, z)$ was prescribed to be $-v(x, y_r, z)$, where y_r is the location of detection. Thus, when fluid was detected at y_r to move towards the wall (sweep) an equally strong blowing velocity was imposed at the wall to "cancel" the sweep event. Similarly, when fluid was detected at y_r moving away from the wall (ejection), an equally strong suction was applied. The initial condition for the calculations was an instantaneous velocity field from fully-developed channel flow. The mass flux was kept constant by adjusting the mean pressure gradient. Thus, any skin-friction reduction would be manifested in the mean pressure gradient necessary to drive

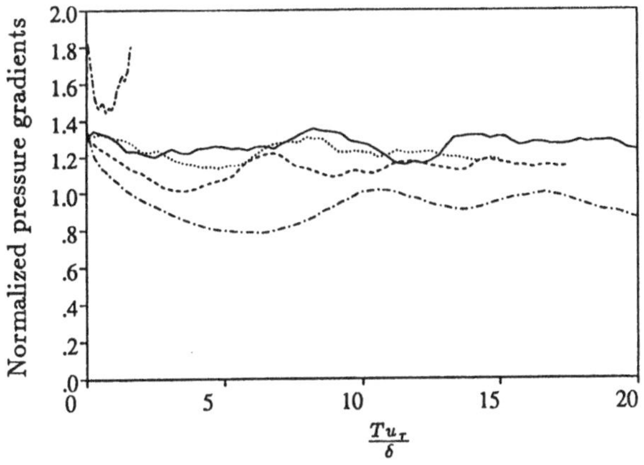

FIGURE 1. Time history of the required pressure gradient to drive the same mass flow rate: ——— , unperturbed channel; ········ , perturbed channel with sensor at $y_r^+ \approx 2$; ---- , $y_r^+ \approx 5$; —·— , $y_r^+ \approx 10$; ——— , $y_r^+ \approx 25$.

the flow with the same mass flow rate. Several computations were performed with the coarse mesh ($32 \times 65 \times 32$) for several different y_r to examine the effect of the location of detection. Using the same initial velocity field, the calculations were continued with the new boundary conditions until a new statistically steady state was obtained or until it became apparent that the drag would increase substantially.

Time histories of the pressure gradient required to drive the same mass flow rate for an unperturbed fully-developed channel flow and perturbed channel flows are shown in Figure 1. Indeed, substantial skin-friction reduction was obtained ($\approx 20\%$ on each wall) with $y_r^+ \approx 10$. For other y_r-locations, either the drag was substantially increased ($y_r^+ \approx 25$) or the reduction is from negligible to small ($y_r^+ \approx 2, 5$).

After the optimum y_r-location was identified with the coarse mesh, a final computation was performed for detail analysis of the modified flow field for the case of $y_r^+ \approx 10$, using $128 \times 129 \times 128$ spectral modes for $Re = 3300$. The mean velocity gradient in the unperturbed fully-developed channel and the perturbed channels are shown in Figure 2(a). The reduction in the velocity derivative in the vicinity of the wall is apparent whereas away from the wall there is a slight increase. The mean velocity profile in a semi-log plot is shown in Figure 2(b). The shift in the profile in the logarithmic layer is due to the reduction in the wall shear velocity, u_τ, in the perturbed channel.

Turbulent intensities and Reynolds shear stress profiles are shown in Figures 3 and 4. Here the velocities are normalized with the shear velocity of the unperturbed channel flow. There is a significant reduction in the intensities throughout the channel. This is in contrast to the experiments with unsteady high-amplitude wavy (but passive) wall motion where the effect of the wall movement was confined to a thin (Stokes-like) layer near the wall.[7] In these latter computations the moving wavy wall was approximated by sinusoidal wall-blowing and suction. Apparently, with the "active" boundary conditions the energy production and transport mechanisms have been significantly altered.

The instantaneous fields have been examined in detail. Contours of constant streamwise velocity fluctuations are shown in Figure 5. It is clear that the structure of the wall-layer streaks has been changed. Their strength has been reduced considerably. Two point correlations of the streamwise velocity component in the spanwise direction

$$R_{11}(r_z) = \langle u_1(x,y,z)u_1(x,y,z+r_z) \rangle$$

show a significant change in the near-wall region (Figure 6). It is well known that the auto-correlation of the streamwise component crosses the r_z axis, and the location of

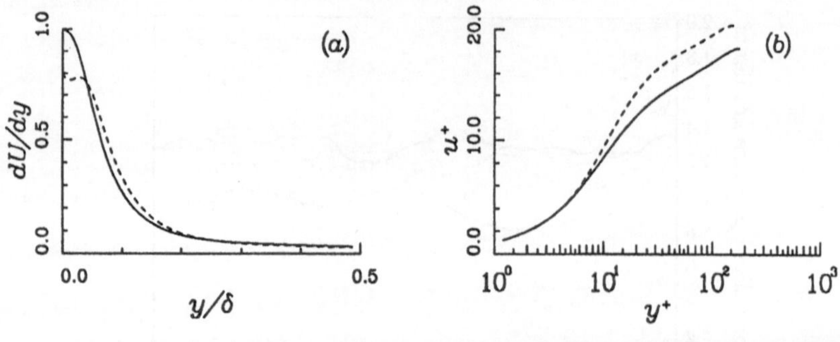

FIGURE 2. Profiles of (a) mean velocity gradient and (b) mean velocity profile normalized by wall variables: ——— , unperturbed channel; - - - - , unperturbed channel.

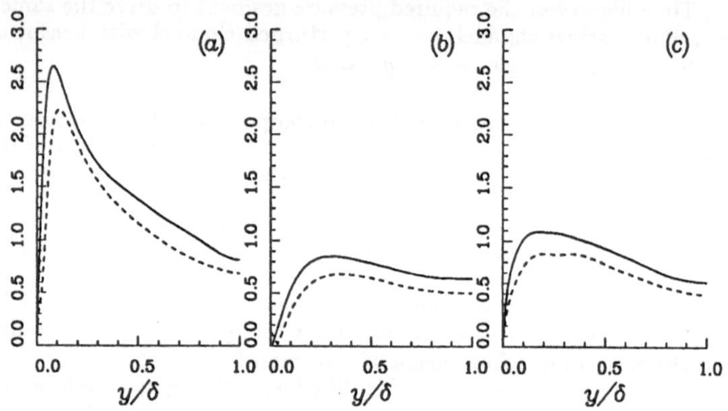

FIGURE 3. Profiles of *r.m.s.* fluctuations of turbulence intensities: ——— , unperturbed channel; ········ , perturbed channel. (a) $< u^2 >^{1/2}$, (b) $< v^2 >^{1/2}$, and (c) $< w^2 >^{1/2}$. All intensities are normalized with the wall-shear velocity of the unperturbed channel.

the (negative) minimum is the mean spacing between low- and high-speed streaks. In the perturbed channel, $R_{11}(r_z)$ only has a weak negative region consistent with the observation that the strengths of the streaks have been reduced. The physical streak-spacing has been increased, whereas their mean-spacing in the wall units has remained approximately the same. Other flow structures also have been affected. For instance, contours of instantaneous streamwise vorticity, ω_x in a (y, z)-plane in the unperturbed and perturbed channels indicated a similar reduction in the intensity of the vortical structures.

The control algorithm just described is probably not optimum. For example, a shift in the streamwise location of the imposed wall velocity relative to the velocity at $y_r^+ \approx 10$ may be more effective in "canceling" local structures or events. Also, one can vary the strength of the blowing and suction at the surface instead of using the same magnitude of v at y_r. Such experiments have not been performed yet. When the normal component of the velocity at the wall was forced *in phase* with the normal velocity at the sensor location, the required pressure gradient to drive the same mass flow rate increased dramatically, indicating that one can also augment turbulence generation and heat transfer from the wall.

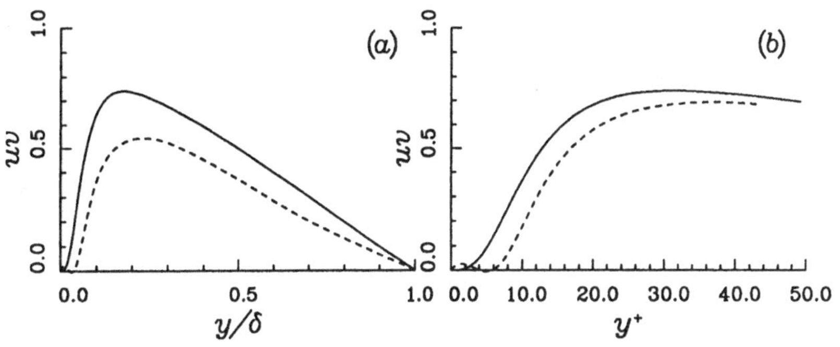

FIGURE 4. Profiles of Reynolds shear stress: ——— , unperturbed channel; ---- , perturbed channel. (a) The shear stress is normalized by the wall-shear velocity of the unperturbed channel; (b) The shear stress and the y-coordinate are normalized by the corresponding wall-shear velocity.

4. ACTIVE CONTROL USING WALL VARIABLES

Although the control algorithm described in section 3 was successful in reducing skin-friction and suppression of turbulence, it is clearly not feasible for practical implementation. Among other things, it is difficult to place sensors within the flow field away from boundaries. In this section we shall investigate the usage of flow variables at the wall for detection of structures above the wall. Only the data from fully-developed unperturbed channel flow will be used. The (non-zero) variables considered are the wall-pressure fluctuations, $p_w(x,z)$, and normal derivative of velocity fluctuations, $\frac{\partial u}{\partial y}|_w$ and $\frac{\partial w}{\partial y}|_w$.

Two-point correlations of the wall-pressure and the velocity $< p(x,z)u_i(x+r_x,y,z+r_z) >$, and vorticity $< p(x,z)\omega_i(x+r_x,y,z+r_z) >$ away from the wall are calculated. Brackets indicate an average over horizontal (x,z) planes and time. Contours of $< p(x,y=0,z)u_i(x+r_x,y,z+r_z) >$ for $(i=1,2)$ on the center plane $(r_z=0)$ are shown in Figure 7. The pattern is that of high-speed fluid impinging on the wall followed by the ejection of low-speed fluid. The pattern is very similar to the conditionally-averaged velocity contours obtained using VISA (spatial version of VITA) by Johansson et al.[4,8] Correlation contours in (y,r_z) planes show the upstream inrush of high-speed fluid followed downstream by a pair of inclined roller eddies pumping low-speed fluid from the wall region. Again, this is very similar to the flow patterns obtained from conditional averages.[8] Wall-pressure appears to be a good detector of the internal shear layers and could be used in a control strategy aimed at manipulating them.

Probability density functions are used to examine the relationship of the wall variables and the flow above the wall. In section 3 we reported substantial drag reduction by imposition of wall-normal velocity opposite to the velocity at $y^+ \approx 10$. Examination of the joint probability density of wall variables and the normal velocity at $y^+ \approx 10$ should reveal to what extent can one reproduce the control experiment of section 3 by placing sensors only *at* the wall. The joint probability density functions of the wall pressure and the normal velocity at different y-locations (not shown here) did not reveal any particular correlations between them, indicating that wall-pressure alone is not an adequate detector of the flow toward the wall or away from it. The joint PDF of $v(x,y^+\approx 10,z)$ and $\partial u/\partial y|_w$ is shown in Figure 8. The streamwise velocity derivative

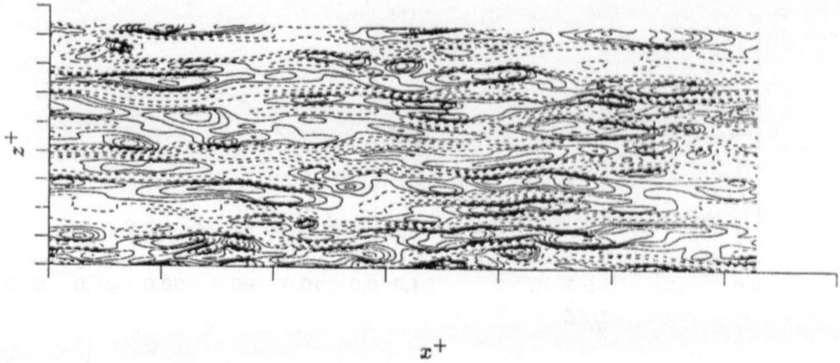

(a)

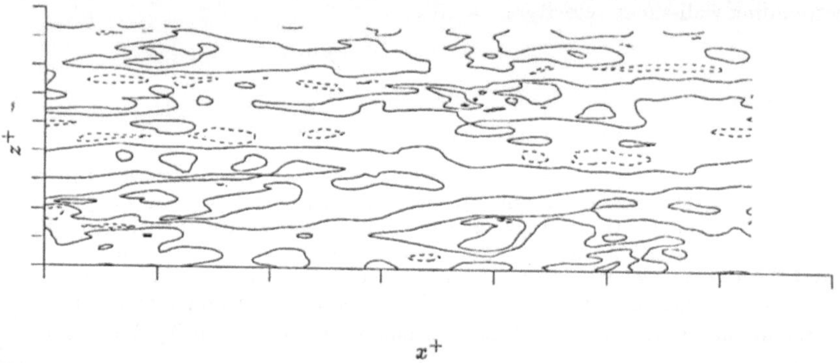

(b)

FIGURE 5. Contours of streamwise velocity fluctuations, u, in an (x, z)-plane at $y^+ \approx 5$: (a) unperturbed channel, u_{max}=6.5 and u_{min}=-3.6; (b) perturbed channel, u_{max}=2.1 and u_{min}=-1.5. The contour levels range from -4 to 7 by increments of 1, and negative contours are dashed. The plot domain extends 2300 wall units in x and 800 units in z.

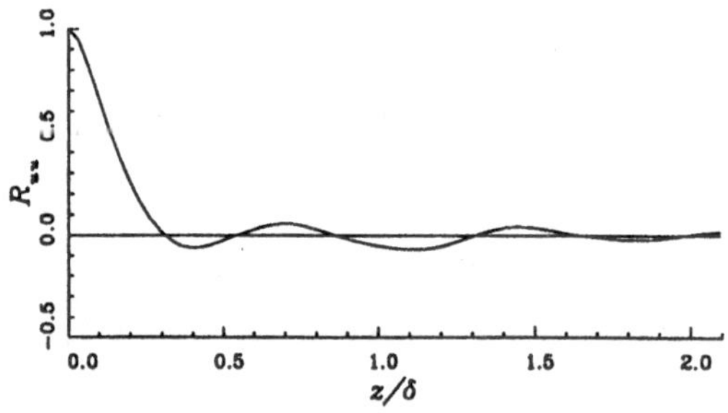

FIGURE 6. Two-point correlation of the streamwise velocity in the spanwise direction, $R_{uu}(r_z)$.

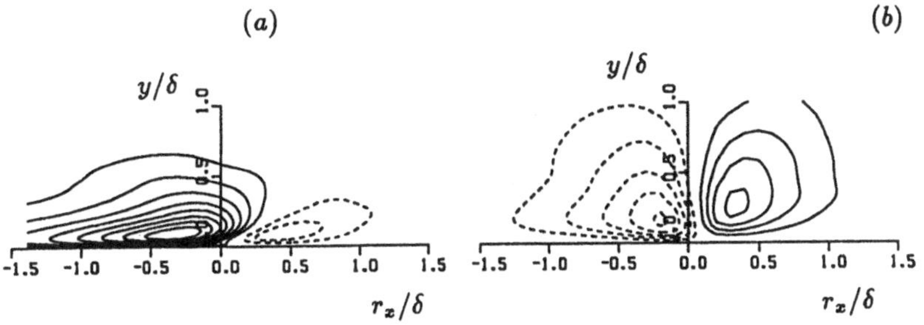

FIGURE 7. Contours of the correlation of wall-pressure fluctuations and (a) streamwise and (b) normal velocity components in the center plane ($r_z = 0$). Negative contours are dashed. All variables are non-dimensionalized with the wall-shear velocity, u_τ, and the channel half width δ. Contour levels are from -0.2 to 0.8 with increments of 0.1 in (a), and from -0.25 to 0.2 with increments of 0.05 in (b). Note that the correlation function is not normalized. For rms values of p and u_i see Kim et al.[6]

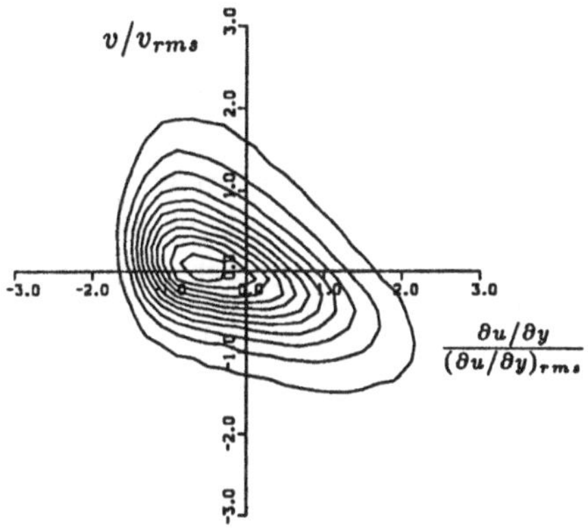

FIGURE 8. Joint probability density of the fluctuating wall-shear stress $\frac{\partial u}{\partial y}$ and the normal velocity, v at $y^+ = 10.5$. The contour levels are from 0.025 to 0.275, with 0.025 increments.

at the wall appears to be a better detector of the events at $y^+ \approx 10$ than the pressure; in particular, high amplitude positive values of $\partial u/\partial y|_w$ are likely to be associated with sweeps. Negative values of $\partial u/\partial y|_w$, however, do not provide adequate discrimination between sweeps and ejections.

Taylor series expansion of the normal velocity component about the wall

$$v(y) = \frac{y^2}{2}\frac{\partial^2 v}{\partial y^2}|_w + \dots \tag{1}$$

suggests that, near the wall, v is highly correlated with its second derivative evaluated at the wall. The accuracy of the first term approximation for v in (1) increases as one approaches the wall. It is of interest to know whether this approximation is acceptable

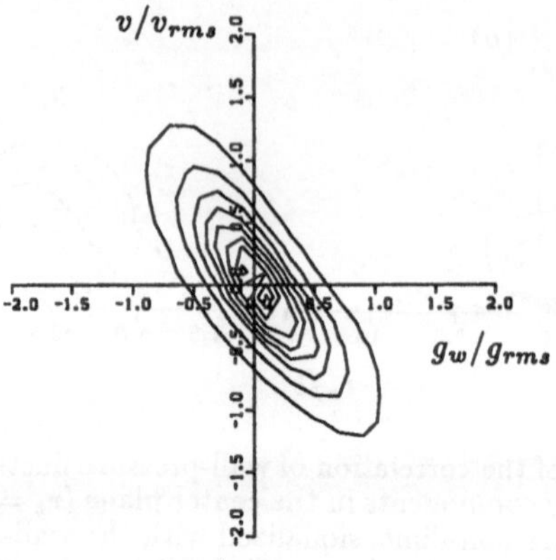

FIGURE 9. Joint probability density of g_w and the normal velocity at $y^+ = 10$. The contour levels are from 0.1 to 0.9 with increments of 0.1.

at $y^+ \approx 10$. From the continuity equation, one can deduce the equivalent relationship

$$v(y) = -\frac{y^2}{2}\left[\frac{\partial}{\partial x}\left(\frac{\partial u}{\partial y}\right)_w + \frac{\partial}{\partial z}\left(\frac{\partial w}{\partial y}\right)_w\right] + \dots \qquad (2)$$

Our numerical tests have shown that the correlation of the first term in the bracket with v is negligible. Joint probability density of v at $y^+ = 10$ and $g_w = \frac{\partial}{\partial z}\left(\frac{\partial w}{\partial y}\right)_w$ is shown in Figure 9. There is a remarkably high correlation between g_w and v. Among the wall-variables considered so far, this is the best detector of the events near the wall. High values of g_w have better correlation with the events of interest at $y^+ \approx 10$. For example, if g_w is greater than its rms value, then the probability that $v < 0$ at $y^+ = 10$ is 0.97, and the most probable value of v at this location is $-1.2 < v^2 >^{\frac{1}{2}}$. If $g_w < -g_{rms}$, then there is an 84% chance for ejection at $y^+ = 10$, with the most probable value of $v = 1.2 < v^2 >^{\frac{1}{2}}$. For large magnitudes, g_w is a better detector of sweeps than ejections.

5. SUMMARY AND CONCLUDING REMARKS

An active control algorithm was tested using direct numerical simulation of turbulent channel flow. The algorithm was based on blowing and suction at the wall exactly opposite to the instantaneous wall-normal velocity at a location near the wall. It was found that the optimum location for matching the transpiration velocity was at $y^+ \approx 10$. Approximately 20% reduction in skin friction was achieved. For practical considerations it is important that the sensor be confined to the wall. Several flow variables were considered for detection of sweep or ejection events at $y^+ \approx 10$. The gradient of wall stress proved to be the most sensitive indicator of these events.

The principle of selective blowing and suction for drag reduction appears to be promising. However, transpiration over the entire surface is far from practical. In the next phase of this study, we plan to explore *selective* blowing and suction targeted at specific structures.

Control of a transitional flow

The same control strategy was applied to a transitional flow to examine the effectiveness of the present algorithm in delaying (or suppressing) transition to turbulence. It was found that one could delay (or completely suppress) all small disturbances at supercritical Reynolds numbers in the plane Poiseuille flow. The change in the boundary condition modified the original eigensystem of the linearized Orr-Sommerfeld equation such that with a proper choice of y_r, the most unstable mode (a 2D Tollmien-Schlichting wave) became stable. This was true for a wide range of Reynolds numbers tested including $Re \approx 50000$, at which a linear disturbance attains its maximum growth rate for the plane Poiseuille flow. The implication of this result for turbulent flows is not completely clear at this point. It does, however, indirectly suggest that in turbulent channel flow, in which about 20% drag reduction was achieved, the present control strategy probably modified the flow instabilities.

We are grateful to Professor W. C. Reynolds for helpful discussions. This paper is a partial reporting of work under progress, and a major portion of it was presented at the AIAA 2nd Shear Flow Conference, March 13-16, 1989, Tempe, Arizona.

REFERENCES

1. Bushnell, D. M. 1984 Proceedings of ASME Conference on Laminar Turbulent Boundary Layers, New Orleans, Louisiana.
2. Kline, S. J., Reynolds, W. C., Schraub, F. A. & Runstadler, P. W. 1967. The structure of turbulent boundary layers. *J. Fluid Mech.*, **30**, 741.
3. Corino, E. R. & Brodkey, R. S. 1969. A visual investigation of the wall region in turbulent flow. *J. Fluid Mech.* **37**, 1.
4. Johansson, A. V., Alfredsson, P. H. & Kim, J. 1987 Shear layer structures in near-wall turbulence. In *Studying Turbulence Using Numerical Simulation Data -bases*, Proceedings of the 1987 Summer Program of the Center for Turbulence Research, NASA-Ames & Stanford University.
5. Kim, J. & Moin, P. 1986 Flow structures responsible for the bursting process. *Bulletin of the American Phys. Soc.*, **31**, 10, 1716.
6. Kim, J., Moin, P. & Moser, R. D. Turbulence Statistics in fully developed channel flow at low Reynolds number. *J. Fluid Mech.*, **177**, 133.
7. Kuhn, G. D., Moin, P. Kim, J. & Ferziger, J. H. 1984 Turbulent flow in a channel with a wall with progressive waves. Proc. of *ASME Symposium on Laminar Turbulent Boundary Layers: Control, Modification and Marine Applications*; New Orleans, 61.
8. Kim, J. 1985 Turbulence structures associated with the bursting event. *Phys. Fluids*, **28**, No. 1, 52.

Part 5
Drag Reduction by Passive Means

Drag Reduction by Passive Devices – a Review of some Recent Developments

By: Dr. A.M. Savill
Rolls-Royce Senior Research Associate
Department of Engineering
University of Cambridge
England

INTRODUCTION

There is currently considerable academic and industrial interest in the possibility of controlling the eddy structure of turbulent boundary layers by passive means (either to reduce skin friction drag, self-noise and mixing or alternatively to increase mixing and heat transfer), both for the practical benefits that may be achieved and for the insight such studies can provide into the mechanisms and dynamics of these flows. A concentrated research programme at NASA Langley [see 1,2] and many other laboratories in the USA, India and Japan [3], together with an increasingly collaborative effort between research groups in Europe (and Canada) [2,4], has established that turbulent skin friction drag can be reduced by suitably scaled modified wall geometries, most notably longitudinal grooves or 'riblets', and thin plate or aerofoil outer-layer-devices (variously also known as LEBUs or manipulators/manipulateurs(Fr)) introduced into the flow itself. Indeed it has now been demonstrated repeatedly that surface drag can be reduced by up to 8% (nett), at least below M=1.2 [see 2], in nominally zero pressure gradient plane flow over either machined/moulded riblets or plastic riblet sheets manufactured by 3M Inc or Hoechst, and there is some evidence that similar benefits might be attainable with aerofoil manipulators.

Both riblets and manipulators have now been tested on bodies of revolution in air and in water, and 3M riblets have been applied, apparently successfully, at full-scale, on rowing shells in the 1984 Olympics and 1987 University Boat Race, and on the 1987 Americas Cup winning yacht, Stars and Stripes (tank testing of alternative U-groove plastic riblet sheets, fabricated by Hoechst, on a 1/3 scale model of the previous winner Australia II, having revealed a 3% total drag saving for the low wave making conditions of the final rounds [see 1,2]). The idea of using such passive techniques to reduce surface drag is thus well established for external flow applications and their application to internal flows through channels and pipes has recently been the subject of a number of studies [see 2,5,6]. Some information is also now available on such practically important effects as the additional influence of pressure gradients, free-stream turbulence [see 1,2] and surface curvature [7]. In addition, although the practical performance of aerofoil manipulators under sea-going and flight-representative conditions is still being evaluated in controlled laboratory experiments, commercial aircraft flight tests of riblets are already being undertaken by Airbus Industry, Boeing, British Aerospace and NASA, as well as Pratt and Whitney [eg.see 8].

At the same time detailed experimental measurements of mean and turbulence quantities, allied to flow visualisation conditional sampling and pattern recognition techniques, have identifieded a number of possible drag reduction mechanisms, and the maximisation of some of these has tentatively been linked to the apparent optimum set of device configuration parameters revealed by parametric drag surveys. Although the exact nature of the processes responsible for any drag saving remain obscured by differences in the interpretation of different experimental findings, such studies have also

A. Gyr (Editor)
Structure of Turbulence and Drag Reduction
IUTAM Symposium Zurich/Switzerland 1989

provided valuable test case data sets for turbulence modelling. Significant progress has already been made in the modelling of such drag reducing flows [7] and it is hoped that future parametric computations will aid the optimisation of device parameters for both high Reynolds number external flow and fully developed internal flow under a variety of practical flow conditions.

The search is now underway for alternative passive techniques with larger drag reduction potential. Already it has been shown that a wide range of apparently rough surfaces may, like riblets, in fact exhibit drag reducing properties when of sufficiently small scale [eg.see 9] and a number are being studied in greater detail with a view to optimising their performance. One way in which larger drag reduction may already be obtained is to apply riblets and manipulators in combination [see 10] or to combine either with various other existing control techniques. Several such combinations have already been studied with beneficial results [6,10].

Since a number of other review publications have already appeared on this subject of drag reduction by riblets and manipulators (covering the research conducted at NASA Langley, and the USA in general, as well as European research efforts and those of ONERA in particular [11]) the scope of this paper has been restricted to consider primarily research conducted either at Cambridge or in collaboration with other research groups. By relating the results to those obtained elsewhere the purpose is to establish the state of the art and probable directions for future work. All of the above developments are therefore reviewed with the emphasis being placed on the latest results obtained for both riblets and aerofoil manipulators (particularly at high

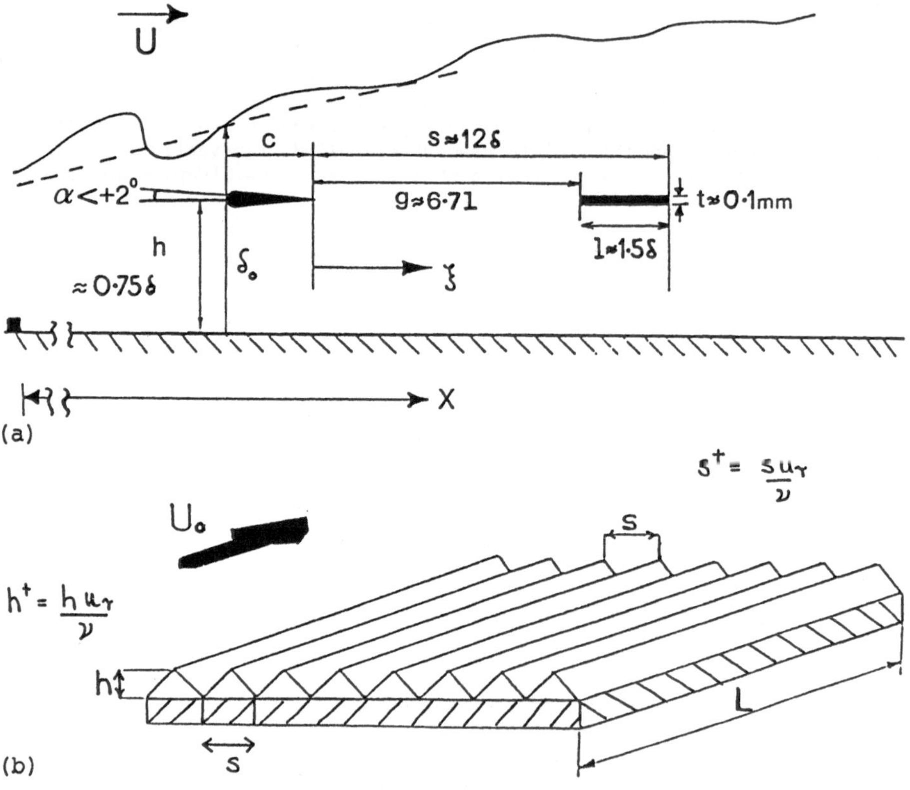

Fig.1: Definition of configurational parameters for: a)manipulators & b)riblets

Reynolds numbers), their combination, studies of alternative surface geometries, and other previously unpublished findings.

IN-FLOW DEVICES

The use of manipulators or LEBUs appears to have its origins in work on grids, screens and honeycombs to control free-stream wind tunnel turbulence [12]. Such 'management' techniques were first applied to boundary layers by reserachers at NAL Bangalore [13] and further studies [14-17] led rapidly to the definition of single or tandem thin elements as the preferred means of achieving drag reduction. Subsequent more detailed parametric surveys [18-22] in nominally zero pressure gradient low Reynolds number ($Re\theta<7000$) boundary layers identified the following optimum parameters for tensioned plates (in terms of boundary layer thickness at the device leading edge - see Fig.1), of: chord length $l=1.3\delta$, thickness $t=0.1mm$, and tandem spacing (leading to leading edge) $s=10-12\delta$. These are all to an extent compromise values since any nett reduction results from a balance of the inherent device drag introduced into the flow and the integrated Cf reduction produced downstream. The latter (but not so much the former) also depends critically on the height at which the devices are mounted within the flow. Typical results are illustrated by Fig.2a-d from which it can be estimated that $h=0.75\delta$ is optimum, although the value deduced naturally depend on the integration length assumed and becomes considerably less when this is shorter. (There is also some evidence that h should be reduced as $Re\theta$ increases [17]). The Cf reduction does not appear to be greatly affected by t [23], the optimum value of this parameter being that required to minimise device drag, but increases with l (there being a marked increase between $l<\delta$ and $l>\delta$) and only the compromise with device drag produces an optimum value (see Fig.3). Tandem spacing also has some affect on both drag elements since the second manipulator benefits from sitting in the wake of the first. The figure quoted for s has emerged from a number of different studies, but it is not clear this is the most appropriate parameter; the gap g between trailing and leading edge perhaps being a better choice [see 24]. In fact comparison of a wide range of data for local, integrated and nett reductions suggests that there may not be a unique optimum choice of g (Fig.4a). Although $g=7.7\delta$ appears slightly favoured, a finding recently independently verified for plates with $l=1.6-2.2\delta$ (when $s\sim10\delta$) [21], for plates of $l<\delta$ a gap $g<2\delta$ appears preferable and nett reduction has been reported for $g<\delta$ [16]. It would therefore appear that g and l should be optimised together and replotting the data in terms of g/l (see Fig.4b, suggested by Bonnet [private communication], by analogy with galloping instabilities for tandem bodies in external flow) reveals two optima: $g=6.7l$ (which for optimum length plates corresponds to $s=10\delta$) and $g=3l$ for a range of $l=2-3\delta$, for which again s is approximately 10δ, most clearly evidenced by the recent EPFL data [21]. A different trend is again suggested for $g<2\delta$ and $l<\delta$, and flow visualisation has suggested that the main effect of such small gaps may simply be to increase the effective length of plate.

The suggestions that there might be greater advantage to be had from introducing the device drag/momentum defect in smaller stages [3] or from utilising small plate gaps to gain similar benefits from shorter lower drag plates, prompted some attempts by the author to further improve manipulator performance by employing triplet, quadruplet or other plate arrangements and some results are presented in Fig.5. Although larger Cf reduction was obtained with three plates rather than two there was no nett advantage when the additional device drag was taken into account (as also found by V.D.Nguyen and by R.Blackwelder [private communications], but a better nett performance was achieved by replacing each element of a tandem by a close pair of plates. In view of the rather small magnitude of the additional benefit it is unlikely such complex configurations would be of practical interest, however it is interesting to note that larger reductions were obtained when the manipulator plates were replaced by meshes and serrated leading and trailing edges were

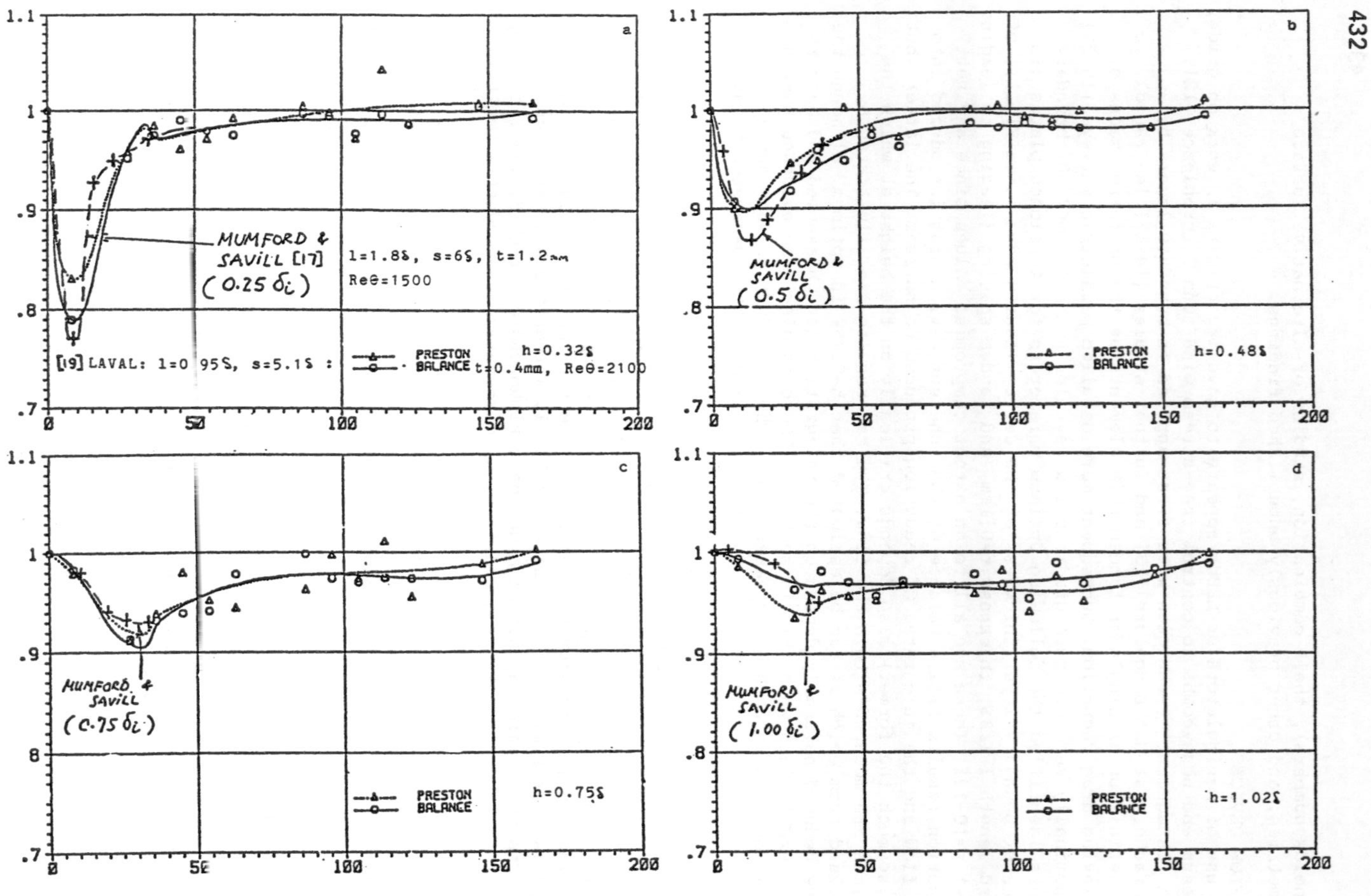

Fig.2: Comparison of tandem manipulator performance at different heights

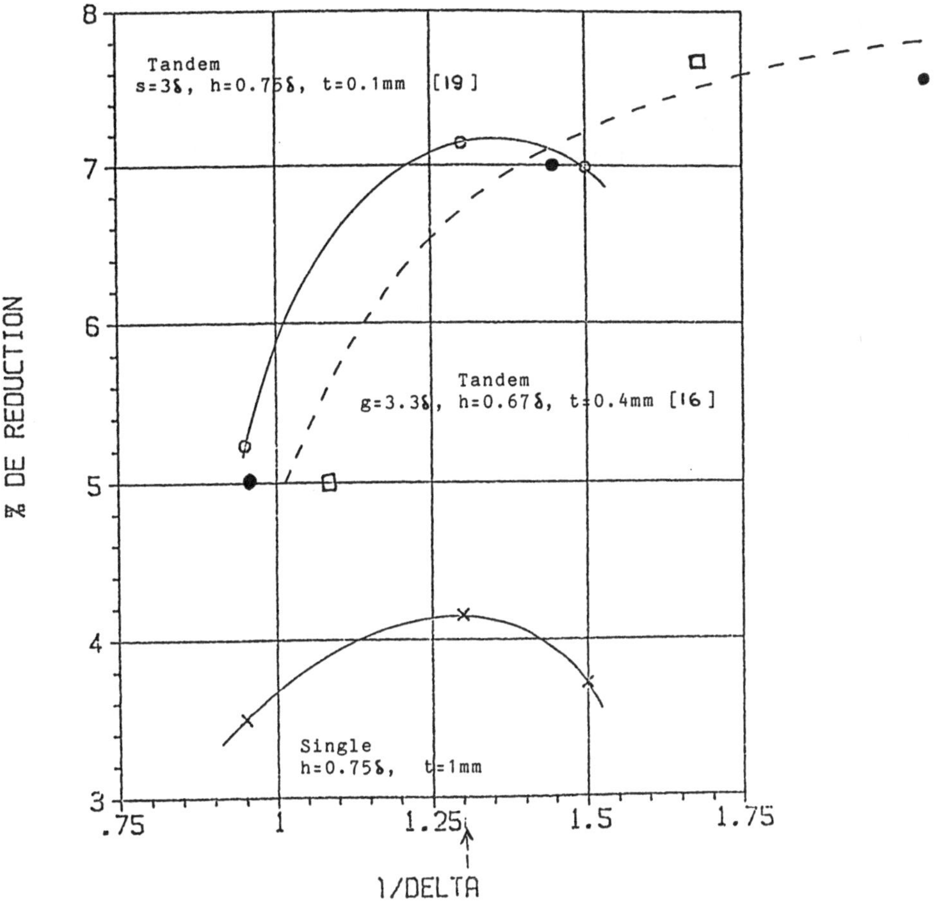

Fig.3: Influence of plate length on drag reduction for single & tandem plates

introduced in an effort to introduce more persistent longitudinal vorticity into the wake.

Studies of aerofoil section manipulators [25,26] have suggested essentially the same optimum values of s,l and h for the rather more robust devices which will be required in any practical applications. The replacement of thin plates by aerofoils introduces two other parameters: profile shape and angle of attack into the optimisation problem. By common consent initial studies have been restricted largely to symmetrical NACA 0009 aerofoils, although some assymetric sections have also been tested, and small positive angles of attack (of less than 2°), have been shown to offer additional benefits.

Unfortunately very few direct measurements have so far been made of device drag [17,27] and there is considerable disparity between the Cf, and hence nett, reduction recorded in different experiments even with optimised tandem arrangements. These differences have been attributed to the transitional nature of the flow on the devices themselves, leading or trailing edge separations with consequent detrimental vibration, and the sensitivity of the flow to fine details of the profile at the low chord Reynolds numbers (Rec<120,000) which

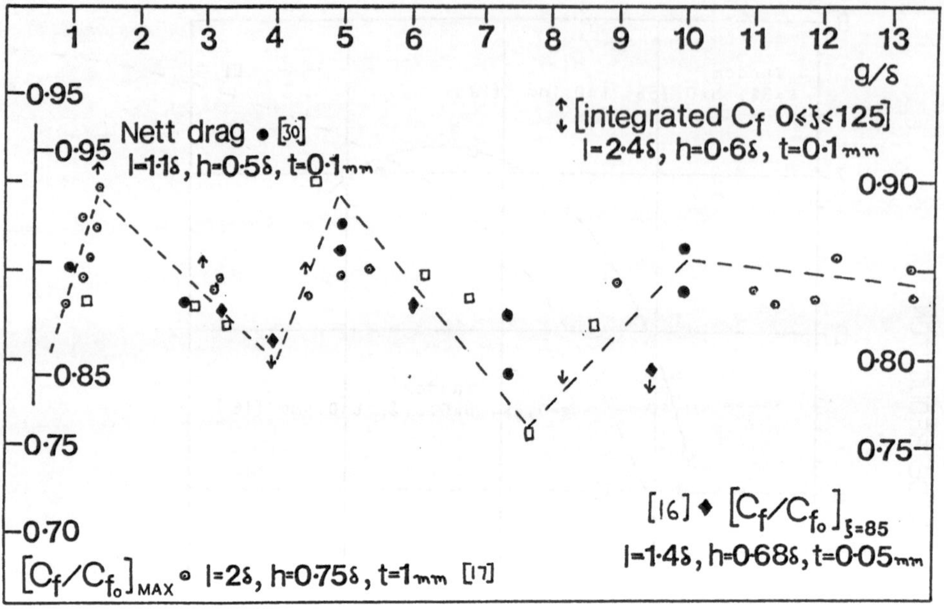

(a)

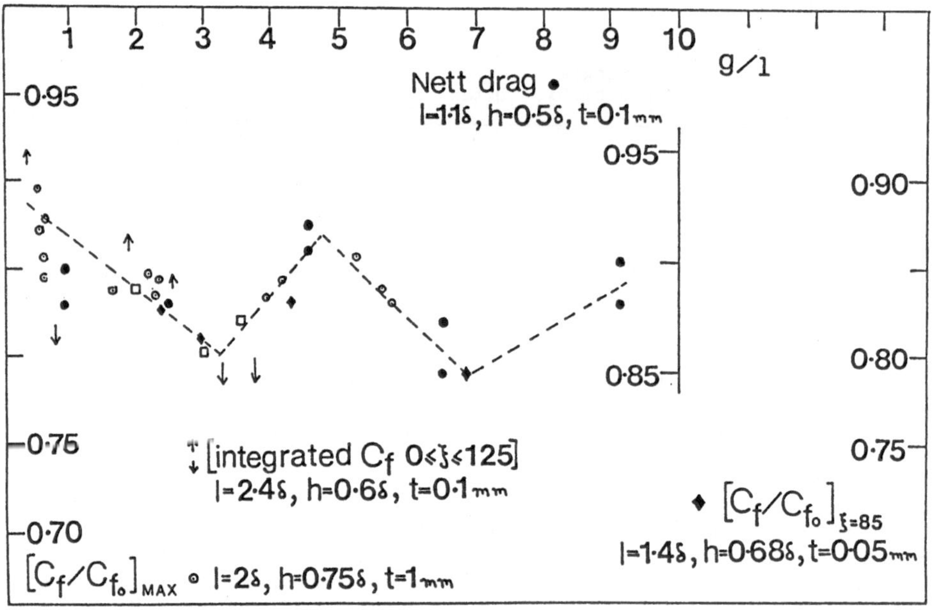

(b)

Fig.4: Variation of tandem performance with plate gap: a) g/δ & b) g/l

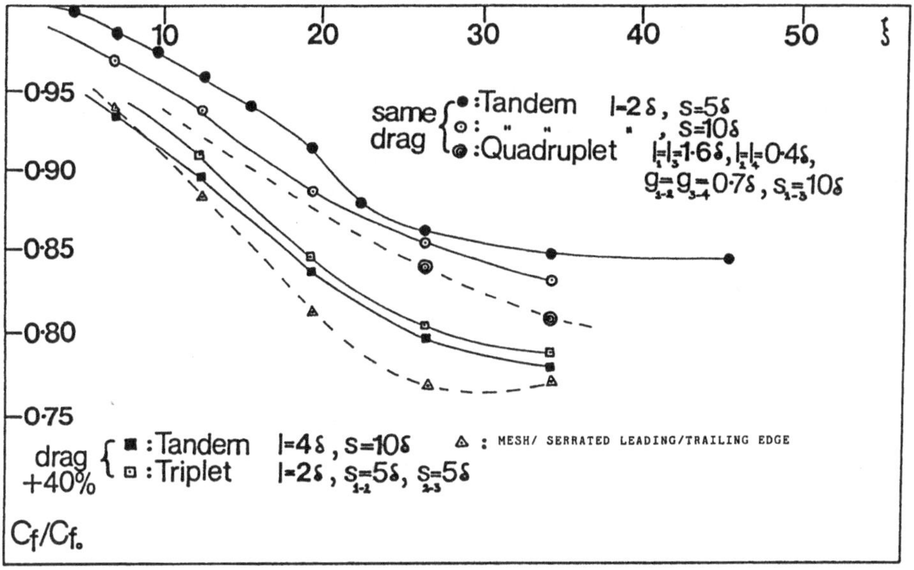

Fig.5: Comparative performance of various multi-plate manipulators

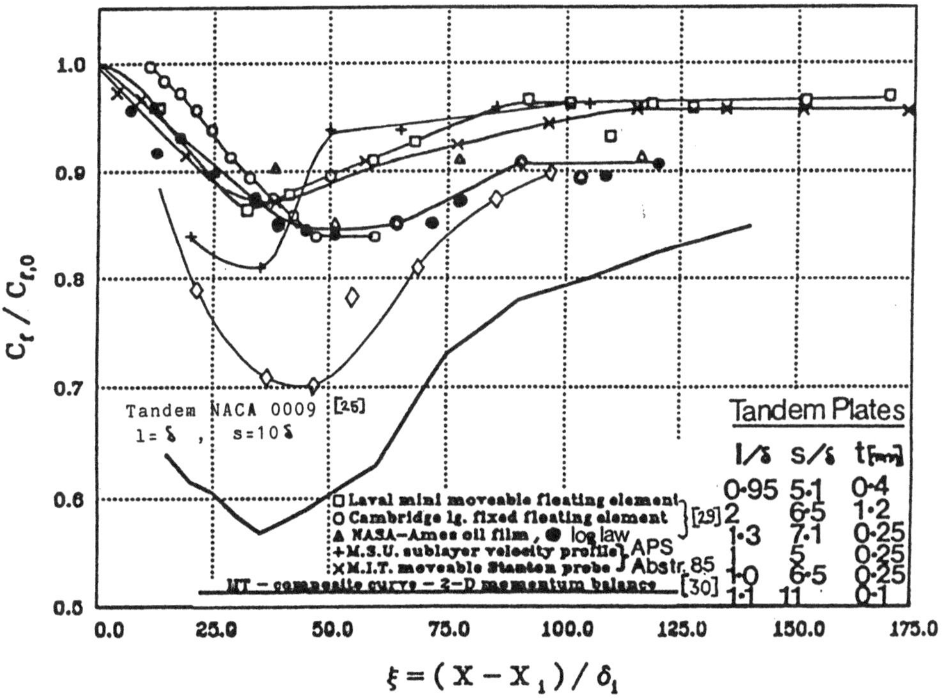

Fig.6: Comparison of different data sets for 'optimised' tandem devices

have so far been examined [28]. However it would appear that substantially larger effects than have been reported in the majority of such experiments [eg.29] can be achieved with very carefully machined, set-up and tensioned flat plates [30] or aerofoils [25] providing they have very sharp trailing edges (see Fig.6).
Manipulators have also been tested on an axisymmetric body [31] and on a towed canoe [see 32] with similar C_f, if not, nett results. Considerably less attention has so far been paid to internal flow applications despite the potential practical benefits in pipelines or ducting. An early study conducted in a fully developed channel flow showed no nett reduction [33] leading to suggestions that the absence of a free interface may provide a critical limitation on performance under such conditions. More recently a series of pipe flow manipulation experiments have been conducted at ETH Zurich [34] at Re_D=180,000. A limited parametric study revealed that when ring manipulators of $l=1.5\delta$, with t tapering from 0.3 to 0.15mm, were mounted within the initial developing boundary layer region $h=0.36$ & 0.71δ proved better than $h=1.1\delta$, as for external flow operation. However even then only half of the pressure drop due to the device drag was recovered within 200δ (or 30D) of the device, a similar recovery distance to that observed in the channel experiments, and the overall drag deficit was simply doubled for two such rings in tandem with $s=12\delta$ (See Fig.7a). In the fully developed region the single devices, then with $l=0.4R$ ($D=2R$), performed equally well when mounted at $h=0.1, 0.2$ or $0.3R$ from the pipe wall, and for the case of the tandem device ($s=3.1R$) mounted at $h=0.2R$ recovery was still continuing at $\xi=60R$ downstream, the limit of the experimental rig (Fig.7b). However it is still unclear what the optimum device parameters should be once these have to be scaled on R rather than δ.

On the basis of more detailed mean flow and single or multi-point turbulence measurements, including very extensive data sets obtained at EPFL [21], CEAT Poitiers [23,35], ONERA-CERT [22], USC [37] and Cambridge [20,23,24] (where in the latest extension phase of this work Pattern Recognition Analysis has been used to confirm changes to the turbulence structure previously deduced from correlation data [38]) allied to conditional sampling and flow visualisation, a general concensus has emerged regarding possible drag reduction mechanisms and their relation to the optimum configurational parameters. The proposed mechanisms may be separated into: 'immediate' effects of the devce:

(i) Suppression of large scale motions (for which an individual plate, (or two plates each $l<\delta$), separated by $g<2\delta$, (with an 'effective ' plate length greater than half the typical length of bulge scale motions, of order 2δ) appear more efficient) by restricting vertical fluctuations (an effect considerably amplified for plates with $s>10\delta$ [39,40])

(ii) 'Downwashing' of the momentum defect towards the wall due to circulation about the device (enhanced for lifting aerofoils at small angles of attack)

and 'persistent' effects of the device wake:

(i) Introduction of small scales which promote an enhanced energy cascade from larger scales.

(ii) Interactions between wake vortices of opposite sign to the bounday layer vorticity leading to damping of near wall fluctuations and associated reduction in momemtum transfer (h determining the rapidity with which this affects the wall region, and hence subsequent relaxation occurs).

(iii) 'Blocking' by the wake vorticity of incursions of external fluid and outer layer fluctuations from respectively either perturbing the wall region directly or increasing near wall fluctuations via interaction-at-a-distance effects of irrotational velocity fields.

(all three effects being proportional to the strength of the wake which seems in turn to depend on l, and in a weak oscillatory manner on s,g as illustrated by Fig.4, perhaps due to the type of 'beating' of the two wakes envisaged by Falco [41]). However no clear evidence has so far emerged for any kind of 'vortex unwinding' [15] , and any reverse transfer of energy [14] appears to

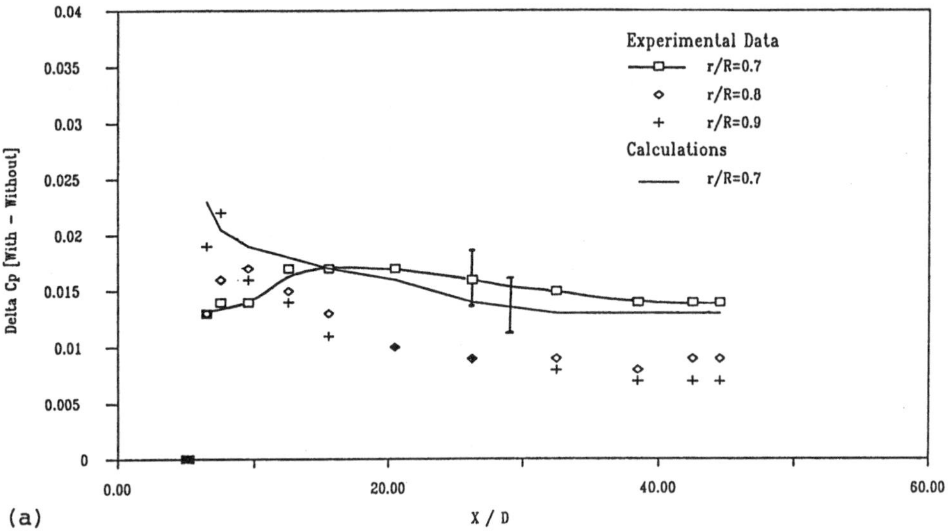

(a)

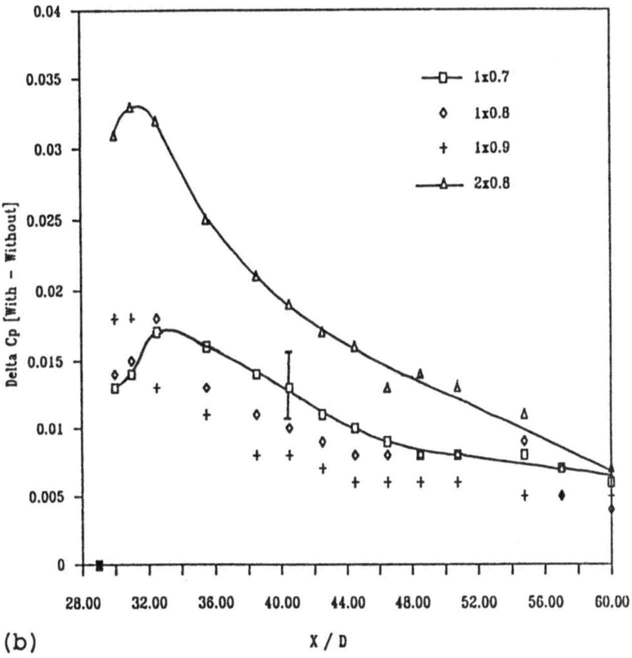

(b)

Fig.7: Effect of manipulators on Cp in: a)developing & b)developed pipe flow

be restricted to the near wake region of thicker devices which do not offer nett drag savings.

SURFACE MODIFICATIONS

The use of longitudinal v-groove riblets has its origins in the inovative turbulent boundary layer skin friction reduction programme initiated at NASA Langley in 1972 (also reponsible for much of the work on manipulators). The present 'standard' aspect ratio one, triangular grooves (see Fig.1b) have evolved from initial studies of similar, but physically larger, ribs which were found to offer approximately 15% increase in heat transfer for little drag penalty [42]. Further investigations [43] revealed that smaller scale with h+,s+ <30 offered nett drag savings, with a maximum reduction of 8% (essentially per unit area) for h+=s+=10; a finding confirmed by subsequent detailed parametric surveys conducted at Langley [44] and a large number of other laboratories in North America and Europe [1,2,4,22,45]. During the course of these it was suggested that a slight additional benefit might be achieved by rounding the grooves and reducing the include angle of the peaks to less than 25 [46]. In fact the use of such U-profile grooves had emerged almost independently from a different line of research pursued by researchers in Germany, where interest in these was initially stimulated by the suggestion (first noted in 1967 [47] according to Dinkelcaker [private communication]) that shark scales, which have a similar cross-section profile, non-dimensional size and are flow aligned, might also act in a similar manner. Discrepancies between various data sets can be attributed largely to experimental uncertainty in measuring such small effects and to the apparent sensitivity of riblet performance to fine details of their geometry, possibly amplified by machining or moulding processes. The latter problem has largely been overcome by the development of adhesive-backed vinyl riblets (size .0004-.006in.)in the USA by 3M Inc, and the manufacture of similar plastic semicircular U (and more recently V also; both with s=0.25mm) groove plastic sheets in Germany by the Hoechst company. Availability of these has led to far more consistent results, confirmed earlier findings including drag reduction for very small h+ [48], and indicated that s+ is the more appropriate parameter for collapsing data for different riblets (eg.Fig.8). However the effect of the ratio h/s on the extent of the drag reduction range still needs to be examined further since any widening of this could be of practical benefit in flows where conditions are changing. Rounding, blunting and notching of riblet peaks have already been found to extend the drag regime to h+ 45-50 [43], but at the expense of a smaller maximum reduction, and it would seem blade or L shaped riblets may offer advantages [52].

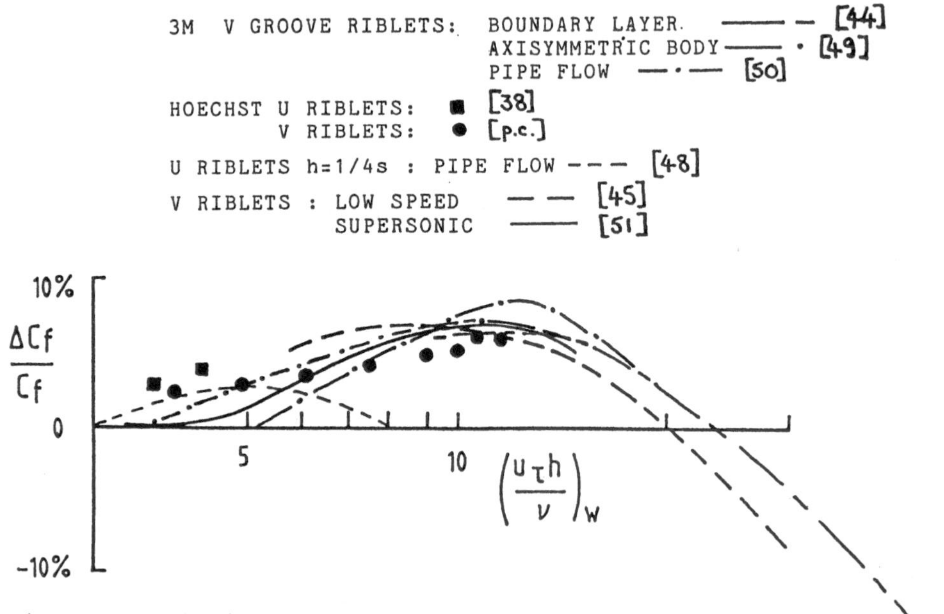

Fig.8: Drag reduction due to V & U groove riblets versus h+

Riblets have also been applied to a 2D aerofoil [22], a glider aerofoil [see 2], axisymmetric bodies [2,49,53] and various 3D bodies including a 1/11th scale A320 aeroplane model [22] and a 1/3 scale model of an Americas cup yacht [54]. In all cases the measured drag reduction was consistent with the Cf reduction recorded on flat surfaces with the same non-dimensional scale riblets, and similar levels of performance have also now been recorded in internal flow application to both pipes [48,50,55] and full developed duct flow [56].

Although a number of very detailed investigations have been conducted into possible modifications to the flow structure both within and above riblets (including measurements of mean, single and multipoint turbulence [4,57] quantities, VITA/burst statistics [58,59], quadrant analysis [4,60], and conditional sampling allied to flow visualisation [61,62]) it is still not entirely clear how they function. There is considerable disparity between the findings of different research groups largely it would seem due to differences in the techniques and thresholds applied. However it would appear that both bursts and sweeps are modified on riblet surfaces and the remarkable similarity and regularity of burst signatures (sufficient in fact for this to be used as a trigger for conditional flow visualisation) that has been recorded in a number of cases [58,59,62] points to the existence of a more stable and more ordered streak structure. The general concensus of opinion seems to be that riblets act primarily to increase the sub-layer thickness, but at the same time influence the near-wall turbulent structure and production directly by inhibiting spanwise motions. Sublayer velocity traverses [57] have suggested that the outer flow sees a virtual wall origin raised to an effective height of 0.75s (peak origin=h; groove origin=0.5h), so that only the top 0.25s of the riblet protrudes. This estimate is found to provide the best collapse of a range of mean velocity data, is close to the figure of 0.2D found for close-packed hemispherical roughness of diameter D

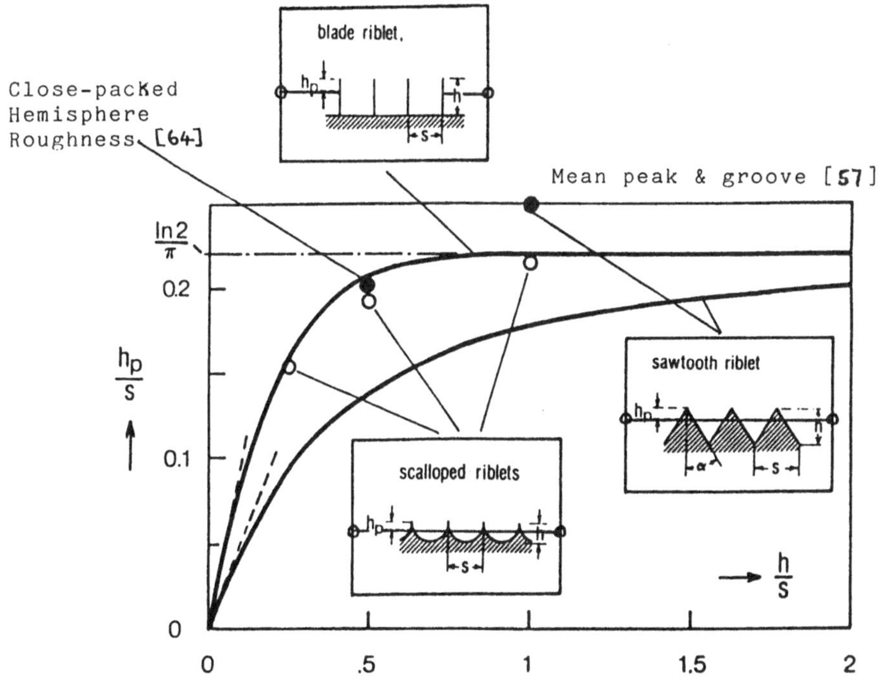

Fig.9: Protrusion heights of different riblets above virtual wall origin

[64], and agrees closely with estiamtes of 'protrusion height' derived for U, V and L riblets from an entirely viscous analysis due to Bechert et al. [63], (see Fig.9). It seems likely therefore that about three quarters of the riblet effect is due simply to the response of the flow to this change in wall condition, although both Choi [61] and Bechert [private communication] have independently pointed out that the secondary effect of the riblets on wall shear stress fluctuations should not be underetsimated since these are continually 'uncovered' by, and hence can interact more directly with, bursts of turbulent activity penetrating the viscous sublayer.

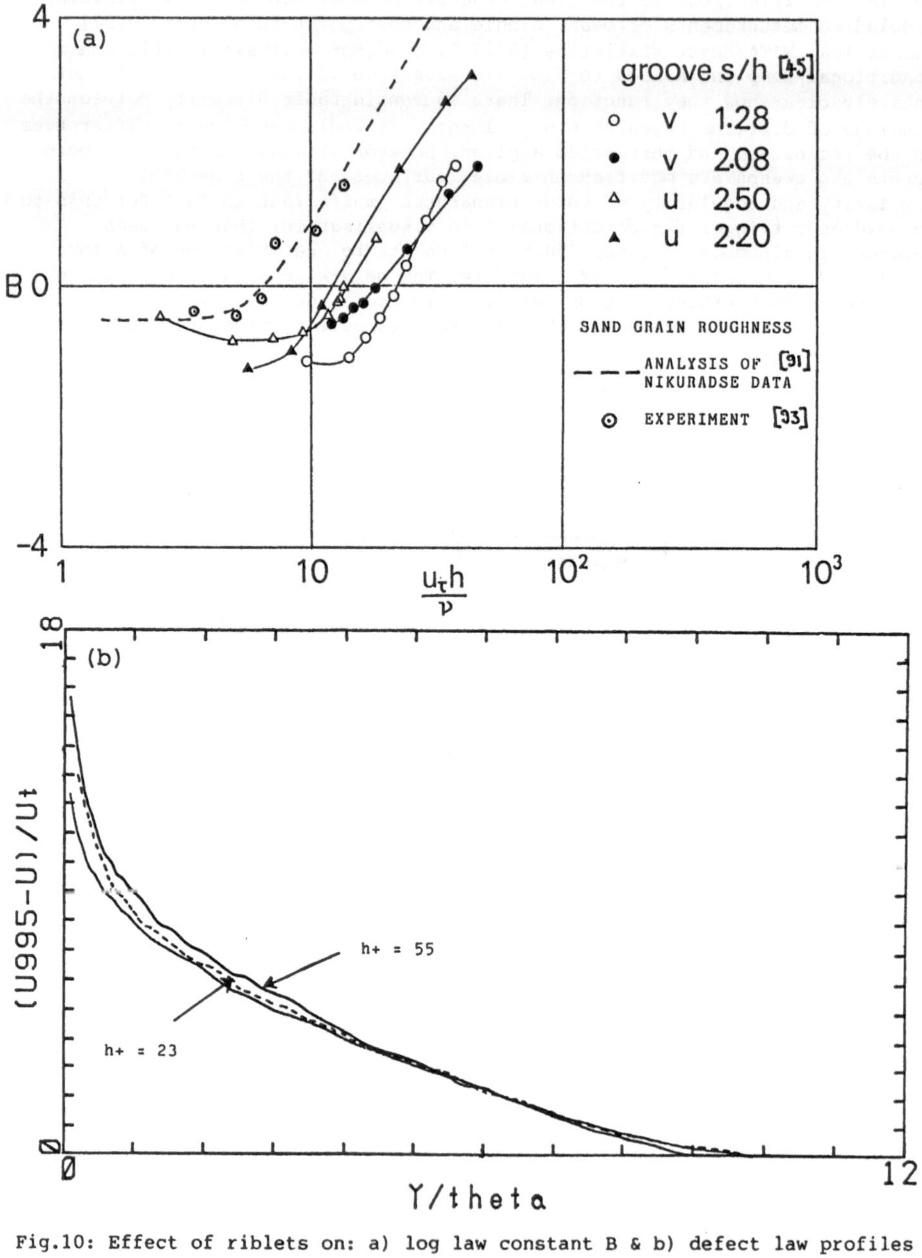

Fig.10: Effect of riblets on: a) log law constant B & b) defect law profiles

When plotted in logarithmic form velocity profile data indicate that riblets act as a kind of 'negative roughness', in the sense that the roughness shift to the additive constant B is positive within the drag reducing range and only changes sign when they are sufficiently large scale to be drag increasing (see Fig.10a). A similar shift has been found in the recovery region of manipulated boundary layers [29] where there is also evidence from correlation maps for a more stable eddy structure [23]. Although the shape factor appears to be unaffected by drag reducing riblets, there is clear evidence for a change in the wake factor and defect law (see Fig.10b) suggesting that the effects of the modified wall geometry are felt throughout the flow.

HIGH SPEED TESTS (a) Aerofoil manipulators:
During the last 3 years the first series of high speed wind tunnel tests of both 3M riblets and NACA 0009 aerofoil manipulators have been conducted in a transonic blow-down tunnel at the University Engineering Department. The tunnel and the techniques employed in operating it have been described in detail elsewhere [65].The working section consisted of either a continuous test plate or a number of separate blocks, one of which contained a floating element balance which could thus be located at a number of positions along the test surface (ξ=-15-200δ). Velocity traverses were conducted at two positions 0.82 and 1.30m downstream of the origin (ξ=90 or 130δ)

The tests were made at Mach numbers of 0.88, 0.77 and 0.5, corresponding Re1 2.20×10^7 per metre at M = 0.88 to 1.50×10^7 per metre at M =0.50. Reθ at the manipulator location was approximately 7500 at M = 0.5 rising to 9000 at M=0.77 and 11,000 at M = 0.88. The floating element drag balance used to measure local Cf was of a type designed by colleagues at Laval University [66], but had been modified by them for such high speed use, and was supplied by Les Industries Fanny Inc. of Quebec. Velocity profiles were measured by means of a flattened pitot tube linked to a traverse gear driven by a stepper motor and automatically controlled by a microcomputer.

For the manipulator tests a NACA 0009 aerofoil manipulators was machined from soid aluminium in the Department Workshops (in spite of its small size the profile was in very close agreement with the NACA 0009 ordinates). This spanned the tunnel, and was mounted at h=0.5, 0.7 or 0.9δ and α=0°, 35cm downstream of the start of the smooth or riblet test section. The chord length c=1=6mm was chosen to provide the same value of chord/boundary-layer thickness as in a complimentary series of tests to evaluate the device drag of such a manipulator under similar conditions conducted at CEAT Poitiers [67] (Relevant aerofoil and boundary-layer parameters are compared in Table 1).

The skin friction distributions recorded behind the manipulator are shown in Fig.11. These show first that the device had no influence on Cf at a station 15δ upstream, a finding consistent with the conclusion from Cp data from both the CEAT experiments and low-speed studies that the upstream influence is limited to only 5δ (see Fig.12). They also indicate that in each case recovery took place around 120δ from the device, in close agreement with low speed measurements [15,29], but the maximum local reduction occured further from the device than expected. Otherwise the M=0.77 & 0.88 results were broadly in line with both the low speed data and the only other available transonic data, preliminary flight test results, obtained at the FFA in Sweden [69]. The largest integrated reduction (average 7%) occured for M=0.88 and and this was found to approximately balance the device drag as estimated from the $C_D$measured at CEAT for their highest M=0.8. This is a remarkable result in view of the non-optimal nature of the NACA 0009 profile aerofoil and the fact that its boundary layers must have been transitional. In fact subsequent oil flow visualisation at CEAT has suggested that CD, which was approximately twice that recorded for the same aeofoil in the free-stream, may have been affected by flow separation on the device, and this has led Falco [private communication] to suggest that the use of an inverted cambered aerofoil may be preferable. Low speed studies conducted earlier by J.Lemay and the author have

TABLE 1.

Comparison of aerofoil manipulator parameters at M= 0.76.

CEAT Poitiers	Cambridge
$\alpha = -1^\circ, 0 +1^\circ$	$\alpha = 0$
M = 0.75	M = 0.77
c = 12.44mm	c = 6mm or 12.7mm
Rec = 150,000	Rec = 120,000 or 260,000
$Re\theta$ = 11,000	$Re\theta$ = 11,000
$\delta \doteq$ 10mm	$\delta \doteq$ 5mm
c/δ = 1.2	c/δ = 1.2 or 2.5
h/δ = 0.3-0.9	h/δ = 0.5, 0.7 & 0.9
t/δ = 0.11	t/δ = 0.11 or 0.22
NACA 0009 profile	NACA 0009 profile
Carbon-fibre (moulded at ONERA)	Metal (Machined at CUED)
Tension 4000N	Tension 50N or 100N

already indicated that NACA 4415 aerofoils may have other benefits in terms of increased Cf reduction (Fig.13) provided these are inverted. Since NASA have obtained similar poorer performance from NACA 4409 aerofoils in their normal operating mode it may be that the improvement results from an amplification of opposite sign vorticity, as obtained by Goodman with a 'one-sided vortex generator [70]. It is interesting to note that similar devices, although of rather cruder shape, were used to achieve substantial Cf reductions in flight.

However the average Cf reduction at M=0.77 amounted to only 4% and comparison with CD measured by CEAT for the same M indicated that only half the device drag was recovered at this condition. Although the most recent results from CEAT suggest CD may be higher for M=0.88 it is interesting that momentum thickness measurements from a series of traverses close behind (ξ =2-30δ) their aerofoil also suggest a larger reduction at M=0.8 compared to M=0.7 [71].

Further Cf results are presented in Fig.11b,c for the aerofoil mounted either higher at h=0.9δ, or lower at h=0.5δ , which suggest that h=0.7δ remains optimum for Cf reduction (and hence nett performance since the CEAT results indicate h=0.6-0.7δ is required for minimum device drag under these conditions). However both lower (expected) and higher (unexpected) locations apparently led to a more rapid recovery, and the maximum Cf reduction for the low position may not have been captured.

Momemtum thickness measurements surprisingly indicated larger nett reductions which could only be consistent with the Cf data if the device drag were reduced to zero (or less) by a Katzmayr effect [72]. The only other possible explanation seems to be that the device must have had some influence on the opposite tunnel wall boundary layer so that the total momentum change

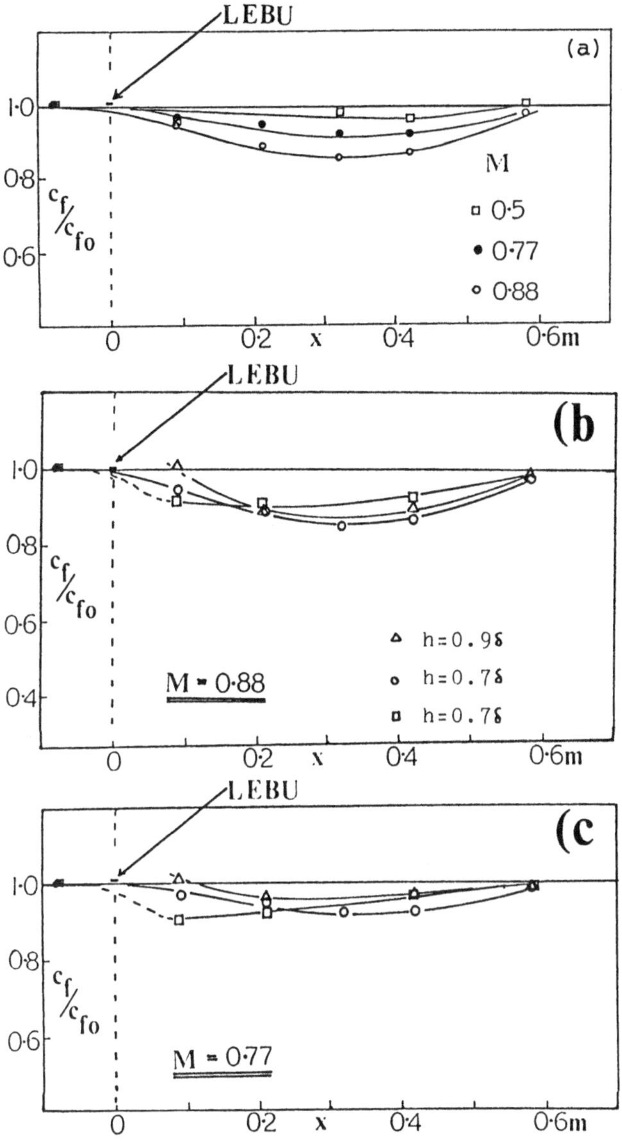

Fig.11: Cf reduction due to a NACA 0009 manipulator at high subsonic Mach No.

was not captured by the boundary layer traverses. This conjecture is supported by Schlieren photographs for the ONERA carbon fibre aerofoil taken at CEAT, which show evidence of pressure waves moving upstream from the aerofoil, extending as far as the opposite tunnel wall, and also suggest the existence of an expansion fan emanating from the device at this condition (see Plate 1a). However this effect could not explain the different behaviour recorded in the only other set of LEBU experiments so far conducted at such high Re_θ (but not M) at NASA Langley on an axisymmetric body in water [73]. In contrast these suggest that the Cf reducing performance drops off sharply as Re

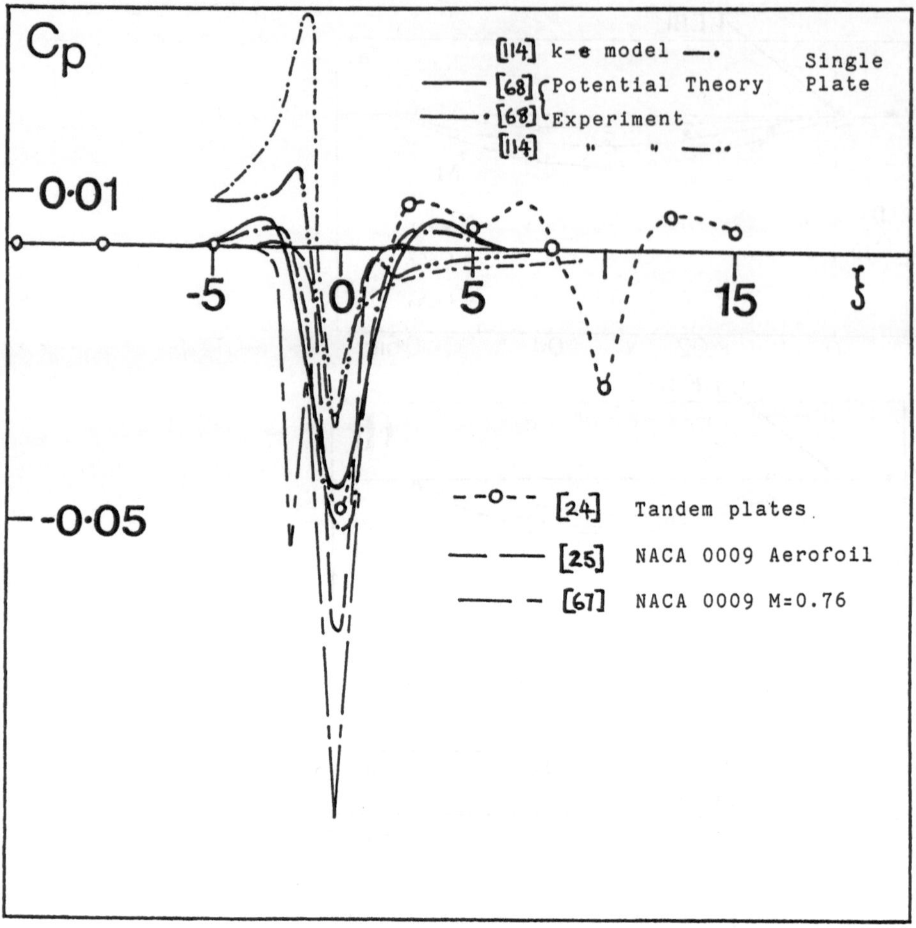

Fig.12: Cp distributions beneath single and tandem manipuators

increases above 7000. One possible explanation for this apparently contradictory finding might perhaps be that their results were affected by vibration (or cavitation) since this is known to have been a problem in other studies of manipulators in water. Similar loss of performance at low speed has been attributed to vibration or fluctuating separation at the leading edge, and some limited drag balance results obtained with the exact same carbon fibre aerofoil manufactured by ONERA for the CEAT tests, showed that no Cf reduction occured as long as the device was vibrating. (Indeed in tests at Cambridge the tension had to be increased to within 1% of the specified ultimate tensile strength, well above that used at CEAT in order to avoid such detrimental behaviour, and the trailing edge was eventually destroyed by vibration). It would be interesting in future to reevaluate CD and the Cf distribution for a truncated profile in order to assess the importance placed on the trailing edge influence by the NASA researchers [28].

(b) 3M Riblets:

Six sizes of 3M riblets (ranging in size from .0007 to .006in and covering a range of h+=10-110) have also been tested in The Cambridge facility at M=0.5 & 0.88, the drag reduction (or increase) being deduced in this case solely from

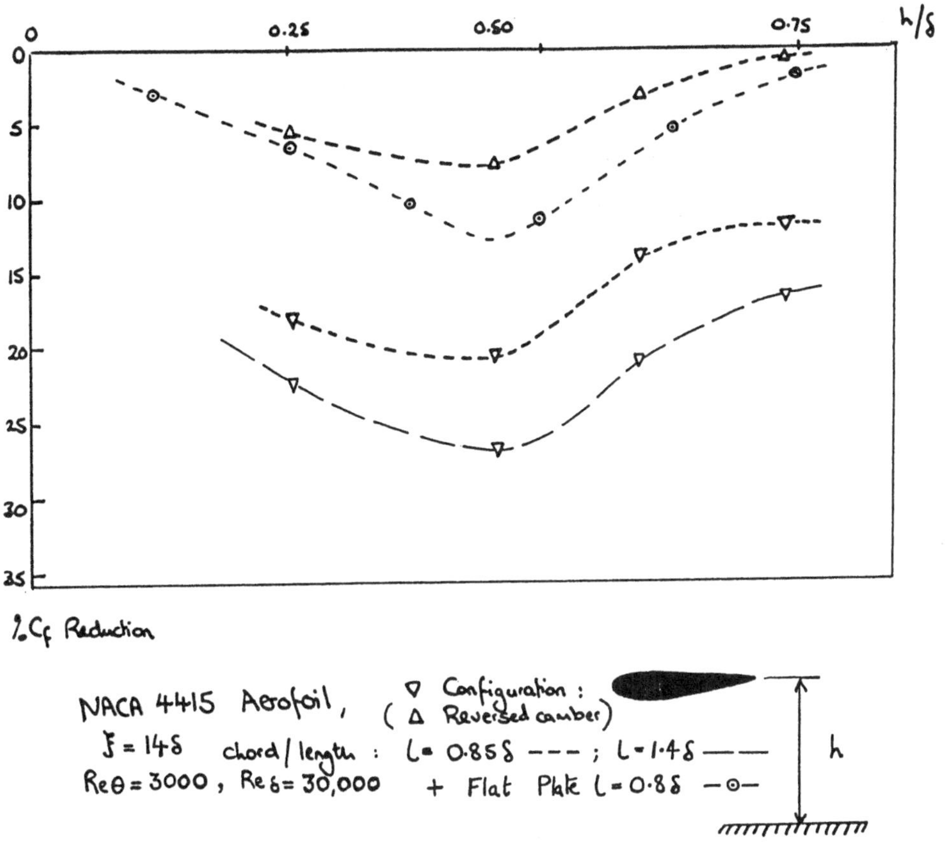

Fig.13: Effect of aerofoil manipulator profile shape on Cf reduction

momentum thickness measurements from traverses onto the downstream end of various lengths of riblet material. Results for 0.28, 0.56, 0.82 and 1.30m lengths (Rel=0.8-3x10^7) have shown that the first of these was insufficient to achieve a nett reduction of more than 2-3% (see Fig.14) whereas the results for the other three development lengths collapsed and indicated a maximum 6% reduction, rather less than that found at low speed, but in line with a theoretical estimate [74]. The data for the longest riblet lengths are compared in Fig.15 with a similar set of controlled laboratory measurements, performed during the course of this study at ONERA/CERT [22]. The ONERA experiments were conducted over the Mach number range M=0.4-0.8 in the T2 tunnel at Toulouse and involved direct drag measurements for .0009,.0013,.002 and .003in. 3M riblets mounted on an ogive. The agreement between the two sets of data is excellent. Similar results have also now been obtained at M=0.53 and 0.87 by Gaudet [see 2] using a drag balance set into the wall of the RAE Bedford 8' tunnel and he has more recently reported the first set of such drag measurements for riblets acting at supersonic speeds up to M=1.2 [51]. These are also plotted on Figs.15 and fall within the scatter band of the transonic tunnel data. Furthermore Figs.8 & 15 shows that the present high speed results follow the broad trend of extensive low-speed investigations performed by NASA [76] (as well as those from a large number of other research groups), and are also in good agreement with flight test data from similar pitot traverses by Boeing [77] and the later NASA flight data, some of which was recorded using a drag balance [78].

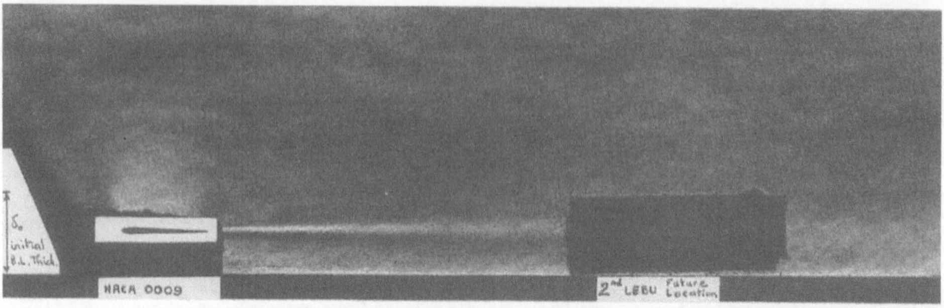

(a)

(b)

Plate 1: Photographs of: a)CEAT LEBU at M=0.7 by Bonnet & b)shock on riblets

In order to investigate further the effect of riblet development length on skin friction reduction a number of other drag balance measurements have also made at Cambridge using the drag balance [65]. With riblets on the balance head it was found that the measured skin friction was virtually independent of the length of the riblet material ahead of the balance provided this length was greater than 5 boundary-layer thicknesses, and a similar investigation of the recovery length behind the ribbing revealed a relaxation length again of order 5δ.

The floating element balance was also used to investigate the effect of yawing the riblet material on the head of the balance relative to the surrounding, flow-aligned ribbing. For yaw angles less than 15° this effect

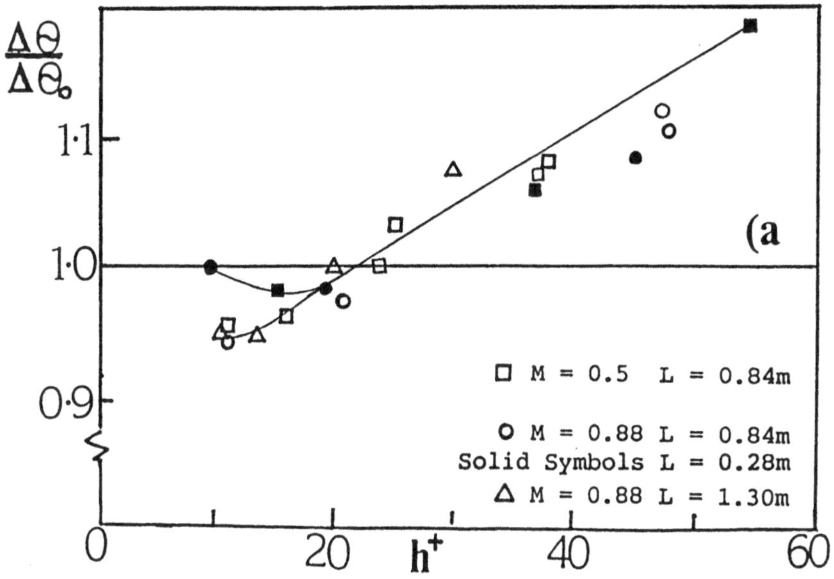

Fig.14: Drag reduction due to riblets at high subsonic Mach No. [75].

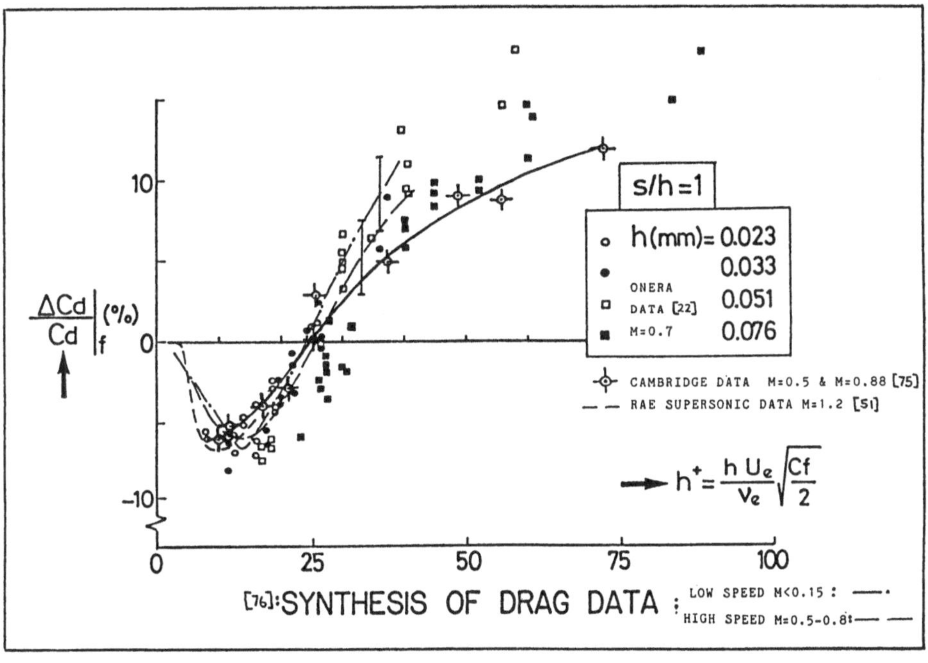

Fig.15: Drag data for riblets at low, high subsonic and transonic speeds

appeared to be insignificant, and despite scatter in the measured results, these were generally consistent with the results of previous NASA low-speed studies [43] and ONERA and RAE high-speed tests [22,51], as illustratrd by Fig.16

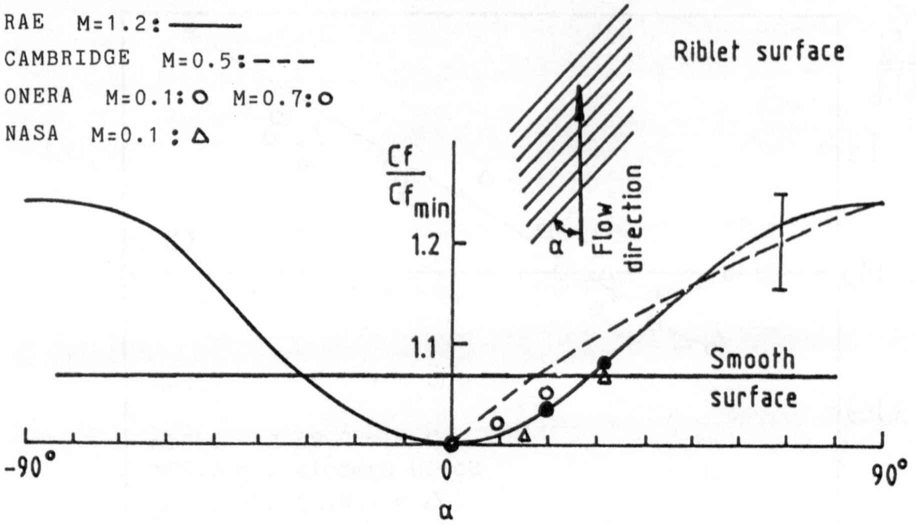

Fig.16: Effect of yaw on riblet performance

Velocity profiles plotted in law-of-the-wall coordinates for four riblet surfaces with $15 < h^+ < 54$ at M=0.5 (assuming that the relevant Cf is the skin-fiction coefficient consistent with the overall drag changes shown in Figs.14) confirmed that the slope, A is the same as the standard wall law, but the additive constant, B is increased for drag reducing riblets (indicating a corresponding increase in sub-layer thickness) as found at lower speed. Despite the uncertainty introduced by the Cf assumption the changes in B are in very close agreement with those deduced by Gaudet [51,74] from the low speed results of Winter and Sawyer [45] and his own high speed experiments, as illustrated by Fig.17.

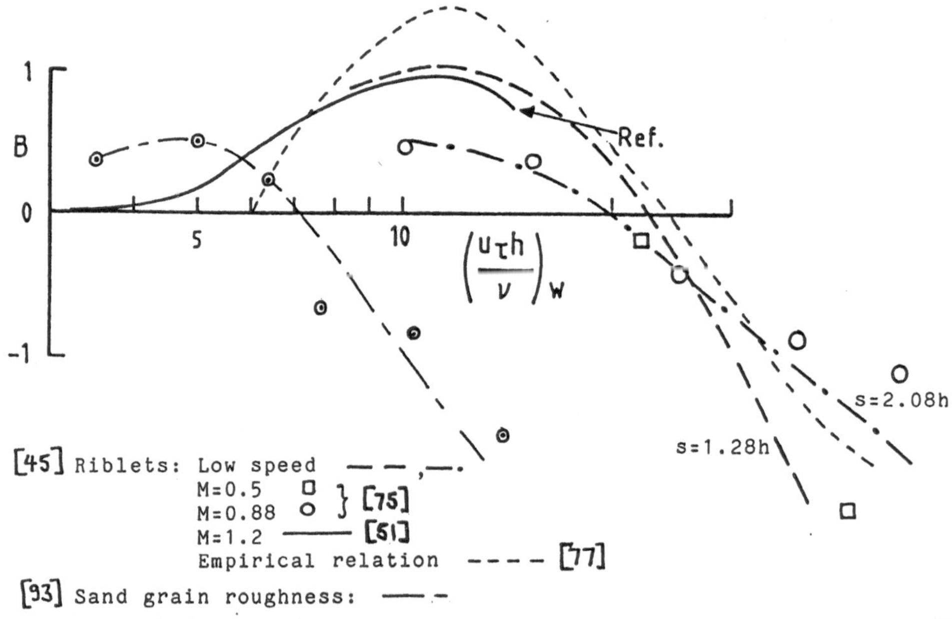

Fig.17: Changes in B at high speed

The effect of adverse pressure gradients on riblets operating at such Re,M was investigated by inclining the top liner of the tunnel test section, by a predetermined angle, and then controlling the choking by varying the angle of inclination of the aerofoil in the diffuser. In this way it was possible to generate two relatively uniform adverse gradients along the lower liner test surface; the Mach number dropping from approximately 0.88 at a position near the origin for the riblet tests to 0.72 (β=0.2) or 0.62 (β=0.5) respectively at the traverse location. All the tests were conducted on a 0.92m riblet length and the effect of the riblets on Cf isolated from the change in θ due to pressure gradient after estimating the latter contribution from computations employing Green's lag-entrainment method for the measured pressure distributions. The results are compared with those obtained in zero pressure gradient in Fig.18. Despite the progressively larger experimental uncertainty due to run-to-run variations in tunnel operating conditions, the evidence is that similar levels of drag reduction can be obtained in mild pressure gradients up to β=0.2 (over a somewhat larger range of h+) and possibly β=0.5, but there may be some loss of efficiency in more severe gradients. Similar results for mild adverse (and favourable) pressure gradients at ONERA by Coustols [private communication] and EPFL data from stronger gradients in a diffuser [79] tend to support these findings, although in both cases proximity to separation at higher B may have been a significant factor.

The first investigation of the effect of riblets in supersonic flow (on shock/boundary layer interaction) was also conducted in the Cambridge tunnel, using this in the conventional mode with a pair of nozzles designed to give a uniform Mach number of 1.5 with the 3M material stuck along the whole length of the lower nozzle surface. Spark shadowgraph photographs were taken with the riblet material attached in the stream direction for h+=9,20,30, 45 and 60 and at yaw angles of 45° and 90° for h+=45. The flow separated at the shock with a significant increase in boundary-layer thickness through the interaction. For h+ = 9 and 20 there appeared to be very little change in the nature or scale of the interaction, in particular the height of the slip line above the surface was unchanged, but at h+ = 30 the scale of the interaction had increased significantly with the height of the slip line being 25% greater than on the smooth surface (see Plate 1b). (This increase became progressively more marked at h+=45 and 60 and for other yaw angles). Since the upstream influence takes place through the subsonic part of the boundary layer it is not suprising that the scale of the interaction increased when riblets with a height of more than 20 wall units are attached to the surface, particularly as Fig.17 indicates these would then have been drag increasing.

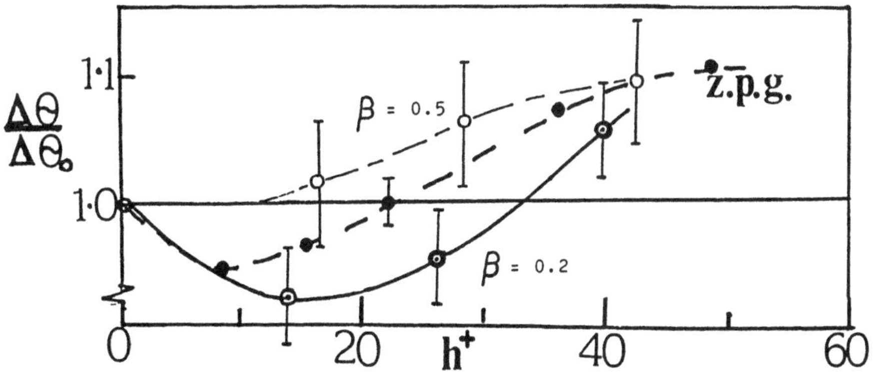

Fig.18: Effect of increasing adverse pressure gradient on riblet performance

SOME PRACTICAL CONSIDERATIONS

In considering possible practical applications to real engineering flows, either on the external surfaces of aircraft or ships or the internal passages of turbo-machinery, ducting and pipe work, a number of studies have been made of manipulator performance under disturbed flow conditions including those associated with flow over sandgrain [80], d-type [80,81] or k-type [80,82] roughness and riblets [44]. The influence of small free-stream turbulence intensity (Tu<3%) and length scale (l<1.5δ) [83] on manipulator preformance and the effects of similar mild [16,21] and strong adverse pressure gradients [24,84] (β=0.2-0.5) as studied for riblets, have also been investigated

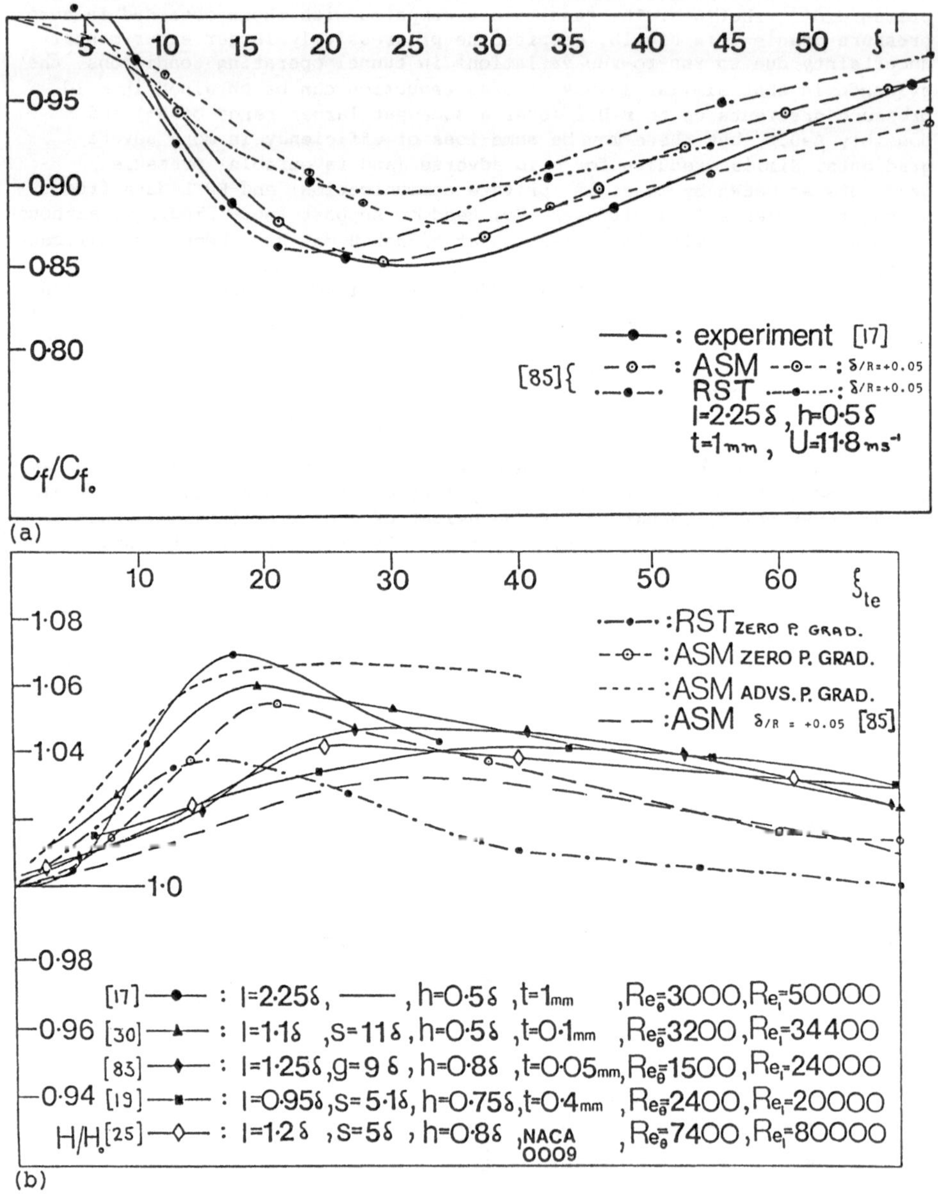

Fig.19: Effect of manipulators on a) Cf & b) H under disturbed flow conditions

experimentally, while turbulence modelling has been used to estimate the effect of longitudinal curvature on a manipulated layer [85]. As Fig.19a shows little effect is predicted for mild curvature, and in general it appears that such extra influences or combinations have little effect on the absolute Cf reduction due to manipulation (see Fig.20) although more extensive computer predictions have suggested that manipulator performance is diminished for very high rates of curvature [85], but enhanced for larger turbulence intensities and particularly length scales [84], both of which may be encountered in the internal passages of aero-enginess. As we have seen, there is also some evidence that the efficiency of riblets may be diminished in more severe pressure gradients. However mimimum Cf may not be the prime consideration under such conditions. Instead van den Berg [86] has shown that it is the associated increase in H which is the more important factor. The increase in H (see Fig.19b, which mirrors the Cf reduction of Fig 19a) found in manipulated flows therefore places an additional restriction on nett drag performance in such flows, and the observation that manipulation may promote separation suggests it may even be necessary to increase Cf before this is approached.

In the case of adverse pressure gradient flow over riblets H was either not affected or very slightly increased, in contradiction to the reduction recorded in the earlier BMT/Cambridge tests of Hoechst riblets on a model Americas Cup yacht [54] which was therefore attributed to three-dimensional effects in the latter case. In fact no information is yet available regarding the performance of either riblets or manipulators in 3D flow except that provided indirectly by such studies and flight tests where incidental 3D effects do not appear to have had much effect. Equally little is known about the influence of riblets or manipulators on transitional flows although detailed hot-wire anemometry studies within larger riblets in water have suggested drag reductions of up to 4% even in laminar flow over riblets [87],

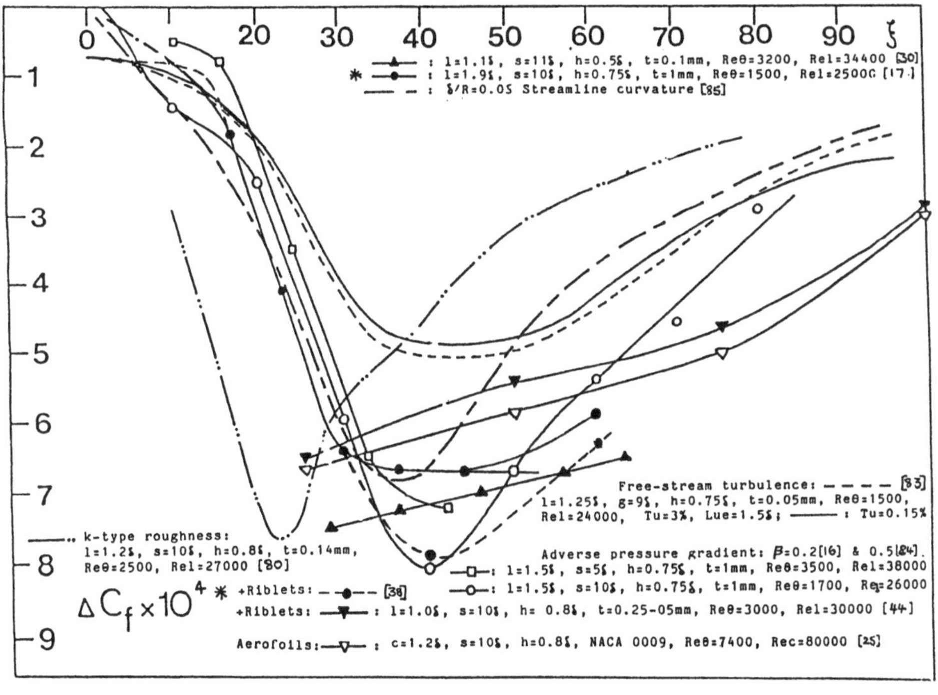

Fig.20: Absoluted Cf reduction due to manipulation under various conditions

and riblets have also been shown to delay transition under the influence of free-stream turbulence, although not by so much as some other surface modifications [88].

In addition no attempt has so far been made to optimise either riblets or manipulators for other than zero pressure gradient smooth wall flow despite the fact that there is already evidence to suggest that changes to manipulator configuration are either required or beneficial when these are applied to flow over rough or riblet surfaces, as discussed below. It seems likely that changes will be required to riblet scales to achieve optimal behaviour at least in flows subjected to wall curvature, adverse pressure gradient, and or manipulation, where rapid changes in Cf occur.

OTHER APPROACHES AND COMBINATIONS OF TECHNIQUES

Already it has been demonstrated that similar drag reduction can also be achieved with compound riblet surfaces exibiting spanwise variations in height and spacing [2,89] and by 3-D arrays of similarly scaled short fin-like ribs [90] suggesting that there may be a wide range of 'deterministic roughnesses' with drag reducing properties (see Fig.21). At the same time a new analysis of Nikuradse's pipe flow data by the same Japanese researcher group has suggested that some sand grain roughness may also exhibit drag reducing properties for $k^+<7$ [91], a maximimum 2-3% reduction being indicated. This finding has been questioned because the skin friction was deduced from the velocity profile records using a different log law constant to that employed by Nikuradse, and indeed when his value was subsequently adopted the drag reduction was rather less [92]. However a very recent boundary layer experiment in which an attempt was made to copy Nikuradse's roughness from original photographs appears to have confirmed the existence of a Cf reduction, reaching as much as 4-5% for $k^+=3$-5 (in fact a figure of 8% was indicated by the anlysis - as indicated on Fig.21 - but when applied to riblet data this over estimated their effect), with the reduction falling to zero at $k^+=7$, and larger sandgrains again acting as a true roughness [93]. Similar reduction may also be deduced from earlier

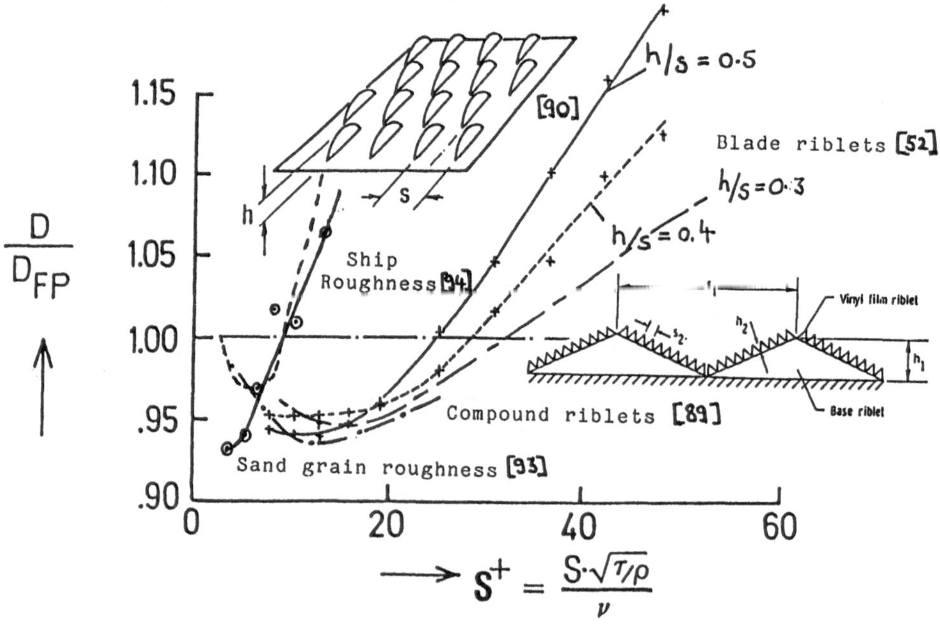

Fig.21: Drag reduction due to modified riblets and other surfaces

studies of ship roughness [94] as also shown on Fig.21 and if correct this is clearly an important finding with considerable practical implications.

It has long been known that the drag of d-type surfaces is anomalously low compared to similar scale k-type roughness and recently Japanese researchers have shown that sparsely spaced d-type transverse cavities, with spacings 10 and, preferably, 20 times their width or depth (considerably larger than previously studies) may also offer drag savings over certain low Reynolds number ranges [95]. Experimental mesurements of mean and turbulent flow quantities, coupled with the first flow visualisation studies of such surfaces [9,96], performed as part a collaborative research project between British Maritime Technology and Cambridge, have lent some support to the Japanese findings and suggested that any drag reduction is due partly to the establishment of stable vortices within the cavities, but primarily to a modification of the small scale structure in the sub-layer region between these. The cavity vortices appear to act as trapped air bearings for the outer flow, thereby reducing Cf, but also seem to absorb, replenish and reorganise the near-wall structure (although the changes in flow structure could equally be considered to be a response to local stress relievance at the cavity location, as pointed out by R.E.Britter [private communication]). The flow visualisation and subsequent image enhancement suggest that additional small-scale vortices are introduced into the sublayer which help maintain a lower skin friction because they tend to act as additional semi-rolling, air-bearings. The extent of the Cf reduction and recovery of turbulence quantities downstream of an isolated cavity appeared to be closely related to the persistence of this modified sub-layer structure and were consistent with an apparent optimum spacing of 20-30w between successive cavities indicated by more extensive studies recently conducted by the Japanese group (private communication). In the cases studied this corresponded to approximately 10δ, the typical relaxation length for inner layer disturbances.

Tani et al.[95] have already shown that drag measurements for such sparse d-type surfaces follow an empirical relationship correlating roughness shift with Cf for a wide range of other d- and k-type roughnesses roughnesses, and that riblets also appear to conform to this (see Fig.22). The wake parameter is also increased, but by a factor similar to that found in many manipulated layers [80,97-99] rather than the more modest increase determined for riblet surfaces (see Fig.23). It therefore appears that a wide range of wall surface geometries may exhibit low drag properties under certain conditions (although they can also act as excressences depending on their size). There is thus now a need to develop an overall roughness concept which can encompass all of these drag reducing surfaces alongside the more conventional random,

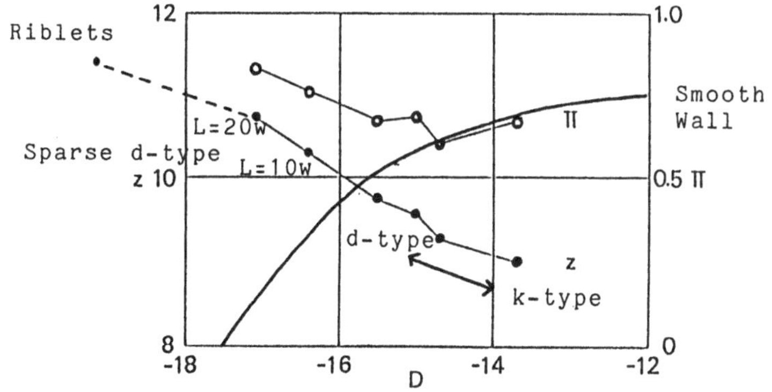

Fig.22: Variations of Π & $z(=\sqrt{2/Cf})$ versus U shift

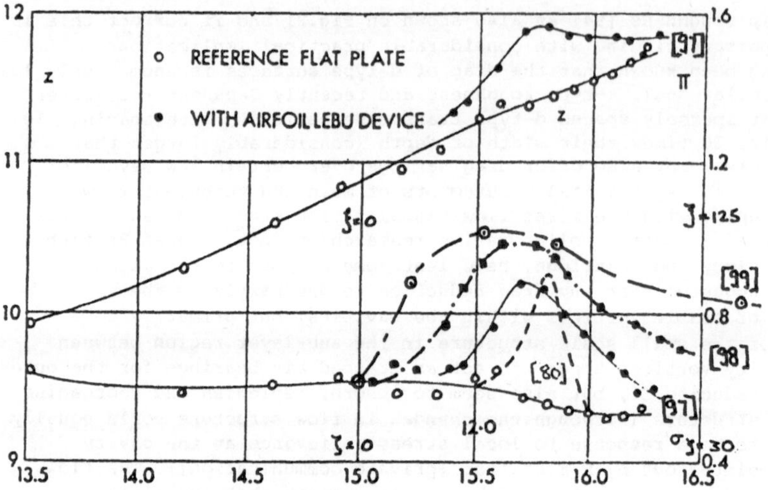

Fig.23: Π & Z for manipulators

stochastic and deterministic roughnesses encountered in practical engineering flows. Further parametric studies are also required to investigate the appropriate scaling laws and Reynolds number range for optimised d-type surfaces. These should be complemented by more detailed quantative multi-point measurements of mean and turbulent, flow, and wall, properties on both random and deterministic rough surface geometries to examine the effect on the flow structure, including its three-dimensional organisation, in greater detail, and to verify the magnitude of the attainable maximum nett drag reduction.

One way in which larger drag reductions can already be achieved is to combine such passive devices with other existing control techniques. It has already been demonstrated that manipulators can reduce the drag of d-type [80,81] or 'micro-air-bearing' [100] surfaces (and been suggested that riblets introduced between adjacent cavities might offer similar benefits [101]), probably by limiting the interactions between the cavity and external flows [9]. Furthermore it appears that following manipulation the flow over such surfaces may recover to a new equilibrium condition, potentially prolonging the recovery and hence drag reduction [97]. Since recent research has also suggested that certain 'sparse d-type' surfaces (with cavity separations an order of magnitude larger than their dimensions) may offer inherent nett drag savings over certain Reynolds number ranges, attention is now also being directed to the combination of these with aerofoil manipulators [102] and studies should soon be conducted to see if they are also amenable to other proposed means of control [103].

A number of other combinations (including: polymers and riblets [50,53,104]; polymers and manipulators [105]; tangential slot injection with manipulators [106] or riblets [63]; riblets with peak suction or blowing [89]; riblets plus surfactants and/or injection [107]; and manipulators with acoustic forcing [108,109]) have been studied to-date and some beneficial results obtained. In particular synergisic results have been obtained for both the combination of riblets with manipulators and with polymers.

Since riblets scale on wall variables and act directly on the sublayer, while manipulators scale on δ, act on the outer layer stucture, and therefore only indirectly influence the wall region, it should not be surprising to find that these could be combined constuctively, and integrated drag measurements performed by NASA have indeed indicated that their overall effects on skin friction were simply additive [44,110]. Subsequent flow visualisation studies conducted at Cambridge [111] have suggested that this is because riblets compliment both the immediate effect of the manipulator and the persistent effect of its wake (in suppressing wallward motions and hence reducing burst activity) by introducing an additional degree of spanwise order, and possibly also a second 'shield' of longitudinal eddies, to the near wall region, but also pointed to the possibilty of some additional benefit given an appropriate matching of scales. In order to investigate this possibility further the first set of local direct drag measurements have recently been made utilising the same floating element balance, wind tunnel flat plate devices, and experimental techniques as for the earlier parametric studies of such manipulators in smooth wall flow ([17] & see Fig.2). For these latest experiments Hoechst U-groove riblets were stuck along the whole of the 2 metre test section wall, starting immediately behind the trip bar, and also onto the drag balance itself, great care being taken to align the riblets on the floating element with those on the surrounding surface and to ensure that the gap between this and the balance was unaltered. The resulting data for riblets alone have been included on Fig.8.

The combined results tend to support the conclusion, based on the NASA earlier integrated drag data, that the effects of riblets and manipulators are broadly additive, at least at low Reynolds numbers. But they have also revealed interesting local departures from this purely additive response, which shed further light on the drag reduction mechanisms involved, and suggest that some beneficial re-optimisation of device parameters may indeed be feasible when the two techniques are used in combination (see [10] for Figures and detailed results).

For an optimised ($l=1.5\delta$, $h=0.75\delta$) single plate it was found that the effect of the riblets was simply to shift the Cf distribution for the manipulated smooth wall layer down by an almost constant amount corresponding to the contribution made by the riblets. However when a longer plate ($l=3.2\delta$) was introduced into the flow over the riblets their additional benefit seemed to be some what reduced in the vicinity of the maximum reduction position. These findings were consistent with conclusions reached by R.E.Falco (see [2]) on the basis of more fundamental studies of the effect such devices have on the near wall structure that riblets should be more efficient close behind a manipulator where the flow angle towards the wall is reduced, but less efficent in the region where the device wake impinged on the wall region. It has been independantly demonstrated that the perturbation introduced by a manipulator travels along characteristic lines and reaches the wall at a point coincident with the maximum reduction location [21]. It therefore seems likely that the poorer performance of the riblets when combined with the larger plate could be attributed to the increased strength and scale of its wake, leading to an earlier, stronger interaction with the wall region.

This conjecture is also supported by the earlier flow visualisation study [11] as well as the finding that with a tandem manipulator of similar device drag ($l=1.5\delta$, $s=5\delta$) the effect of the riblets in the vicinity of the maximum reduction location was found to be rather more than additive, and when s was increased to the optimum value for smooth wall flow of 10δ it appeared that the overall benefit might be slightly larger than the sum of the individual contributions from the manipulator and riblets alone.

One might expect that the height of the manipulator could be increased advantageously in such flow over riblets because this should extend the initial region over which the effect of these is at least additive; since the wake/wall interaction would then be delayed. However when this was tested it appeared that an earlier recovery was then initiated, in some cases perhaps even before the delayed maximum reduction occured, so that no overall benefit resulted. Reducing h from $3/4\delta$ to $2/3\delta$ also had little effect on the manipulator performance, and again a more rapid recovery then occured as also observed in the manipulation of k-type rough walls. However when the riblets were employed in conjunction with a manipulator plate of only 0.8 chord their effect was rather more than additive. This was surprising in view of the step down in manipulator performance which occurs when l is reduced to less than δ for smooth wall flow, and has been attributed to a dimunition in the ability of the device to interact with large, order δ , scale motions. It is possible either that such motions play a less significant role in flow over riblets or that only a weaker manipulator wake/momentum deficit is required to reduce Cf in such a layer. If so the implication is that smaller manipulators, having less inherent device drag, may be sufficient to produce the same integrated Cf reduction on a riblet surface as on a smooth wall; thus making larger nett drag savings possible. There may thus be some advantage in also reducing the spacing of shorter tandem plates since this appears to be related to l (in the manner indicated by Figs.4 & 5). Certainly in future attention should be directed towards varying other device parameters, and it seems likely further benefits will come from re-optimising riblet geometry for manipulated wall layer conditions. However attempts to extend the present work to the NACA 0009 aerofoil manipulator in combination with 3M V-groove riblets at higher Reynolds numbers and sub-sonic Mach numbers representative of flight applications have so far indicated little benefit although as pointed out earlier the mometum thickness measurements employed are in some doubt. These studies are continuing and further more detailed investigations of the combination at low speed are now being undertaken by V.D.Nguyen and J.Dickinson at Laval University (private communication).

The effect of combining riblets with polymers has also been the subject of several recent investigations. In an extension to the towing tank tests of riblets on a 1/3 scale model of the Americas Cup yacht Australia II [54] a polymer solution was sprayed onto the riblet film in situ and the drag measurements repeated. Comparison with the baseline tests for the smooth hull and riblets or polymer coating alone showed that the sum of the two effect was nearly equal for the optimum s+ operating conditions of the riblets. The overall effect was rather less than additive at smaller s+, but rather larger at higher s+ and Froude numbers. Such a synergistic response has also been indicated in pipe flow studies of riblets [50], where the polymer was injected into the flow upstream, and has been attributed primarily to the the reduction in wall shear stresses by the riblets delaying degradation of the polymer additive, although other effects may have contributed to the reduction. It has been suggested that the poorer effect at low S+, F might may have resulted from the near wall shear stress being sufficiently reduced at low speed to inhibit the action of the polymers, although entirely contradictory findings have emerged from studies conducted of a riblet-covered axisymmetric body drooped through a tank of polymer solution [53]. In contrast this study suggested that, by reducing the friction velocity, riblets may lower the threshold for polymer activity, although apparently only in some cases, and also raised the possibility that riblets may provide nucleation sites for beneficial microbubble generation [112] if the solution is saturated. Further studies are clearly needed to resolve these apparently contradictory findings, but it does appear that useful practical benefit are obtainable.

MODELLING

The need to avoid further expensive experimental parametric optimisation studies for internal flows, and also for external flows at high Mach and Reynolds numbers under a variety of flow conditions, has stimulated an interest in the prediction of such drag reducing flows. From a modelling standpoint manipulated boundary layers are of inherent interest because they represent a step-up in complexity from the plane boundary layers subjected to additional straining considered at the 1980-81 Stanford Conference and highlight weaknesses in current Pressure-strain, Diffusion and Dissipation closure approximations as well as near wall modelling approaches in general.

Initial attempts to model manipulated boundary layers with simple mixing length and k-e models have led rapidly to more sophsticated treatments. Early computational efforts were restricted to caculations of the effect of changes in length scale [15] and/or imposed momentum deficits on boundary layer development [113], but with the provision of test case quality detailed mean and turbulence profile measurements as input data and for comparison with output, computations have now been performed using a wide range of models and model closures: Integral method [86]; Mixing length [H.M.Nagib & R.Westphal p.c.,113,M.Veuve p.c]; Prescribed eddy viscosity [M.Dhanak p.c.]; Υ-1 (Method of characteristics [21]); k-ϵ [15,114]; low-Re k-ϵ [L.Johnston p.c.,34]; ASM [20,24,84,85]; 3-equation [114] and 5-equation [22,114] Differential Stress Transport (DST) model calculations, and full RST schemes [84,85] together with Rapid Distortion Theory (RDT) [39,40]; Point vortex [115]; and a vorticity/stream function approach [116].

Within the framework of parabolic turbulent transport approaches the ASM scheme appears in some ways the most promising, not least because it represents the simplest upgrade from the k-ϵ level of closure which is widely used in aero-industry design codes. It has the advantages that it can provide predictions for all the Reynolds stress components and contains most of the physics while avoiding the extra approximations for diffusion which have to be made in higher order routines. The inclusion of ϵ, pressure-strain, and the fact that it reduces to an eddy viscosity prescription, make it amenable to structural modifications based on experimental results [117,118], numerical simulation 'data' and RDT calculations [85,119]. The ASM scheme of Launder and colleagues [120] performed generally at least as well as full RST or reduced-equation DST approaches, and considerably better than k-ϵ models, on the simpler Stanford Test Cases, including the effect of free-stream turbulence on a zero pressure gradient, flat plate boundary layer, and the adverse pressure gradient cases to which it was applied. The same model, with essentially the same set of 'standard' constants has subsequently been found to perform remarkably well on a range of confluent wake/boundary layer interactions in both zero and adverse pressure gradients [121] and with modifications to account for counter-gradient transport by the eddy structure has been shown capable of handling 'negative production' regions [118]. Since there is interest in modelling such 'extra strain effects' and a manipulated flow might be regarded as a category of wake-boundary layer interaction where the wake generating body resides within rather than outside the boundary layer, this model was therefore selected for the first attempts by the author at modelling manipulated boundary layers with this level of closure. Perhaps not surprisingly in view of the above comments, good results were again achieved, particularly after adjustments had been made to the manner in which the initial dissipation profile was specified (from a mixing length description) to take account of the influence of the external turbulence on the device wake [see 84]. The same model was then applied with equal success to the case of the manipulated layer subjected to such strong adverse pressure gradient that the flow separated, and has subsequently been used to predict the effect of streamline curvature (see Fig.19) and free-stream turbulence intensities/length-scales representative of conditions within the internal passages of aero-engines [84,85]. Since there are still as yet no corresponding experimental data the latter were TRULY PREDICTIVE results. However to perform computational parametric optimisation elliptic schemes, with calculations starting upstream of the device are needed. Such a scheme, employing a k-ϵ, 3- or 5-equation turbulence model has been developed at ONERA [114] and this has now also been applied to a number of test cases producing results in excellent agreement with experimental data for a range of flat plate manipulators obtained earlier at ONERA [22] (eg.see Fig.24), CEAT [35], Rolls-Royce [83] and Cambridge [17]. Further prametric predictions confirm the optimum tandem spacing, $s=12\delta$ for maximum integrated Cf reduction (see Fig.25a), and optimum manipulator height of $h=0.3\text{-}0.6\delta$, depending on the integration length (see Fig.25b) found experimentally. (It is interesting to note that RDT calculations tend to confirm the conjecture that close gaps icrease the effective length of plates so far as the flow is concerned, but

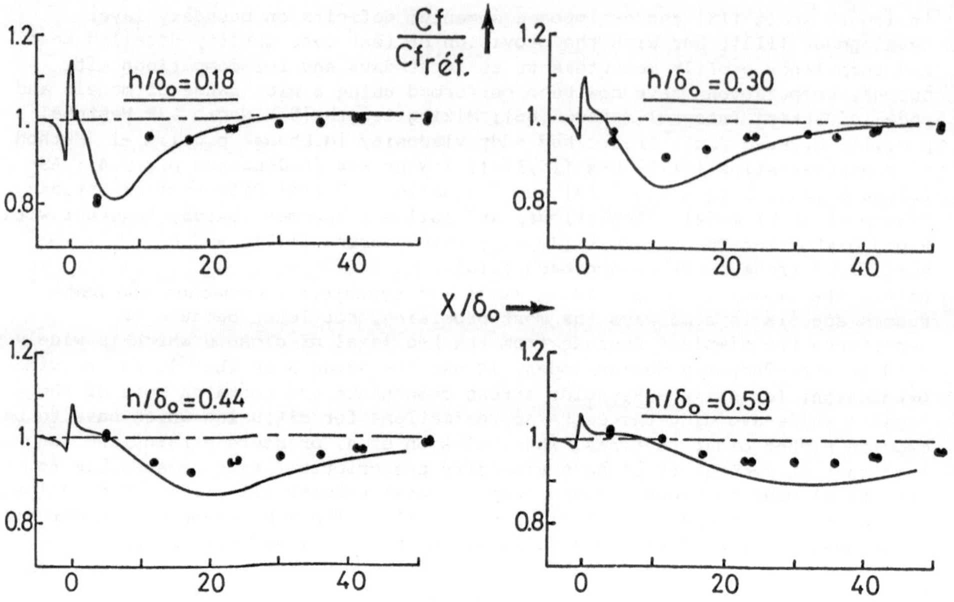

Fig.24: Comparison of ONERA manipulator computations & experiment

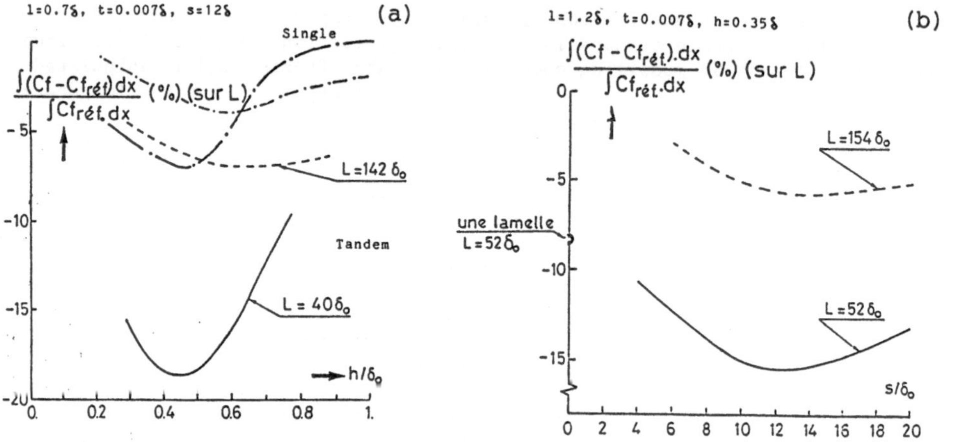

Fig.25: ONERA predictions for optimum manipulator h & s (integration over L)

that s>10δ is to be prefered since the second plate then has a mutiplicative rather than additional effect [39], while the method of characterisics approach when suitably modified for manipulated flow more accurately predicts the growth of an overshoot in $\overline{uv}$ and $\overline{v^2}$ as the flow relaxes. As illustrated by Fig.26a-c this effect is reminiscent of the 'stress bore' observed in boundary layers recovering from convex curvature [122] and may reflect similarities due to supression of large scales).

As part of a continuing collaborative effort between ETH Zurich, Cambridge and Queens University, Canada a similar elliptic code, containing a low-Re k-e

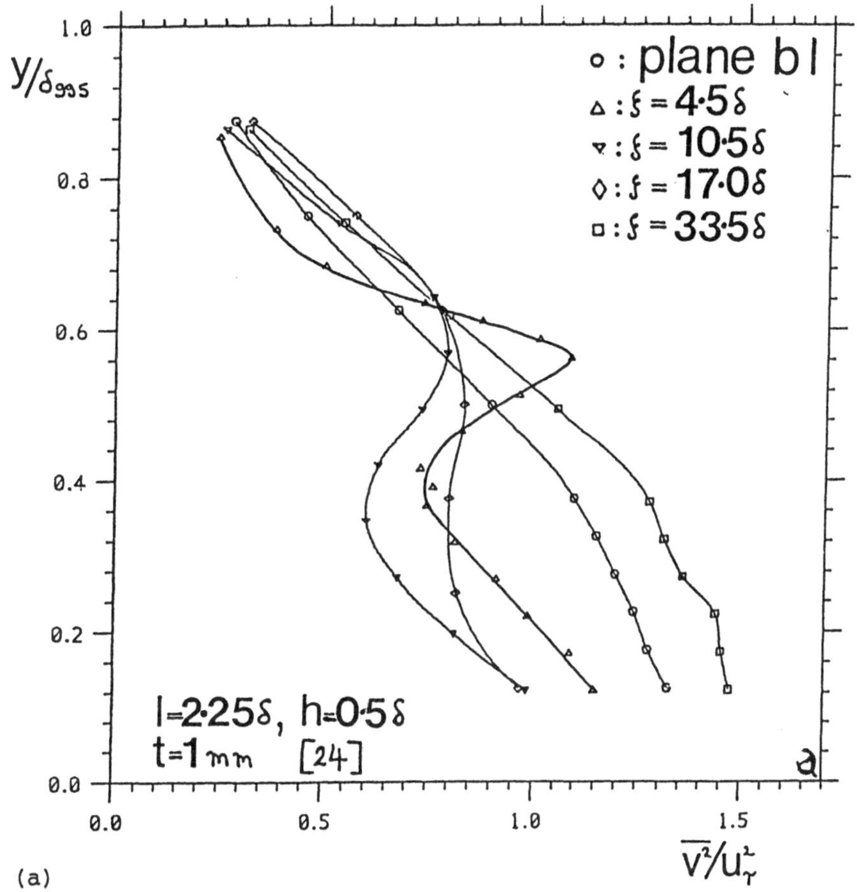

1.0
y/δ_{995}
0.8
0.6
0.4
0.2
0.0
○ : plane bl
△ : ξ = 4·5δ
▽ : ξ = 10·5δ
◇ : ξ = 17·0δ
□ : ξ = 33·5δ
l=2·25δ, h=0·5δ
t=1mm [24]
a
0.0
0.5
1.0
1.5
$\overline{v^2}/u_\tau^2$

(a)

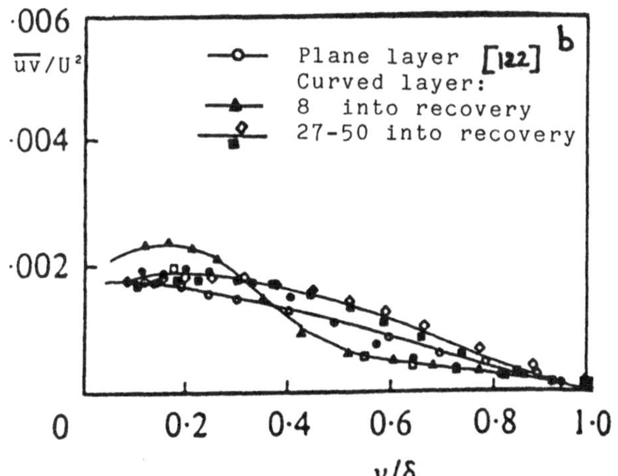

·006
$\overline{uv}/U^2$
·004
·002
0
0·2
0·4
0·6
0·8
1·0
y/δ
Plane layer [122]
Curved layer:
8 into recovery
27-50 into recovery
b

(b)

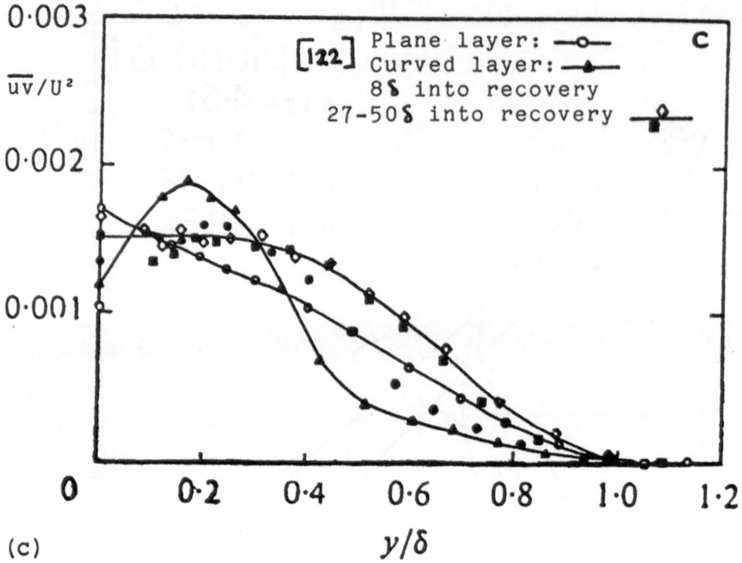

(c)

Fig.26: Recovery $\overline{v^2}$ (and $\overline{uv}$) profiles in: a) manipulated & b) curved layers

model, has also recently been be applied with some success to the pipe flow manipulator experiments studied at ETH. Two cases were considered: first with a single ring device mounted at h=1.1δ in the developing flow region and then at h=0.2R in the fully developed region. In general the agreement with (Fig.7) experiment was good except that in the first case the model tended to overpredict the initial disturbance close behind the device probably due to the larger trailing edge thickness although a similar defficiency is apparent in k-ε computations for an external layer [114]. Further computations are now being performed with an alternative turbulence model and for the tandem manipulator ring case [123]. Comparisons will be made later with recent ETH data data for higher ReD, and other fully developed pipe flow experiments now being conducted by V.D.Nguyen & J.Dickinson at Laval University [124], in an effort to determine whether any nett reduction can in fact be obtained under such conditions, and it is hoped that further parametric computations for both single and tandem devices will indeed allow optimum parameters to be defined for subsequent experimental studies of both the developing and fully developed flow situations.

A similar range of modelling techniques have now also been applied to compute the effect of riblets in a laminar boundary layer at IMST [125], Marseille and ONERA [126] as well as a channel flow at UMIST [127]. The IMST group have recently reported some forward marching, finite volume, parabolic Navier-Stokes computations, performed on a Cartesian grid assuming W=0, for laminar smooth wall flow stepping down onto V-groove riblets [125]. The predicted velocity profiles were in excellent agreement with their own LDA data; the computed shear stress falling to almost zero in the trough, but increasing to a level 50% above the smooth wall value near the peak. Parametric computations showed h/s was more important than h/δ, with h=0.8s proving optimum and resulting in a predicted 4% nett drag reduction, rather larger than the 1% gain indicated by similar compuatations performed earlier by Khalid [128]. Coustols [126] has reported that computations for U- and L-groove riblets, obtained with the same computer code, indicated a similar reduction, but that subsequent grid refinement particularly in the vicinity of the trough and peak reduced this to zero. Comparable laminar flow computations for L-groove riblets in a fully developed duct flow by Launder & Li Shoaping [127] have indicated that the minimum drag (determined from mass flow rate) occured in limit of h tending to zero and it is by no means clear that any drag reduction can be expected under laminar conditions. Preliminary turbulent flow calculations with mixing length and k-e models (which did not admit the development of turbulence driven secondary vortices) have shown, as expected, that the predicted drag depended critically on the near wall damping assumptions employed, although Khalid has obtained results consistent with experimental findings using an ASM scheme [128]. Turbulent flow computations for riblets in diffuser flow employing a mixing length plus damping function model, modified to account for the expected roughness shift using an empirical relation (see [77] & Fig.17) have proved to be in remarkably good agreement with experiment [79].

Present discrepancies between computations and experiments can be largely attributed to the lack of a sufficiently accurate near wall or manipulator model and the need to account correctly for the damping effect of manipulator wakes. However it seems likely these defficiencies can be overcome by taking advantage of contemporary developments in the modelling of turbulence for the two-dimensional or two-component limit. The extension to curvilinear cordinates for aerofoil manipulator optimisation, and the compilmentary modelling of turbulent flow over riblets, are now the subjects of a collaborative research effort by groups from England, France and Switzerland, centred on the ERCOFTAC Pilot Centre at EPF Lausanne.

In order to predict more complex turbulent flows where combinations of strain rates are acting simultaneously, and in particular to make use of the alternative 'curvature' (or extra strain rate) concept for drag reduction [129] in the aerodynamic design, it will be necessary to model both the "instantaneous" response of turbulent flow to general curvature and also the "memory" effects of strain "history" on mean and turbulent quantities. The possiblity of predicting such effects is presently being examined. By writing the equations in general orthogonal curvilinear coordinates, in which the curvature terms are expressed in terms of the pricipal radii of curvature of the coordinate surfaces any ambiguities associated with body geometries having compound curvature are removed and, in some flow situations, the analysis simplifies the analysis to one of geodesic curvature (K) effects. Such a formulation also allows for a unique physical identification of the different curvature terms as they appear in the stress transport equations and provides a convenient framework in which to examine the various types of curvature. Calculations have already been performed for both the 2D and 3D flow cases in the limit where production and pressure-strain redistribution dominate over any diffusion effects [130], a condition satisfied in the outer region of turbulent boundary layers and in turbulent wakes, using the closure approximations from the well-known Launder, Reece and Rodi RST model since these appears to provide the correct form of response to simple curvature.

It has been found that the assymetric response and relaxation effects observed between experiments on convex and concave surfaces can be predicted provided that all relevant terms are retained in the full set of transformed transport equations. A similar assymetry of response is also predicted for the case of concave followed by convex curvature which is of particular relevance to the application of curvature for drag reduction (see Fig.27a). The 'unexpected' effect on turbulent shear stress and hence Cf of combining lateral divergence (K31:destablising in isolation) with convex curvature (K12:stablising) to produce an overall beneficial reduction has also been modelled. Surprisingly the analysis was found to correctly predict the anomalous sign, although not magnitude, of the shear stress under such combined influences (see Fig.27b), and although it is clear that further attention needs to be paid to the modelling of strain history effects it does appear such a general 3D curvature approach offers the potential for predicting the effects of such combinations and sequences of strain rates and thus exploiting them for practical benefits.

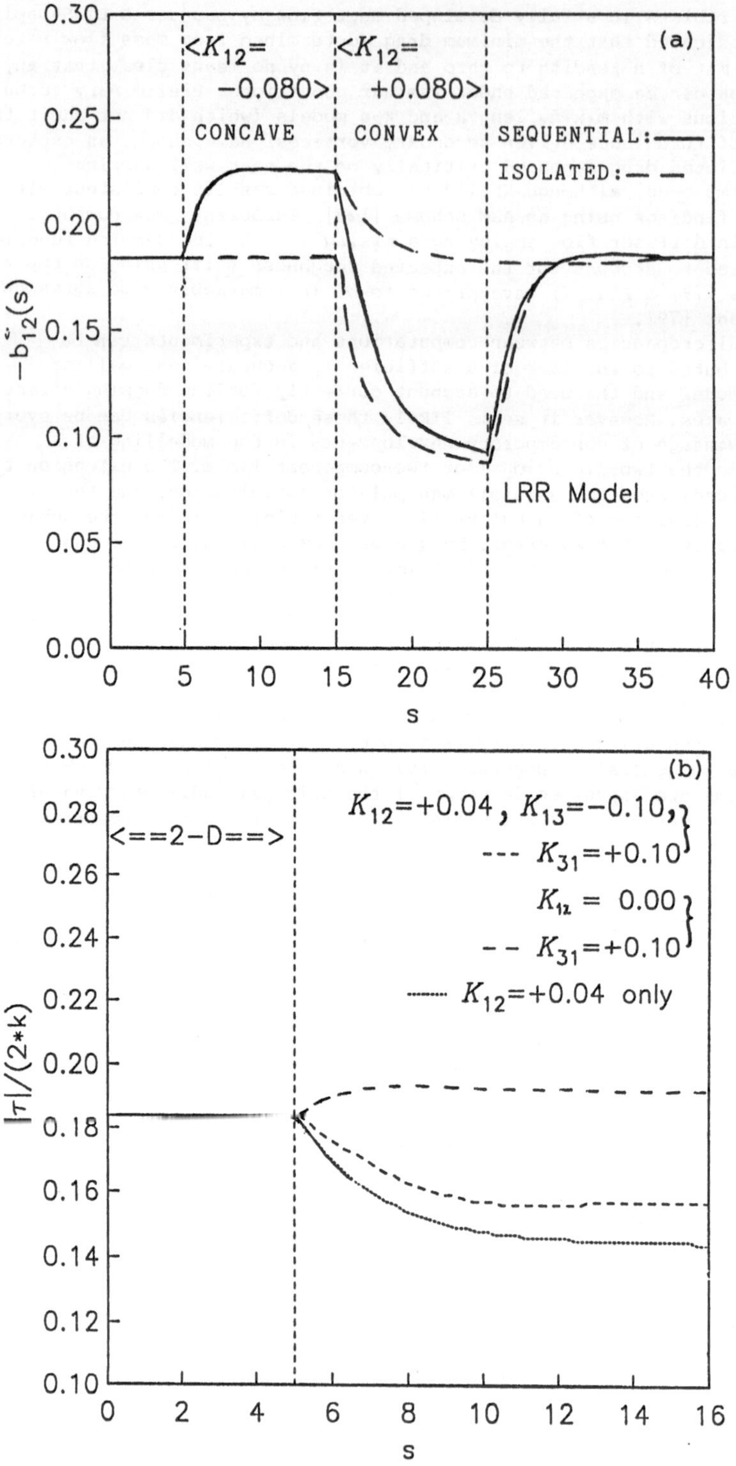

Fig.27: Convex curvature: a) after concave & b) with lateral divergence

Acknowledgements: Thanks are due to all those who have assisted in the preparation of this review through the provision of material and results often as private communications. Much of the work discussed in the review was funded by Rolls-Royce plc and the author gratefully acknowledges their continuing support under Mater Brochure PVA3-02D. Special thanks are also due to Frank Marentic of 3M and Dr.D.W. Bechert of DFVLR together with Dr.G. Meinel of Hoechst for making riblet material freely available and to Professor John Dickinson for his assistance with the drag balances. The high speed experiments were also supported by Airbus Industrie Contract 30124.5. Additional equipment and running costs were provided by Rolls-Royce plc via Brochure PVA3-05D and by an SERC Equipment Grant GR/E/0682.4. Further funding for specific projects by BACC,BAe,BMT and the British Council is also acknowledged. The work has benefited greatly from collaboration with UMIST(ASM and RST codes provided by Professor B.E.Launder & Dr.B.A.Younis), BMT,CEAT,EPFL,ETHZ,ONERA-CERT,NASA, Laval & Queens University, Canada.

References:

1. Proc. Roy. Aero. Soc. Int. Conf. Turbulent Drag Reduction By Passive Means, London, Vol.1 & 2, RAeS (1987)
2. Appl. Sci. Res. Special Volume 46, No.3 (1989).
3. Narasimha, R. & Sreenivasan, K.R.: AIAA-85-0517 & IIS FM 4 (1986)
4. Savill, A.M., Truong, T.V. & Ryhming, I.L.: J. de Mecanique, 7, No.4, p.353 (1988).
5. This Conference Volume.
6. Proc. DRAG REDUCTION 89, Davos, Ellis Horwood Ltd (1989).
7. Savill, A.M.: "Modelling Approaches to Drag Reducing Flows". Invited Review for Prog. Aero. Sci. (1989).
8. Drewer, C. Down with Drag. Flight International, 134, p.26 (1988).
9. Savill, A.M.: Proc. FLUCOME 88, Sheffield, p.436, H.S.Stephens & Associates (1988).
10. Savill, A.M., Kourey, E., Sebastian, & Squire, L.C.: see [6] (1989)
11. Cousteix, J., Coustols, E. & Arnal, D.: AGARD CP Fluid Dynamics of Three Dimensional Turbulent Shear Flows and Transition, Turkey (1988).
12. Loerke, R.I. & Nagib, H.M.: AGARD Rep 598 (1972)
13. Yajnik, K.S. & Acharya, M.: Lecture Notes in Physics 76, p.249 (1977)
14. Corke, T.C, Nagib, H.M. & Guezennec, Y.: NASA CR 165861 (1982)
15. Hefner, J.N., Anders, J.B. & Bushnell, D.M.: Prog Aero & Astro 72, p.110 (1979)
16. Bertelud, A., Truong, T.V. & Avellan, F.: AIAA-82-1370 (1982)
17. Savill, A.M. & Mumford, J.C.: ASME Fluids Eng. Div. 11, p.41 & JFM 191, p.389 (1988).
18. Hefner, J.N., Anders, J.B. & Bushnell, D.M.: AIAA-84-0345 (1984).
19. Lemay, J., Provencal, D., Gourdeau, R., Nguyen, V.D. & Dickinson, J.: AIAA-85-0521, 1985.
20. Savill, A.M.: Advances in Turbulence, p.533, Springer (1986)
21. Veuve, M. Ph.D. thesis No.768, EPFL, Switzerland, 1988.
22. Coustols, E.: AIAA-89-0963 (1989)
23. Lemay, J., Savill, A.M., Bonnet, J.-P. and Delville, J.: TSF6, Springer (1987).
24. Savill, A.M. (1987) IUTAM Symp. Turbulence Management and Relaminarisation, Bangalore Springer p.89 (1987).
25. Anders, J.B. & Watson, R.D.: AIAA-85-0520 (1985)
26. Coustols, E.: Proc. 16th ICAS, Jerusalem (1989)
27. Govindaraju, S.P. & Chambers, F.W.: AIAA-86-0284 (1986)
28. Anders, J.B.: SAE Tech paper 861769 (1986)
29. Nguyen, V.D., Savill, A.M. & Westphal, R.V.: AIAA J. 25, p.498 (1987)
30. Plesniak, M.W. & Nagib, H.M.: AIAA-85-0518 (1985)
31. Poddar, K. & van Atta, C.W.: TSF5, Springer (1985)
32. Bertelrud, A., Drougge, G. & lamdahl, M.T.: FFA TN 60 (1984)

33. Prabhu, A., Kailas Nath, P., Kulkarni, R.S. and Narasimha, R., "Blade manipulators in channel flows", Proc. IUTAM Symp. Turbulence Management and Relaminarisation, Bangalore, Springer, p.97 (1987).
34. Pollard, A., Savill, A.M., Thomann, H.: see [2 & 6] (1989)
35. Lemay, J., Bonnet, J.-P. & Delville, J.: see [1] (1987)
37. Chang, S.I. Ph.d thesis University of Southern California (1987)
38. Savill, A.M. & Ferre, J.A.: Proc. 10 ACFM, Melbourne (1989)
39. Gebert, G. & Atassi, H.: AIAA-89-0212 (1989)
40. Balakumar, P. & Widnall, S.E.: Phys. Fluids 29, p.1779 (1986)
41. Falco, R.E.: AIAA-83-0377 (1983)
42. Walsh, M.J. & Weinstein, L.M.: AIAAJ 17, p.770 (1979)
43. Walsh, M.J.: AIAA-82-0169 (1982)
44. Walsh, M.J. & Lindemann, A.M.: AiAA-84-0347 (1984)
45. Winter, K.G. & Sawyer, W.: see [1] (1987)
46. Bechert, D.W., Hoppe, G. & Reif, W.-.E.: AIAA-85-0546 (1985)
47. Applegate, S.P.: In Sharks, Skates and Rays, John Hopkins Press (1967)
48.Dinkelacker, A., Nitschke-Kowsky, P. and Reif, W.-E.: Proc. IUTAM Symp. Turbulence Management and Relaminarisation, Bangalore, Springer, p.109 (1987).
49. Neumann, D & Dinkelacker, A.: see [6] (1989)
50. Rohr, J.J., Reidy, L.W. & Anderson, G.W. see [6] & AIAA-88-0138 (1989).
51. Gaudet, L: see [2] (1989)
52. Wilkinson, S.P. & Lazos, B.S.: Proc. IUTAM Symp. Turbulence Management and Relaminarisation, Bangalore, Springer (1987)
53. Beauchamp, C.H & Phillips, R.B.: Symp. Hydodynamic performance Enhancement for Marine Applications, Newport RI (1988)
54. Choi, K.-S., Pearcey, H.H. & Savill, A.M.: see [1] (1987)
55. Liu, K, Christodoulou, C, Riccius, O and Joseph, D.D.: see [5] (1989)
56. Lowson, M.V., Jewson, A.D. & Bates, J.H.J.: see [6] (1989)
57. Hooshmand, A., Youngs, R.A. & Wallace, J.M.: AIAA-83-0230 (1983).
58. Bacher, E.V. & Smith, C.R.: AIAA-85-0548 (1985)
59. Choi, K.S.: Advances in Turbulence, Springer (1986)
60. Prasad, K.K., Nieuwstadt, F. & Pulles, C.: see [2] (1989)
61. Choi, K-S.: see [5] (1989)
62. Clarke, D.: 4th European drag Reduction Meeting Report to appear in ERCOFTAC Bulletin.
63. Bechert, D.W., Bartenwerfer, M & Hoppe, G.: Proc. 15th ICAS, London (1986)
64. Coleman, H.W.: Mississipi State Univ. Rep. MSSU-TFD-84-1 (1983)
65. Squire, L.C. & Savill, A.M.: see [1] (1987)
66. Nguyen, V.D., Dickinson, J.D., Jean, Y., Chalifour, Y., Anderson, J., Lemay, J., Haeberle, D. & Larose, G.: AIAA-84-0346 (1984).
67. Bonnet, J.-P., Delville, J. & Lemay, J.: Proc. 16th ICAS, Jerusalem (1988)
68. Bandyopadhyay, P.R. & Watson, R.D.: Expts. in Fluids 5, (1987)
69. Bertelrud, A.: AGARD CP 365 (1984)
70. Goodman, W.L.: AIAA-85-0550 (1985)
71. Bonnet, J.P., Poirier, D. & Delville, J.: see [6] (1989)
72. Katzmayr, V.: see [15]
73. Anders, J.B.: AIAA-89-1011 (1989)
74. Gaudet, L.: see [1] (1987)
75. Squire, L.C. & Savill, A.M.: see [2] (1989)
76. Walsh, M.J. & Anders, J.B.: see [2] (1989)
77. McLean, J.D., George-Falvy, D.N. & Sullivan, P.P.: see [1] (1987)
78. Walsh, M.J., Sellers, W.L. & McGinley, C.B.: AIAA-88-2554 (1988)
79. Pulvin, P. & Truong, T.V.: see [2] (1989)
80. Bandyopadhyay, P.R.: AIAA-86-1126 & J. Fluids Eng. 108, p.247 (1986)
81. Sakamoto, M. & Osaka, H.: see [1] (1987)
82. Pineau, F., Nguyen, V.D., Dickinson, J. & Belanger, J.: AIAA-87-0357 (1987)
83. Roach, P.E. & Brierley, D.H.: Submitted to Appl. Sci. Res. (1989)
84. Savill, A.M.: see [1] (1987) and Int. J. Heat & Fluid Flow, June (1989)
85. Savill, A.M.: Proc.3rd Int. Symp. Refined Flow Modelling and Turbulence Measurements, Tokyo, Universal Academy Press Inc. (1988).

86. van den Berg, B.: NLR MP 86060 U (1985)
87. Djenidi, L., Anselmet, F. & Fulachier: see [1] (1987)
88. Kozlov, V.E., Kuznetsov, V.R., Mineev, B.I. & Secundov, A.N.: Proc. Int. Sem. Ser. Near Wall Turbulence, Dubrovnik (1988)
89. Wilkinson, S.P., Anders, J.B., Lazos, B.S. and Bushnell, D.M.: see [1] (1987)
90. Bechert, D.W.: see [1] (1987)
91. Tani, I: Proc. Japan. Acad. 64, Ser.B (1988)
92. Tani, I: Proc. 2nd European Turbulence Conf., Berlin, Springer (1988)
93. Abe, K., Matsumoto, A, Munakata, H. & Tani, I.: see [5] (1989)
94. Karlsson, R.I.: ASME Conf.Paper 81-FE-35 (1981)
95. Tani, I., Munakàta, A., Matsumoto, A. & Abe, K.: Proc. IUTAM Symp. Turbulence Management and Relaminarisation, Bangalore, Springer, p.161 (1987).
96. Choi, K.-S., Fujiasawa, N . & Savill, A.M.: Proc. %th Int. Symp. Flow Visualisation, Prague (1989)
97. Tani, I. & Motohashi, T.: Proc. Japan Acad. 61, Ser.B p.333 (1985)
98. Westphal, R.V.: AIAA-86-0283 (1986)
99. Krishnan, V. & Yajnik, K.S.: NAL Tech. Mem. FM 8802 (1988)
100. Eidson, T.: Abstr. ONR-NSSC-AFOSR-NASA Symp. Drag Reduction and Boundary layer Control, Washington (1985).
101. Khalid, M.: NAE Canada Aero Note NAE-AN-39, NRS No.26163, 1986.
102. Fanourakis, P. & Savill, A.M. 4th European Drag Reduction Meeting Report, to appear in ERCOFTAC Bulletin (1989).
103. Bandyopadhyay, P.R.: AIAA-88- (1988)
104. Choi, K.-S., Gadd, G.E., Pearcey, H.H. & Savill, A.M., Svenson, S.: see [2] (1989).
105. Park, J.T. & Johnson, J.E.: Abstr. ONR-NASA-AFOSR-NSSC Drag Reduction Conference, Washington (1985).
106. McInville, R.M. & Hassan, H.A.: AIAA-85-0541 (1985).
107. Reed, J.C. & Weinstein, L.M.: AIAA-89-0962 (1989).
108. Papathanasiou, A.G. & Nagel, R.T.: AIAA-86-1954 (1986).
109. Bardakhanov, S.P., Kozlov, V.V. & Larichkin, V.V.: see [6] (1989)
110. Walsh, M.J. & Anders, J.B. Abstr ONR-NSSC-AFOSR-NASA Symp. Drag Reduction and Boundary layer Control, Washington (1985).
111. Savill, A.M.: Flow Visualisation IV, Hemisphere Publishing Corporation, Washington, p.303 (1986).
112. Merkle, C.L. & Deutch, S.: AIAA-85-0537 (1985)
113. van den Berg, B.: AGARD CP 365 (1985)
114. Tenaud, C.: Ph.D. thesis, ONERA-CERT (1987)
115. Dowling, A.P.: JFM 160, p.447 (1985)
116. Kinney, R.B., Taslim, M.E. & Hung, S.C. Univ. Arizona Coll. Eng. Rep. NAG-1-141 (1984)
117. Savill, A.M.: Proc. IMA/SMAI Conf. Computational Methods in Aeronautical Fluid Dynamics, OUP (1987)
118. Savill, A.M. & Zhou, M.D.: $th Asian Congr. Fluid Mech., Hong Kong (1989)
119. Savill, A.M.: Ann. Rev. Fluid Mech. 19 (1987)
120. Launder, B.E., Leschziner, M.A. & Sindir, M.: Proc. AFOSR-HTTM-Stanford Conf. Complex Turbulent Flows, Vol.II (1981)
121. Agouropoulos, D.: Ph.D thesis, University of Cambridge, 1986.
122. Smits, A.J, Young, S.B. & Bradshaw, P.: JFM 94 (1979)
123. Pollard, A., Savill, A.M. & Thomann, H.H.: Proc. 10th AFCM, Melbourne (1989)
124. Nguyen, V.D. & Dickinson, J.: 4th European Drag Reduction Meeting Report to appear in ERCOFTAC Bulletin (1989)
125. Djenidi, L., Liandrat, J., Anselmet, F. & Fulachier, L.: see [2] (1989)
126. Coustols, E.: 4th European Drag Reduction Meeting Report to appear in ERCOFTAC Bulletin (1989)
127. Launder, B.E. & Shaoping, L.: see [2] (1989)
128. Khann, M.M.S.: AIAA-86-1127 (1986)
129. Bandyopadhyay, P.R.: AIAA-88-0135 (1988)
130. Gatski, T.B. & Savill, A.M.: AIAA-89-1014 (1989)

Relation Between Outer Structures and Wall-layer Events in Boundary Layers with and without Manipulation

C. E. Wark and H. M. Nagib
Fluid Dynamics Research Center
Illinois Institute of Technology
Chicago, IL 60616 USA

Abstract

The scaling of structures related to high Reynolds stress production in turbulent boundary layers was investigated over a wide range of regular and drag reducing conditions. Consistent with the development of their integral size, which is proportional to the outer scale, a model for the growth of these vortical structures based on joining of neighboring pairs is conjectured from the measurements.

Introduction

Experimental investigations have been carried out in our laboratories over the past several years to examine the link between outer-layer structures and wall-layer events, in regular and manipulated turbulent boundary layers, over the Reynolds number range $1500 < Re_\theta < 5200$. While a link was documented between them (e.g. Guezennec 1985, Nagib et al. 1987 and Wark 1988), the recent results of Wark et al. (1989) suggest that the outer structures play only a partial role in the wall process and that this production process can be incipiently generated and self sustained. The key to the understanding of these relations has been provided by three-dimensional, pseudo-instantaneous information regarding the character of coherent structures associated with the wall-layer production process, as well as their time and space evolution. In particular, the existence of a hierarchy of scales (consistent with ideas of Perry et al. 1982 & 1986) have been quantitatively documented by Wark (1988).

An important aspect in understanding the production cycle is to be aware of the nature of the events identified using the various detection techniques. Figure 1 represents a comparison between three of the common detection schemes, VITA, u-level and Quadrant. The events related to Reynolds-stress production ($uv < 0$) are those falling into quadrants 2 and 4. Both the VITA and u-level schemes recognize events not related to Reynolds-stress production. A thorough comparison of the various detection techniques can be found in Bogard and Tiederman (1986) and Luchik and Tiederman (1987). It must be remembered that these are single-point comparisons and differences may be attributed to detection at different phases (i.e., spatial points) of the production cycle.

Based upon a wall shear-stress detection method (i.e. T- and T+ events), Guezennec (1985) concluded that ensemble-averaged roller-like coherent structures are associated with high levels of perturbation in the streamwise wall-shear stress. Figure 2 represents the roller-like structure associated with a streamwise wall-shear stress less than 1.9 times the long time rms value ($\tau_x < -1.9\,\tau_{x_{rms}}$). These large counter-rotating structures, at a shallow inclination angle, extend several δ in the streamwise direction, and have a spanwise and normal extent of approximately $0.5\,\delta$. Later, Wark (1988) documented the three-dimensional structure based on the shear-stress and quadrant detection (i.e. Q2 and Q4 events) methods and obtained information regarding the distribution of sizes (i.e. hierarchy of scales) of the instantaneous structures. That is, she mapped the scales associated with the individual realizations of the ensemble-averaged picture. Both investigations looked at the coherent structures associated with ejection of fluid away from the wall (T- and Q2) and sweeps of fluid to-

A. Gyr (Editor)
Structure of Turbulence and Drag Reduction
IUTAM Symposium Zurich/Switzerland 1989

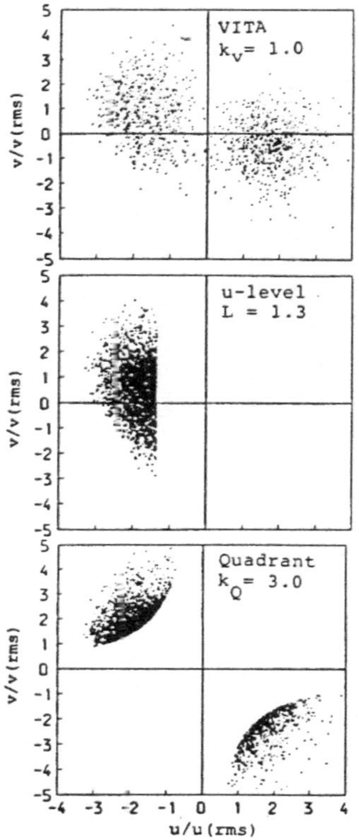

Figure 1. Scatter Plot of Distribution of u and v Associated with VITA, u-level and Quadrant Detection Techniques.

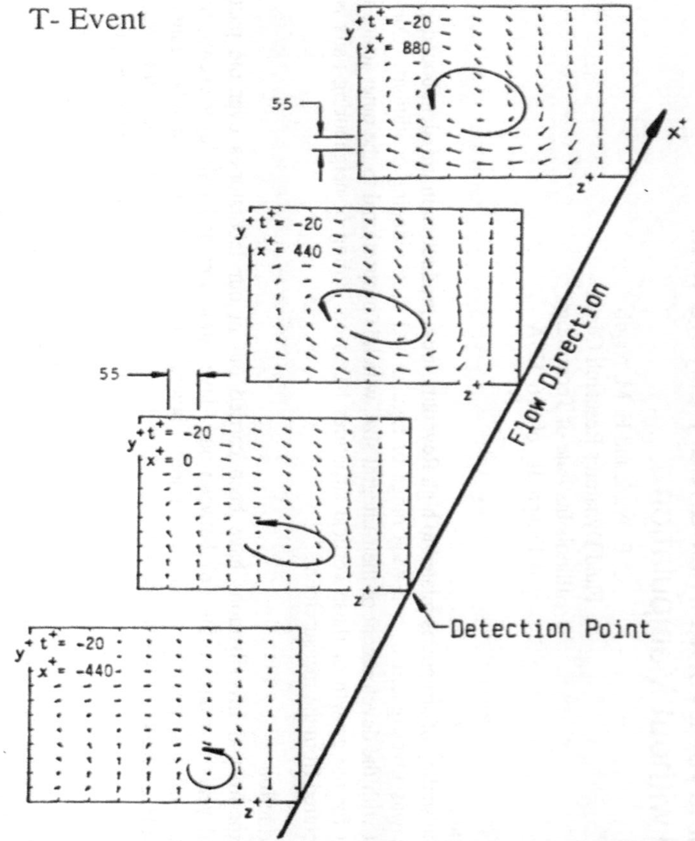

Figure 2. Streamwise Evolution of v-w Vector Maps Associated with Wall-Detected T- Events, from Guezennec (1985).

wards the wall (T+ and Q4). Also investigated were the convection speeds for both the ejection and sweep type coherent motions.

The objective of the present work is not only to further clarify the relation between outer structures and wall-layer events, but also to document and attempt to explain the generation and development of the structures within the hierarchy. Manipulated boundary layers producing drag reduction were used to focus on the early stages of their regeneration.

Results

The interaction between the outer and inner layer is studied based on the scaling of the frequency of detected events and also on the scaling of an integral-length measure of the ensemble-averaged roller-like structure. Figure 3 represents the Reynolds number dependence of the detection frequency at $y^+=55$ using inner scaling. The measured frequency is corrected using the scheme of Blackwelder and Haritonidis (1983) and non-dimensionalized using u_τ and ν / u_τ. The solid line is the best linear fit to the data points while the dashed lines represent +- 10% to this fit. The slope of this solid line is a measure of the Reynolds number dependence and is shown for all three scaling laws as a function of y^+ in Figure 4. For $y^+ < 110$, inner scaling provides the greatest degree of Reynolds-number independence as compared with mixed and outer scaling. This was also seen for the u-level, VITA and Q4 detection frequencies (see Wark and Nagib 1988). For the Q2 and VITA results, mixed scaling provides the best collapse with Reynolds number for the higher y^+ positions investigated. This may be caused by the fact that the frequency correction scheme has not been validated for the higher y^+ positions and for x-wire probes. Nevertheless, it is fairly well agreed upon that the frequency of the Reynolds-stress production process scales with inner layer variables suggesting that the process is initiated by the wall region.

The degree to which the various schemes (quadrant, shear-stress, u-level and VITA) identify the same 3-D structures was analyzed. This was done based on both the ensemble-averaged structure and the instantaneous events. An integral-length scale (Λ_z) associated with the ensemble-averaged structure is measured from the spanwise distribution of the streamwise velocity associated with a detection (see Nagib et al. 1987). Briefly, it is a measure of the spanwise extent of the ensemble-averaged roller-like structures. The variation of Λ_z (normalized using both inner, Λ_z^+, and outer scaling, Λ_z / θ), with Reynolds number is shown by the results of Figure 5. In addition to the present experimental results two points at lower Reynolds numbers from the NASA-Ames computational work have been added from Guezennec et al. (1987). It is clear that inner scaling does not provide a collapse of Λ_z^+ with Reynolds number; however, Λ_z / θ appears to be insensitive to the Reynolds number. It is important to note that Λ_z was found to be equivalent for the quadrant, shear-stress and u-level schemes. A VITA detection resulted in a significantly lower value for Λ_z.

The hierarchy of scales is depicted by the results of Figure 6. The gray level represents the probability of a Q2 event to occur at a given point in the sampling volume when detecting a u-level event at (x^+,y^+,z^+) = (0,35,0). The distribution of scales is easily visualized from this representation. By comparison with the results of Wark et al. 1989, it was concluded that the shear-stress, quadrant and u-level detection schemes lead to approximately equivalent distribution or hierarchy of scales. However, the VITA technique was found to not only mix ejection and sweep events, as Figure 1 suggests, but also to detect only the smaller of the ejections and sweeps in the distribution. The measurements regarding convection speeds of the sweep and ejection events displayed similar trends.

To further study the interaction between the outer and inner layer, the effect of manipulator blades on the ensemble-averaged 3-D structure was investigated (Nagib et al. 1987). The blades were placed in the outer region of the boundary layer and Λ_z was measured at various

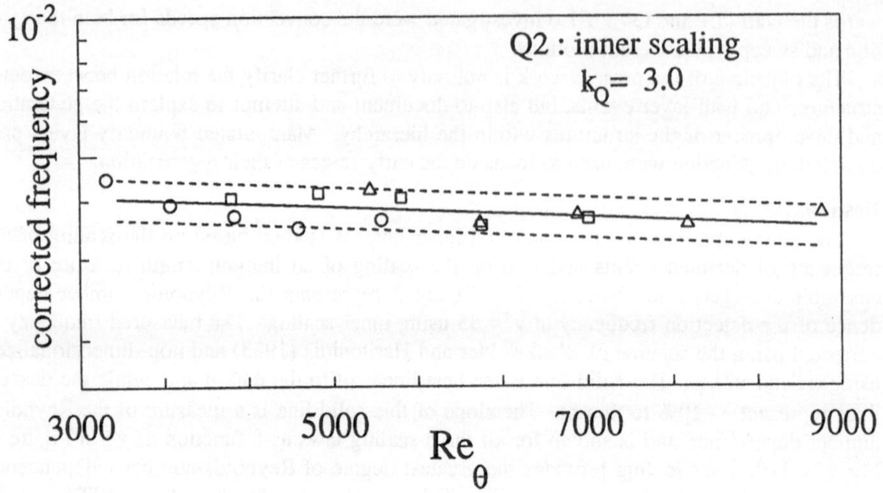

Figure 3. Corrected Frequency of Occurrence of Q2 Events for a Regular Boundary Layer at $y^+=55$ using Inner Scaling.

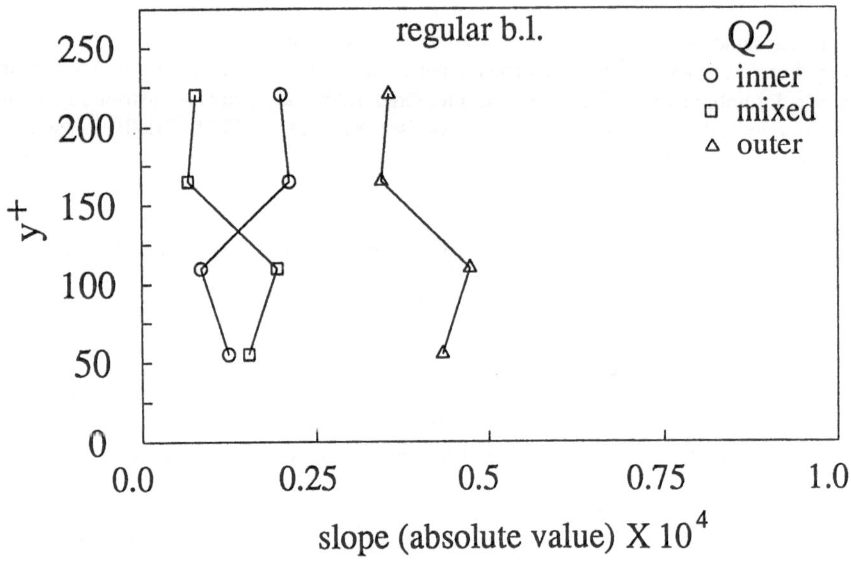

Figure 4. Variation of a Measure of Reynolds Number Independence (slope of line in Fig. 3 for $y^+=55$) With Height for Corrected Q2 Frequencies, Using All Three Scaling Laws.

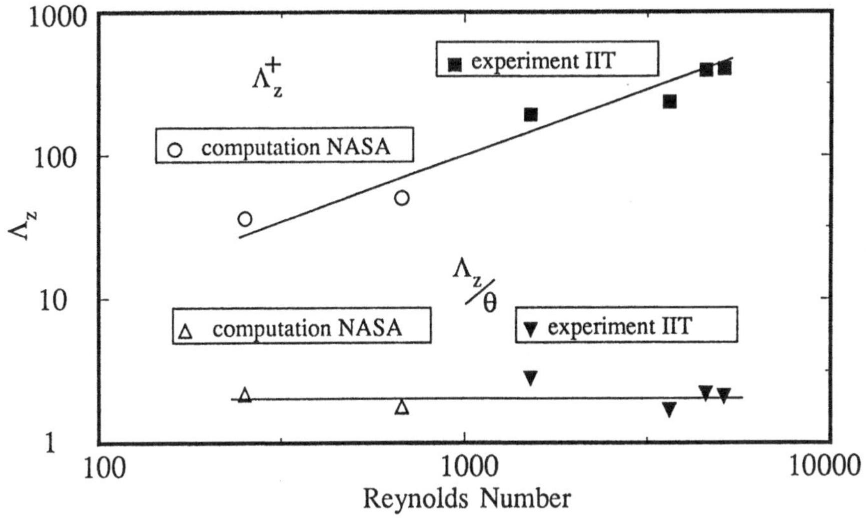

Figure 5. Scaling of an Integral Length Scale of Quadrant Detected Events.

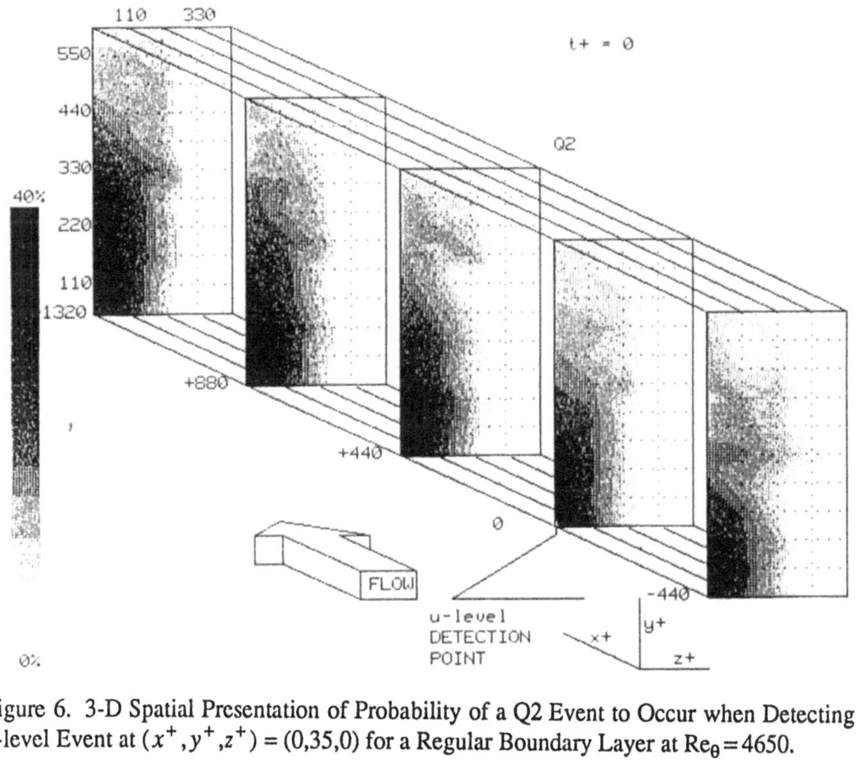

Figure 6. 3-D Spatial Presentation of Probability of a Q2 Event to Occur when Detecting a u-level Event at $(x^+, y^+, z^+) = (0,35,0)$ for a Regular Boundary Layer at $Re_\theta = 4650$.

distances downstream of the blades. Figure 7 illustrates the return of Λ_z to conditions equivalent to a regular boundary layer by 96 boundary layer thicknesses downstream of the blades. It was concluded that manipulator blades placed at relatively high positions in the boundary layer ($h/\delta=0.7$) affect only the largest of the events associated with the production process. A "relaxation" back to conditions similar to a regular boundary layer occurs over ≈ 100 boundary layer thicknesses downstream of the manipulators. To achieve significant alteration of the ensemble-averaged coherent structure, the blades had to be placed at relatively low heights ($h/\delta=0.1$) in the boundary layer; see Wark et al. 1989. In contrast to the slow redevelopment of the larger scales in the hierarchy, the smaller scale structures exhibit a very rapid return over a distance less than 7 δ.

Five significant conclusions may be reached from our results: 1) The wall layer is responsible for the *initiation* of coherent structures of all sizes; 2) a hierarchy of similar scales exists for high Reynolds numbers; 3) the integral scale of the hierarchy is proportional to the outer scale of the boundary layer; 4) while the development of the smaller scales of the hierarchy is very rapid a redevelopment of the full hierarchy after manipulation is gradual; and 5) the larger scales of the hierarchy (so called "outer scales") play only a partial role in the wall process since the smaller scales of the hierarchy appear to be incipiently generated and self sustained.

A model was proposed by Nagib et al. (1987) in an attempt to bring together some of these ideas. A big unanswered question in that model is related to the development of the larger scales. However, the model postulated that the ejection (Q2 or T-) events are related to upright lambda structures and that the sweep (Q4 or T+) events are associated with inverted lambda structures. A process similar to that occurring during transition may continue *indefinitely* in the wall region of turbulent boundary layers. The rollers (Townsend's typical eddies) are the legs of these vortical structures.

In a model based on a hierarchy of geometrically similar eddies, Perry and Chong (1982), propose "vortex pairing of two eddies in one hierarchy to form an eddy in the next hierarchy", as one possible mechanism for the growth of structures (e.g., lambda or hairpin). Later Perry et al. (1986) conclude that the work of Acarlar and Smith (1987) had "shown that the speculated pairing process does in fact occur, at least for hairpin-type vortices" artificially introduced into laminar boundary layers. The interaction documented by Acarlar and Smith (1987) is of eddies occurring sequentially in space and therefore is of a similar type to the well understood "pairing" in free-shear flows. A model for the growth of eddies within the hierarchy, more consistent with the original ideas of Perry and Chong (1982), is outlined in Figure 8. Joining of neighboring structures, and destruction of their adjoining legs through cancellation of opposite sign vorticity, would lead to the required doubling of the transverse scale. Similar interaction, depicted in the central part of Figure 8, between eddies of dissimilar size can lead to the non-symmetric structures found in computational results at the lower Reynolds numbers. A similar scenario can also be drawn for the inverted (sweep) events.

One inconsistency between our ideas and those of Perry et al. (1982 & 1986) is the inclination of the dominant vortical structures. They require a 45 degree inclination based on earlier flow visualization while a much shallower angle is clearly revealed by our measurements at high Reynolds numbers. The "neck" of the lambda (or hairpin) structures is more likely to be at the 45 degree angle and may be more visible and active at lower Reynolds numbers.

Conclusions

While the frequency of events producing large Reynolds stresses in the wall region scales with inner variables of the boundary layer, an integral measure of the hierarchy of scales in-

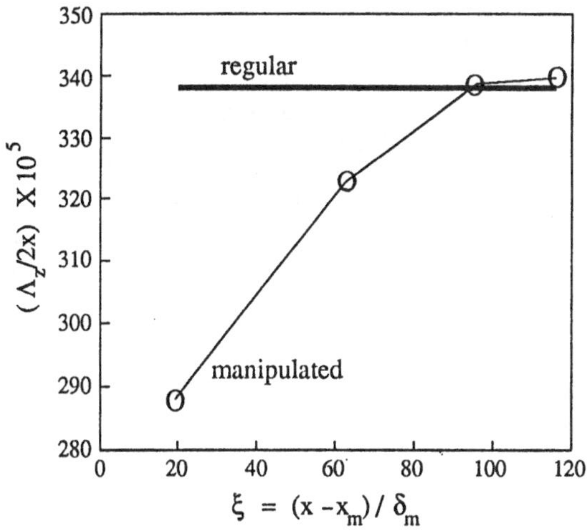

Figure 7. Variation of Integral Length Scale of Fig. 5 as a function of Downstream Distance from Manipulator Blades.

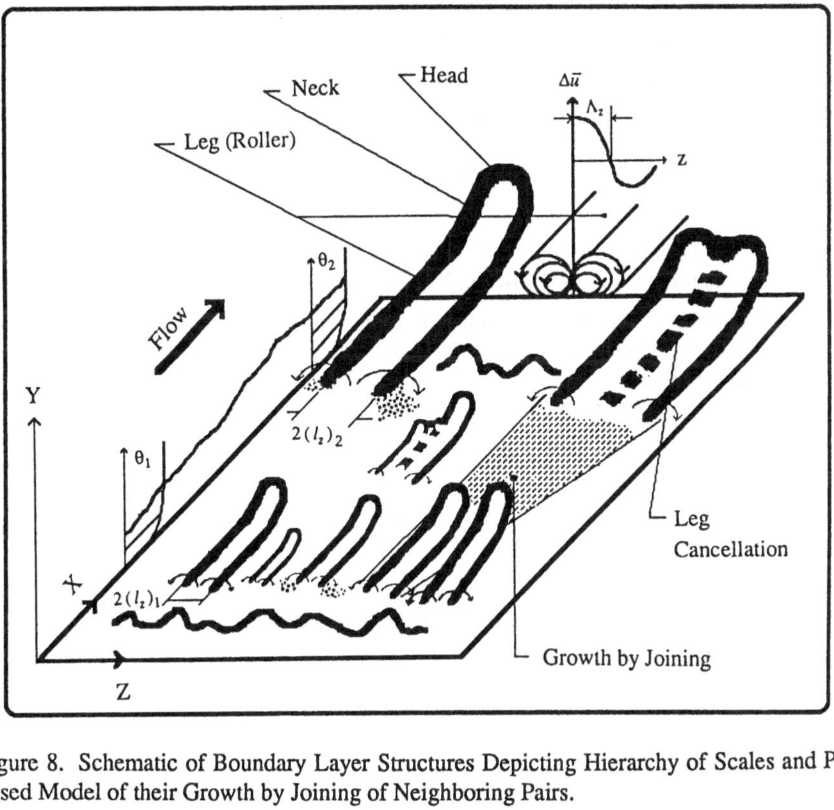

Figure 8. Schematic of Boundary Layer Structures Depicting Hierarchy of Scales and Proposed Model of their Growth by Joining of Neighboring Pairs.

volved is found to be proportional to the outer scales. This conclusion is based on analysis of the experimental data from IIT and computational data from NASA-Ames using the same measure of the scales over one decade of Reynolds number variation. Experiments downstream of drag reducing manipulator blades and measurements of the distribution of scales within the hierarchy confirm this result and suggest the existence of a *rapid* or geometric growth mechanism for the coherent vortical structures. A model is proposed for this growth based on the joining of neighboring transversely aligned pairs of similar vortical structures. The conjectured model is consistent with the ideas of Perry and Chong (1982) and may also provide an explanation for the existence of non-symmetric structures which have been captured by the computation.

Acknowledgment

This work was supported by AFOSR under contract number F49620-86-C-0133 monitored by Dr. J. McMichael.

References

Acarlar, M.S. and Smith, C.R. 1987. A study of Hairpin Vortices in a Laminar Boundary Layer. Parts 1 and 2. J. Fluid Mech., Vol. 175, pp. 1-83.

Blackwelder, R. F. and Haritonidis, J. H. 1983. Scaling of the Bursting Frequency in Turbulent Boundary Layers. J. Fluid Mech., Vol. 132, pp. 87-103.

Bogard, D.G. and Tiederman, W.G. 1986. Burst Detection with Single Point Velocity Measurements. J. Fluid Mech., Vol. 162, pp. 389.

Guezennec, Y. G. 1985. Documentation of Large Coherent Structures Associated with Wall Events in Turbulent Boundary Layers. Ph.D. Thesis, Ill. Inst. of Tech., Chicago, IL.

Guezennec, Y.G., Piomelli, U. and Kim, J. 1987. Conditionally-Averaged Structures in Wall-Bounded Turbulent Flows. Proceedings of the 1987 Summer Program, Center for Turbulence Research, Stanford, CA. pp. 263-272.

Luchik T.S. and Tiederman, W.G. 1987. Timescale and Structure of Ejections and Bursts in Turbulent Channel Flows. J. Fluid Mech., Vol, 174, pp. 529-552.

Nagib, H. M., Wark, C. E. and Guezennec, Y. G. 1987. Documentation of Turbulence Producing Structures in Regular and Manipulated Turbulent Boundary Layers. Proceedings of the IUTAM Symposium on Turbulence Management and Relaminarisation. Bangalore, India. Springer-Verlag, Paper#1.

Perry, A.E. and Chong M.S. 1982. On the Mechanism of Wall Turbulence. J. Fluid Mech., Vol. 119, pp. 173-217.

Perry, A.E., Henbest, S. and Chong, M.S. 1986. A Theoretical and Experimental Study of Wall Turbulence. J. Fluid Mech., Vol. 165, pp. 163-199.

Wark, C.E. 1988. Experimental Investigation of Coherent Structures in Turbulent Boundary Layers. Ph.D. Thesis, Illinois Institute of Technology, Chicago Illinois.

Wark, C.E. and Nagib, H.M. 1988. On the Character of Turbulence Producing Events in Near-Wall Turbulence. Proceedings of the Zoran Zaric Memorial International Seminar on Near-Wall Turbulence, May 16-20, 1988, Dubrovnik Yugoslavia.

Wark, C.E., Naguib, A.M. and Nagib, H.M. 1989. Effect of Flat-Plate Manipulation on the Coherent Structures in a Turbulent Boundary Layer. Proceedings of the 2nd AIAA Shear Flow Control Conference, Tempe, Arizona, 1989.

Boundary Layer Manipulators at High Reynolds Numbers

J. B. Anders
NASA Langley Research Center
Hampton, VA 23665-5225

Summary

Airfoil large-eddy breakup (LEBU) devices were tested on an axisymmetric body in the Langley Tow Tank up to speeds of 50 ft/sec. NACA-0009, NACA-2412, E-193, and Clark Y contours were tested in single and tandem configurations. At the higher Reynolds numbers local skin friction downstream of the devices showed minimal reductions (O(10 percent)) and total body drag was increased 1 to 3 percent. At lower Reynolds numbers skin-friction reductions as large as 25 percent were measured and total body drag tended toward net reductions. The loss of effectiveness with increasing Reynolds number of conventional, outer layer devices suggests a decoupling of the outer and inner scales in high Reynolds number turbulent boundary layers.

Introduction

Embedded flow-aligned plates can be employed to reduce skin friction under turbulent boundary layers (Refs. 1-2), and by careful control of the device geometry, net-drag reductions of 5 to 10 percent can be obtained (Refs. 3-9). The exact mechanisms whereby the devices reduce wall shear are still somewhat unclear; however, it is clear that they they produce measurable structural changes in the boundary layer (Refs. 9-13) and the region of reduced skin friction can persist over 100 boundary-layer thicknesses downstream. The passive nature of these add-on devices make them a potentially attractive method of reducing turbulent skin-friction drag on aircraft if their effectiveness at flight conditions can be verified.

Applying large-eddy breakup (LEBU) devices to flight vehicles requires operation at transonic speeds and flight-level Reynolds numbers, i.e., device chord Reynolds numbers (R_c) of 3-5 x 10^5 and host boundary layer momentum thickness Reynolds numbers (R_θ) of 3-7 x 10^4. Unfortunately, with one exception, the existing LEBU data base is confined to $R_c < 1 \times 10^5$ and $R_\theta < 7 \times 10^3$. Reference 3 pointed out the influence of low chord Reynolds numbers on the drag of LEBU devices, and suggested that unsteady laminar separation on the devices was responsible for much of the variability in reported net drag reductions. However, an equally important issue is the low momentum thickness Reynolds numbers of the existing data. Significant structural changes are known to occur within the turbulent boundary layer as Reynolds number is increased (Refs. 14-19) and these findings suggest that a high Reynolds number turbulent boundary layer may interact differently with the LEBU devices.

A. Gyr (Editor)
Structure of Turbulence and Drag Reduction
IUTAM Symposium Zurich/Switzerland 1989

Simplistically, one may argue that at low Reynolds numbers the wall events that produce most of the Reynolds stress are roughly only a factor of 10 smaller than the largest eddies. At these conditions, techniques which alter the large eddies (LEBUs, wall curvature) have a dramatic effect on wall shear indicating a strong relationship between the two scales. However, as Reynolds number increases, the scale mismatch between wall events and the large-scale structures can be of the order of 100 or more, and a much-weakened relationship could exist. Thus, one might expect that conventional LEBU devices, which presumably operate on the large eddies, may be much less efficient in reducing wall shear due to this decoupling of the scales.

The experiments of Sahlin et al (Ref. 20) are the single published LEBU results at high Reynolds number ($R_c \simeq 2.6 \times 10^5$, $R_\theta \simeq 2 \times 10^4$). They reported average skin-friction reductions downstream of the devices of only 8 percent (at best), and net drag increases of 0 to 3 percent, even though device drag was low. They concluded that net drag reductions at high Reynolds numbers seem implausible, and speculate that LEBU effectiveness decreases with increasing Reynolds number because of the decrease in size of the most energetic turbulent scales relative to the boundary-layer thickness. This implies that the scale mismatch between outer and inner scales disables the LEBU mechanism. Further evidence of this effect has been noted in unpublished, ultra-high Reynolds number results obtained in transonic wind-tunnel experiments at Langley Research Center ($2.9 \times 10^5 < R_c < 8.6 \times 10^5$; 3.2×10^4 $R_\theta < 6.8 \times 10^4$). These preliminary results indicated that C_f reductions downstream of conventional outer layer devices were virtually nonexistent.

The preceding discussion raises serious doubts as to whether conventional, low Reynolds number optimized, outer-layer devices will reduce skin friction under high Reynolds number turbulent boundary layers. The present research addresses that question.

Facility, Model, and Instrumentation

The present tests were conducted in a fresh water towing tank at Langley Research Center, leased and operated by the Naval Underwater Systems Center of Newport, Rhode Island. The tank is 24 feet wide, 12 feet deep, and 3,000 feet long and is equipped with an electrically-driven carriage supporting a water-piercing strut/sting assembly (Ref. 21). The sting-mounted axisymmetric model used for these tests was approximately 12 feet long with a maximum body diameter of about 21 inches (Fig. 1). It was positioned approximately 4.5 feet beneath the water surface. A non-separating tailcone closed the body at the rear and the entire body/tailcone assembly was mounted on a free-floating balance mechanism with axial drag forces measured by a load cell.

Fig. 1. Axisymmetric model in the dry dock of the Langley Tow Tank.

The model was designed for a near-zero pressure gradient in the region $1 < x < 10$ ft. A trip wire was located one foot downstream of the nose at the end of the favorable pressure gradient region to ensure fully turbulent flow at the device location. Injected dye and liquid crystals were used to verify that transition was fixed at the trip location for all speeds of the current study. Test conditions at the LEBU device location are shown in Table 1.

Table 1. Test Conditions

U, ft/sec	R_∞, ft^{-1}	R_c	R_θ	h/δ	c/δ	δ, inches
6	3.8×10^5	3.2×10^4	2.5×10^3	.57	1.14	.88
10	7.6×10^5	6.3×10^4	4.5×10^3	.63	1.25	.80
20	1.5×10^6	1.3×10^5	8.1×10^3	.69	1.39	.72
30	2.3×10^6	1.9×10^5	1.2×10^4	.74	1.47	.68
40	3.0×10^6	2.5×10^5	1.5×10^4	.77	1.54	.65
50	3.8×10^6	3.2×10^5	1.8×10^4	.79	1.59	.63

Local skin friction was measured using modified Preston tubes (Ref. 22) at selected locations on the model surface. A 24 tube boundary-layer rake was located at a station 10 feet downstream of the nose to provide velocity profiles and integral thicknesses.

The LEBU devices tested were stainless steel, airfoil-section rings mounted at a station 5 feet downstream of the nose. Both single and tandem arrays of NACA-0009, NACA-2412, E-193, and Clark Y airfoils (Ref. 23) were tested at zero angle of attack. Figure 1 shows a tandem array of devices mounted on the body. The chord of the airfoils was 1 inch and the height above the model surface was 0.5 inches. Table 1 shows the chord (c) and height (h) of the devices normalized by the local boundary layer thickness (δ). The airfoils were supported by three airfoil-shaped (NACA-0009 section) struts spaced 120° apart. A flutter analysis similar to that reported in Reference 7 was used to determine the maximum unsupported length allowable for a maximum test velocity of 50 ft/sec in water. The spacing between devices for the tandem configurations was 4δ. Lift coefficients at zero angle of attack for the NACA-0009, NACA-2412, E-193, and Clark Y airfoils were 0, 0.23, 0.28, and 0.33, respectively, in the Reynolds number range of the present tests.

Results and Discussion

Model drag without manipulators is shown in Figure 2. The large increase in drag coefficient at low free-stream Reynolds number per foot (R_∞) reflects the significant wave-drag component present at low speeds. The magnitude of this component is a function of the model size, submergence depth, and finite dimensions of the tank (Ref. 24).

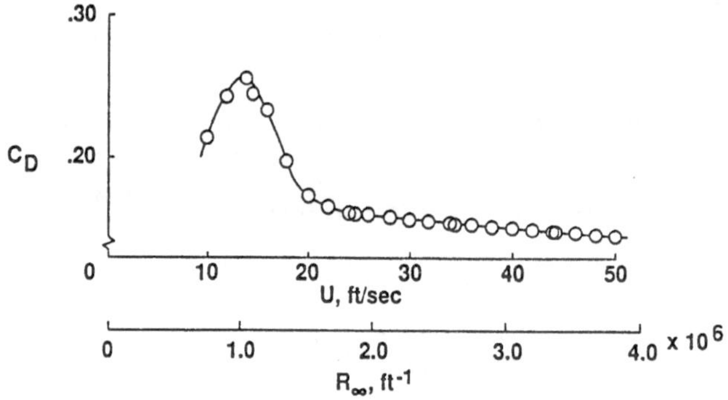

Fig. 2. Reference body total drag (without LEBU devices).

Unfortunately, this large wave-drag component limited the ability to resolve small drag changes incurred by the manipulators in this Reynolds number range. In addition, at the lowest speeds (less than 20 ft/sec), the load cell output was less than 5 percent of range and carriage vibration contributed significantly to the measured signal. As a result, repeatability of total body drag at U = 10 ft/sec was typically no better than ± 4 percent and no reliable reading could be obtained at U = 6 ft/sec. Above U = 20 ft/sec the repeatability was better than ± 1 percent. Corrections to total body drag due to incomplete body closure at the rear of the sting-supported model were typically 1 percent, indicating that at the higher speeds most of the body drag was due to skin friction.

Figure 3 shows the normalized total body drag with single and tandem devices. Except for a few points at the lowest Reynolds numbers, all airfoil devices increased the total drag from 1 to 5 percent. The low Reynolds number data in Figure 3 showing net drag reductions should be viewed with some caution in light of the earlier comments on the variability of the low Reynolds number drag measurements. Typical repeat measurements are shown in each figure. No clear trend with airfoil contour is apparent in these data

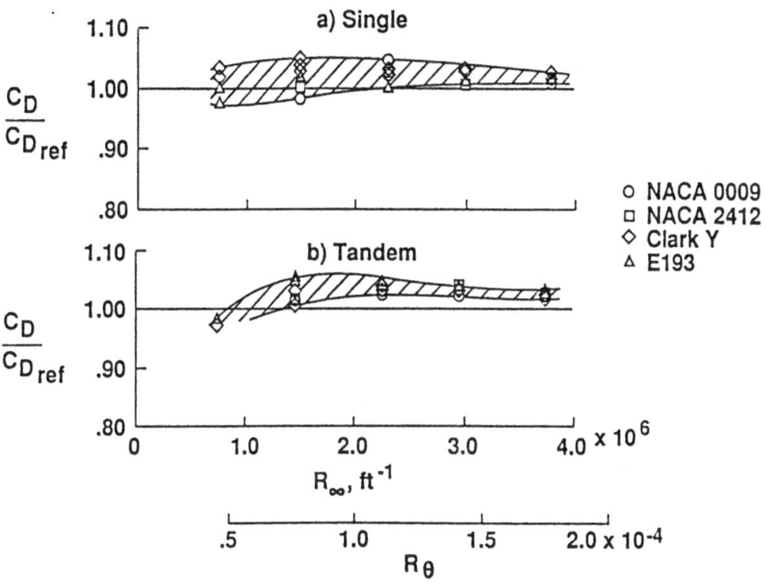

Fig. 3. Total body drag with airfoil LEBU devices.

and the faired data bands are meant to direct attention to broad trends rather than detailed behavior. The faired bands for both single and tandem devices give the distinct impression that net drag increases are low (perhaps even small net reductions) at low Reynolds numbers but increase with increasing Reynolds number.

Local C_f measurements from the Preston tubes are shown in Figures 4 and 5. It is immediately obvious from these data that, for $R_\theta > 8.1 \times 10^3$ the maximum skin-friction reductions achieved are less than 10 percent and average reductions are 5 percent or less. These maximum C_f reductions are significantly less than those reported in the low Reynolds number experiments of References 3, 4, 5, 7, 25, 26 and others, but in agreement with the high Reynolds number experiment of Reference 20.

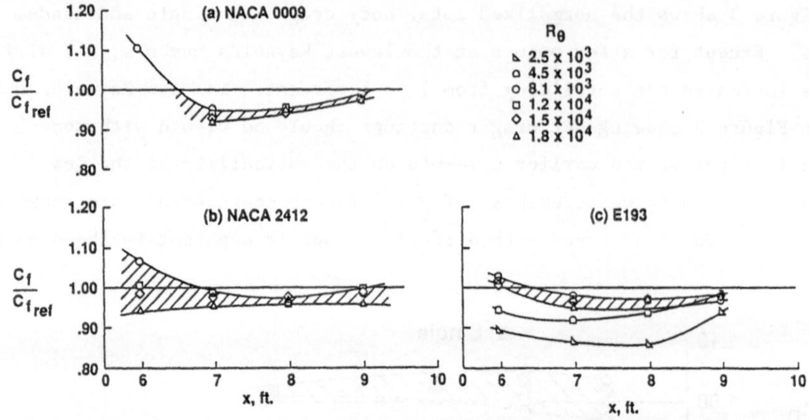

Fig. 4. Skin-friction distribution downstrean of single LEBU devices.

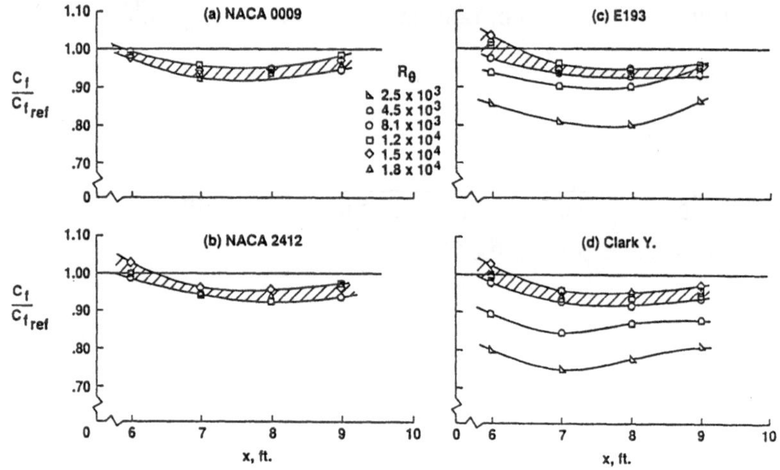

Fig. 5. Skin-friction distribution downstream of tandem LEBU devices.

Figures 4c, 5c, and 5d show that at very low Reynolds numbers ($R_\theta < 8.1 \times 10^3$) larger skin-friction reductions were found. Maximum C_f reductions were approximately 15 percent for single devices and 25 percent for tandem devices. Unfortunately, inadvertant damage to the NACA-0009 and NACA-2412 airfoils during testing prevented low Reynolds number C_f measurements with either of these two devices. However, the Clark Y and single E-193 airfoil devices exhibited a clear increase in effectiveness with decreasing Reynolds number. Estimated average C_f reductions appear to be of the order of 15 to 20 percent, equaling the best results obtained in low Reynolds wind-tunnel tests.

The larger reductions obtained with the Clark Y devices (18 percent higher lift coefficient) may be indicative of the beneficial effects of device lift. With this level of average skin-friction reduction, a net drag reduction might be expected for the two lowest Reynolds number test conditions. Total body drag measurements for the tandem E-193 and Clark Y LEBUs did show a slight net reduction of approximately 2 percent (Fig. 3) but as mentioned previously, repeatability of the load-cell measurements at $R_\theta = 4.5 \times 10^3$ was poor and limits any firm conclusion. Also, since the LEBU device chord Reynolds numbers for these lowest Reynolds number test conditions were 3.2×10^4 and 6.3×10^4 it is likely that device drag was high due to laminar separation, as described in Reference 3.

Conclusions

The present tests indicate a sharp degradation of the LEBU drag-reduction mechanism with increasing Reynolds number. Some evidence of this is also seen in the high Reynolds number results of Reference 20, unpublished LaRC transonic tests, and even in the low Reynolds number results of Reference 27. The most likely source of this degradation is a decoupling of the inner and outer scales at higher Reynolds numbers. As a result of this decoupling, breakup of the large structures by outer-layer devices has minimal influence on the near-wall, shear-producing scales. This suggests that smaller devices, closer to the wall may be required to regain effectiveness at high Reynolds numbers (in agreement with the conclusions of Ref. 12).

References

1. Falco, R. E.; and Rashidnia, N.: What Happens to the Large Eddies When Net Drag Reduction is Achieved by Outer Flow Manipulators. Presented at International Conference on Turbulent Drag Reduction by Passive Means, Royal Aeronautical Society, London, September 15-17, 1987.
2. Wilkinson, S. P.; Anders, J. B.; Lazos, B. S.; and Bushnell, D. M.: Turbulent Drag Reduction at NASA Langley, Progress and Plans. Presented at International Conference on Turbulent Drag Reduction by Passive Means, Royal Aeronautical Society, London, September 15-17, 1987.
3. Anders, J. B.: Large-Eddy Breakup Devices as Low Reynolds Number Airfoils. SAE Paper No. 86-1769, presented at the Aerospace Technology Conference and Exposition, Long Beach, CA, October 13-16, 1986.
4. Corke, T. C.; Guezennec, Y.; and Nagib, H. M.: Modifications of Drag of Turbulent Boundary Layers Resulting from Manipulation of Large-Scale Structures. NASA CR-3444, July 1981.
5. Lynn, T. B.; and Sreenivasan, K. R.: Measurements in Manipulated Turbulent Boundary Layers. Yale University, Department of Mechanical Engineering, Report 1-FM-85.
6. Bertelrud, Arild: Full Scale Experiments into the Use of Large-Eddy Breakup Devices for Drag Reduction on Aircraft. AGARD Symposium on Improvement of Aerodynamic Performance Through Boundary Layer Control and High Lift Systems, AGARD CP No. 365, May 1984.

7. Anders, J. B.; and Watson, R. D.: Airfoil Large-Eddy Breakup Devices for Turbulent Drag Reduction. AIAA Paper No. 85-0520, March 1985.
8. Coustols, E.; and Cousteix, J.: Reduction Du Frottement Turbulent: Moderateurs De Turbulence. 22 EME Colloque D'Aerodynamique Appliquee, Lille, France, November 13-15, 1985.
9. Rashidnia, N.; and Falco, R. E.: Changes in the Turbulent Boundary Layer Structure Associated with Net Drag Reduction by Outer Layer Manipulators. Report TSL-86-1, Dept. Mech. Eng., Michigan State University, May 1986.
10. Falco, R. E.: Correlation of Outer and Passive Wall Region Manipulation with Boundary Layer Coherent Structure Dynamics and Suggestions for Improved Devices. AIAA Paper 89-1026, March 1989.
11. Chang, S. I.: Modification of Large Eddies in a Turbulent Boundary Layer. Ph.D. Dissertation, University of Southern California, Los Angeles, California, August 1987.
12. Savill, A. M.; and Mumford, J. C.: Manipulation of Turbulent Boundary Layers by Outer Layer Devices: Skin Friction and Flow-Visualization Results. Journal of Fluid Mechanics, Vol. 191, 1988, pp. 389-418.
13. Westphal, R. V.: Skin Friction and Reynolds Stress Measurements for a Turbulent Boundary Layer Following Manipulation Using Flat Plates. AIAA Paper No. 86-0283, January 1986.
14. Head, M. R.; and Bandyopadhyay, P.: New Aspects of Turbulent Boundary-Layer Structure. Journal of Fluid Mechanics, Vol. 17, 1981, pp. 297-338.
15. Robinson, S. K.: Space-Time Correlation Measurements in a Compressible Turbulent Boundary Layer. AIAA Paper No. 86-1130, May 1986.
16. Zakkay, V.; Barra, V.; and Hozumi, K.: Turbulent Boundary Layer Structure at Low and High Subsonic Speeds. AGARD CP-271, September 1979.
17. Robinson, S. K.: An Experimental Search for Near-Wall Boundary Conditions for Large Eddy Simulation. AIAA Paper No. 82-0963.
18. Andreopoulos, J.; Durst, F.; and Jovanovic, J.: The Influence of Reynolds Number on Characteristics of Turbulent Boundary Layers, In: Structure of Turbulence in Heat and Mass Transfer, Zoran P. Zaric', editor, Hemisphere Publishing Corporation.
19. Antonia, R. A.; Subramanian, C. S.; Rajagopalan, S.; and Chambers, A. J.: Reynolds Nusmber Dependence of the Large Structure in a Slightly Heated Turbulent Boundary Layer. In: Structure of Turbulence in Heat and Mass Transfer, Zoran P. Zaric', editor, Hemisphere Publishing Corporation.
20. Sahlin, A.; Johansson, A. V.; and Alfredsson, P. H.: The Possibility of Drag Reduction by Outer Layer Manipulators in Turbulent Boundary Layers. Physics of Fluids, Vol. 31, October 1988.
21. Brown, D. A.: Langley Sea Water Tow Tank Facility: Description and Operation. Naval Underwater Systems Center Technical Document, June 1985.
22. Bertelrud, A.: Total Head/Static Measurements of Skin Friction and Surface Pressure. AIAA Journal, Vol. 15, No. 3, pp. 436-438, March 1977.
23. Miley, S. J.: A Catalog of Low Reynolds Number Airfoil Data for Wind Tunnel Turbine Applications. Department of Aerospace Engineering, Texas A&M University, College Station, Texas, RFD-3387, February 1982.
24. Hoerner, S. F.: Fluid Dynamic Drag. Hoerner Fluid Dynamics, Brick Town, New Jersey, 1965.
25. Lemay, J.; Provencal, D.; Gourdeau, R.; Nguyen, V. D.; and Dickinson, J.: More Detailed Measurements Behind Turbulence Manipulators Including Tandem Devices Using Servo-Controlled Balances. AIAA Paper 85-0521, March 1985.
26. Plesniak, M. W.: Optimized Manipulation of Turbulent Boundary Layers Aimed at Net Drag Reduction. M.S. Thesis, Illinois Institute of Technology, Chicago, Illinois, December 1984.
27. Bandyopadhyay, P. R.: Drag Reducing Outer-Layer Devices in Rough Wall Turbulent Boundary Layers. Experiments in Fluids, Vol. 4, pp. 247-256, 1986.

Large-scale Turbulence Structures in a Manipulated Channel Flow

R. Friedrich and H. Klein
Lehrstuhl für Strömungsmechanik
Technische Universität München
Arcisstraße 21, 8000 München 2
Federal Republic of Germany

SUMMARY

Large-eddy simulation (LES) is a proper means to show the effect of large-eddy break-up devices (LEBUs) on turbulence structures in a high Reynolds number channel flow. Three different types of simulations are performed on the basis of the filtered Navier-Stokes equations: 1) a LES of fully developed channel flow with periodicity boundary conditions (PBC) in order to generate inflow boundary conditions, 2) a LES of the same flow using inflow/outflow boundary conditions (IOBCs) with the aim of testing their influence and 3) a LES of manipulated channel flow (MCF) with inflow boundary conditions from 1). Characteristics of the numerical method are a leapfrog-scheme for time integration and the capacitance matrix technique to solve the Poisson equation for the pressure. Results for a 64×32×32 grid are presented. They clearly exhibit the effect of breakup of large scale instantaneous structures and of decrease of turbulence intensities and Reynolds stress due to the primary plate effect.

1. INTRODUCTION

It is generally known that turbulent shear flows (either free or wall-bounded) contain so-called coherent structures. These are typically of large scale, comparable to the global flow dimension, have characteristic life-spans and a high degree of organization in structure as well as in dynamics [1]. They contain most of the turbulence energy and do most of the turbulent transport. As a consequence the wall-shear stress, the noise-production and -emission etc. essentially depend on these large scale structures. These facts have opened up the possibility of controlling turbulence in two different directions, namely towards an amplification (in chemical engineering applications) or towards a suppression (relevant in aerodynamics). We will focus on the latter aspect in this paper. It has been noted by several authors, e.g.[2,3,4,5], that flat plate manipulators are promising means of reducing skin friction. They do not solely act as large-eddy break-up devices, but change the flow structure in several ways [2],[6],[7]. There is the primary plate effect which leads to a reduction of the turbulence intensities and the Reynolds stress by either restricting the burst events near the wall or suppressing the entraining eddies in the outer region of a boundary layer [7]. Especially the wall normal fluctuations are inhibited by the LEBU and this effect is felt in the whole space

A. Gyr (Editor)
Structure of Turbulence and Drag Reduction
IUTAM Symposium Zurich/Switzerland 1989

between LEBU and wall. Another consequence of the plate effect has been noted in [8]. It is the damping of large scale fluctuations due to an acceleration of the flow in the gap between LEBU and wall. Finally, there is the secondary wake effect which is shown to be far more important than had previously been expected. The wake acts as a shield that prevents incursions of high-speed fluid and blocks the interaction between turbulent fluctuations in the outer part of the boundary layer and the near wall region [7]. Besides this, it introduces small scales into the flow which partly stimulate the energy cascade and partly transfer energy from the turbulence back to the mean flow [2]. It was observed in [7] that the maximum local skin-friction reduction always occured close to the position where the wake reached the wall. This fact explained why the height of a LEBU is an important parameter. In a boundary layer it should be varied inversely with the Reynolds number. Now, for the present investigation being concerned with manipulated channel flow (MCF), entrainment of potential outer flow is not relevant, but, the obstruction of wall normal fluctuations due to LEBU devices and an acceleration resulting from growing thin boundary layers are clearly discernible effects. The spatial resolution of the present simulations does not suffice to resolve details of the wake nor is the flow field long enough in order to include the position of minimal local skin friction. Irrespective of these drawbacks which are due to limitations of the available computer, instructive results could be obtained which reflect the right tendencies of important drag reducing mechanisms.

2. NUMERICAL MODEL

The time-dependent and three-dimensional Navier-Stokes equations for an incompressible fluid are low-pass filtered via integration over the finite volume of a cartesian equidistant grid [9]. Using the divergence theorem to convert volume integrals into surface integrals and defining cell-surface averaged velocity components $\overline{v_i}$ and a cell-volume averaged pressure $\overline{p}$, we obtain the following nondimensional form of the filtered equations:

$$\delta_j \, \overline{v}_j = 0 \ , \tag{1}$$

$$\frac{\partial \, \overline{v}_i}{\partial t} + \delta_j(\overline{v}_j^i \overline{v}_i^j + \overline{v_j{}' v_i{}'} + \overline{p} \, \delta_{ji} - \overline{\tau}_{ji}) = 0 \ . \tag{2}$$

Since the averaged or grid-scale (GS) quantities, $\overline{v_i}$, $\overline{p}$ are related to fixed positions in the grid ($\overline{p}$ is related to the cell center and $\overline{v_i}$ to the cell surfaces), spatial discretization of (1) and (2) has resulted naturally and is reflected in the central difference operator

$$\delta_j \, \psi = \frac{\psi(x_j + \frac{\Delta x_j}{2}) - \psi(x_j - \frac{\Delta x_j}{2})}{\Delta x_j} \ . \tag{3}$$

Evaluation of the momentum transport (2) in a staggered grid needs velocity components at positions where they are not defined. Simple algebraic averaging of the form

$$\overline{\psi}^j = \frac{\psi(x_j - \frac{\Delta x_j}{2}) + \psi(x_j + \frac{\Delta x_j}{2})}{2} \tag{4}$$

removes this difficulty. A serious closure problem is left over in (2) in the form of the subgrid scale (SGS) stresses, $\overline{v_j'v_i'}$, which describe the momentum transport through the cell-surfaces due to deviations of velocity components from the GS variables. We adopt Schumann's algebraic model [9] for $\overline{v_j'v_i'}$ which contains a statistical and a fluctuating part, viz:

$$\overline{v_j'v_i'} = -\mu_{inh} <\overline{D}_{ji}> -\mu_{iso}(\overline{D}_{ji} - <\overline{D}_{ji}>) \ . \tag{5}$$

The eddy viscosities contain length scales which are related to the grid spacing (in case of μ_{iso}) or are empirical. The velocity scales are either computed from the fluctuating or from the statistically averaged ($< \ldots >$) deformation tensor

$$\overline{D}_{ji} = \delta_i\overline{v}_j + \delta_j\overline{v}_i \ . \tag{6}$$

(6) is also used to define the 'molecular' shear stress in (2):

$$\overline{\tau}_{ji} = \frac{\overline{D}_{ji}}{Re} \ . \tag{7}$$

The statistical part in (5) is designed to take inhomogeneities in the SGS turbulence into account which are certainly important in cases of strong shear and coarse mesh systems. For more details of this model see [9,10].

Time-integration of (2) proceeds in cycles of explicit steps consisting of 49 leapfrog steps and an averaging step to avoid $2\Delta t$-oscillations. The use of Fortin's version of the projection method [11] leads to a Poisson equation for the pressure which is solved directly. Since the flow considered is statistically two-dimensional, a fast Fourier transformation can be applied to convert the 3D Poisson problem into a 2D problem in a multiply connected domain. The capacitance matrix technique serves to solve the remaining problem. It essentially consists in a twofold application of a cyclic reduction algorithm. A detailed description of this technique is given in [12,13].

Boundary conditions deserve special attention. Let us start discussing *wall boundary conditions* first. The physical no slip condition cannot be satisfied since the velocity component tangential to the wall is only defined in a cell-surface perpendicular to the wall. Instead, the instantaneous wall shear stress is specified taking it in phase with the longitudinal velocity component [9]. Only the coefficient of proportionality in this relation is computed from empirical laws of the wall. The validity of this assumption is supported by experimental findings [14]. New approximate boundary conditions will be discussed in a paper to appear [15]. The velocity component normal to the wall can be prescribed exactly. Such type of b.c.s are used not only at the channel walls but also on the surface of the LEBU. In the spanwise direction ***periodicity boundary conditions*** (PBCs) do a good job. If the flow is fully developed PBCs are preferably used in the main flow direction as well. In the case of a MCF this is however not possible, unless the computational domain can be made long enough. The flow field is no longer homogeneous in planes parallel to the channel walls, but develops downstream. The LES, therefore, needs ***inflow/outflow boundary conditions*** (IOBCs). They consist in specifying the 3D velocity vector in the I/O-planes as a function of time during the whole simulation process. Boundary conditions for $\overline{p}$ follow from a Neumann condition and are thus expressed in terms of velocity components. This is a special feature of the explicit version of the projection method used, see [16]. In order to provide the necessary inflow data, a separate simulation for fully developed channel flow with PBCs in streamwise and spanwise directions is performed from which the instantaneous 3D velocity vector in a cross section is obtained and stored on magnetic tape over 6000 successive time steps. A special treatment of the velocity vector in the outflow plane has proved successful. It uses linear extrapolation of the mean streamwise component, constant extrapolation of the remaining two components, but a convection equation for all the velocity fluctuations with $<\overline{u}>$ as proper transport velocity. The conditions are discussed in more detail in [17].

3. SPECIFICATION OF THE SIMULATED FLOWS AND RESULTS

We briefly describe the geometry of the flow field and the relevant fluid dynamical parameters. The size of the computational domain in a (x,y,z)-coordinate system is $8 \times 4 \times 2$ in terms of the channel half-width h. Two infinitely thin LEBU-devices are mounted at a distance 2h downstream of the inflow plane. The wall distance of the symmetrically grouped plates is 0.5 and their length amounts to 1.25 which is too short to yield the maximal plate effect. The Reynolds number based on friction velocity at the entrance and on h is 3240 or based on channel width 2h and bulk velocity, 1.5×10^5. A $64 \times 32 \times 32$ equidistant grid has been chosen which of course cannot properly resolve neither the boundary layers on the plates nor their wakes. Nevertheless, encouraging results are obtained which give a glimpse of the power of the LES technique.

Two graphs may suffice to demonstrate the quality of inflow data taken from a separate LES of fully developed flow with PBCs. Fig.1 shows profiles of primitive variables and of the different contributions to the shear stress. All quantities in this and in subsequent figures are non-dimensionalized with the friction velocity and the channel half-width. The pressure p represents the difference between the local variable $<p(x,y,z)>$ and the arbitrarily chosen reference value $<p(x=4,z=0)>$. Being completely determined by the velociy field, it varies with the intensity of the vertical velocity fluctuations. The

Reynolds stress is composed of its grid scale part (TAU GS) and a contribution from the unresolved small scales (TAU INH) which is important only in the near wall region where the viscous stress (TAU VISC) complements the Reynolds stress to form the total shear stress (TAU SUM). The latter attains its linear distribution perfectly which is an indication of fully developed simulated flow. Fig.2 is meant to demonstrate the effect of IOBCs which are described at the end of chapter 2. From bottom to top we have plotted contour lines of rms-velociy fluctuations in spanwise direction, of rms-pressure fluctuations and of the mean pressure itself. The results indicate a weak effect of our outflow conditions. Further improvements seem to be difficult to achieve. We found quadratic extrapolation of $<u>$ and linear extrapolation of $<v>$, $<w>$ to produce similar results.

The following figures give an impression of the plate effect of LEBU devices. The LES was performed over 6000 time steps out of which 4000 have been used for statistical averaging besides averaging in the homogeneous direction. The mean longitudinal velocity in fig.3 shows the plate effect and the formation of two wakes. The acceleration of the flow due to the boundary layer displacement effect is obvious from the plot to the right with the enlarged scale. The numbers 1 to 65 indicate locations of grid volumes in x-direction. I=18 and 27 correspond to positions slightly downstream of the plate leading edge and one grid volume upstream of the trailing edge, respectively. The displacement of the flow is also reflected in the mean vertical velocity (Fig.4). The last profile (I=65) should be considered with care because it may be affected by the outflow conditions. Fig.4 also contains rms-profiles of longitudinal velocity fluctuations. The main effect of the plates is the suppression of these fluctuations and of those in the two remaining directions (Fig.5). The results for u- and v-components seem to confirm what is experimentally found [4], namely a surplus of turbulence intensity generated by the device. We do not find such an effect for the fluctuations normal to the plates, instead they show the strongest suppression among the three components. The no-slip condition on the plates increases the wall shear stress there, while the total stress in the wakes of the plates is again reduced. The latter is also true for the resolved part of the Reynolds stress (TAU GS), see fig.6. A comparison of the mean pressure contour lines in fig.7 with those for the unmanipulated flow in fig.2 is worth-while. Their shape in the wake region mirrors the effect of reduced vertical fluctuations. From rms-pressure fluctuations, fig.8, we draw the conclusion that the plate leading edges form the primary obstruction to the turbulent flow field which breaks up the eddies and realigns the velocity vectors. The effect of the remaining part of the plate is much less dramatic. The instantaneous turbulence structure is represented in fig.9 in terms of contour lines of the spanwise velocity fluctuations and of fluctuating velocity vectors in a (x,z)-plane at J=8 (y=h). Among the three velocity components, v reveals best the typical inclination of large scale structures with respect to the walls. The lower plate is about to cut a large structure into two. The whole flow field consists of fluctuating velocity vectors which in the lower half point mostly backwards upwards and forwards downwards (and the reverse in the upper half). These events contribute most to the Reynolds stress. In fig.10 again vectors and contour lines of fluctuating velocity components are chosen in planes perpendicular to the main flow in order to illustrate manipulated large scale structures. The plots to the left correspond to planes shortly upstream (I=27) and those to the right to planes shortly downstream of the plate trailing edge (I=30). In the wake region the (u,w)-vectors are still aligned due to the plate effect and the contour lines of w-fluctuations exhibit the breakup of large eddies.

CONCLUSION

LES of high Reynolds number turbulent channel flow manipulated by infinitely thin LEBU devices using a fairly coarse grid reveals some of the important plate effects. The spatial resolution of a 64×32×32 grid in a 8×4×2 computational domain does however not suffice to resolve the boundary layers growing along the plates and consequently it cannot predict the new fine scale turbulence generated by the plates. Much finer grids will be needed in order to reproduce such details. The channel geometry has been chosen as a test case because of the possibility to generate inflow boundary conditions via LES of fully developed flow with periodic boundary conditions. The more complicated case of a manipulated boundary layer will be treated next with a refined grid.

ACKNOWLEDGMENT

The work is supported by Stiftung Volkswagenwerk.

LITERATURE

[1] FIEDLER, H.E. (1987): – In: Advances in Turbulence, G.Comte-Bellot, J.Mathieu (Eds.), Springer Verlag, pp.320

[2] CORKE, T., NAGIB, H.M., GUEZENNEC (1982): NASA CR 165861

[3] BUSHNELL, D.M. (1985): AGARD Report No. 723

[4] COUSTOLS, E., TENAUD, C. and COUSTEIX, J. (1989): In: Turbulent Shear Flows 6, Springer Verlag, pp.164

[5] LEMAY, J., SAVILL, A.M., BONNET, J.-P. and DELVILLE, J. (1989): In: Turbulent Shear Flows 6, Springer Verlag, pp.179

[6] SAVILL, A.M. (1987): – In: Advances in Turbulence, G.Comte-Bellot, J.Mathieu (Eds.), Springer Verlag, pp.533

[7] SAVILL, A.M. and MUMFORD, J.C. (1988): J.F.M. 191, pp.389

[8] COUSTOLS, E., COUSTEIX, J. and TENAUD, C. (1987): In Savill et al. (eds.), A review and report on the 1st European Drag Reduction Meeting, T-87-3, Lausanne.

[9] SCHUMANN, U. (1975): J.Comp.Phys., 18, pp.367

[10] FRIEDRICH, R. (1988): In: Computational Methods in Flow Analysis, Vol.2, H. Niki and M. Kawahara (Eds.), Okayama Univ.Press, Japan, pp.833

[11] FORTIN, M., PEYRET, R.,TEMAN, R. (1971): Lecture Notes in Physics, 8, pp.337, Springer Verlag.

[12] KLEIN, H. (1987): Lehrstuhl für Strömungsmechanik, TU München, Report No. TUM-LSM-87/24.

[13] SCHMITT, L. and FRIEDRICH, R. (1988): In: Notes on numerical fluid mechanics, Vol.20, Vieweg Verlag, Braunschweig, pp.355

[14] RAJAGOPALAN, S. and ANTONIA, R.A. (1979): Phys. Fluids 22, pp.614

[15] PIOMELLI, U., FERZIGER, J., MOIN,P. and KIM, J. (1989): New approximate boundary conditions for large eddy simulations of wall-bounded flows. To appear in Phys. Fluids.

[16] PEYRET, R. and TAYLOR, T.D. (1983): Computational Methods for Fluid Flow. Springer Series in Computational Physics. pp.160.

[17] FRIEDRICH, R. and RICHTER, K. (1989): In: Notes in Numerical Fluid Mechanics, Vol.25, pp.108, Vieweg, Braunschweig.

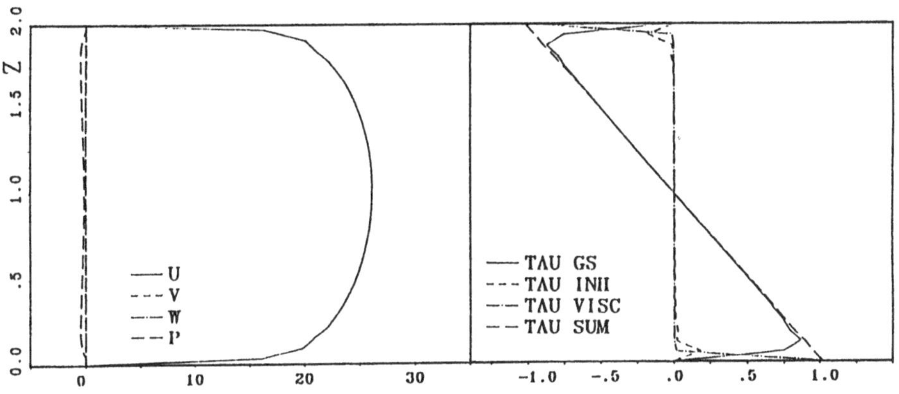

Fig.1 Mean primitive variables and shear stress from LES with PBCs.

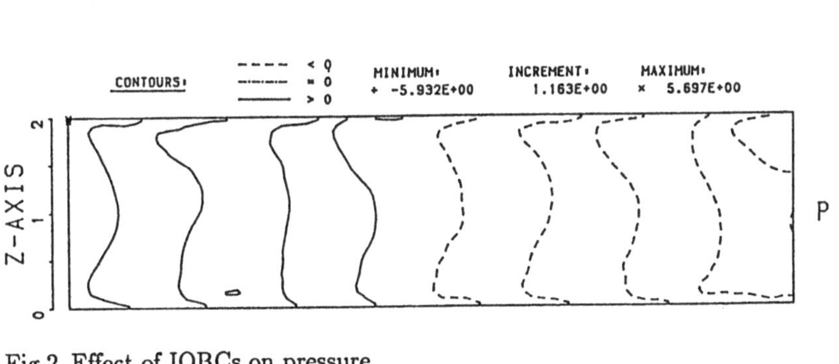

Fig.2 Effect of IOBCs on pressure.

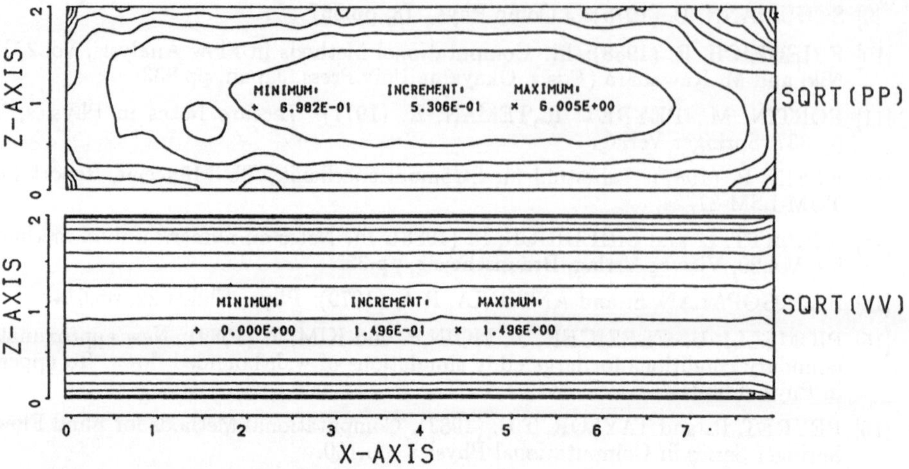

Fig.2 (cont.) Effect of IOBCs on rms values of pressure and velocity fluctuations.

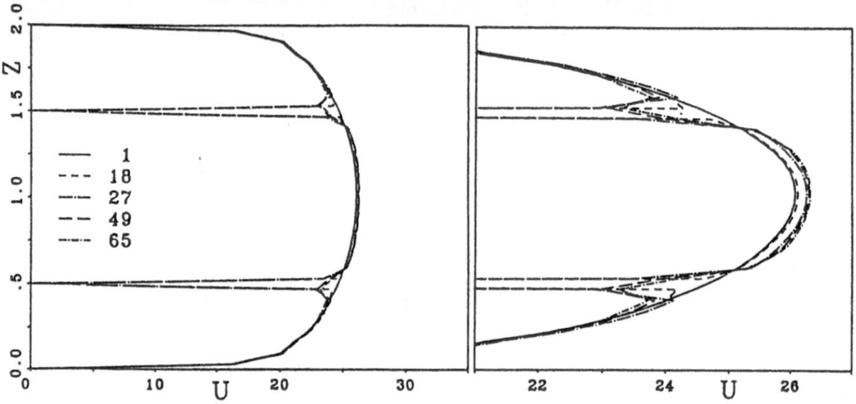

Fig.3 Mean longitudinal velocity on two different scales.

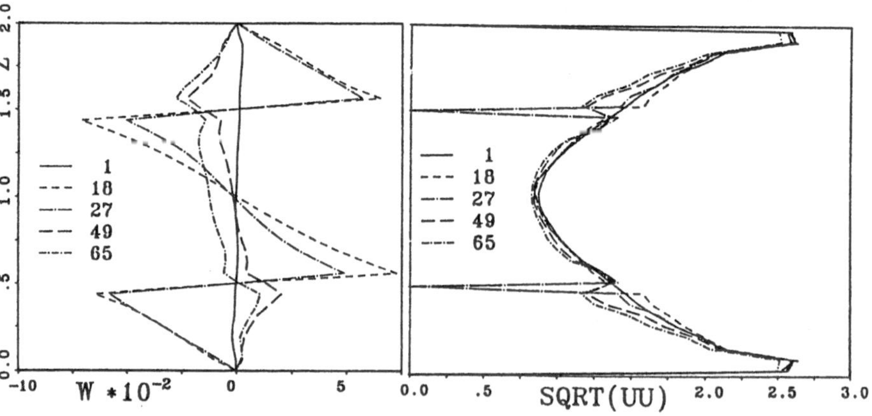

Fig.4 Mean vertical velocity and rms longitudinal velocity fluctuations.

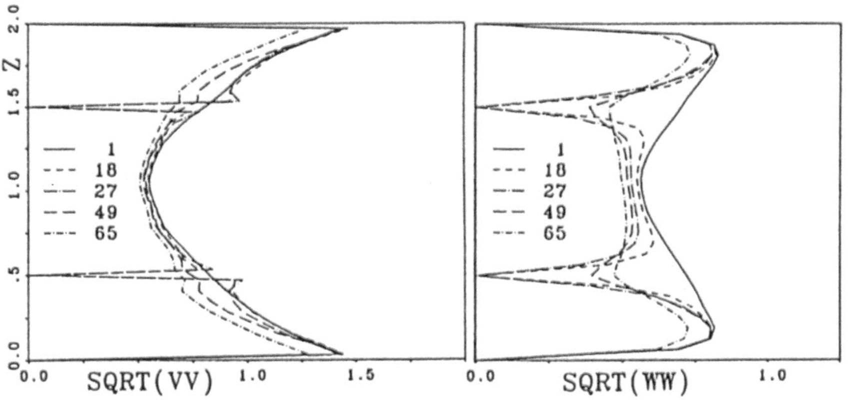

Fig.5 RMS velocity fluctuations in the spanwise and vertical directions.

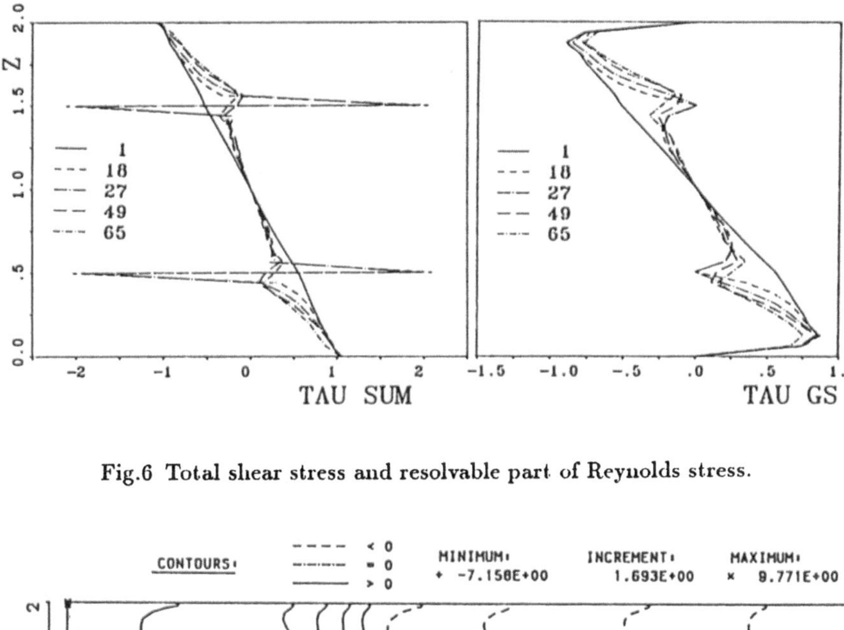

Fig.6 Total shear stress and resolvable part of Reynolds stress.

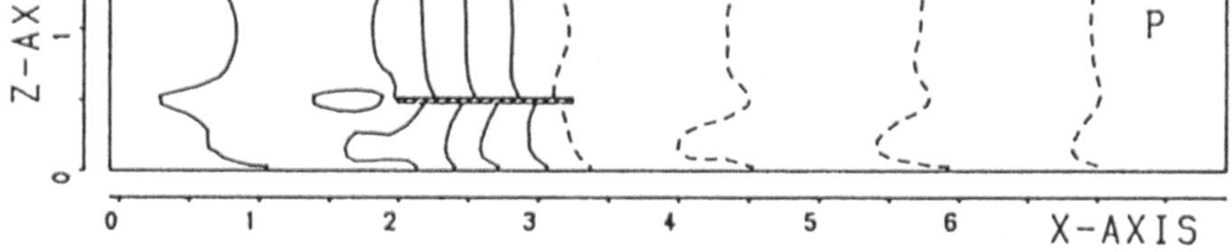

Fig.7 Mean pressure

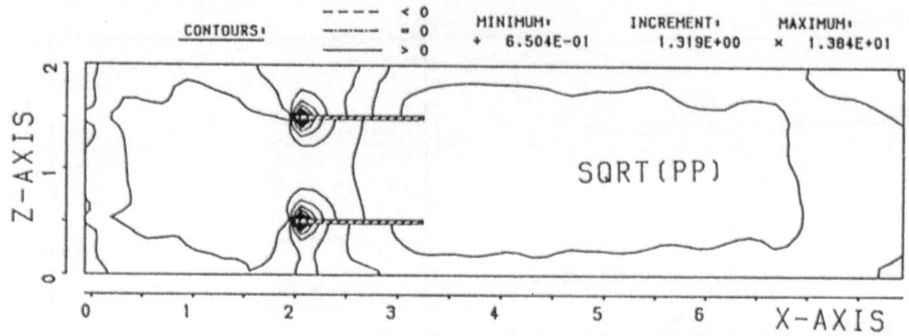

Fig.8 RMS pressure fluctuations

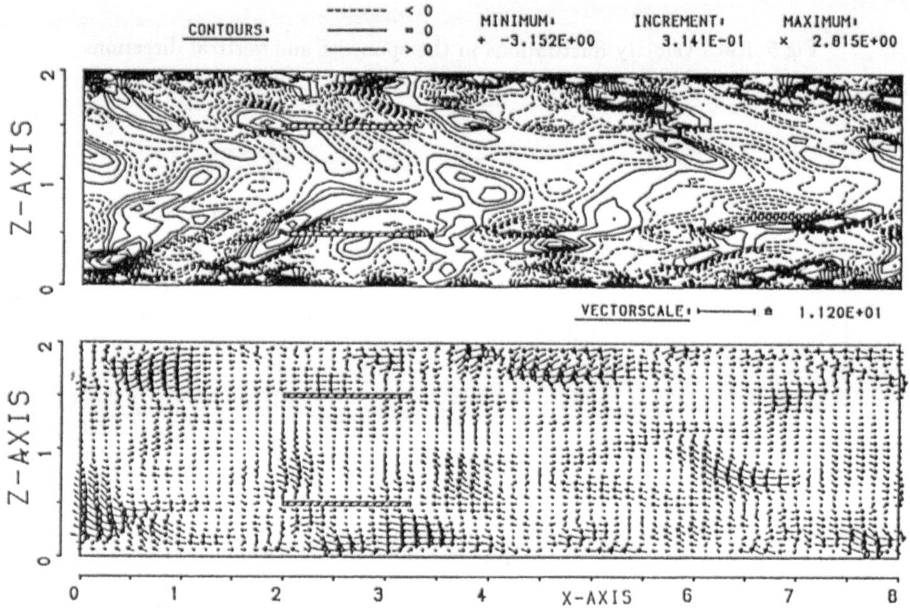

Fig.9 Contour lines of spanwise velocity fluctuations and fluctuating velocity vectors in (x,z)-planes at J=8.

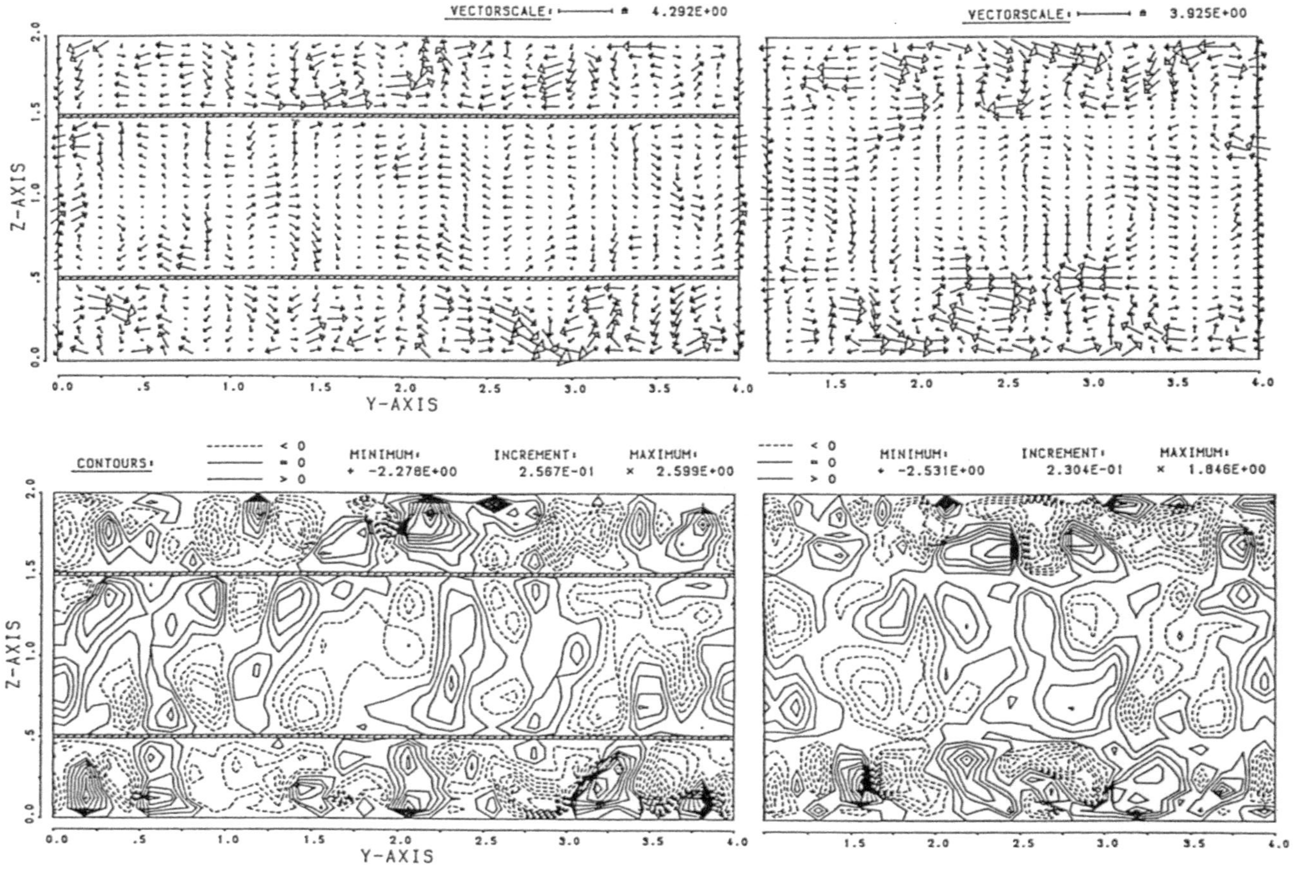

Fig.10 Fluctuating velocity vectors and contour lines of vertical fluctuations in (y,z)-planes at I=27 and 30.

Analysis of the Wake of an Outer Layer Manipulator

J. DELVILLE, J.P. BONNET, J. LEMAY*
Centre d'Etudes Aérodynamiques et Thermiques
LEA, UA CNRS 191, Poitiers, France
*Dept. de Génie Mécanique, Université Laval,
Québec, Canada

I. INTRODUCTION.

Among the different ways of manipulating turbulent boundary layers to obtain drag reduction, external manipulation (blades or aerofoils profiles) has been intensively investigated. If the Cf reduction downstream of the manipulator is somewhat comparable whatever the configuration, it seems that the net drag reduction that can be achieved is strongly related to the manipulator by itself, or more precisely to the interaction between the turbulent structures of the boundary layer and the manipulator. For example, the unsteadiness of the large scale structures of the boundary layer or Large Scale Motions (L.S.M.) as called by Falco / 1 / may create separated regions on the manipulator and by this way increase dramatically its drag. On the other hand, the incoming L.S.M. may modify the coupling between the wake structures of the manipulator and the surrounding flow. In this context, we first analyse the coherence of longituninal velocity in the wake; second, a slight heating of the manipulator is used in order to get a passive marker of the flow passed on it.

II. EXPERIMENTAL SET UP AND CONDITIONS.

The experiments are performed in a closed loop wind tunnel (test section: $300\text{x}300\ \text{mm}^2$, length: 2 m). The external velocity is Ue = 25 m/s. A cartesian coordinate system is defined as follows: X is the streamwise direction, starting at the manipulator trailing edge with $\zeta=X/\delta_0$, Y normal to the floor (Y=0 at the wall). At the manipulator location, the boundary layer thickness δ_0 (at U/Ue = 0.995) is 23.5mm and the Reynolds number based on the momentum thickness is 4000.The manipulator is a thin steel flat blade with square

A. Gyr (Editor)
Structure of Turbulence and Drag Reduction
IUTAM Symposium Zurich/Switzerland 1989

edges (length 25mm, thickness 0.1mm), located at 10mm of the wall. The chord Reynolds number is then 36000. A slight heating of the blade (15°C greater than the ambiant) can be performed by applying a voltage between its ends. A description of the experimental arrangement may be found in / 2 / and / 3 /. The following table summarizes the overall features of the manipulated boundary layer at the locations where the measurements are performed.

Streamwise Location ζ	B.L.Thickn. δ (mm)	Wake Axis Y_0/δ	Wake Width $2b/\delta$	Wake Edges Lower	Upper
0.2	23.6	0.43	0.10	0.39	0.48
0.4	23.6	0.43	0.20	0.33	0.53
2.1	24.2	0.43	0.31	0.27	0.59
4.0	25.2	0.43	0.42	0.22	0.64
7.5	26.7	0.43	0.60	0.13	0.73
13.0	27.8	0.43	0.79	0.04	0.83

The velocity measurements are performed by Constant Temperature hot-wire Anemometry with no heating of the blade. When the plate is heated, the temperature measurements are performed by use of a specially designed Constant Courant Anemometer, providing a 300 μA courant to a DISA 55P31 probe (wire dia.: 1 μm). A dynamical calibration involving direct heating of the wire by use of a chopped Laser beam has shown that, under the present experimental conditions, the cutoff frequency of the probe is about 6 kHz. Considering this good frequency response, neither analogic nor numerical compensation are used. All the signals are sampled at 50 kHz.

III. STREAMWISE VELOCITY COMPONENT COHERENCE FUNCTION ANALYSIS.

The coherence function $\gamma(f)$ between the streamwise component of velocity, at the two points Y_1 and Y_2, is defined by $\gamma^2(f)=|Ey_1y_2(f)|^2/Ey_1(f)/Ey_2(f)$ where Ey_1y_2 is the cross spectral density function and Ey_1 and Ey_2 are the power spectral density functions at locations Y_1 and Y_2. The coherence functions have been measured at 2 streamwise locations: ζ=0.2 close to the trailing edge of the blade and ζ=4 in the region of the wake expansion located upstream of the maximum drag reduction position. In all the following figures, dashed lines correspond to the coherence for the unmanipulated boundary layer and a schematic drawing shows the relative positions of the two probes within the flow.

a) In the very near wake region (ζ=0.2), the most striking feature is the strong disconnection between the wake and the boundary layer. When the fixed probe is located on the wake axis (Y_1=0.43δ), an important decrease of the coherence is found for probes separations greater than the wake extent (fig. 1a, b). For frequencies scaling with the boundary layer thickness ($f\delta/Ue<1$), the coherence is then always less than 0.2. For small probes separations (i.e. when the two points remain in the wake), a sharp peak of coherence located at $f\delta/Ue \approx 8.5$ can be detected (fig. 2a). In this case, the two points are located on both sides of the wake axis. Peaks at the same frequency have been previously observed on the u' or T'- spectra / 2 /, in the wake region for streamwise locations $\zeta<1$. The frequency of this peak corresponds to a typical spatial size of $\delta/8.5$ which is close to the local vertical width of the wake (0.1 δ) and can then be related to the presence of well defined eddies generated by the blade. For this probes configuration, a large decrease of coherence for $f\delta/Ue<1$ is also noticeable. This lack of coherence, observed when the two probes are inside the wake, may be interpreted not as the result of large scales reduction, but rather in terms of a decorrelation between the upper and lower part of the wake for this range of scales. This feature is confirmed by the asymmetrical behaviour of the coherence in these two regions. Figure 1b shows that, relatively to the wake axis, the lower part of the wake (facing to the wall) seems to be more affected; for given separation and frequency values, the coherence is lower than in the upper part. However this difference in coherence reduction may also be explained as the superimposition of two phenomena. On one hand, the wake is disconnected of the boundary layer and for small separations, only the scales matching with those of the wake are correlated, the other ones being almost fully damped (fig. 2a). On the other hand, the upper part of the wake, where a greater coherence level is obtained in the low frequency range, can be thought as beeing preferently submitted to a large scale excitation coming from the L.S.M. of the incoming boundary layer. This feature is also exhibited when one analyses the coherence between points located on both sides of one edge of the wake: both for the upper edge and lower one. In the first case (fig.2b), up to $f\delta/Ue \approx 0.5$, the coherence in the manipulated boundary layer keeps a lower value than in the natural case but evolves in the same way. When the frequency increases, the coherence remains constant up to $f\delta/Ue \approx 2$. After this plateau, the natural coherence is recovered. In the lower part of the wake (fig. 2c) the evolution is different. The coherence is always lower than in the natural case, but more reduced when compared with the corresponding upper part of the wake and no plateau can be detected.

b) The same asymmetrical behaviour of the two parts of the wake can be found further downstream (ζ=4). Like in the very near wake region, the coherence in the upper part of the wake flollows a plateau region from $f\delta/Ue \approx 0.5$ until it recovers the natural coherence level (fig. 3.a, b, c). In the lower part of the wake, the same feature as in the very near wake region can be found (fig. 3.d, e , f). The characteristic plateau begins at the same reduced frequency (0.5) for ζ=0.2 and ζ = 4. Between these two locations, the width of the wake increases in a ratio of 4, indicating that this characteristic traduces a direct influence rather of the L.S.M. of the boundary layer (which typical scales slowly evolves between these location) than of the wake itself.

This asymmetry between the upper and lower part of the wake may be explained by the skewed behaviour of the L.S.M. / 1 /. In the outer part of the boundary layer, the positive angles of attack of the incoming flow on the manipulator (corresponding to outward motion) are found to be slowly varying for continuous periods of time; these periods being greater than for the negative angles which are related to more sudden variations.

IV. SLIGHTLY HEATED BLADE TEMPERATURE FIELD ANALYSIS.

The temperature marker allows to visualize the wake up to ζ=15 (fig. 4). Downstream this location, the wake reaches the wall but the upper part of the wake can still be detected. The general distribution of means and fluctuations of the temperature agrees very well with the usual description of turbulent wakes. From these profiles, the wake axis Y_0 and the width of the wake can be found, the wake edges being then defined by $Y_0 \pm b$ (fig. 4a). The manipulator wake appears to be remarkably symmetrical in the Y direction relatively to a well defined axis. This axis follows the layer expansion and remains at the same distance from the wall Y/δ=0.43 corresponding to the height of the blade. However the RMS temperature profiles show that up to ζ=7 the maximum of fluctuation is more important in the lower part of the wake (fig. 4b). The greater fluctuation level in this part of the flow is associated with a more skewed and intermittent temperature, as may be seen from the iso-probability density contours plot (fig. 5a, b). An overall positive skewness, (shift toward the left), and intermittent feature, (size and number of the iso-probability contours) can be observed on all the plots of figure 5. This is related to the passage of spots of hot fluid, generated in the blade vicinity, within the ambiant colder flow. On these plots the probability maxima are located on the wake edges. For the two last streamwise locations tested $\zeta=7.5$ (fig. 5c) and $\zeta=13$ (fig. 5d), the upper part of the wake still exhibits the intermittent feature whilst in the lower part this tendancy vanishes more quickly. This behaviour may be due to the wall vicinity. At the location $\zeta=13$, close upstream of the maximum drag reduction location, the well defined iso-probability contours in the upper edge of the wake, show that individual spots may still be detected there inferring the persistance of the upper part of the wake in this region of the boundary layer. In order to get a more precise idea of the intermittent feature of the wake, we show on figure 6 typical temperature signals corresponding to the upper edge of the wake, for several streamwise locations. Only the upper part of the wake has been choosen in order to avoid the wall interference with the scales measured. Two different time scales are used on these plots: a first one is based on the local width of the wake (upper signal of each sketch), a second one is based on δo. All along the wake expansion, when the wake time scale is used, individual spots of hot temperature can be detected. Their mean duration τ_B evolves following the wake development. For these spots, the dimensionless value $\tau_B.Ue/2b$ remains close to 1.2 (fig. 7). These spots can be referred to as eddies of the wake. On the other hand, when the time scale based on to δ is used, it can be observed that sets of eddies close to each other are separated by quieter periods of time. This may be interpreted as the intermittent passage of strings of eddies, corresponding to a modulation by the L.S.M.. The average separation time T_A between such trains of eddies has been evaluated for various Y locations in the vicinity of the upper edge. These separations remain nearly constant up to $\zeta=7.5$ and then seem to be independent of the wake evolution; the average separation between sets of eddies beeing $Ta.Ue/\delta o$ of the order of 5 and then close of the average spatial separation of L.S.M. given by Falco / 1 /.

In the same way as the behaviour of the coherence of the velocity, the evolution of the thermal wake is then consistent with the idea of a strong modulation of the wake by the L.S.M. of the turbulent boundary layer.

REFERENCES

/ 1 / Falco R.
Correlation of Outer and Passive Wall Region Manipulator with Boundary Layer Coherent Structure Dynamics and Suggestions for Improved Devices, 2nd Shear Flow Conference, March 1989 (AIAA 89-1026)

/ 2 / Bonnet J.P., Delville J. and Lemay J.
Study of LEBUS Modified Turbulent Boundary Layer by Use of Passive Temperature Contamination. In Proc. Inter. Conf. on Turbulent Drag Reduction by Passive Means. R.A.S., London, Sept. 15-17, 1987.

/ 3 / Lemay J., Savill M.,Bonnet J.P., Delville J.
Some Similarities Between Turbulent Boundary Layer Manipulated by Thin and Thick Flat Plate Manipulator"
TSF Toulouse, Springer 1989, pp.179-193

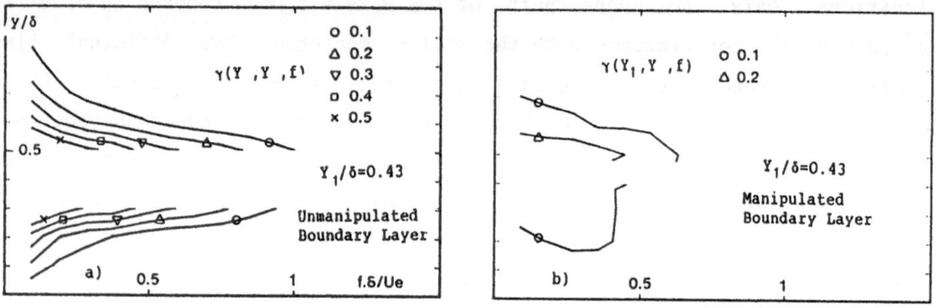

Fig. 1 Coherence Function of U contour plot. Separations normal to the wall. Fixed Probe on the wake axis: ζ=0.2

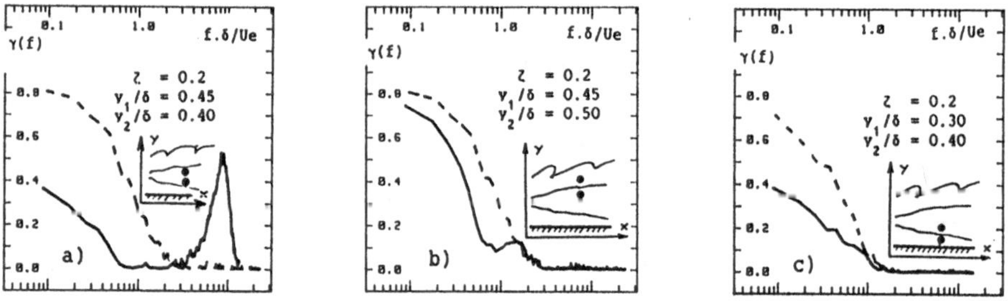

Fig. 2 Coherence Function of U. Separations normal to the wall: ζ=0.2

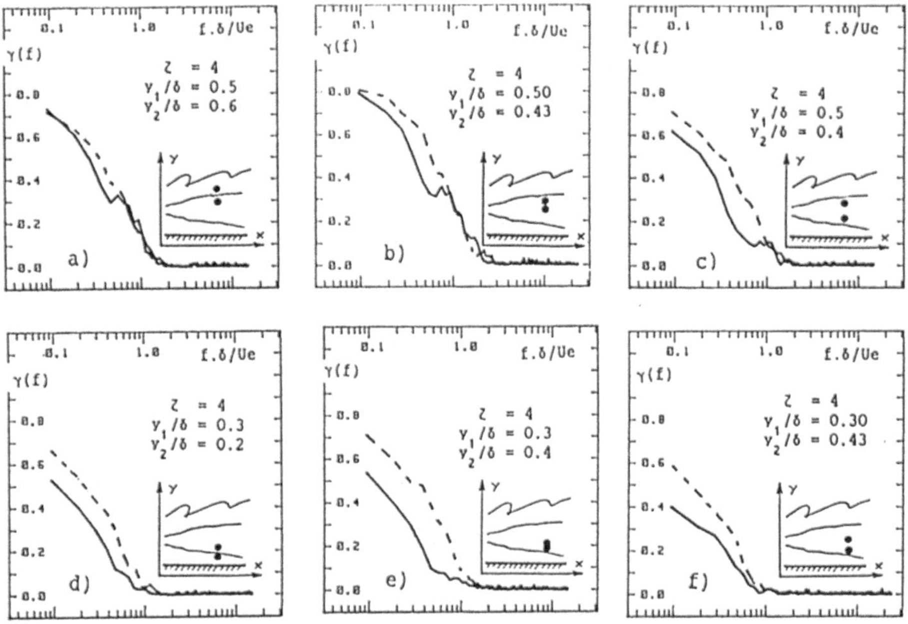

Fig. 3 Coherence Function of U. Separations normal to the wall: ζ=4.0

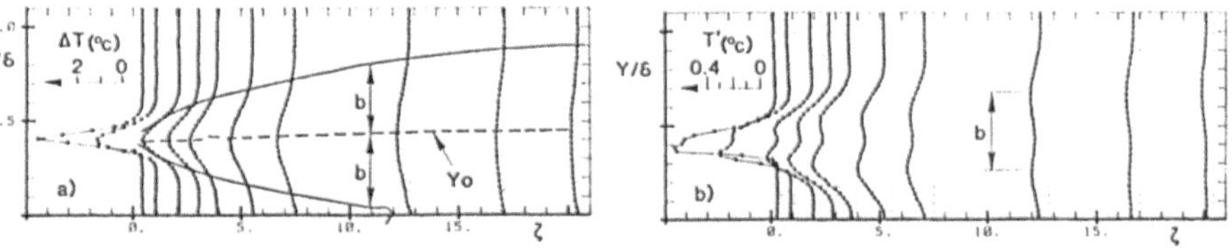

Fig. 4 Slightly heated blade temperature fields:

a) Mean temperature $\Delta T = \overline{T}\text{-}T_e$

b) RMS value of temperature fluctuations,

- - - wake axis location : Y_o, ——— wake edges locations $Y_o \pm b$.

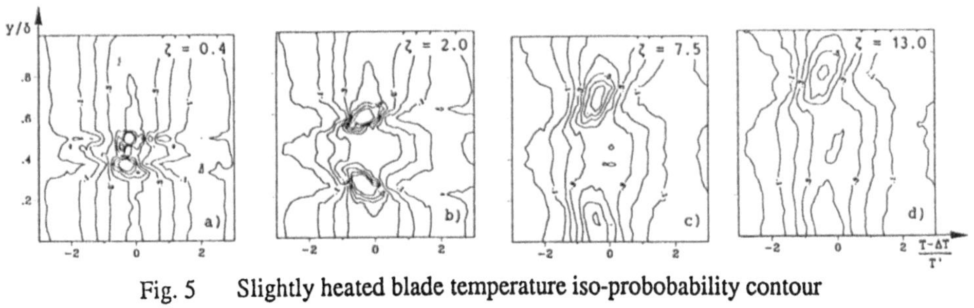

Fig. 5 Slightly heated blade temperature iso-probobability contour plots. Streamwise evolution

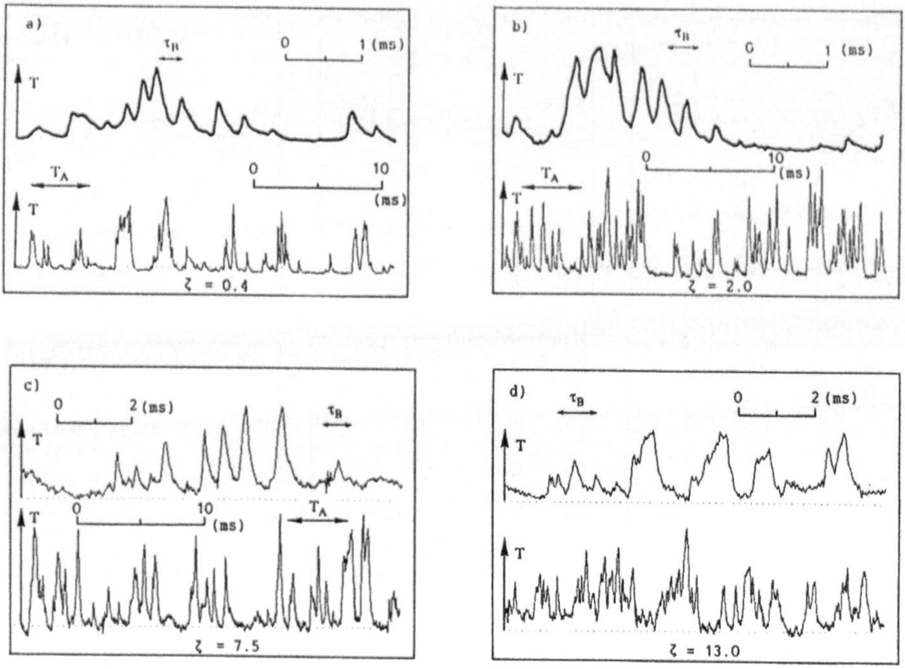

Fig. 6 Typical temperature signals in the upper edge of the wake.

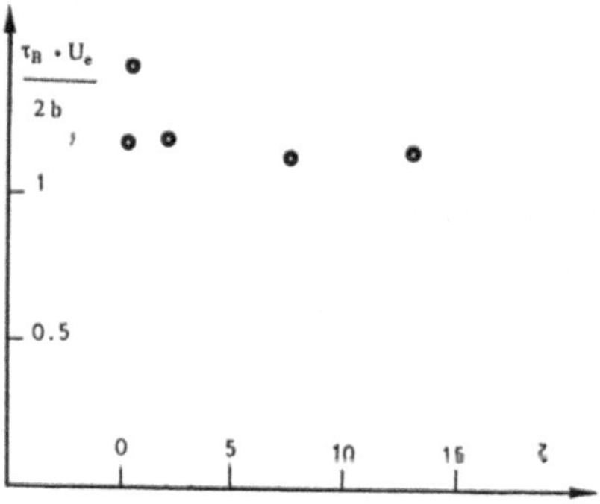

Fig. 7 Streamwise evolution of the average eddies duration.

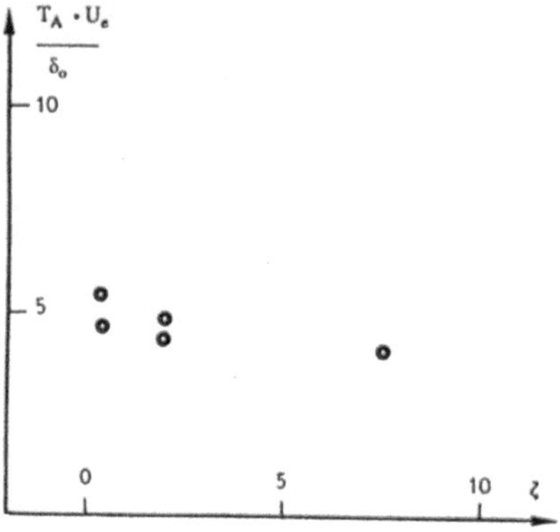

Fig. 8 Streamwise evolution of the average separation between strings of eddies.

Influence of an Acoustic Field on the Flow Structure Behind a LEBU in a Turbulent Boundary Layer

S. P. BARDAKHANOV; V. V. KOZLOV; V. V. LARICHKIN

Institute of Theoretical and Applied Mechanics,
Siberian Branch of USSR Academy of Sciences,
Novosibirsk, USSR

Summary

Coherent structures in the turbulent wake behind thin plates in a turbulent boundary layer have been studied including the influence of an acoustic field.

The reduction of the surface friction of different objects moving in gaseous and liquid media is one of the key problems in aerodynamics. Until the 1970s, attempts to solve this problem were connected with studies of the control of laminar flows. At the same time it is common knowledge that on the surface of the majority of real objects a turbulent flow region is formed. When trying to lower the skin friction for the turbulent boundary layer it was naturally considered that the smoother the surface, i.e. the fewer bulges it has and the less uneven it is, the lower the friction would be. But recently a new concept has been developed which presupposes that a certain type of surface unevenness or certain additional surfaces introduced into the turbulent boundary layer can considerably lower the skin friction [1]. It is from this concept that the LEBU (large eddy breakup devices) originated. In the majority of cases a profiled surface located at some distance from the surface is used to lower the skin friction. The mechanisms of the influence of a LEBU on the skin friction are still unknown. Nevertheless, using them one can lower the friction by about 20%. Up to now LEBUs have been studied only in laboratories, although in several cases they have been tested on a real aircraft. The peculiarity of these studies is that the results obtained not only by different scientists, but also in the same laboratory, were not strictly repeatable. One of the reasons for this, according to some authors, is that the degree of the LEBU influence on the turbulent boundary layer strongly depends on external background disturbances.

The object of this paper is to show that acoustic disturbances can affect the flow structure behind the LEBU. The studies were carried out in the aerodynamic tunnel T-324 ITPM SO AN SSSR. In Fig.1 a sketch of the experiment is given. The LEBU models consisted of steel plates 1mm thick, with a rounded front and sharpened edges. The chord of the models equalled 50mm. The models could be mounted separately or together at different distances one from another. When the models were mounted in tandem, the distance between them equalled 255 mm. The dimensions of the models, their location in a boundary layer and their reciprocal location were chosen according to the data presented in [2].

A. Gyr (Editor)
Structure of Turbulence and Drag Reduction
IUTAM Symposium Zurich/Switzerland 1989

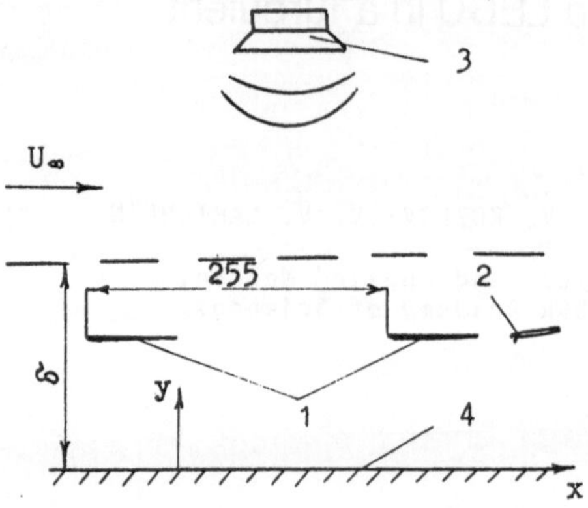

Fig.1: Scheme of experiment. 1 LEBU, 2 hot-wire probe, 3 loudspeaker, 4 wall of the test section

The signals were processed by a hot-wire anemometer installation DISA and a frequency analyzer FAT-1. Fig.2 presents the distributions of the mean velocity and of the r.m.s. value of the fluctuation velocity u in the turbulent boundary layer. The boundary layer thickness at the point of measurement was about 60 mm. A typical turbulent wake is formed in the wake behind a single LEBU, but the mean velocity distribution was not affected beyond the LEBU wake. At the same time it becomes apparent that the presence of the wake with a typical turbulence level causes the pulsations to decrease in the boundary layer.

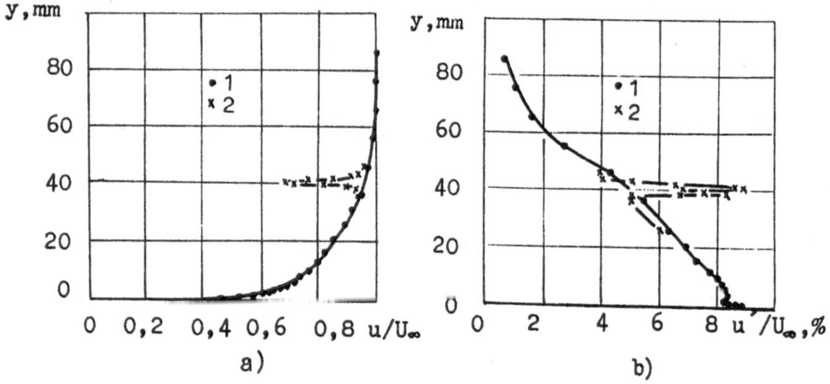

Fig.2: Distribution of mean velocity (a) and integral intensity of velocity pulsations (b) in the wake behind a single LEBU. U_∞ = 20 m/s, x = 10 mm.
1 without model, 2 with model

Fig.3 shows analogous distributions in a boundary layer when a tandem LEBU is used. It becomes apparent that in this case the wake is wider and this is quite natural, since the first model is placed in the flow at a larger distance upstream from the second.

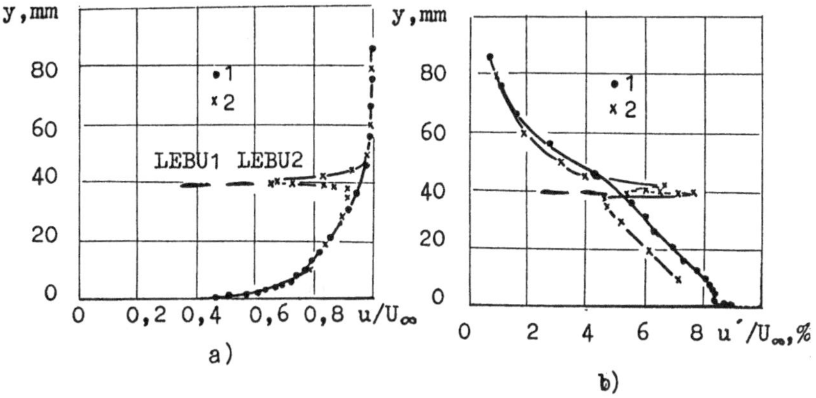

Fig.3: Distribution of mean velocity (a) and integral intensity of velocity pulsations (b) in wake behind two LEBU. U_∞ = 20 m/s, x= 10 mm.
1 without models, 2 with models

At the same time, the influence on the velocity fluctuation is considerably greater than in case of the single LEBU. Now, as far as the mechanism of the LEBU influence is concerned, one can say that the eddies (or coherent structures) developing in the wake impede the formation of large eddies in the boundary layer, and that this causes the change in skin friction.

The flow with a LEBU was subjected to acoustic disturbances from a loudspeaker with the frequency f = 827 Hz and an intensity of 120 dB. Fig.4 presents the distributions of the velocity fluctuations in the band 4Hz at the frequency of the acoustic influence measured in the wake. In Fig.4a: in the wake behind a single LEBU, in Fig.4b: behind two LEBUs. One can see that in the wake behind the models coherent structures appear under the acoustic influence as in

a turbulent wake behind a plate or an aerodynamic profile in a homogeneous flow [3-5]. Their excitation at the given level of the acoustic field intensity did not affect integral characteristics. At the same time one can see that at rather high sound intensities these coherent structures in the wake can interact with boundary layer coherent structures and thus change the efficiency of the LEBU.

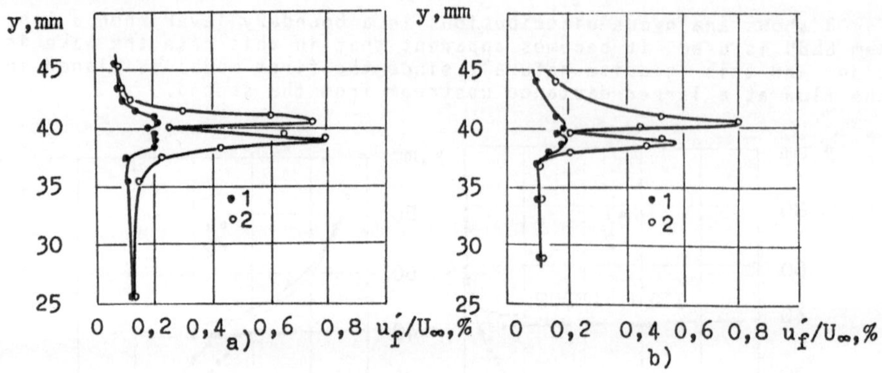

Fig.4: Distribution of velocity pulsation intensity in 4 Hz band. (a) in wake behind single LEBU, (b) in wake behind two LEBU.
1 without acoustic influence, 2 with acoustic influence on frequency f = 827 Hz, U_∞ = 20 m/s,
x = 10 mm from trailing edge LEBU

References

1 Wilkinson S.P., Anders J.B., Lazos B.S., Bushnell D.M. (1987), Proc. International Conference on Turbulent Drag Reduction by Passive Means, London, 15 - 17 September

2 Bertelrud A., Watson R.D. (1987), Proc. IUTAM Symposium on Turbulence Management and Relaminarization, Bangalore, 19 - 23 January, pp. 16 - 35

3 Bardakhanov S.P., Kozlov V.V. (1987), Perspectives in Turbulence Studies, Springer Verlag, pp. 154 - 187

4 Yanenko N.N., Bardakhanov S.P., Kozlov V.V. (1984), DAN SSSR, t. 274, N° 1, pp. 50 - 53 (in Russian)

5 Bardakhanov S.P., Kozlov V.V., Yanenko N.N. (1984), IFZh, t. 47, N° 4, pp. 533 - 536 (in Russian).

Analysis of the Structure of a Turbulent Boundary Layer, with and without a LEBU Using Light Sheet Smoke Visualizations and Hot Wire Measurements

M. STANISLAS - M.C. HOYEZ

Office National d'Etudes et de Recherches Aérospatiales,
Institut de Mécanique des Fluides
5 Boulevard Paul Painlevé
59000 LILLE FRANCE

1. INTRODUCTION

The concept of random turbulence, which has led to the statistical approach of scientists like BACHELOR, TOWNSEND and VON KARMAN, has been reconsidered since the paper by BROWN and ROSKO (1) on shear layers which has put into view the existence of large scale coherent structures in turbulent flows. Since then, coherent eddies have been observed in other types of flows such as, for example, jets and wakes. The boundary layer has also been the subject of intensive and numerous studies. Experiments in water, initiated by KLINE *et al.* (2) using hydrogen bubble techniques and those of CORINO and BRODKEY (3) using dye, have shown the existence of long streaks in the near wall region and of phenomena of sweeping and bursting associated with the streaks. Much work has been done since on this part of the boundary layer, but the reasons for the sweeping and bursting process are still not well understood, and the eventual coherent structures associated with it have not been put in evidence.

Concerning the outer part of the boundary layer, the results are fewer and are due essentially to two research groups. The first one, whose results are presented in two papers by HEAD and BANDYOPADHYAY (4-5), is at Cambridge. The second group is at Michigan States University and the results are in several papers by FALCO *et al.* (6-10). We won't discuss here in detail the results of these authors, this discussion is reported to section 4.

The aim of our own studies was to use the specificity of our illuminating system, composed of pulsed ruby lasers, to make light sheet visualizations of an equilibrium turbulent boundary layer, in order to try to get a better understanding of its physical behaviour. Our work deals essentially with the outer part of the boundary layer.

2. EXPERIMENTAL SET UP

We study the turbulent boundary layer developing on the vertical wall of the wind tunnel presented on figure 1. The tunnel has a cross section of 300x300 mm2 and is 2.5 m long. It has been especially designed for visualizations. The boundary layer is visualized at the rear part of the test section (2.55 m from the tripping which is in the convergent 20 cm upstream of the test section). The velocity outside the boundary layer is 22 m/s. The main characteristics of the boundary layer at this station are

X mm	δ mm	$\delta 1$ mm	θ mm	H	R θ
2555	36.99	5.75	4.28	1.34	6278

A. Gyr (Editor)
Structure of Turbulence and Drag Reduction
IUTAM Symposium Zurich/Switzerland 1989

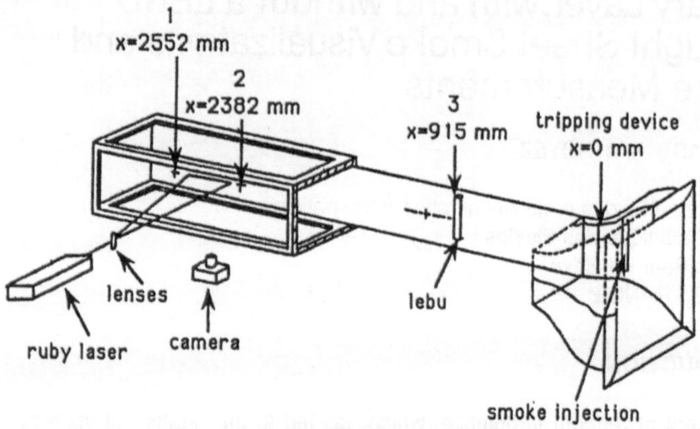

FIG.1 EXPERIMENTAL SETUP

Profiles of the Reynolds stresses measured with hot wires are presented in figure 2. To measure the instantaneous wall shear stress, a hot-film probe can be installed at positions 1 or 2 in figure 1. At position 1 the instantaneous velocity can also be recorded with two hot-film probes situated respectively at 4.45 mm and 19.65 mm from the wall. The first hot-film is straight above the wall shear stress probe, the second is 15.2 mm downstream.

To manipulate the boundary layer, a LEBU can be fixed at x = 915 mm from the tripping device, in position 3 in figure 1. This LEBU is a flat plate of 20 mm chord, 0.1 mm thickness, and is at 6 mm from the wall. The boundary layer thickness at this station is 13.6 mm. We present in figure 3 the mean velocity and turbulence intensity profiles with and without LEBU at $x/\delta o$ = 2,9. These results agree quite well with those of CERT-DERAT (11), who proposed us the LEBU configuration studied here.

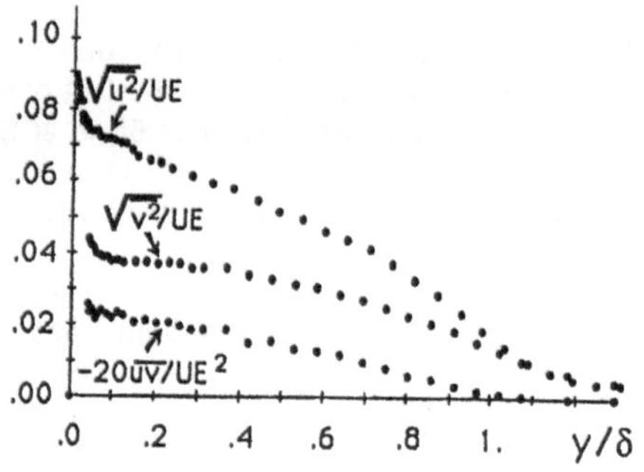

FIG.2 FLUCTUATING VELOCITY PROFILES AT X = 2555 MM

The visualizations are realized using a system of 4 pulsed ruby lasers mounted parallel on a bench. Each cavity emits a pulse with an energy of 100 mJ and a duration of 20 ns. The time interval between two pulses can be varied from 5 µs to 10 s. The laser beams pass through a system of spherical and cylindrical lenses to generate light sheets normal to the wall and parallel to the mean velocity. These light sheets are 15 cm long and 0.5 mm thick, and they are introduced in the test section through a plexiglas window.

The boundary layer is seeded with oil smoke which is injected by a slit underneath the laminar layer. We have made sure by hot wire anemometry that this injection does not disturb the flow significantly.

The images are recorded, using a drum camera, on 35 mm film. The laser beams can be triggered in synchronism with the camera. This trigger can be conditioned by an outer signal (a hot-wire signal for example).

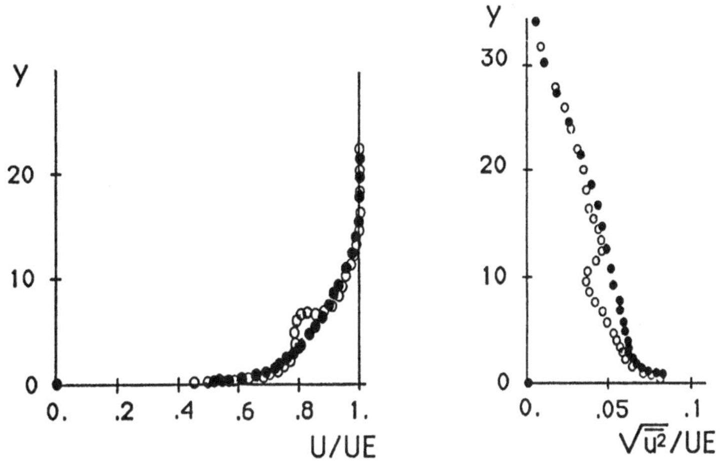

FIG.3 MEAN VELOCITY AND TURBULENCE INTENSITY PROFILE WITH AND WITHOUT LEBU AT X/δ0 = 2.9

Two types of views have been realized. In a first configuration, the 4 light sheets follow each other in the longitudinal direction to obtain a light field about 25 cm long. The interest of such views is to follow the evolution of the large scale outer structures with the time on a relatively important distance. In this case the wall shear stress probe and the two velocity probes are placed just downstream of the last light sheet.

The second configuration consists in superimposing the 4 light sheets. This reduces the longitudinal field but allows a close up view to get more details inside the boundary layer.

These two types of views have been applied both to the free and the manipulated boundary layer.

3. RESULTS

Concerning the equilibrium boundary layer, the images realized in the first configuration show the existence of two types of large scale motions. Some of them evolve little and seem to lose their coherence progressively, the others develop with a general motion upstream and outwards. Concerning the simultaneous signals of the wall shear stress and the instantaneous velocity, no coherence between them has been noted, even through digital correlation.

With the close up views, we are able to see smaller scale structures present in the boundary layer. The analysis of these images suggests that ring-like coherent structures are produced near the wall. They have a mushroom shape and a mean size of 150 ν/u_τ (as defined by FALCO (6)). It seems that, according to the moment and the position where they are emitted, and according to their respective vorticity, these structures interact with each other differently. Sometimes they move upwards forming a straight front inclined of 15 to

FIG.4 SMALL SCALE STRUCTURES MOVING ON STRAIGHT FRONT
a) Typical Eddy
b) Structure rotating in the opposite sense to the natural vorticity

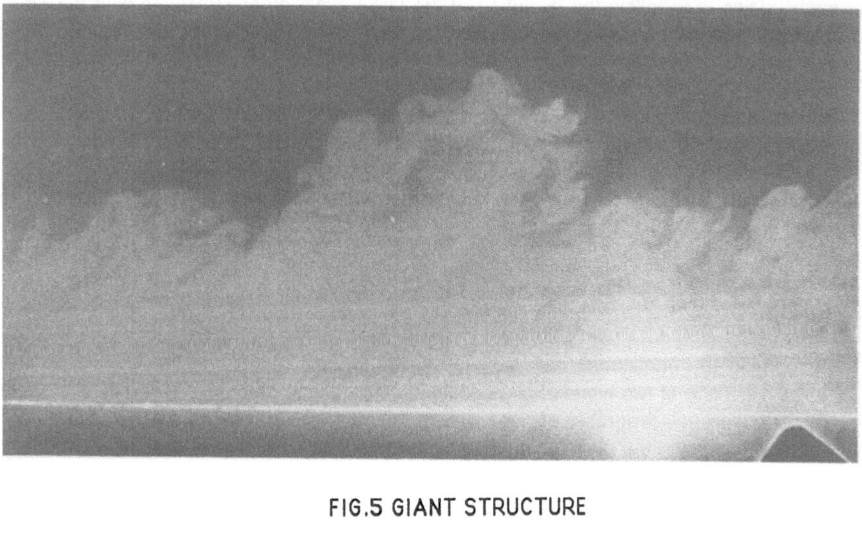

FIG.5 GIANT STRUCTURE

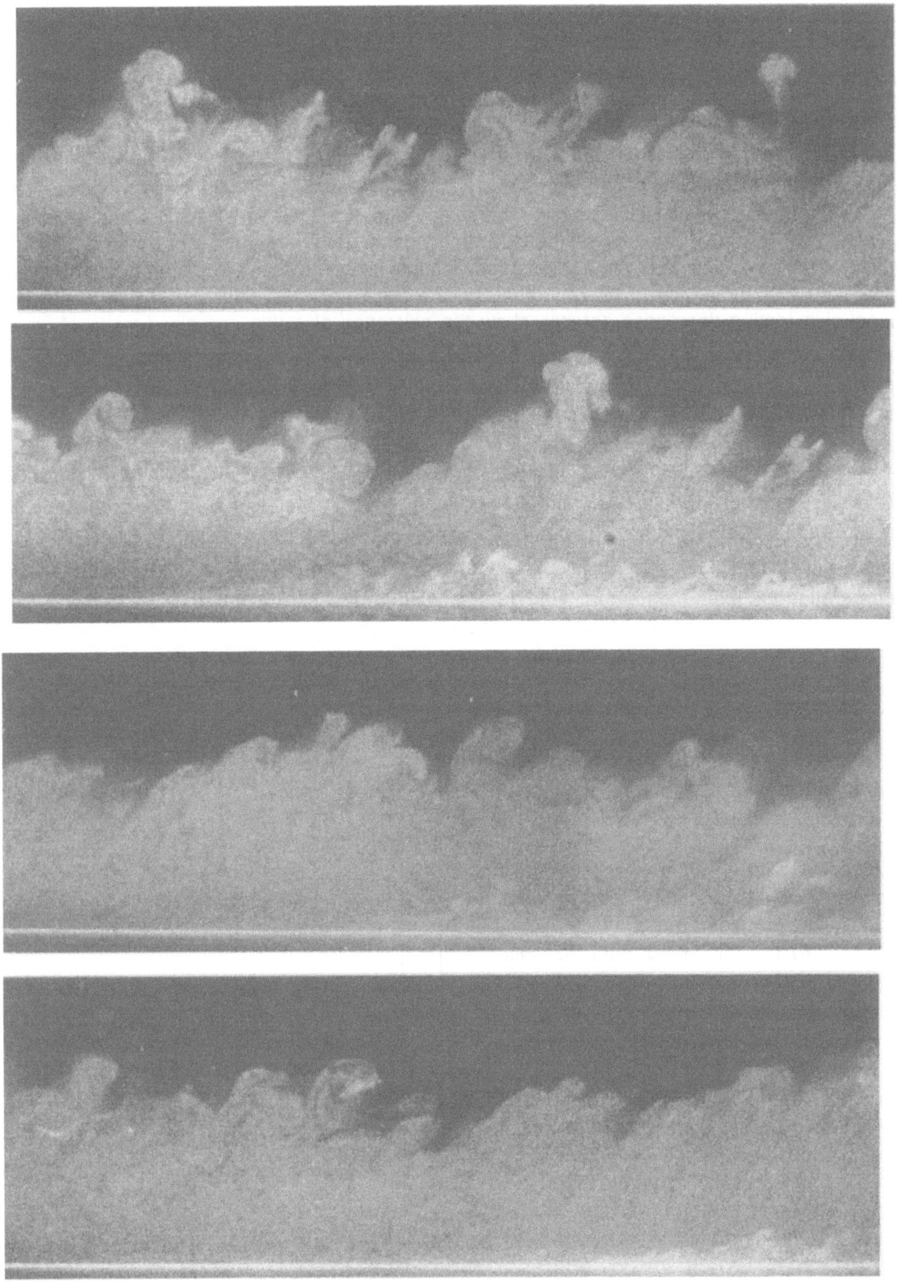

FIG.6 EXAMPLE OF A SEQUENCE OF VIEWS JUST BEHIND THE LEBU
UE = 22 m/s – Δt between two views = 1,9 ms

25 ° (figure 4); sometimes they organize as a big rotating structure which can become giant (figure 5). This interaction is also detectable at the upper part of large scale motions by the existence of structures of the same dimension as the mushroom eddies and rotating in the opposite sense to the natural boundary layer vorticity (figure 4).

For the manipulated boundary layer, the two types of views correspond to a zone between 5 and 25 δo for the large field views and between 0 and 6 δo for the close up views (δo being the boundary layer thickness at the trailing edge of the LEBU). Visualizations has been realized at the same stations without the manipulator. By comparaison of these images, it has not been possible to put in evidence a noticeable influence of the LEBU on the large scale motions of the boundary layer. Nevertheless, the close up views near the LEBU, where the local skin friction decreases notably, show the existence of a thin organized wake of small eddies which persist until about 5 δo and seem then to reorganize themself with the boundary layer vorticity. Anyway, the wake vortices are quite difficult to detect due to the fact that we seed the whole boundary layer with smoke.

Visualizations realized with a thin film of oil on the wall, which is vaporized by the first laser pulse, show that the bursting phenomenon is still persistant in the zone 0-6 δo (figure 6). The only hypothesis that can be proposed on the basis of these results is that the wake of the flat plate, maintaining its coherence for about 5 δo, acts as a screen for the penetration of high momentum fluid to the wall, reducing the intensity of the vortices produced at the wall but not suppressing them.

4. DISCUSSION

Two types of organization have been proposed yet for the outer part of the boundary layer. HEAD and BANDYOPADHYAY (5) have suggested the existence of hairpin vortices created near the wall and extending at an angle of about 45° across the boundary layer. FALCO (6) has observed what he calls "typical eddies" of annular form, the size of which is of the order of 100 $\nu/u\tau$ and which move on the fronts of large scale motions. The results obtained here weaken the hypothesis of HEAD and BANDYOPADHYAY and appear to bo in quite good agreement with those of FALCO. In particular, they clearly show the existence of mushroom shape typical eddies which seem to be produced near the wall.

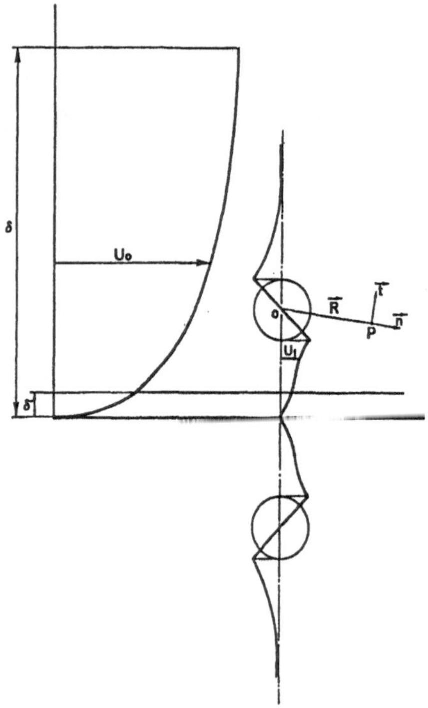

FIG.7 VISCOUS VORTICES MODEL IN A TURBULENT BOUNDARY LAYER

To try to get a better understanding of what is observed on our images, we can use a simple phenomenological model to simulate the vortices behaviour. In the main part of the boundary layer , the

REYNOLDS stresses which find their origin in the non linear terms of the NAVIER-STOKES equations dominate the viscous stresses. We propose to simulate the boundary layer using the model presented in figure 7. The mean velocity profile is that of an equilibrium turbulent boundary layer (12). It is perturbed by one or several vortices. The perturbation is supposed small enough to use a superposition principle. Due to the presence of the wall, we use the image method to verify the wall condition on the normal velocity. So, the local velocity at a point of the flow is the sum of the mean velocity of the boundary layer at this point and of the velocities induced by the vortices and their images. Each vortex is supposed to move with the velocity of its center. The viscosity has to be taken into account at two places:

- very near the wall where the REYNOLDS stresses go to zero. The mean velocity profile verifies already the no slip condition. For the vortices we use a damping function which acts in the viscous sublayer and brings the induced velocity to zero at the wall.

- in the vortices, by using a viscous vortex model as presented in figure 7.

Using this model, it is possible to compute the displacement of vortices in the boundary layer and their mutual interaction. Of course, this is a very crude model. It has serious limitations which must be kept in mind (the main being that it is two dimensional and that it doesn't take the stretching into account). But it has been found helpfull to interpret our visualizations. Figure 8-a presents a simulation based on our experimental configuration. The smoke is represented by discrete points spread uniformly at the beginning of the computation. This case presents two pairs of counter-rotating vortices representing two typical eddies (n°1, n°2) inclined at 45° and staggered downstream and outward. The computation shows that, if the initial distance between the two eddies is enough (depending on their production time), they move slowly outwards on trajectories the angle of which depends on their own vorticity, and they get organized on something which looks like straight fronts (our experimental results show also that the angle of the trajectories is smaller than the angle of the fronts).

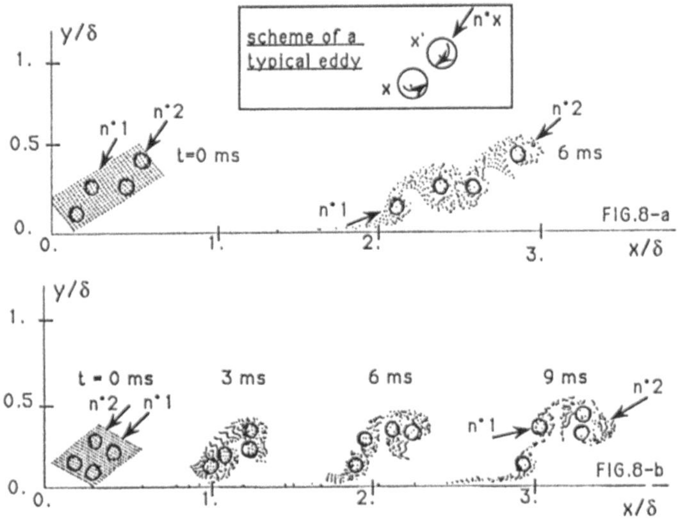

FIG.8 SIMULATION OF TYPICAL EDDIES INTERACTION

In figure 8-b, we present another sequence of production of the typical eddies. One can observe a strong interaction of the two eddies, the upper one sucking the lower one up and blowing it outwards. If the computation is continued further, it appears that the eddy n°1 rotates around the eddy 2' and turns upside down.

5. CONCLUSION

We have presented some experimental results obtained concerning the structure of the turbulent boundary layer. On the basis of the "typical eddies" model and using a simple simulation model, the main features observed on our visualizations can be explained. But these results don't explain in fact the way these eddies are produced near the wall and how they are associated with the sweeping and bursting phenomena. A simulation of hydrogen bubble visualizations using our model is presented in figure 9. The comparison with the results of KIM, KLINE and REYNOLDS (13) seems to indicate that typical eddies have to be taken into account very close to the wall. Some interesting results and ideas have been proposed on this subject by FALCO (14).

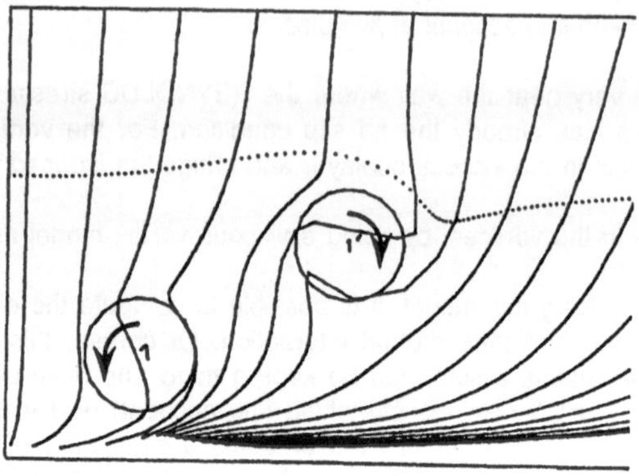

FIG.9 SIMULATION OF HYDROGEN BUBBLE VISUALIZATION

We are thankfull to the French Ministery for Defense which, through the D.R.E.T. , has supported this work.

REFERENCES

(1) G.L. BROWN - A. ROSKO : "On density effects and large structures in turbulent mixing layers."
J.F.M. (1974) vol.64 p.775

(2) S.J. KLINE - W.C. REYNOLDS - F.A. SCHRAUB - P.W. RUNSTADLER : " The structure of turbulent boundary layers."
J.F.M. (1968) vol.30 pp.741-773

(3) E.R. CORINO - R.S. BRODKEY : "A visual investigation of the wall region in turbulent flows."
J.F.M. (1969) vol.37 pp.1-30

(4) M.R. HEAD - P. BANDYOPADHYAY: "Combined flow visualization and hot-wire measurements in turbulent boundary layers."
(1978) Workshop on Coherent Structure of Turbulent Boundary Layers

(5) M.R. HEAD - P. BANDYOPADHYAY : "New aspects of turbulent boundary layer structure."
J.F.M. (1981) vol.107 pp.297-338

(6) R.E. FALCO : "Some comments on turbulent boundary layer structure inferred from the movements of a passive contaminant."
AIAA paper n°74-99 (1974)

(7) R.E. FALCO : "Coherent motions in the outer region of turbulent boundary layers."
Phys. Fluids (1977) vol.20 ptII pp.124-132

(8) R.E. FALCO : "A structural model of the turbulent boundary layer."
Proceedings of the 14th Annual Meeting of SES, Lehigh University, Nov.1977

(9) R.E. FALCO : "The role of outer flow coherent motions in the production of turbulence near a wall."
in Coherent Structure of Turbulent Boundary Layers (1978)
ed. by C.R. SMITH and D.E. ABBOTT, AFOSR/Lehigh

(10) R.E. FALCO : "Structural aspects of turbulence in boundary layer flows."
in Turbulence in Liquids (1979) ed. PATTERSON and ZAKIN, University of Missouri Press pp.1-14

(11) E. COUSTOLS - J. COUSTEIX : "Réduction du frottement turbulent: Modérateurs de trubulence."
La Recherche Aérospatiale (mars-avril 1986) n°2 pp.145-160

(12) T CEBECI - A.M.O. SMITH : "Mean velocity and shear-stress distributions in incompressible flows on smooth surfaces."
in Analysis of Turbulent Boundary Layers (1974) pp.113-128
ed. by F.N. FRENKIEL and G. TEMPLE, Academic Press

(13) H.T. KIM - S.J. KLINE - W.C. REYNOLDS : "The production of turbulence near a smooth wall in a turbulent boundary layer."
J.F.M. (1971) vol.50 ptI pp.133-160

(14) R.E. FALCO : "New results, a review and synthesis of the mechanism of turbulence production in boundary layers and its modifications."
AIAA paper n°83-0377 (1983)

The Effects of Longitudinal Roughness Elements and Local Suction upon the Turbulent Boundary Layer

J.B. Roon and R.F. Blackwelder
Department of Aerospace Engineering
University of Southern California
Los Angeles, California

ABSTRACT:

To directly interact with the bursting process, microgeometry in the form of longitudinal roughness elements was combined with local suction in an attempt to inhibit low speed streak lift-up. The roughness elements consisted of long cylindrical strings aligned in the streamwise direction. The size and spacing were based upon the optimal geometry of Johansen and Smith[1] for aligning the low speed streaks over the roughness elements. Suction was applied under the roughness elements for a small streamwise extent of $\Delta x^+ = 400$. Measurments were performed with a spanwise rake of twelve hot-wires which extended over three of the roughness elements. A detection algorithm was developed to study the low speed streaks and the effects of suction upon them. Statistics of their duration and frequency of occurence have been calculated with and without suction. The addition of the roughness elements increased the number of streaks above them while keeping the streak length nearly constant. When suction was applied, the number and length of the streaks decreased as expected. The VITA algorithm was used to detect the strong shear layers associated with the bursting process. The frequency of VITA detections also decreased with increasing suction. These changes confirm that the selective suction concept is a valid one, but large suction rates of $C_q > 0.005$ were needed to obtain significant structural changes. If the suction were applied over a longer streamwise extent, similar results may be obtained with lower suction rates.

INTRODUCTION:

In the last two decades, substantial effort has been focused upon understanding the coherent structures present in the turbulent boundary layer. The level of knowledge has reached the point where it is now possible to design ways of interacting with these coherent structures. Along these lines, Gad-el-Hak and Blackwelder[2] suggested an approach called selective suction in which small amounts of suction could be applied below low speed streaks (LSS) to reduce the intensity of ejections or remove the streak altogether. The net result should be a reduction of the frequency of such bursting events. In their work, LSS were artificailly generated in a laminar boundary layer by the sudden withdrawl of fluid through two small holes ($d^+ = 4$) with a spanwise separation of $\Delta z^+ = 100$. Suction was applied through a single streamwise slot that was one viscous length unit wide and 1500 units long, located between and directly downstream of the two holes that made up the burst generator. This artificially created eddy structure was

A. Gyr (Editor)
Structure of Turbulence and Drag Reduction
IUTAM Symposium Zurich/Switzerland 1989

similar to the bursting phenomena found in bounded turbulent shear flows. They found that a suction rate of only $C_q = 0.0006$ was successful in eliminating the artifical bursting events (where the suction coefficient, C_q, is defined to be the ratio of the average velocity through the wall to the freestream velocity). To apply selective suction in a turbulent flow, one must first decrease or eliminate the spatial randomness of the LSS. Johansen and Smith[1] were able to greatly reduce the spatial randomness of the LSS by introducing longitudinal roughness elements (LRE) at the wall. Their histograms of streak spacing obtained from hydrogen bubble visualizations showed a marked peak and minimal scatter when the LRE spacing coincided with the low speed streaks most probable spacing of $\Delta z^+ = 80$. Gad-el-Hak and Blackwelder[2] suggested incorporating these roughness elements into the selective suction scheme. The effects of adding just the LREs to a turbulent boundary layer were studied earlier by Blackwelder and Roon[3] and those results are included in the discussion and graphs below.

EXPERIMENTAL SETUP:

Measurments were taken on a flat plate in a zero pressure gradient boundary layer 5 m downstream of the plate's leading edge. The freestream velocity was 5 m/s and the friction velocity, u_τ, was 0.2 m/s. The details of the wind tunnel and flat plate were reported by Blackwelder and Haritonidis[4]. The roughness elements consisted of long cylindrical strings aligned in the streamwise direction (figure 1). Their non-dimensional diameter was $d^+ = 5$ and the spanwise spacing was $\Delta z^+ = 80$. The overall length of the LREs was 1.5 m corresponding to $\Delta x^+ = 20,000$. The measurment station was at the end of the suction screen, 30 mm upstream from the end of the strings, where the momentum thickness Reynolds number was 3000. The spanwise rake of hot-wires was located at the elevation where the rms velocity reaches a maximum for a flat plate ($y^+ \cong 12$). The hot-wire sensors were each 1 mm long and spaced 1.5 mm apart. Measurements were also taken at $y^+ \cong 9$ and showed the same trends so were not included for the sake of brevity. Suction was appied by a separate pump through a plenum behind the porous plate. This provided uniform suction through a square screen lying in the plane of the plate as shown in figure 1. The screen was 30 mm x 30 mm with 0.6 mm diameter holes ($d^+ = 8$). Six LREs spanned the screen. Tape was placed on the backside of the screen so the suction was concentrated under the LREs and thus under the LSS. The tape had a width of $\Delta z^+ = 40$ and was centered between every pair of LREs.

The algorithm developed by Blackwelder and Roon[3] to detect LSS was used in the present work. It consists of a simple u-level scheme where the instantaneous hot-wire signal triggers

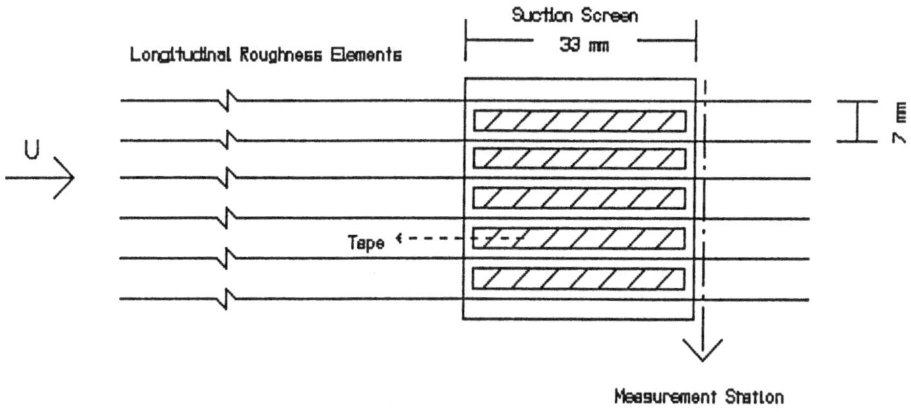

Figure 1: Experimental Setup of LREs and Suction Screen.

a possible detection if the velocity drops one rms below the mean. Since the LSS are very elongated in the streamwise direction, a streak detection is only accepted if the signal stays below this threshold for a time $t^+ = 20$ corresponding to a length of $x^+ \cong 200$ using Taylor's hypothesis. This allowed measurement of the frequency of occurence of streaks as well as their duration. Each hot-wire in the rake was processed separately; no correction was applied for two or more neighboring hot-wires detecting the same streak. Both this LSS algorithm and the VITA technique for measuring the bursting frequency depend upon the mean and rms velocity values that are applied. Each signal could be processed with the value obtained at the same height in a normal flat plate boundary layer or the local values particular to each hot-wire. The flat plate reference values are used in the following because they provided clearer results. The calculations were also performed with the local values and the trends for the LSS statistics were the same, but with more scatter.

RESULTS:

As a first approximation, one might assume that the effect of suction would be to pull the entire boundary layer downwards. So as the suction is increased the measured velocity should also increase. This is seen in the spanwise variation of the mean velocity plotted in figure 2. In all figures, the variables have been normalized with kinematic viscosity, ν and with the friction velocity, u^*, measured with a Coles[5] fit of the velocity profiles, $U(y)$. Directly above each of the strings, there is a retardation of fluid; that is, the velocity above each string is significantly lower than its flat plate value. With no suction, the velocity between the strings relaxed back to

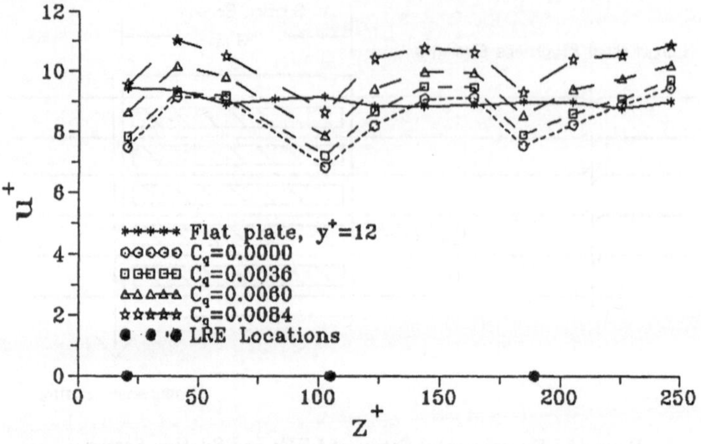

Figure 2: Mean Velocity Profile Across the Span with LREs and Suction.

the natural state. As suction was applied and increased, the velocity at all locations increased as expected.

The rms velocity distribution is plotted in figure 3. The strings alone without suction slightly lowers the fluctuation level. A large amount of suction is required to change the rms levels significantly. Since the data in figure 3 was obtained near the peak of the rms velocity profile, $u_{rms}(y)$, the sensitivity to perturbations is small. Similar data at $y^+ \cong 9$ showed the rms velocity decreased over the strings but remained near the natural values between the strings. The values at $y^+ \cong 9$ were also relatively insensitive to the suction levels. This contrasts the idea that suction is only moving the boundary layer profiles towards the wall. The rms velocity profiles at both locations were more sensitive to the introduction of the LREs than to the applied suction. This may be due to the fact that roughness elements were applied over a much longer streamwise extent (a factor of 50 longer) than the suction. Because the events that create velocity fluctuations are highly intermittant, only events that occured over the screen would be drastically changed by suction. Those that occured upstream and were convected over the screen would only feel slight changes. Thus it could be argued that the rms velocity might be insensitive to local suction since all of the fluctuations measured were not equally effected.

The frequency of occurence of streaks, N, is plotted non-dimensionally as $N^+ = N\nu/u_\tau^2$ in figure 4. After adding the strings, a substantial increase in N^+ occured over the LREs while only slight decreases are seen between them. The increased values of N^+ suggest that there are more low speed streaks with the strings in place. However, the detection algorithm is triggered

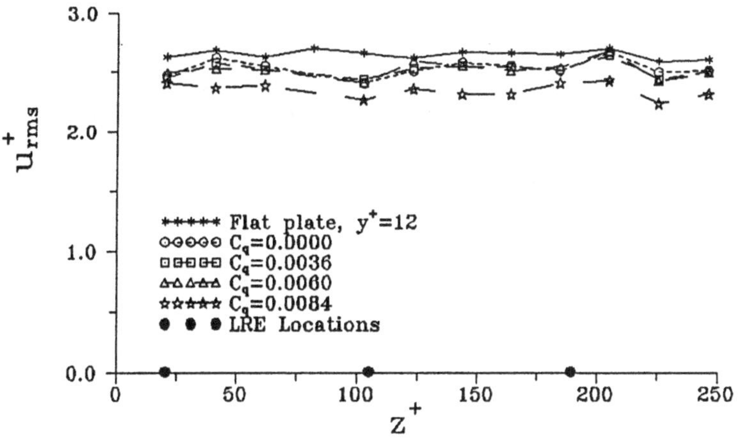

Figure 3: Rms Velocity Distribution Across the Span with LREs and Suction.

when the instantaneous velocity drops one rms below the mean *based upon the flat plate values.* Refering back to figure 2, the mean velocity over an LRE is less than the flat plate reference value; thus the algorithm is triggered more easily by the fluctuations. Increasing suction drastically reduced the number of streaks detected. The strings are effective in aligning the LSS as seen by the decreasing number of streaks between the LREs as the suction becomes very large. Since the suction is applied selectively under the LREs, streaks that occured between them should not have been changed drastically. But the number of LSS between the LREs tends to zero with increasing suction. This result agrees with the conclusion of Johansen and Smith[1] that the LREs tend to spatially align the LSS.

The mean duration of the streaks, T^+, is plotted in figure 5. This represents the mean non-dimensional time of the low speed streaks detected by each hot-wire. The mean streamwise length could be obtained by the use of Taylor's hypothesis, but the results were left in the time domain because the convection velocity changes with applied suction levels. The addition of the strings increased the "length" relative to the flat plate values. This indicates that the strings tend to inhibit spanwise movement of the streaks; thus a streak appears for a longer time over the LREs. The addition of suction decreases this "length" everywhere across the span. As fluid is removed from the streaks, they must decrease in size; thus the number seen at a fixed elevation as well as the length of individual streaks would decrease as seen in figures 4 and 5.

The goal of the selective suction concept is to slow or disrupt the bursting process. The VITA algorithm of Blackwelder and Kaplan[6] was used to detect the strong shear layers associated with

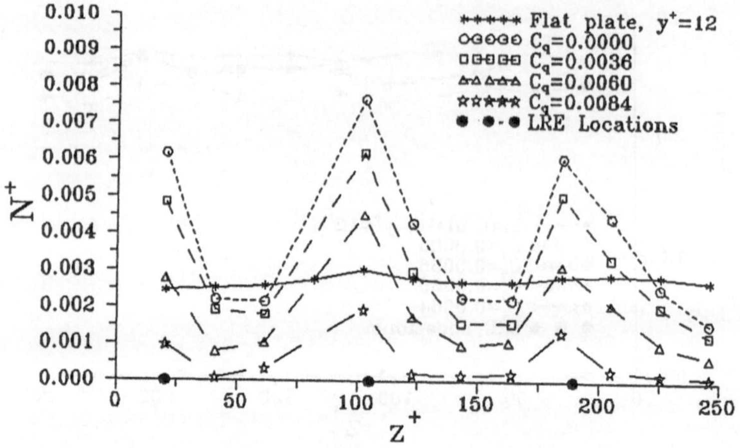

Figure 4: Frequency of Occurence of Low Speed Streaks.

the bursting process. Figure 6 is a plot of the bursting frequency across the span as detected with the VITA algorithm. The bursting frequency is clearly reduced with increased suction is spite of the fact that the suction was applied over a limited streamwise extent. In order to identify an ejection, Bogard and Tiederman[7] preferred a streamwise distance of 200 - 350 viscous scale lengths in the streamwise direction. The small streamwise range of $\Delta x^+ = 400$ used in the present investigation may have had some end effects associated with it which could have reduced the effectiveness of the suction. The selective suction concept assumes one can interact with the ejection stage of the bursting process, but unless the entire ejection occured over the suction screen, smaller changes would be obtained. This may explain why the mean velocity and LSS statistics appear greatly changed, while the rms and bursting frequency are less sensitive. That is, the streaks exist over a large spatial extent, thus the probability of their being effected by local suction is high. But ejections, and thus turbulence production, occur over a relatively smaller spatial extent and thus the chance of interacting with them with local suction fixed at one spatial location is lower.

CONCLUSIONS:

The selective suction concept has been investigated and found to be effective for interacting with low speed streaks. The longitudinal roughness elements have been found to be successful in reducing the spatial randomness inherent in low speed streaks in confirmation of the results of Johansen and Smith[1]. The addition of local suction was seen to decrease the frequency of the

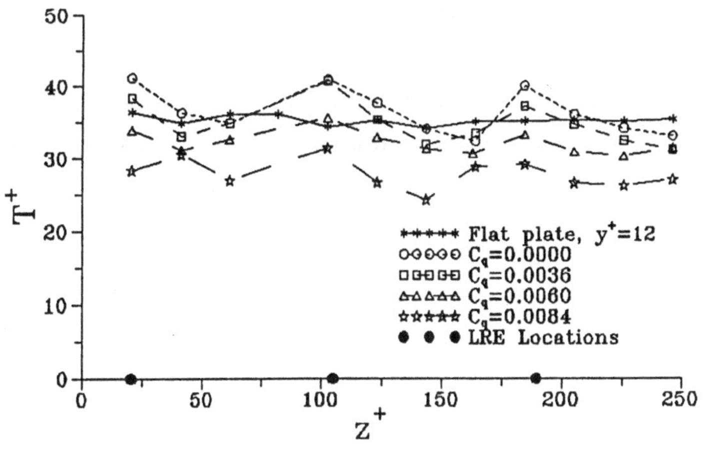

Figure 5: Duration of Low Speed Streaks.

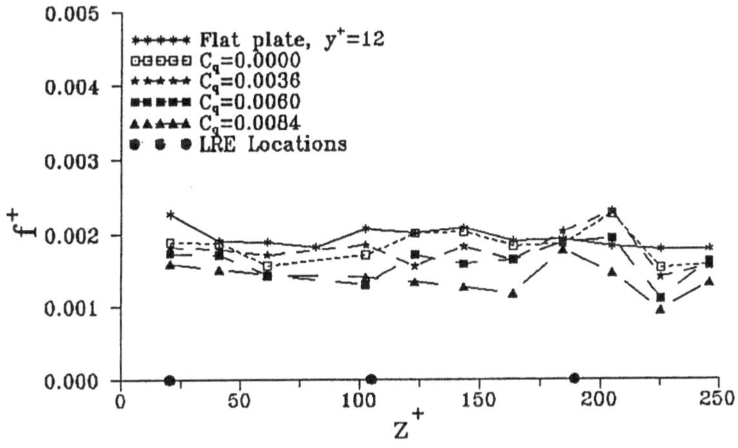

Figure 6: Bursting Frequency Across the Span with LREs and Suction.

low speed streaks as well as their duration. The bursting frequency was also seen to generally decrease with increasing suction, but the scatter in the data is larger than for the LSS statistics. The roughness elements do decrease the spatial randomness in the spanwise direction of the low speed streaks, but ejections still occur randomly in the streamwise direction along the LRE and in time.

ACKNOWLEDGEMENTS:

This research was funded by the Office of Naval Research under the University Research Initiative Contract N00014-86-K-0679 monitored by Mike Reishman. Their support is gratefully acknowledged.

REFERENCES:

[1]Johansen and Smith, 1986, "The Effects of Cylindrical Surface Modifications on Turbulent Boundary Layers", J. AIAA, 24, 1081.

[2]Gad-el-Hak and Blackwelder, 1988, "Selective Suction for Controlling Bursting Events in a Boundary Layer", J. AIAA, 26.

[3]Blackwelder and Roon, 1988, "The Effects of Longitudinal Roughness Elements upon the Turbulent Boundary Layer", AIAA-88-0134.

[4]Blackwelder and Haritonidis, 1983, "Scaling of the Bursting Frequency in Turbulent Boundary Layers", JFM, 132, 87.

[5]Coles, 1968, "The Young Person's Guide to the Data", Proceedings of the Computations of Turbulent Boundary Layers - 1968, AFOSR-IFP-Stanford Conference, 2, Mechanical Engineering, Stanford, CA.

[6]Blackwelder and Kaplan, 1976, "On the Wall Structure of the Turbulent Boundary Layer", JFM, 76, 89.

[7]Bogard and Tiederman, 1087, "Characteristics of Ejections in Turbulent Channel Flow", JFM, 179, 1.

Turbulent Drag Reduction by Nonplanar Surfaces – A Survey on the Research at TU/DLR Berlin

D.W. Bechert, M. Bartenwerfer, DLR, Abt. Turbulenzforschung
Müller-Breslau-Str. 8, 1000 Berlin 12, F.R.G., and
G. Hoppe, Hermann-Föttinger-Institut für Thermo- und Fluiddynamik,
TU-Berlin, Straße des 17. Juni 135, 1000 Berlin 12, F.R.G.

The operation of riblets as a means for drag reduction is explained. It is shown theoretically, that the viscous flow on riblet surfaces creates a hampering effect for the lateral fluctuations near the surface. The analysis of the cross flow constituent reveals that: (i) the origin of the viscous cross flow profile is located farther away from the surface than the origin of the longitudinal flow, and (ii) flow separation and recirculation of the cross flow occurs down to vanishing Reynolds numbers. The solution of the biharmonic equation for the stream function of the viscous cross flow is obtained by novel elastic plate simulation measurements. The results are verified with a low Reynolds number Couette flow experiment. For various riblet shapes, the anisotropic properties of the viscous flow on the riblets are determined. This information can be utilized for riblet optimization as well as an input for the computer simulation of the complete turbulent flow field.

Besides riblet surfaces, other still hypothetical concepts are discussed which are derived from shark skin. In order to test these new concepts, new facilities are necessary, such as a large oil channel which is to be completed in summer 1989. This channel has a differential shear stress balance with which drag reducing surfaces and smooth surfaces can be compared directly. The engineering design and the construction of suitable new experimental facilities and the production of exotic surfaces represents the greatest part of the effort of this scientific program. In an appendix (supplied at the conference) we provide the necessary backing for our theoretical findings. There, the general unsteady viscous fluid motion on a flat surface is worked out and discussed.

1. Introduction

This survey will deal mainly with recent ideas on theory. We shall not consider the subtleties and challenging difficulties of experimental research, which is, nevertheless, the backbone of our research. On the other hand, without new, theoretically supported ideas we see little chance for an improvement of existing drag reducing surfaces. From painful learning through failures in numerous experiments we have extracted a strategy for drag reduction research. It is the "no penalty strategy". The idea is that *turbulent shear stress reduction can be achieved only if the device introduced does not produce any parasitic drag by itself.* Excluded from this conceptual statement are drag reducing schemes in which the whole status of the flow is changed, e.g., from separated to attached flow, or from tur-

A. Gyr (Editor)
Structure of Turbulence and Drag Reduction
IUTAM Symposium Zurich/Switzerland 1989

bulent to laminar flow. The "no penalty strategy" strongly suggests that drag reducing surfaces should be imbedded in the viscous sublayer, with protrusions smaller than $y^+ \approx 5$. Thus, parasitic drag is avoided. On the other hand, this concept opens the door widely to the use of comparatively simple *linear* theories to deal with the unsteady flow in the viscous sublayer. In this way we shall derive a simple model for the drag reduction mechanism of riblet surfaces. Interestingly, even the *separation* at riblets under cross flow conditions can be modeled by *linear* theory. On the other hand, new ideas on possible drag reduction mechanisms can be screened with the "no penalty strategy". Amongst the ideas which could not be rejected thus far, are two additional, but still hypothetical mechanisms derived from shark skin.

2. Basic equations

Our considerations shall be mainly confined to the viscous flow in the vicinity of the wall of a turbulent boundary layer. As we will show later, the drag reducing effect on the turbulence is created by a change in the near wall viscous flow boundary conditions for the turbulent flow. In the viscous sublayer, it is admissible to omit the convective terms of the Navier-Stokes equations. Thus we have

$$\Delta u - \frac{1}{\nu}\frac{\partial u}{\partial t} = \frac{1}{\mu}\frac{\partial p}{\partial x} ; \quad \Delta v - \frac{1}{\nu}\frac{\partial v}{\partial t} = \frac{1}{\mu}\frac{\partial p}{\partial y} ; \quad \Delta w - \frac{1}{\nu}\frac{\partial w}{\partial t} = \frac{1}{\mu}\frac{\partial p}{\partial z} . \tag{1}$$

2.1. Longitudinal mean flow on a riblet surface

A fairly simple case is the calculation of the mean flow distribution on riblets aligned with the flow direction x. Obviously, the time derivative becomes zero here, and the mean flow pressure gradient on a flat plate can be also set to be zero. Thus, we end up with

$$\frac{\partial^2 u}{\partial y^2} + \frac{\partial^2 u}{\partial z^2} = 0 . \tag{2}$$

With this Laplace equation in two dimensions the mean flow velocity u on a riblet surface can be calculated. In fig. 1 an example from our previous work [1] can be seen. Obviously, this calculation is only valid in the near vicinity of the riblets. A typical riblet spacing s expressed in wall units is $s^+ \approx 10$-15 for optimal drag reduction [2,3]. Thus, at about half a riblet spacing above the rib tips deviations from our calculation are to be expected. In a previous paper [1] we expressed our feeling, that the protrusion height h_{pL} in fig. 1, the distance between rib tip and average origin of the velocity profile, may have a special importance for the drag reduction. In the following sections we will return to this particular issue.

2.2. Stokes waves on a flat surface

We consider a fluctuating shear flow on a flat surface. This flow problem is amenable to a more or less complete analysis and thus it provides some useful insights. The fluctuating shear flow represents the flow field close to the wall, induced by some event farther away in the "active" region of a turbulent boundary layer. As a first simplistic approach we will confine our-

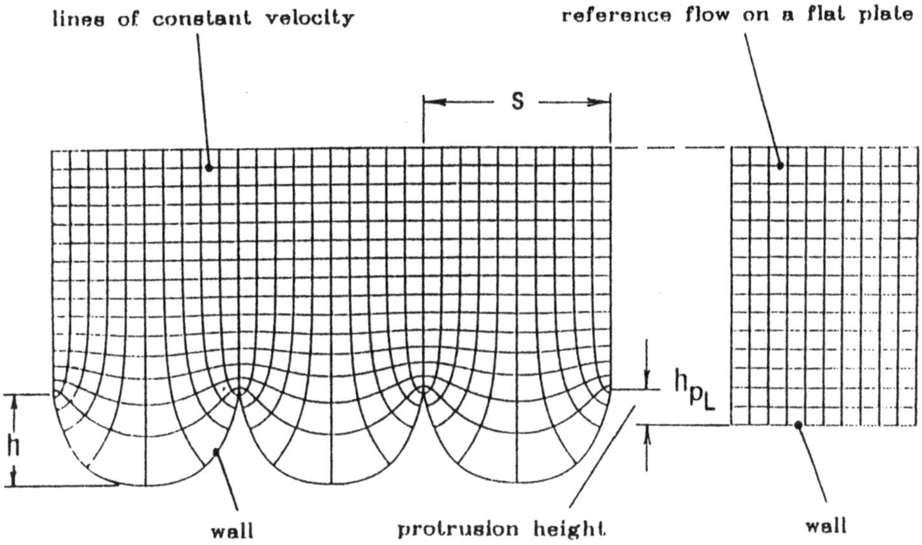

Fig. 1 Mean flow velocity distribution on a riblet surface

selves to a shear wave of infinite wavelength. An analysis with finite wavelength can be also carried out (see appendix), but it will give very similar results as long as the perturbation wavelength is much larger than the typical size of, e.g., a riblet. In fact, this ratio is typically about 10. This model may be applied to longitudinal (u) and transversal (w) shear fluctuations. For infinite wavelength, the pressure gradient can be neglected. In addition, the vertical velocity component (v) vanishes here. Assuming periodical motion $\propto e^{-i\omega t}$ (or taking a Fourier transform) we obtain for cross flow fluctuations

$$\frac{\partial^2 w}{\partial y^2} + \frac{i\omega}{\nu} w = 0 . \tag{3}$$

The solutions of this equation are

$$w = A\, e^{+(1-i)\sqrt{\omega/2\nu}\, y} + B\, e^{-(1-i)\sqrt{\omega/2\nu}\, y} \tag{4}$$

Introducing the boundary condition $w = 0$ at $y = 0$ yields $A = -B$. For comparison, a steady Couette flow would be represented by $w = \Omega_0 \cdot y$, with Ω_0 being the vorticity of the Couette shear flow profile. In order to produce comparable conditions for the fluctuating shear flow, we set

$$w = \frac{\Omega_0 (1+i)}{4\sqrt{\omega/2\nu}} \left[e^{+(1-i)\sqrt{\omega/2\nu}\, y} - e^{-(1-i)\sqrt{\omega/2\nu}\, y} \right] . \tag{5}$$

Then, the real part of w becomes $\Omega_0 \cdot y$ for $\omega \rightarrow 0$, as in a steady Couette flow. Fig. 2 shows real and imaginary part of eq. (5) as well as modulus and phase. w^* is a nondimensional form of w with $w^* = w \cdot \sqrt{\omega/2\nu}/\Omega_0$. For large distances y the phase gradient becomes a constant. In fig. 2 also values for

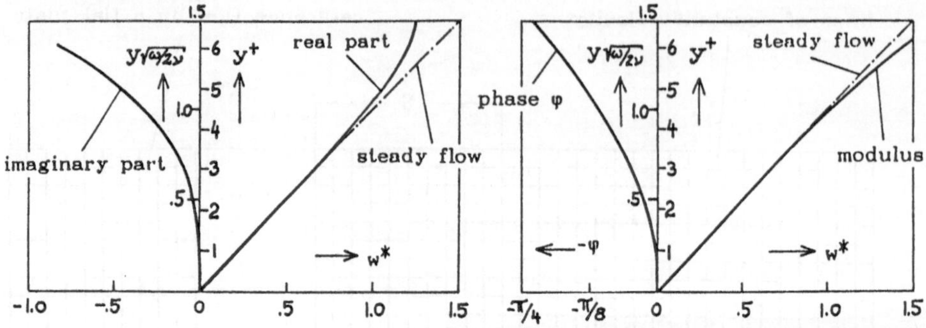

Fig. 2 Fluctuating shear flow

y^+ are plotted. For that, we have assumed $\omega^+ = 0.1$, the frequency of the maximum of the viscous sublayer oscillations, according to Morrison et al. [4] and Jang et al. [5]. The conclusion is, that at $y^+ < 3$, the unsteady shear flow behaves as if it were a steady shear flow oscillating without spatial phase delay.

2.3. Riblets with cross flow

According to the previous section and to the more detailed calculations given in the appendix, the shear flow around one single riblet resembles more a plane Couette flow than a general three-dimensional wavy flow. Hence, we assume that the cross flow on riblets is two-dimensional. Then, our basic equations can be simplified considerably. Eliminating the pressure from equations (1) yields an equation for the vorticity $\Omega = \partial w/\partial y - \partial v/\partial z$

$$\Delta\Omega - \frac{1}{\nu}\frac{\partial\Omega}{\partial t} = 0 . \qquad (6)$$

For a two-dimensional flow we can introduce a stream function ψ with $w = \partial\psi/\partial y$ and $v = -\partial\psi/\partial z$. Inserting this into the vorticity equation gives with $\psi \propto e^{-i\omega t}$

$$\Delta\Delta\psi + \frac{i\omega}{\nu}\Delta\psi = 0 . \qquad (7)$$

It is certainly possible to solve this linear equation by superimposing singularities (Green's functions). The appropriate singularities are Kelvin functions and their derivatives. At lower frequencies the simpler biharmonic equation

$$\Delta\Delta\psi = 0 \qquad (8)$$

should be sufficient. This is also supported by the considerations of the previous section, because equation (5) is indeed a solution of eq. (7). However, it is unlikely, that this simplification may work well beyond $\omega^+ \approx 0.1$, the maximum of the fluctuation spectrum. In order to produce solutions of eq. (8), the general methods of Green's function superposition or of Fourier transform/series expansion are available. In addition, the Goursat approach utilizing functions of complex variables is possible [6]. On the other hand, one is inclined to look into other fields of physics for solutions of the biharmonic equation (8), because fluid dynamicists seem to have lost interest in purely viscous flows. One such field is the elasticity of plates. There, an enormous amount of work has been spent to solve eq. (8) for various boundary conditions [7]. Unfortunately, to our knowledge, our problem does not seem to have been solved there either. Nevertheless, a plate bending *experiment* shall be used in section 2.5. to provide data for our viscous flows. Before doing that, however, we shall look into the problem with a direct (albeit less accurate) flow experiment with riblets in a viscous cross flow.

2.4. A viscous flow experiment

Our simple viscous flow experiment can be seen in fig. 3. The setup shown in this schematic view is immersed in glycerine in a small container (aquarium). A canvas belt runs over two cylinders. These cylinders consist

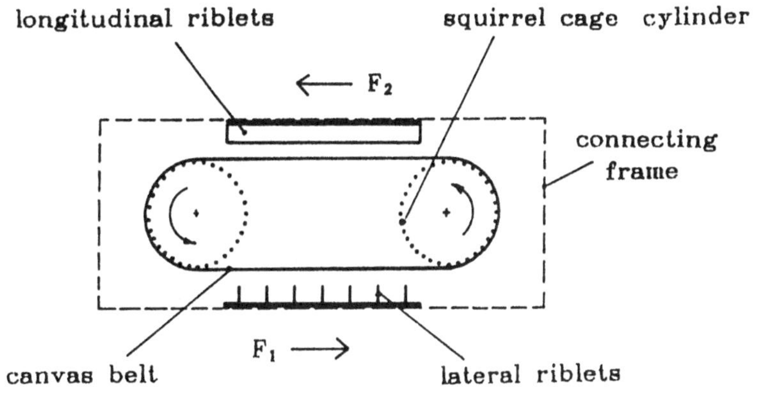

Fig. 3 Couette flow experiment

of an array of rods, a squirrel cage, to allow the fluid to enter and exit through the cylinder surface (it had turned out, that normal tubular cylinders produce a disastrous buildup of glycerine between cylinder and belt). The belt is moved with a crank. One riblet plate is aligned with its ribs in the direction of the Couette flow and the other plate has its ribs oriented laterally to the flow direction. The surface of the plates and the total length of the riblet wedges are equal on both sides. In addition, the distances of the rib tips to the belt surface are the same on both sides. The riblet plates are attached to a frame which is suspended with blade springs. Thus, a small shift of the frame in the flow direction is possible but not perpendicular to it. The forces exerted on the frame are the shear forces F_1 and F_2 (see fig. 3), acting in opposite directions. Consequently, a movement of the frame to the right hand side indicates that the shear force on the lateral riblets is greater than the one on the longitudinal riblets. As a matter of fact, that is what is observed in the experiment. An increased shear force also indicates that the origin of the shear flow profile is closer to the rib tip. Thus, the protrusion height for riblets with lateral flow is smaller than that with longitudinal flow. Indeed, this is the mechanism of riblet surfaces in a real situation: *The shear resistance to cross flow is higher because the origin moves farther away from the surface.* We suggest that this is a *linear flow phenomenon,* because eq. (8) is linear. In our low Reynolds number Couette flow experiment we see, however, flow *separation* between the ribs under cross flow conditions, see fig. 4. The streamlines in fig. 4 are drawn according to the visual observation of the paths of small aluminium flakes in the fluid. The dimensions of the experiment are also given (in mm) in fig. 4. A typical belt velocity is 5-100 mm/sec and the fluid viscosity is $\nu = 1.25\times10^{-3}\ m^2/s$. This leads to Reynolds numbers in the range of Re = 0.04-0.8. Interestingly, the flow separation seems to persist to even lower Reynolds numbers and does not appear to be changed by small flow unsteadiness. Indeed, the whole flow separation pattern looks like being rigidly linked to the belt motion, like a tooth wheel. This suggests that this particular separation is a property of the solution of the *linear* viscous equation (8) under these particular boundary conditions. As fluid dynamicists we are inclined to think that the separation might be rather due to the existence of convective effects close to the rib tips. Note, however, that the separation does not seem to occur directly at the rib tips. In addition, the flow is completely symmetrical in the flow direction. In order to decide whether or not this is a linear effect, it would be desirable to have an experiment at Re → 0. This is exactly what our elastic plate simulation experiment will do.

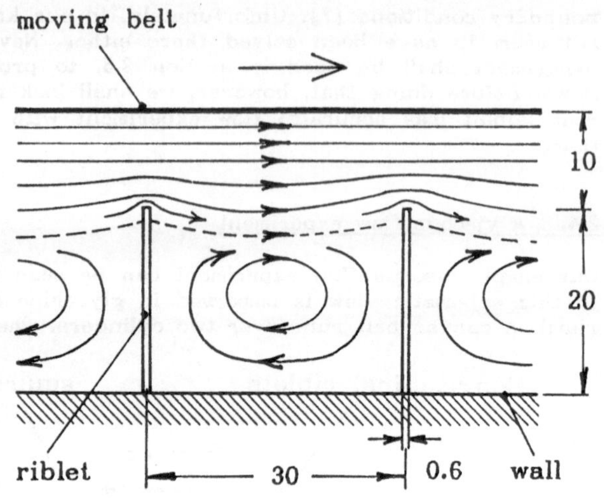

Fig. 4 Cross flow streamlines

2.5. Viscous flow simulation with an elastic plate experiment

The biharmonic equation (8) for plane viscous fluid motion is also valid for small deflections of elastic plates [7]. In spite of the fact that in both fields the equation has been known for more than a century, to our knowledge experimental simulations of viscous flow fields by plate bending have not been tried before. The value of the deflection perpendicular to the plate surface corresponds to the value of the stream function ψ in a two-dimensional viscous flow. The boundary conditions for the fluid on the riblet surface, with stream function and velocity being zero, correspond to a plate with deflection and its gradient being zero in a "built-in edge". This condition is easy to generate in an experiment. Furthermore, for large distances y from the riblet surface, a Couette flow can be simulated by a bending moment M applied continuously to the other edge of the plate, see fig. 5. There are, however, some subtleties to the experiment: Strictly speaking, the deflection is a circle at large y rather than a parabola assumed in the theory. In addition, the plate has to end somehow at the sides, preferably at or near a location of zero lateral bending moment. The configuration in fig. 5 seems to be a suitable compromise there. The origin of the bending problem (= of the velocity profile) is determined by a large number of measurements at various locations on the plate, not too close to the rib tips. The limits to the accuracy of the measurements are given by (i) local variations of the elastic properties of the metal plate and (ii) by problems with the lateral finiteness of the plate.

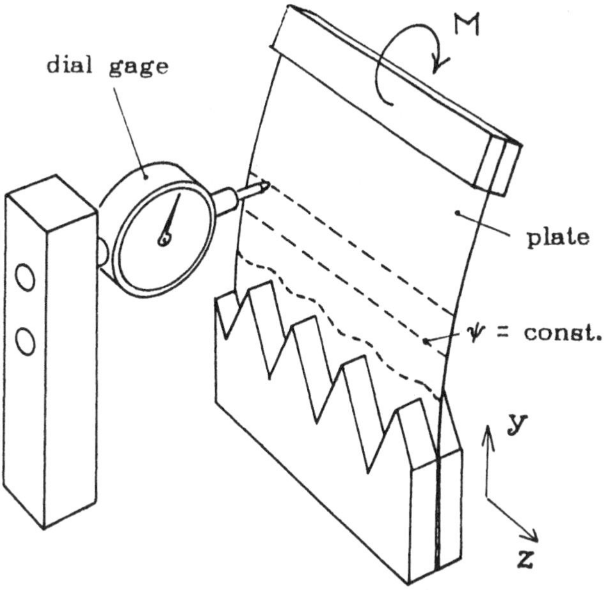

Fig. 5 Plate bending simulation experiment

Figure 6 shows the configurations which we investigated experimentally. Configuration 6(a) is a "blade riblet" with a blade thickness of 2.1% of the lateral riblet spacing. The distance h_{pc} marks the location of the origin for cross flow conditions. The quantity h_{pc}, the cross flow protrusion height, is determined. For $h/s = 0.5$ we find $h_{pc}/s = 0.08$; the same value is found for $h/s \rightarrow \infty$, i.e., without "floor" between the ribs. For configuration (b), scalloped riblets with semicircular shape, we find also $h_{pc}/s = 0.08$. The third configuration, sawtooth riblets with 60° angle, produces again a very similar value of $h_{pc}/s = 0.075$. The fourth configuration, convex ribs, produces $h_{pc}/s = 0.05$. The accuracy of the measurements is estimated to be ≈3% of the rib spacing s.

Finally, by our dial gage measurements, we find that the plate exhibits negative deflection between the ribs which proves the existence of closed streamlines and thus clearly shows that the *flow separation occurs down to vanishing Reynolds number. Hence it is a linear effect.* The hampering effect of the riblets to cross flow is proportional to the difference between longitudinal protrusion height h_{pl} minus the cross flow protrusion height h_{pc}. For the configurations 6(a)-(c) we have an "impedance" $(h_{pl}-h_{pc})/s$ of 0.10 to 0.14, whereas for convex riblets this is only about 0.02. Configurations (a)-(c) did exhibit turbulent drag

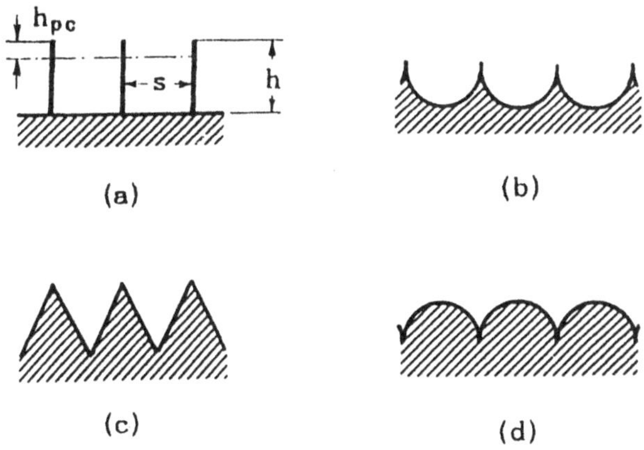

Fig. 6 Various riblet configurations

reduction in the previous experiments [2,3,8], whereas convex riblets did not [2].

2.6. The anisotropic boundary condition for riblet surfaces

The directivity of the riblets is equivalent to the difference of the longitudinal protrusion height h_{pL} and the cross flow protrusion height h_{pc}. The location of the origin is given at

$$hp = \frac{h_{pL} + h_{pc}}{2} + \frac{h_{pL} - h_{pc}}{2} \cos 2\alpha . \qquad (9)$$

Because the dependence on the cross flow angle α is a cosine law, the low sensitivity of the drag reduction to cross flow misadjustment is obvious, which has been found previously [9]. On the other hand, any structure having grooves perpendicular to the flow direction is predicted to produce an enhanced turbulent drag. By contrast, two intersecting groove arrays at ±45° angle have *no* negative effect. This is in fact the overall scale pattern of shark skin. Our three-dimensional staggered riblets also had such a structure. Their performance of 6% drag reduction matched the best conventional riblet surfaces [10].

According to our present data, riblets having a high h_{pL}/s produce also a high rectifying effect $(h_{pL}-h_{pc})/s$. This confirms our previously rather tentative suggestion [1] that a high h_{pL}/s should be optimal for drag reduction. Finally, it should be mentioned, that eq. (9) may be plugged into a full Navier-Stokes code for turbulent boundary layer computation. In this way, a quantitative numerical prediction of riblet drag reduction may be obtained.

3. Hypothetical mechanisms derived from shark skin

In a previous paper [11] we have suggested a drag reduction mechanism with slits oriented into the streamwise direction, derived from observations on shark skin. Our "no penalty strategy" suggests that also this configuration should be imbedded in the viscous sublayer. The slits should be oriented preferably at a cross flow angle of about ±45°. In the following, we will discuss first a new *linear* streak compensation mechanism with these slits. We simplify real shark skin (see [11]) to an array of guide vanes with slits above a cavity. The viscous shear flow will produce a pressure distribution as shown in fig. 7. Note that these pressure differences are created by viscous forces alone, and they are of the order of the mean shear stress τ_0 on the surface. Thus, the cavity under the guide vanes will be at a pressure level being by about τ_0 lower than the ambient pressure. Normally, the velocity profile has the shape (o) in fig. 7. If the shear stress is locally reduced (i), say, in a low speed streak, the lower pressure in the cavity causes local suction through the slits. This produces an increased local shear stress and partly cancels the low speed streak. On the other hand, if the local shear stress raises above the average value (ii), the opposite effect occurs and fluid is (linearly) added to the shear layer. Again this tends to compensate the deviation. Thus far, fluctuations of the static pressure have not been accounted for (see appendix). The static pressure fluctuation coupled with the shear stress fluctuation produces a phase shift of this dynamic suction control. A compensation of this phase shift by a pressure-difference controlled angle of the individual scales seems conceivable in liquids. It should be mentioned, however, that the phase coupled pres-

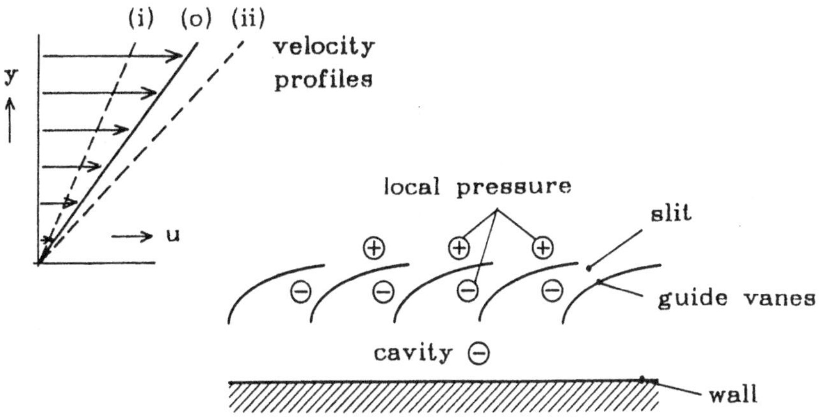

Fig. 7 Simplified shark skin configuration

sure fluctuation does not eliminate the effect alltogether. Besides this new *linear* control concept, we have also refined the previous slit ejection concept [11], which is genuinely *nonlinear*. It is based on the idea, that the pressure fluctuations in a real turbulent boundary layer are strong and by an order of magnitude greater than the shear stress fluctuations. In fact, the RMS level of the pressure fluctuations is 3-4 times greater than the mean shear stress τ_0. Hence, it is conceivable to feed back energy from the pressure fluctuation to the mean flow. We think, that a slit surface should be capable of doing that. At higher fluctuation levels (and Reynolds numbers), the slits are operating like orifices. An inflow produces a potential sink flow pattern, whereas an outflow produces a jet. Thus, the fluctuating in- and outflow produces a thrust which is felt as a drag reduction. The efficiency of this mechanism is bound to be poor, but since the pressure differences are so much larger than the mean shear stress, even a poor efficiency could cause an appreciable drag reduction. The limits to the efficiency of this effect are the following:

i) The small negative bias pressure in the cavity under the guide vanes, being of the order of 1/10 of the RMS pressure fluctuation level, is expected to be somewhat detrimental.
ii) The mechanism, by definition, has an efficiency of less than 50%, because thrust is generated only in the ejection phase.
iii) Low Reynolds number reduces the effect, but our own glycerine tank simulations have shown, that an (albeit reduced) thrust is still generated at $Re_{slit} \approx 1$.
iv) Another general performance limit is due to the restricted ratio of slit opening area to total projected surface area.

It is interesting, that elastically anchored scales would reduce the deteriorating influences (iii) and (iv), because the slits would open up during the ejection phase. Again, this latter compliant scale mechanism is only feasible for liquids. For the speculative ideas displayed in this section, experimental confirmation is badly needed. The facilities for this purpose are discussed in the following section.

4. Experimental facilities

Most of the time invested in this project went into design and construction of suitable experimental facilities. Most of our previous data have been collected with a plate installed in a wind tunnel [11]. This flat plate with a turbulent boundary layer has a variable length between 2.5 and 10 m, producing maximal Reynolds numbers of $Re_x \approx 3\times10^7$. The turbulent shear stress is measured with a large (0.6×0.75 m) floating element balance. Suitable riblet widths for this facility lie between s = 0.15 and s = 0.5 mm. The accuracy has been improved to better than ±1% error for the shear stress measurements.

It is obvious, that the suggested complicated surfaces of section 3 would be extremely difficult to build for a wind tunnel experiment. Thus, it is desirable to have a facility where the surface elements would be ten times larger in size. Therefore, we have designed and built an *oil channel* (see fig. 8). It is an advanced version of the Reichardt/ Eckelmann facility in Göttingen. It has a higher Reynolds number range and it has an incremental shear stress balance which compares the shear stress on plates at both sides of the channel. Thus, smooth plates and riblet surfaces can be compared directly.

OIL CHANNEL		
cross section	:	0.25 × 0.85 m
max. velocity	:	2 m/s
viscosity ν	:	1.35×10^{-5} m²/s

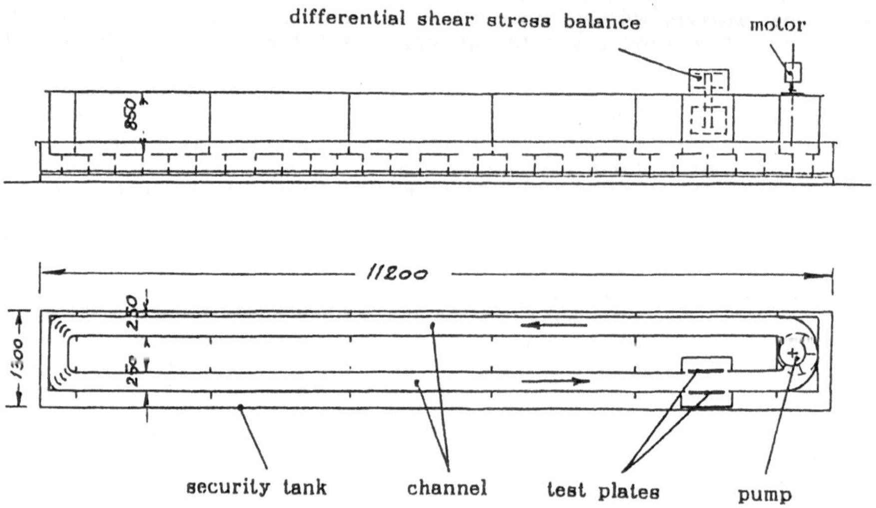

Fig. 8 DLR/HFI oil channel

Acknowledgement
Funding for this project has been provided by the Deutsche Forschungsgemeinschaft.

Appendix
The appendix dealing with "waves in the viscous sublayer above a flat surface" will be supplied at the meeting.

References

[1] Bechert, D.W. and Bartenwerfer, M., The viscous flow on surfaces with longitudinal ribs. J. Fluid Mech., 1989 (in press).
[2] Walsh, M.J., Turbulent boundary layer drag reduction using riblets. AIAA-paper 82-0169 (1982).
[3] Walsh, M.J., Optimization and application of riblets for turbulent drag reduction. AIAA-paper 84-0347 (1984).
[4] Morrison, W.R.B., Bullock, K.J. and Kronauer, R.E., Experimental evidence of waves in the sublayer. J. Fluid Mech. (1971), 47, pp. 639-656.
[5] Jang, P.S., Benney, D.J. and Gran, R.L., On the origin of streamwise vortices in a turbulent boundary layer. J. Fluid Mech. 169, pp. 109-123.
[6] Müller, W., Einführung in die Theorie der zähen Flüssigkeiten. Akad. Verl.-Ges., Leipzig 1932.
[7] Timoshenko, S.P. and Woinowsky-Krieger, S., Theory of plates and shells. McGraw-Hill, New York 1959.
[8] Wilkinson, S.P. and Lazos, B.S., Direct drag and hot-wire measurements on thin-element riblet arrays. IUTAM Symposium on Turbulence Management and Relaminarization. Bangalore, India, Jan. 1987. Eds. Liepmann & Narasimha, Springer-Verlag, Berlin 1988.
[9] Bechert, D.W., Gerich, D. and Hoppe, G., Short report on measurements with sawtooth riblets (3M plastic riblet film). DFVLR/HFI Berlin, 1987.
[10] Bechert, D.W., Experiments on three-dimensional riblets. Proc. "Turbulent drag reduction by passive means", Royal Aeronaut. Soc. 15-17. Sept. 1987, London.
[11] Bechert, D.W., Hoppe, G. and Reif, W.-E., On the drag reduction of the shark skin. AIAA-paper 85-0546 (1985).

IUTAM–SYMPOSIUM ON STRUCTURE OF TURBULENCE AND DRAG REDUCTION, ZÜRICH 25.–28. JULY 1989

Appendix of the paper:

TURBULENT DRAG REDUCTION BY NONPLANAR SURFACES. A SURVEY ON THE RESEARCH AT TU/DLR BERLIN.

D.W. Bechert, M. Bartenwerfer, DLR, Abt. Turbulenzforschung, Müller–Breslau–Str. 8, 1000 Berlin 12, F.R.G., and G. Hoppe, Hermann–Föttinger–Institut für Thermo– und Fluiddynamik, TU Berlin, Straße des 17. Juni 135, 1000 Berlin 12, F.R.G.

Waves in the viscous sublayer above a flat surface:

The viscous flow equations

$$\left.\begin{aligned} \Delta u - \frac{1}{\nu}\frac{\partial u}{\partial t} &= \frac{1}{\mu}\frac{\partial p}{\partial x} \\ \Delta v - \frac{1}{\nu}\frac{\partial v}{\partial t} &= \frac{1}{\mu}\frac{\partial p}{\partial y} \\ \Delta w - \frac{1}{\nu}\frac{\partial w}{\partial t} &= \frac{1}{\mu}\frac{\partial p}{\partial z} \end{aligned}\right\} \qquad (A1)$$

$$\frac{\partial u}{\partial x} + \frac{\partial v}{\partial y} + \frac{\partial w}{\partial z} = 0 \qquad (A2)$$

together with a wave motion approach

$$u,v,w,p \propto e^{i(\alpha x+\beta z)-i\omega t} \qquad (A3)$$

are used to develop more specific equations for the fluctuating quantities in the viscous sublayer. By taking derivatives of eq. (A1) and by utilizing the continuity equation (A2) one can find the Laplace equation being valid for the pressure

$$\Delta p = 0 \qquad (A4)$$

All these equations are, of course, linear. Thus, combinations of solutions are also solutions. Let us start with the spatial distribution of the pressure. Equation (A4) becomes with eq. (A3)

$$\frac{\partial^2 p}{\partial y^2} - (\alpha^2+\beta^2)\, p = 0 \tag{A5}$$

The determination of the distribution of the pressure is straightforward

$$\left.\begin{array}{l} p_1 = p_{10}\, e^{+\sqrt{\alpha^2+\beta^2}\, y} \cdot e^{i(\alpha x+\beta z) - i\omega t} \\ p_2 = p_{20}\, e^{-\sqrt{\alpha^2+\beta^2}\, y} \cdot e^{i(\alpha x+\beta z) - i\omega t} \end{array}\right\} \tag{A6}$$

The equations (A1) become then

$$\left.\begin{array}{l} \dfrac{\partial^2 u}{\partial y^2} - (\alpha^2+\beta^2 - \dfrac{i\omega}{\nu})\, u = \dfrac{i\alpha}{\mu} \cdot p_{1,2} \\ \dfrac{\partial^2 v}{\partial y^2} - (\alpha^2+\beta^2 - \dfrac{i\omega}{\nu})\, v = \pm \dfrac{\sqrt{\alpha^2+\beta^2}}{\mu} \cdot p_{1,2} \\ \dfrac{\partial^2 w}{\partial y^2} - (\alpha^2+\beta^2 - \dfrac{i\omega}{\nu})\, w = \dfrac{i\beta}{\mu} \cdot p_{1,2} \end{array}\right\} \tag{A7}$$

There are nonhomogenous and homogenous solutions to these equations

$$\left.\begin{array}{l} u_1 = \dfrac{\alpha}{\rho\omega}\, p_1 \;; \qquad u_2 = \dfrac{\alpha}{\rho\omega}\, p_2 \\ u_3 = u_{30}\, e^{+\sqrt{\alpha^2+\beta^2 - i\omega/\nu}\, y} \cdot e^{i(\alpha x+\beta z) - i\omega t} \\ u_4 = u_{40}\, e^{-\sqrt{\alpha^2+\beta^2 - i\omega/\nu}\, y} \cdot e^{i(\alpha x+\beta z) - i\omega t} \end{array}\right\} \tag{A8}$$

where u_1 and u_2 are the nonhomogenous and u_3 and u_4 are the homogenous solutions, respectively. In a completely analogous way we obtain equations for v

$$\left.\begin{array}{l} v_1 = -\dfrac{i\sqrt{\alpha^2+\beta^2}}{\rho\omega}\, p_1 \;; \qquad v_2 = \dfrac{i\sqrt{\alpha^2+\beta^2}}{\rho\omega}\, p_2 \\ v_3 = v_{30}\, e^{+\sqrt{\alpha^2+\beta^2 - i\omega/\nu}\, y} \cdot e^{i(\alpha x+\beta z) - i\omega t} \\ v_4 = v_{40}\, e^{-\sqrt{\alpha^2+\beta^2 - i\omega/\nu}\, y} \cdot e^{i(\alpha x+\beta z) - i\omega t} \end{array}\right\} \tag{A9}$$

and for w

$$\left.\begin{aligned} w_1 &= \frac{\beta}{\rho\omega} p_1 \ ; \qquad w_2 = \frac{\beta}{\rho\omega} p_2 \\ w_3 &= w_{30}\, e^{+\sqrt{\alpha^2+\beta^2-i\omega/\nu}\; y} \cdot e^{i(\alpha x+\beta z)-i\omega t} \\ w_4 &= w_{40}\, e^{-\sqrt{\alpha^2+\beta^2-i\omega/\nu}\; y} \cdot e^{i(\alpha x+\beta z)-i\omega t} \end{aligned}\right\} \qquad \text{(A10)}$$

The solutions with indices 1 and 2 are related to an exponentially growing and to an exponentially decaying pressure wave. The solutions with the indices 3 and 4 are exponentially growing and exponentially decaying vorticity waves. The magnitudes of each group of the latter are connected by the continuity equation (A2). We obtain for the amplified vorticity waves

$$i\alpha u_{30} + \sqrt{\alpha^2+\beta^2-i\omega/\nu}\; v_{30} + i\beta w_{30} = 0 \qquad \text{(A11)}$$

and for the decaying vorticity waves

$$i\alpha u_{40} - \sqrt{\alpha^2+\beta^2-i\omega/\nu}\; v_{40} + i\beta w_{40} = 0 \qquad \text{(A12)}$$

At this point a short discussion of the different contributions to the viscous sublayer motion is necessary. If we assume that the sublayer is driven by events in the active region above the sublayer, we can make a sensible choice of the solutions to be used. The flow constituents which drive the viscous sublayer are the ones which decay in the $-y$ direction, i.e., are amplified in the $+y$ direction. There are, however, two such amplified solutions, one being a pressure wave (p_1) and one being a shear wave (u_3,v_3,w_3). The events in the active region are likely to produce both influences at once. On the other hand, it is not prohibited to study both influences separately since linear superposition is possible.

A.1 Pressure waves

First we consider a viscous sublayer being excited by a pressure wave of the form p_1 (eq. (A6)). The influence of the boundary condition, i.e., the presence of the wall at $y=0$ does, however, also produce contributions of the other wave forms (p_2 and a shear wave u_4,v_4,w_4) decaying in the $+y$ direction. At the wall, all velocities must be zero. Thus, for u,v,w we have the conditions

$$\left.\begin{aligned} &\frac{\alpha}{\rho\omega}(p_{10}+p_{20})+u_{40}=0 \\ &\frac{-i\sqrt{\alpha^2+\beta^2}}{\rho\omega}(p_{10}-p_{20})+v_{40}=0 \\ &\frac{\beta}{\rho\omega}(p_{10}+p_{20})+w_{40}=0 \end{aligned}\right\} \qquad (A13)$$

In addition, we have equation (A12), the continuity equation, linking the velocity components of the shear wave constituent. For p_1, as an excitation, is given, we have now four equations for four unknowns, $p_1, u_{40}, v_{40}, w_{40}$. With a little algebra we find:

$$\left.\begin{aligned} &p_{20}=-\frac{1+\sqrt{1-i\bar{\omega}}}{1-\sqrt{1-i\bar{\omega}}}\cdot p_{10} \\ &u_{40}=\frac{2\alpha}{\rho\omega}\cdot\frac{\sqrt{1-i\bar{\omega}}}{1-\sqrt{1-i\bar{\omega}}}\cdot p_{10} \\ &v_{40}=\frac{i\cdot 2\sqrt{\alpha^2+\beta^2}}{\rho\omega}\cdot\frac{1}{1-\sqrt{1-i\bar{\omega}}}\cdot p_{10} \\ &w_{40}=\frac{2\beta}{\rho\omega}\cdot\frac{\sqrt{1-i\bar{\omega}}}{1-\sqrt{1-i\bar{\omega}}}\cdot p_{10} \end{aligned}\right\} \qquad (A14)$$

with the abbreviation

$$\bar{\omega}=\frac{\omega}{\nu(\alpha^2+\beta^2)}$$

Thus, with the equations (A8÷10) the complete fluctuation distribution for the pressure and the velocities is given, for this case of a pressure induced motion.

Another interesting aspect in connection with measurable quantities and with the application to drag reducing surfaces is the relation between pressure and shear stress. Given the motion of the viscous layer under a pressure wave, these relations can be worked out. We obtain after some intermediate calculations for the fluctuating quantities at the wall (index w)

$$\left.\begin{aligned} \tau_{x_w} &= -\,i p_w \cdot \frac{\cos\varphi}{\sqrt{1-i\bar{\omega}}} \\ \tau_{z_w} &= -\,i p_w \cdot \frac{\sin\varphi}{\sqrt{1-i\bar{\omega}}} \end{aligned}\right\} \qquad \text{(A16)}$$

In these equations, we use the abbreviation φ for the angle of the wave number vector. Explicitly we have for this abbreviation

$$\cos\varphi = \alpha/\sqrt{\alpha^2+\beta^2} \;; \qquad \sin\varphi = \beta/\sqrt{\alpha^2+\beta^2} \qquad \text{(A17)}$$

A2 Shear waves

In this case, we assume that $p_1=0$ and that there exists a (shear) wave field decaying in the $-y$ direction which has velocity components u_3, v_3, w_3. Due to the existence of the wall also this shear wave field will produce pressure fluctuations p_2, decaying in the $+y$ direction as well as a secondary shear wave field with velocity components u_4, v_4, w_4 also decaying in the $+y$ direction. The condition that all velocities must be zero establishes the relation between the different constituents

$$\left.\begin{aligned} &\frac{\alpha}{\rho\omega}\, p_{20} + u_{30} + u_{40} = 0 \\ &\frac{+i\sqrt{\alpha^2+\beta^2}}{\rho\omega}\, p_{20} + v_{30} + v_{40} = 0 \\ &\frac{\beta}{\rho\omega}\, p_{20} + w_{30} + w_{40} = 0 \end{aligned}\right\} \qquad \text{(A18)}$$

As we see from the continuity equation (A11), v_{30} is determined once we have chosen u_{30} and w_{30}.

$$v_{30} = \frac{-i\,(\alpha u_{30}+\beta w_{30})}{\sqrt{\alpha^2+\beta^2-i\omega/\nu}} \qquad \text{(A19)}$$

In addition to equations (A18) we have equation (A12) from continuity which links v_{40} to u_{40} and w_{40}. Thus we have again 4 equations for the 4 unknown quantities p_{20}, u_{40}, v_{40} and w_{40}. We obtain after a little algebra

$$\left.\begin{aligned} p_{20} &= -\rho\omega\vartheta \\ u_{40} &= -u_{30} + \alpha\vartheta \\ v_{40} &= -v_{30} + i\sqrt{\alpha^2+\beta^2}\,\vartheta \\ w_{40} &= -w_{30} + \beta\vartheta \end{aligned}\right\} \tag{A20}$$

with the abbreviation

$$\vartheta = \frac{2(\alpha u_{30}+\beta w_{30})}{(\alpha^2+\beta^2)\cdot\sqrt{1-i\bar{\omega}}} \tag{A21}$$

The excitation is by u_{30} and w_{30}. We can, however, condense this into one single exciting quantity by introducing a proportionality coefficient σ between both quantities

$$w_{30} = \sigma u_{30} \tag{A22}$$

In general σ can be a complex quantity to allow for a phase shift between the two contributions. We can establish relations between wall pressure and wall shear stresses as we did for the pressure waves. For the shear waves we obtain slightly more complicated expressions

$$\tau_{xw} = -ip_w \cdot \frac{\left(\frac{\sqrt{1-i\bar{\omega}}}{\cos\varphi+\sigma\sin\varphi} + \cos\varphi\right)}{1 + \sqrt{1-i\bar{\omega}}} \tag{A23}$$

$$\tau_{zw} = -ip_w \cdot \frac{\left(\frac{\sigma\sqrt{1-i\bar{\omega}}}{\cos\varphi+\sigma\sin\varphi} + \sin\varphi\right)}{1 + \sqrt{1-i\bar{\omega}}} \tag{A24}$$

A3 Discussion

To provide a feeling for what these equations mean we insert some data of velocity fluctuation measurements from Morrison, Bullock & Kronauer [1]. They found $\omega^+ = 0.1$, $\alpha^+ = 0.01$ and $\beta^+ = 0.047$ for the maxima of their frequency and wave number spectra. The "+" quantities are defined as $\omega^+ = \omega\nu/u_\tau^2$ and $\alpha^+ = \alpha\nu/u_\tau$, $\beta^+ = \beta\nu/u_\tau$ with u_τ being the shear stress velocity $u_\tau = \sqrt{\tau_0/\rho}$, where τ_0 is the (mean) wall shear stress.

The expression $\overline{\omega}$ then becomes $\omega^+/(\alpha^{+2}+\beta^{+2})$. With the quantitative values of Morrison et al. [1] we find $\overline{\omega}$ =43.3. Thus, the quantity $\left|\sqrt{1-i\overline{\omega}}\right| = 6.6$. The wave vector components become $\sin\varphi = \beta^+/\sqrt{\alpha^{+2}+\beta^{+2}} = 0.98$ and $\cos\varphi = \alpha^+/\sqrt{\alpha^{+2}+\beta^{+2}} = 0.21$. We estimate σ being of order 1 and thus, for shear waves, the shear stress fluctuations would be approximately

$$|\tau_{xw}| \approx |p_w| \approx |\tau_{zw}| \qquad \text{(A25) (shear waves)}$$

For two-dimensional shear waves, for comparison, this would be the exact result. For pressure waves, the relations would be different

$$\left.\begin{aligned} |\tau_{xw}| &= 0.032\ |p_w| \\ |\tau_{zw}| &= 0.15\ \ |p_w| \end{aligned}\right\} \qquad \text{(A26) (pressure waves)}$$

provided that the wavelengths of both waves are the same. It is likely, however, that the typical wavelength would be rather larger [2], and this would make the shear stresses generated by pressure waves even smaller. Experimental results show that the pressure wave RMS values are one order of magnitude higher that the RMS values of the shear stress fluctuations [2]. Thus, the pressure fluctuations are clearly dominated by our pressure mode waves, the p_1 type waves of eq. (A6). There is no phase shift in the y direction, leading to phase planes perpendicular to the surface ("pressure potatoes"). On the other hand, one cannot decide clearly from the present calculations that the shear stress is dominated only by shear waves. The shear waves do have a significant phase shift in the y direction originating from the "heat conduction" type of the original equations. There is evidence for this phase delay from experiments by Eckelmann [3].

Finally, we want to show, that the Stokes–wave approximation is valid for an estimation of the cross flow distribution as assumed in section 1.2. We have an exact equation for the shear wave case to compare with

$$\begin{aligned} w = \ & w_{30}\left(e^{+\sqrt{\alpha^2+\beta^2-i\omega/\nu}\,y} - e^{-\sqrt{\alpha^2+\beta^2-i\omega/\nu}}\right) + \\ & +\vartheta\beta\left(e^{+\sqrt{\alpha^2+\beta^2-i\omega/\nu}\,y} - e^{-\sqrt{\alpha^2+\beta^2}\,y}\right) \end{aligned} \qquad \text{(A27)}$$

with ϑ given by eq. (A21). For simplicity, the wave coefficient $e^{i(\alpha x+\beta z)-i\omega t}$ is omitted. With the data from Morrison et al. [1] we obtain

$$\vartheta\beta = (1+i)\,(0.044\,u_{30} + 0.20\,w_{30}) \tag{A28}$$

and $\omega/\nu \gg \alpha^2+\beta^2$ or $\omega^+ \gg \alpha^{+2}+\beta^{+2}$, namely $\omega^+=0.1$ and $\alpha^{+2}+\beta^{+2}=0.0023$. Thus, a Stokes–type flow with

$$w = w_{30}(e^{+\sqrt{-i\omega/\nu}\,y} - e^{-\sqrt{-i\omega/\nu}\,y}) \tag{A29}$$

(compare eq. (6)) is a reasonable approach.

References

[1] W.R.B. Morrison, K.J. Bullock and R.E. Kronauer: "Experimental evidence of waves in the sublayer". J.F.M. 47 (1971), pp. 639–656.

[2] W.K. Blake: "Mechanics of flow–induced sound and vibration". Academic Press, Orlando, 1986.

[3] H. Eckelmann: "Experimentelle Untersuchungen in einer turbulenten Kanalströmung mit starken viskosen Wandschichten". Mitteilungen aus dem Max–Planck–Institut für Strömungsforschung und der AVA Göttingen, Nr. 48 (1970).

Drag Reduction in Pipes Lined with Riblets

K. N. Liu*, C. Christodoulou, O. Riccius**, D. D. Joseph*****
The University of Minnesota, Minneapolis, MN

Theme

In the present paper, experiments are reported establishing a maximum drag reduction of five to seven percent in fully developed turbulent flow of water through 25.4mm and 50.8mm diameter pipes lined with a film of grooved equilateral triangles of base 0.11mm. The maximum reduction occurs when the height of the riblets is 11 to 16 wall units. This correlates well with the Taylor microscale of the fluctuating velocity gradient.

Contents

There is a large literature about drag reduction using riblets in turbulent boundary layer flow over flat plates. Some of the earliest and more important results were obtained by Walsh[1,2,3]. He showed that drag reduction could be obtained when the height of the riblet structure expressed in wall units $S^+=Su^*/\nu$ is below 30; the maximum of 7–8% occurred when S^+ is about 15. Here S is the height and base of the riblets, u^* is the friction velocity and ν is the kinematic viscosity. He also found that triangular grooves are among the most effective in reducing drag.

---

* Visiting scholar from the People's Republic of China (Assoc. Prof.)

** Graduate student, Department of Aerospace Engineering and Mechanics

*** Professor, Department of Aerospace Engineering and Mechanics

A. Gyr (Editor)
Structure of Turbulence and Drag Reduction
IUTAM Symposium Zurich/Switzerland 1989

Less is known about the effect of riblets on drag reduction in pipe flow. Nitschke[4] studied air flow in a pipe with rounded peaks and flat valleys machined into the pipe surface. A maximum drag reduction of 3% was measured using pressure drop measurements over a length of 120 pipe diameters. Drag reduction was obtained when the riblet spacing was between 8 and 23 wall units with the maximum in the neighborhood of 11 to 15.

The test section of our experimental apparatus consisted of two pipes in series: the test pipe and the control pipe. The test pipe was lined with 0.11mm riblet film while the control pipe was either a smooth PVC pipe or a pipe lined with smooth film. The flow was fed by a gravity feed tank which maintained a constant head of 11.6m. The water flowed from the head tank first through a 7.6cm (3") diameter pipe, then turned in a 15.2cm (6") elbow toward the test sections. The large elbow helped to damp unwanted eddying before the flow entered the test section. The distance to the test section was 2.13m (7 ft.) or 84 d, where d is the pipe diameter. This large L/d ratio appears to suffice for achieving fully developed flow. We say that a flow is fully developed if it gives rise to a linear pressure gradient and passes the interchange tests discussed in the paragraph following equation (3) below.

The two pipes that constitute the test section were in series, each equipped with 4 pressure taps at equal distances. A flowmeter was placed downstream from the test section and a gate valve, to control the flow, followed the flowmeter. Both the flowmeter and the gate valve were located far enough from the test section to avoid any backflows or any effects on the pressure measurements.

Flowrates, pressure drops and temperatures of the water were measured during the experiments. The pipes used were PVC, smooth 50.8mm (2") and 25.4mm (1") diameter and 3.05m (10 ft.) long. For technical reasons, only 1.5m were lined with film in the 25.4mm (1") case and 2.4m in the 50.8mm (2") case. The fabrication of good pressure holes was the most demanding part of the project. Poor holes lead to incorrect measurements. The pressure holes made in the unlined pipe have sharp corners and are free of burrs. It was more difficult to get good holes in the lined pipes. Counter pressure from inside the pipe was applied when drilling to prevent the film from separating from the PVC. After drilling with an end mill, the holes were trimmed of film debris and reamed with a dentist's end reamer. They were repeatedly trimmed with the reamer until constant pressure gradients were achieved.

We shall designate the Darcy friction factor by

$$f = \frac{\Delta P}{\rho g} \frac{2g}{U^2} \frac{d}{L}. \tag{1}$$

where ΔP is the pressure drop over the length L of pipe, g is gravity, d is the pipe diameter and U is the average flow velocity. An effective riblets diameter was defined by

$$d_r = \sqrt{\frac{4A}{\pi}} \tag{2}$$

where A is the cross-sectional area of the pipe lined with riblets.

The measured values of the friction factor for the smooth unlined pipes and the pipes lined with smooth film were compared with the values given by

$$f = \left[1.8 \log_{10} \left(\frac{Re}{6.9}\right)\right]^{-2} \tag{3}$$

where the Reynolds number $Re = dU/\nu$, which is an excellent approximation of Prandtl's formula for the range of our experiments. In computing Re, the measured values of the volume flow rate Q and the various diameters were used. There was found to be a quite good agreement with average differences of about 1 percent with a maximum difference of about 3 percent.

During the experiments one section was smooth (lined or unlined) and the other lined with riblets. Each experiment was carried out twice but with test sections interchanged. In other words, the first time the smooth pipe was downstream with the riblet pipe upstream, and the second time the riblet pipe was downstream and the smooth pipe upstream. The purpose of the interchange was twofold. First, it ensured that our data were repeatable. Second, it ensured that both sections were in the fully-developed-turbulent-flow region, since the pressure drop was not affected by the position of the pipe.

The comparison between drag in smooth pipes and pipes lined with riblets is shown in Figure 1. In these experiments S is fixed but S^+ varies. The data do not determine a lower limit for drag reduction, although it suggests that it must be around

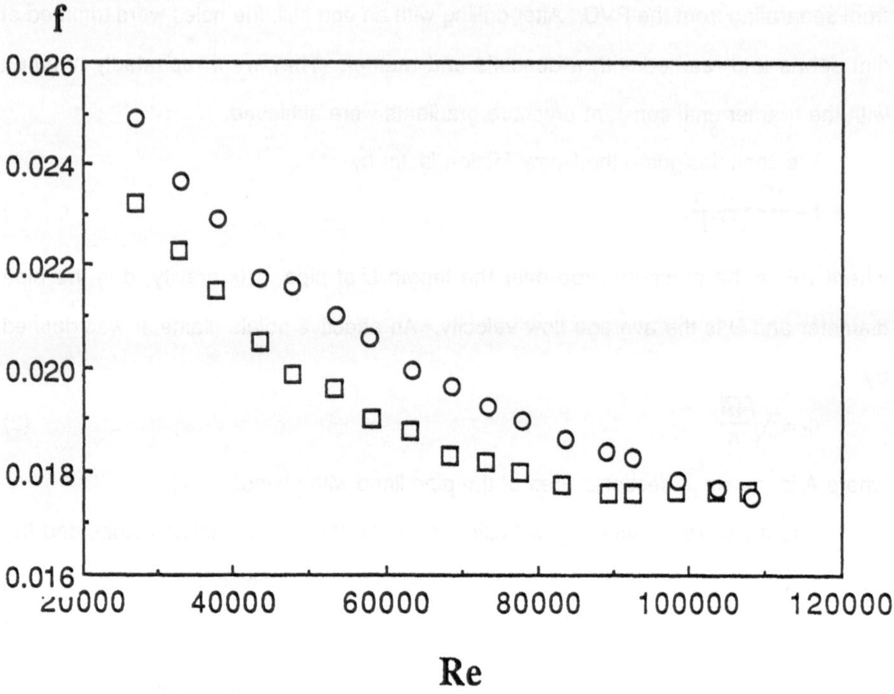

Figure 1: Comparison of the measured values of the friction factor in the section with the smooth film with the 25.4mm diameter pipe lined with riblets.

□ Riblet film

○ Smooth pipe

$S^+=3$. The largest value of S^+ for which drag reduction was achieved is approximatoly 23. After this, at larger speeds with $S^+>23$, riblet linings lead to a drag increase. Nearly identical results have been reported by Nitschke[4] in a study of air flow in pipes with grooved walls.

The maximum drag reduction occurs for $S^+\approx 11\sim 13$ (see Figure 2), in excellent agreement with previous investigations for pipe flow and boundary layers. The maximum drag reduction was between 5 and 7%.

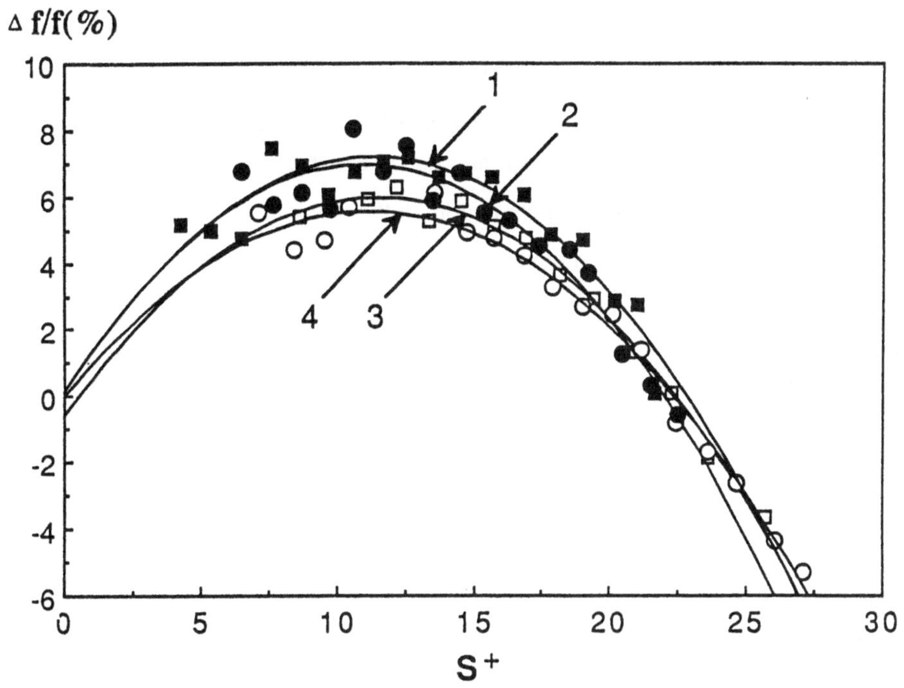

Figure 2: Percent drag reduction due to riblets:

Line 1:	25.4mm pipe lined with smooth film	●
Line 2:	50.8mm pipe lined with smooth film	■
Line 3:	25.4mm smooth unlined pipe	○
Line 4:	50.8mm smooth unlined pipe	□

It seems not to have been noted before* that the Taylor microscale λ in the spanwise direction, determined from the quadratic approximation of the correlation function

$$R_{xx}(z) = 1 - z^2 / 2\lambda^2 \qquad (4)$$

where x is the streamwise and z the spanwise coordinate, gives rise to $\lambda \sim S^+$ where

$$\lambda^2 = \overline{u_y^2} / \left(\overline{\frac{\partial u_y}{\partial z}}\right)^2 . \qquad (5)$$

---

* Falco[5] appears to be the only other reference to mention Taylor microscales and drag reduction. He relates the microscales to pocket scales (his Fig. 17) and the pocket to riblet scales, without recognizing the importance of the spanwise microscale or the values $S^+=12\pm2$. The pocket scales do not correlate with drag reduction.

This microscale can be viewed as a spanwise correlation length for the fluctuating wall shear stress on smooth walls. Finnicum and Hanratty[6] have shown that the data from the experiments of eight different authors give rise to $\lambda=12\pm2$.

Remarkably, this λ is also near the value of S^+ which maximizes drag reduction with riblet linings in our experiments and in all the many other experiments on drag reduction due to streamwise grooves. Perhaps this is a striking result both for drag reduction and the determination of important scales for sublayer turbulence. Certainly, the appearance of the same correlation length, about 12, in two groups of many experiments of very different types ought not to be dismissed out of hand. It is also of interest that maximum production of turbulent energy $-\overline{uv}dU/dy$ peaks at $y^+\sim15$ in fully-developed pipe and channel flows[7] and that the minimum spanwise distance between sensors which is required to sense structure in turbulence is reported to be 11 wall units[8].

Acknowledgements

We would like to express our thanks and appreciation to Mr. F. Marentic of the 3M Company for his valuable assistance on riblet technology. We also are grateful to K. R. Sreenivasan for useful comments on an earlier draft of this paper.

This work was supported by the Department of Energy, the National Science Foundation, and the U. S. Army. The work of K. N. Liu was supported by the People's Republic of China. The work of O. Riccius was supported by the Graduate School of the University of Minnesota.

References

1 Walsh, M. S., "Drag characteristics of V-groove and transverse curvature riblets", in "Viscous flow and drag reduction", G. R. Hough, editor, Progress in Astronautics and Aeronautics, Volume 72, presented at the AIAA Symposium on Viscous Drag Reduction, Dallas, Texas, November 1979.

2 Walsh, M. S., "Riblets as a viscous drag reduction technique", AIAA Journal **21**, No. 4, p. 485, 1983.

3 Walsh, M. S., and Lindemann, A. M., "Optimization and application of riblets for turbulent drag reduction", AIAA Paper 84-0347, presented at the 22nd AIAA Aerospace Science Meeting, Reno, Nevada, January 1984.

[4] Nitschke, P., "Experimental investigation of the turbulent flow in smooth and longitudinal grooved pipes", Max-Planck Institute für Strömungsforschung, Göttingen, West Germany, 1983.

[5] Falco, R. E., "New results, a review and synthesis of the mechanism of turbulence production in boundary layers and its modification", AIAA Paper 83-0377, January 1983.

[6] Finnicum, D. S., and Hanratty, T. J., "Turbulent normal velocity fluctuations close to a wall", Phys. Fluids **28**, No. 6, p. 1654, 1985.

[7] Sreenivasan, K. R., "A unified view of the origin and morphology of the turbulent boundary layer structure", Proceeding of the Symposium on Turbulence Management and Relaminarisation, Bangalore, India, 1981, pp. 37–61.

[8] Simpson, R. L., "An Investigation of the Spatial Structure of the Viscous Sublayer", Max-Planck Institute für Strömungsforschung, Göttingen, **118**, August 1976.

Drag Reduction Mechanisms and Near-wall Turbulence Structure with Riblets

Kwing-So Choi

British Maritime Technology
Teddington, Middlesex, TW11 8LZ, UK.

Abstract

Hot-wire and film measurements were carried out in low-speed wind tunnels to study the mechanisms of turbulent drag reduction by riblets. The results seem to suggest that the drag reduction is mainly due to the "structural" changes in the near-wall boundary layer caused by the riblets, which inhibit the sideways movement of the hairpin legs during the "near-wall bursts". The turbulence intensities in all components and Reynolds stress were reduced by up to 10 and 20%, respectively over the riblet surface.

1. Introduction

It has been known for some time that riblets are effective in reducing skin-friction drag of the turbulent boundary layer and there have been a number of studies carried out to identify some of the benefits under different flow conditions. For example, compressibility and Reynolds number effects on the riblets were investigated in flight (McLean et al., 1987) and in a high-speed tunnel (Squire and Savill, 1987). Demonstration of the effectiveness of the passive device over a double-curvature surface with non-zero pressure gradient was given by Choi et al. (1987) using Hoechst U-grooves in a towing tank. Later on, they extended the experiment to test a combination of two drag-reducion methods, the riblets and polymer coating (Choi et al., 1989,a). Studies of the effects of pressure gradient are presently being carried out at several institutions across Europe (Choi and Johnson, 1988; Coustols and Savill, 1989). For a summary of the recent development in riblets for turbulent drag reduction, readers are referred to the review papers by Wilkinson et al. (1987) and Savill et al. (1988).

As for possible mechanism of drag reduction by the riblets, there are basically two schools of thought. One school advocates that it is merely a viscous phenomenon (Vukoslavcevic et al., 1987), linking the drag reduction with the change in mean velocity profile. Choi (1989,b), on the

A. Gyr (Editor)
Structure of Turbulence and Drag Reduction
IUTAM Symposium Zurich/Switzerland 1989

other hand, presented experimental results to suggest that the drag reduction is due to the changes in the near-wall turbulence structure. He argued that the restriction of spanwise movement of longitudinal vortices by the riblets would have a prime responsibility for the drag reduction. Bacher and Smith (1985) and Bechert et al. (1986) also suggested similar mechanisms independently, emphasising an importance of impeding the cross-flow by the longitudinal ribs for the drag reduction.

The purpose of this paper is to reinforce the argument of the latter school and to develop it further by presenting some of the new experimental evidence that the turbulence drag reduction by the riblets essentially stems from the structural changes in the near-wall boundary layer.

2. Experiments

The experiments were carried out at BMT's environmental wind tunnel at Teddington and RAE's boundary layer tunnel at Bedford. The tests were conducted at zero-pressure gradient condition with a free-stream velocity of 3m/s. Details of the experimental set-up and procedure together with the information on the riblets used for the tests can be found in Choi (1989,b) and Choi and Johnson (1988). Basically, two sets of experiments were carried out over a smooth and riblet surface, and the structural differences between the two cases were documented. For most of the measurements, DISA miniature hot-wire probes and TSI surface-flush mounted hot-film sensors were used in a constant-temperature mode. The Reynolds number of the tests based on the momentum thickness was 4.6×10^3 and 2.7×10^3, respectively.

3. Results and Discussion

Figure 1 compares the mean velocity profiles over the smooth and riblet surface plotted in a log-law format. The thickening of the viscous sublayer by the riblets is evident from the upward shift of the log-law, which is coupled with the reduction of turbulence activities near the wall region. Figure 2 shows a ten percent reduction of the longitudinal turbulence intensity over the riblet surface. Vertical (u') and spanwise (w') turbulence intensities are also reduced over the riblet surface by a similar amount in the near-wall region, say within 70 wall units, as shown in Figure 3. The Reynolds stress is reduced by nearly twice as much as the turbulence intensities (see Figure 4). This may be expected since the Reynolds stress is proportional to the velocity squared, while the turbulence intensities are linear in velocity.

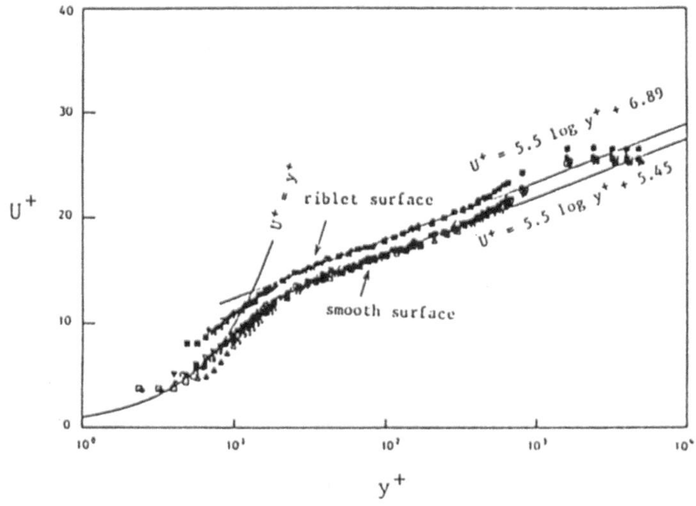

Fig. 1 : The log-law plot
of mean velocity profiles

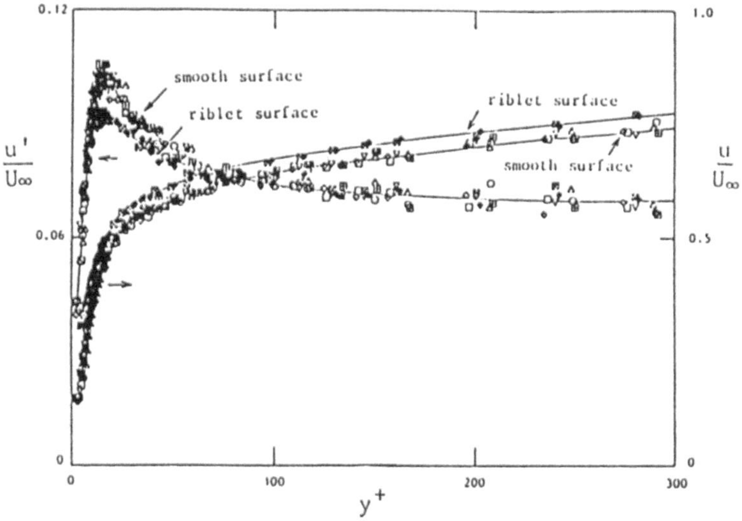

Fig. 2 : The longitudinal turbulence
intensity and mean velocity profiles

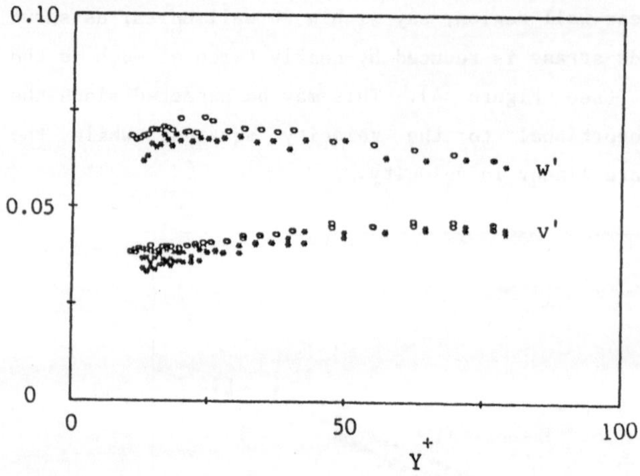

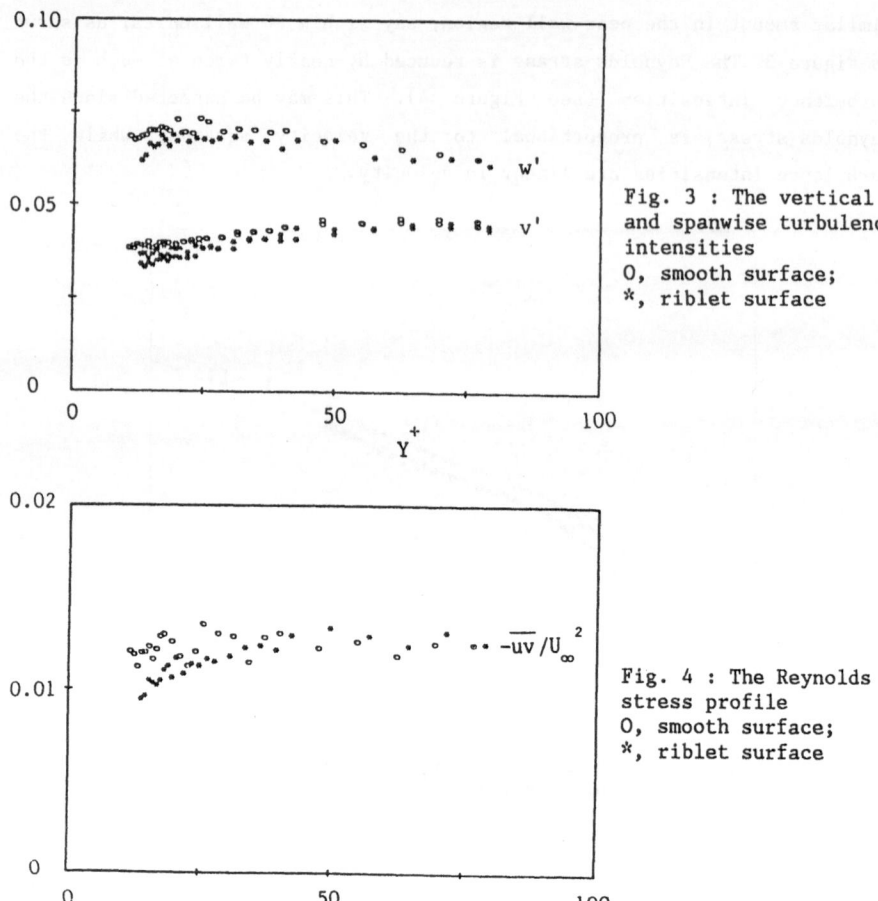

Fig. 3 : The vertical and spanwise turbulence intensities
O, smooth surface;
*, riblet surface

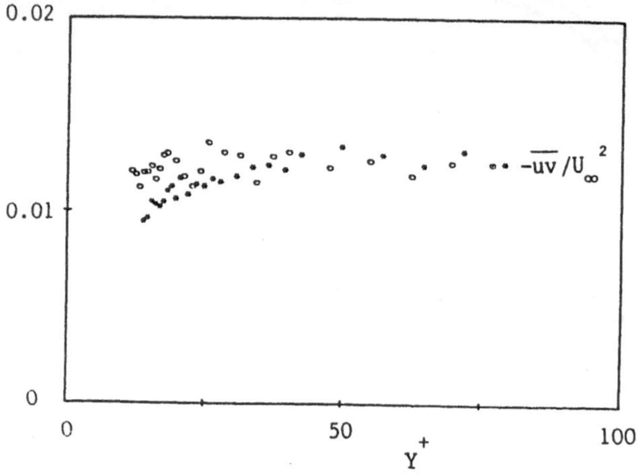

Fig. 4 : The Reynolds stress profile
O, smooth surface;
*, riblet surface

The changes in the near-wall turbulence structure are also shown in the probability density of longitudinal velocity fluctuation (u) in Figures 5(a)-(c), and of wall-shear stress fluctuation (τ_w) in Figure 5(d). It is clear from these figures that the riblets seem to increase the probability in the positive tail, effectively making the skewness more positive. It should be noticed that the probability densities over the smooth and riblet surface are already nearly identical at and above y^+=16, which is very close to the position of maximum turbulence intensity (Choi,1989,b) and the maximum reduction. Also noticed is a systematic change in the shape of probability density with height. The skewness of u seems to become from positive to negative with height irrespective of the surface condition. As is expected that the probability density of τ_w is very similar to that of u close to the wall surface (y^+=7).

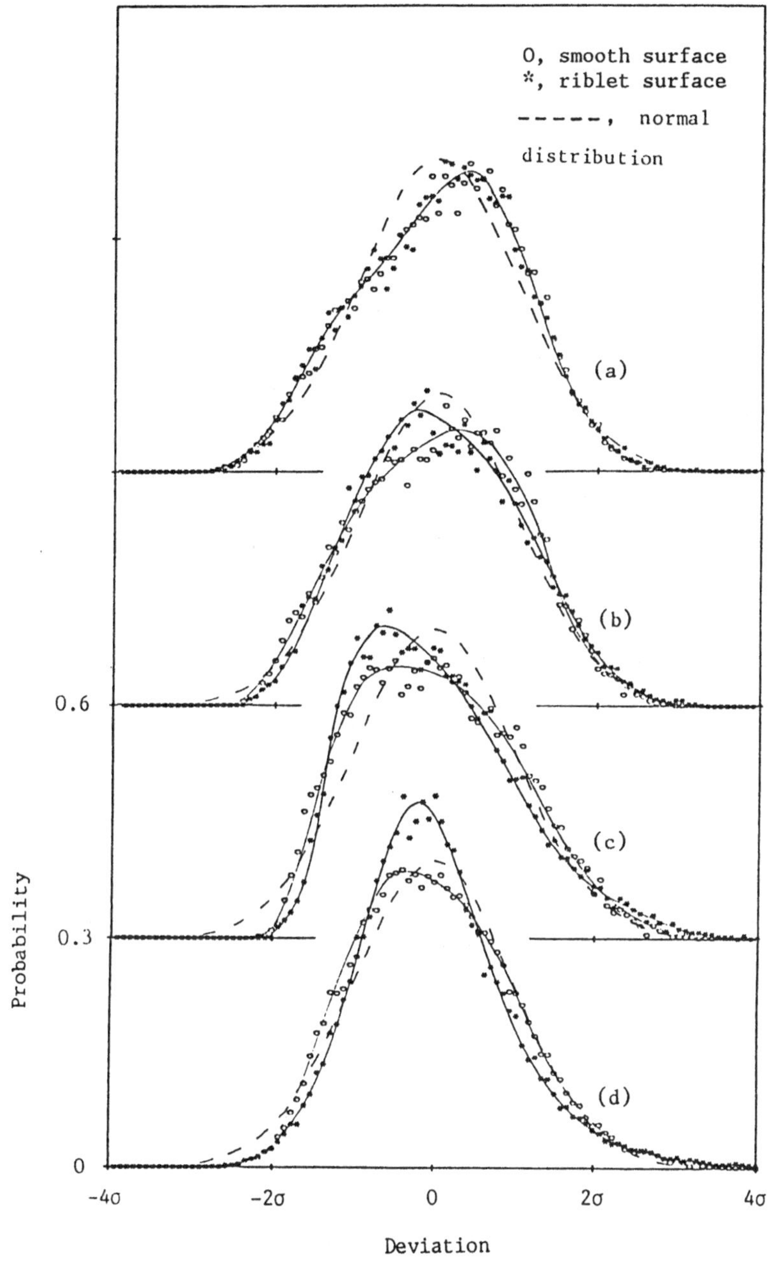

Fig. 5 : The probability density of u at y^+=16(a), 11(b) and 7(c), and of τ_w(d)

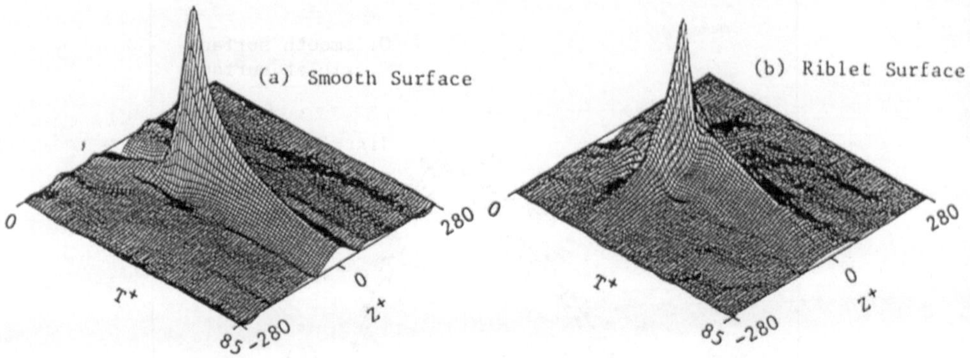

Fig. 6 : The conditionally averaged signatures of wall-skin friction during the near-wall bursts over the smooth (a) and riblet surface (b)

Figures 6(a) and (b) are the conditionally sampled and ensemble averaged signatures of wall-skin friction during the "near-wall bursts" over the smooth and riblet surface, respectively, where the "near-wall bursts" are the events taken place in the near-wall region of the turbulent boundary layer which are associated with a sharp increase of wall-skin friction fluctuation (Choi, 1989,b). As it is believed that a large part of the turbulent skin friction is produced during the near-wall bursts, the changes in the turbulence structure by the riblets during the events should reveal a clue to the mechanisms of the drag reduction. It is observed in the figures that the duration of the bursts is shorter and the spanwise correlation length larger for the flow over the riblet surface. The frequency of the near-wall bursts, which is indicated by the tail probability of the density function, is increased by the riblets as shown in Figure 7. The energy spectrum of the wall-skin friction fluctuation measured at the valley of the riblets (Figure 8) indicates a substantial reduction of the energy by the riblets at medium to low frequency range. This suggests that the "strength" of the near-wall bursts is reduced substantially.

All of these results and those of conditional flow visualisation by Choi (1986) suggest the following conceptual model of the burst sequence which is shown in Figure 9. In the Stage 1 of the model, the concentrated vortex filament imbedded in the boundary layer is deformed by the locally large velocity field created by the near-wall bursts. The vortex filament then

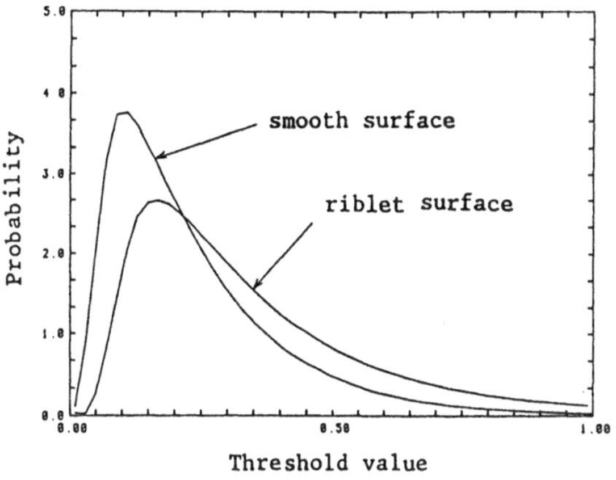

Fig. 7 : The probability density of the occurrence of the near-wall burst

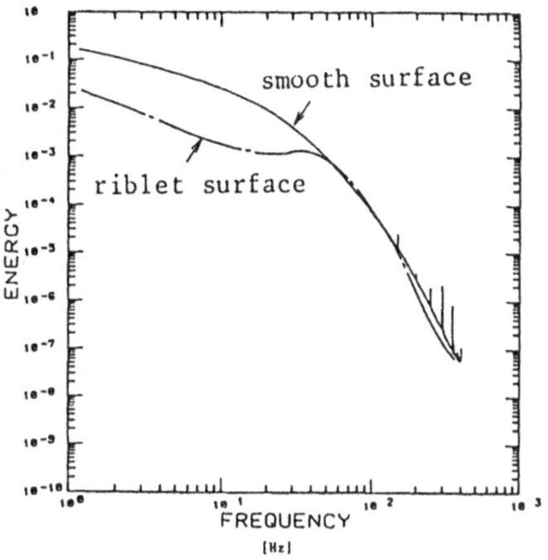

Fig. 8 : The energy spectrum of the wall-skin friction fluctuation

develops by the self-induction mechanisms (Stage 2) into hairpin loops, which then eject away from the wall surface (Stage 3). At this last stage, the neighbouring legs of the loops form pairs of counter-rotating longitudinal vortices. Downwash of high-momentum fluid then takes place between each of these pairs, producing a large wall-shear stress momentarily. This is the event which we call the near-wall burst. The vortex pairs are also shown on the left of Stage 1, thus completing a burst cycle.

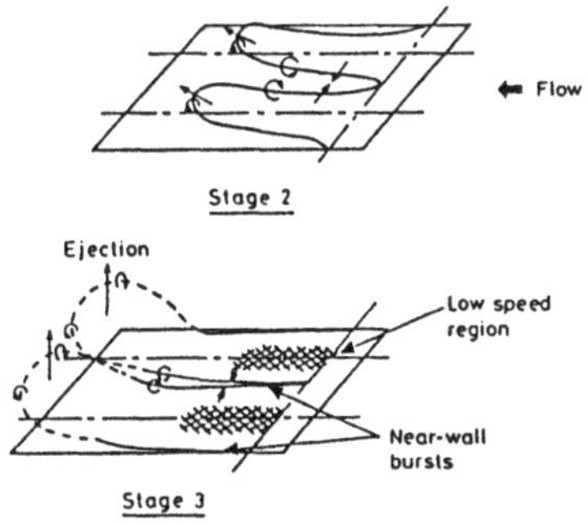

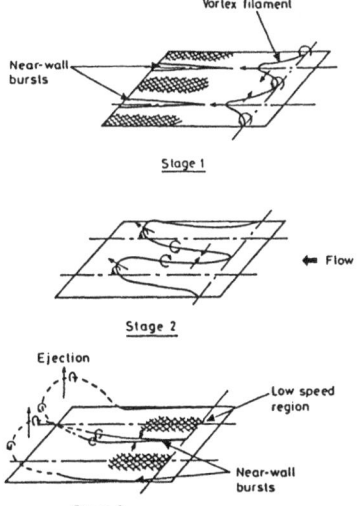

Fig. 9 : Conceptual model for the sequence of the burst event

4. Concluding Remarks

It has been shown that the effects of riblets are to thicken the viscous sublayer and to reduce the turbulence intensities in all components and the Reynolds stress. The skewness of U and τ_w are also affected by the riblets to become more positive. However, the height of influence is different for different quantities. For example, the turbulence intensities and Reynolds stress are affected up to about 70 wall units, while the skewness does not seem to be changed at over 16 wall units.

As far as the drag reduction mechanisms are concerned, riblets would work just like small fences to inhibit the lateral turbulence activities near the wall surface. The sideways movement of the hairpin legs during the near-wall bursts are prevented by the riblets leading to "premature" bursts. When these happen, the intensity of the bursts becomes less, the duration shorter, and the spanwise correlation length greater as the bursts take place prematurely before the legs come very close to each other. The frequency of the bursts is increased over the riblet surface.

This work has been supported by British Aerospace plc and the Department of Trade and Industry. U.K.

5. References

1) Bacher, E.V. and Smith, C.R. (1985) AIAA Paper No. 85-0548.
2) Bechert, D.W. et al. (1986) 15th Congress, Int. Council Aero Sci., London. Paper No. 86-1.8.3.
3) Choi, K.-S. (1986) In: Advances in Turbulence, ed. by Mathieu, J. and Comte-Bellot, G., Springer-Verlag.
4) Choi, K.-S. et al. (1987) Proc. Int. Conf. Turbulent Drag Reduction by Passive Means, London.
5) Choi, K.-S. and Johnson, R. (1988) Proc. 2nd European Turbulence Conf., Berlin.
6) Choi, K.-S. et al. (1989, a) To appear in Appl. Sci. Res.
7) Choi, K.-S. (1989, b) To appear in J. Fluid Mech.
8) Coustols, E. and Savill, A.M. (1989) To appear in Appl. Sci. Res.
9) McLean, J.D. et al. (1987) Proc. Int. Conf. Turbulent Drag Reduction by Passive Means, London.
10) Savill, A.M. et al. (1988) J. Theor. Appl. Mech. 7, 353.
11) Squire, L.C. and Savill, A.M. (1987) Proc. Int. Conf. Turbulent Drag Reduction by Passive Means, London.
12) Vukoslavcevic et al. (1987) Proc. Int. Conf. Turbulent Drag Reduction by Passive Means, London.
13) Wilkinson, S.P. et al. (1987) Proc. Int. Conf. Turbulent Drag Reduction by Passive Means, London.

The Bursting Process over Drag Reducing Grooved Surfaces

A.D.Schwarz–van Manen, J.H.H.Thijssen, C.Nieuwvelt and K.Krishna Prasad.
The Laboratory for Fluid Mechanics and Heat Transfer, Faculty of Technical Physics, Eindhoven University of Technology, Eindhoven, The Netherlands.
F.T.M.Nieuwstadt
Laboratory of Fluid Dynamics, Faculty of Mechanical Engineering and Maritime Technology, Technical University of Delft, The Netherlands.

1 Introduction

Drag reduction by microgrooves in the streamwise direction has been confirmed by measurements in several laboratories in the last ten years (Pulles 1988). However, the mechanism for this behavior is as yet unclear. Various heuristic arguments have been proposed to explain the drag reducing behaviour of these surfaces. In this paper we shall restrict our attention to one argument based on the hypothesis that the bursting process, which is assumed to contribute to a significant proportion of turbulent production, is less vigorous near the grooved wall than that near a smooth wall. We present results on this aspect over a grooved and a smooth wall. In addition we look at the statistical properties of the different quantities that are used to describe a burst and also conditional averages of the fluctuating signals during the bursting process, since these are expected to be more relevant.

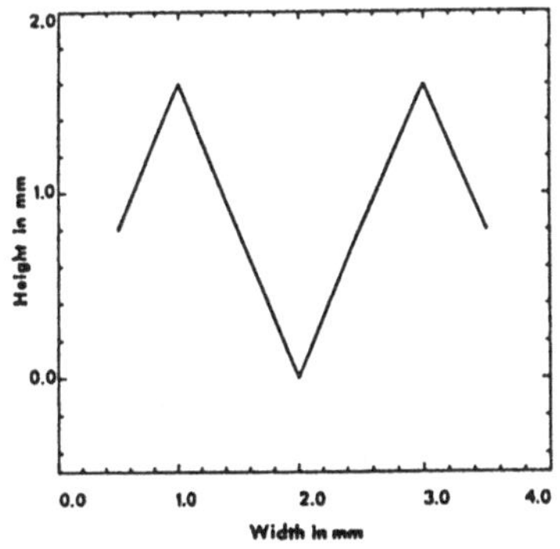

Fig.1

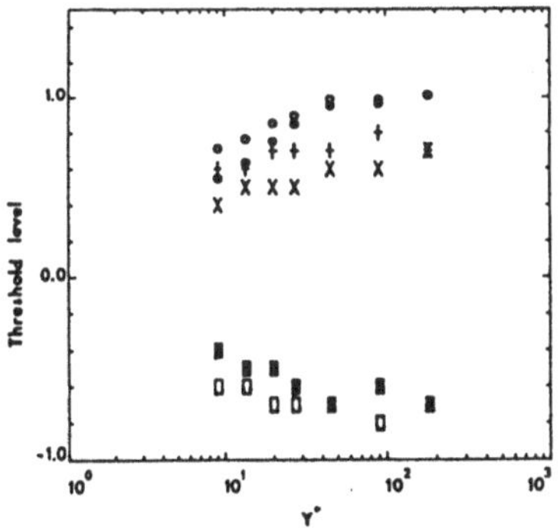

Fig.2

Fig.1 The profile of the grooved wall
Fig.2 The threshold levels versus the height y^+ above a smooth and a grooved wall. H_2 smooth o , grooved ● ; h_u smooth □ , grooved ■ ; h_v smooth + , grooved × .

A. Gyr (Editor)
Structure of Turbulence and Drag Reduction
IUTAM Symposium Zurich/Switzerland 1989

2 Equipment and measurement method

The experiments were done in the low speed water channel of the Laboratory for Fluid Mechanics and Heat Transfer, Faculty of Physics at the Eindhoven University of Technology. A detailed discription of the instrumentation and the measurement method has been given by Pulles (1988). The advantage of this channel is that the boundary layer is quite thick, so that it is easy to do experiments at low values of y^+. The measurements were done at several heights above a smooth and a streamwise grooved wall (Fig.1),at a distance of 2.90 m from the tripping wire. The flow conditions were: free stream speed $U_0 = 0.20$ m/s; at a temperature of 20 ^{0}C, the Reynolds number $Re_\theta =$ 1340; the length scale $l^+ = 0.11$ mm;and the time scale $t^+ = 0.0125$ s. A Laser Doppler Anemometer measured the streamwise (U) and the normal (V) components of the velocity.

3 Results

For both walls the profiles of mean quantities (mean velocities, rms values of the fluctuating components and the Reynolds stress) showed small differences in agreement with the results of Pulles(1988).

The initial efforts were concentrated on the analyses of u and v signals by means of the VITA detection technique (Blackwelder and Kaplan 1976). This did not give a satisfactory result. We therefore chose the quadrant detection technique of Lu and Willmarth (1973) refined by Comte–Bellot et al. (1978) and Bogard and Tiederman (1986). This technique was chosen because it had a high probability of detecting a correct event and a low probability of detecting a wrong event as demonstrated by Bogard and Tiederman. The absolute values of the fractional distribution for all the quadrants increased with decreasing height for both the walls. The values for both walls are the same for $y^+ > 40$. Further discussion in this paper will be restricted to the second quadrant (ejections).

3.1 Threshold levels

The threshold levels for the second quadrant were determined from the data according to the procedure described by Comte–Bellot et al. The procedure involves three separate threshold levels: the amplitude of uv to be larger than an arbitrarily fixed threshold H; the amplitude of u to be larger than a threshold h_u; the amplitude of v to be larger than a threshold h_v; the latter two being used while investigating the amplitude distributions of u and v signals during ejections. Application of these ideas to the signal led Comte–Bellot et al to define a characteristic threshold level according to

$$H_2 = \frac{1}{n} \sum_{i=1}^{n} \frac{(uv)_{2,i}}{uv} \tag{1}$$

In Fig.2 these threshold levels are plotted as a function of height y^+ for both the walls. (Fig.5a shows the threshold levels in the second quadrant in a qualitative manner). The threshold level H_2 increases as y^+ increases till $y^+ = 40$ and thereafter it is constant. The threshold level H_2 for the measurements above the grooved wall is smaller than that above the smooth wall for $y^+ < 40$. For higher values of y^+ there is no difference between the grooved and the smooth wall. The results for the smooth wall are in agreement with those of Comte–Bellot et al except they did not report any results for $y^+ < 70$. Thus the present results show that in general the ejections are weaker closer to the wall and in particular they are much weaker above the grooved wall than above the smooth wall. The absolute values of the threshold levels h_u and h_v for the u and v signals are smaller for the grooved wall than for the smooth wall for every height.

With these threshold levels one is able to determine the ejections, their duration and the period between two successive ejections (T_e).

3.2 Determination of τ_{max}

In this paper we use the word " burst" to denote the breakup of a streak. Since the breakup of a streak can involve either a single ejection or mutiple ejections which are closely grouped together Bogard and Tiederman introduced the concept of a cut–off time τ_{max}. Ejections separated by less than τ_{max} are taken to be from the same burst while those separated by greater than τ_{max} are from different bursts. τ_{max} Is determined by

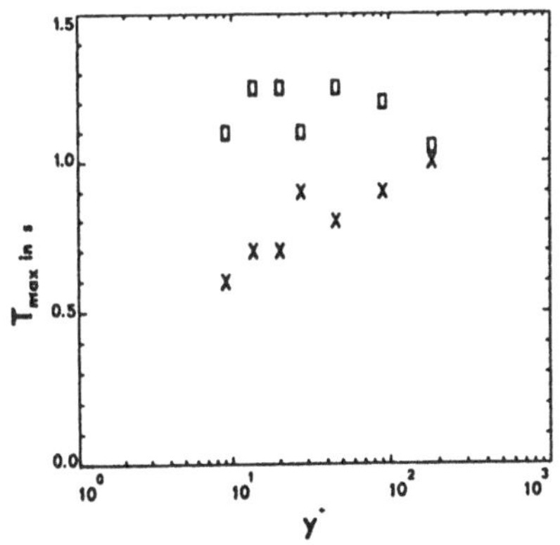

Fig.3

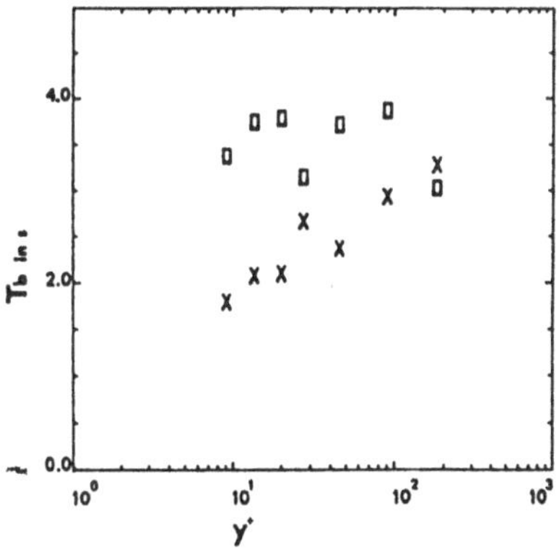

Fig.4

Fig.3 τ_{max} versus the height y^+ for a smooth × and a grooved □ wall.
Fig.4 The burst period T_b versus the height y^+ for a smooth × and a grooved □ wall

that point where the probability distribution function of T_e crosses the exponential distribution function modified for $T_e \geq T_0$;

$$\left[1 - \exp\left\{ (T_0 - T_e) / (\overline{T}_e - T_0) \right\} \right] \tag{2}$$

where $\overline{T}_e$ is the average period between ejections and T_0 is the average duration of an ejection.

τ_{max} and the burst frequency (T_b^{-1}) are plotted against y^+ for both the walls in figures 3 and 4. We see that τ_{max} for the grooved wall is constant with increasing height, while τ_{max} for the smooth wall increases but still is lower than the τ_{max} for the grooved wall. Now we are able to classify whether a burst is a single ejection burst or a group ejection burst involving two or more ejections. We will discuss all the bursts in general with respect to the burst frequencies and the conditional averages.

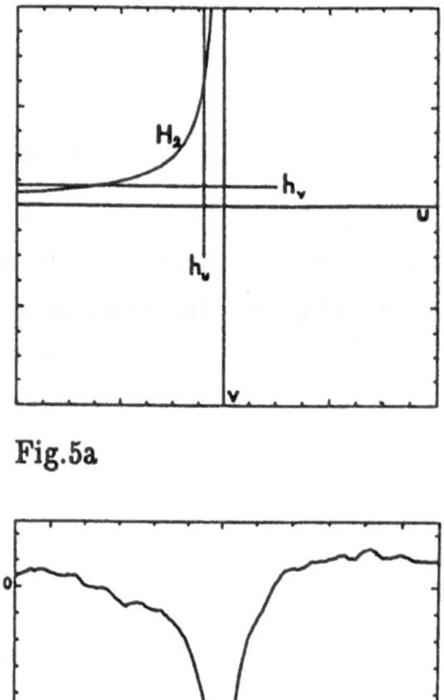

Fig.5a

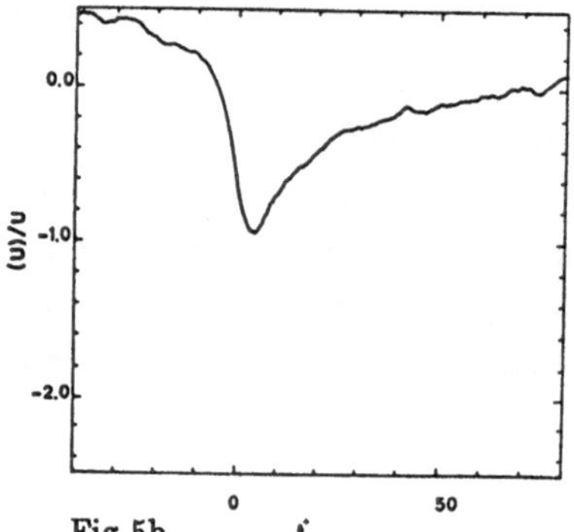

Fig.5b

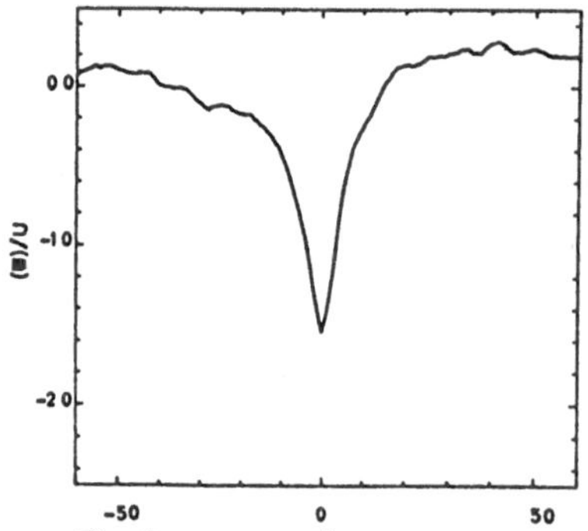

Fig.5c

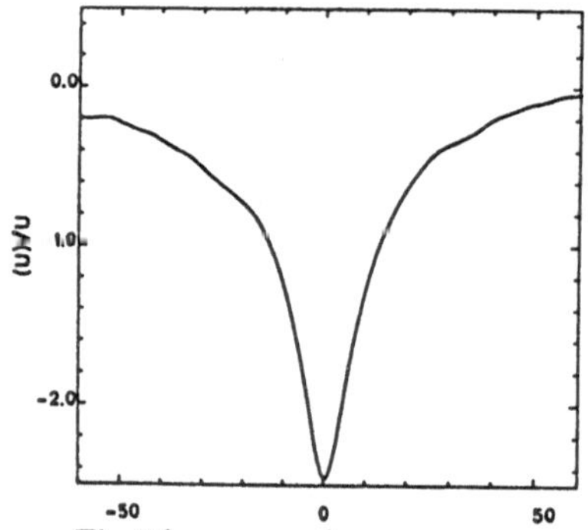

Fig.5d

Fig.5a The threshold levels H_2, h_u and h_v in the second quadrant in a qualitative manner. Fig.5b The peak triggering conditional average. Fig.5c The starting point triggering conditional average. Fig.5d The Euler correlation.

Figure 4 shows that the burst frequency above the grooved wall is reduced compared with that above the smooth wall.

3.3 Conditional averages

To characterize the *uv*, *u* and *v* signals one must form the ensemble averages of the detected events. The first step was to store the starting and the peak positions of the the detected events of the *uv* signal so that conditional averages could be computed by aligning the detected events either at the peak or the starting point of the event (Fig.5b and 5c give examples of this procedure for the *u* signal) .Fig.5d is the Euler correlation (Nakagawa & Nezu 1981) for the *u* signal, which is *independent* of the Euler correlations for the *v* and *uv* signal. It is remarkable to notice the similarity between 5b and 5d thus validating the choice of the peak position of the *uv* signal as the trigger–point.

The conditional averages with the starting position as trigger point do not have any comparison with other detection methods to the authors' knowledge. The conditional averages, which are triggerd at the end position, show some slight

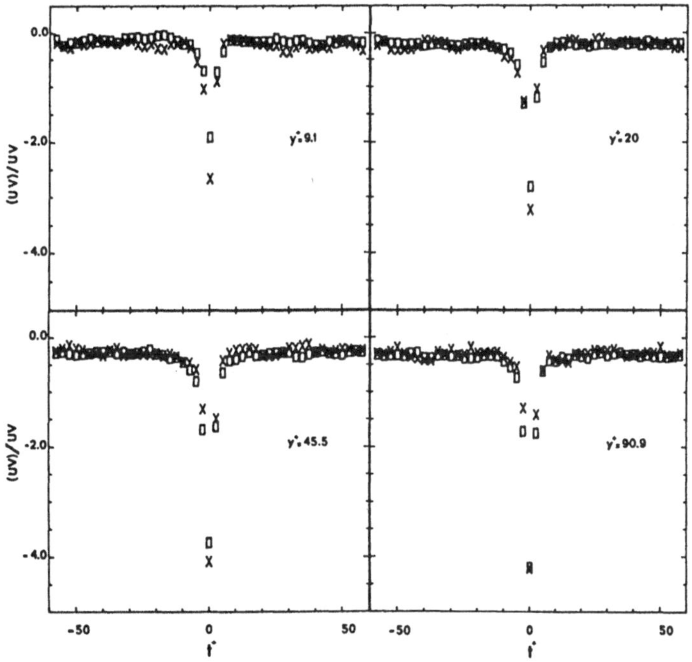

Fig.6a

Fig.6 The peak triggering conditional average for the smooth × and the grooved □ wall for several heights. Fig.6a The *uv* conditional average. Fig.6b The *u* conditional average. Fig.6c The *v* conditional average.

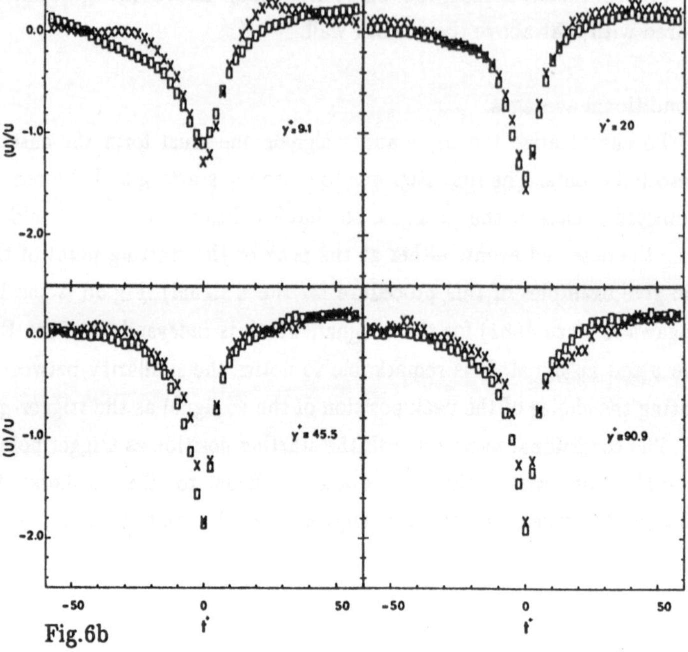

(u)/u
0.0
-1.0
-2.0
y+=9.1
y+=20
y+=45.5
y+=90.9
-50
0
50
t+

Fig.6b

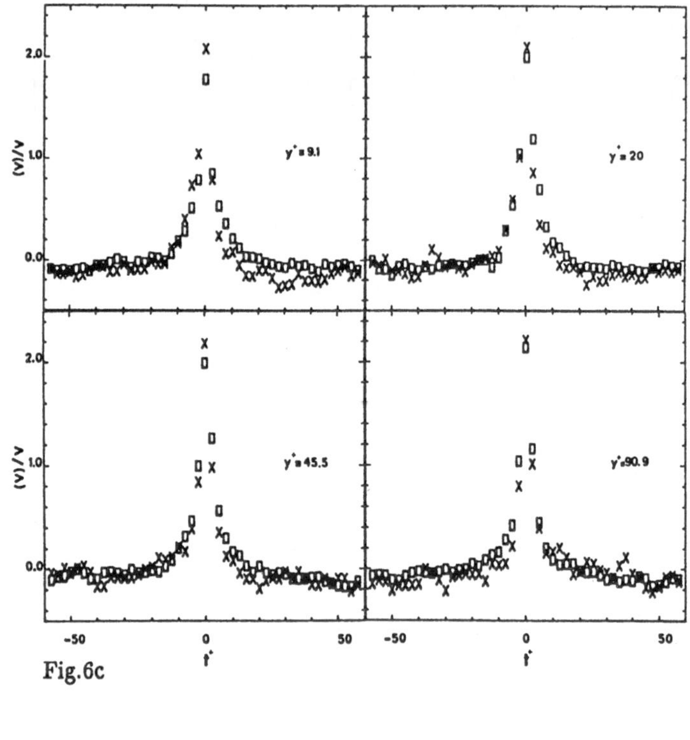

(v)/v
2.0
1.0
0.0
y+=9.1
y+=20
y+=45.5
y+=90.9
-50
0
50
t+

Fig.6c

comparisons with the conditional averages found by Blackwelder and Kaplan; this is currently under investigation.

In Fig.6a–c the conditional averages with the trigger located at the peak position of uv value are plotted for the three signals of both walls for the heights $y^+ = 9.1$, 20.0 , 45.5 and 90.9. These heights were chosen because $y^+ = 9.1$ and 20 are very near the wall and $y^+ = 45.5$ and 90.9 cover a reasonable part of the log–law region.

The uv averages clearly show smaller peaks for the grooved wall compared with those above the smooth wall. These differences gradually reduce as we move away from the wall with virtually no difference being observed at $y^+ = 90$. The v averages show similar tendencies except at $y^+ = 20$ the difference between the two surfaces is smaller. By far the striking differences are noticed with the u averages. Over the grooved plate the deceleration and the acceleration are much more gradual than is the case with the smooth wall particularly near the wall. As we move away from the wall these differences become less pronounced.

For calculating the peak triggered conditional averages for the u, v and uv signal, the peak position of the uv signal is used. Remarkable is that the peaks of these three averages are all at the same time $t^+ = 0$. It could have been possible that the peaks of the u and v signals were shifted to the left or right. Furthermore we draw the same conclusion as Nakagawa and Nezu : "The Reynolds stress shows a pulse–like behavior and the momentum transfer in the vertical direction may be done in a short time".

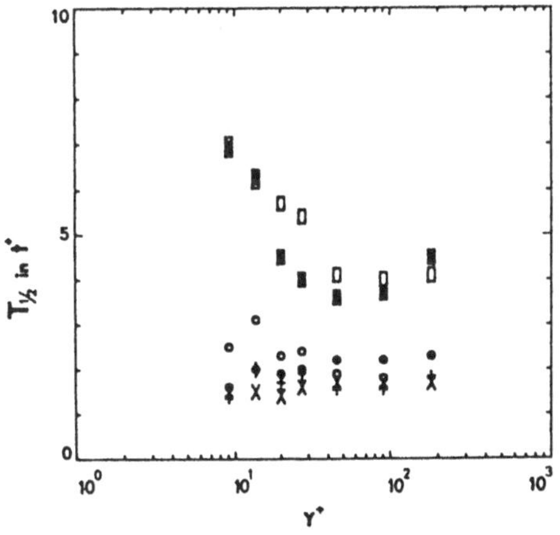

Fig.7a

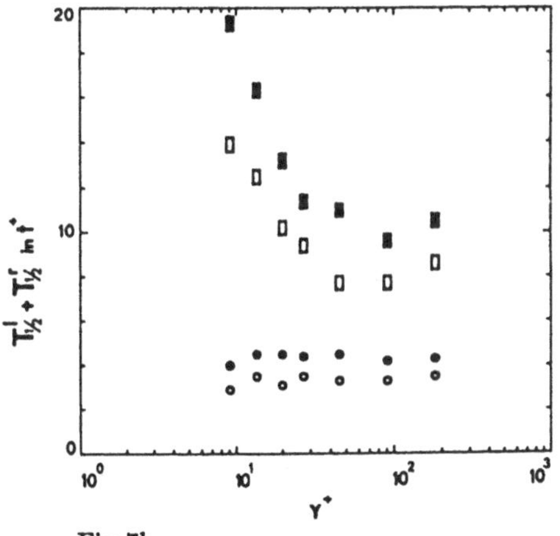

Fig.7b

Fig.7a The width of the left (open symbol) and right (closed symbol) side of the peak for the u □ , v o and uv ×(right) +(left) signals for a smooth wall. Fig.7b The width of the peak for the u □ and uv o signal for the smooth (open symbol) and the grooved (closed symbol) walls.

3.4 Characteristic width of the conditional averages

For the conditional averages we determined the width of the peak by looking at what time the signal is half the size of its peak value. This was done for the left side of the peak ($T_{1/2}{}^{l}$) and for the right side of the peak ($T_{1/2}{}^{r}$). In Fig.7a $T_{1/2}{}^{l}$ and $T_{1/2}{}^{r}$ are plotted for the u, v and uv signals for the smooth wall, and in Fig.7b is plotted the value ($T_{1/2}{}^{l} + T_{1/2}{}^{r}$) for the u and uv signals for both walls.

In Fig.7a one can see that the v and uv signals are sharper than the u signal, and for the uv signal the time of decelaration and accelaration is almost equal.
Fig.7b shows clearly that the duration of an event above the grooved wall is longer than the duration of an event above the smooth wall. For the u signal we see that with increasing height the duration of the event becomes shorter while it stays constant for the uv and v signals which is to be seen in both pictures.

4 Conclusion

The major conclusion to emerge from this study is that the grooved wall in general weakens the bursting process. It not only reduces the frequency of bursting but also weakens the accelerations and decelerations associated with the event. The effect of the grooves does not penetrate much beyond the buffer layer. The τ_{max} and T_b at $y^+ = 27$ seem to show distinctly different behaviour for the grooved and smooth walls. This point is half way the critical layer of Sreenivasan(1988). More measurements are necessary to charracterize the flow in the neighbourhood of this point.

5 References

Blackwelder,R.F. & Kaplan,R.E. 1976 J. Fluid Mech. 76 , 89–112.

Bogard,D.G. & Tiederman,W.G. 1986 J. Fluid Mech. 162 , 389–413.

Comte Bellot,G., Sabot,J. & Saleh,I. 1978 Proc. of the Dynamic Flow conference 1978.

Lu,S.S. & Willmarth,W.W. 1973 J. Fluid Mech. 60 , 481–511.

Nakagawa,H. & Nezu,I. 1981 J. Fluid Mech. 104 , 1–43.

Pulles,C.J.A. 1988 Ph.D. Thesis 1988 , Eindhoven University of Technology.

Sreenivasan,K.R. 1988 Proc. of the Turbulence Management and Relaminarisation, IUTAM Symposium Bangalore, India, 1988.

Riblets in Internal Flows with Adverse Pressure Gradients

Ph. Pulvin and T.V.Truong
Swiss Federal Institute of Technology-Lausanne / IMHEF, Switzerland

Abstract
An experiment has been performed to investigate the influence of the riblets in internal flows with adverse pressure gradients. The skin friction reduction effect due to the riblets was evident through the velocity profiles, the integral parameters and the pressure recovery in weak negative up to large positive pressure gradients. The efficiency domain of the riblets was evidenced in term of their non-dimensional geometric parameters.

1. Introduction

Longitudinally ribbed surfaces have been extensively investigated in the context of drag reduction during the last decade. Most of the published experiments were conducted in a zero pressure gradient flow and have shown a common value of 8% in drag reduction with some defined values for the riblet geometric parameters. The interaction between the ribbed surfaces and the turbulence production is not yet clearly established but different investigators have demonstrated that the rate of turbulent bursts was reduced with the riblets. This stabilization of the wall layer leads to an increase in the thickness of the sublayer and thus to a decrease in the skin friction. The purpose of the present investigation is to extend these findings into the internal flows with longitudinal pressure gradient and in particular in a planar diffuser flow with different geometric configurations; in this diffuser flow, the Clauser pressure gradient parameter is not constant along the channel but, in the other hand, the geometry is perfectly defined.

2. Experimental facility

The experimental set-up shown in figure 1 is a closed loop wind tunnel specially designed to give a stable uniform flow with low turbulence level. The contraction preceded by one honeycomb and two screens has a contraction ratio of 18 and an exit section of $B \times W_1 = 800 \times 117$ (mm^2). A transition trip device (sand paper) installed at 0.5 hydraulic diameter of the inlet section is used all along this investigation.The working section is a two dimensional channel, 2 (m) long with 2 fixed transparent side walls and 2 aluminum flat plates with hydrodynamically smooth surface and adjustable in angle. These plates were replaced by

A. Gyr (Editor)
Structure of Turbulence and Drag Reduction
IUTAM Symposium Zurich/Switzerland 1989

identical ones where different riblets are mounted. The half angle is adjustable symmetrically or non symmetrically up to 10° covering a large working domain. Different traversing mechanisms have been installed at the entrance and the exit of the working section to investigate the flow field with pressure probes and hot wire anemometers; they are equipped with step-by-step motors and controlled by the computer used for the data acquisition, the data reduction and to adjust the upstream flow velocity to a prescribed inlet Reynolds number. Along the diffuser walls and in the spanwise direction, pressure taps are installed to investigate the pressure distribution. Different riblet combinations were tested but only the relevant riblet results will be discussed mainly for the combination 60/60 and 45/100 [3M symmetric "V" shapped riblets with h = s = 0.0060" and 0.0045" and Hoecht riblets with h = s = 0.0100"], the first figure for the first half diffuser lenght and the second for the remaining diffuser.

3. Results and discussion

The smooth walls in the working section are investigated throughout the whole working domain to set up the reference data for the riblet comparison. This investigation covers the pressure distribution, the overall pressure recovery, the flowfield at the entrance and the exit of the working section with around 50 stations in the spanwise direction for different inlet Reynolds numbers and diffuser total angles. This condensed station mesh is necessary since it is not unusual for a so-called 2-dimensional boundary layer to contain a spanwise variation in the momentum thickness θ of the order of the magnitude drop in θ induced by the riblets.

In the figure 2, the mean velocity U/U_e and the rms value of the longitudinal fluctuating velocity $[u_{rms}/U_e]$ profiles are shown as function of the non-dimensional distance from the wall $(y/\frac{W_2}{2})$ at the exit of the working section for 3 diffuser total angles ($2\theta = 0.25^o, 2^o, 4^o$) and the inlet Reynolds number of 3×10^5. At high adverse pressure gradient (4°), the potential core flow is reduced and the turbulence level is slightly affected at this station. In the figure 3, the pressure distribution defined as $C_p(x) = [p(x) - p_{in}] / \frac{1}{2}\rho U_{in}^2$ is plotted as function of the non-dimensional streamwise distance x/L for 3 values of 2θ ($2^o, 4^o, 6^o$) and the same inlet Reynolds number (3×10^5). No difference can be detected between the upper and lower diverging walls.

To examine the riblet effect in this particular pressure gradient flow, their geometric spacing is non-dimensionalized by the following relation $s^+ = \alpha s^+_{inlet} + (1-\alpha)s^+_{outlet}$ to take account of the friction velocity u_τ variation along the channel. α is a weighted coefficient obtained by numerical simulation, and $s^+_{in,out}$ scaled with the values $u_{\tau,in,out}$ measured with smooth walls at the inlet and exit stations of the working section.

In weak pressure gradient flows (negative and positive), the riblet effect is presented in term of the ratio $\Delta\theta_{rib}/\Delta\theta_{smooth}$ as function of the inlet free stream velocity in the figure 4a for the riblet combination 60/60, in the figure 4b for the riblet combination 45/100 and as function of the riblet height h^+ (= s^+) in the figure 4c for both combinations. The value $\Delta\theta$ defined as the difference (θ_{exit} - θ_{inlet}), is mainly due to the skin friction reduction at these pressure gradients and the pressure gradient parameter $\beta = \frac{\delta^*}{\tau_p} \cdot \frac{dp}{dx}$ is considered constant along the working section. The positive effect of the riblets is clearly evident and their maximum efficiency is about 6% with a value of 13 for h^+ ($=s^+$). This skin friction reduction effect is less pronounced at lower values of h^+ down to 6 and vanishes at values h^+ above 20 - 25. In the figure 5, the ratio of the displacement δ^*, the momentum θ thicknesses and the shape parameter H with and without riblets (60/60, 45/100) are plotted as function of h^+ for these weak pressure gradient flows. The shape parameter H is not affected by the riblets.

In the figure 6, the mean velocity profiles with the riblet combinations 60/60 and 45/100 are compared with the smooth wall profiles in term of the non-dimensional difference $[U_{rib} - U_{smooth}] / U_c$ for 3 total diffuser angles 2θ (0°, 0.25^o, 0.5^o) and the inlet Reynolds number of 3×10^5. Close to the wall, the riblet velocity profiles have a deficit corresponding to a decrease in the boundary layer momentum thickness and thus to a decrease in the skin friction. In the external flow, these profiles exhibit an overshoot corresponding to the decrease in the boundary layer displacement and momentum thicknesses.

With increasing positive pressure gradients, the effect of the riblets is still present as shown in the following figures. The ratio of the integral parameters δ^*, θ, H with and without riblets (60/60) are plotted as function of the inlet Reynolds value in the figure 7 for the diffuser total angles of 0°,2°, 4°. With the riblets (60/60), all these parameters are reduced confirming the riblet potential in pressure gradient flows.

For the mentioned riblet combinations, the pressure recovery defined by $C_p = [p_{exit} - p_{inlet}] / \frac{1}{2} \rho U_{in}^2$ is plotted as function of the inlet Reynolds number in the Figure 8a for small pressure gradients [$2\theta = 0^o, 0.25^o, 0.5^o$] and for large pressure gradient in the Figure 8b. The positive effect of the riblets is evident in the figure 8a and the influence of the Reynolds number is small. There too, the maximum efficiency of the riblets is for the Reynolds number roughly of 2.5×10^5 corresponding to a value s^+ of 13-15. The combinations 60/60 and 45/100 are quite similar for these weak pressure gradient flows. For low Reynolds number flows, the riblet effect is still apparent though the flow is not really laminar due to the transition trip device installed. With increasing pressure gradient, the riblets still have a positive contribution at $2\theta = 2^o$ and the combination 45/100 is more effective than the 60/60 one due to a growing geometric parameter more appropriate to the diverging flow. At $2\theta = 4^o$, the combination 45/100 is still superior to the 60/60 one but the overall effect of the riblets is extremely small. The Reynolds number dependance is not evident. The pressure recovery increase is larger the less the skin friction and the shape parameter H are low [Ref. 2] and the magnitude order is comparable with the variation of the boundary layer displacement thickness. The decreasing effect of the riblets with adverse pressure gradient is not surprising as evidenced by the von Karman's equation: the order of magnitude of the skin friction contribution on the momentum increase are respectively 100%, 30%, 10% compared to the pressure gradient 0%, 70%, 90% for the values $2\theta = 0.25^o, 2^o, 4^o$.

4. Concluding remarks

In pressure gradient flows (negative and positive), the positive effect of the riblets is evident in the skin friction reduction up to well defined values of their non-dimensional geometric parameters. This beneficial contribution is apparent through the boundary layer thicknesses, the shape parameter and the blockage effect but it is masked in large pressure gradients by the pressure term. The riblets with geometric parameters adapted with the flow exhibit better efficiency. The lower limit of the positive effect is not yet clearly well defined.

Acknowledgements

The authors are grateful to Prof. I.L.Ryhming, Head of the Fluid Dynamics Laboratory of the EPFL, for many fruitful discussions and suggestions. The Hoecht riblets were kindly supplied by Dr. A.M.Savill.

Nomenclature

B	width of the working section	C_p	non-dimensional pressure coefficient
L	lenght of the adjustable (upper or lower) plate	p(x)	static pressure at station x
H	shape parameter	p_{in}	static pressure at the inlet of the working section
U	mean velocity in the streamwise direction	u_τ	friction velocity
U_e	mean velocity in the inviscid flow and in the streamwise direction	h^+	non-dimensional height of the riblet
U_{in}	mean velocity at the inlet of the working section	s^+	non-dimensional spacing of the riblet
W_1	height at the inlet of the working section	δ^*	displacement thickness
W_2	height at the exit of the working section	θ	momentum thickness
x	streamwise distance	θ	diffuser half angle
y	distance normal to the wall	τ_p	wall shear stress

References

1. A.M.Savill, T.V.Truong and I.L.Ryhming (1988): Turbulent drag by passive means: a review and report on the first European drag reduction meeting. Journal of theoretical and applied mechanics, Vol. 7, No: 4, 353-378
2. T.V.Truong and Ph.Pulvin (1989): Influence of wall riblets on diffuser flow", *accepted for publication in* Applied Scientific Research - Special Volume on Turbulent Drag Reduction, Kluwer Academic Publishers, July
3. Ph.Pulvin : Ph.D.Dissertation (*in preparation*) EPFL-Lausanne, Switzerland

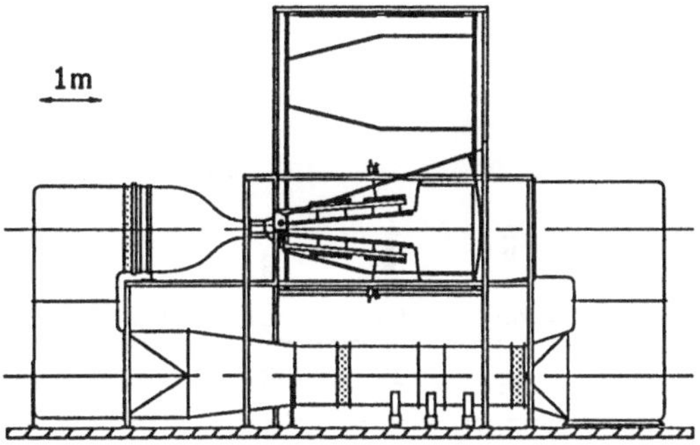

Figure 1: Experimental facility

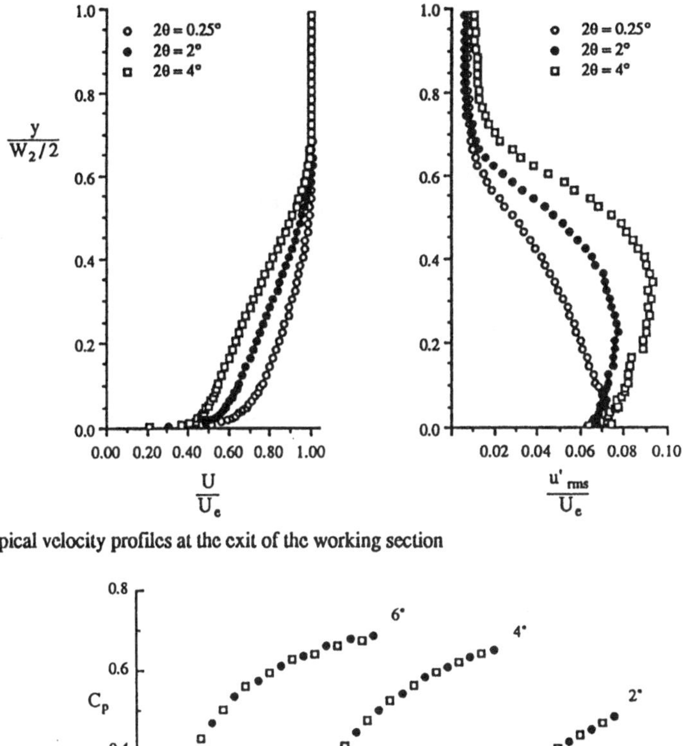

Figure 2: Typical velocity profiles at the exit of the working section

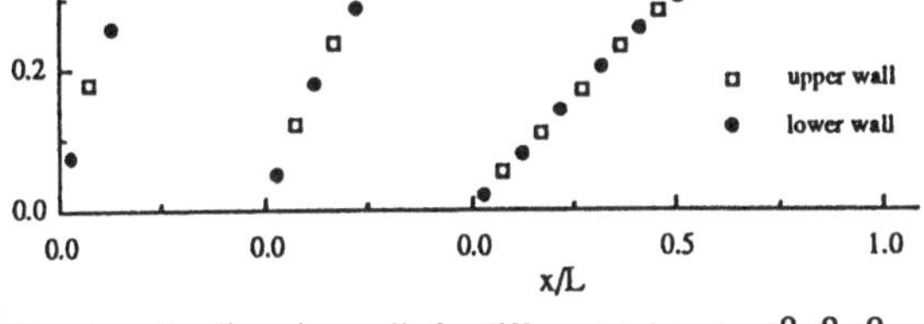

Figure 3: Pressure distribution along the diverging walls for diffuser total angles $2^o, 4^o, 6^o$

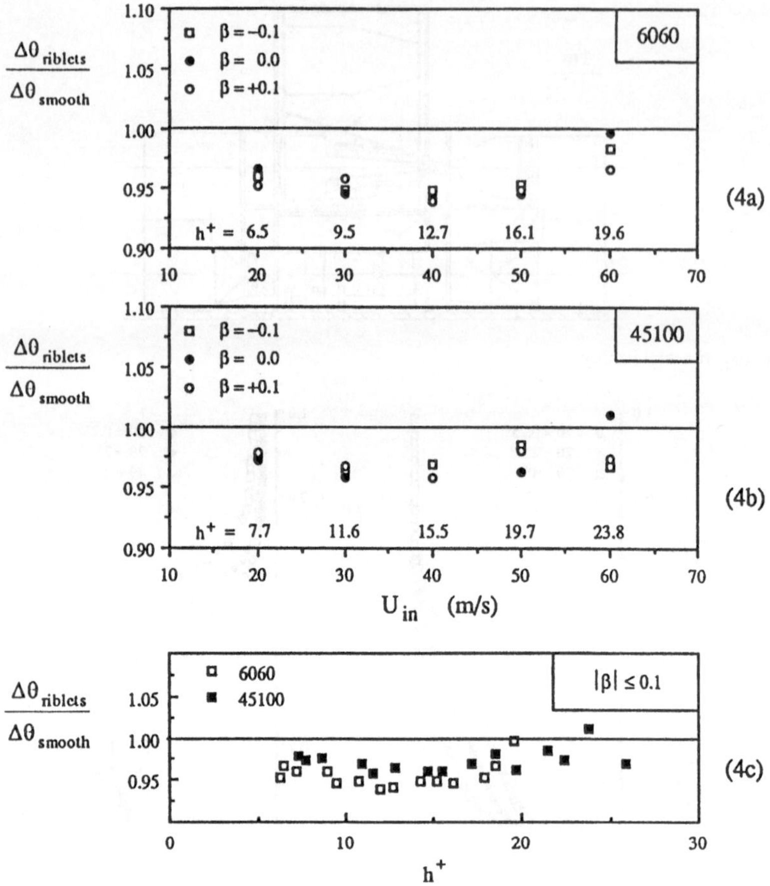

Figure 4: Ratio of momentum thickness variation with and without riblets for weak pressure gradients

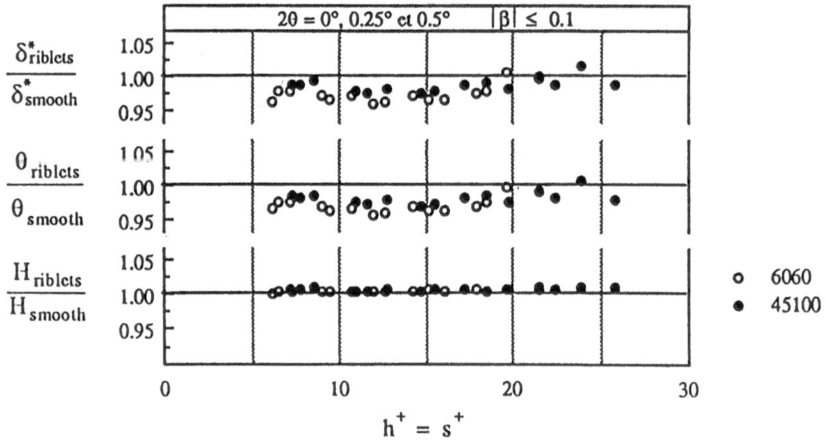

Figure 5: Ratio of the integral parameters with and without riblets at the exit working section

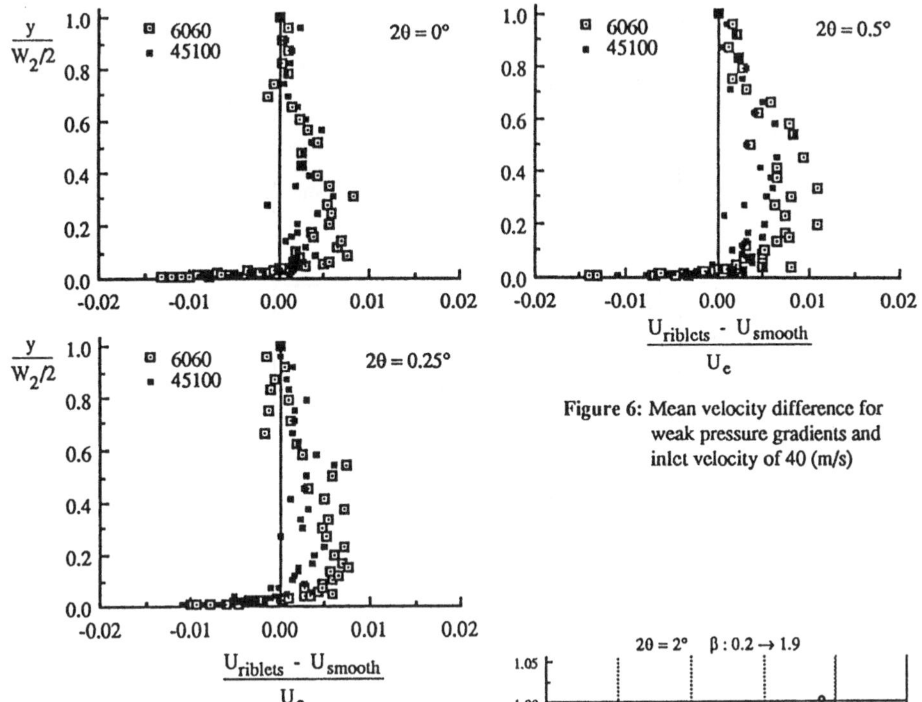

Figure 6: Mean velocity difference for weak pressure gradients and inlet velocity of 40 (m/s)

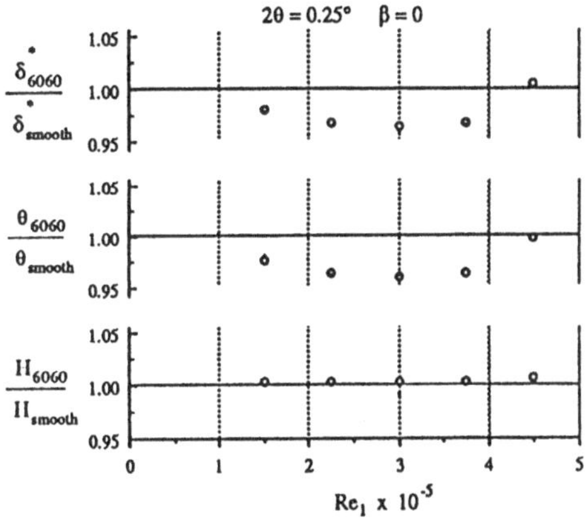

Figure 7: Ratio of the integral parameters without and with riblets (60/60) as function of the inlet Reynolds number

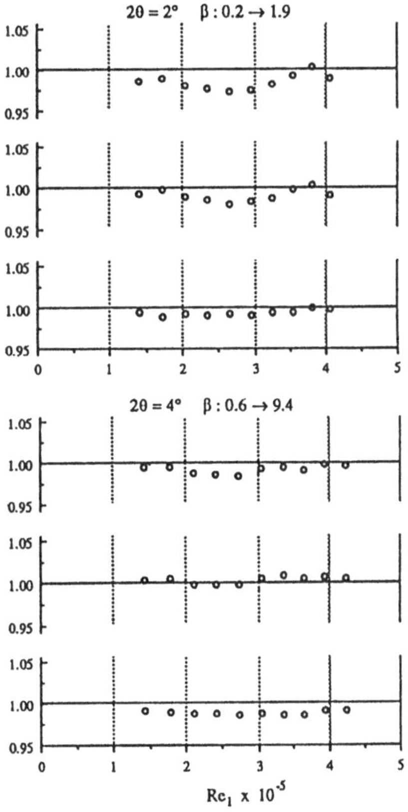

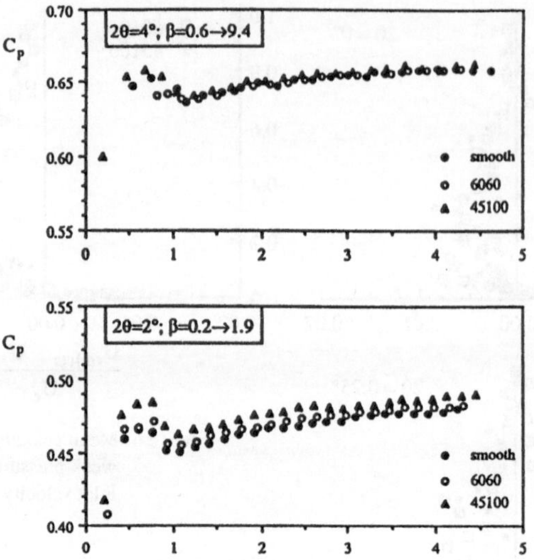

(8b)

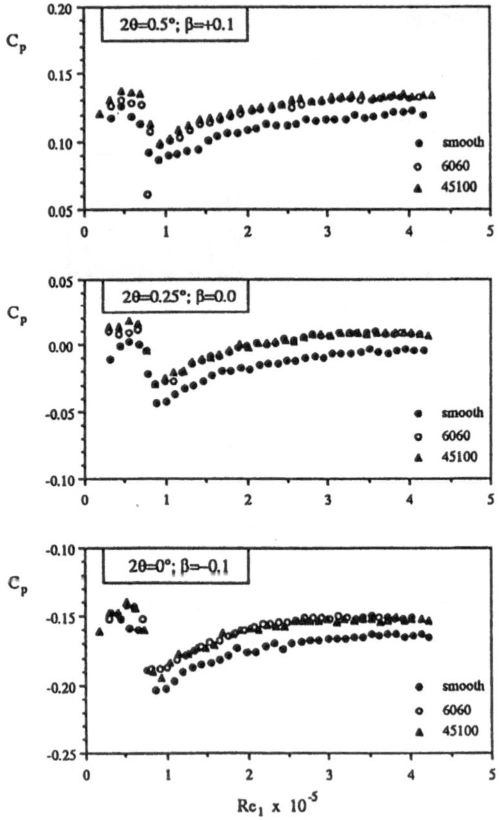

(8a)

Figure 8: Pressure recovery as function of the inlet Reynolds number for riblet combinations 60/60 and 45/100

Experimental Investigation of Turbulent Boundary Layers Manipulated with Internal Devices: Riblets

E. Coustols & J. Cousteix – ONERA / CERT - Aerothermodynamics Department - 2, Avenue Edouard Belin - 31055 TOULOUSE CEDEX - FRANCE

ABSTRACT – The purpose of this paper is to summarize the current status of the experimental research, as regards internal devices, carried out in low-speed wind tunnel when manipulating two-dimensionnal boundary layers. The main objective is to provide an estimate of attainable drag reduction performances in zero- as well as adverse pressure gradient flows. Moreover, the effect of combining internal and external devices has been checked in order to control whether or not the overall skin-friction reductions are additive.

Introduction

Interest in reducing the drag of aerodynamic surfaces has led to the control of the eddy structure of turbulent boundary layers. By assuming that the turbulence production arises from a cyclic process, it is rather evident that, whatever method for turbulent drag reduction is concerned, its action will be closely involved with a modification of that specific process. Furthermore, new skin-friction reduction techniques have been largely stimulated by the recognition of identifiable flow patterns, the so-called "coherent structures", providing mechanisms for turbulence production. Among the existing passive methods "which work" as it was written by Wilkinson et al, [1], two of them have been considered in detail at ONERA/CERT :

– Insertion of aerofoil section devices within the external part of the turbulent boundary layer : external manipulators ("LEBU's,BLADE's,OLD,...").

– Modification of the wall geometry with small streamwise surface grooves : internal manipulators commonly called ribs or "riblets".

The present paper summarizes the status of the experimental research conducted in low-speed wind tunnel. Emphasis is made upon the performances of such internal devices under either zero or adverse pressure gradient flows. Besides the fact that, throughout the literature, "riblet" data are consistent enough to affirm their potential for turbulent drag reduction, it has been decided to explore the effects of these manipulators on the mean and fluctuating quantities of a two-dimensional boundary layer. Thus, these results might help in the understanding of the mechanisms involved in such a drag reducing process. Let us add that the main conclusions will only be resumed since more detailed information could be found in [2], [3] and [4], for instance.

A rather new area of investigation has started by looking at the behaviour of these devices when set together ; indeed, could they be combined successfully for increasing local and overall skin-friction reductions?

A. Gyr (Editor)
Structure of Turbulence and Drag Reduction
IUTAM Symposium Zurich/Switzerland 1989

Experimental Apparatus

All the results discussed below have been obtained in the same wind tunnel at low subsonic free-stream speeds with nominally two-dimensional flows developing on the lower floor of the test section. The turbulence level in the external flow is rather constant and of the order of 0.25% for outer velocity range : 18 – 36 ms^{-1} ; it slightly increases up to 0.4% at about 12 ms^{-1}. The transition is tripped ahead of the test section inlet ; the virtual origin of the turbulent boundary layer is set at the tripping location whatever the value of the free-stream velocity is, [3], [4]. The cross-section is rectangular : 0.3m high, 0.4m wide ; its height allows to consider rather thick aerofoils without being worried with any blocking effect.

Experiments in zero-pressure gradient flows

The boundary layer developing along the lower wall of the test section is manipulated by altering the wall geometry. Several machined aluminium surfaces as well as "riblet" films made from thin vinyl sheets having an adhesive backing have been tested. The examined models, the length of which L is close to 0.64m, allow to evaluate the effect of rib spacing (s), rib height (h) and eventually rib cross-section shape. The performances of these devices have been judged from differences between momentum thicknesses obtained through hot-wire surveys and evaluated just behind the trailing edge of the ribbed surface. One has to be aware that the amount of drag reduction or increase might be subject to discussion since the momentum balance technique is very depending upon a lot of parameters, which might not be completely controlled. Then, the results have to be discussed in terms of "tendency".

For instance, it turns out that for different symmetric V-shaped machined models (s/h=1, 2 or 3) nett drag reductions were achieved when $h^+ < 13$ where h^+ denotes the rib height scaled with the inner variables of the turbulent boundary layer, [3], [4]. Looking at the effect of the aspect ratio (s/h) on riblet performances is rather difficult since it has been shown the extreme sensitivity to the quality of the machined surface. One could think that this dependency might be cancelled by considering symmetric grooves made from vinyl sheets having an adhesive backing, supplied from most of them by the 3M Company.

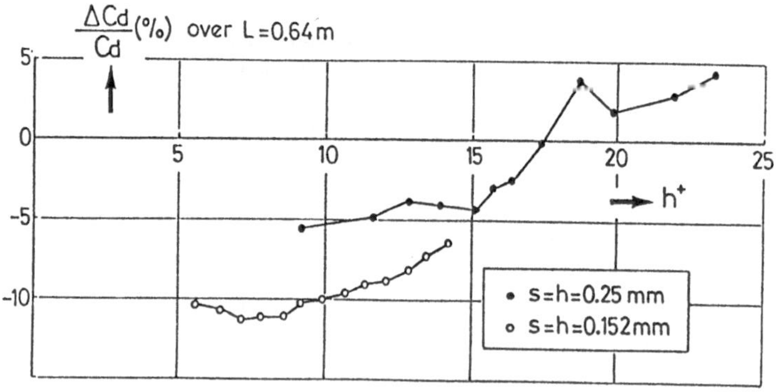

Fig. 1 Drag data for two V-grooved models (o vinyl, • machined models)

Such an observation is confirmed by looking at results plotted on Figure 1 for a vinyl and a machined models having the same assumed triangular shape (s/h=1), but different heights. As a consequence, the explored range of h^+, obtained by varying the external free-stream velocity, is not the same. It appears that a maximum of nett reduction occurs for h^+ close to 7–8. The level of drag reduction is not the same for the two models though one could guess that the zero drag reduction cross-over point would be roughly constant : $h^+ \sim 17 - 18$, which is consistent with experiments performed by Walsh et al, [5], on the same type of grooves.

Moreover, experiments revealed that peak curvature has negative effect on devices performances, [3], and that grooved models keep a beneficial effect for angles of yaw up to 20° ; in that case, the ribs are aligned at a given angle to the direction of the mean external flow, [3], [4]. The average skin-friction reductions are lower than those recorded without any angle of yaw.

In order to look at the response of the boundary layer to such a manipulation, detailed hot-wire measurements have been undertaken. The aim is to investigate whether skin-friction reductions are due to wall geometry arrangement (the average local Cf can be lower than over a smooth plate, because it might be tremendously decreased within the valleys) or to some re-organisation of turbulence structure, [6], [7]. For that, streamwise fluctuations spectra were made at different altitudes within the viscous sub-layer and the buffer layer, upon the "riblet" model (Figure 2).

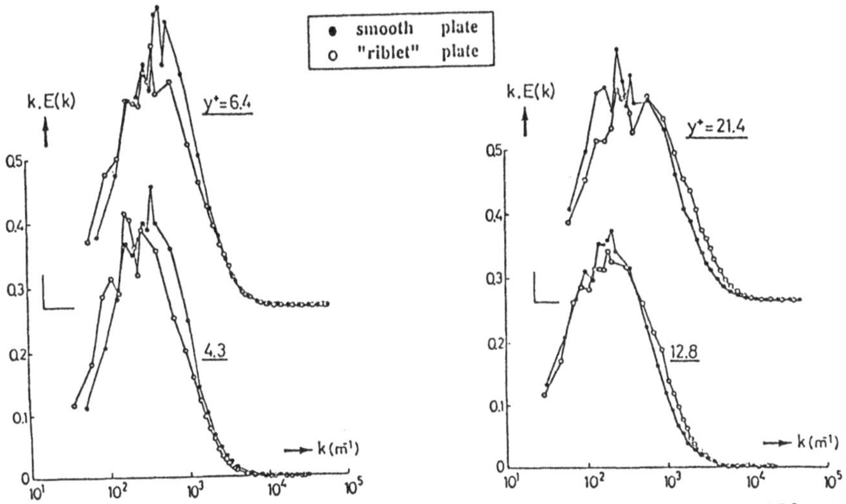

Fig. 2 u'^2-spectra upon the "riblet" model s = h = 0.152 mm

Let us mention that y= 0 corresponds to the crests plane. Some modifications exist : the grooves would induce a deficit of small scale structures at the upper part of the viscous sub-layer and, on the other hand, an excess in the high frequency range in the middle of the buffer layer ; thus, in the region of intense turbulence production ($y^+ \sim 10 - 15$), ribs would induce smaller structures. Because of the wires probe geometry and the dissipative structures lengths, such a spectral analysis has to be considered as a comparative study between smooth and grooved walls and analysed in such a way. Considering small U-type ribs, Wilkinson et al, [1], have shown that thin fins reduce the turbulence intensity, diminish the skin-friction-velocity correlations and act on the turbulence structure ; this would confirm the preceding remarks.

Experiments in adverse pressure gradient

In order to apply this drag reducing process to most of the transport aircraft, it is useful to study the behaviour of such devices when the turbulent boundary layer grows under adverse pressure gradients.

The chosen aerofoil is a LC100D one, the chord length of which, c, is 0.4m. The experimental set-up allows to modify its position within the test section and its angle of attack, α ; its rotation axis is located at 0.12m from the lower floor in order to clear out the upper side of the aerofoil which will be covered with "riblet" films, [8].

Two mean free-stream velocities have been considered : 20 ms^{-1} and 30 ms^{-1} which lead to chord Reynolds numbers, Rc, close to 5.30 10^{5} and 7.9510 ^{5}. The transition is tripped on the upper side at about 2.5% chord length from the leading edge, with a cylindrical wire (ϕ = 0.6mm). Different "riblet" models, made from thin vinyl sheets having an adhesive backed film, have been applied between x/c = 0.2 and 0.95. On the other hand, upwards to the leading edge as well as downwards to the trailing edge, a smooth vinyl sheet has been used (Figure 3). Thus, the reference configuration will consist of an aerofoil the upper side of which is entirely covered with smooth vinyl film from the tripping wire down to the trailing edge. Let us mention that the thicknesses of the ribbed and smooth films are almost identical ; the former depending in fact upon the ribs height.

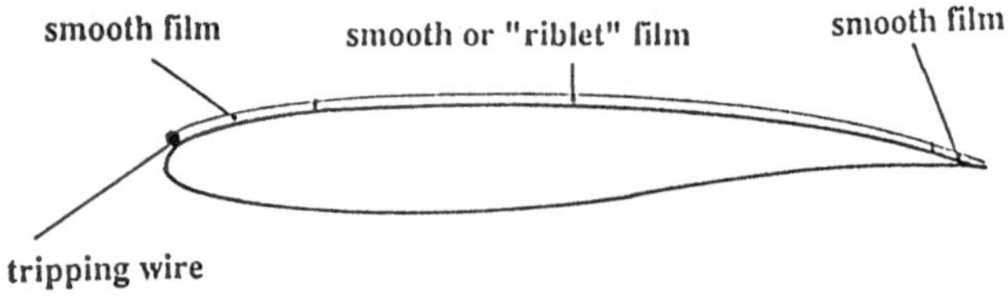

Fig. 3 LC100D aerofoil covered with smooth and "riblet" film

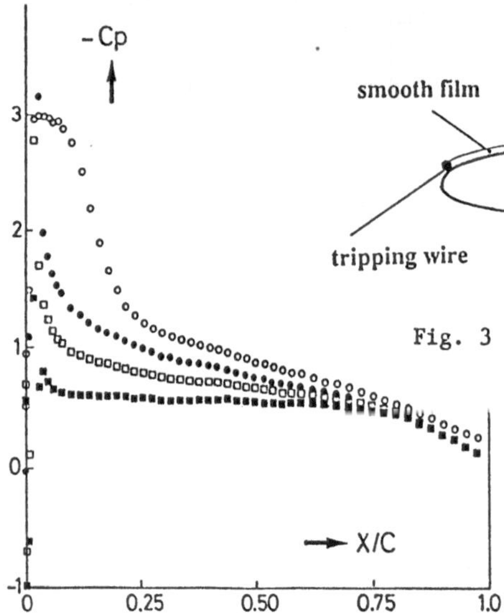

Fig. 4 Pressure distribution on the upper side of the aerofoil

On Figure 4 is plotted the Cp distribution on the upper side of the aerofoil, without any vinyl film, for $\alpha = 0°, 2°, 4°$ and $6°$ at Rc=5.310 [5] ; this distribution does not change by increasing Rc. At $\alpha = 0°$, one can observe a zero-pressure gradient condition over more than 50% of the chord length. The velocity peak increases with α up to 4°. However, at $\alpha = 6°$, a short separation buble occurs in the first 10% of chord length. For the reference case, when the aerofoil is covered with smooth film, the uncorking of the pressure taps reveals that some discrepancies exist in the leading edge region ; indeed, the "backward step effect", behind the tripping wire, is weaker since the film leads on the wire. For instance, at $\alpha = 6°$ (resp. 0°) there is a pressure drag increase of 16% (resp. 9%) at Rc=7.9510 [5]. On the other hand, removing smooth film with "riblet" film does not bring noticeable modification on the pressure field, [8].

In order to quantify the intensity of the pressure gradient, the parameter β – $\beta = -(\delta/u_\tau) \times (dUe/dx)$, where δ denotes the physical boundary layer thickness and u_τ the friction velocity – has been computed through boundary layer code. As β varies with the streamwise abscissa, its average value along the manipulated length (i.e between 20% and 95% of c) has been considered. Then, $\overline{\beta}$ =0.018 (for $\alpha = 0°$), 0.048 (2°), 0.118 (4°) and 0.255 (6°).

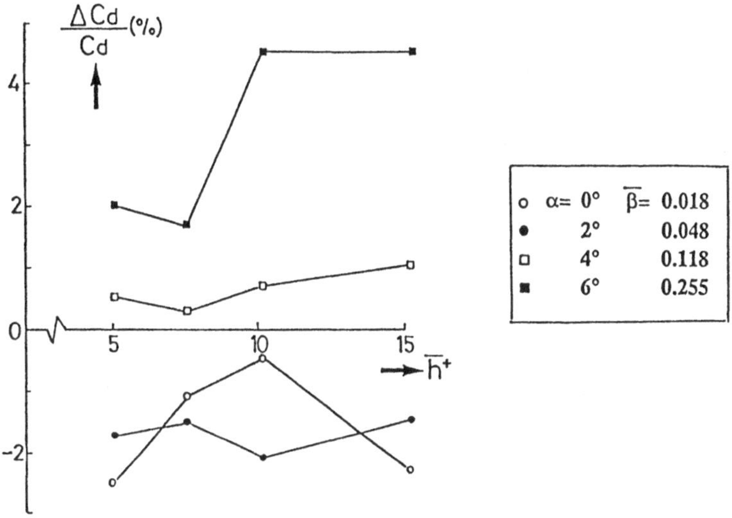

Fig. 5 Drag coefficient variations with adverse pressure gradients

The effect of grooved surfaces has been determined through Pitot tube surveys in the aerofoil wake. The streamwise location corresponds to x/c=1.5 since measurements revealed that the momentum thickness was constant downstream this location. Two "riblet" films, the cross section of which is triangular, have been considered : s=h=0.152mm and s=h=0.076mm. Boundary layer computations have shown that the h^+ parameter had a smooth evolution with x, except along the rear part of the aerofoil because of its camber ; moreover, its average value along the manipulated length was independent of α at least up to 4°. We will assume it will keep the same value at 6°. Then, considering two rib heights and two Rc made possible to cover the $\overline{h^+}$ range : 5.1 – 15.2, [8]. On Figure

5 are plotted the variations of the total drag coefficient (estimated from wake surveys) versus $\overline{h^+}$ for different pressure gradients. For the lowest angles of attack, significant nett drag reductions ($2 - 2.5\%$) have been recorded. At α=2° and for Rc=5.30 10^5, by integrating the computed local skin-friction coefficient one could estimate the contribution of the friction drag from the upper side : 40% of the total drag. Then, since 75% of the suction side is covered with "riblet", a 2.1% reduction of the total drag at $\overline{h^+} \sim 10$ corresponds to a 7% reduction of the friction drag over the manipulated length. This is consistent with the usual advertised values for zero-pressure gradient flows.

For higher angles of attack, there is no more drag reduction ; the interpretation of the results is very difficult since a separation bubble occurs close to the leading edge and, moreover, flow visualisations have shown that the boundary layer separates along the rear part of the aerofoil. There is, also, no noticeable modification of the shape parameter very close to the trailing edge.

In non-zero-pressure gradient flows, the influence of ribbed surfaces has raised up very little interest. Let us point out that experimental results obtained by Pulvin, [9], in a subsonic diffuser ring have shown that "riblets" keep a beneficial effect as long as β is rather small : ma- ximum skin-friction reductions of the order of $5 - 6\%$ for $\beta < 0.2$ at $\overline{h^+} \sim 13$.

Combined effect of external and internal manipulators

This effect is studied under the manipulation of the turbulent boundary layer developing along the lower floor of the afore-mentionned low speed wind-tunnel. The considered manipulators are :

– external device : a NACA0009 aerofoil section device, made from carbon fiber. It is mounted between two supports, each of them having a transverse degree of freedom, allowing then to apply some tension during the tests. At the manipulator location, 0.5m from the tripping wire, the natural boundary layer thickness is close to 17mm and the momentum thickness Reynolds number is about 2900. The free-stream velocity is $32 ms^{-1}$ which leads to a chord Reynolds number Rc $\sim$ 42000 (device chord length = 20mm).

– internal device : a V-shape "riblet" model, made from vinyl sheet having an adhesive backing, s=h=0.152mm. That model starts at 0.4m from the tripping wire and covers about 0.9m of the lower floor of the test section.

Hot-wire surveys have been performed at the station $X/\delta_o = 49$, where X= 0 refers to the trailing edge of the external device ; that streamwise abscissa is located at $3.4\delta_o$ downstream of the grooved surface trailing edge. The comparison of momentum thickness with and without manipulators at that location allows to judge of an increase or decrease in the friction drag coefficient. Indeed, at this rather far downstream station, the pressure variations induced by the external device can be neglected. However, such an estimation of Cd changes is not realistic since it takes the upstream non-manipulated region into account. It is, then, more appropriate to evaluate ΔCd/Cd over the manipulated length, L, i.e "riblet" model length (L $\sim 53\delta_o$).

Without the internal device, the evolution of the variations of the friction drag coefficient over L are plotted versus the height of the manipulator within the turbulent boundary layer, h_L (Figure 6). Whatever the value of h_L is, drag increases have been recorded. The available manipulated length downstream of the external device is not long enough so that the device drag penalty balance the observed local Cf reductions, [2].

Without the external device, the grooved wall parameter h^+ decreases very slowly with the steamwise abscissa ; it is then possible to characterize that device with the average value $\overline{h^+} \sim 12$. Measurements revealed a 5% nett skin-friction drag reduction over L.

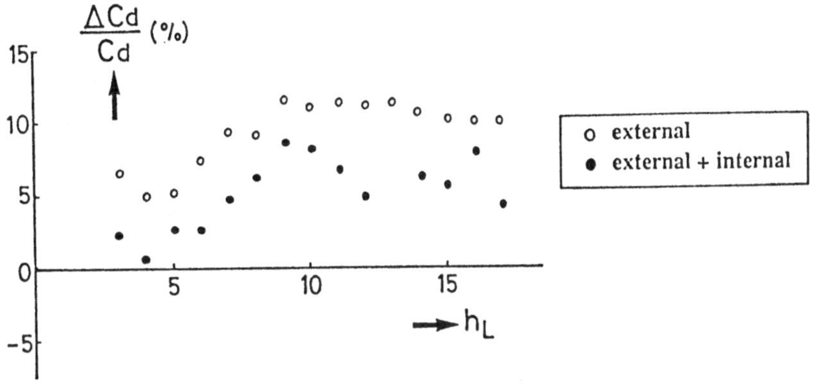

Fig. 6 Combined effect of internal and external manipulators

The question which arises, now, is : what happens when these two devices are associated ? Looking at Figure 6, one can notice the beneficial effect of combining the manipulators ; indeed, the level of drag increase, obtained with only the external device, diminishes. Behind the aerofoil section device, rather large local Cf reductions have been recorded depending, of course, of the value of h_L. For instance, a 25% reduction on Cf would induce a 13% reduction of the h^+ parameter ; that means that the ribbed surface might not be optimised in some areas. On the other hand, when the manipulator is moved away from the wall, the downstream local skin-friction reductions are weaker and one can observe that the "riblet" drag reduction seems to be additive.

This result is consistent with those recorded by Walsh et al, [1], for subsonic zero-pressure gradient flows. In their experiments, the external device was made from three stuck small plates, which led to a rather large wake. Since the external device acts on the large scale structure and indirectly upon the wall structure through its wake, whereas the "riblet" model operates directly upon the wall turbulence structure, one might wonder what happens when the wake reaches the wall... According to Falco, [10], "grooved surfaces should perform better in the region close behind outer layer manipulators, but worse near the maximum reduction location, when the device wake interacts strongly with the wall".

Conclusion

The main objective of this experimental investigation was to give evaluations of passive drag reduction techniques through internal manipulators, at low subsonic speeds for zero- and adverse pressure gradient flows. These experiments have allowed to verify with laboratory measurements (boundary layer or wake surveys), that carefully optimized "riblet" models could end up with nett overall drag reductions. The level of recorded skin-friction reduction is narrowly dependent upon the "crucial" parameter : h^+ (rib height scaled with the inner variable of the turbulent boundary layer). Maximum gains occured for V-groove models such that h^+ was less than 18.

Concerning the application of some ribbed surfaces on the upper side of an LC100D aerofoil, over 75% of the chord length, maximum total drag reductions of the order of $2-2.5\%$ have been obtained through wake surveys, when the average pressure gradient parameter, $\overline{\beta}$, was less than 0.12. It turned out that grooves, scaled such as $\overline{h^+} \sim 10$, keep a beneficial effect at least until the turbulent boundary layer goes to separation.

The coupling effect of NACA0009 aerofoil section device and symmetric V-groove "riblet" has been investigated for zero-pressure gradient flows. The first set of experiments has revealed that the "riblet" drag reduction was approximately additive to the drag increase recorded behind the external manipulator. This observation is very attractive if one has in mind transport aircraft applications. Though some experiments showed that internal devices lead to significant drag reductions $(6-8\%)$ in transonic flows, [5], [11] or [12], the behaviour of such external devices just began to be under investigation in that specific flow conditions, [13], [14] or [15].

References

[1] Wilkinson S.P., Anders J.B., Lagos B.S., Bushnell D.M. : Int. Conf. on Turbulent Drag Reduction by Passive Means (London, 1987)

[2] Coustols E., Cousteix J., Belanger J. : Int. Conf. on Turbulent Drag Reduction by Passive Means (London, 1987)

[3] Coustols E., Cousteix J. : 16th ICAS Congress (Jerusalem, 1988)

[4] Coustols E. : AIAA Paper 89 – 0963 (1989)

[5] Walsh M.J., Lindemann A.M. : AIAA Paper 84 – 0347 (1984)

[6] Coustols E., Gleyzes C., Schmitt V., Berrue P. : 24^{ieme} Colloque AAAF (Poitiers, 1987)

[7] Cousteix J., Coustols E., Arnal D. : AGARD Symposium (Cesme, 1988)

[8] Coustols E. : Technical Report CERT/DERAT (1989)

[9] Pulvin P. : Ph D dissertation (in preparation) (1989)

[10] Coustols E., Savill A.M. : To appear in Applied Scientific Research, Vol.46, (1989)

[11] Douglas McLean J., George-Falvy D.N., Sullivan P.P. : Int. Conf. on Turbulent Drag Reduction by Passive Means (London, 1987)

[12] Walsh M.J., Sellers W.L.III, McGinley C.B. : AIAA Paper 88 – 2554 (1988)

[13] Bonnet J.P., Delville J., Lemay J. : 16th ICAS Congress (Jerusalem, 1988)

[14] Bonnet J.P, Poirier D., Delville J. : Drag Reduction 89 Conf. (Davos, 1989)

[15] Savill A.M. : IUTAM Symposium (Zurich, 1989)

3. What particular flow conditions are necessary for the formation of coherent structures?

   To understand the flow and its internal stresses we have to know the scales in space and time and their interactions. Therefore, it is necessary that we know what is behind the formation of turbulent structures (Falco).

4. What is the importance of interactions between coherent structures (e.g. vortex pairing)?

   If we are to understand the formation, we must, in another step, attempt to understand the size distribution of the structures, the shear stress they produce, and how frequently they appear. Ho explained that this can only be achieved if we understand the interaction between them and their interaction with the walls.
   Accordig to Ho, the free shear layer is a good object for studying the interactions between the structures since this flow configuration is mainly two-dimensional and has no vorticity source after the trailing edge. As the main result of such studies, he mentioned the change in wavelength of the spanwise and streamwise vortices of the mixing layer with distance downstream of the trailing edge.
   In spanwise direction, the ratio of the frequency times the layer thickness, divided by the convectional velocity, is constant. The wavelength in streamwise direction changes and the local ratio of the two wavelengths is a constant of 2/3.
   The situation is quite different for boundary layer flows. The mixing layers are dominated by regular pairing processes, whereas boundary layer flows are highly three-dimensional and the wall is a continuous source of vorticity for them. This vorticity is transported by a bursting mechanism, and we do not even know how its frequency scales.
   The interactions of the structures in the boundary layer produce a continuous change in their population. Ho explained these ideas with two examples:
   1. The so-called hairpin vortices in the boundary layer close to the wall produce secondary vortices and their legs are capable of coalescing and so reducing the number of structures in the outer regions. The interaction of the wall-near structure with outer ones is still not understood. We neither know in which way the population is changed nor how to scale the frequency of the outer flow since this frequency

Panel Discussions

Introduction

Tapes of about three hours of lively discussions can hardly be reproduced on a few pages. However, since it was the opinion of most of the participants that the panel discussions were very valuable sessions, we will here try to present the main questions and replies on the different subjects. Before doing so, however, I must apologize for the fact that in many cases it could no longer be determined who raised a question or who contributed which comments to the discussion. These records of the panel discussions are therefore a summary. A lot of spontaneity is lost, but I hope that the participants, in any case, will find in these pages an aid in refreshing their memory.

1 Coherent Structures in Turbulence
(R.W. Falco, C.M. Ho & C.R. Smith)

Their main questions for discussion were:

1. Why are we studying coherent structures?

   For C.R. Smith, who raised this question, the ultimate goal of such studies is a better predictability and control of the turbulent flow by a better understanding of its dynamic. He claimed that drag reduction as well as inhibition and enhancement of mixing would be elements to be controlled and that the use of coherent structures for modelling this should improve the prediction.

2. How completely must we be able to characterize coherent structures in order that they be of utility?

   Falco explained this question by discussing the visualization of a turbulent boundary layer. He shows that single point measurements are by far too restrictive to give information on the flow field and that, on the other hand,the information gained by visualization through passive contaminants is not unequivocal. We are submerged in data. Therefore, the question of how much information we need is an essential one.

normalized with δ is not constant, probably because of the interaction of the structures with the wall.
2. Ho described how a jet interacts with the wall and showed that the shear layer produces a counterrotating vortex which is responsible for a local pressure distribution at the wall leading to separations, which in turn produce high shear rates.
To conclude, the main mechanisms in mixing and boundary layers are fundamentally different.

As a consequence the fifth question is obviously:

5. What are the implications and what is the importance of coherent structures with regard to turbulence control and drag reduction?

This question remained unexplained.

General discussion.

Wygnansky opened the discussion with a comment on visualization. He insisted that we must all be very careful in our interpretations of visualized flow fields and referred to Hama's early work, in 1962, on mixing layers, which showed how difficult it is to interpret simple flows produced by sinuisodal oscillations. For reliable visualizations, Wygnanki recommended particle tracking with a constant density of particles. But even if we get a good Lagrangean view of the flow field, this may not be sufficient for describing its vorticity.

Virk stressed a similar point in two questions addressed to those working in the field of flow visualizations. 1. Has a visualization ever been predicted before the experiment was carried out? 2. How accurate can a quantitative description based on visualization be? He raised the questions because he is convinced that by visualization the change in structures through riblets, which produce a 5% of drag reduction, cannot be visualized. Falco agreed with this statement on visualizations with passive contaminants, but pointed out that, with his techniques, which allow for two-point vorticity correlations to be measured, such small differences should be measurable.

Kline answered Virk's question by stating that they had qualitatively predicted a drag reduction by riblets some years ago, but that they had not publish this since, in their view, the effect of 2-3 % was too small. He went on to the third and fourth questions by remarking that one should probably rather ask how the coherent structures can be turned off at high enough Reynolds numbers, as they are produced by any disturbance. If the shear stresses are high enough, the structures are always there. They are necessary to maintain these stresses. But the question is again, how can we turn them off? Falco stressed the need for more knowledge on the formation of coherent structures and believes to have contributed some new ideas in his paper. He agreed that we should also know about the turn-off process, but that it is easier to start with the formation. Kline contradicted him by claiming that transitional processes are too simple to allow for the understanding of the full process in the developed turbulent flow.

Kim anwered the first three questions from a more general viewpoint by saying that we are too much concerned with details and still do not realise what we see. We recognize interesting structures, but their real characteristics, like their vorticity, are still unknown. This can probably be understood from our common view that we need a wall to produce coherent structures in a boundary layer and that we have to start with the streaky structure to interpret the whole assembly of vortices. However, as Kline already pointed out, we do not need the wall to produce structures; a high enough shear is a sufficient condition.

What do dyes or markers really make visible? Blackwelder commented on this fundamental question by recalling that, during this first session, we had all seen very different visualizations, by different authors under approximately the same conditions, and he felt that the differences probably have to do with the amount of dye the different authors had injected into the flow at different levels. He therefore suggested that we quantify these parameters, as Tiederman had done some years ago. Blackwelder asked Falco why he had seen streaks one time and pockets another, under the same flow conditions. Falco agreed that visualizations depend on the amount of dye or markers injected. With a very small amount, only streaks can be seen since the regions where the dye concentrates are the ones with the slowest motion. With a medium amount of dye, we can see pockets and with concentrated dyes wavy systems. Although we see a whole variety of structures and substructures, the visualizations give a good insight into the flow field. The ambiguity of the pictures is not hypercritical for the experienced scientist.

If the interpretation of structures is so subjective, Virk's question is still unanswered. Bechert remarked that predicted flow structures had been visualized, such as Tolmien-Schlichting waves, and that in regard to riblets we lag only two million years behind nature. This was the key word for Wygnanski, who felt that visualizations with and without riblets differ to the extent, that, without riblets, the streaks meander in z-direction with an unaltered spacing. However, with riblets, the streaks remain in the grooves; they stop meandering. Whether this is the reason for the small drag reduction observed or not, is another question. Virk suggested doing pressure measurements.

Reischman agreed with Kline by insisting that the control of the flow should be the main goal of our investigations, which does not necessarily mean a complete characterization of the involved mechanisms, but one that is sufficient to control them. Kline put these ideas into an even more general statement. He claimed that the history of science tells us that the understanding of a phenomenon usually requires three steps: (1) A qualitative description, (2) a taximometrical ordering and (3) a quantitative understanding leading to a mathematical description. To return to the problem of visualizations, their purpose is to help understand the phenomena in a qualitative sense and not to make quantitative final decisions. Visualization should help in evaluating what we can later measure quantitatively. Taximometrical ordering supports the engineer by answering the questions he raises in order to solve a problem.

Virk said that he has nothing against visualizations, but on the contrary, finds them very useful. He liked Kline's description of a burst as involving three different stages. However, where Kline uses five or eight elements, he is in doubt wether a description with so many elements can finally be reduced to something useful. Kline said that he does not care for the simplification of the problem from an academic standpoint, but does not care for complexity from the practical standpoint either.

Anselmet maintained that since the consequences of the coherent structures are still unknown, in the absence of a description of the turbulent shear stresses by a model based on coherent structures, it is problematic to conclude anything from a change in the structures. From his work with riblets he believes to have evidence that the interactions are more complicated than what visualizations would suggest. Choi postulated that the main changes of the flow by riblets happen in a layer of less than 15 viscous units' thickness, and that the main changes induced by an alteration of the coherent structures only appear at higher moments.

Wygnanski asked whether anybody would like to comment, not only on the reduction of the skin friction, but also on pressure drops due to changes in the separation behaviour since a change in the coherent structures should also either delay a separation or promote reattachment.

There was no response to this question and Falco then asked the computing community whether they could incorporate passive contaminants into their direct computation schemes. Such calculations would allow for a comparison of simulations with visualizations. Kline replied that this had already been done, however, probably not to the extent Falco suggested. Falco stressed that he had never found a useful calculation of this kind in literature. Several interruptions made clear how controversial this point is. It seemed that the different wishes were quite individual. Falco, for example, would have liked to have had more evidence than given in the presentations in the morning. He agreed that the possibility of calculating a pressure distribution is very valuable, but demanded particle path lines in a Lagrangean sense. Kim pointed out that they had checked the bursting processes seen with the hydrogen bubble technique and found good agreement with the pictures produced by simulation. Falco asked that further efforts be made.

Tiederman hoped to settle this controversy by saying that he believes that the time averaged models of the structures should be ultimately sufficient to produce a good description based on coherent structures. However, this does not mean that we can give up investigating the problem in its whole complexity, since there are different mechanisms of drag reduction and also interactions between structures wich can only be distinguished by considering their whole complexity.

Kline defended the computational simulation by saying that ho had compared pictures of a direct simulation of the turbulent boundary layer with laboratory experiments for the same flow conditions, and could put them side by side without being able to tell the difference. Falco argued that this was only a qualitative statement, but that what we need are calculations for the passive contaminants so that we can interpret what we see. Kline accepted this remark and asked the experimentalists to make a catalogue of specific simulations for our assistance.

The discussion ended in a comment on question two. The speaker claimed that by a better understanding of the structures not only drag reduction, but also mixing and heat exchange can be controlled. Preliminary experiments on mixing in pipe flows have shown promising results in this respect.

2 Drag Reduction in Dilute Polymer Solutions and by Inhomogeneous Injection of Concentrated Polymer Solutions

(B. Gampert, M.M. Reischman & W.G. Tiederman)

The conductors of this panel discussion also raised some questions. The first one seemed to be the main issue.

1. What is the general mechanism of the drag reduction produced by injected polymer solutions of high concentration in the centre of a pipe flow (heterogeneous drag reduction)? and
2. In which scope does it differ from the one producing drag reduction by polymers of the same concentration injected at the wall?

Bewersdorff pointed out that between central and wall injected polymer threads there exists no difference as soon as drag reduction has fully developed. However, the ratio between the injection and the local mean flow velocity has to be the same constant and the polymer thread has to remain intact.

Reischman demanded a more mechanistic interpretation of how the outer flow communicates with the wall. Falco gave an explanation by using the model he had presented the first day: in the development of hairpins with a viscous time scale of 5 there must exist areas with a strain rate of about 1000 s^{-1}. An attempt should therefore be made at getting the molecules into this area. Hanratty speculated that one possibility would be an enhancement of the effective dissipation in the core region. Since the wavelength in the wall region has to adjust to this dissipation, the structures close to the wall have to grow in size, which results in drag reduction.

Virk saw no discrepancy between the homogeneous and the heterogeneous kind of drag reduction. In his opinion, drag reduction appears when stretched molecules produce a non-Newtonian buffer or elastic layer. It is not important, Virk argued, whether it is fed with polymers in a diffusion process out of the thread or by premixing. Concerning the elastic layer, it is not crucial whether it starts to grow from the core or the wall side. However, the development of such a layer requires a layer of sufficient strength in the flow. Virk summarized that entangled polymer molecules probably diffuse out of the thread and, due to their length, produce a drag reduction at an even lower concentration, but that this is not necessarily a new effect. Reischman wanted to know if Virk believes that the interaction of the polymers with the production process of turbulence is necessary. Virk feels that this is the case, but in the sense of an overall process. He argued that the main mechanism is a redistribution of the kinetic energy from the elastic layer into the core, where a net deficit exists. Reischman asked if anyone would like to comment on Virk's conception since it contradicts the common belief that heterogeneous drag reduction is a new effect.

Bewersdorf commented that, if Virk's idea is correct, there are several observations in cases of heterogeneous drag reduction. For example, drag reduction produced by a thread has a periodicity in flow direction. The frequency formed by the wavelength of this periodicity and the mean axial flow velocity are equivalent to the main frequency of the axial velocity fluctuations found in the autocorrelation of the time sequence of these fluctuations in the near wall region.

Tsninober made clear that we have to know if there is a diffusion process or not. If the thread, with a diameter which is only a small amount of the diameter of the tube, were embedded in a flow of a Newtonian fluid, we would be confronted with an incredibly interesting problem. Its solution would also clarify our understanding of a turbulent flow. Virk insisted that there must be a diffusion process. However, the polymers don't have to reach the wall to produce a drag reduction. It is sufficient that they form an elastic layer, for example close to the thread. There the diffusion time can be as low as 10^{-5} second. Falco confronted Virk with the experimental results of Berman, who showed that no concentration outside the thread could be measured.

In an injection type experiment, one speaker said, the polymer concentration in the thread is of the order of 6-8000 ppm and therefore the viscosity is higher than in the ambient fluid. He therefore suggested testing if the drag reduction was not a viscous effect by using a non-elastic fluid of higher viscosity. Bewersdorff reported that they had carried out this experiment by using glycerol as a fluid of higher viscosity. They had found a small effect just behind the injection point, but 50 diameters downstream there had been only a small increase in drag due to the higher viscosity of the injected fluid. Virk therefore suggested repeating the experiment, but this time, dissolving a very dilute amount of polymer additives in the glycerol. This would put his diffusion hypothesis to the test.

Reischman posed the next question:

3. When documenting an experiment, what must we pay attention to so that information can be exchanged and compared?

Several participants spontaneously demanded precise information on shear and wallshear stresses. Zakin, from his long experience, asked for a better description of the polymer solutions used. This could perhaps be done by giving the intrinsic viscosity before and after the experiment. Reischman asked if a characterization by molecular weight and by the intrinsic viscosity is sufficient.

Berman was very critical since, as he claimed, there is no way to reproduce a mixed solution, not even in one and the same laboratory. He believes that the best comparison would be achieved if the authors published their friction-Reynolds number relation. The next speaker agreed with Berman, but suggested concentrating on drag reduction experiments made at a constant shear rate instead of aiming at too many experiments. These experiments would reveal more about how a change of flow conditions at the wall influences the flow. The speaker also said he would appreciate it if everyone published the diameter of their tubes. With this wish he made clear that there exist, even in regard to the most simple data, too many insufficiently documented publications.

Falco suggested that the NBS supply the community world-wide with a standard, well caracterized polymer at a reasonable price. Reischman was in support of this idea, provided we can get rid of the aging problem.

Fruman felt it was of minor importance whether we experiment with different materials. Even under different hydromechanic conditions, it is important that we understand the physics behind the general phenomenon. However, exactly in this respect, we are no further than we were five years ago. Today everyone, even without understanding polymers or turbulent flows, can produce drag reduction. Therefore, we should not pay too much attention to details, but stress the physics. Falco protested by claiming that we have made considerable progress in understanding the turbulent structures, though not enough yet to explain the heterogeneous drag reduction. He therefore suggested studying and characterizing the turbulent flow in this particular case of drag reduction. The main goal should be a better understanding of the interaction of the polymer thread with the large eddies. But he also suggested making these experiments in a channel flow rather than in a pipe, although we do not know if drag reduction appears in such flows or if heteregeneous drag reduction is restricted to the rotational symmetric case.

Reischman posed the next question:

4. What is the imput of rheology? Can rheologists supply us with a constitutive equation?

James said that, for shear flows, we have good constitutive equations. However, whenever strain is important, we run into a problem since for characterizing the extension, we need the extensional viscosity behaviour, which, in the case of dilute polymer solutions, is unknown. To underline this statement he reported that at their latest international conference rheologists had had to present their elongational viscosity measurements made with a standard sample of a polymer solution. These measurements had been done by a half a dozen different methods. This had resulted in the extensional viscosity as a function of the elongational strain differing by a factor of 1000!

Gampert argued that, if we had a plot of the elongational viscosity as a function of the elongational stress, we would not be able to use it since we do not know the strain in the turbulent flow field. Falco argued that we do know the strain. James argued that with the numerical simulations of the turbulent flows we have a test flow from which we can calculate the strain in the flow. Reischman remarked that in this case we need more knowledge of the elongational viscosity behaviour, and he asked James if no trends had been found in the data he had referred to. James agreed that most of the measurements differed only by a factor of 10. The method which had produced the extreme high data is the fibre spinning rheometer, which should not be used for the characterization of the elongational viscosity. (The very high elongational viscosity values in literature, e.g. in Lumley's review article, are probably based on this fibre spinning method.)

Bewersdorff remembered that James too had found that drag reduction also depends on the pre-shear of the fluid. Therefore, the situation is much more complicated since what is needed, is the deformation history or a three dimensional plot of the elongational viscosity due to strain and shear rate. Reischman agreed that history is crucial.

Virk encouraged the rheologists to continue their work although they have not yet been successful in formulating a useful constitutive equation. The physical relations of how a molecule expands in dependence of a flow strength, for example, are very useful.

Reischman asked what part the numerical simulations play in finding a constitutive equation, and if someone had already used one or a variety of constitutive equations in his simulations. Gampert supplemented this question by one addressed to James. What kind of constitutive equation would he put into a simulation?

James suggested an approach in three steps. In a first one, particles would be put into the simulation of a flow which, in a physical experiment, would produce drag reduction. A calculation of this sort would show where the particles are elongated and to which extent. A second step would be the characterization of the areas of the flow field by the interaction of the molecules with the flow. In a third step we could then hope to develop a constitutive equation.

Hanratty said that we have the techniques to follow particles in a simulation scheme, and so follow the particles' history ; e.g. if we take a flow condition which starts to produce drag reduction, we can study what kind of deformation of the particle is needed to initiate the effect. Calculations of this kind could be compared to a simulation for which we use a constitutive equation, e.g. based on a dumbell model; or we could develop a new polymer model to enter the constitutive equation.
" No more models, please", was the reaction of the audience.

Falco referred to an argument he had already put forth in the first panel discussion. He suggested that, whenever a simulation with drag reducing particles is to be made, a passive contaminant should also be added because what we are looking for is an answer to the question: What causes the extension when we see an extension?

Reischman said that the discussion about constitutive equations showed that the main information gained so far is rather a byproduct of this branch of research, comparable to the very useful knowledge on the influence of strain and strain history. He wanted to encourage especially the rheologists to try to supply us with a reliable description of the elongational viscosity. With these summarizing remarks he continued the discussion and explained the next question.

5. How will the turbulence due to additives be modified and how is this ultimately related to the structural turbulence?

Tsinober said that one of the most important issues related to this question is to find an answer in the controversy on heterogeneous drag reduction. Therefore, he again asked if the fluid is around the string Newtonian or not. Virk claimed that this was initially the case but, as the diffusion process went on, did not remain the case.

Bewersdorff reported on an experiment with isobutene as a polymer dissolved in petroleum di-chloretan, wich is insoluble in water. Injected into water, it produces about 20% drag reduction. However, the polymers are at the limit of the flow rate in their set-up and only just above onset of the effect. Therefore, Bewersdorff felt that much higher drag reduction could be achieved. Virk was not convinced and reckoned that about 5ppm could still be dissolved.

Nychas commented on Reischman's question by asking why practically all the turbulent kinetic energy is produced in flow direction, as Tiederman had shown in his review lecture. Judging from our knowledge of the coherent structures, this would not have to be so. Tiederman made clear that he had shown time averaged values and that Nychas' argument was one on instantaneous values. Nychas agreed, but stressed that his comment was still valid. Tiederman felt that this problem asks us to investigate how the structures, as described by Kline, differ for drag reduction and if we are to add new specific elements to this list.

Falco said that this was what disappointed him most. He complained that he had not been offered any new information on structural parameters in drag reducing flows that day, such as: Measurements analysed with conditional sampling to evaluate the large eddies, or information on the bursting process. There had been no LDA measurements on w, not even on streak spacing differences at higher Reynolds numbers, in other words, on the usual structural characterization of a turbulent Newtonian flow. In his opinion, their is just not enough overlap between flows with and without drag reduction to make progress in the interpretation of the effect.

Hanratty felt that more progress could be achieved if we first studied the structures in Newtonian flows and then proposed experiments based on physics, instead of chasing eddies under drag reducing conditions.

Gyr disagreed with the statement that no progress had been made in the past years. He stressed that the discovery that drag reduction goes hand in hand with an enhanced anisotropy of the turbulent flow field was a promising result. Therefore, he argued, it should be a subject of study, which classes of anisotropic flows have the effect of being drag reducing, and only then, if and in which way the additives contribute to an altered turbulent flow field belonging to such an anisotropic class. He was convinced that the drag reduction in MHD flows, as presented by Tsinober, give a hint in the direction of such investigations of the anisotropy.

Virk suggested turning Reischman's question around. He felt that we have all learned a lot from the investigations of a turbulent flow of Newtonian fluid for the understanding of the drag reducing effect. However, the goal still is to improve our insight into the turbulent flows based on experiments made with drag reducing liqiuds. Does such an experiment exist? Tsinober said that this was the reason he insisted on clarifying the situation in which heterogeneous drag reduction occurs. This could well be a key experiment of the kind Virk is looking for.

Bewersdorff referred to Gyr's argument by commenting on the results Gampert had shown. They indicate that, independent of the Reynolds number and drag reduction, the w component of the velocity fluctuations remains nearly constant. The increasing drag reduction is therefore linked with the anisotropy of the flow, and it is this quantity which increases with increasing drag reduction.

Falco suggested studying the deformation of the string in a heterogeneous drag reducing experiment due to the viscoelasticity of the material by adding dye of alternative color just at the nozzle. He suggested this experiment since experiments with a thread, instead of the viscoelastic string, do not show any drag reduction.

Tiederman hoped that a new imput would be the result of the numerical simulation. Such calculations would help define structures in an Eulerian scheme. Tiederman argued that this information is the missing link in the study of the specific structures under drag reducing conditions. Hopefully, these structures will provide us with some informative differences. With this outlook, Reischman closed the engaged discussion session.

3 Drag Reduction by Surfactants, Implication of Stability and Direct Computation of Turbulent Flows Making Use of the Concept of Coherent Structures (N.S. Berman, J. Kim & J.L. Zakin)

Zakin explained the first question:

1. Is the mechanism of homogeneous drag reduction the same for polymers and surfactant solutions?

Zakin summarized the main features of surfactant solutions. They are only drag reducing if the molecules form rod-like mycells. The mechanism which produces the mycells is based on shear fields in the flow. The process is temperature and concentration dependent. At a critical shear stress the drag reducing effect gets lost. Zakin concluded that a plot of the friction factor versus Reynolds numbers already shows that the mechanism of the two drag reducing effects must be different.

One participant wanted to know whether any experiments had been made with drag reduction in external flows. The purpose is to extend our knowledge on flows with a decrease of the shear velocity in flow direction.

Bewersdorff announced that Riediger would present results of an experiment with surfactant solutions in a plane mixing layer in Davos. He summarized that the coherent structures are altered by the surfactants, but also differ from the coherent structures present in dilute polymer solutions.

Virk argued that the mechanism does not mean anything since it should be indipendent of the additives. In his opinion, the flow scarcely depends on what exact additives produce the drag reduction. Zakin replied that Virk himself had described a different behaviour of the drag reduction when he introduced type A and B drag reduction, although he had not explicitly spoken of different mechanisms. Virk insisted on his ideas. In his opinion, the gross flow behaviour can be different although the mechanism is the same. This means, drag reduction is based on the same interaction in site, e.g. produced by extended molecules. At first, only few polymers contribute to the drag reduction, whereas in the process of the flow field getting stronger, more molecules get elongated and the gross flow behaviour looks different although the mechanism remains the same. Zakin stressed this point by asking if that implies that we only need to increase the concentration for the type A material to get type B. In a certain sense, yes, Virk answered and postulated that type A is a subset of type B.

A speaker remembered that Smith had shown flow visualizations of the near-wall region of a drag reducing flow with surfactants. It is remarkable that no streaks can be observed in these pictures. This is a unique situation, which nobody had previously observed. Whether this implies a change of mechanism, is open for speculation.

Bewersdorff argued that, based on the rheology induced by the two different additives, the mechanisms must also be different. Surfactants always produce drag reduction when the shear rate is high enough for the molecules to build up shear induced structures. Therefore, the viscosity in the buffer zone increases. This is in opposition to dilute polymer solutions where the shear viscosity is not affected.

Zakin asserted a disagreement in the answer to the first question, but closed the subject.

Berman formulated the second question:

2. What measurements are needed to test constitutive equations for drag reducing additive systems in order to simplify the formal description of an internal flow?

He explained the question by saying that our main goal is to set up a model for the momentum flux in drag reducing flows. For a gross model we would primarily need the viscosity at zero shear rate and the viscosity of the solvent and, in addition, the number of backbones of the molecule, perhaps supplemented by a type of spring constant or many spring constants which give a model of the polymer. However, the problems arise in the general description, which includes the history, and so makes things quite difficult for computation. Even if we took a limited model, say, for a low, elongational flow and we could so consider separately in our rheological description the Newtonian and the non-Newtonian part, we would have to model the non-Newtonian part, and we would have to do so in the simplest model, which has an infinite series for what Berman found is the first term alone of importance, even for high drag reduction. In general, that first term contains the products of the shear rate tensor and some rotation tensor. Berman said that even under this condition he does not know if it would be possible to calculate all the products that occur in the equation. Another point is that if we take the time average of the production and the dissipation for a drag reducing polymer solution, the difference is not as much as one would think.

Landahl suggested making a simulation with a very simple model with rod-like elements, as used by Batchelor. This, he argued, would elucidate the mechanism of Virk's type B model. Berman disagreed since he felt the rod model would miss the elastic term. Landahl stressed that this is the problem to be investigated since elasticity may or may not be of importance.

Banerjee inquired what the difference between fully extended molecules and rigid rods is. Berman agreed that the constitutive equation looks the same and is characterized by an anisotropic viscosity. But it remains open if changes in the shear viscosity only produce drag reduction, as this may be the case for surfactants, or if the elastic properties of the elongated molecules are of importance.

Virk went one step further when he said that rheology has given no insight whatsoever into drag reduction. He feels that computation is making such fast progress that rheology loses its actuality since rheology is implicitly incorporated in a simulation. Berman replied that rheology can help in answering a lot of questions, e.g. explaining the onset of the effect. But for Berman the main remaining question is whether the elastic part is necessary or not for the description of the effect. Virk replied that the question could be reduced to: Is it the extended molecules that cause drag reduction or is it the actual extension that does it? He claimed that the actual extension is not of importance. Berman closed this part of the session by agreeing with Virk on the fact that the actual extended molecule is of importance.

Kim proceeded to the next questions.

3. What can we learn from flow instabilities?

Kim stressed that we do not understand the flow instabilities and, therefore, cannot understand how the flow structures form. This situation also holds for the most basic structures, such as streaks or streamwise vortices. But, without understanding their formation, he argued, it will be hard or impossible to manipulate them. To him, the most troublesome result is that most stability studies are based on the mean turbulent flow. He believes this approach is invalid. However, these calculations provide good results and, therefore, we should really know more about what occurs in the flow. Another most intriguing fact is the drag reduction by a polymer thread, especially, since in Kim's opinion, it is hard to imagine that such a thread interacts with the wall-near structures.

According to Kim, another question to be discussed was:

4. What is the role of coherent structures?

Kim said that we all believe that, if we understand the coherent structures, we can come up with a better turbulent model with which to predict the flow due to a Reynolds type simulation. This means that we have to combine two incomparable ideas, the Reynolds averaging and the idea of coherent structures. How can we join two approaches, which are so different, Kim asked.

Nagib agreed with Kim that we do not understand enough about the stability of a turbulent flow and he reminded the audience that an instability of streaky type can also be formed in laminar boundary layer flows. He said that Landahl showed a new approach, but believes that there are still many others to be considered. He disagreed with Kim on the second question. He argued that we do not necessarily have to use the coherent structures in conjunction with a model that is based on Reynolds averaging. He stressed that there are a number of things going on, like an analysis of the stability of dynamic systems which aims at certain features. An analysis along these lines was set up by Lumley and his coworkers. We do not necessarily have to face comparability; for building a model it is of importance to understand coherent structures.

Kim explained his standpoint. We are looking for a better situation to simulate the turbulent flow, he said. But this cannot be done by a combination of the two concepts, as they result from two different approaches. Nagib said he would agree with this, but stressed that the engineering community is waiting for better methods for simulations. He also noted that it is very difficult to incorporate the new ideas into today's approaches.

Falco offered another piece of information. Our work with LEBUs shows us that the local skin friction can be lowered by 15-20 % due to the interaction of the large eddies with the wake of the plates in the outer part of the boundary layer. Something similar happens in heterogeneous drag reduction, Falco noted. The large eddies are somehow affected. Falco concluded that the large eddies and the skin friction are definitely connected with each other. Kim inquired if Falco believes that the thin string affects the large scale structures. Falco confirmed this.

Kline agreed with Falco, although he had no explanation of how it works. He said that he could observe as much as 20 % interaction. He reminded us that at convex surfaces with a flow of around 20° around, the production stops, which means we have only decay, a result predicted by the rapid distortion theory. This explains several points, Kline said. There is an interaction, though of 20 % only; the other 80 % are still kept going. There is an outer interaction, but we must take care not to think only in models of mixing type of interaction since the mechanisms are more complicated than that.

In answer to the question whether the structures are important or not, Kline said that if we omit all structures, we have a laminar flow by definition, and if we define coherent structures as the features which control the Reynolds shear stress, they are obviously important. Whether we can get a close model, is quite a different question. There are hints that a certain kind of vortex dominating the outer flow and another dominating the inner flow exists. This gives us quite a different picture than the assumption that one model dominates the whole thing.

Tsinober reminded us that, at least for polymer solutions, the power spectra show that for drag reduction the remaining spectra are dominated by large scale features.

Nagib figured that we should not try to fit all drag reducing appraoches to the same mechanism. He reminded us that we have clear evidence that we may achieve drag reduction by reducing the large scale structures or by altering the small scale structures. The richness of the structures of the boundary layer provides a wide field of interaction and we can have drag reduction even for a grid turbulence by either altering the small or the large scale structures.

Gustavsson's statement had to be reduced on account of the quality of the recording. In short, he said that whether we look at inviscid instability in viscous cases, as he had done, or calculate second instabilities, the dt disturbances are the ones to focuse on. Gustavsson also expressed his disappointment with the fact that we have made practically no progress in understanding drag reduction in the past. This is, in his opinion, due to the lack of a constitutive equation which is necessary for investigating the stability of the flow.

Kim pointed out that we should pay more attention to the stability behaviour. Such studies, he argued, are also required for numerical simulations. The evolution of the structures is the most important feature for a better understanding. We should be aware that what we usually get are only sequences of kinematical pictures. From them we may derive an idea of the dynamical process itself, which is necessary for a consistent model, but not more.

Banerjee inquired if any measurements like the streak spacing in the case of heterogeneous drag reduction had been made and if such measurements had been compared to measurements of homogeneous drag reduction. Virk answered that all we know is that for homogeneous drag reduction the streak spacing increases at least in case of low drag reduction. This was confirmed by a series of speakers. Therefore, Banerjee insisted on having more information on the case of heterogeneous drag reduction. Berman stressed that there are no reliable measurements on this. But the situation is even more crucial as nobody even knows if heterogeneous drag reduction exists at all in a channel flow. So far it has been recognized only in pipe flows, but for reliable measurements and comparisons we would require streak spacings in a channel flow.

Landahl suggested studying the structures and their interactions in a more concise form. Possibly, a laminar flow would be suitable. In a laminar flow, disturbances could be a lot better controlled and we would be able to evalidate the cause and effect in much more detail than for turbulent flows with their background disturbances. He stated that we can produce streaks in a laminar flow since they seem not to be part of the turbulence as such. They are part of the interaction of the fluctuations with the background shear. This is one approach; another possibility is to do selective experiments like the one presented by Smith, who forced the low speed streaks.

Virk suggested investigating (1) how instabilities are restored and (2) how the laminar turbulent transition is affected by the additives. His experience shows that these transition happens sooner, which puzzles him. This would justify a study along the lines Landahl had suggested. Kim noted that the presented papers show that polymer affects the transition through the second instabilities, which are delayed and produce bigger vortices.

Bechert started a discussion on the question if streaks also occur in free shear layers and, provided they exist, if they too are altered. Kim stressed that for laminar mixing layers they had not found any streaks, which was obvious since they cannot be produced under the shear existing in such flows. The streaks in laminar boundary layers are part of the high shear existing close to the wall. Nagib said that streaks are possible in mixing layers, that he disagreed with Bechert and agreed with Kim, but added that what is necessary is a high gradient in the transversal velocity as shown in Bernal's experiment. This experiment showed that if there is some reminence of a high gradient at the splitter plate, we get streaks in the mixing layer.

Nychas suggested adding a further question: What are the overall structures? But Kim decided to end this session here.

4 Drag Reduction by Passive Means
(D. Bechert, H.M. Nagib & A.M. Savill)

Bechert opened the final panel discussion by introducing Professor Clark, who showed a series of slides with pictures of the visualization of a channel flow over a floor of transparent riblets. The riblets produce a partition of the flow into a flow near and between the roughness with rollers in the grooves and the outer flow region, which could be altered by the rollers. Clark's presentation resulted in the main panel discussion.

1. How do the riblets relate to the structures of the flow, more precisely when does the maximum of drag reduction occur?

Clark stressed that the rollers which build up in the grooves are presumably the main changes in the structures. They provide the flow with a mechanism which enhances the exchange of fluid visible at the wall. To what extent the fluid is recirculated cannot be defined quantitatively yet. Nagib added that the observed drag reducing effect is only of the order of 4-7%. Therefore, he speculated that the riblets act only by an enhancement of the exchange and not by an additionally altered structure. Bechert, however, reminded us that in the crossflow experiment he had presented, the fluid at low Reynolds numbers is definitely caught in rollers sitting in the grooves. At high Reynolds numbers, however, he found an asymmetric transport between the two ranges, which could be interpreted as an additional type of structure. Clark agreed that his experiments too had been made at the lower limit of the Reynolds number for a turbulent boundary layer flow. Without tripping, the flow would have produced some turbulent spots only occasionally. With tripping, he thought that he had produced the lowest possible steady state turbulent boundary layer flow.

Someone mentioned that we should always keep in mind that riblets only produce a drag reduction of 6-7%,for which an improvement of not more than 1% can be expected. Therefore, the question arises if it is advisable to put a lot of effort and time into the investigation of conventional riblets. Bechert made clear that riblets work, but that the decision to apply them is not a scientific, but an economic one, in other words, determined by the price of oil . Falco was more optimistic in his opinion because he relies on our knowledge of how riblets work. The riblets only rearrange the structures to a small degree. The streaks are less strong, the shear layer is a little bit weaker and, therefore, the second instabilities too all result in a drag reduction of about 5%. If we wanted to improve the riblets, we would have to take into account the pressure drag. Based on this concept, Falco suggested reducing the number of ribs to acquire a surface which looks more like a d-type roughness. The redistribution of the pressure drag should show up in a drag reduction which should be about twice as much as the conventional one. However, this enhancement would work only as long as the spacing is less than 100 viscous wall units.

Blackwelder reminded us that Falco's suggestion could be tested by a simulation with an active condition, as the one presented by Kim for the v fluctuations. Kim mentioned that Blackwelder had said that he thought of also investigating the influence of an active manipulation of the w component, but that he had renounced the idea for practical reasons. For the understanding of the physics, however, Blackwelder stressed, it would well be of practical interest.

Nychas commented that the physics of suctional loading is the enhancement of the interaction and not an alteration of the production. It would therefore be interesting to know how riblets woul affect the flow just above the crests of the riblets. So he asked if anyone could present results of such an experiment. Blackwelder had made such an experiment. He had not found any, or at least not any significant, improvement of the net drag reduction by a combination of suction with the riblets.

Nagib continued with the next question:

2. Did any new passive drag reducing techniques come up in the last years or, possibly, new applications of a combination of techniques already known?

For fun someone proposed heated riblets. Another participant asked if there exist new results on a manipulated porous wall by which the pressure fluctuations at the wall could be suppressed by a compensational process, a process which could be enhanced by active suction or blowing. Bechert replied that NASA, as well as they themselves, had attempted such an experiment with a disastrous result; possibly, because the porosity consisted of holes with openings perpendicular to the flow instead of being tangential. However, riblets do work, he said, and an improvement should be possible.

Blackwelder pointed out that for the shark skin there exists a drag reduction wich is associated with a roughness, which can be characterized by a scale of its configuration which is much larger than the spacing of the riblets. Bechert stressed that the shark skin problem is far too complicated to be explained in the frame of such a discussion and reminded us that the main results can be found in his full paper. Someone, however, insisted on the influence of porous walls by remarking that we find in nature series of roughnesses combined with a porosity which probably produces drag reduction, and that we should therefore see if we cannot reproduce nature and even improve it by active suction and blowing, for example. Bechert reminded us that an example of this kind already exists: It is the bypass of fluid from the stagnation line of a wing through the wing into the separation zone, where the separation bubble can be manipulated by this mechanism. However, this is not a reduction of the skin friction, but rather a change in the pressure drag.

Falco suggested that we prevent the formation of some hierarchies of the so-called lamda-vortices, perhaps, by adding other roughnesses instead of streamwise fixed regular roughnesses only. The full explanation cannot be given here since Falco explained his ideas on the transparencies he had already used to explain the main structural features close to the wall.

Kim was surprised that nobody had mentioned compliant walls. The rough simulation they had done some years ago had been positive where the flexibility of the wall had been adequately chosen. Flows with moving wall boundary conditions were another promising simulation they had tried. They had found that for particular wave numbers of the applied boundary condition they had got drag reduction where the motion coincided with the phase speed. Kim speculated that combinations of compliant walls with moving wall boundary conditions would still be an attractive subject of study. Blackwelder reported that studies in this direction had been made, though mainly with the aim to manipulate the acoustic part of the problem. For drag reducing applications, he said, we would first have to know more on the interaction of the structures with the mean flow.

Savill mentioned that there exist some studies on transitional flows over subgels.

But the main subject he promoted was the discussion on LEBUs or manipulators. He considered himself the right man to do this since he is much more optimistic than most of the audience, who would rather like to "bury" the subject. In his opinion, the riblets act as drag reducers by two different mechanisms: (1) by shifting away the turbulence from the wall, and (2) by influencing the spacing between the structures. Savill said that there exist several fairly different opinions on how manipulators work. This is obvious, he went on, since manipulators interact much more with the flow than riblets do. In his view, there exist at least five different mechanisms involved in this kind of drag reduction. He had mentioned them in his review lecture.

Since the main attitude towards the future of manipulators is very negative, he decided to provoke accordingly he said:

"The manipulators are dead!" Let's discuss this!

The audience's reaction was silence. And then someone remarked that yes, they were dead, at least in regard to internal or channel flows. Nagib argued that before being so critical, the reaction of the flow to a change in the angle of attack on the manipulators should be examined since the drag is very sensitive to this angle. To accentuate this statement, Nagib reported that they had not been able to find any features in the boundary layer which are different, with the exception of the so-called downwash case. Someone confirmed Nagib's comment and added that negative angles of attack enhance the large scale structures. He also argued that we should not study manipulators from the standpoint of drag reduction only, but also discuss them as manipulators of the delayed separation or as mixing elements. Nagib insisted that Savill's optimism concerned drag reduction, and that this was the subject of discussion.

There was agreement on the fact that the manipulators are not applicable for internal flows. However, since there is no conceptual reason why they should not work here when they work in other cases, we would have to thoroughly examine whether the negative result is only due to the angle of attack.

Nagib confessed that his enthusiasm for manipulators had been drastically reduced by the fact that they, at least in simple configurations, are ineffective at high Reynolds numbers. Therefore he could not share Savill's optimism.

One speaker claimed that manipulators have only been studied in quasi two-dimensional cases instead of in three-dimensional situations, which correspond much better to flows in practical applications.

Savill replied that his optimism stems from two remaining questions which, if answered, could really improve the effectiveness of manipulators. (1) A better knowledge of how the separation works, especially when interacting with a boundary layer. (2) The effect of manipulators on transitional flows, which is a question of how the stability can be altered.

Choi commented that they had experimented with three-dimensional boundary layer flows without having arrived at any results differing from the one for a more or less two-dimensional boundary layer flow.

Bechert added that the same questions apply to riblets. They reduce separations and their highest drag reducing effect appears in transitional flows. Savill said that he would show an experiment on a ship hull, which would confirm what Bechert said, in Davos.

One participant commented that the argument that three-dimensional flow probably does not interact with manipulators by reducing the drag should be taken very seriously, since at high Reynolds numbers every boundary layer is three-dimensional, and this may be the reason why manipulators do not work under these conditions.

Nagib summarised some points of the controversy with the aim that we take the questions back home with us to ponder over them in peace.

Can drag reduction for polymers, complient walls, riblets and manipulators be described by one mechanism only?

In Nagib's view, the richness of the boundary layer is large enough to allow for more than one approach. Therefore, what works for polymers, need not work for passive devices. A hint in this direction is their different response to high Reynolds number flows. (Editor's remark: This question has to be relativized. One of the goals of the symposium was to discover what the different drag reducing mechanisms have in common , especially with respect to the characteristics of the turbulent structures.)

How should the scales of the structures be compared under different flow conditions?

We should be very careful, Nagib stressed, when comparing the range of structures in outer parameters which are Reynolds number dependent with structures scaled with inner parameters, which, in turn, have not to be Reynolds number dependent. In other words, we must examine the mechanisms in how they differ, but also the balances of the forces.

Nature, Nagib reminded us, does not allow us to simply turn something on or off and get an in- or decreasing effect in the drag.

Choi agreed that there must exist different mechanisms for the different methods of achieving drag reduction. He stressed that we should carefully compare the basic statistic features of the coherent structures for the different methods by comparing them one by one. In addition, he said, we should treat the effect like a black box and learn more about the specific mechanisms by varying the initial and boundary conditions.

Weiss asked why there was no contribution on additives in gas flows, especially in dense gas flows. The physics should be essentially the same. Zakin answered that studies of this kind had been reported in 1974, but, to his knowledge, nobody had treated this problem since because such experiments are extremely difficult. Nagib said that from the basic aspects involved we should encourage the scientific community to pick up this subject again. In numerical simulation there is no difference as long as compressibility needs not be taken into account.

Tsinober, as the final speaker, aimed at provoking and supplementing Nagib's two questions. He said that to impose some restriction on some important scale, for example on the energy containing one, is less important than a constraint on the interaction. This had been the message of his review. To him, the interaction of the three well known scales is the key to all effects: the energy containing scales, the Taylor microscale and the dissipative Kolmogorov scale. To him, the polymers are so efficient because they interact with all scales and the coupling mechanisms between them, whereas passive devices are so inefficient, if drag-reducing at all, because they interact with one scale only.

Instead of a final overall discussion, Gyr closed the session and the Symposium with the hope that communication between the scientists of the various branches had been successful and would bear fruition for future conferences.